2014 IEEE 21st International Symposium on the Physical and Failure Analysis of Integrated Circuits

(IPFA 2014)

Singapore
30 June – 4 July 2014

IEEE Catalog Number: CFP14777-POD
ISBN: 978-1-4799-3911-4

Copyright © 2014 by the Institute of Electrical and Electronic Engineers, Inc
All Rights Reserved

Copyright and Reprint Permissions: Abstracting is permitted with credit to the source. Libraries are permitted to photocopy beyond the limit of U.S. copyright law for private use of patrons those articles in this volume that carry a code at the bottom of the first page, provided the per-copy fee indicated in the code is paid through Copyright Clearance Center, 222 Rosewood Drive, Danvers, MA 01923.

For other copying, reprint or republication permission, write to IEEE Copyrights Manager, IEEE Service Center, 445 Hoes Lane, Piscataway, NJ 08854. All rights reserved.

This publication is a representation of what appears in the IEEE Digital Libraries. Some format issues inherent in the e-media version may also appear in this print version.

IEEE Catalog Number: CFP14777-POD
ISBN 13:   978-1-4799-3911-4

Additional Copies of This Publication Are Available From:

Curran Associates, Inc
57 Morehouse Lane
Red Hook, NY 12571 USA
Phone: (845) 758-0400
Fax: (845) 758-2633
E-mail: curran@proceedings.com
Web: www.proceedings.com

2014 IEEE 21st International Symposium on the Physical and Failure Analysis of Integrated Circuits (IPFA 2014)

Singapore
30 June - 4 July 2014

IEEE Catalog Number: CFP14777-POD
ISBN: 978-1-47993-911-4

Proceedings of the

21ᵗʰ International Symposium on the

Physical and Failure Analysis of Integrated Circuits (IPFA)

30 June – 4 July 2014
Marina Bay Sands, Singapore

Organised By	Technically co-sponsored By		Supported by

IEEE Singapore Reliability/CPMT/ED Chapter			

Copyright and Reprint Permission: Abstracting is permitted with credit to the source. Libraries are permitted to photocopy beyond the limit of U.S. copyright law for private use of patrons those articles in this volume that carry a code at the bottom of the first page, provided the per-copy fee indicated in the code is paid through Copyright Clearance Center, 222 Rosewood Drive, Danvers, MA 01923. For other copying, reprint or republication permission, write to IEEE Copyrights Manager, IEEE Operations Center, 445 Hoes Lane, Piscataway, NJ 08854. All rights reserved. Copyright ©2014 by IEEE.

Technical support & Inquiries
Research Publishing Services
Blk 12 Lorong Bakar Batu,
#02-11, Singapore 348745.
t: +65-6492 1137
f: +65- 6747 4355
e:enquiries@rpsonline.com.sg

Welcome Message

Welcome to 21st IEEE International Symposium on Physical and Failure Analysis of Integrated Circuits - IPFA 2014, back in Singapore!

On behalf of the Organising Committee, we thank you for your participation at IPFA 2014. Over the next 3 days, 49 oral presentations and 39 poster presentations will be delivered and displayed during the technical symposium and poster session respectively. Despite the increasing presence of related conferences, we appreciate your continuous support in selecting IPFA as the platform to publish your latest research in the areas of microelectronics reliability and failure analysis (FA). Besides the submitted papers, there are 6 invited papers and 2 exchange papers from International Symposium for Testing and Failure Analysis (ISTFA) and European Symposium on Reliability of Electron Devices, Failure Physics and Analysis (ESREF). Also, there are 2 keynote papers by very prominent experts in their fields. **Prof. Michael Pecht** from University of Maryland, USA will present "Advances in the Qualification of Microelectronic Devices". **Mr. Raj Nair**, from GLOBALFOUNDIRES, Singapore will present "Failure Analysis Techniques for More than Moore Semiconductor Technologies".

In addition, a dedicated tribute session is organized in remembrance of the late Professor Jacob Phang in recognition of his prolific contributions to the FA and Reliability communities. You are cordially invited to join us in this meaningful event. Notwithstanding the challenging economic outlook, IPFA continues to receive strong support from our sponsors. FEI Company has remained as the Platinum sponsor for the 7th consecutive year and is co-sponsoring the banquet as well. Leica Microsystems, Digit Concept and Synopsys have also taken up the Silver, lanyard and tea break sponsorships respectively. We would like to take this opportunity to thank them, as well as Singapore Tourism Board, for their financial contribution, without which this symposium will not be possible.

The 3 days equipment exhibition that is held in conjunction with the symposium remains an important activity of IPFA, both financially and technically. There are 28 exhibitors from both local and international companies this year. We thank them for choosing IPFA as the conference to showcase their latest equipment and technologies. Many of themare taking a leading role in the microelectronics characterization and FA market, especially in advanced technologies such as 3D integrated systems and green electronics. Please visit their booths for more information during the tea and lunch breaks.

The "Art of Failure Analysis" Photo Contest which is an integral part of IPFA provides FA enthusiasts a platform to showcase their best FA photos. It continues to receive good participation from the FA community and the top 10 photos will be published in IEEE Spectrum. Also, these photos will be exhibited over the next 3 days and the top 3 winners will be determined through the votes of IPFA participants.

The success of IPFA is attributable to the contributions from every stake holders. Besides the sponsors, exhibitors, authors and participants, we would like to thank all the organising committee members, who have carried out their roles diligently. Special thanks to the members of Technical Programme Steering Committee (TPSC). They are experts from all over the world and are essential in soliciting quality papers for the symposium, and reviewing them despite their busy schedules. Last but not least, we would like to acknowledge the immense contribution by our secretariat, whom has been our constant source of information and reminder since the planning of IPFA 2014.

IPFA has been held outside of Singapore in alternate years since 2004. It is our pleasure to take this opportunity to announce that IPFA 2015 will be held in Taiwan. We hope everyone will have a rewarding experience over the next 3 days at IPFA 2014.

Yeow Kheng Lim
General Chair

Diing Shenp Ang
Technical Programme Chair

Committees

IPFA Board

CHIN JiannMin (Chair), *Advanced Micro Devices*
Alastair TRIGG (Immediate Past Chair), *Institute of Microelectronics*
GAN Chee Lip, *Nanyang Technological University*
LIMYeow Kheng, *STATS ChipPAC*
PEY Kin Leong, *Singapore University of Technology and Design*
John THONG, *National University of Singapore*

Organising Committee

General Chair**Lim Yeow Kheng**, *STATS ChipPAC*

Technical Program Chair
Ang Diing Shenp, *Nanyang Technological University*

Technical ProgramCo-Chair
Vinod Narang, *Advanced Micro Devices*

Tutorial Chair
Alan Street, *Apple*

Tutorial Co-Chair
Gan Chee Lip, *Nanyang Technological University*

Art of FA
Zhao Siping, *GLOBALFOUNDRIES*

Art of FA Co-Chair
Zeng Xu, *GLOBALFOUNDRIES*

Finance Chair
Tan Soon Huat, *Semicaps*

Exhibitions/Sponsorship Chair
Chin Jiann-Min, *Advanced Micro Devices*

Exhibitions/Sponsorship Co-Chair
Hua Younan, *WinTech Nano-Technology Services*

Publicity Chair
Sanjay Kumar Thakur, *Rolls-Royce*

Publicity Co-Chair
Goh Szu-Huat, *GLOBALFOUNDRIES*

Publications Chair
Tang Lei Jun, *Institute of Microelectronics*

Publications Co-Chair
Aaron Chin, *Systems on Silicon Manufacturing Co*

Banquet & VIP Dinner
Tan Kok-Tong, *T2 Integrated Solutions*

AV / Student Volunteers
Nagarajan Raghavan, *Singapore University of Technology and Design*
Wardhana A. Sasangka, *Singapore-MIT Alliance for Research and Technology*

Technical Program Committee

A. Novel Gate Stack/Dielectrics and FEOL Reliability and Failure Mechanisms
Gennadi Bersuker, *SEMATech, USA (Chair)*
Nagarajan Raghavan, *Singapore University of Technology and Design, Singapore (Co-Chair)*
Diing Shenp Ang, *Nanyang Technological University, Singapore*
Tibor Grasser, *TU Wien, Austria*
Jim Stathis, *IBM, USA*
C. K. Maiti, *IIT Kharagpur, India*
Chadwin Young, *University of Texas at Dallas, USA*
Jianfu Zhang, *Liverpool John Moores Univeristy, UK*

B. Advanced Interconnects and BEOL Reliability and Failure Mechanisms
Jeffrey Gambino, *IBM, USA (Chair)*
Chee Lip Gan, *Nanyang Technological University, Singapore (Co-Chair)*
Ehrenfried Zschech, *Fraunhofer Institute for Non-Destructive Testing IZFP, Germany*
Kristof Croes, *IMEC, Belgium*
Galor Zhang, *GLOBALFOUNDRIES, Germany*
Timothy Turner, *College of Nanoscale Science and Engineering, USA*
Tony Oates, *TSMC, Taiwan*

C. Novel Devices, LED Reliability and Failure Mechanisms
Guoqiao Tao, *Philips (Chair)*
Mingxiang Wang, *Soochow University, China*
Zung-Sun Choi, *BSAF, Korea*
Hwang Nam, *Korea Photonics Technology Institute, Korea*

D. Product Reliability Evaluation and Approaches
M. K. Radhakrishnan, *NanoRel, India (Chair)*
Alan Lek, *GLOBALFOUNDRIES, Singapore (Co-Chair)*
Frank Huang Weidong, *Microsoft, China*
Zoran Stamenkovic, *IHP, Germany*

E. Die/Package-level Failure Analysis Case Study and Failure Mechanisms
Susan Li, *Spansion (Chair)*
Alastair Trigg, *Insitute of Microelectronics, A*STAR, Singapore (Co-Chair)*
Jiann Min Chin, *AMD, Singapore*
Aaron Chin, *SSMC, Singapore*
Jean-Jacques Hajjar, *Analog Devices*
Frank Wei, *DISCO, Japan*

Daniel Rhee Min Woo, *Institute of Microelectronics, A*STAR, Singapore*
Thomas Rupp, *Infineon Technologies Austria AG, Austria*

F. Advanced Failure Analysis Techniques
Christian Boit, *Technical University Berlin, Germany (Chair)*
Vinod Narang, *Advanced Micro Devices, Singapore (Co-Chair)*
Phillippe Perdu, *CNES, France*
David Su, *Taiwan Semiconductor Manufacturing Co, Taiwan*
Terence Kane, *IBM, USA*
Alan Street, *IRIS, Singapore*
Hobert Christian, *GLOBALFOUNDRIES, Germany*
Goh Szu Huat, *GLOBALFOUNDRIES, Singapore*

G. Sample Preparation, Metrology and Material Characterization
Ralf Heiderhoff, *Bergische Universitat Wuppertal, Germany (Chair)*
John Thong, *National University of Singapore , Singapore (Co-Chair)*
Chih Hang Tung, *Taiwan Semiconductor Manufacturing Co, Taiwan*
Sam Subramanian, *Freescale, USA*
Hans-Jürgen Engelmann, *GLOBALFOUNDRIES, Germany*
Leijun Tang, *Institute of Microelectronics, A*STAR, Singapore*

Conference Secretariat	On-LineWeb & Conference Management System
Jasmine Leong (Ms.) Blk 121 Paya LebarWay #03-2801, Singapore 381121 Tel: (65) 6743 2523 Email: ieee_ipfa@singnet.com.sg	Ra. Sankaran (Mr) iTEK CMSWeb Solutions Blk 12 Lorong Bakar Batu, #02-11 Singapore 348745. Tel: (65) 6492 1137, Fax: (65) 6748 2556 Email: sankaran@itekcms.com Web: www.itekcms.com

Keynote Speaker 1

Title : Failure Analysis Techniques for "More than Moore" Semiconductor Technologies

Raj Nair
GLOBALFOUNDRIES
Singapore

Raj Nair is the Vice President for Technology Development at GLOBALFOUNDRIES Singapore. The Singapore TD team develops embedded NVM, RF, Power Management and MEMS solutions for a wide range of products and customers. Raj has over 25 years of Operations and R&D experience in the semiconductor and advanced battery industries. He has held executive positions at Johnson Controls, Maxim, Motorola and On Semiconductor. At Johnson Controls he was the Vice President for Advanced Manufacturing and Quality and launched the world's largest automotive Li ion battery plant in Holland, MI. He is a pioneer in the development of Smartpower integrated circuit technologies. Raj has an MSEE degree from NTU and an MS ChE from Syracuse U. He has also written a series of books of the electronics industry in China, Korea, Japan and India. He consults for 22 international companies. He has written over 700 technical articles and has 5 patents.

Abstract

Due to the capital and development costs of new semiconductor processes, the number of companies offering commercial leading edge logic technologies continues to drop. At the same time technologies such as embedded non-volatile memories (eNVM), MEMS, Power Management, RF, Display and TSV are attracting greater interest driven primarily by the mobility markets. This talk will begin with a review of industry trends in these "More than Moore" technologies. It will then discuss the Failure Analysis techniques required for these applications. The ability to identify the fault location and the nature of the fault is always important in diagnosing integrated circuit failures and implementing corrective action for all the semiconductor technologies.

Scanning laser microscopy, photoemission microscopes with Si CCD and InGaAs, and thermal emission microscopes with InSb and MCT detector all play a critical role in isolating the fault location. In additional to optoelectronics based fault isolation techniques, e-beam based fault isolation techniques, such as EBIC and EBAC, were also developed to meet the fault isolation demand at the circuit level. A breakthrough in nanoprobing techniques in the past 10 years has enabled great strides in the fault isolation capability at circuit and transistor levels. Backside fault isolation plays an important role in detecting the failing locations due to ESO, ESD, and latchup which are observed in high voltage analog circuitry.

Other than the destructive techniques, X-ray microscopy and scanning acoustic microscopy can inspect the internal structures of a device or a package without removing the covering layers or materials. They are critical in the failure analysis on MEMS products. Moving forward, due to the special requirements of the more than Moore processes, further developments in failure analysis techniques are expected. These include increasing the maximum voltage during nanoprobing, improving the leakage current sensitivities for reliability failure, X-ray resolution in nano defect detection, multi-layer analysis capability in TSV and non-destructive analysis for MEMS.

Keynote Speaker 2

Title : **Fusion Prognostics-based Qualification of Microelectronic Devices**

Michael Pecht
University of Maryland
USA

Prof Michael Pecht is a world renowned expert in strategic planning, supply chain management, design, test, IP and risk assessment of electronic products and systems. In 2013, he was awarded the University of Wisconsin-Madison's College of Engineering Distinguished Achievement Award. In 2011, he received the University of Maryland's Innovation Award for his new concepts in risk management. In 2010, he received the IEEE Exceptional Technical Achievement Award for his innovations in the area of prognostics and systems health management. In 2008, he was awarded the highest reliability honor, the IEEE Reliability Society's Lifetime Achievement Award. Prof. Pecht has an MS in Electrical Engineering and an MS and PhD in Engineering Mechanics from the University of Wisconsin at Madison. He is a Professional Engineer, an IEEE Fellow, an ASME Fellow, an SAE Fellow and an IMAPS Fellow. He has previously received the European Micro and Nano-Reliability Award for outstanding contributions to reliability research, 3M Research Award for electronics packaging, and the IMAPS William D. Ashman Memorial Achievement Award for his contributions in electronics analysis.

He is the editor-in-chief of IEEE Access, and served as chief editor of the IEEE Transactions on Reliability for nine years, chief editor for Microelectronics Reliability for sixteen years, an associate editor for the IEEE Transactions on Components and Packaging Technology, and on the advisory board of IEEE Spectrum. He is the founder and Director of CALCE (Center for Advanced Life Cycle Engineering) at the University of Maryland, which is funded by over 150 of the world's leading electronics companies at more than US$6M/year.

The CALCE Center received the NSF Innovation Award in 2009, as well as the National Defense Industrial Association award (2009) for demonstrating outstanding achievement in the practical application of Systems Engineering principles, promotion of robust systems engineering principles throughout the organization, and effective systems engineering process development. He is currently a Chair Professor in Mechanical Engineering and a Professor in Applied Mathematics at the University of Maryland.

He has written more than twenty books on product reliability, development, use and supply chain management. He has also written a series of books of the electronics industry in China, Korea, Japan and India. He consults for 22 international companies. He has written over 700 technical articles and has 5 patents.

Abstract

The rapid evolution of electronic products has resulted in numerous choices for customers. This has made for intense competition between manufacturers to reduce costs and minimize the time to market for their products. One bottle-neck in getting products to market is the qualification process, which has traditionally been time-consuming and often inadequate to prevent failures in field. In particular, in the past decade, there have been significant numbers of microelectronic devices that have passed qualification tests but failed in the field. The resulting costs of these failures have been in the billions of dollars. Thus, there is a need to develop approaches to qualification methodologies that quicken the development time but also prevent product failures in the field. This paper discusses the current state of qualification practices in the electronics industry. Then, an alternative approach, called fusion prognostics, for qualification is presented that can make the process more efficient and cost-effective. This approach involves an in-situ qualification process that incorporates a fusion of machine learning techniques and physics-of-failure based prognostics. The machine learning techniques are used to monitor the degradation behavior during testing. On the other hand, the physics-of-failure techniques identify critical failure mechanisms and the acceleration factors.

Invited Speakers

Invited Speaker 1

Title : Fault Backside Failure Analysis Techniques: What's The Gain Of Silicon Getting Thinner?

Christian Boit
Technical University of Berlin
Germany

Christian Boit holds the chair of semiconductor devices at TUB (Berlin University of Technology), Germany, focussing on simulation, characterization analysis and debug of Integrated Circuits (ICs) and other specific devices and their respective operating modes. After gaduating from TU Berlin, he joined Siemens Research Laboratories, Munich, Germany, in 1986. From 1990 to 1993, he participated IBM-Siemens 64M DRAM project at East Fishkill, NY. Later, he has been Director of Failure Analysis at Siemens Semiconductor and Infineon Technologies, respectively. In 2002 he accepted the call to TUB. Founded on optical interaction techniques and Focused ion Beam applications for ICs, security related issues belong to his research areas as well. Christian has more than 100 publications and 6 international publication awards. He cofounded the worldwide unique society for electronic device failure analysis EDFAS in 1998 and the European FA network EUFANET. In 2002, he served as General Chair of the International Symposium of Testing and Failure Analysis of electronic devices ISTFA. He is member of IEEE, VDE and acatech, the German Academy of Science and Engineering. He is General Chair of the European Symposium on Reliability of Electronic Devices, Failure Physics and Analysis ESREF in 2014, held at Berlin.

Abstract

Failure analysis (FA) of electronic devices today is mostly conducted through the device backside. Advanced silicon (Si) backside preparation for this purpose has developed over the past years, with a final Si thickness from around 300 µm to 100 µm down to around 20 µm to 10 µm. This paper discusses what to expect if Si can be processed a little thinner, from 20 µm down to a few µm. Improvement of optical imaging, spectral extension of photon emission and expanded optical interaction for stimulation techniques are investigated. Further, Si thickness is coming close to the penetration depth of particle beams. The interaction potential for device analysis is discussed, preliminary results are presented.

Invited Speaker 2

Title : **Bias Temperature Instability Investigation of Double-Gate FinFETs**

Chadwin D. Young
The University of Texas at Dallas
USA

Chadwin D. Young is the Assistant Professor for Materials Science & Engineeringt at The University of Texas at Dallas USA. He received his Ph.D. in Electrical Engineering from the North Carolina State University at 2004. He received his M.S. in Electrical Engineering from the North Carolina State University at 1998. He received his B.S. in Electrical Engineering from the University of Texas - Austin at 1996.

Invited Speaker 3

Title : SRAM Failure Analysis Evolution Driven by Technology Scaling

Zhigang Song
SRDC, IBM
USA

Zhigang Song received Bachelor of Science from Fudan University, China, 1988, Master of Engineering from Beijing Institute of Chemical technology, China, 1990, and Ph. D degree from National University of Singapore, 1998. He started failure analysis career in AMD, Singapore as a device analysis engineer in 1997. In 1999, he moved to Chartered Semiconductor Manufacturing Limited, where he had held senior engineer, principal engineer, senior principal engineer and member of technical staff positions in failure analysis lab. In 2007, he joined Semiconductor Research and Development Center (SRDC) of IBM as a senior engineer in PFA Lab. Since then, he had led failure analysis team to support 45nm, 32nm, 28nm and 20nm bulk technology development and also been responsible for failure analysis of logic failures (scan chain and ATPG) for 32nm and 22nm SOI technology development. Currently, he is involved in failure analysis for 14nm SOI FinFET technology development. He has authored or coauthored more than 50 papers published in Journals and Conferences, included ISTFA and IPFA.

Abstract

Demand for high speed and more function microelectronic devices has driven semiconductor industry to continue developing technologies with ever-shrinking geometry. During technology development, Static Random Access Memory (SRAM) is often chosen as the process qualification and yield learning vehicle. Thus SRAM failure analysis is the major activity in any microelectronic device failure analysis lab. Conventional physical failure analysis in old technology nodes has achieved high success rate since the SRAM bitcell failures can be precisely localized by functional test and the defect causing such failures is within the failing bitcells. However, As SRAM feature size decreases with technology scaling down, the size of the defect causing SRAM failure also scales down. Some of the defects are so tiny that they are invisible in ultra-high resolution SEM. On the other hand, the SRAM bitcell number greatly increases, and thus the SRAM design, especially address decoder scheme becomes more complex. More and complicated SRAM logic type failures arise. Therefore, the conventional physical failure analysis has faced increasing challenges and encountered low success rate. This paper will talk about how SRAM failure analysis evolves to maintain high success rate.

Invited Speaker 4

Title : **Hourglass Concept for RRAM: A Dynamic and Statistical Device Model**

Robin Degraeve
IMEC
Belgium

Robin Degraeve received the M.Sc. degree in electrical engineering from the University of Ghent, Belgium, in 1992 and the Ph.D. degree from the Catholic University of Leuven, Belgium, in 1998. In 1992, he joined imec, Leuven, in the Device Reliability and Characterization Group, where he is currently a Principal Scientist. His work has been focusing on the reliability aspects of thin insulating layers under electrical stress. His current research interests include the physics of degradation and breakdown phenomena in gate oxide films, the reliability of flash memory devices, the reliability of ultrathin oxide and oxynitride layers for VLSI technologies, and the characterization and the reliability of high-k materials as gate insulators for future CMOS generations and memory applications. Recently he has been working mainly on Resistive RAM memory development and modeling.

Abstract

In this paper we review a dynamic device model for filamentary RRAM in HfO-based dielectrics. We summarize its transient modeling features and its statistical properties. The model explains with satisfactory quantitative resolution all main features of the RRAM switching, not just the voltage, time and temperature dependence, but also statistical fluctuations resulting from atomistic motion and their resulting LRS and HRSdistributions.

Invited Speaker 5

Title : Material Characterization and Failure Analysis of Through-Silicon Vias

Rui Huang
University of Texas at Austin
USA

Rui Huang received his Bachelor degree in Theoretical and Applied Mechanics from the University of Science and Technology of China in 1994 and his PhD degree in Civil and Environmental Engineering, with specialty in Mechanics, Materials, and Structures, from Princeton University in 2001. He joined the University of Texas at Austin as an Assistant Professor in September 2002 and was promoted to Associate Professor with tenure in 2008. He currently holds the endowed position of Mrs. Pearlie Dashiell Henderson Centennial Fellowship in Engineering. His research interests include mechanics of integrated materials and structures at micro- and nanoscales, reliability of advanced interconnects and packaging for microelectronics, and mechanical instability of thin films and nanostructures. He has published over 80 journal articles, 40 conference papers, and 3 book chapters.

Abstract

In this paper, the effects of Cu microstructure on the mechanical properties of TSV and via extrusion are studied using two types of though-silicon vias (TSVs) with different grain size distributions. A direct correlation is found between the Cu grain size and the mechanical properties of the TSVs. An analytical model is used to explore the relationship between the mechanical properties and via extrusion. The results show that small and uniform grains in the Cu vias led to smaller via extrusion. Such grain structures are effective for reducing via extrusion failure to improve TSV reliability.

Invited Speaker 6

Title : **One Die Logic Analysis through the Backside**

Mike Bruce
Semicaps
USA

Dr. Bruce is a consultant with over 17 years experience in the semiconductor industry applying optical based fault isolation techniques. He has a BS and PhD in Physics from the University of Texas at Austin and conducted post doctoral research in chemical physics at Indiana University. He worked 14 years at Advanced Micro Devices, Inc. in reliability and failure analysis. At AMD, he has developed many optical based tools and techniques for debug of microprocessors including Soft Defect Localization (SDL) for isolating speed paths in IC's, single element Time Resolved Emission (TRE), and Resistive Interconnect Localization (RIL). Mike holds 74 patents and has published numerous papers in referred journals and conferences, including a best paper and outstanding paper at ISTFA for RIL and SDL, respectively. He is past editor of the Electronic Device Failure Analysis magazine and has served on numerous technical committees at Sematech, ISTFA, IPFA, and IRPS

Abstract

On Die Logic Analysis (ODLA) uses a scanning optical microscope (SOM) to quickly determine logic timing patterns, and then uses this information to identify logic pattern matches/mismatches on-the-fly from the backside. In this paper, the ODLA system and methodology will be described along with how, in one universal method, it can replace a slew of techniques such as Laser Timing Probe (LTP), Frequency Mapping (FM), and Phase Imaging (PI). It will be demonstrated on a chain of scan cells.

Tutorials

Tutorial 1

Session	Tutorial 1
Date	30 Jun 2014 / 08:30 – 12:00 noon
Topic	FIB-Backside Processing for Circuit Edit, Probing and High Resolution Circuit Analysis
Presenter	**Christian Boit**, *Technical University of Berlin, Germany*

Biography

Prof. Christian Boit holds the chair of semiconductor devices at TUB (Berlin University of Technology), Germany, focussing on simulation, characterization analysis and debug of Integrated Circuits (ICs) and other specific devices and their respective operating modes. After gaduating from TU Berlin, he joined Siemens Research Laboratories, Munich, Germany, in 1986. From 1990 to 1993, he participated IBM-Siemens 64M DRAM project at East Fishkill, NY. Later, he has been Director of Failure Analysis at Siemens Semiconductor and Infineon Technologies, respectively. In 2002 he accepted the call to TUB. Founded on optical interaction techniques and Focused ion Beam applications for ICs, security related issues belong to his research areas as well. Christian has more than 100 publications and 6 international publication awards. He cofounded the worldwide unique society for electronic device failure analysis EDFAS in 1998 and the European FA network EUFANET. In 2002, he served as General Chair of the International Symposium of Testing and Failure Analysis of electronic devices ISTFA. He is member of IEEE, VDE and acatech, the German Academy of Science and Engineering. He is General Chair of the European Symposium on Reliability of Electronic Devices, Failure Physics and Analysis ESREF in 2014, held at Berlin.

Abstract

Processes for FIB Backside Access presented in terms of practical techniques will be the core of this tutorial. Backside Circuit Edit will be the starting point. Backside FIB sample preparation will be discussed under the aspect that the circuit is maintaining its full functionality. Supporting invasiveness studies will be included. Based on this aspect, an advantage of specific backside FIB processing is the capability of probing, even nanoprobing and other circuit analysis techniques that can be helpful for circuit functional analysis with high spatial resolution. FIB backside processing is able to speed up devices, so delay trimming is an option as well. Finally, a view is taken on the risk these practises are posing to security applications in smartcards etc.

Tutorial 2

Session	Tutorial 2
Date	30 Jun 2014 / 01:30 – 05:00 pm
Topic	The Practice of High Resolution and Analytical Transmission Electron Microscopy
Presenter	**Tim White**, *Nanyang Technological University, Singapore*

Biography

Prof. Tim White is currently a Professor in the School of Materials Science and Engineering at Nanyang Technological University. Tim White received his Ph.D. in Chemistry from the Australian National University. For the 30 years, he has worked at universities and national laboratories in Australia, Europe and North America as well as Singapore. His primary expertise is materials characterization explored using X-ray, electron and neutron diffraction, directed towards the design of advancedmaterials for environmental remediation, superconductivity, hydrogen storage, catalysis and ion exchange. At NTU, Tim is concerned with developing new pedagogies for materials characterization to enable fresh graduates to completely exploit state-of-the-art instrumentation. He launched the firstMassiveOpenOnline Course (MOOC) for NTU on the topic Beauty, Formand Function: An Exploration of Symmetry that included a brief introduction to SEM and TEM. He is author or co-author of over 200 papers, 3 conference proceedings, and numerous confidential industry and government reports.

Abstract

This tutorial will familiarize new users of transmission electron microscopy (TEM) to best practice in collecting diffraction patterns and images from thin samples, and also the conditions required to accumulate reliable analytical information. The major topics to be covered will be:

- sample orientation: how to use electron diffraction to achieve perfect alignment for imaging;
- basic imaging: dark field, bright field and phase contrast imaging;
- compositional imaging: high angular dark field imaging and bright field scanning transmission electron microscopy (STEM);
- microanalysis: the strengths and limitations of energy dispersive X-ray spectroscopy (EDS) and electron energy loss spectroscopy (EELS); and
- specimen artifacts: problems in specimen preparation and electron beam damage.

Software for simulating electron scattering will be introduced.

Tutorial 3

Session	Tutorial 3
Date	30 Jun 2014 / 08:30 – 12:00 noon
Topic	Silicon Probing for Debug
Presenter	**Mike Bruce**, *Semicaps, USA*

Biography

Dr. Bruce is a consultant with over 17 years experience in the semiconductor industry applying optical based fault isolation techniques. He has a BS and PhD in Physics from the University of Texas at Austin and conducted post doctoral research in chemical physics at Indiana University. He worked 14 years at Advanced Micro Devices, Inc. in reliability and failure analysis. At AMD, he has developed many optical based tools and techniques for debug of microprocessors including Soft Defect Localization (SDL) for isolating speed paths in ICs, single element Time Resolved Emission (TRE), and Resistive Interconnect Localization (RIL). Mike holds 74 patents and has published numerous papers in referred journals and conferences, including a best paper and outstanding paper at ISTFA for RIL and SDL, respectively. He is past editor of the Electronic Device Failure Analysis magazine and has served on numerous technical committees at Sematech, ISTFA, IPFA, and IRPS.

Abstract

This tutorial will cover post-silicon debug and the role of physical probing tools in debug. After discussing the debug process, electrical debug tools and techniques will be discussed. Furthermore, a special emphasis is placed on optical probing from the backside. Techniques such as static and Time Resolved Emission (TRE), Laser Timing Probe (LTP), Laser Assisted Device Alteration (LADA), Soft Defect Localization (SDL), Laser Induced Voltage Alteration (LIVA) and Thermal Induced Voltage Alteration (TIVA)/Optical Beam Induced Resistance Change (OBIRCH) will be discussed. Finally, future challenges will be discussed as devices continue to scale down below 20nm.

Tutorial 4

Session	Tutorial 4
Date	30 Jun 2014 / 08:30 – 12:00 noon
Topic	Memory Reliability: From Fundamental Understanding to Reliability Prediction
Presenter	**Robin Degraeve**, *IMEC, Belgium*

Biography

Dr. Robin Degraeve received the M.Sc. degree in electrical engineering from the University of Ghent, Belgium, in 1992 and the Ph.D. degree from the Catholic University of Leuven, Belgium, in 1998. In 1992, he joined imec, Leuven, in the Device Reliability and Characterization Group, where he is currently a Principal Scientist. His work has been focusing on the reliability aspects of thin insulating layers under electrical stress. His current research interests include the physics of degradation and breakdown phenomena in gate oxide films, the reliability of flash memory devices, the reliability of ultrathin oxide and oxynitride layers for VLSI technologies, and the characterization and the reliability of high-k materials as gate insulators for future CMOS generations and memory applications. Recently he has been working mainly on Resistive RAM memory development and modeling.

Abstract

This tutorial aims at making the link between fundamental understanding of intrinsic and extrinsic material imperfection to final degradation issues of a memory device. Some examples:
1) How does the learning on defect charging, defect-assisted conduction and defect creation result in understanding and predicting the retention behavior of flash memories?
2) How do stress-induced leakage currents affect SRAM characteristics?
3) What material parameter is crucial for a filamentary RRAM retention?
4) How do defects in the blocking oxide of a TANOS affect program and retention?
Using this basics-to-end-characteristic insight, we also demonstrate the scaling properties and make a comparison between different memory concepts - both mature as well as emerging concepts.

Tutorial 5

Session	Tutorial 5
Date	30 Jun 2014 / 01:30 – 05:00 pm
Topic	Static Random Access Memory Failure Analysis
Presenter	**Zhigang Song**, *SRDC, IBM, USA*

Biography

Dr. Zhigang Song received his Bachelor of Science from Fundan University, China, 1988, Master of Engineering from Beijing Institute of Chemical technology, China, 1990, and Ph. D degree from National University of Singapore, 1998. He started failure analysis career in AMD, Singapore as a device analysis engineer in 1997. In 1999, he moved to Chartered Semiconductor Manufacturing Limited, where he had held senior engineer, principal engineer, senior principal engineer and member of technical staff positions in failure analysis lab. In 2007, he joined Semiconductor Research and Development Center (SRDC) of IBM as a senior engineer in PFA Lab. Since then, he had led failure analysis team to support 45nm, 32nm, 28nm and 20nm bulk technology development and also been responsible for failure analysis of logic failures (scan chain and ATPG) for 32nm and 22nm SOI technology development. Currently, he is involved in failure analysis for 14nm SOI FinFET technology development. He has authored or coauthored more than 50 papers published in Journals and Conferences, included ISTFA and IPFA.

Abstract

Static Random Access Memory (SRAM) is often chosen to be the process qualification vehicle during technology development and yield monitor vehicle during manufacturing, thus failure analysis of SRAM accounts for the large part of work in any microelectronic failure analysis Lab. This tutorial will cover 3 parts. The first part will talk about SRAM basics, including what is the SRAM architecture? How does the SRAM operate? How to test the SRAM functionality and what are the SRAM failure types and failure patterns. The second part will describe the SRAM failure analysis methodology and processes, including sample preparation, fault isolation with passive voltage contrast technique, defect imaging technique selection and FIB cross-section. The third part will show various case studies for SRAM bitcell type failure, simple column and row type failure and logic type failure, demonstrating how to use the SRAM basic knowledge and failure analysis methodology to find the root causes. Finally, electrical probing techniques will be discussed with several cases of SRAM soft failures.

Tutorial 6

Session	Tutorial 6
Date	30 Jun 2014 / 01:30 – 05:00 pm
Topic	Thermomechanical Reliability of TSV Structures in 3D Interconnects: A Material and Processing Perspective
Presenter	**Rui Huang**, *University of Texas at Austin, USA*

Biography

Dr. Rui Huang received his Bachelor degree in Theoretical and Applied Mechanics from the University of Science and Technology of China in 1994 and his PhD degree in Civil and Environmental Engineering, with specialty in Mechanics, Materials, and Structures, from Princeton University in 2001. He joined the University of Texas at Austin as an Assistant Professor in September 2002 and was promoted to Associate Professor with tenure in 2008. He currently holds the endowed position of Mrs. Pearlie Dashiell Henderson Centennial Fellowship in Engineering. His research interests include mechanics of integrated materials and structures at micro- and nanoscales, reliability of advanced interconnects and packaging for microelectronics, and mechanical instability of thin films and nanostructures. He has published over 80 journal articles, 40 conference papers, and 3 book chapters.

Abstract

The 3D integration has emerged as a potential solution to overcome the wiring limit imposed on chip performance, power dissipation and packaging form factor beyond the 22nm technology node. In the 3D ICs, the through silicon via (TSV) is a critical element connecting die-to-die in the integrated stack structure. The high aspect ratio and the thermal mismatch between TSV and Si can induce complex stresses to drive interfacial crack growth and TSV protrusion, degrading the performance and reliability of 3D interconnects. This presentation will first analyze the effect of process-induced stresses on the thermomechanical reliability of TSV structures. The analysis will be based on the Griffith criterion by evaluating the crack driving force and the reliability impact on the Cu TSV structure. Recent results from measurements of thermal stress and plasticity characteristics of Cu TSV structures will be reviewed, including wafer curvature, micro-Raman spectroscopy and synchrotron X-ray micro-diffraction techniques. The stress effect on reliability and the mechanism for Cu extrusion in TSV structures will be discussed. We will explore the potential of material and processing optimization to build reliable TSV structures for 3D interconnects.

Program-At-A-Glance

Wednesday, 2 July 2014	
8:45 - 10:15	**Session 1** Opening Ceremony
10:15 - 10:40	TEA BREAK
10:40 - 12:20	**ISTFA 2013 / ESREF 2013 Best Paper**	**Session 2** Die/Package Level Failure Analysis Case Study and Failure Mechanisms I
12:20 - 13:45	**Lunch**
13:45 - 15:10	Special Tribute Session in Memory of Prof Jacob Phang
15:10 - 15:35	**Tea Break**
15:35 - 17:15	**Session 3** Advanced Failure Analysis Techniques I

Thursday, 3 July 2014			
8:30 - 10:35	**Session 4** Die-/Package-Level Failure Analysis Case Study and Failure Mechanisms II		
10:35-11:00	**Tea Break**		
11:00-12:10	**Session 5** Novel Devices, Led Reliability and Failure Mechanisms		
12:10-14:00	**Session 6** Poster Session / Luncheon / "Art Of Fa" Photo Contest		
14:00-15:10	**Session 7** Sample Preparation, Metrology and Material Characterization I		
15:10-15:35	**Tea Break**		
15:35-17:00	**Session 8** Novel Gate Stacks/Dielectrics and Feol Reliability and Failure Mechanisms	**Session 8A** Novel Gate Stacks/Dielectrics and Feol Reliability and Failure Mechanisms	**Session 8B** Die-/Package-Level Failure Analysis Case Study and Failure Mechanisms III
17:00	**Conference Banquet**		

xxiii
Proceedings of the 21th International Symposium on the Physical and Failure Analysis of Integrated Circuits (IPFA)

Friday, 4 July 2014		
08:30 - 09:30	**Session 9A** Sample Preparation, Metrology and Material Characterization II	
09:30 - 10:15	**Session 9B** Die-/Package-Level Failure Analysis Case Study and Failure Mechanisms IV	
10:15 - 10:40	Tea Break	
10:40 - 12:20	**Session 10** Advanced Interconnects and Beol Reliability and Failure Mechanisms	
12:20 - 13:45	Lunch	
13:45 - 15:15	**Session 11A** Product Reliability Evaluation and Approaches	**Session 11B** Die-/Package-Level Failure Analysis Case Study and Failure Mechanisms V
15:15 - 15:35	**Tea Break**	
15:35 - 16:35	**Session 12** Advanced Failure Analysis Techniques II	
16:35 - 17:00	**Closing Ceremony**	

Technical Papers

Session 1	Keynotes
Date/Time	Wednesday, 2 July 2014 / 08:45 – 10:15
Venue	Roselle 4712 & 4713
Chair	**J. M. Chin**, *Advanced Micro Devices, Singapore*

Failure Analysis Techniques for "More than Moore" Semiconductor Technologies N/A
Raj Nair, GLOBALFOUNDRIES, Singapore

Fusion Prognostics-based Qualification of Microelectronic Devices 383
Michael Pecht, University of Maryland, USA

Session	ISTFA 2013 / ESREF 2013 Best Paper
Date/Time	Wednesday, 2 July 2014 / 10:40 – 12:20
Venue	Roselle 4712 & 4713
Chair	**Alan Street**, *Apple, Singapore*

Two - Photon - Absorption - Enhanced Laser - Assisted Device Alteration and Single - Event Upsets in 28nm Silicon Integrated Circuits
K. A. Serrels, K. Erington, D. J. Bodoh, D. T. Reid, C. Farrell, N. Leslie, T. R. Lundquist and P. Vedagarbha

Effective and Reliable Heat Management for Power Devices Exposed to Cyclic Short Overload Pulses
M. Nelhiebel, R. Illing, T. Detzel, S. Wöhlert, B. Auer, S. Lanzerstorfer, M. Rogalli, W. Robl, S. Deckerd, M. Ladurner and J. Fugger

Session 2	Die/Package Level Failure Analysis Case Study and Failure Mechanisms I
Date/Time	Wednesday, 2 July 2014 / 10:40 – 12:20
Venue	Roselle 4712 & 4713
Chair	**Alan Street**, *Apple, Singapore*

Failure Analysis of Off-State Leakage in High-voltage Word-Line Decoder Circuit of Memory Device 1
K. W. Lai, A. S. Teng, C. H. Tu, T. Y. Chang, Julia Hsueh, M.-Y. Lee, Albert Kuo, Y. H. Chao, Scott Hu, U. J. Tzeng and C.-Y. Lu

MOSFET Implant Failure Analysis Using Plane- View Scanning Capacitance Microscopy Coupled with Nano- Probing and TCAD Modeling 5
Hun-Seong Choi, Yong-Woon Han and Il-Sub Chung

Resolving Systematic Voltage Sensitive Soft Failures in 28nm Microprocessor Devices 9
Dnyan Khatri, Soon Huat Lim, Mun Yee Ho, Vinod Narang, Dakshina-Murthy Srikanteswara and Keith Kasprak

A Stain Chemical Etching Rate Study and Used to Detect Implant Defect 13
Miao Wu and Yi Che

Session 3	Advanced Failure Analysis Techniques I
Date/Time	Wednesday, 2 July 2014 / 15:35 – 17:15
Venue	Roselle 4712 & 4713
Chair	**M. K. Radhakrishnan**, *NanoRel, India*

Backside Failure Analysis Techniques: What's the Gain of Silicon Getting Thinner? 17
Christian Boit, Norbert Schäfer, Daniel Abou-Ras, Clemens Helfmeier, Arkadiusz Glowacki and Uwe Kerst

Advanced TEM Applications in Semiconductor Devices 22
A. Y. Du, J. Zhu, Y. K. Zhou, B. H. Liu, Eddie Er, Z. Q. Mo, S. P. Zhao and Jeffrey Lam

Improvement of 3D Current Mapping by Coupling Magnetic Microscopy and X-Ray Computed Tomography 26
Nicolas Courjault, Fulvio Infante, Vincent Bley, Thierry Lebey and Philippe Perdu

Wafer-level Fault Isolation Approach to Debug Integrated Circuits JTAG Failures 30
S. H. Goh, G. F. You, B. L. Yeoh, Hu Hao, N. L. Chung, C. P. Yap and Jeffrey Lam

Advanced Package Circuit Modification by μMilling 35
Christian Hollerith, Bernd Krüger, Gürcan Gezerci, Siegfried Pauthner and Gunnar Zimmerman

Study and Mechanism of Static Scanning Laser Fault Isolation on Embed SRAM Function Fail 39
Changqing Chen, Huipeng Ng, Ghinboon Ang, J. Lam and Zhihong Mai

Session 4	Die-/Package-Level Failure Analysis Case Study and Failure Mechanisms II
Date/Time	Thursday, 3 July 2014 / 8:30 – 10:35
Venue	Roselle 4712 & 4713
Chair	**Zhigang Song**, *IBM, USA*

Material Characterization and Failure Analysis of Through-Silicon Vias 312
Chenglin Wu, Tengfei Jiang, Jay Im, Kenneth M. Liechti, Rui Huang and Paul S. Ho

Failure Analysis for Metal Bridge Defect in Logic Area of Mixed-signal IC 42
Diwei Fan and Winter Wang

Application of Scanning Capacitance Microscopy on SOI Wafer in Die-Level Failure Analysis 46
Seah Pei Hong, Zheng Xin Hua, Chng Kheaw Chung and Aaron Chin

Failure Analysis of Zn-Mo Particle in the Molding Compound Causing Gate-Source Short in Non-Passivated MOSFET Device 50
C. K. Lau and C. H. Tan

Thermal-Electric Activated Ions Diffusion on Printed Circuit Board 54
Lai-Seng Yeoh, Kok-Cheng Chong and Susan Li

Study on the High Via Resistance by TEM Failure Analysis 58
Binghai Liu, Eddie Er , Si Ping Zhao, Changqing Chen, Ang Ghim Boon, Kunihiko Takahashi, Chivukula Subbu and Jeffrey Lam

Failure Analysis of Low-ohmic shorts using Lock-In Thermography 62
Kannu Wadhwa, Rudolf Schlangen, Joy Liao, Tung Ton and Howard Marks

A Case Study on the Defective Contact with Schottky Junction Character 66
Jinglong Li, ChangYan Qi, Yi Che, Quande Zhang, Horse Ma, Jonathon Liu, Motohiko Masuda and Binhai Liu

Session 5	Novel Devices, LED Reliability and Failure Mechanisms
Date/Time	Thursday, 3 July 2014 / 11:00 – 12:10
Venue	Roselle 4712 & 4713
Chair	**Robin Degraeve**, *IMEC, Belgium*

Bias Temperature Instability Investigation of Double-gate FinFETs 70
C. D. Young, A. Neugroschel, K. Majumdar, Z. Wang, K. Matthews and C. Hobbs

Short Localization in a Multi Chip BGA Package 74
Jan Gaudestad, Antonio Orozco, Mark Kimball, Kalon Gopinadhan and Thirumalai Venkatesan

An Improved Coffin-Manson Model for Mid-Power LED Wire-Bonding Reliability 78
Bin Zhang and Guoqiao Tao

Quantitative Analysis for Noise Generated from Share Circuitries within DDR3 DRAM 83
I. Nam, J. Lim, H. Hwang, K. Cho and J. Choi

Session 6	Poster Papers
Date/Time	Thursday, 3 July 2014 / 12:10 – 14:00
Venue	Outside – Roselle 4612 & 4613
Chair	**VINOD Narang**, *Advanced Micro Devices, Singapore*

Microstructural Approach to Failure Analysis of Thin Film Transistors 87
Ju Ho Lee, Sungsoon Choi and Kwan-Hun Lee

Characterization Studies of Fluorine-induced Corrosion Crystal Defects on Microchip Al Bondpads using X-ray Photoelectron Spectroscopy 90
Hua Younan, Xing Zhen Xiang and Li Xiaomin

Characterization of Wet Chemical Etching for Effective Backside Sample Preparation on Devices with Exposed Pads 94
Andrew C. Sabate and Rowin V. Galarce

Library Setup for Epitaxial Layer Dopant Profile using Spreading Resistance Profiling Analysis 98
Lim Saw Sing and Lim Chan Way

Decapsulation of Silver-Alloy Wire-Bonded Devices 102
Francois Kerisit, Matthew J. Lefevre, Bernadette Domenges, Wilfrid Prellier and Michael Obein

High Energy Electron Degradation of the Bonding Connections 106
P. R. Laskowskia, M. Olszackia, M. Al Bahri, P. Pons and M. Jasiorski

An Unique Sample Pretreatment Method for Circuit Edit on WLCSP Devices 110
Sheng-Yu Chen, Chino Wang and Puma Wu

ICP-RIE Platinum (Pt) Sputter Etching 114
R. G. Mendaros and M. T. Marcelo

Spatial Correction in Dynamic Photon Emission by Affine Transformation Matrix Estimation 118
S. Chef, S. Jacquir, P. Perdu, K. Sanchez and S. Binczak

XPS and TEM Studies of Oxidation States on Sn Solder Ball 123
Shen Yiqiang, Chen Yixin, Lee Hwang Sheng, Chow Shue Yin, Xing Zhen Xiang, Hua Younan and Li Xiaomin

Inverted Scan Transducer Mount Technique: A Cost Effective Acoustic Scanning of IGBT Modules for Failure Analysis 127
Em Julius De La Cruz, Sheenel Karl De La Rea, Stephen McDonough and S. F. Chai

ATR-FTIR, DUAL BEAM FIB-SEM, TEM and TOF-SIMS Studies on High Temperature and Moisture Induced "White Haze" Following the Pattern of Electrodes in Touch Panels 131
Chen Yixin, Hao Meng, Shao Jingjing, Lee Esther, Khoo Bing Sheng, Chooi Meailing, Li Kai, Xin Qiuju, Kon Cambridge, Lee Hwang Sheng, Shen Yiqiang, Song Lu, Xing Zhenxiang, Zhou Yongkai, Feng Yang, Fu Chao, Hua Younan and Li Xiaomin

On-Chip Device and Circuit Diagnostics on Advanced Technology Nodes by Nanoprobing 135
M. K. Dawood, T. H. Ng, P. K. Tan, H. Tan, S. James, P. S. Limin, H. H. Yap, J. Lam and Z. H. Mai

Simple, Novel and Low Cost Numerical Aperture Increasing Lens System for High Resolution Infrared Image in Backside Failure Analysis 140
Li Tian

Localized FIB Delayering on Advanced Process Technologies 146
David Donnet, Oleg Sidorov, Pete Carleson, Chad Rue, Roger Alvis and Surendra Madala

An Innovative Method to Overcome Signal Instability during TDR measurement of Power MOSFET 150
S. Y. Tan, K. K. Ng, S. Y. Gan, and C. K. Sin

The In-depth Description for the FA Case with Gate-to-Source or Drain Short by Nanoprobing Analysis 156
LiLung Lai, Oscar Zhang, Ling Zhu, Feng Qian and Mason Sun

A Study of Isolation Test on FullPak Device 160
S. Y. Gan, Lokman Alias and W. Y. Ng

Study on Sensitive Character of Unexpected High Impedance Circuit in VLSI Failure Analysis 165
Gaojie Wen

Case Study of Embedded Memory Failure Analysis for Dislocation Issue 169
Cheng Wei Tang, Shin Chia Lin, Yi Chen Lin, Mei Ying Hsiao, Yau Shan Wu and Chi Lin

Electrical and Physical Analysis of a 28nm FPGA Programmable Delay Circuit Single Tap Delay Failure 173
Chow Yew Meng, Bai Haonan, Grace Tan, Peter F Salinas and Johney Ou Yang

Finding a New Type of In-line Failure Mechanism "Floating Antenna Effect" and its Solution 178
Yutian Zhang, Junzhi Sang ,Yun Xu and Zhimin Zheng

Failure Analysis Methodology for the Localization of Thin and Ultra-Thin Metal Barrier Residue 182
A. C. T. Quah, N. Dayanand, S. P. Neo, G. B. Ang, M. Gunawardana, H. H. Ma, Z. H. Mai and J. Lam

Idss Failure Investigated by SIMS Profiling and TCAD Simulation 186
Lei Zhu, M. B. Bai, X. P. Wang, Y. H. Huang, Kenny Ong, A. B. S. Sumarlina, W. G. Park, Z. Q. Mo, Peck Y. Zheng, S. P. Zhao and Jeffrey Lam

Fault Isolation by Conquering Obstruction Effect of Resistor in Complex Cases Analysis 191
Gaojie Wen

Analysis of Insertion Force of Electric Connector based on FEM 195
Ying Li, Fulong Zhu, Yanming Chen, Ke Duan, Kai Tang and Sheng Liu

Study of RFID Tag Reliability Issue Under Abnormal Temperature Ambient N/A
Po-Ying Chen, Wen-Kuan Yeh, Yukan Chang, Chi-Chang Chen, Chwei-Hsiung Tsai, Yu-Jung Huang, Wei-Cheng Lin, Shao-I Chu, Chung-Long Pan and Shyh-Chang Liu

Observation of Long Term Potentiation in Papain-based Memory Devices 199
A. Bag, M. K. Hota, S. Mallik and C. K. Maiti

Bipolar Resistive Switching in Different Plant and Animal Proteins 203
A. Bag, M. K. Hota, S. Mallik and C. K. Maiti

Reliability Prediction and Real World for LED Lamps 207
G. Mura and M. Vanzi

Hot Carrier Injection on Back Biasing Double-Gate FinFET with 10 and 25-nm Fin Width 211
Wen-Teng Chang, Li-Gong Cin, Wen-Kuan Yeh and Po-Ying Chen

Junction induced Variation and Reliability for Ultra-Thin-Body and Bulk Oxide MOSFETs 215
Wen-Kuan Yeh, Wen-Teng Chang, Po-Ying Chen and Cheng-Li Lin

Palladium-Copper Inter-diffusion during Copper Activation for Electroless Nickel Plating Process on Copper Power Metal 219
Poo Khai Yee, Wan Tatt Wai and Yong Foo Khong

Novel Technique for Deep Vertical Interconnect Access Fault Isolation 268
T. P. Chua, C. H. Chong and K. N. Liew

The Observation of Mobile Ion of 40nm node by Triangular Voltage Sweep 332
Clement Huang, James W. Liang, Alex Juan and K. C. Su

Session 7	**Sample Preparation, Metrology and Material Characterization I**
Date/Time	Thursday, 3 July 2014 / 14:00 – 15:10
Venue	Roselle 4712 & 4713
Chair	**Philippe Perdu**, *CNES, France*

SRAM Failure Analysis Evolution Driven by Technology Scaling 223
Song Zhigang

Fast and Easy Sample Preparation with Reduced Curtaining Artifacts using a P-FIB 231
S. Moreau, D. Bouchu and G. Audoit

Novel Inverted Sample Thinning Method by Ex-situ Lift-out 236
Liew Kaeng Nan

FIB-SEM Investigation and Auto-metrology of Polymer-Microlens/CFA Arrays of CMOS Image Sensor 240
Pradeep Sharma, Tai Shan Chiu, Sajal Biring, Te-Fu Chang, Chih-Hsun Chu and Yong-Fen Hsieh

Session 8	**Novel Gate Stacks/Dielectrics and Feol Reliability and Failure Mechanisms**
Date/Time	Thursday, 3 July 2014 / 15:35 – 16:00
Venue	Roselle 4712 & 4713
Chair	**Chadwin D. Young**, *University of Texas at Dallas, USA*

Hourglass Concept for RRAM: A Dynamic and Statistical Device Model 245
R. Degraeve, A. Fantini, N. Raghavan, L. Goux, S. Clima, Y. Y. Chen, A. Belmonte, S. Cosemans, B. Govoreanu, D. J. Wouters, Ph. Roussel, G. S. Kar, G. Groeseneken and M. Jurczak

Session 8A   **[Parallel] Novel Gate Stacks/Dielectrics and Feol Reliability and Failure Mechanisms**

Date/Time   Thursday, 3 July 2014 / 16:00 – 17:00

Venue   Roselle 4712 & 4713

Chair   **Chadwin D. Young**, *University of Texas at Dallas, USA*

Study of (Correlated) Trap Sites in SILC, BTI and RTN in SiON and HKMG Devices 250
Erik Bury, Robin Degraeve, Moon Ju Cho, Ben Kaczer, Wolfgang Goes, Tibor Grasser, Naoto Horiguchi and Guido Groeseneken

Transient to Temporarily Permanent and Permanent Hole Trapping Transformation in the Small Area SiON p-MOSFET Subjected to Negative-Bias Temperature Stress 254
Z. Y. Tung and D. S. Ang

Evidence for Defect Pairs in SiON pMOSFETs 258
T. Grasser, K. Rott, H. Reisinger, M. Waltl and W. Goes

Understanding of Self-Heating Enhanced Degradation in pLDMOSFETs by MR-DCIV Method 264
Yandong He, Ganggang Zhang and Xing Zhang

Session 8B   **[Parallel] Die-/Package-Level Failure Analysis Case Study and Failure Mechanisms Iii**

Date/Time   Thursday, 3 July 2014 / 16:00 – 17:00

Venue   Roselle 4711

Chair   **Rui Huang**, *University of Texas at Austin, USA*

Back-End Defect Localization for 28nm FPGA 271
Jack Ng Yi Jie, Liew Chiun Ning and Khoo Khai Ling

40nm NAND Flash Reliability Failure Analysis with Identification Tools Combination 274
Mei Ying Hsiao, Yi Heng Chen and Ling Kuey Yang

Detailed Package Failure Analysis on Short Failures after High Temperature Storage 278
Z. Y. Oh, F. J. Foo and W. Qiu

Session 9A   **Sample Preparation, Metrology and Material Characterization II**

Date/Time   Friday, 4 July 2014 / 08:30 – 09:30

Venue   Roselle 4712 & 4713

Chair   **Jeff Gambino** , *IBM, USA*

Structure and Composition of the Cu/Low k Interconnects de-Layered with FIB 283
Dandan Wang, Pik Kee Tan, Yamin Huang, Jeffrey Lam and Zhihong Mai

Identification of Cu-Al Intermetallic Phases of Copper Wire Bonding Using TEM Nano Beam Diffraction Indexing Technique 287
Foo Khong Yong

- Experiments and Results of Raman and FTIR Complementary Vibrational Spectroscopy for IC Reliability Failure Analysis 291
  Huang Yamin, Hao Tan, Dandan Wang, Jeffrey Lam and Zhihong Mai
- 3D EBSD Characterizations on Copper TSV for 3D Interconnections 295
  W. N. Putra, A. D. Trigg, H. Y. Li and C. L. Gan

Session 9B	**Die-/Package-Level Failure Analysis Case Study and Failure Mechanisms IV**
Date/Time	Friday, 4 July 2014 / 09:30 – 10:15
Venue	Roselle 4712 & 4713
Chair	**Jeff Gambino** *IBM, USA*

- Case Study of Wet Chemical Stain to Identify Implant Related Low Yield Issue 300
  Yi- Cheng Lin and Sheng-Min Chen
- Comprehensive Study and Corresponding Improvements on the ESD Robustness of different nLDMOS Devices 304
  Yuan Wang, Guangyi Lu, Lizhong Zhang, Jian Cao, Song Jia and Xing Zhang
- Systematic Methods to Identify and Verify Non-Visible Defects in Silicon Substrate 308
  Hongwei Huang, Winnie Wei, JJ Xin, Candy Liu, Luke Wu, Clieve Dai, Pinglung Liao and Wei Xu

Session 10	**Advanced Interconnects and Beol Reliability and Failure Mechanisms**
Date/Time	Friday, 4 July 2014 / 10:40 – 12:20 hrs
Venue	Roselle 4712 & 4713
Chair	**Chee Lip Gan** *Nanyang Technological University, Singapore*

- Electromigration Reliability of Open TSV Structures 317
  Wolfhard H. Zisser, Hajdin Ceric, Josef Weinbub and Siegfried Selberherr
- Effects of Sidewall Scallops on the Performance and Reliability of Filled Copper and Open Tungsten TSVs 321
  Lado Filipovic, Roberto Lacerda de Orio and Siegfried Selberherr
- Imaging of Through-Silicon Vias using X-Ray Computed Tomography 327
  J. P. Gambino, W. Bowe, D. M. Bronson, S. A. Adderly
- Electromigration Reliability of Solder Bumps 336
  Hajdin Ceric and Siegfried Selberherr

Session 11A	Product Reliability Evaluation and Approaches
Date/Time	Friday, 4 July 2014/ 13:45 – 14:15 hrs
Venue	Roselle 4712 & 4713
Chair	**Mike Bruce***IBM, USA*

★ New Technique for Acquiring Dead Pixel Free and Fine Inspection Image of Advanced LSI Package with Rough Surface Using New Technique for Acquiring Dead Pixel Free and Fine Inspection Image of Advanced LSI Package with Rough Surface Using Scanning Acoustic Tomograph 340
Kaoru Kitami, Masakatsu Murai, Natsuki Sugaya, Osamu Kikuchi and Shigeru Ohno

★ Reliability Analysis from Field Data and Prediction Models for Customer Risk Assessments: Case Studies and Strategy 344
Corinne Bergès, Yves Chandon and Pierre Soufflet

Session 11B	Die-/Package-Level Failure Analysis Case Study and Failure Mechanisms V
Date/Time	Friday, 4 July 2014 / 14:15 – 15:15 hrs
Venue	Roselle 4712 & 4713
Chair	**Mike Bruce** *IBM, USA*

★ Defect Localization by Lock-in IR-OBIRCH on Some Recovered Cases 350
Chunlei Wu, , Grace Song and Suying Yao

★ Single Bit Cell SRAM Failure Due to Titanium Particle 354
Rachel Siew and WF Kho

★ Gate Oxide Rupture Localization by Photon Emission Microscopy with the Combination of Lock-in IR-OBIRCH 358
Chunlei Wu and Suying Yao

★ Non-Destructive Techniques for Internal Solder Bump Inspection of Chip Scale Package-Ball Grid Array Package 362
Jason H. Lagar, Rudolf A. Sia and Marlyn C. Grancapal

Session 12	Advanced Failure Analysis Techniques Ii
Date/Time	Friday, 4 July 2014 / 15:35 – 16:35
Venue	Roselle 4712 & 4713
Chair	**Christian Boit**, *Technical University of Berlin, Germany*

★ One Die Logic Analysis through the Backside 366
M. R. Bruce, L. K. Ross and C. M. Chua

★ Temperature Effect on Reflected Laser Probing Signal of Multiple Elementary Substructures 370
M. M. Rebaï, F. Darracq, J-P. Guillet, D. Lewis, P. Perdu and K. Sanchez

★ Defect Localization Enhancement using Light Induced CI-AFP 375
N. Dayanand, A. C. T. Quah, C. Q. Chen, S. P. Neo, G. B. Ang, M. Gunawardana, Z. H. Mai and J. C. Lam

★ Cluster Matching in Time Resolved Imaging for VLSI analysis 379
S. Chef, S. Jacquir, P. Perdu, K. Sanchez and S. Binczak

Author Index

A

Abou-Ras, Daniel
Adderly, S. A.
Alias, Lokman
Alvis, Roger
Ang, D. S.
Ang, G. B.
Ang, Ghinboon
Audoit, G.

B

Bag, A.
Bahri, M. Al
Bai, M. B.
Belmonte, A.
Bergès, Corinne
Binczak, S.
Biring, Sajal
Bley, Vincent
Boit, Christian
Boon, Ang Ghim
Bouchu, D.
Bowe, W.
Bronson, D. M.
Bruce, M. R.
Bury, Erik

C

Cambridge, Kon
Cao, Jian
Carleson, Pete
Ceric, Hajdin
Chai, S. F.
Chandon, Yves
Chang, T. Y.
Chang, Te-Fu
Chang, Wen-Teng
Chang, Yukan
Chao, Fu
Chao, Y. H.
Chauhan, Preeti
Che, Yi
Chef, S.
Chen, C. Q.
Chen, Changqing
Chen, Chi-Chang
Chen, Po-Ying
Chen, Sheng-Min

Liu, Binghai
Liu, Binhai
Liu, Candy
Liu, Jonathon
Liu, Sheng
Liu, Shyh-Chang
Lu, C. -Y.
Lu, Guangyi
Lu, Song

M

Ma, H. H.
Ma, Horse
Madala, Surendra
Mai, Z. H.
Mai, Zhihong
Maiti, C. K.
Majumdar, K.
Mallik, S.
Marcelo, M. T.
Marks, Howard
Masuda, Motohiko
Matthews, K.
McDonough, Stephen
Meailing, Chooi
Mendaros, R. G.
Meng, Chow Yew
Meng, Hao
Mo, Z. Q.
Moreau, S.
Mura, G.
Murai, Masakatsu

N

Nam, I.
Nan, Liew Kaeng
Narang, Vinod
Neo, S. P.
Neugroschel, A.
Ng, Huipeng
Ng, Jack Yi Jie
Ng, K. K.
Ng, T. H.
Ng, W. Y.
Ning, Liew Chiun

Chen, Sheng-Yu
Chen, Y. Y.
Chen, Yanming
Chen, Yi Heng
Chin, Aaron
Chiu, Tai Shan
Cho, K.
Cho, Moon Ju
Choi, Hun-Seong
Choi, J.
Choi, SungSoon
Chong, C. H.
Chong, Kok-Cheng
Chu, Chih-Hsun
Chu, Shao-I
Chua, C. M.
Chua, T. P.
Chung, Chng Kheaw
Chung, Il-Sub
Chung, N. L.
Cin, Li-Gong
Clima, S.
Cosemans, S.
Courjault, Nicolas
Cruz, Em Julius De La

D

Dai, Clieve
Darracq, F.
Dawood, M. K.
Dayanand, N.
Degraeve, R.
Degraeve, Robin
Domenges, Bernadette
Donnet, David
Du, A. Y.
Duan, Ke

E

Er, Eddie
Esther, Lee

F

Fan, Diwei
Fantini, A.
Filipovic, Lado
Foo, F. J.

O

Obein, Michael
Oh, Z. Y.
Ohno, Shigeru
Olszacki, M.
Ong, Kenny
Orio, Roberto Lacerda de
Orozco, Antonio

P

Pan, Chung-Long
Park, W. G.
Pauthner, Siegfried
Pecht, Michael
Perdu, P.
Pons, P.
Prellier, Wilfrid
Putra, W. N.

Q

Qi, ChangYan
Qian, Feng
Qiu, W.
Qiuju, Xin
Quah, A. C. T.

R

Raghavan, N.
Rea, Sheenel Karl De La
Rebaï, M. M.
Reisinger, H.
Ross, L. K.
Rott, K.
Roussel, Ph.
Rue, Chad

S

Sabate, Andrew C.
Salinas, Peter F
Sanchez, K.
Sang, Junzhi
Schäfer, Norbert
Schlangen, Rudolf
Selberherr, Siegfried
Sharma, Pradeep
Sheng, Khoo Bing
Sheng, Lee Hwang
Sia, Rudolf A.
Sidorov, Oleg

G

Galarce, Rowin V.
Gambino, J. P.
Gan, C. L.
Gan, S. Y.
Gaudestad, Jan
George, Elviz
Gezerci, Gürcan
Glowacki, Arkadiusz
Goes, W.
Goes, Wolfgang
Goh, S. H.
Gopinadhan, Kalon
Goux, L.
Govoreanu, B.
Grancapal, Marlyn C.
Grasser, T.
Grasser, Tibor
Groeseneken, G.
Guillet, J-P.
Gunawardana, M.

H

Han, Yong-Woon
Hao, Hu
Haonan, Bai
He, Yandong
Helfmeier, Clemens
Ho, Mun Yee
Ho, Paul S.
Hobbs, C.
Hollerith, Christian
Hong, Seah Pei
Horiguchi, Naoto
Hota, M. K.
Hsiao, Mei Ying
Hsieh, Yong-Fen
Hsueh, Julia
Hu, Scott
Hua, Zheng Xin
Huang, Clement
Huang, Hongwei
Huang, Maggie Yamin
Huang, Rui
Huang, Y. H.
Huang, Yu-Jung
Hwang, H.

Siew, Rachel
Sin, C. K.
Sing, Lim Saw
Song, Grace
Song, Zhigang
Soufflet, Pierre
Srikanteswara, Dakshina-Murthy
Su, K. C.
Subbu, Chivukula
Sugaya, Natsuki
Sumarlina, A. B. S.
Sun, Mason

T

Takahashi, Kunihiko
Tan, C. H.
Tan, Grace
Tan, H.
Tan, Hao
Tan, P. K.
Tan, S. Y.
Tang, Cheng Wei
Tang, Kai
Tao, Guoqiao
Teng, A. S.
Tian, Li
Ton, Tung
Trigg, A. D.
Tsai, Chwei-Hsiung
Tu, C. H.
Tung, Z. Y.
Tzeng, U. J.

V

Vanzi, M.
Vasan, Arvind
Venkatesan, Thirumalai

W

Wadhwa, Kannu
Wai, Wan Tatt
Waltl, M.
Wang, Chino
Wang, Dandan
Wang, Winter
Wang, X. P.
Wang, Yuan
Wang, Z.
Way, Lim Chan
Wei, Winnie
Weinbub, Josef

I

Im, Jay
Infante, Fulvio

J

Jacquir, S.
James, S.
Jasiorski, M.
Jia, Song
Jiang, Tengfei
Jingjing, Shao
Juan, Alex
Jurczak, M.

K

Kaczer, Ben
Kai, Li
Kar, G. S.
Kasprak, Keith
Kerisit, Francois
Kerst, Uwe
Khatri, Dnyan
Kho, W. F.
Khong, Yong Foo
Kikuchi, Osamu
Kimball, Mark
Kitami, Kaoru
Krüger, Bernd
Kuo, Albert

L

Lagar, Jason H.
Lai, K. W.
Lai, LiLung
Lam, J.
Lam, J. C.
Lam, Jeffrey
Laskowski, P. R.
Lau, C. K.
Lebey, Thierry
Lee, Ju Ho
Lee, Kwan Hun
Lee, M. -Y.
Lefevre, Matthew J.
Lewis, D.
Li, H. Y.
Li, Jinglong
Li, Susan
Li, Ying
Liang, James W.

Wen, Gaojie
Wouters, D. J.
Wu, Chenglin
Wu, Chunlei
Wu, Luke
Wu, Miao
Wu, Puma
Wu, Yau Shan

X

Xiang, Xing Zhen
Xiaomin, Li
Xin, J. J.
Xu, Wei
Xu, Yun

Y

Yamin, Huang
Yang, Feng
Yang, Johney Ou
Yang, Ling Kuey
Yao, Suying
Yap, C. P.
Yap, H. H.
Yee, Poo Khai
Yeh, Wen-Kuan
Yeoh, B. L.
Yeoh, Lai-Seng
Yin, Chow Shue
Yiqiang, Shen
Yixin, Chen
Yong, F. K.
Yongkai, Zhou
You, G. F.
Younan, Hua
Young, C. D.

Z

Zeng, Zhimin
Zhang, Bin
Zhang, Ganggang
Zhang, Lizhong
Zhang, Oscar
Zhang, Quande
Zhang, Xing
Zhang, Yutian
Zhao, S. P.
Zheng, Peck Y.
Zhenxiang, Xing
Zhou, Y. K.
Zhu, Fulong

Liao, Joy
Liao, Pinglung
Liechti, Kenneth M.
Liew, K. N.
Lim, J.
Lim, Soon Huat
Limin, P. S.
Lin, Cheng-Li
Lin, Chi
Lin, Shin Chia
Lin, Wei-Cheng
Lin, Yi Chen
Lin, Yi- Chen
Ling, Khoo Khai
Liu, B. H.

Zhu, J.
Zhu, Lei
Zhu, Ling
Zimmermann, Gunnar
Zisser, Wolfhard H.

Exhibitors

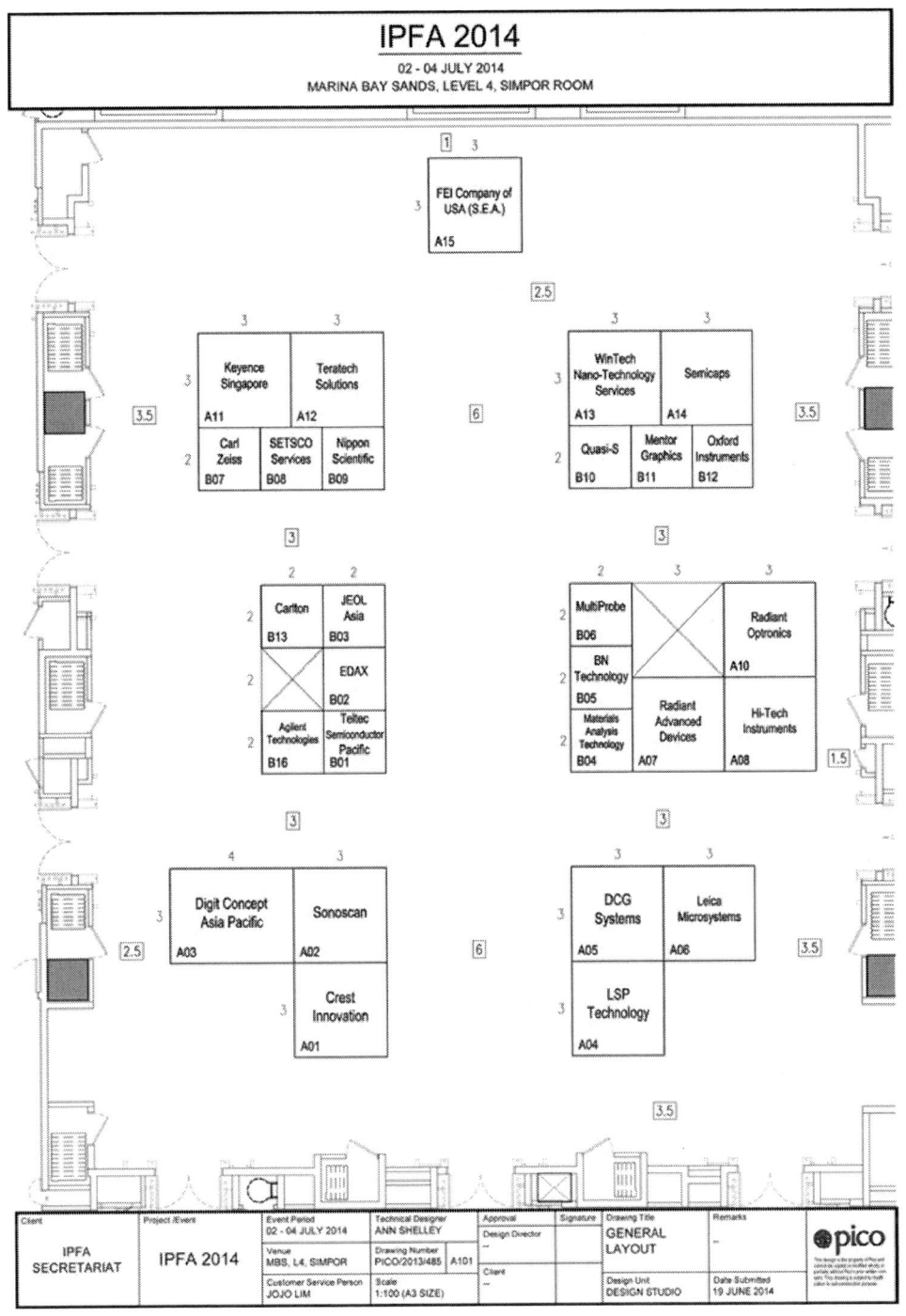

Sponsors

PLATINUM SPONSOR

Booth No: A15
FEI COMPANY OF USA (S.E.A.) PTE LTD
1 Jalan Kilang Timor, #04-02 Pacific Tech Centre
Singapore 159303
Tel: (65) 6272 0050; Fax: (65) 6272 0034
Website: www.fei.com
Contact: Mr. Paul Lawrence
Business Development Manager (Electronics Business Unit - Asia)
Email: Paul.Lawrence@fei.com

FEI's industry leading workflows deliver fast, accurate answers for accelerating IC design and production decisions. In the lab or in the fab, integrated imaging and analysis provide superior images, rich feature sets, cross-sectional metrology, and automation to speed process defect identification, enable root-cause analysis, reduce yield loss, and accelerate time-to-market for new products. Our experts engage with your applications, engineering, and manufacturing teams to address today's challenges, while our leadership and significant R&D commitment are paving the way to 10-nanometer technologies and beyond. FEI is the brand trusted by all the major semiconductor manufacturers to provide imaging and metrology tools for their next-generation computing and storage products.

SILVER SPONSORS

Leica Microsystems is a world leader in microscopes and scientific instruments for life science, industry, and medical applications. Founded as a family business in the nineteenth century, the company's history was marked by unparalleled innovation through its historically close cooperation with the scientific community. Leica Microsystems' tradition of innovation draws on users' ideas and creates solutions tailored to their requirements. The company is represented in over 100 countries with 6 manufacturing facilities in 5 countries, sales and service organizations in 20 countries, and an international network of dealers. The company acts globally and is headquartered in Wetzlar, Germany.

About Singapore

Singapore, a dynamic city rich in contrast and color, is where you'll find a harmonious blend of culture, cuisine, arts and architecture. Singapore has grown into a thriving centre of commerce and industry. Located in the heart of fascinating Southeast Asia, Singapore is the busiest port in the world with over 600 shipping lines. Brimming with unbridled energy and bursting with exciting events, the city offers countless unique, memorable experiences waiting to be discovered.

Singapore's location on the major sea route between India and China. The strategic position has made it grow into an excellent harbor for trade and tourism. Its geographical location is 136.8 km north of the equator, between latitudes 103°38'E and 104°06'E.

Singapore's Central Business District actually spreads across both the central and southern parts of the island (you'll know when you're there - it boasts striking high-rise structures). You can get a good visual orientation to the city as you cross the Benjamin Sheares Bridge on the East Coast Parkway, which links the airport to the city center.

Four Reasons to visit Singapore

1. Latest attractions - Be WOW by the World's Biggest Trees

Visit the state-of-the-art Gardens by the Bay, located in the heart of Marina Bay that will define Singapore as the world's premier tropical garden. It's a mega showcase, where visitors can experience and enjoy flora across the board from Mediterranean, Tropical Montane and temperate annual plants and flowering species. Be WOW by the world's biggest trees, 'Supertrees', ranging from 25 to 50 meters, made of steel structures with several functions; as unique, vertical tropical gardens; as the engine room for the conservatory environmental systems; as receptacles to collect rainwater, to provide shade, to be lit up at night and to house an exclusive bar or F&B outlet. Some of the supertrees will also be connected by aerial walkways.

2. Resorts World Sentosa

Resorts World Sentosa is the first ever integrated resort on Sentosa island, bringing with it highly anticipated and exciting new attractions such as Universal Studios Singapore®, FestiveWalk™ and, Voyage de la Vie™. Live The Movies™ at the many exclusive attractions found only at the region's first Hollywood movie theme park, Universal Studios Singapore®. Featuring movie-themed rides and attractions, which are unique to Singapore, thrill seekers and families will be the first to experience many new rides and shows based on blockbuster hits. Feast your senses on a world of non-stop entertainment, signature shopping experiences and culinary adventures at FestiveWalk™, and take pleasure in the comfort and luxury of the resort's unique world-class hotels. Brace yourselves for Voyage de la Vie™, an original rock circus spectacular comprising of world-renowned circus stars. Playing at the Festive Grand™ theatre, this engaging spectacle is told through pulsating music, fabulous costumes, awe-inspiring sets and death-defying stunts. Created by an international team, Voyage de la Vie™ will be an entertainment experience like no other. With so much more in store, your adventure at Resorts World Sentosa is only limited by your imagination. This world-class integrated resort is the ideal resort destination where everyone can come together for moving experiences and lasting memories. Come discover a million truly rewarding moments, all in one world.

3. Savour Singapore's Melting Pot of Flavours - Join us in Singapore's favourite pastime

Singapore offers an incredible variety of colourful cuisine and scrumptious dinning choices. Whether in hawker centres, open-air food centres or fine-dining restaurants, you will find an eatery at everycorner that suits your taste-bud. Try our local delights such as Chicken Rice, Laksa, Chilli Crab and Satay!

4. Indulge in Singapore's Shopping Paradise

With the second most favourite pastime next to eating, Singapore is passionate about shopping and is fondly called the paradise of shopping. Many tourists come here for the sole purpose of shopping. This small island country has around 250 shopping malls that offer excellent shopping places which will leave you with an enriching buying experience at the end of the day. For more information please click to visit Your Singapore — Singapore Tourism Board's official tourism website.

Key Information at a Glance

Climate

Singapore is known for its hot and humid weather, with little variation throughout the year. The average daytime temperature is 31°C (88°F), dropping to around 24°C (75°F) in the evenings.

Population

Singapore's total population was 5.31 million as at end June 2012. There were 3.82 million Singapore residents, comprising 3.29 million Singapore citizens and 0.53 million permanent residents, and 1.49 million non-residents. In 2012, the Chinese formed the majority at 74.2 per cent of the resident population, followed by the Malays at 13.3 per cent and the Indians at 9.2 per cent. The minority Caucasians, Eurasians and Asians of different origins formed the remaining 3.3 per cent.

Language

English is the main working language in Singapore. Other official languages used are Mandarin, Malay and Tamil.

Getting Around

Singapore's public transport system is well-developed. The network of Mass Rapid Transit (MRT) trains, buses and taxis serves to shuttle its population across the city state every day, at relatively inexpensive fares;

Getting around Singapore is a breeze with the MRT. The four main lines move commuters along the North-South, East-West, North East Lines, and around the city fringe through the Circle Line.

SBS Transit runs 250 bus services with a fleet of more than 3,000 buses while SMRT buses operate a fleet of 1,050 buses. In total, both bus companies move about 3 million people to their destinations daily in clean, air conditioned comfort. For bus routes and timetables, go to the SMRT or SBS Transit websites.

If you prefer to take the taxi, just flag one down by the road (or at any taxi-stand if you are in the Central Business District), call the common taxi booking line at 6-3425-222, or tap out the alphabetical prompter 6-DIAL-CAB. 10 different taxi operators run over 26,000 air-conditioned taxis in Singapore and all are dedicated to providing smooth, comfortable rides for commuters. Cab fares start at SGD3 at flag down, and on average rise by SGD0.20 after travelling every 400m or so - See more at: http://app.singapore.sg/society/transportation#sthash.WslkRxrL.dpuf

Currency

The currency used in Singapore is the Singapore dollar (S$). Money changing services can be found not only at the Singapore Changi Airport but also most shopping centres and hotels around the island. You can also access the automated teller machines (ATMs) located everywhere in Singapore, that accept most of the main credit cards such as Visa, Master Card and American Express.

Electricity

Singapore uses the "Type G" (British 3-pin rectangular blade) electrical plug. Voltage is 230V, 50Hz.

Shopping

Brace yourself for a world-class shopping experience. There are stretches of shopping malls, centres and outlet stores in Singapore that offer you a wide array of shopping choices. You only need to take a walk down Singapore's iconic shopping districts to find out what's in store for you. One thing's for sure, you won't be leaving empty-handed, but with bag loads of great bargains and gifts.

For the latest in fashion trends, check out Orchard Road which is Singapore's main shopping district. But if you're looking for more cultural shopping, then the districts of Kampong Glam, Little India and Chinatown are most ideal. Here, you'll find an assortment of ethnic products, jewellery, textiles, antiques and more.

Most shops and shopping centres open at 1000hrs and close between 2100hrs to 2200hrs, seven days a week.

Tax Refund

A 7% Goods and Services Tax (GST) is levied on most goods and services. Tax can be refunded if you spend a minimum amount of SGD300. Receipts of SGD100 or more from different shops participating in the tax refund scheme can be pooled to meet the minimum requirement. Ask for tax free shopping cheques to be completed by the shop and present the goods to customs at the airport who will then stamp your shopping cheques. Refunds are either by cheque through the mail, deposited into your credit card account or cashed at the Global Refund Counters at the airport before your departure.

Cell Phone Usage

Singapore's international dialing code is +(65). While in Singapore and if you have international roaming service on your cell phone, you don't have to press +(65) as it will automatically connect you to the local numbers here.

Orientation

Welcome to the IPFA 2014 on Web. This proceedings is best viewed with Internet Explorer 7.0 or later, Firefox 8.0 or later, Safari 5.1 or later in Microsoft® Windows™, Linux and MacOS. If you are using Firefox, you are required to use the adobe acrobat or reader to view pdf files instead of the built-in PDF viewer (To change settings in Firefox, go to Tools > Options > Applications & define the default action for the Portable Document Format (PDF)).

Getting Started

The end-user needs to install Adobe® Reader to view the pdf files. If you do not have acrobat reader in your system or finding difficulties to access the PDF, you may wish to obtain from http://www.adobe.com/.

For technical support and enquiries, please contact:

Singapore:
Research Publishing Services
Blk 12 Lorong Bakar Batu,
#02-11, Singapore 348745.
t:+65-6492 1137
f:+65-6748 2556
e: enquiries@rpsonline.com.sg

Failure Analysis of Off-state Leakage in High-voltage Word-line Decoder Circuit of Memory Device

K. W. Lai, A. S. Teng, C. H. Tu, T. Y. Chang, Julia Hsueh, M.-Y. Lee, Albert Kuo, Y. H. Chao, Scott Hu, U. J. Tzeng, C.-Y. Lu

Macronix International Co., Ltd.
No. 16, Li-Hsin Road, Science-Based Industrial Park, Hsin-chu city 300, Taiwan
Email: kuoweilai@mxic.com.tw

Abstract-After 500-hour HTOL reliability test on memory device, off-state leakage was found in nMOSFETs of word-line decoder. According to electrical and physical failure analysis on IC and device level, we found that holes were trapped in SiN of STI edge, which lowered threshold voltage of nMOSFETs and lead to off-state leakage from drain to source. The hole-traps came from anode gate low current flow and were trapped in SiN layer. The hole-traps could be annihilated by electron beams from SEM, and this phenomenon might mislead judgment during failure analysis. Detailed failure analysis, failure mechanism and corresponding improvement of circuit and process are presented in this paper.

I. INTRODUCTION

As the dimension of memory cell scales down, higher cell threshold voltage (Vt) was required to suppress cell punch-through [1]. Besides, to distinct each level of multi-level cell (MLC) memory, the cell was programmed even much higher. For these reasons, the operational bias must increase to read each level properly. Therefore high voltage stress on accessing circuit, like charge pumping and word-line (WL) decoder, becomes an important topic on reliability.

In this paper, we investigated in single WL read failure after long-term (500 hours) high temperature operation life (HTOL) test. Electrical failure analysis (EFA) showed that abnormal leakage was on WL decoder (or X-decoder) path and caused read failure. Physical failure analysis (PFA) discovered that the leakage came from off-state current of nMOSFET in X-decoder circuit, but it disappeared after exposing to scanning electron microscope (SEM) during nano-probing measurement. Further device level analysis revealed that there were hole trapped in the Silicon-Nitride (SiN) liner of shallow trench isolation (STI) edge after long-term high gate bias stress on nMOSFET (Fig. 1a) at high temperature. The hole-traps induced leakage of off-state nMSOFET in selected WL decoder circuit (Fig. 1b) and caused read failure. Further research showed that the holes trapped in SiN liner came from anode gate low current flowing through STI edge [2]. The low current came from high operational bias applied on anode gate to substrate across thick STI. The hole-traps mechanism explained the recovery phenomena observed after high temperature baking and SEM nano-probing. Finally, the corresponding circuit and process modification was proposed to reduce voltage stress and prevent hole traps generation respectively.

II. FAILURE ANALYSIS

A. Electrical Failure Analysis

After 500-hour HTOL reliability test on MLC memory IC, a number of single WL read failures and failed bits were located only in one of physical banks (Fig. 2a). Each bank had its own WL decoder (Fig. 3), and the WL decoder was used for accessing the operational bias to selected WL. The Vt-distribution (Fig. 2b) of memory cells for failed WL address showed the abnormal tails comparing to normal one. With external voltage accessing WL path directly, an unexpected leakage of microampere for failed WL address was measured. The failure symptoms indicated that less voltage was applied on failed WL, because of the leakage pulling down voltage value in the WL path. Therefore the abnormal Vt-distribution appeared higher than normal one.

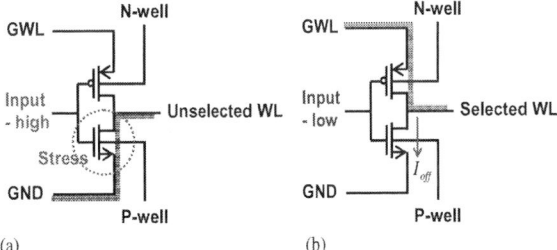

Figure 1. (a) Unselected state: in input-high state, nMOSFET suffers high gate bias stress. (b) Selected state: in input-high state, WL voltage pass from GWL. Abnormal *Ioff* occurred for the failed chip.

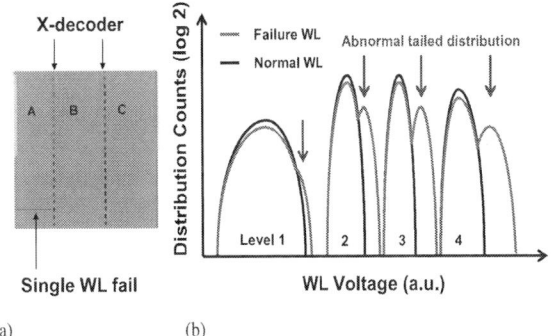

Figure 2. (a) The failed bitmap. Failed bits were most located in one physical bank for a single WL. (b) The schematic illustration of *Vt*-distribution for one WL and the abnormal tailed distribution from bank A.

978-1-4799-3911-4/14 $31.00 © 2014 IEEE

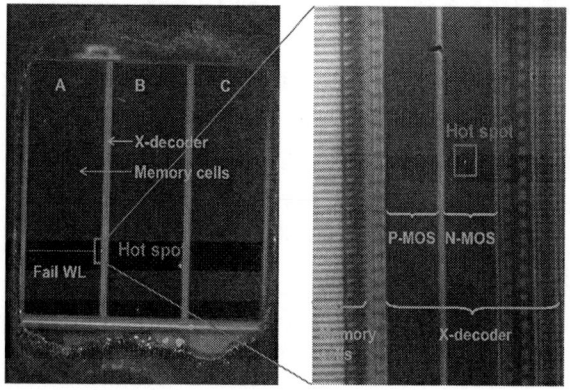

Figure 3. The hot spot was located on the nMOSFET in WL decoder (or X-decoder).

B. Physical Failure Analysis

To find out the leakage path, photoemission electron microscope (PEM) was preformed and showed hot spot on nMOSFETs in WL decoder while failed WL was applied high bias 10V (Fig. 3). Leakage might come from three parts of the circuit, which made hot spot appear at the same nMOSFETs decoder in direct or indirect way (Fig. 4). There were three possibilities to this hot spot: (1) two WLs short in memory array, (2) two metal-lines short in WL decoder and (3) leakage from MOSFET in WL decoder. Using focused ion beam (FIB) to separate WL decoder from memory array (Fig. 5a), the same WL leakage was still measured. Hence WLs in memory array were fine and the leakage came from WL decoder circuit. To check possibility of two metal-lines short, the failed chip was backside de-processed to contact layer by polish and chemical etch [3] (Fig. 5b), but no leakage was shown by nano-probing. This result showed backend interconnect was fine. Thus, the leakage was attributed to the frontend MOSFET in WL decoder. After FIB cutting off nMOSFET source side metal line (Fig. 5a), the leakage through nMOSFET was further identified because no leakage was measured. In summary, we presumed that the leakage was from nMOSFET itself rather than WL or metal line bridge, and hot spot just showed it directly.

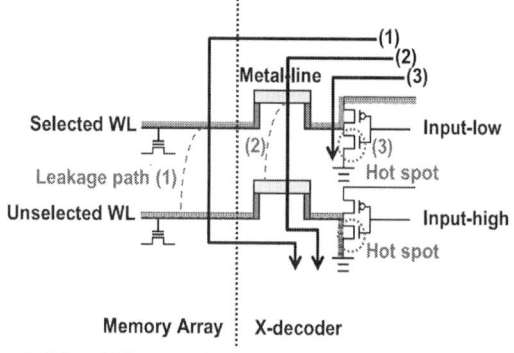

Figure 4. Selected WL was applied high bias 10V, and unselected one was 0V. There are three possible leak path (1), (2), and (3), each path could result in voltage drop at nMOSFET. Because PEM hot spot resolution is smaller than our condensed WL decoder, we can not distinguish two adjacent nMOSFETs by one hot spot.

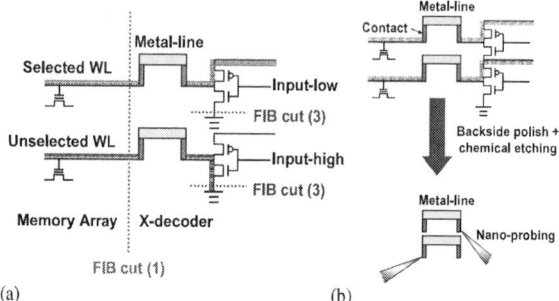

Figure 5. (a) FIB cut (1) to separate WL decoder from memory array, and FIB cut (3) cut off nMOSFET source side metal line. These two adjacent nMOSFETs' source side links to common metal line, so FIB cut (3) cut both of them. (b) backside de-processed to contact layer with polish and chemical etch, and then nano-probing to check leakage.

To further investigate nMOSFET leakage behavior, the failed chip was polished from topside to contact layer and current of nMOSFET was measured by nano-probing. However, the leakage disappeared just after polishing. This strange result impelled us toward other probing method, which was to fabricate probing pads from nMOSFET terminals to passivation without polishing and nano-probing. The gate, source and P-Well terminals of nMOSFET in WL decoder were isolated and linked out with Pt probing pads by FIB hole-drilling and cutting from top passivation (Fig. 6a). Then memory tester controlled failed chip and accessed failed WL in operation mode, meanwhile external source/measure unit (SMU) controlled the bias of Pt pads by micro-probing (Fig. 6b). Fig. 7 shows that the measured leakage from terminal GWL matched that from nMOSFET's source side (terminal B). It meant that the leakage was major from nMOSFET punchthrough without other leak source. The curve was similar to Id-Vd curve of MOSFET and the leakage could be controlled by P-Well bias, which implied that the leakage from nMOSFET was induced by lower Vt .

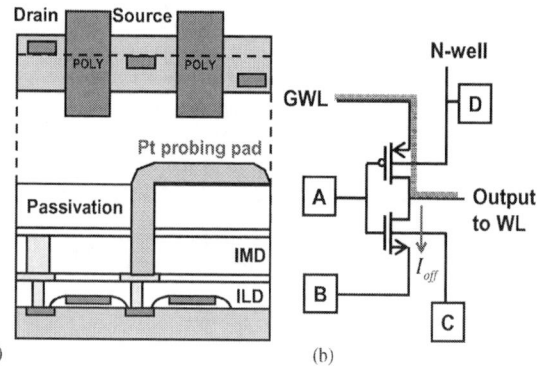

Fig. 6. (a) FIB-made Pt probing pad on passivation. (b) Node of GWL voltage is supplied from internal circuit. Node of A, B, C, D are isolated from circuit and biased from external source supplier. Node D is for voltage reference.

C. Device Level Analysis

Since off-state leakage of nMOSFET was suspected, we investigated in the device-level reliability test of single nMOSFET with high voltage and temperature stress Fig. 8 shows that after accelerated gate bias stress, the off-state leakage increases with time, and the leakage current has the same order

978-1-4799-3911-4/14 $31.00 © 2014 IEEE

as the failed chip. Moreover, PEM demonstrated that leakage path of stressed device was at the STI-edge (Fig. 9a). The holes trapped in the SiN layer of STI edge explained the STI-edge leakage (Fig. 9b) [2], which made channel edge Vt lower and resulted in nMOSFET punchthrough. The generation of the hole-traps was discussed later.

Then, in order to double check the previous strange disappeared leakage on chip level, the stressed device was measured by nano-probing system for comparison. Again the leakage became much smaller even on device level (Fig. 10). In fact, leakage disappearance is recovery of off-state current. Because the current recovered about four orders, we though it disappeared.

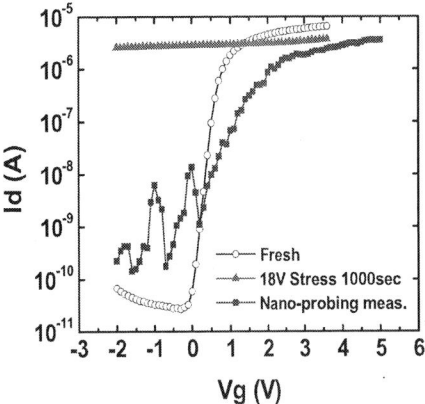

Figure 10. IV-curves of nMOSFET (measured with Vd = 0.1V) for fresh state (black), after 18V gate bias for 1000sec stress by tester (red) and followed by nano-probing system (blue). The recovery effect with lower leakage was found.

D. Recovery of Off-state Current

To investigate recovery phenomenon, the leaked nMOSFET with different thickness of top passivation was exposed under SEM for different time and then measured on ordinary probing system (Fig. 11). The results indicated that electron beams from SEM penetrated the STI [4] and recombined the hole-traps in the SiN, so the leakage reduced (Fig. 12a). They not only explained the recovery phenomena for both chip and device level during nano-probing in SEM system, but also implied that the hole-trap in SiN liner was the failure root cause.

E. Circuit Level Analysis

Since off-state leakage of nMOSFET was confirmed, we preformed HTOL test on partial array (Fig. 12b) to discover from where high-V long term stress was during circuit operation. The experiment result showed that the failed WLs were spread in whole chip even though the unselected array was not read. It meant that nMOSFET suffered high gate bias stress during unselected state for a long time (Fig. 1a), which lead to nMOSFET off-state leakage when reading the WL (Fig. 1b).

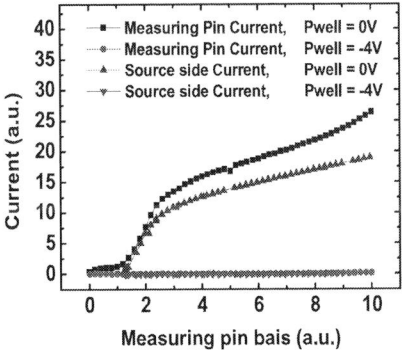

Fig. 7. The measured leakage from terminal GWL matched that from nMOSFET's source side (terminal B) and controlled by P-Well bias.

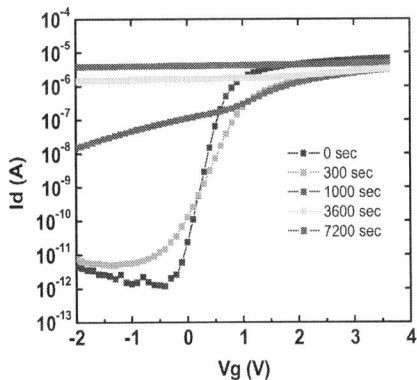

Figure 8. The degradation of IV curve with time (measured with Vd = 0.1 V) for a test key of nMOSFET under high gate voltage stress.

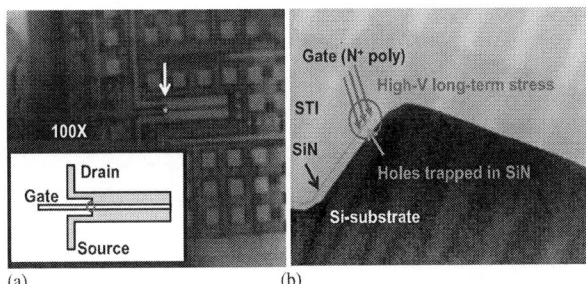

Figure 9. (a) PEM of stressed nMSOFET and leakage path is shown at the STI-edge. (b) TEM cross section of STI edge and schematic illustration of stress-induced hole traps. The degradation of IV curve with time (measured with Vd = 0.1 V) for a test key of nMOSFET under high gate voltage stress.

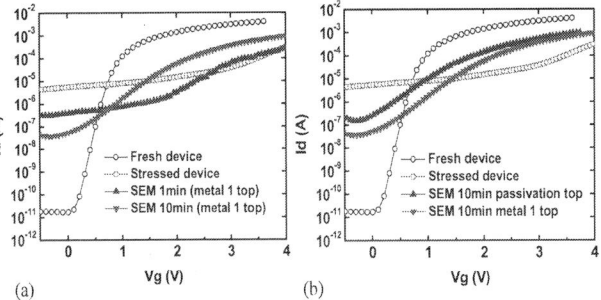

Figure 11. (a) Longer exposure time, more recovery of off-state current (measured with Vd = 3.8 V). (b) Thinner layer above the STI edge, more recovery of off-state current.

978-1-4799-3911-4/14 $31.00 © 2014 IEEE

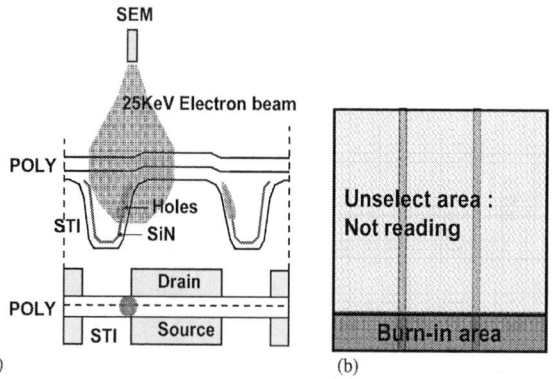

(a)   (b)

Figure 12. (a) 25KeV electron beams from SEM penetrated top passivation and recombined the hole trapped in SiN layer of STI edge. (b) Partial burn-in experiment: the failed WLs were spread in whole chip even though the unselected array was not read (but under high gate voltage for nMOSFET in WL decoder).

III. FAILURE MECHANISM

A simple two carrier model points out the generation of hole-traps after positive gate bias-temperature stress. When high gate voltage is applied, electric field on the center gate oxide are high electric field, however, it becomes low electric field in STI-edge due to the thick STI, and low electric field provides low current through SiN (Fig. 13). In SiN layer with low current flow, the electrons travel toward the anode-gate electrode whereas the holes injected from anode gate, and both of them get forward-trapped in SiN gradually. However, they stop getting forward-trapped when they come across to each other, because they recombine each other at common boundary (Fig. 14). The common boundary become away from anode-gate due to higher hole mobility than electron in SiN layer. It is found that hole mobility in the SiN layer is greater than electron mobility due to shallow hole traps as opposed to deep electron traps [5][6]. Under this situation, hole traps outnumber electron traps and dominate the negative Vt shift of STI-edge and cause the off-state leakage [2].

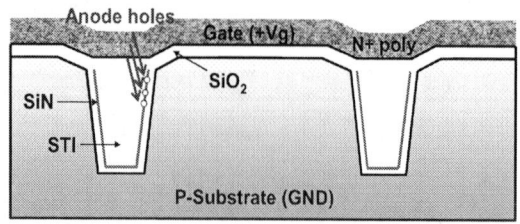

Figure 13. The holes trapped in SiN liner came from anode gate low current flowing through STI edge

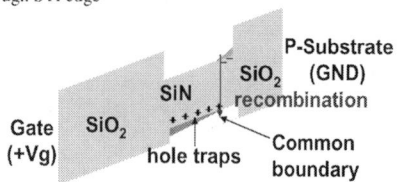

Figure 14. Low current flow situation, electrons and holes recombine each other at common boundary, and hole traps outnumber electron traps and dominate the negtive V_t shift.

IV. IMPROVEMENT

This failure root cause is hole-traps at the SiN occurred under long term high voltage stress, and high voltage came from input-high gate bias because of MLC operation bias. Therefore we came up with two solutions for process and circuit aspects. Solution for process was to remove the SiN liner process from STI. However, SiN liner in STI is indispensable in our device, because it is used to prevent HDP oxide drilling in silicon substrate. Solution for circuit was to prevent nMOSFETs decoder from long-term high gate bias stress. The method was simply to supply 0V through pMOSFET in unselected state rather than nMOSFET (Fig. 15) in order to reduce stress time.

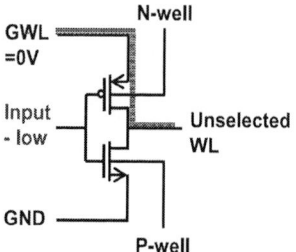

Figure 15. Modified unselected state: in input-low state, the ground voltage passing through from pMOSFET rather than nMOSFET.

V. CONCLUSION

In summary, EFA and PFA are performed on both chip and device level. The root cause of gate bias-temperature-stress leakage was identified to the hole-traps at the STI-edge. And the mechanism of the hole-traps generation is anode gate low current flowing through STI and holes were trapped in SiN liner. The hole-traps will be annihilated by electron beams from SEM. This phenomenon might misguide judgment during failure analysis. Once an IC has high operational bias, process with SiN liner in STI must be avoided; otherwise some unexpected leakage or Vt shift may occur under long term high-V stress. If SiN process is indispensable, circuit ought to be modified to reduce high gate bias stress.

REFERENCES

[1] S. Ogura et al., "A Half Micron MOSFET Using Double Implanted LDD" IEEE IEDM Technical Digest, pp.718-721, 1982.

[2] A. S. Teng et al., "Gate Bias Temperature Stress-Induced Off-state Leakage in nMOSFETs: Mechanism, Lifetime Model and Circuit Design Consideration" IEEE IRPS, accepted..

[3] S. Prejeanm et al., "Special Techniques for Backside Deprocessing" Microelectronics Failure Analysis: Desk Reference, pp.496-500, 2004.

[4] P. J. Potts, A handbook of silicate rock analysis, Chapman & Hall, p. 336, June 1987.

[5] M. Aminzadeh, Shinji Nozaki, R. V. Giridhar, "Conduction and charge trapping in polysilicon-silicon nitride-oxide-silicon structures under positive gate bias," IEEE Tanns. on Elec. Dev., vol. 35, Issue 4, pp. 459–467, April 1988.

[6] S. Manzini and F. Volonté, "Charge transport and trapping in silicon nitride-silicon dioxide dielectric double layers," J. Appl. Phys., vol.58, Issue 11, pp. 4300–4306, December 1985.

978-1-4799-3911-4/14 $31.00 © 2014 IEEE

MOSFET Implant Failure Analysis Using Plane- View Scanning Capacitance Microscopy Coupled with Nano- Probing and TCAD Modeling

Hun-Seong Choi[1,2], Yong-Woon Han[1], Il-Sub Chung[2]

1) Analysis Science & Engineering Group, System LSI Division, Samsung Electronics
Giheung-Gu, Youngin-City, Gyeonggi-do, 449-711, Republic of Korea
2) School of Information and Communication Eng., Sungkyunkwan University
Chunchun-dong, Jangan-gu, Suwon, Gyeonggi-do, 440-746, Republic of Korea
phone: +82-10-5649-1232, fax: +82-31-209-7136, Email: hshs.choi@samsung.com

Abstract - This paper presents MOSFET implant failure analysis using plane view scanning capacitance microscopy (PVSCM) at silicon substrate level. Failing transistors are characterized by nanoprobing (NP) at contact level. The cause of failure was deduced from the combination of PVSCM, IV characteristics from NP measurement, and TEM cross-section analysis. Technology computer aided design (TCAD) simulation was implemented for failure modeling and SCM data verification. Failure analysis case studies of samples manufactured by 65 nm and 45 nm process are presented.

I. INTRODUCTION

In modern integrated circuit fabrication, ion implantation (IIP) is commonly used at a variety of doping processes. Implant related failures, however, are hard to be revealed by physical failure analysis (PFA) since they are invisible in conventional observation tools such as scanning electron microscope (SEM) or transmission electron microscope (TEM) [1].

Recently, scanning capacitance microscopy (SCM) is one of widely used techniques for carrier profiling. PVSCM is available at either contact or silicon substrate level. Contact level SCM provides useful information to detect abnormal contacts as SEM passive voltage contrast (PVC) does [1,2]. However, due to its complex nature, contact level SCM data lacks substantial clue for implant failures, whereas SCM at silicon substrate directly extracts local doping information beneath the prober tip. Contact level SCM has following drawbacks; 1) SCM data consists of various capacitive components. They are from both silicon substrate and contact: global depletion capacitance in connected junction area, gate capacitance, or parasitic capacitance from defective contacts. Moreover, 2) a contact on multi-finger gate includes multiple transistors' gate information. Some special contacts can be shared by both gate and diffusion layer. For those cases, mixed SCM signals can affect sensitivity of detection. Despite the advantage of direct doping information from silicon level SCM, a drawback of this technique is difficulty of leaving shallow junction

region of MOSFET intact [2]. In this paper, we performed SCM at around source, drain and well region without considering shallow junction region such as source drain extension (SDE). By scanning the sample, the variance of carrier density was mapped as SCM images [3]. In our experiment, image anomaly was observed at failed MOSFET and was verified by TCAD simulation with the combination of transistor's NP characterization results.

II. EXPERIMENTAL DETAILS

Failed transistors were measured by NP at contact level to characterize their electrically failing symptoms. And then the samples were polished down to silicon substrate layer with diamond suspension. Polishing was stopped at around source, drain, and well region. The backside was also polished to be conductive for biasing. A schematic diagram of our experimental setup is depicted in Fig.1. We applied DC bias voltage 0 V and AC modulation bias amplitude 0.4~0.8 V with frequency 40~80 KHz. The UHF (~1 GHz) resonant capacitance sensor measured the change of probe tip while scanning the surface of the sample. In advance, the capacitance sensor was tuned to maximize sensitivity by setting the operation point at the frequency of largest slope in sensor tuning curve. The lock-in amplifier extracted *dC/dV* signals [4] and then the *dC/dV* signals were mapped to a SCM image.

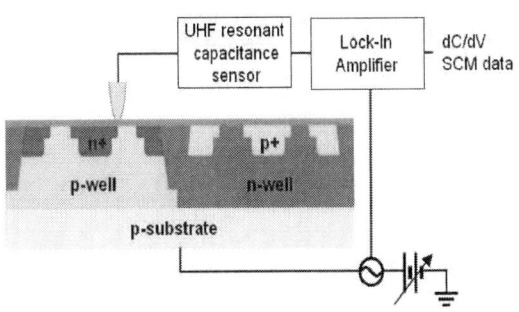

Fig. 1. Schematic diagram of experimental setup

978-1-4799-3911-4/14 $31.00 © 2014 IEEE

III. RESULTS AND DISCUSSIONS

To identify the status of sample surface, SCM imaging was performed at SRAM bit-cell region. In Fig.2, SCM images are listed with a SEM reference image. Comparing Fig.2.(b) and (c), well regions are discontinuous in (b). Source and drain region is not visible while the strong image is observed at only well region. This is because *pn* junction depletion region of diffusion layer blocks the sample bias to modulate capacitance between tip and silicon. In contrast, in the further polished sample, well regions are continuous: Fig.1.(c) as source and drain *pn* junction were removed.

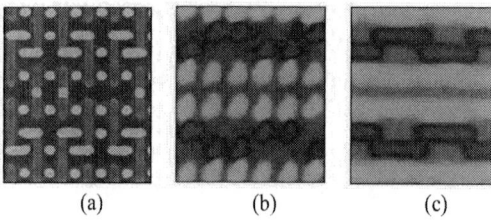

(a)   (b)   (c)

Fig. 2. (a) SEM image at contact level. SCM *dC/dV* image, (b) at source, drain, and well (c) at well level, green: *p* well, blue: *n* well

Case Study 1

A 65 nm node SRAM was failing with a cluster of bits. In SEM PVC inspection, abnormal contacts were observed at *n* channel transistors of the failing bits: Fig.3. (a). Contacts on both gate and *n+* diffusion and gate showed darker contrast than references. In general, the darker contrast represents the higher resistivity [5]. TEM cross-section was performed to investigate darker contacts and no defect was found as shown in Fig.3. (b).

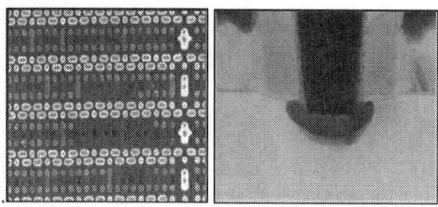

Fig. 3. (a) SEM PVC image, contacts on n FET showed darker contrast than reference. (b) No defect was found in TEM cross-section.

The abnormal transistors were measured by NP. Compared with the reference, low on-current, high threshold voltage, and punch-through current from source to drain was observed: Fig.4. The measured sample was polished down to silicon substrate and SCM imaging was performed. In SCM image, channel regions of dark contrasted contacts

were not clearly observed: Fig.5.(a). To clarify the distinction, SCM *dC/dV* signals along the markers are shown in Fig.5.(b). The channel region of failing transistor did not produce SCM signal. Since the strength of the doping is qualitatively related to SCM signal intensity, the weak signal postulates low channel doping density [4]. For the more detailed analysis, a new sample with the same failing symptom was prepared for TEM cross-section analysis. The sample was chemically stained to inspect junction regions clearly. As a result, source and drain extension region was not observed in failing transistors: Fig.6. The missing SDE region can result from masking of ion implantation step. Since SDE and halo IIP are the same manufacturing process steps where IIP mask is shared, it is reasonable to suspect that channel doping density can be low as well.

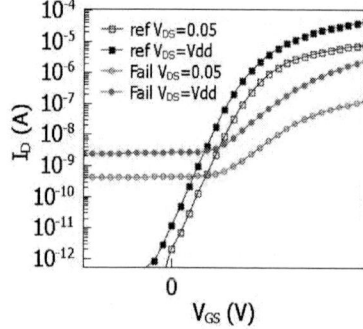

Fig.4. IV curve of SRAM pass gate transistors.

(a)

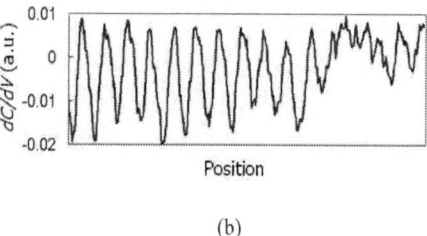

(b)

Fig.5. (a) PVSCM image at source, drain, and well (b) SCM signal along the makers

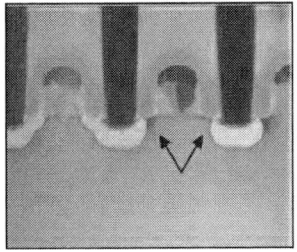

Fig.6. TEM cross-section of chemically stained sample, SDE region is not observed in failing region.

Case Study 2

A 45 nm node memory was failing with a group of columns. Through the diagnostic test and circuit analysis, the fault candidates were narrow downed to the specific periphery circuit. In PFA, there found no visible defects. The suspect transistors were measured by NP and characterized as they have high punch-through current from source to drain with dropped on-current: Fig.7. This behaviour is similar as was observed in case study 1. The suspect area was scanned by SCM and the abnormal transistors of high punch-through current were not visible: Fig.8.(b).

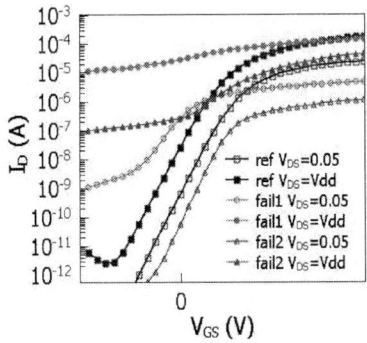

Fig.7. IV curve of transistors in the suspect area

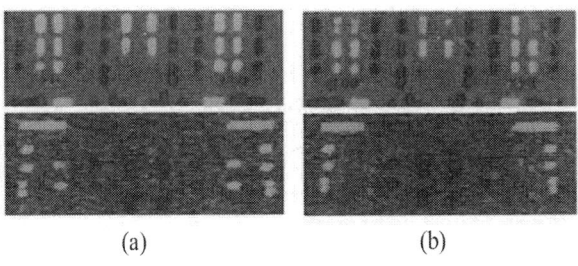

(a)   (b)

Fig.8. SCM images at silicon substrate level: source and drain remains (a) reference (b) suspect region, arrows indicate weak or none SCM signal.

TCAD Modeling

A MOSFET structured was established for TCAD simulation. The structure has source/drain, SDE and halo region. The simulation was performed by adjusting the thickness of IIP blocking layer during halo and SDE IIP process. We assumed that the remaining photoresist (PR) after development worked as an IIP blocking layer. As the simulation results, IV of MOSFET and CV to derive SCM dC/dV signal by prober tip are described: Fig.10. (a), (b). The CV curves in Fig.10.(b) explicitly shows that significant difference of SCM signal can be drawn at a proper bias window of V_{AC}. Considering probe tip material, flat-band shift should be taken into account. For this case, DC offset voltage can play a role in optimizing the amplitude of sample bias. From the simulation results, the SCM image where failing transistors of low channel doping is invisible can be explained by comparing the case of thk03 with nominal case in Fig.10.(b). The amplitude of AC voltage less than 0.6 V does not produce dC/dV signal at around zero V_{DC} while nominal case does. In the worst case, the signal of negative phase can be induced under high amplitude bias condition when halo IIP is perfectly blocked. Considering NP results, TEM cross-section, and TCAD simulation together, the cause of failures in case studies were deduced as low or blocked doping at halo and SDE IIP step. Low halo doping might increase depletion width, which caused punch-through from source to drain. And the missing SDE might cause insufficient overlap between gate and source/drain, thereby increasing threshold voltage significantly accompanying on-current drop. The effect of IIP blocking is described in Fig.10.(c).

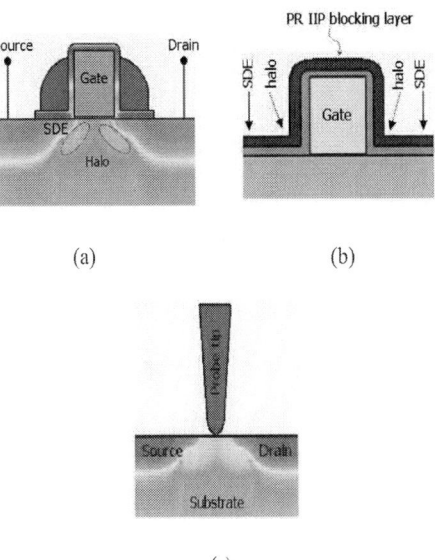

Fig.9. (a) MOSFET structure used for TCAD simulation. (b) IIP blocking layer was adapted at halo and SDE IIP step. (c) a simulation of CV measurement at silicon substrate level.

978-1-4799-3911-4/14 $31.00 © 2014 IEEE

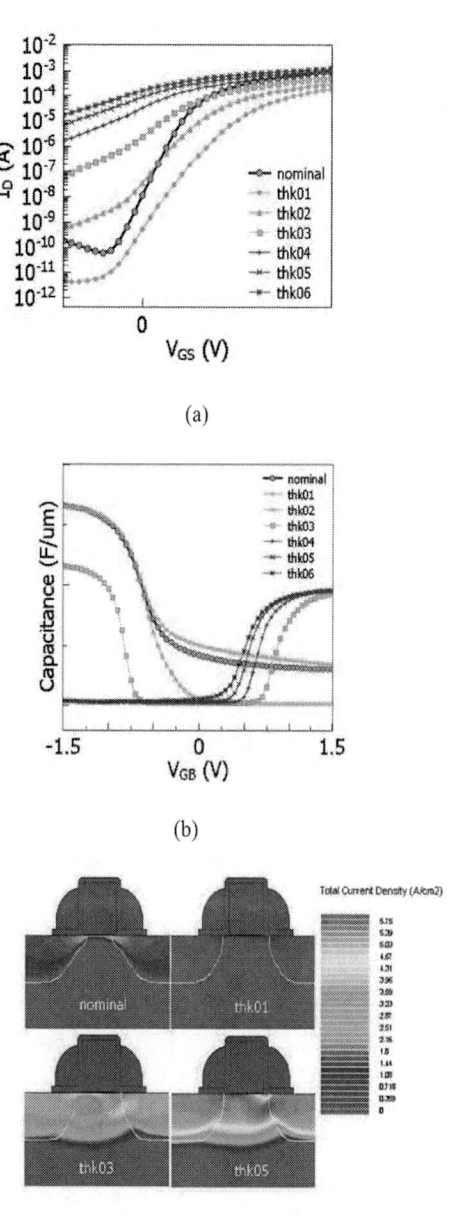

(a)

(b)

(c)

Fig.10. TCAD simulation results as the thickness variation of implantation blocking layer (a) IV curve where $V_{DS} = V_{DD}$ (b) high frequency CV plot at channel region seen by SCM probe tip (c) off-state current density where $V_{GS}=0$, $V_{DS}= V_{DD}$. Metallurgical *pn*-junctions are designated by white lines.

IV. CONCLUSIONS

Implant failure analysis using PVSCM at silicon substrate level is presented. The combination of SCM, NP characterization results, and TCAD simulation is useful to establish failure model. Additionally, SCM *dC/dV* signal was studied in terms of SDE and halo implantation

blocking as a variance of blocking layer thickness. Although SCM at plane-view does not provide direct doping profile, it has capability of fast fault isolation to detect doping related issues across wide area.

REFERENCES

[1] Cha-Ming Shen, Shi-Chen Lin, et al. "Couple passive voltage contrast with Scanning Probe Microscope to identify invisible implant issue," Proc. ISTFA 2005, pp. 212-216.

[2] Kartik Ramanujachar, "Scanning Capacitance Microscopy at transistor contact level," Proc. ISTFA 2004, pp. 482-486.

[3] Edwards, Hal, et al. "Scanning capacitance spectroscopy: An analytical technique for pn-junction delineation in Si devices," Applied Physics Letters 72, pp. 698-700, 1998.

[4] Williams, C. C., "Two-dimensional dopant profiling by scanning capacitance microscopy," Annual review of materials science 29.1, pp. 471-504, 1999.

[5] Zhang, Hai-Bo, Ren-Jian Feng, and Katsumi Ura. "Utilizing the charging effect in scanning electron microscopy," Science progress 87.4, pp. 249-268, 2004.

Resolving Systematic Voltage Sensitive Soft Failures in 28nm Microprocessor Devices

Dnyan Khatri[1,*], Soon Huat Lim[1], Mun Yee Ho[1], Vinod Narang[1], Dakshina-Murthy Srikanteswara[2], Keith Kasprak[3]

[1] Device Analysis Laboratory, 508, Chai Chee Lane, AMD Singapore 469032
[2] Product Development Engineering, 508, Chai Chee Lane, AMD Singapore 469032
[3] Silicon Design MHDC, AMD Austin, 7171 Southwest Parkway, Austin TX 78735
*Ph: (+65)82280656, E-mail: dnyan.khatri@amd.com

Abstract- **With rapid developments in semiconductor manufacturing technologies, new and more complicated challenges emerge in the Failure Analysis space. It has been a challenge to perform failure analysis for voltage-sensitive soft failures, especially those occurring in SRAM circuitries. However, fault localization in sub-micron devices is successful if existing FA techniques are innovatively and extensively leveraged during physical fault isolation. This paper emphasizes the use of SEM-based nano-probing followed by advanced TEM techniques to successfully identify the root cause of failure, thus enabling the wafer fab to take appropriate corrective measures to mitigate such failures. A successful case study involving these techniques will also be discussed.**

I. INTRODUCTION

Semiconductor industry is rapidly scaling down devices to nanometer range to pack more transistors in the chip and get better device performance. During wafer fabrication of such complex devices, there are several kinds of killer defects introduced at various manufacturing stages and processes. SRAM circuitries in modern day microprocessors are densely packed to save chip space and thus are highly susceptible to fab process marginalities. They also have very stringent operating voltage requirements. Any defects introduced in SRAM cells can thereby prove to be fatal. Fab process marginality related defects do not always cause hard failures (open/short) in SRAM circuits. In many cases like the one discussed in this paper, failures occur only above or below certain voltages at which these circuits are tested. Finding the root cause and determining the correct failure mechanism prove to be critical in enabling process fixes, resulting in yield improvement and monetary loss prevention. In this paper, we will demonstrate how the use of SEM-based nano-probing coupled with advanced TEM techniques can be used in solving a systematic wafer fab issue that resulted in SRAM memory failures for 28nm microprocessor devices. We will also discuss simulation studies done to show how SRAM cell stability is affected by the presence of defects. We will also show a process flow snapshot to indicate how layout-process interaction can cause defects and what steps can be taken to correct them.

II. EXPERIMENTS

During product qualification, devices were reported to have double bit memory failures occurring only at higher voltages. These devices passed at lower voltages. One such failing unit was used for analysis.

Bitmapping helped to narrow the failing location down to two failing bits adjacent to each other. Since the failures were voltage dependent (soft failures), fault isolation was performed using Soft Defect Localization (SDL) to confirm the failing location. SDL is an established fault isolation process for localizing soft defects using laser heating [1]. Results from SDL confirmed the failing site to match with the bitmapping results as shown in Fig. 1(a).

Fig.1 (a) When the laser hits the failing bit location the number of "pass" increases significantly, highlighted in blue. SDL site matches bitmap results

Following fault isolation, PFA using parallel polishing was performed. Since the unit was suffering a double bit failure, analysis was focused on metal 2 layer and below. SEM inspection, via Passive Voltage Contrast and Conductive-AFM at via layers were undertaken at the failing bit location but no obvious anomaly could be seen. The unit was further prepared for nano-probing to check for any abnormal transistor behavior at the failing bits [5]. Nano-probing is a well-established technique to characterize failing transistors and identify possible failure mechanisms [2]. The unit was de-processed to contact layer and nano-probing was performed using SEM-based DCG n-prober system on the failing bits as well as a reference bit to obtain the transistor characteristics of all 6 individual transistors in the failing SRAM bitcells. From the transistor characteristics seen in Fig. 2 (a), Fig. 2 (b) and Fig. 2(c), we can see that NMOS transistors for both failing bits have a lower drive current as compared to the reference bit.

978-1-4799-3911-4/14 $31.00 © 2014 IEEE

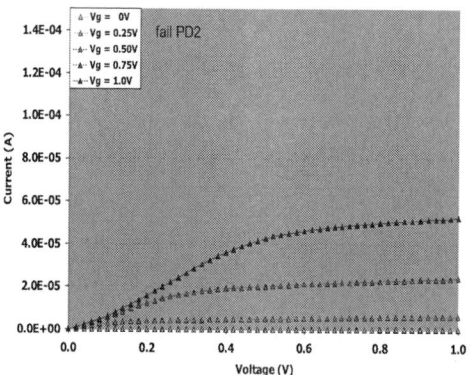

Fig. 2 (a) Id vs. Vd for failing pull down transistor of first failing bit (drain sweep)

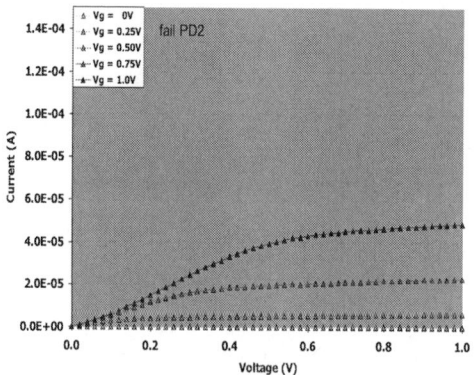

Fig. 2 (b) Id vs. Vd for failing pull down transistor of second failing bit (drain sweep)

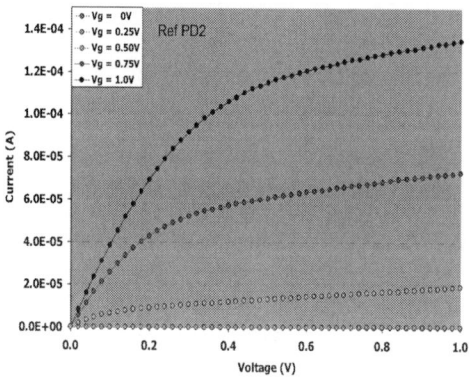

Fig. 2 (c) Id vs. Vd for failing pull down transistor of reference bit (drain sweep)

Since this transistor behavior showing lower drive current points to a possibility of higher resistance at the nodes, further experiments were necessary to find the source of high resistance. Single probe Source vs. Substrate resistance measurements were thus performed for the failing and reference bits as seen in Fig. 3(a) and Fig. 3(b) and they showed that the current values were three orders higher for the reference bit (E-5 range) as compared to the failing bit (E-8 range).

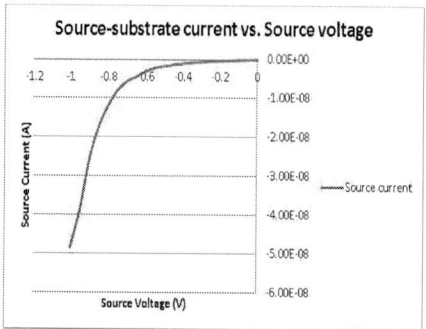

Fig. 3 (a) Source current values for one of the failing bits (E-8 range)

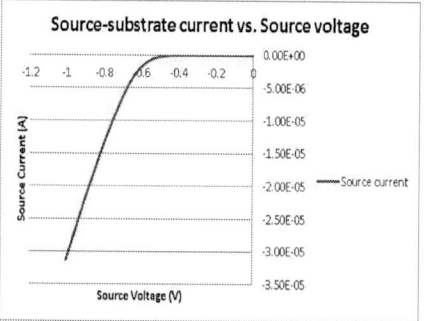

Fig. 3 (b) Source current values for the reference bit (E-5 range)

III. ANALYSIS AND DISCUSSION

From analysis of the nano-probing data, we concluded that there is a high resistance [3] between Source and Drain of pull down transistors resulting in the failure of two adjacent bits. Since the only contact common to both these failing bits is a source contact, as seen in Fig. 4, further analysis was needed to understand the failure mode. TEM lamella was prepared as shown in Fig. 4. From the conventional STEM image [4] shown in Fig. 5(a), we can see that the suspected source contact has a slightly shallower contact etch as compared to the adjacent drain contacts. However, this observation is not sufficient to explain high resistance and the nickel silicide profile is also unclear.

In order to see the silicide profile underneath the contacts, a new TEM analysis technique was put to test. The lamella was intentionally tilted off-axis so that regions underneath the contacts could be made visible, as shown in Fig. 5(b).

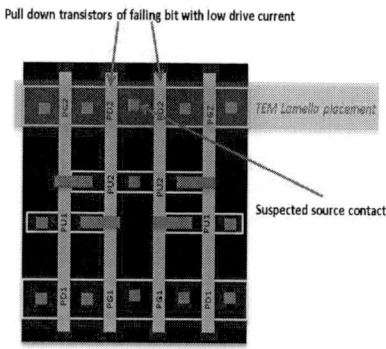

Fig. 4 Rough schematic of suspected common source contact between two failing bits

978-1-4799-3911-4/14 $31.00 © 2014 IEEE

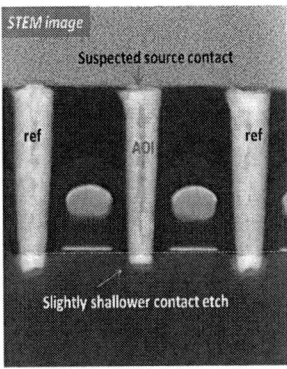

Fig. 5(a) Conventional STEM image showing suspected source contact having a slightly shallower contact etch as compared to adjacent contacts

It can be seen that there is barely any silicide underneath the suspected source contact while all adjacent contacts have enough silicide, as shown in Fig. 5(c). Fig 5(d) shows the EELS element mapping at the defect region and the absence of Ni can be clearly seen. From Fig. 4 we can see that the TEM lamella is along the X-axis. This has been done deliberately in order to get silicide profile of the defective contact as well as adjacent contacts so as to make a comparison. A lamella along Y-axis allows us to inspect only the defective contact and no silicide profile comparison can be made.

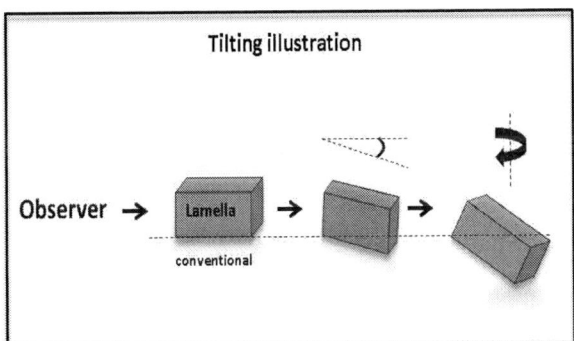

Fig. 5(b) TEM lamella deliberately tilted for a better view of silicide profile

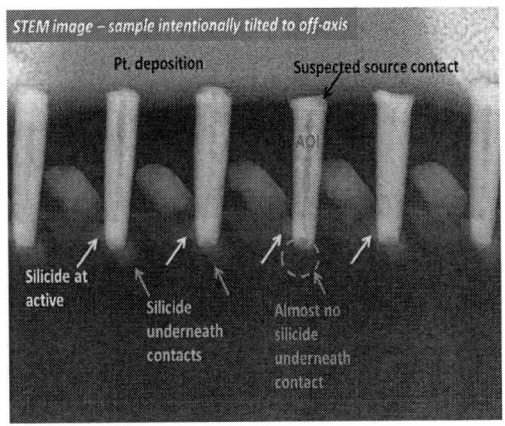

Fig. 5(c) Off-axis tilted STEM image to view the silicide profile better. Suspected source contact has barely any silicide underneath

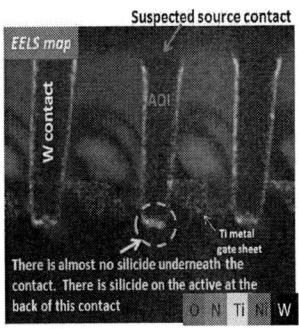

Fig 5(d) EELS elemental map showing absence of Ni underneath the defective contact

From all the evidence gathered, we can say that partially missing nickel silicide (NiSi) results in the high resistance [6] at the common source contact, which in turn results in failing of two bits adjacent to each other. Following a feedback of FA results to the wafer fab, investigation work was carried out to find and fix the process resulting in the missing NiSi. Fig. 6 shows that due to peculiar layout-process interaction, insufficient spacer etch resulted in nitride residue at the bottom of the contact that could not be cleared before silicide formation. This initially led to missing NiSi at the location as the deposited Ni was blocked from reacting with the (expected) active Si. The residual nitride also resulted in the contact depth at the defective location being marginally lower than others. The combination of a shallow contact combined with missing NiSi underneath the contact landing eventually led to a high resistance contact. To fix this issue, the wafer fab increased the spacer over-etch timings and results showed an improvement in the residue removal under sensitive contacts. Further investigations are on to determine why only common source contacts are impacted by this process marginality.

IV. SIMULATION STUDIES

From the electrical and physical FA data, we can see that missing NiSi can result in a higher source contact resistance resulting in lower transistor drive strength. The SRAM bitcell failures reported were voltage dependent and were occurring at higher voltages (cells were still stable at lower voltages). It is essential to know how this particular defect can cause an SRAM cell to fail at higher voltages while remaining stable at lower voltages [7]. Fig. 7 shows a simulated graph of SRAM cell stability versus its operating voltage. For these studies, we have used a reference base cell and incrementally added a source resistance to a Pull Down transistor to observe its effect on the cell stability. Cells with stability values greater than 1 are expected to be stable (with varying operating voltage).

As we can see, a base cell is found to be stable across a wide range of operating voltage values, but as we start adding a source resistance, we begin to see a roll-off in the curves. In some cases, the cell stability value falls below 1, resulting in cell failure. It is also interesting to see how at lower voltages the high source resistance does not cause any drift in cell stability, explaining why even defective cells are able to pass at low test voltages.

978-1-4799-3911-4/14 $31.00 © 2014 IEEE

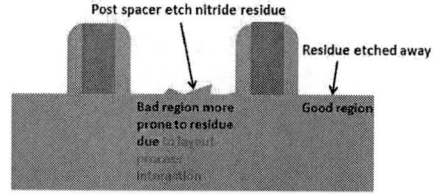

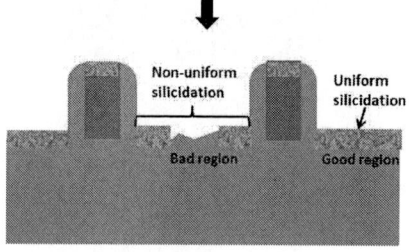

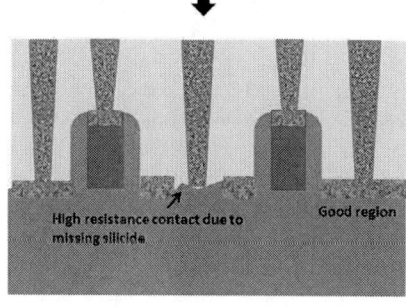

Fig. 6 Process flow showing the effect of nitride residue on contact formation resulting in a high resistance contact

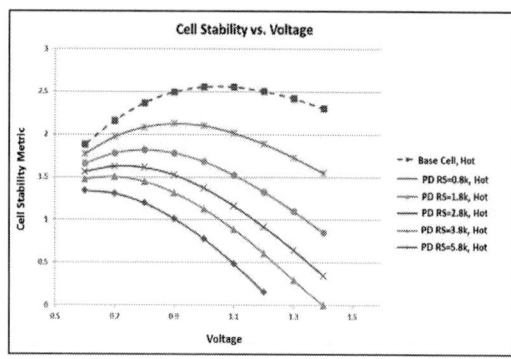

Fig. 7 Plot showing SRAM cell stability versus operating voltage for an SRAM cell

V. SUMMARY

From this case study, we can see the significance of extensive transistor level nano-probing in assisting to find the root cause of memory failures. We have also observed limitation of conventional TEM for uncovering soft defects. A new way of tilting the TEM lamella has proved to be crucial in successfully determining the cause of failure. Understanding which process steps are the source of the defect as well as detailed simulation studies can help us get a much clearer understanding of the failure mechanism and why a specific type of defect can result in a particular type of failure. We can thus also see that for complex soft failures in microprocessors, it is imperative for failure analysis team to work collaboratively with design and process technology teams in order to resolve the issue.

ACKNOWLEDGMENT

The authors would like to thank colleagues from the Device Analysis Lab at AMD, Singapore and Anser Muhammad from Yield Engineering for their contributions to this work. We also express gratitude to Foundry Operations and Product Development and Engineering teams for their valuable assistance.

REFERENCES

[1] V.K. Ravikumar, S.L. Phoa, V. Narang, J.M. Chin, "Understanding Soft Defect Localization Set-Points for Reducing Cause-Not-Founds in Integrated Circuits," *IEEE Proceedings of 15th IPFA – 2008*

[2] Caho-Chin Wu, Jon C. Lee, Jung-Hsiang Chuang, Tsung-Te Li; "Single Device Characterization by Nano-probing to Identify Failure Root Cause", *Proceedings from the 31st ISTFA 2005*

[3] Ma Yinzhe, Oh Chong Khiam, Nyi Ohnmar, Chuan Zhang, Donald Nedeau, Lim Seng Keat and King Ming Chu, "Nanoprobing As An Essential and Fast Methodology in Identification of Failure's Root Cause for Advanced Technology", *Proceedings from the 39th ISTFA 2013*

[4] Naoko I. Kato, Akira Nishikawa, Juichi Matsuzawa, Wonjin Moon, and Yoshiteru Kohno, "Nano-scale Defect Investigation by Site-Specific Transmission Electron Microscopy and Electron Energy Loss Spectroscopy", *2001 IEEE International Symposium on Semiconductor Manufacturing*

[5] S. L. Toh, P. K. Tan, Y. W. Goh, E. Hendarto, J. L. Cai, H. Tan, Q. F. Wang, Q. Deng, J. Lam, L. C. Hsia, and Z. H. Mai, "In-Depth Electrical Analysis to Reveal the Failure Mechanisms With Nanoprobing", *IEEE Transactions On Device And Materials Reliability, Vol. 8, No. 2, June 2008*

[6] John Foggiato, Woo Sik Yoo, "Integration of Nickel Silicide: Minimizing Defect Generation During Formation", *Advanced Semiconductor Manufacturing Conference and Workshop, 2005 IEEE/SEMI*

[7] Xiaojun Li, Jin Qin, Bing Huang, Xiaohu Zhang, and Joseph B. Bernstein, "SRAM Circuit-Failure Modeling and Reliability Simulation With SPICE", *IEEE Transactions On Device And Materials Reliability, Vol. 6, No. 2, June 2006*

978-1-4799-3911-4/14 $31.00 © 2014 IEEE

A Stain Chemical Etching Rate Study and Used to Detect Implant Defect

Miao Wu, Yi Che

Product Analysis Engineer of Quality Department in Freescale Semiconductor (China) Limited

No.15 Xinghua Avenue, Xiqing Economic Development Area, Tianjin, China 300385

Phone: 86-22-85684614   Email: B28231@freescale.com

Abstract

In this paper, a chemical is used to stain the implant profile. Its etching rate to different dope type and density is studied by performing cross-section on three typical devices and then dipping in the chemical for staining. Due to etching rate difference, the method is shown to be effective in localizing implant profile in the Si active area. Meanwhile real failure analysis cases are presented, and the stain chemical is proved effective in localizing non-uniform implantation.

Key words- implant; etch rate; stain chemical; failure analysis

I. INTRODUCTION

Semiconductor manufacturing process is kept innovating to scale down CMOS device to meet the requirement of higher density, higher performance and lower power consumption. Consequently, leakage mechanism in deep submicrometer regimes is becoming a significant research topic [1].

In wafer process, the steps before CT layer are usually called as front-end and remaining steps are called as back-end. Implant process is a very important key step. It helps to form the MOS, Resistors, Bipolars and Diodes. All devices cannot exist without implant process.

Completing implant needs cooperation of lithography and implant process. Excursions during these two steps may cause serious leakage in devices. Furthermore, normal physical analysis methods cannot find any visual hints unless wet chemical methods adopted, such as Wright Etch.

II. ETCH RATE FOR DIFFERENT DOPING

Wright Etch is generally used for implant issue analysis [2, 3]. Doping profile can be visualized because of the wet chemicals behave different etch rate for different doping agent. In this research, the chemical is a little different with Wright etcher. Its major components include H_2O, HNO_3, HF and oxalic acid. Some devices are chosen to do the etch rate experiment. The chosen devices include PLD45, PNPL and DZPDN_LV.

A convenient notation is given as following. For example, N13 and P13 are used to express doping type and density. N13 means the doping is N type and density is on the order of thirteenth power of ten. Same principle can be used to understand P13, N15 and so on.

PLD45 is a P type LDMOS. It consists of PSD (P15), HVPW (P12), NSD (N15), NBL (N15) and NHV+LVNW (N13). The device includes N and P type doping agent, and appropriate density range to compare etch rate. The cross-section diagram of PLD45 is shown in Fig 1.

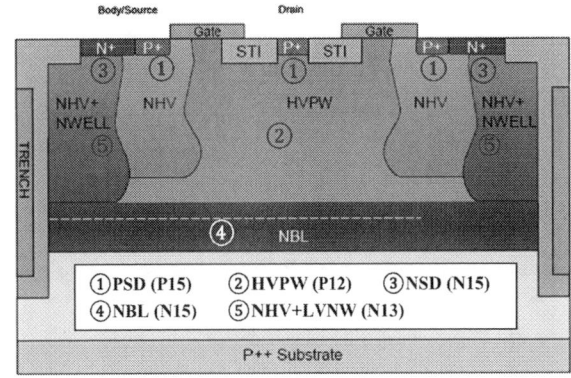

Fig. 1. The cross-section diagram of PLD45

After dipping in the chemical for 5 seconds, the implant profile is seen as Fig 2. From the cross section image, we can roughly get the qualitative conclusion due to the image resolution limitation.

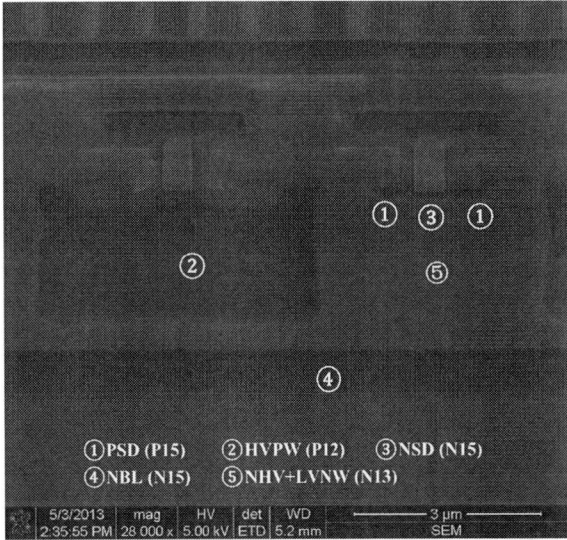

Fig. 2. The cross-section of PLD45, after chemical etching 5 seconds

978-1-4799-3911-4/14 $31.00 © 2014 IEEE

Firstly, for P type doing, obvious profile is seen if density higher than P12, such as HVPW (P12).

Secondly, for N type doping, no profile can be seen if density lower than N13, such as NHV+LVNW (N13). But profile is visible if density higher than N15.

Thirdly, for same doping type, etch rate of high density is higher than that of low density according to the depth of the color, such as NBL (N15) with obvious profile which means part of the area etched away, versus NHV+LVNW (N13) with almost zero etch rate.

Lastly, for same density, etch rate of P should be higher than that of N, such as HVPW (P12) with high etch rate, also versus NHV+LVNW (N13).

PNPL is a Bipolar. It consists of PSD (P15), HVPW (P12), NSD (N15), NBL (N15) and NEXT (N12). The device also includes N and P type doping agent, and appropriate density range to compare etch rate. The cross-section diagram of PNPL is shown in Fig 3.

After 5 seconds etching, the implant profile is seen as Fig 4. The conclusion from Fig 2 is double confirmed by the result of PNPL. Especially, the last conclusion is confirmed accurately. HVPW (P12) and NEXT (N12) get the density on a same order. The former gets high etch rate, but zero for the latter.

To analyze with more density, the third device is chosen. DZPDN_LV is a Diode. It consists of PSD (P15), NSD (N15), NBL (N15), DPN (N14), HVNW (N12) and NHV+LVNW (N13). The device includes DPN which is on the order of fourteenth power of ten. The cross-section diagram of DZPDN_LV is shown in Fig 5.

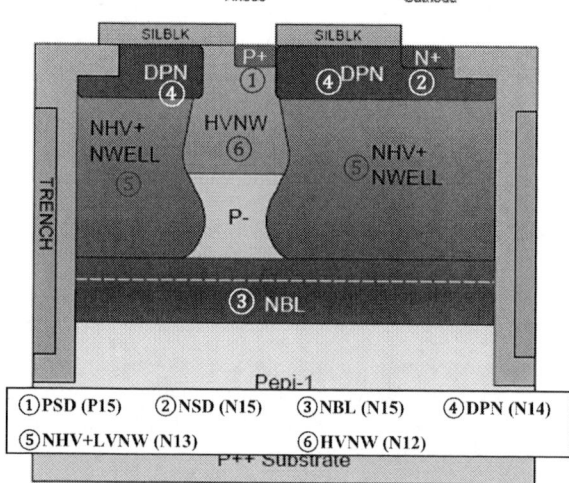

Fig. 5. The cross-section diagram of DZPDN_LV

Fig 6 reveals the result of DZPDN_LV after 5 seconds chemical etching. The more accurate conclusion is that, for N doping agent, profile can be seen if density higher than N14. The other conclusions from Fig 2 are confirmed again.

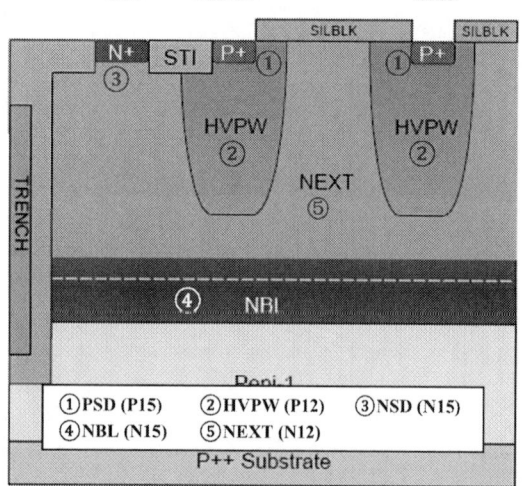

Fig. 3. The cross-section diagram of PNPL

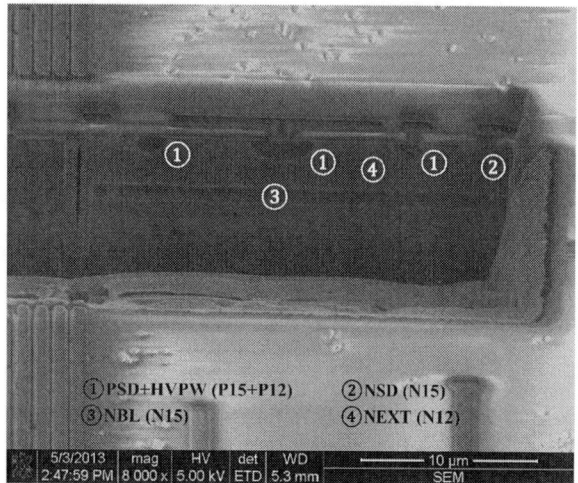

Fig. 4. The cross-section of PNPL, after chemical etching 5 seconds

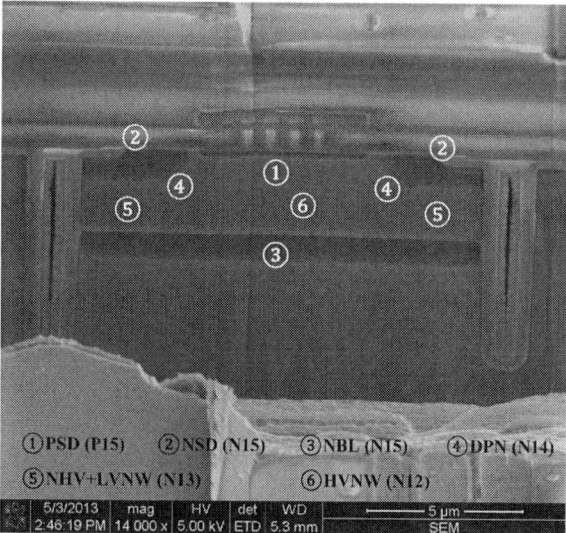

Fig. 6. The cross-section of DZPDN_LV, after chemical etching 5 seconds

Based on above three devices, we emphasize again the conclusion as Fig 7. The "0" etch rate does not mean the rate is really zero, but almost same with the etch rate of substrate. The "0" is a relative value to substrate. In our case, the substrate is P-doping. Additionally, no obvious stacking effect is observed. For example, NHV (N13) and LVNW (N13) in same area are also zero etch rate. It is supposed that etch rate can be obviously increased only if stacking doping increases density by an order of magnitude.

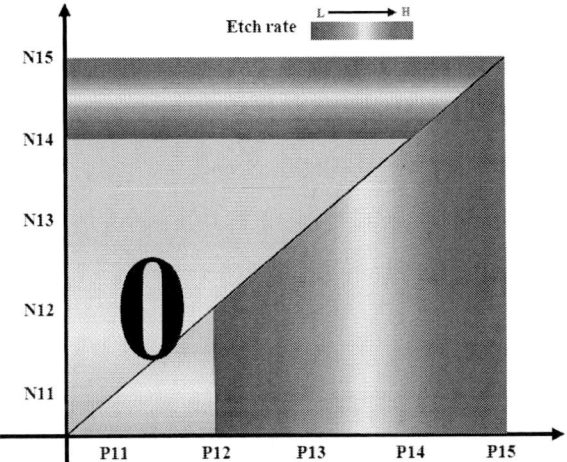

Fig. 7. The etch rate diagram, based on chemical etching 5 seconds

III. REAL CASE SHARING

A. Resistance value affected by implant defect

The failure of the case is resistance lower than normal. From curve trace, the value is 100K ohms, half of normal 200K ohms.

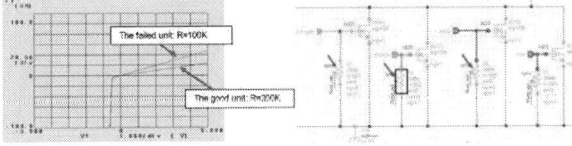

Resistor PHV

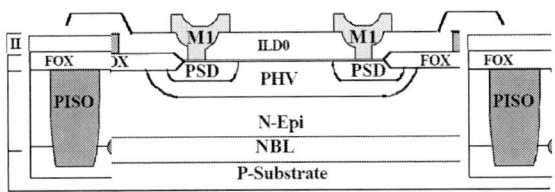

Poly ring to cut parasitic MOS

Fig. 8. The failure mode, related circuit and cross-section diagram of the PHV resistor

After a series of analysis, including PEM (Photon Emission Microscopy) and TLS-OBIRCH (Thermal Laser Stimulation -Optical Beam Induced Resistance Change), the failure was isolated to a resistor. According to schematic and layout analysis, related circuit and device are shown in Fig 8. The resistor is formed by PHV (P13) implant, which is picked up by PSD (P15), and isolated by PISO (P15).

However, physical analysis did not find any visual hints after delayer to AA by face lapping. Chemical etching has to be used, and duration is 25 seconds. Some abnormal color is revealed at PHV implant area. The color of the anomaly is deeper than adjacent PHV area, which means the etch rate for the area is higher than PHV (P13), as shown in Fig 9.

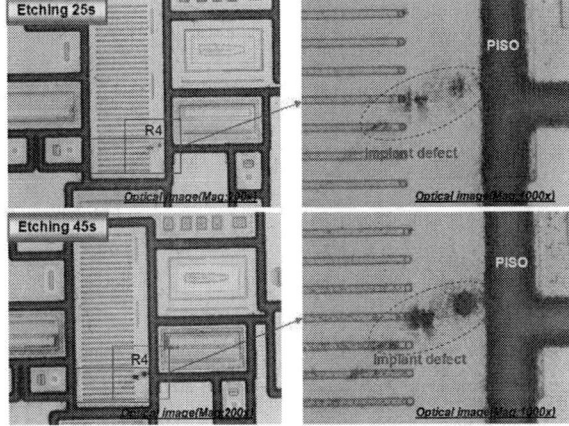

Fig. 9. The top view of PHV resistor after etching

According to layout, PHV resistor is isolated by PISO ring. Etching result reveals the etch rate for the defect is same with PISO (P15), higher than PHV (P13). It is reasonable to believe the defect is PISO. Product manual indicates that PISO gets sheet resistance of 13 ohms per square, much lower than that of PHV, 700 ohms per square. Furthermore, the direction of defect shortens almost half of effective length of PHV resistor. The failure mechanism diagram is shown in Fig 10.

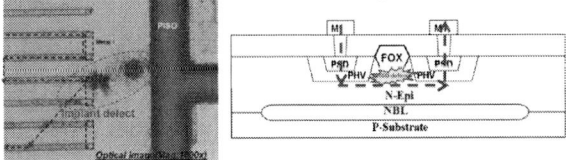

Fig. 10. Failure mechanism diagram

B. Leakage induced by implant defect

The failure mode of the failure is resistance short between HS1 and GND, as shown in Fig 11 (a). TLS-OBIRCH reveals resistance change spot on a MOSFET "MLD3", as shown in Fig 11 (b). But schematic analysis does not find relationship between HS1-GND through MLD3. According to layout, Fig 11 (d), MLD3 is surrounded by PISO (P15), which was connected with GND. Its function is isolating MLD3 with adjacent devices.

Similar with previous case, physical analysis did not find any visual hints after delayer to AA by face lapping. Chemical etching has to be used.

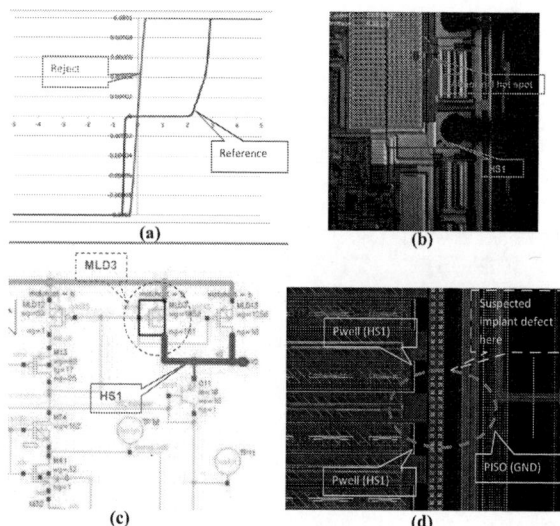

Fig. 11. (a) I-V curves of HS1-GND for Reject and Refenect parts; (b) OBIRCH images; (c) Schematic of related circuit; (d) Top-view and cross-section Diagram of MLD3

Chemical etching result reveals the etch rate for the defect is same with PISO (P15), higher than PWELL (P13). It is reasonable to believe the defect is PISO. The failure mechanism is PISO implant defect creating a link between MLD3 Drain (PWELL) and GND (PISO), as shown in Fig 12.

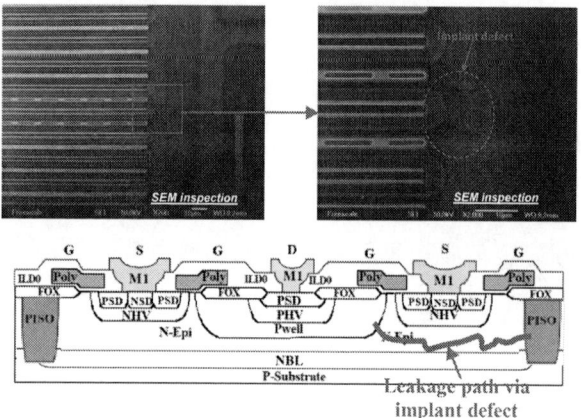

Fig. 12. The top view of MLD3 after etching and failure mechanism diagram

IV. SUMMARY AND CONCLUSION

This research studies a stain chemical etching rate for different doping. By choosing typical devices including appropriate doping agent and density, etch rate is collected. Qualitative conclusion is drawn as following.

N14 and P12 is the starting density to see obvious profile for N and P respectively. For same density, P has higher etching rate than N. For same doping type, higher density gets higher etch rate.

According to the conclusion, real case is presented. Etch rate difference causes obvious difference of color depth. Based on the color depth difference, defect composition can be supposed and then failure mechanism is easily setup.

ACKNOWLEDGMENT

The author would like to thank Yi Che, for his support on sample preparation and helpful suggestions. Meanwhile I sincerely appreciate my colleagues in Tianjin product analysis laboratory of Freescale semiconductor.

REFERENCES

[1] KAUSHIK ROY, SAIBAL MUKHOPADHYAY, "Leakage Current Mechanisms and Leakage Reduction Techniques in Deep-Submicrometer CMOS Circuits", PROCEEDINGS OF THE IEEE, VOL. 91, NO. 2, FEBRUARY 2003, pp. 305-327.

[2] Lee, W.F., Chin, A., Seah, P.H, "Application of Wright Etch in failure analysis on localized abnormal implant profile in wafer fabrication", IPFA 2011, pp. 1-4.

[3] Neo, S.P., Phong, O.L., Song, Z.G., "Study and application of wright etch in sub-quarter-micron technology", Semiconductor Electronics, 2004. ICSE 2004. IEEE International Conference on.

Backside Failure Analysis Techniques: What's The Gain Of Silicon Getting Thinner?

(Invited Paper)

Christian Boit[1], Norbert Schäfer[2], Daniel Abou-Ras[2], Clemens Helfmeier[1], Arkadiusz Glowacki[1], Uwe Kerst[1]

[1]TUB Berlin University of Technology
Einsteinufer 19, Sekr E2, 10587 Berlin, Germany
christian.boit@tu-berlin.de

[2]HZB Helmholtz-Zentrum Berlin für Materialien und Energie
Hahn-Meitner-Platz 1, 14109 Berlin, Germany

Abstract—**Failure analysis (FA) of electronic devices today is mostly conducted through the device backside. Advanced silicon (Si) backside preparation for this purpose has developed over the past years, with a final Si thickness from around $300\,\mu m$ to $100\,\mu m$ down to around $20\,\mu m$ to $10\,\mu m$. This paper discusses what to expect if Si can be processed a little thinner, from $20\,\mu m$ down to a few μm. Improvement of optical imaging, spectral extension of photon emission and expanded optical interaction for stimulation techniques are investigated. Further, Si thickness is coming close to the penetration depth of particle beams. The interaction potential for device analysis is discussed, preliminary results are presented.**

I. Introduction

With the complexity of interconnect routing and frontside preparation, electronic device failure analysis (FA) mostly has to access the device through the chip backside. Then, transmission through homogeneous silicon (Si) is the criterion whether interaction with the active devices is possible. On the sample preparation side, Si is conductive, so does not charge up much and can usually be removed without penalty.

The paper discusses the optimization of backside thickness and back surface reflection properties that allow the application of a variety of techniques. Especially the thickness of the remaining Si plays an important role when modern technology nodes are analyzed. In a second step these two important parameters for backside FA are discussed with respect to optical and particle beam techniques.

II. Sample Preparation

For a successful backside FA approach, sample preparation is a key factor. Without proper preparation, access to active devices is almost impossible. Furthermore, for a good interaction with active devices, homogeneous preparation is as essential as choosing the right target Si thickness. In this work, we targeted remaining Si thicknesses of one to twenty micrometers. MAX V complex programmable logic devices (CPLD) from Altera with on-die non-volatile memory storage fabricated in a 180 nm technology were used [1]. The samples were configured to build a scan chain with externally driven data and clock inputs and data output.

A. Mechanical preparation

Mechanical preparation allows to thin down Si comfortably to a remaining thickness between fifty and ten micrometers. The samples were thinned using the UltraTec Inc. ASAP-1 chemical-mechanical polishing machine down to a target Si thickness of $20\,\mu m$ with more than $90\,\%$ success rate. Processing such a thin sample requires knowledge about the samples in use and frequently checking the remaining Si thickness using optical techniques (e.g. a photometer) is mandatory.

Due to various factors, pure mechanical preparation is insufficient when the bulk Si should be thinned down to single digit micrometer thicknesses. For reaching further down, additional preparation must be employed or very advanced mechanical preparation may be used. For subsequent Focused Ion Beam (FIB) preparation, the samples were spin-coated with the UltraTec Inc. FIB-friendly anti-reflective coating (ARC) fluid for better chemical resistance.

B. Focused Ion Beam (FIB) preparation

A FIB tool can be used to thin bulk Si, too. Though different suggestions were made to exclusively use a FIB tool for backside Si removal [2], it may not be the preferred way. A FIB beam current of 20 nA was applied for 7 min to 14 min with xenon-di-fluorine (XeF_2) gas assistance, regularly rotating the sample. By this special procedure, a trench planarity better than 500 nm was achieved for areas as large as $200\,\mu m \times 200\,\mu m$, even with a single jet system. Apart from obviously reducing valuable FIB operating time, this approach also allows the controlled homogeneous thinning of the sample down to one micrometer without the risk of losing the device functionality. If full thickness samples were used for FIB thinning, inhomogeneous etching parameters and difficult endpointing would greatly increase the risk for partial over-etching and subsequently losing the sample.

The authors were capable of identifying the endpoint through a variety of means. Apart from the often used passive or active voltage contrast when reaching well level, a coaxial optical system allows to estimate the etching speed from observing the pattern and strength of interference fringes. For

978-1-4799-3911-4/14 $31.00 © 2014 IEEE

this approach, a coaxial optical system with a narrow spectral bandwidth in the near infrared (IR) regime is used during FIB operation.

C. Summarizing sample preparation

With processes adapted to sample preparation, successful and reliable thinning of backside Si is possible. Though thinning to shallow trench isolation (STI) level with only several hundred nanometer Si left has been shown [3], [4], conducting such approaches on large areas in every day work may be difficult. Instead of working at STI level, this paper discusses means to deal with Si on a micrometer thickness basis. Micrometer thickness can be reliably achieved using standard FA equipment available in every FA lab.

III. NEAR-IR OPTICAL TECHNIQUES

Optical techniques applied from the backside of the die suffer from light absorption in the bulk Si. Backside bulk Si thinning is significantly improving the performance of the techniques. Advantages of thinner Si require a spectral sensitivity of the detector for photon energies considerably higher than Si band gap, significant for Si-based and negligible for InGaAs detectors. Laser-based techniques, in turn, gain from thinner Si mainly in cases when laser light wavelength is shorter than the one corresponding to the Si bandgap. In this section we will focus on the improvement of reflected light microscopy (RLM) and photon emission microscopy (PEM) using Si-CCDs.

A. Reflected light microscopy (RLM)

The reflected light imaging microscopy performed from the backside of the die takes advantage of the light transmitted through the bulk Si. A halogen lamp illuminates the sample. A portion of light is reflected from the back surface (wavelength dependent phenomenon) and the remainder is transmitted into the bulk Si. There, depending on the doping and thickness of the sample, a portion of light is absorbed. The remainder is reflected in part from the structural features on the device frontside (the imaging target) and again absorbed on its way to the back surface. Again, a portion is reflected at the back surface (multiple reflections phenomenon). The transmitted portion outside the silicon is used for imaging. The light which is absorbed in the bulk Si represents essentially lost information. The light that is reflected from back surface at the first incidence is parasitic noise. Therefore, the goal in backside RLM is to maximize the intensity ratio of the light coming out of the sample to the light reflected from the back surface at the first incidence (used later in the text as *Ratio*).

The transmission into the Si and reduction of parasitic reflection is optimized by low roughness back surface coated with ARC. The bulk Si is thinned to reduce the absorption. All the related phenomena are wavelength dependent. Thinning Si and a low roughness back surface enables the use of shorter wavelengths for imaging, enhancing the optical resolution limit. As a result, the mentioned Ratio is enhanced and contrast of the images improved. Because Si-based CCD detectors

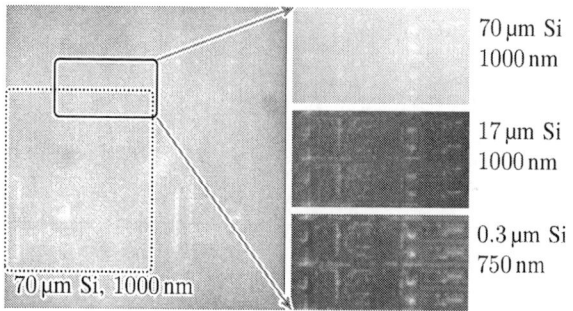

Fig. 1. Reflected light images acquired through bulk Si of various thicknesses and using various spectral filtering. The dashed area corresponds to the approximate position of Figure 6 for later use.

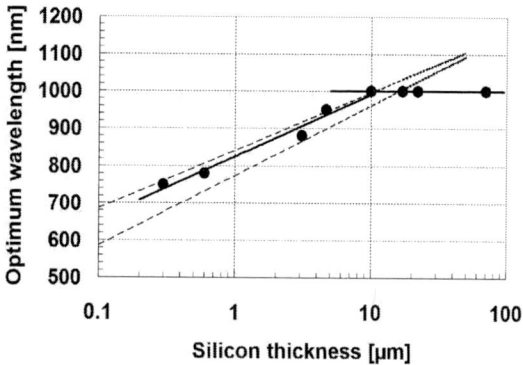

Fig. 2. Optimum wavelength for the reflected light imaging as a function of the bulk Si thickness.

are sensitive to exactly those wavelengths which are also absorbed in bulk Si, only a very narrow spectral detection window remains for regular die thicknesses. The bandwidth of the window depends on the doping of the Si material and the actual thickness of the bulk Si. Therefore, wideband imaging (using the entire available spectrum of the Si-CCD) is not the best choice for backside imaging as there is always light reflected from the back surface. Narrowing down the observation bandwidth by using interference bandpass filters and proper selection of the central wavelength enhances the Ratio.

Narrow-band backside imaging has then been performed for numerous samples having various bulk Si thicknesses from 70 μm down to STI level (≈ 0.3 μm). For each case a set of bandpass filters have been used for imaging to determine the optimum observation wavelength. The combined effect of the bulk Si thinning and proper selection of observation wavelength can be seen in Figure 1. The improvement of the image quality and resolution is apparent as the bulk Si is thinned down. From all acquired images, the dependence between the optimum observation wavelength (best contrast, best resolution) and the thickness of the bulk Si layer could be established and is presented in Figure 2. For samples thicker than ≈ 10 μm the optimum solution is observation at 1000 nm. This limitation is due to the use of Si-CCD (sensitivity

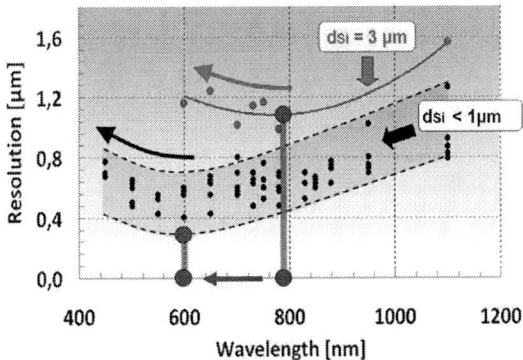

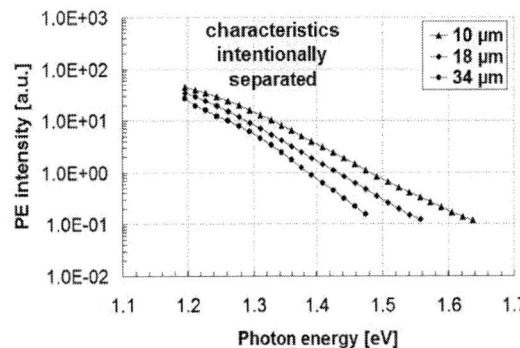

Fig. 3. Improvement of the spatial resolution of reflected light imaging performed using Si-CCD and interference bandpass filters for samples of various bulk Si thicknesses [5].

Fig. 4. Impact of the backside bulk Si thinning on the PE spectral range detectability. Measurement data obtained using non-intensified Si-CCD.

rapidly decreasing beyond ≈ 1000 nm) and is marked by the black bold line. Below $10\,\mu m$ thickness, the sample opens up significantly to transmit shorter wavelengths. Therefore, the optimum point of observation is shifting towards shorter wavelengths. The two dashed lines represent the simulated data obtained for two different values of the Ratio assumed to correspond to best image quality.

In addition, another study has been previously carried out [5] where simple rectangular shape test structures were used in order to be able to calculate the image resolution by edge response functions. The calculated resolution as a function of the observation wavelength for two different Si layer thicknesses is presented in Figure 3. Again, most importantly, there is an optimum wavelength for each sample thickness that gives the best resolution. While decreasing the wavelength of observation, the image quality and resolution improve (from basic diffraction-limitation law) until a wavelength specific to each Si thickness. Beyond that, absorption reduces the Ratio, resulting in poorer contrast and resolution. This approach of optimum wavelength estimation gives slightly different numbers compared to the pure qualitative approach presented before (Figure 1). Nevertheless, the results are consistent and the differences are probably due to the method of the resolution determination.

B. Photon emission microscopy (PEM)

The detection of photon emission signals can also be significantly improved by bulk Si thinning due to the decreased absorption in the remaining bulk Si layer. However, the exact gain depends also on the spectrum of the light emitted by the devices being imaged. By thinning the Si the transmission characteristic of the remaining layer is shifting towards shorter wavelengths leaving a wider spectral regime available for detection. As a result, the integral PE intensity will increase due to reduced absorption in a thinner layer for all wavelengths in varying degrees, but may additionally increase due to the detection of previously masked wavelengths.

Recently it has been demonstrated that a Si-CCD can also be successfully used for backside spectroscopic PE analysis if

performed through backside thinned samples. The detectable spectral regime is dependent on the remaining thickness of the bulk Si. Figure 4 demonstrates PE spectrum acquisition for an n-type field effect transistor (FET) operated in saturation for various bulk Si thicknesses. PE spectrum of saturated FET is an exponential function of PE intensity versus wavelength. Thinning down bulk Si layer enables the transmission of lower wavelengths and effective acquisition of a wider spectral regime.

Using recent intensified Si-CCDs allows the acquisition of useful PE spectra for $50\,\mu m$ and thicker samples. Vice versa, if samples are thinned further, the spectral range can be easily extended better filling out the full Si-CCD sensitivity bandwidth and/or the acquisition time can be significantly shortened. For instance, for thin samples ($10\,\mu m$ or less), photons with an energy in the range of $1.7\,eV$ to $1.9\,eV$ can also be detected resulting in a detectable spectral range of $0.5\,eV$ to $0.7\,eV$ as compared to the full Si-CCD spectral sensitivity range of $1.6\,eV$ (from $1.1\,eV$ to $2.7\,eV$).

IV. ELECTRON BEAM TECHNIQUES

The currently most common electron beam based techniques for physical analysis are electron beam probing (EBP), electron beam induced current (EBIC) and electron beam absorbed current (EBAC). These are considered with respect to the backside approach in the following sections.

A. Electron Beam Probing (EBP)

The electron beam is directed to the site of interest, causing emission of secondary electrons (SE). Beside material, orientation and topology contrast, the SE emission intensity also depends on the surface potential at the specific emission location. Thus, for known (or homogeneous) other parameters, the electrical potential at the targeted surface is reconstructed from the SE intensity.

For backside analysis, EBP could be conducted on the body or the drain of a transistor. For the observation of the body potential, the bulk material must be removed until the transistor body is separated from other transistors' body areas, by exposing STI in FIB preparation. Subsequently,

978-1-4799-3911-4/14 $31.00 © 2014 IEEE

the transistor body exhibits signal signatures of its drain voltage [6]. EBP can only be conducted on samples that expose STI, increasing the likeliness of over-etching the device and permanently damaging the sample. The technique allows unique performance but at quite a risk of losing the sample.

B. Electron Beam Induced Current (EBIC)

The analysis of carriers in semiconductor devices through electron beams began as early as 1958 [7]. The basic understanding is build upon the creation of electron hole pairs in the semiconductor [8], [9]. Different theoretical works have suggested models for the carrier generation and resulting carrier concentration [9]–[11]. The devices collect carriers and the resulting current is observed on a terminal. From literature [8], a penetration depth of below $1\,\mu m$ results from beam energies below $7\,keV$ when Si is targeted. In order to achieve high resolution with the EBIC technique, low beam energies should be used. The resulting carrier generation is limited to the electron range and thus can be more precise for small ranges [8].

Though low beam energies are not as intensively studied as energies of several keV, recent simulation works suggest, that the penetration depth depends exponentially on the beam energy in this range [12]. This makes $100\,nm$ ranges accessible for energies below $2\,keV$ in Si. Even at such low energies, electrons can generate substantial amounts of free carriers. For Si, the number of carriers generated per incident electron N is $N = {}^{E}/3.65$ [13], allowing for attenuation of the primary beam current by factors between 100 and 1000 for useful beam energies E.

Applying EBIC to the backside of active electronic devices thus has some limiting factors. Device currents in the µA range are possible with nA beam currents but resolution must be investigated from case to case. Once the actual setup of the sample is understood, a suitable beam energy can be chosen to reach the pn-junction of interest. Consequently, the knowledge of the exact sample thickness is not of first importance as the beam energy is a parameter that can be easily adjusted. Though diffusion is of importance for EBIC application [8], the relatively thin samples do not allow for high diffusion length due to surface recombination. This actively limits the spatial distribution of the injected current making the technique easier to apply on smaller dimensions.

C. Electron Beam Absorbed Current (EBAC)

EBAC is very similar to EBIC. Instead of generating excess carriers inside a semiconductor material, the primary beam is absorbed by the sample. Depending on the material and physical connectivity of the irradiated region, the current can be detected externally [14]. This technique has been adopted to FA for backend analysis of opens and shorts [15]–[17]. The resolution is limited in a similar way as that of the EBIC technique by electron scattering inside the solid. Thinning could be required to achieve nm scale resolution [17].

In order to absorb the primary beam without carrier generation, metals and field free insulators should be targeted. For

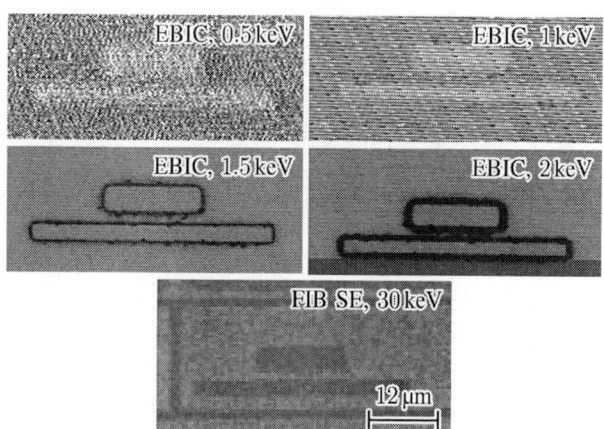

Fig. 5. Electron beam induced current (EBIC) images at different beam energies together with a FIB image at 30 keV showing surface potential from IO pad. For low electron beam energies, the carriers are absorbed inside the well whereas for higher electron beam energies, the electrons generate additional carriers inside the space charge region. The trench thickness is $0.8\,\mu m$, exposing the wells.

backside application, this is only possible through STI or when the full semiconductor material is removed [16]. This either inevitably removes the full circuit functionality, or requires yet unknown details on electron hole pair creation in insulators.

V. PRACTICAL RESULTS OF ELECTRON BEAM ON BACKSIDE PREPARED SAMPLES

The prepared devices were connected in situ to external driving and acquisition systems for signal feed-out. A Raith beam blanker was installed on a Zeiss Gemini electron microscope to minimize the collateral influence on the sample from the electron beam. We present the results of two experiments: EBAC and EBIC on IO-structures and EBADA on switching devices.

A. IO-Structure EBIC and EBAC

Using the electron beam, both diffusions and wells could be detected. With low beam energies, the current was absorbed in the exposed layer. The well contrasts in the two upper images of Figure 5. As the beam energy is increased, the current is able to generate electron hole pairs that are most efficiently collected directly inside the space charge region (SCR). For higher beam energies, the well SCR is stronger in contrast, see the next two images of Figure 5. For best absorption of the primary beam current, the structure should be exposed and subsequently, a beam energy below 1.5 keV should be used. For carrier generation, a slightly higher beam energy is necessary.

B. Electron Beam Assisted Device Alteration (EBADA)

Once the electron beam can be precisely controlled to cause additional current inside pn-junctions, it can be used to inject faults. When an inverter is loaded with an additional current in its drain diffusions, the performance will alter. Thus, the electron beam can be used to shmoo a soft fault, similar to the

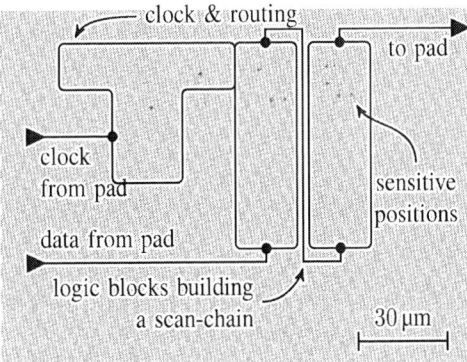

Fig. 6. The electron beam causes the introduction of additional high values during scanning a low-to-high transition at sensitive sites (arrow). Data and clock are sensitive to the electron beam bombardment. Frames are approximate, the visible area corresponds to the dashed frame in Figure 1.

process done when performing light assisted device alteration (LADA). On the chosen device, bits could be flipped along the scan chain, very similar to the approach taken in [18], [19] though the chosen setup did not yet allow reliable operation, see Figure 6.

During the course of this work, we observed failing devices due to electron beam irradiation of various beam energies down to 2 keV. A similar observation was made for frontside EBIC experiments on MOS structures [8] and backside electron beam micromachining on ring oscillators [20]. The effect was very quickly saturated in terms of typical electron beam operating times. The quantitative study [20] showed beam dose values comparable to charges to breakdown for Fowler Nordheim tunneling [21] and direct tunneling [22] in a FET gate. Today, oxide stress can be controlled for reliable product functionality and the same is expected to apply to electron beam irradiation. Improving the electron beam work-flow for backside irradiation of MOS devices may create beneficial conditions where device degradation is limited and the desired effect is still observed. The background and mechanisms are subject to ongoing research.

VI. ACKNOWLEDGMENTS

The authors would like to thank DCG Systems for support on our OptiFIB tool and Hamamatsu KK for support on our PHEMOS 1000 system. Special thanks are due to Harald Stapel (HZB) for technical assistance with the EBIC measurements. Daniel Abou-Ras and Norbert Schäfer would like to acknowledge support with the EBIC system by point electronic GmbH. We thank Andreas Eckert (TUB) for the meticulous and experienced sample preparation.

REFERENCES

[1] A. Corporation, *MAX V Device Handbook*, 2011.

[2] C. Rue, S. Herschbein, and C. Scrudato, "Backside circuit edit on full-thickness silicon devices," in *Conference Proceedings from the 34th International Symposium for Testing and Failure Analysis.* ASM International, Nova 2008, pp. 141–150.

[3] R. Schlangen, R. Leihkauf, K. U., T. Lundquist, P. Egger, and C. Boit, "Physical analysis, trimming and editing of nanoscale IC function with backside FIB processing," *Microelectronics Reliability,* vol. 49, pp. 1158–1164, 2009.

[4] R. Schlangen, R. Leihkauf, T. Lundquist, P. Egger, U. Kerst, and C. Boit, "Trimming of IC timing and delay by backside FIB processing - comparison of conventional and strained technologies," in *Electron Devices Meeting, 2008. IEDM 2008. IEEE International,* 2008, pp. 1–4.

[5] A. Glowacki, C. Helfmeier, U. Kerst, and C. Boit, "Improvement of optical resolution through chip backside using FIB trenches," in *Proceedings of the 36th International Symposium for Testing and Failure Analysis (ISTFA 2010),* vol. 36. ASM International, 2010, pp. 176–180.

[6] R. Schlangen, R. Leihkauf, U. Kerst, C. Boit, R. Jain, T. Malik, K. Wilsher, T. Lundquist, and B. Kruger, "Backside e-beam probing on nano scale devices," in *Test Conference, 2007. ITC 2007. IEEE International,* 2007, pp. 1–9.

[7] T. Everhart, *Contrast Formation in the scanning electron microscope,* 1958.

[8] H. Leamy, "Charge collection scanning electron microscopy," *Journal of Applied Physics,* vol. 53, no. 6, pp. R51–R80, Jun 1982.

[9] J. I. Hanoka and R. O. Bell, "Electron-beam-induced currents in semiconductors," *Annual Review of Materials Science,* vol. 11, no. 1, pp. 353–380, 1981.

[10] T. E. Everhart and P. H. Hoff, "Determination of kilovolt electron energy dissipation vs penetration distance in solid materials," *Journal of Applied Physics,* vol. 42, no. 13, pp. 5837–5846, 1971.

[11] D. Kyser and D. Wittry, "Spatial distribution of excess carriers in electron-beam excited semiconductors," *Proceedings of the IEEE,* vol. 55, no. 5, pp. 733–734, May 1967.

[12] O. Kurniawan and V. K. S. Ong, "Investigation of range-energy relationships for low-energy electron beams in silicon and gallium nitride," *Scanning,* vol. 29, no. 6, pp. 280–286, 2007.

[13] L. Reimer, *Scanning Electron Microscopy.* Springer-Verlag, 1985.

[14] C. Smith, C. Bagnell, J. Cole, E.I., F. DiBianca, D. Johnson, W. V. Oxford, and R. H. Propst, "Resistive contrast imaging: A new sem mode for failure analysis," *Electron Devices, IEEE Transactions on,* vol. 33, no. 2, pp. 282–285, Feb 1986.

[15] W. P. Lin and H. J. Chang, "Physical failure analysis cases by electron beam absorbed current amp; electron beam induced current detection on nano-probing sem system," in *Physical and Failure Analysis of Integrated Circuits (IPFA), 2010 17th IEEE International Symposium on the,* July 2010, pp. 1–4.

[16] K. Dickson, G. Lange, K. Erington, and J. Ybarra, "Electron beam absorbed current as a means of locating metal defectivity on 45nm soi technology," in *International Symposium for Testing and Failure Analysis, Proceedings from the 36th (ISTFA),* 2010.

[17] M. Simon-Najasek, J. Jatzkowski, C. Große, and F. Altmann, "A new technique for non-invasive short-localisation in thin dielectric layers by electron beam absorbed current (ebac) imaging," in *38th International Symposium for Testing and Failure Analysis, Conference Proceedings from the (ISTFA),* 2012.

[18] T. Kiyan, C. Brillert, and C. Boit, "Timing analysis of scan design integrated circuits using stimulation by an infrared diode laser in externally triggered pulsing condition," *Microelectronics Reliability,* vol. 48, no. 8–9, pp. 1327–1332, 2008, 19th European Symposium on Reliability of Electron Devices, Failure Physics and Analysis (ESREF 2008).

[19] T. Kiyan. "Dynamic analysis of tester operated integrated circuits stimulated by infra-red lasers." Ph.D. dissertation, 2010.

[20] Y. Greenzweig, Y. Drezner, A. Raveh, O. Sidorov, and R. H. Livengood, "E-beam invasiveness on 65 nm complementary metal-oxide semiconductor circuitry," *Journal of Vacuum Science Technology B: Microelectronics and Nanometer Structures,* vol. 29, no. 2, pp. 021 202–021 202–5, Mar 2011.

[21] Y.-B. Park and D. Schroder, "Degradation of thin tunnel gate oxide under constant fowler-nordheim current stress for a flash eeprom," *Electron Devices, IEEE Transactions on,* vol. 45, no. 6, pp. 1361–1368, Jun 1998.

[22] S. Lombardo, J. Stathis, B. Linder, K. L. Pey, F. Palumbo, and C.-H. Tung, "Dielectric breakdown mechanisms in gate oxides," *Journal of Applied Physics,* vol. 98, no. 12, pp. 121 301–121 301–36, Dec 2005.

Advanced TEM Applications in Semiconductor Devices

[1]A. Y. Du, [1]J. Zhu, [2]Y. K. Zhou, [1]B. H. Liu, [1]Eddie Er, [1]Z. Q. Mo, [1]S. P. Zhao and [1]Jeffrey Lam

[1]TD-FA, GLOBALFOUNDRIES Singapore, 60 Woodlands Industrial Park D, Street 2,
Singapore 738406

[2]Wintech Nano-technology Services Pte Ltd. 10 Science Park Road #03-28, The Alpha, Singapore Science Park II Singapore
117684

Phone: (65) 6360 3469. Fax: (65) 63622938. Email: anyan.du@globalfoundries.com

Abstract-There is increasing demand of advanced TEM techniques for modern IC failure analysis. Some practical issues of using TEM holography in studying MOSFETs P-N junction, channel strain and magnetic domains are discussed in this paper. It is shown that salicide/contact have significant effect on the phase diagram of shallow S/D P-N junction and hinders its application in shallow junction devices. In holography strain measurement, it is found that intensity of diffraction beam depends strongly on the sample thickness and crystallography orientation. A proper sample thickness should be chosen to maximize the nearly two-beam diffraction beam intensity and therefore the measurement sensitivity. Magnetic domain imaging is much affected by the FIB damage during sample preparation and low energy milling should be employed. At last, it is demonstrated that numerical de-convolution of EELS spectrum provides improved energy resolution for more reliable dielectric bonding chemical analysis.

I. INTRODUCTION

In the past two decades, Transmission Electron Microscopy (TEM) has become an indispensable tool in the rapid-advancing semiconductor industry for process evaluation and failure analysis. However, with continuous scaling in device feature size, introduction of new materials and more sophisticated structures, there is increasing demand of advanced TEM techniques for analyzing materials properties such as P-N junction profile, channel strain, and magnetic domains with nanometer scale spatial resolution. TEM holography technology is a powerful technique for such applications [1-3]. In this work, we report the use of TEM holography to study P-N junction profile in Si MOSFETs, distribution of compressive channel strain induced by S/D eSiGe and magnetic domain imaging.

TEM is commonly used in failure analysis because of its (sub-) nanometer spatial resolution for structure characterization and chemical analysis. Among many TEM techniques, electron holography is a powerful technique to retrieve information of electron wave phase which can be used to evaluate P-N junction profile, strain distribution and magnetic domains. There are more stringent requirements on TEM sample for holography P-N junction analysis as compared with sample used in the conventional imaging. In order to avoid abrupt electron wave phase change caused by local thickness variation, sample thickness must be kept as uniform as possible in the area of interest. Usually, back-side FIB milling is used to obtain sample with uniform thickness since front-side FIB milling shows thickness variation due to FIB curtain effect. In addition,

interface charging should be minimized by coating conductive layer, e.g., a thin Pt layer can be deposited at the top surface of sample by electron beam assisted deposition. In this work, two samples with different processing history were used for P-N junction study. One sample is processed after S/D ion implantation followed by rapid thermal annealing, and the other one is processed after S/D contact formation.

Strain engineering is a cost-effective method used in modern VLSI fabrication in order to improve MOSFETs speed. Carrier's mobility can be enhanced by applying either tensile or compressive stress onto crystalline Si channel which reduces the effective mass of electrons or holes, respectively. Different TEM techniques have been reported for measuring channel strain including, convergent electron beam diffraction (CBED), nano-beam diffraction (NBD), HRTEM geometric phase analysis and dark-filed off-axis holography (DFOH). Among these techniques, DFOH attracts much attention since it provides 2-D strain mapping with nano-meter spatial resolution. By replacing Lorentz lens with objective lens, we've demonstrated sub-nanometer scale strain mapping of Si fin structure. There is less stringent requirement on thickness uniformity for strain measurement comparing with P-N junction analysis; hence TEM sample used in this work is prepared with normal front-side FIB milling. However, a proper sample thickness should be chosen in order to maximize diffraction beam intensity as beam intensity is a function of sample thickness as predicted by electron dynamical diffraction theory. Sample thickness should also be thick enough to avoid strain relaxation. Since DFOH is carried out in dark-field mode, one practical limitation is the low intensity of diffraction beam which makes experiment difficult. Therefore, an optimum sample thickness with high beam intensity and minimum strain relaxation should be chosen.

Electron holography is very useful in studying magnetic materials since electron wave phase change can be easily imaged at presence of magnetic field. When FIB is used for TEM sample preparation, it is critical to minimize ion beam damage to the materials crystal structure in order to retain the original materials property. There are many reports on FIB milling-induced amorphous layer and ion beam implantation damage of the compound semiconductor materials. With the conventional TEM technique, there may not be clear evidence on the magnetic property change induced by FIB damage. Very

978-1-4799-3911-4/14 $31.00 © 2014 IEEE

thin TEM specimen is necessary for good quality HRTEM and chemical composition analysis of magnetic layer with heave metallic elements. Electron holography which is very sensitive to the material's magnetic property provides direct evidence on magnetic property change caused by FIB milling.

Electron energy loss spectroscopy (EELS) in conjunction with mono-chromator and probe corrector is a very powerful technique for nano-structure chemical bonding analysis, but the high cost of such equipment is unaffordable for most of the FA labs. How to improve the EELS energy resolution of old TEM system without use of mono-chromator is the key challenge. We have set-up in-house software to improve the energy resolution of EELS spectra so that it is possible to analyze the chemical bonding information which usually requires mono-chromator. The chemical bonding study from such analysis provides more information for process team to fine tune manufacturing parameters and improve device performance.

II. EXPERIMENTAL AND RESULTS

A. P-N junction 2D profile with electron holography

FEI Titan 80~300kV TEM equipped with Lorentz lens and biprism is used for off-axis electron holography study in this work. Two samples were selected for P/N junction profiling, one sample was processed after S/D dopant implantation and RTA annealing while the other sample was processed after S/D metal contacts formation. TEM specimens were prepared with FIB backside milling in order to obtain uniform substrate thickness at P/N junctions. With same Si thickness across P-N junction, electron wave phase change is caused by built-in electrostatic potential in the depletion region of P-N junction. Fig. 1 & 2 show the reconstructed amplitude and colour phase images of two samples. In both samples, apparent phase changes can be observed around P/N junctions. Fig. 3 & 4 show the plots of phase change across P/N junctions in two samples. As shown in Fig. 3b for the sample processed after S/D implantation annealing, the total phase change ($\Delta\varphi$) across P/N junction is 1.1 radian corresponding to built-in potential of 0.59V which is reasonable for P/N junction doping concentration used in modern MOSFETs device. Built-in voltage (V_{bi}) can be calculated using equation:

$$\Delta\varphi = C_E \cdot V_{bi} \cdot t \qquad (1)$$

where, $\Delta\varphi$ is the wave phase change measured from holography phase image, C_E is a constant value for TEM operated at certain voltage, V_{bi} is the built-in potential across P/N junction, and t is the sample thickness. Sample thickness is estimated from reconstructed holography amplitude image with vacuum area as the reference and 180nm as inelastic scattering mean free path of Si material at 300kV TEM voltage. However as shown in Fig. 4b for the sample processed after S/D contact formation, the total wave phase change is 6.5 radian which is about 6 times greater than the phase change observed in previous sample. Clearly, such abrupt phase change can't be merely from built-in potential across P/N junction. In this case, P/N junction is so close to metallic salicide and contacts that the measured phase

change is mainly attributed to difference in materials inner potentials instead of built-in potential. Such effect can also be observed in the amplitude image in Fig. 2a, where abrupt contrast change across P/N junction is due to big difference in electron scattering factor between Si and metallic materials.

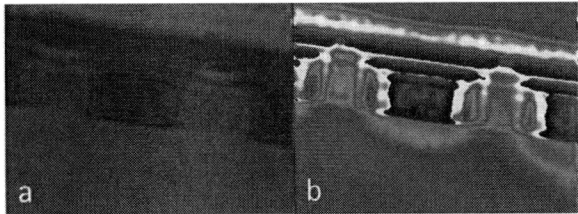

Fig. 1 (a) holography amplitude image and (b) color phase image of P/N junction from sample processed after S/D annealing.

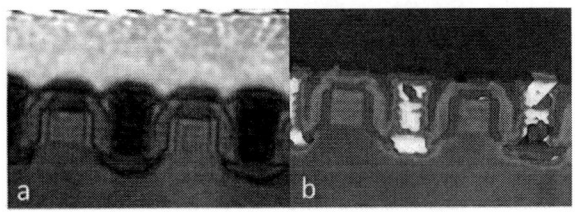

Fig. 2 (a) holography amplitude image and (b) color phase image of P/N junction from sample processed after S/D contact formation.

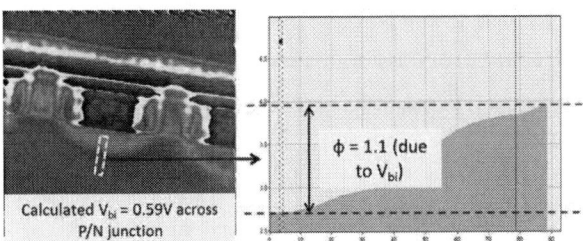

Fig. 3 electron wave phase plot across P/N junction of sample processed after S/D annealing. Phase change is due to built-in potential within depletion area.

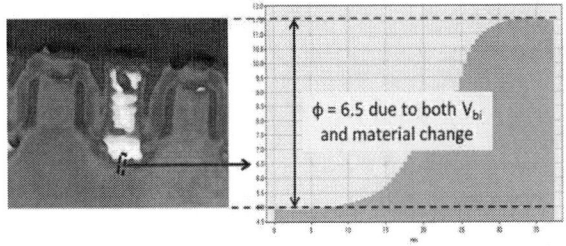

Fig. 4 electron wave phase plot across P/N junction of sample processed after S/D contact formation. Phase change is mainly attributed to the large difference in mean inner potential between Si and metals.

To support continuous gate length scaling while reducing channel leakage current (I_{off}), shallow S/D ion implantation is used to reduce drain-induced barrier lowering (DIBL) which results in V_t roll-off, commonly known as short channel effect. As a result, the physical space between the bottom of salicide/contacts and metallurgical P/N junction is very close. Hence, the reconstructed phase image of dopant profile overlaps with that from metals and therefore V_{bi} is difficult to be measured accurately. In the case where salicide and P/N

978-1-4799-3911-4/14 $31.00 © 2014 IEEE

junction is far enough, e.g., in 0.18um device, electron holography is still applicable for P/N junction dopant profiles study (Fig. 5). Again, an abrupt phase change is observed in salicide area which supports our above discussion.

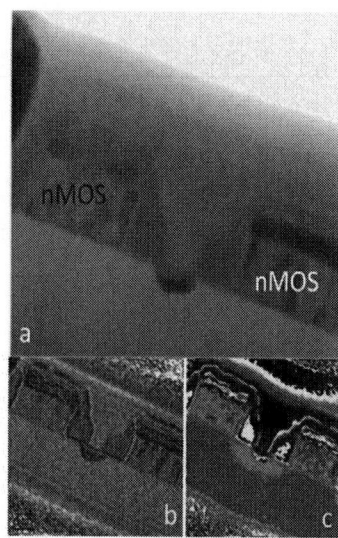

Fig. 5 (a) TEM bright field image, (b) TEM hologram, and (c) colour phase image of P/N junction from a sample processed after salicide formation.

B. 2-D strain profile using dark field off-axis holography (DFOH)

Embedded SiGe source/drain deposited by epitaxial growth is widely used in 65nm and below technology node to improve pMOSFETs performance. Direct measurement of channel strain is important in understanding the effectiveness of strain engineering which can be tuned by optimizing fabrication parameters. The dark-field off-axis holography provides 2D strain distribution of Si channel, as shown Fig. 6. During DFOH, TEM specimen is tilted to strong two beam diffraction condition in order to increase diffraction beam intensity and reduce dynamical different effects. The intensity of diffraction beam affects the sensitivity of strain analysis in DFOH. According to electron diffraction theory, simple prediction shows that maximum intensity is expected from sample with thickness $(n+1/2)\cdot\xi_g$, and minimum beam intensity from specimen thickness of $n\cdot\xi_g$, where n is 0 or integer and ξ_g is the extinction distance of Si which depends on material and TEM voltage. For 200kV TEM, extinction distances of 220 and 004 diffraction beam of Si material are listed in Table 1 [4]. Based on the values shown in table 1, one can make a rough prediction that DFOH is better for g_{004} but poor for g_{220} since the active PMOS in SRAM for 40nm is about 100nm (from beam intensity point of view). Similarly if sample thickness is 150nm, DFOH is better for g_{220} but poor for g_{004} DFHO strain analysis. In practice, beam intensity change is more complicated due to dynamical multi-beam coupling effect since a real two-beam condition is not attainable. In addition, variations in excitation errors also exist depending on sample tilting. We've studied correlation between diffraction beam intensity, sample thickness and TEM voltage using a wedge-shaped Si sample and found the optimum

thicknesses for g_{220} and g_{004} that provide highest beam intensity for DFOH study while considering both strain relaxation and FIB beam damage (results will be published later).

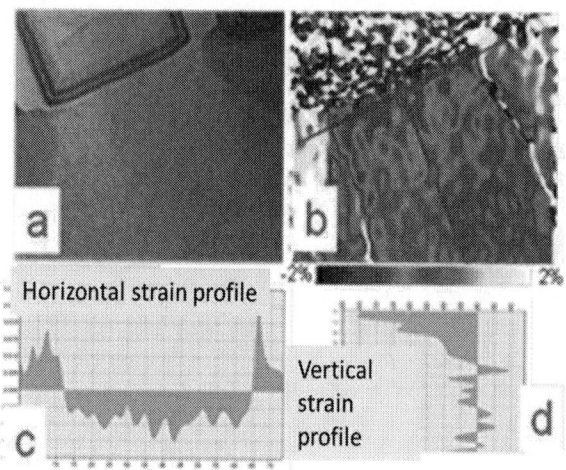

Fig. 6 (a) TEM bright field image of transistor with eSiGe S/D structure, (b) 2-D DFOH strain mapping at channel region, (c) strain profile along channel direction, and (d) strain profile along direction perpendicular to channel.

Table 1 lists extinction distance of Si 220 and 004 diffraction beam at 200kV

ξ_g (nm)	g_{220}	g_{004}	remark
	100	160	
$1/2\xi_g$	50	80	Maximum for diffraction beam
ξ_g	100	160	Minimum of diffraction beam
$3/2\xi_g$	150	240	Maximum for diffraction beam
$2\xi_g$	200	320	Minimum of diffraction beam

C. Magnetic domain imaging with holography

TEM holography is commonly used to study magnetic property of materials such as magnetic domains. In holography experiment set-up, Lorentz lens is used instead of normal objective lens since strong magnetic field generated near the objective lens can affect magnetic domain structure change in the sample. When electron wave passes through magnetic material, electron wave phase is changed by the magnetic field. Fig. 7 is TEM electron holography image of hard-disc magnetic materials with phase image showing the magnetic domain. Due to the presence of heavy elements in the magnetic materials, mean-free patch of the magnetic layers is very short. In order to obtain good quality HRTEM image, TEM specimen needs to be very thin. FIB sample preparation usually introduce amorphous layer on sample surface and thickness of amorphous layer is mainly dependent on ion beam voltage, e.g., ~25nm amorphous layer is formed on each side of Si material when ion beam voltage is 30kV. Such beam damage will certainly affect materials crystalline structure and therefore the magnetic domain profiling. The upper left corner of TEM sample represents a thin area (circled area in Fig. 7a) and is good to show the effect of beam damage to magnetic domain structure.

978-1-4799-3911-4/14 $31.00 © 2014 IEEE

In holography phase image as shown in Fig 7d, it is clear that the magnetic domains in this thin area (circled region) can't be imaged when comparing with the domain profile on the right side. In order to minimize beam damage to the crystalline structure of magnetic materials, low energy FIB milling is necessary.

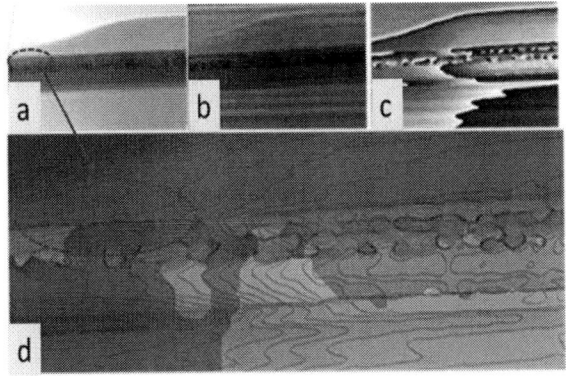

Fig. 7 (a) TEM bright field image of hard-disc sample prepared by FIB milling, (b) TEM hologram of sample, (c) reconstructed phase image, and (d) circled area shows absence of magnetic domains due to FIB damage.

D. EELS chemical bonding analysis with R-L de-convolution

Chemical bonding of Si in ultra-low-k (ULK) dielectric layer is one of the key factors affecting k-value of the material. The high k value of the dielectric layer increases time delay in BEOL and results in poor performance of the device. There is transition layer at ULK layer near the interface of ULK with metal line edge introduced by plasma etching and cleaning process during metal trench formation process. The transition layer may be SiO_2 type chemical bonding resulting high k-value (~3.4 for SiO_2 and ~2.4 for ULK). Here, EELS spectrum was de-convoluted with Richardson-Lucy method which effectively improves the energy resolution of spectrum which allows chemical bonding analysis of ULK dielectric layer The EELS spectra were taken at both edge and center of ULK for comparison. EELS analysis positions in the TEM specimen are shown in Fig.8a as indicated by the red dots. Both the original and de-convoluted EELS spectra are shown in Fig 8b. The purpose of using numerical de-convolution is to sharpen and spectrum and reveal the minor edges that are difficult to be differentiated in raw spectrum. In Fig. 8d, de-convoluted EELS spectrum of Si clearly shows sp2 hybridized state at ULK dielectric edge as indicated by the circle. The Si sp2 hybridized state is due to SiO_2 type chemical bonding formed during plasma cleaning process at edge of ULK dielectric, which is not present in the normal low-k dielectric layer as shown in the spectrum form center. In Fig. 8c, Oxygen K-edge shows typical shape of SiO_2 chemical bond at the ULK edge implying there should be higher k-value at edge location.

III. CONCLUSION

In summary, electron holography has been applied as an advanced failure analysis method for semiconductor failure analysis. The salicide/contacts effects in shallow junction on P/N junction profiling was discussed. Because of small distance between salicide/contacts and P/N junction in shallow junction device, holography is not suitable for junction profiling

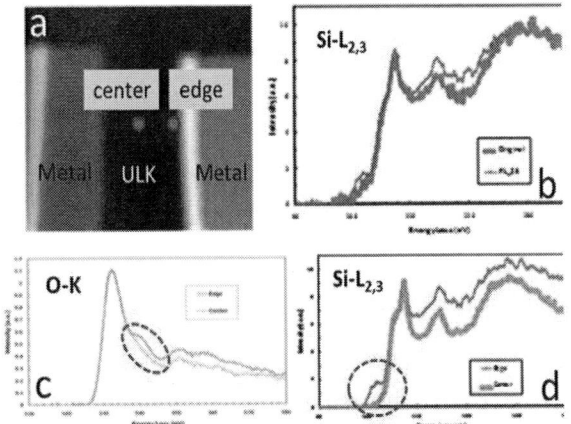

Fig. 8 (a) STEM image of BEOL structure with ULK dielectric, (b) comparison of EELS raw spectrum (blue) and de-convoluted spectrum (red), (c) EELS spectra of de-convoluted spectra of O-K edge, and (d) EELS spectra of de-convoluted spectra of Si-$L_{2,3}$ edge.

in 65nm or below devices. The extinction distance of Si affects the diffraction beam amplitude and therefore the sensitivity of DFHO strain analysis. The DFHO is very sensitive to the sample thickness at different sample orientation. Low energy FIB milling should be used to avoid FIB induced damage to the crystalline magnetic material in order to obtain reliable domain profiles. Application of numerical de-convolution method is presented here to improve energy resolution of EELS spectra for chemical bonding analysis in ULK. This method is particularly useful for those FA labs that can't afford to purchase very expensive mono-chromator for TEM.

ACKNOWLEDGMENT

We are very grateful to the support and encouragement from GLOBALFOUNDRIES Singapore TD-FA management team for this project development.

REFERENCES

[1] W. D. Rau, P. Schwander, F. H. Baumann, W. Höppner, and A. Ourmazd. "Two-Dimensional Mapping of the Electrostatic Potential in Transistors by Electron Holography" Phys. Rev. Lett. 82, 2614–2617 (1999).
[2] M. Hÿtch, F. Houdellier, F. Hüe and E. Snoeck. "Nanoscale holographic interferometry for strain measurements in electronic devices". Nature 453, 1086-1089 (19 June 2008).
[3] T. Matsuda, S. Hasegawa, M. Igarashi, T. Kobayashi, M. Naito, H. Kajiyama, J. Endo, N. Osakabe, A. Tonomura, and R. Aoki. "Magnetic field observation of a single flux quantum by electron-holographic interferometry" Phys. Rev. Lett. 62, 2519–2522 (1989).
[4] R. Uemichi, Y. Ikematsu and D. Shindo. "Precise Evaluation of Specimen Thickness by Convergent–beam Electron Diffraction Technique and Electron Energy–loss Spectroscopy" J. Japan Inst. Metals, Vol. 65, No. 5 pp. 427–433(2001).

Improvement of 3D Current Mapping by Coupling Magnetic Microscopy and X-Ray Computed Tomography

Nicolas Courjault (1), Fulvio Infante (1), Vincent Bley (2), Thierry Lebey (2) and Philippe Perdu (3),

(1) Intraspec Technologies, 3 Avenue Didier Daurat, 31400 Toulouse, France

(2) CNRS UMR 5213 LAPLACE, Bat 3R3 118 Route de Narbonne 31062 Toulouse Cedex 9

(3) French Space Agency, CNES DCT/AQ/LE bpi 1414, 18 Avenue E. Belin, 31401 Toulouse, France

Phone: +33 (0)5 61 27 45 54 ; Fax: +33 (0)5 61 27 47 32 ; E-mail: nicolas.courjault@intraspectechnologies.com

Abstract - **Magnetic Microscopy has demonstrated all its functionality for 2D component thanks to its ability to image current density distribution from magnetic field. At the "More than Moore" age, we need to improve our capabilities to detect and localize failure in 3D components. Unfortunately, it is not possible to directly image 3D current density from a magnetic field scan. 3D conductive path information, that could come from design, and failure assumptions are also needed. In this paper, a new approach based on X-Ray Computed Tomography that bypasses the need of design information and failure site assumptions is presented and its results are discussed.**

I. INTRODUCTION AND MOTIVATIONS

Magnetic Microscope is a failure analysis tool (figure 1) able to measure the magnetic field on an electronic system or component. Usually, the vertical component of the magnetic field is scanned along a horizontal plane parallel to the sample. In static or very low frequency, there is no or few electric field E variations so Magnetic Field B and Current density J are linked by the simplified Maxwell-Ampère equation (1)

$$\overrightarrow{rot}\,\vec{B} = \mu_0 \vec{J} + \frac{\delta \varepsilon_0 \vec{E}}{\delta t} \approx \mu_0 \vec{J} \quad (1)$$

Therefore, it is possible to image a current density from a magnetic scan. This technique named Magnetic Current Imaging (MCI) [1], works well on 2D samples where all the currents are in the same plane. In this case, $J_z=0$ (and $\frac{\delta J_z}{\delta z} = 0$) so there is a direct relation between the other components of J from the equation of continuity (2) when there is no charge (ρ) modification:

$$div\,\vec{J} = -\frac{\delta \rho}{\delta t} \approx 0 \quad (2)$$

$$div\,\vec{J} = \frac{\delta J_x}{\delta x} + \frac{\delta J_y}{\delta y} + \frac{\delta J_z}{\delta z} = \frac{\delta J_x}{\delta x} + \frac{\delta J_y}{\delta y} \approx 0$$

Because $\frac{\delta J_x}{\delta x} \approx -\frac{\delta J_y}{\delta y}$ it is possible to reverse B to J in this case [2]. Then, by mainly making a comparison between a good and a "bad" component, or system, we are able to localize defects like short and resistive open.

Unfortunately, this approach is no longer working for 3D devices because $\frac{\delta J_z}{\delta z}$ is no longer equal to zero and multiple parallel planes can be involved. Specific 3D cases have been solved by mapping magnetic field in two orthogonal planes [3] but this solution works only for simple current paths in 3D cubes. However, a more generic alternate approach has been setup few years ago.

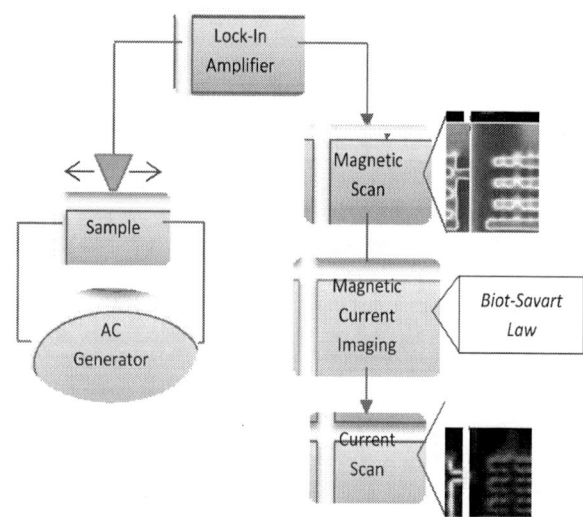

Fig 1. Synoptic of the Magnetic Microscope

This approach is based on comparison of a magnetic field 2D map acquired by magnetic microscope and a similar map obtained from simulation of magnetic fields generated by a current path. This current path comes from design information and assumptions about where the short takes place. This approach has been validated on simple devices [4] and then on more complex 3D via chain [5].

Nevertheless, this approach faces various challenges. First, we do not have necessarily the design of the 3D sample and second, it is time consuming to check numerous defect site assumptions in complex devices. The need of having a precise and accurate 3D current path on failed and golden devices is mandatory.

On the other hand, constant improvement of X-Ray Computed Tomography and advanced pieces of software to implement specific applications have opened the door to the extraction of these conductive paths from complex devices.

Fig. 2. Example of RX Computed Tomographic system

The objective of our work is to couple X-Ray Computed Tomography and magnetic microscopy to setup a generic approach for 3D current analysis in complex devices.

II. COUPLING DESCRIPTION

By using X-Ray Computed Tomography, we would like to obtain the current path inside the 3D component and introduce it into the magnetic simulator in order to obtain the theoretical magnetic field. Then, we compare magnetic field measured by magnetic microscopy and magnetic field generated by the introduced current path. Both results are fitted to visualize the 3D current path inside the real device. This coupling is illustrated by the following schematic (figure 3)

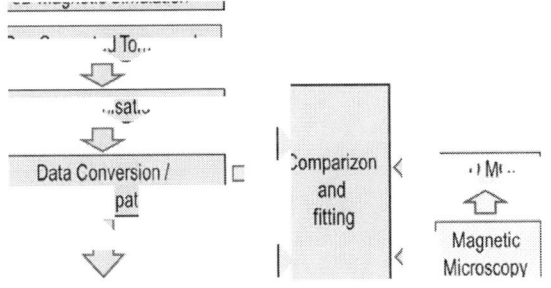

Fig. 3. Schematic of the X-Ray CT and Magnetic Microscopy coupling

III. EXPERIMENTAL RESULTS

In this paper, experiments were done on a Magnetic Microscope Magma C30 with a SQUID sensor, a X-Ray Computed Tomography V-Tome Xs240 and the sample is a two layer TSV chain structure. Each via has a 45 μm diameter, distance between two TSVs is 125μm and distance between layers is 120 μm (figure 4).

The tomography was realized with a resolution of 4.5μm at a voltage of 90kV and 50μA of current. A set of 1500 images has been captured in order to represent this TSV as a volume. The magnetic microscopy used an AC current of 1mA at a frequency of 5333Hz. A 2.5μm step on X and Y coordinates and a 30μm working distance between the top layer of the sample and SQUID window had been implemented. That corresponds to a distance of 180μm between sensor and the sample.

Skeletonisation is then obtained in order to find the theoretical path of the current. It uses a special algorithm, named *Distance-ordered Homotopic Thinning*, which was realized especially for 3D digital images by C. Pudney [6].

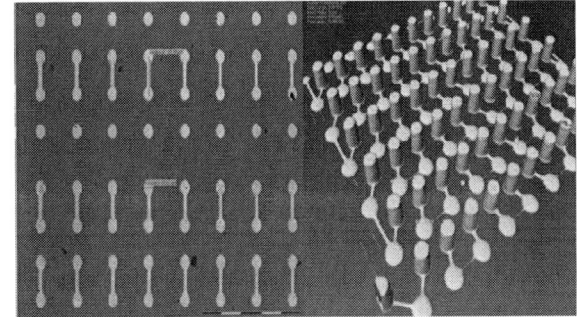

Fig. 4. (Left) Photography of the TSV. (Right) 3D X-Ray CT picture with 3D observation software (*Courtesy of CEA LETI*).

This algorithm allows to recover a spatial graph which represents the 3D current way in the component (Figure 5).

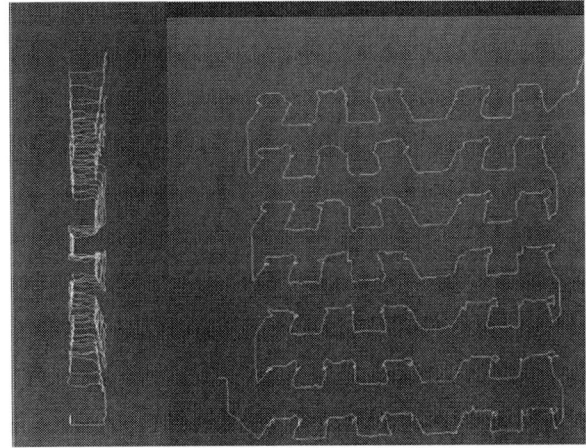

Fig. 5. Spatial graph obtained after making skeletonisation. (Left) Side view, (Right) Front view.

978-1-4799-3911-4/14 $31.00 © 2014 IEEE

The resulting skeleton file contains data that have been converted to be compatible with magnetic field simulation software. Finally, the current path is properly introduced in the magnetic simulator. Comparisons between simulation and measurement are then performed and final fitting is achieved. Superimposed results are shown in Figure 6.

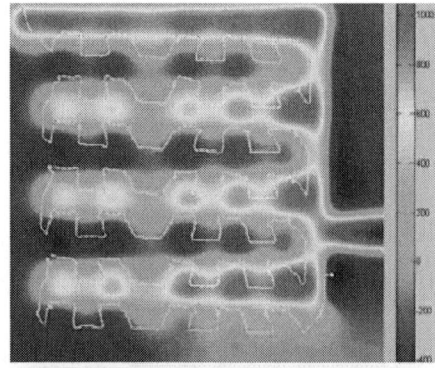

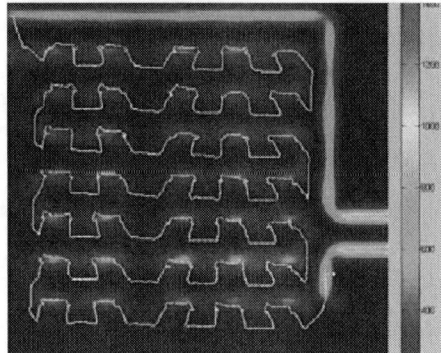

Fig. 6. (top) Alignment between magnetic scan and current path obtained by skeletonisation. (Bottom) Alignment between 2D current scan and current path.

An algorithm of the nearest neighbor has been used to allow a good representation of the current path in the magnetic simulator.

Then, physical alignment and scaling has been done. A two point alignment technics is used to do it. We take two points of the magnetic scan and the same points on the current path. Of course, we need to be sure that these points are identical on both magnetic scan and current path. Otherwise, the alignment would be bad.

These results, presented in figure 5, allow to conclude on the coupling procedure. It shows wearable to link information which comes from RX Computed Tomography and Magnetic Microscopy.

Now, what can be useful for failure analysis is to apply the same procedure on a failed component. On a one hand, a magnetic scan would be realized on the failed sample. And, on the other hand, a RX tomography would be performed on a good one. Then we can bring face to face the current path of the good sample and the magnetic scan of the "bad" sample, and let the simulator localize the short circuit.

IV. DISCUSSIONS

This approach needs a good X-Ray contrast to compute the tomography database (3D). It can be challenged by conductive layer thickness or X-ray "transparent" material (Aluminum for instance). Another issue is related to short circuits inside components that are not necessarily contrasted by X-Ray. In this case, we don't have any other choice that to presuppose a part of the current path. Then, the simulator is supposed to modify the current path in order to concur with the magnetic scan of the failed sample to finally obtain the position of the 3D defect.

This coupling procedure needs a certain time to be performed and currently, it only has been realized on a good device. It can be easily extended to faulty device. As is common to find in Failure Analysis, we need to realize a comparison between the "bad" and the good sample. In this case, we can move back to the old process but we will focus only on magnetic field differences.

Another issue concerns the simulator. We need to perform a procedure to prevent the current path from forming crossing lane or return trip as illustrated in figure 7. Colors indicate the two different layers of this sample. The path corresponds to the skeleton obtained after making RX computed tomography. We can observe that the path is not perfectly straight. This is due to limited resolution of the tomography and also because of the hollow TSV.

Some crossing lanes or return trip can cause malfunctions to the simulator and have caused artefacts in the simulation results.

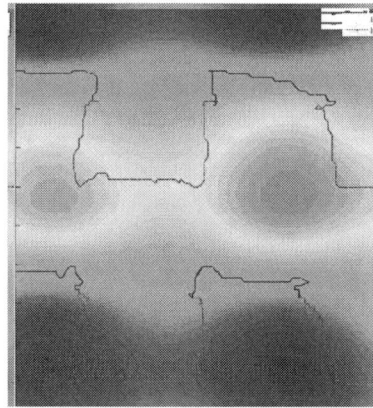

Fig 7. Observation of crossing lane in the current path

V. CONCLUSION AND PERSPECTIVES

In 3D devices, Magnetic Microscopy can achieve short and leakage defects localization. In case of 3D current path, there is no unique solution of the reverse problem therefore it is mandatory to use simulation (on the current and the magnetic field) and then to compare with magnetic field measurements.

This comparison is based on defect site assumption and can be time consuming for complex devices with many assumptions. In addition, design data are not always available to start with.

In this paper, we have proposed a new approach based on X-Ray computed tomography to extract the 3D current path with or without defects. The result can be directly used in the simulation and superimposed to the magnetic field scan after basic data conversion and alignment. This generic approach is promising and has been demonstrated on a TSV chain.

ACKNOWLEDGMENT AND DISCLAIMER

I want to thank Alex Jeffers from Maryland University who generously help me to solve various problems by using the 3D magnetic simulator. Alex Jeffers is the software programmer of the 3D magnetic simulator.

Part of this work has been supported by the Intelligence Advanced Research Projects Activity (IARPA) via Air Force Research Laboratory (AFRL) contract number FA8650-11-C-7101. The U.S. Government is authorized to reproduce and distribute reprints for Governmental purposes notwithstanding any copyright annotation thereon.

Disclaimer: The views and conclusions contained herein are those of the authors and should not be interpreted as necessarily representing the official policies or endorsements, either expressed or implied, of IARPA, AFRL, or the U.S. Government.

REFERENCES

[1] L. A. Knauss, S. I. Woods, and A. Orozco, "Current Imaging using Magnetic Field Sensors," *Microelectronics Failure Analysis Desk Reference Fifth Edition*, pp. 303–311.

[2] J.P. Jr Wikswo, "The Magnetic Inverse Problem for NDE", H. Weinstock (ed.), SQUID Sensors: Fundamentals, Fabrication and Applications, Kluwer Academic Publishers, pp.629-695, 1996.

[3] F. Felt *et al.*, "Construction of a 3-D Current Path Using Magnetic Current Imaging," *Proceeding of the 33rd International Symposium for Testing and Failure Analysis*, San Jose, November 2007.

[4] F. Infante et al, "A new Methodology for Short Circuit Localization on Integrated Circuits using Magnetic Microscopy Technique Coupled with Simulations", *Proceedings of the IEEE 16th International Symposium on the Physical and Failure Analysis of Integrated Circuits*, pp 208-212, July 2009

[5] H. B. Kor et al, "3D Current Path in Stacked Devices: Metrics and Challenges", *Proceedings of the IEEE 18th International Symposium on Physical and Failure Analysis of Integrated Circuits (IPFA)*, July 2011

[6] C. Pudney, "Distance-Ordered Homotopic Thinning: A Skeletonization Algorithm for 3D Digital Images", *Computer Vision and Image Understanding*, December 1998, pp. 404-413

978-1-4799-3911-4/14 $31.00 © 2014 IEEE

Wafer-level Fault Isolation Approach to Debug Integrated Circuits JTAG Failures

SH Goh, GF You, BL Yeoh, Hu Hao, NL Chung, CP Yap, Jeffrey Lam

GLOBALFOUNDRIES, Technology Development, New Technology Prototyping, Singapore

Email: SzuHuat.Goh@globalfoundries.com

Abstract- **Boundary scan test failures in the early phase of integrated circuit device yield engineering suggest fundamental manufacturing weaknesses and require fast response to fix the I/O connectivity, without which, chip functionality cannot be validated further. This paper presents a complete wafer-level workflow for JTAG-based boundary scan debug. We also show how a tester-based fault isolation technique called frequency mapping can be extended to JTAG data registers shift failures with the help of basic JTAG test methodology knowledge.**

I. INTRODUCTION

Boundary Scan is a widely used structured Design-for-Test (DFT) technique in integrated circuits (IC) design to test I/O connectivity on a final product. There are 2 main boundary scan architecture that are implemented in modern board, system and chip design - the IEEE standard 1149.1 Test Access Port also known as JTAG (Joint Test Action Group) and the NandTree [1, 2]. JTAG is focused in this work.

Originally, JTAG was developed as a solution to PCB manufacturing test challenges from increasing chip logic complexity and board functional density, to detect IC assembly defects. The primary advantage of JTAG lies in its ability to test nets connectivity without the influence of IC complexity. This is achieved with the introduction of strategic boundary scan cells or data shift registers, additional control signals and Test Access Port (TAP). The TAP operates a TAP controller which is basically a state machine to enable different test modes via Test Clock (TCK), Test Mode Select (TMS) pins, data and instruction registers [1]. Fig. 1(a) and 1(b) shows a typical TAP control architecture and an application of JTAG to validate chip I/O integrity on a system board. Thereafter, JTAG is so successful because of its automated testing feature and flexibility for application to other test features that it is adopted at chip level to test logic and memory I/O interconnects [3].

In almost all IC device production sort test, JTAG usually precedes AC functional tests. Any JTAG fail occurrences signal a fundamental manufacturing line weakness and therefore, considered critical. Without good I/O connectivity, the chip will likely not yield at all and it is meaningless to continue chip functionality validation. Therefore, it is crucial that JTAG failures be diagnosed and defects isolated for root cause learning in the shortest time. Any delay will affect product yield ramp which directly impacts time-to-market, hence, profitability.

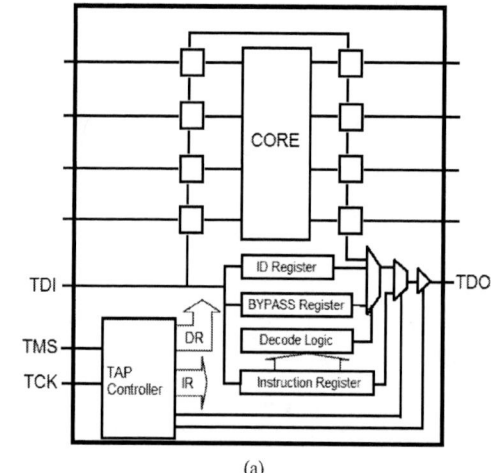

(a)

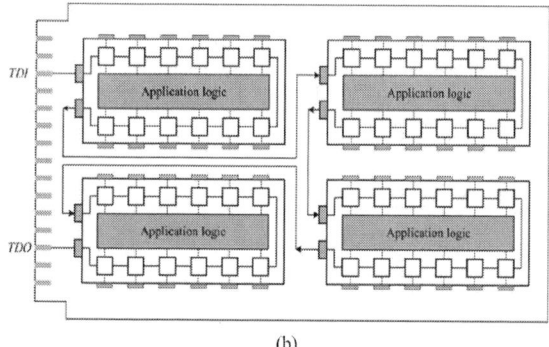

(b)

Fig. 1 (a) Typical TAP control architecture and (b) an application of JTAG to validate chip I/O integrity on a system board.

Following hardware-based fault isolation (FI) approach, JTAG failure FI has been realized before using conventional static electrical FI techniques like thermally induced voltage alteration (TIVA) [4], also known as optical beam induced voltage alteration (OBIRCH). However, these techniques require the defects to be related to either the I/O cell circuitry supplies or power planes such that a simple few pins bias is able to reveal the fault electrically. For defects within the logical nets or JTAG design architecture, such methods become less effective. Although tester-based FI techniques such Picosecond Imaging Circuit Analysis (PICA) have been demonstrated to be useful for localization of abnormal gates switching in logic operational faults [5], there have been no published applications on resolving JTAG failures in specific. To make debugging more challenging, JTAG FI is also not possible using software

978-1-4799-3911-4/14 $31.00 © 2014 IEEE

fault simulation approach, unlike scan chain integrity and scan logic. DFT Scan diagnosis does not apply to JTAG [6].

To address these challenges, this paper presents a complete wafer level debug approach to diagnose JTAG failures by introducing failure signature characterization on Automatic Test Equipment (ATE) [7] and hardware-based FI tools. It offers a more systematic approach as will be evident in the next sections as compared to traditional practice where several fail dies are brought for curve trace analysis and the dies with the highest leakages are identified for FI. Volume curve tracing approach is known to be tedious and it is not efficient because most JTAG fails that occur in advanced technology node products are functional failures, with similar IV characteristics to a reference die. The true failure mechanism could be missed by just focusing on the static FI feasible dies. As part of the proposed diagnostic methodology, we also show how a tester-based FI technique called frequency mapping which is commonly used for broken scan chain FI [8] can be extended for JTAG data registers debug with the help of basic JTAG architecture knowledge. In principle, frequency mapping uses a continuous near-Infrared (NIR) laser to raster scan the die region of interest. The reflected laser beam modulates according to the change in surface optical reflectivity, predominantly due to free carrier absorption effects, at locations where the transistors in the scan cell switches. In this way, the faulty cell can be isolated.

This paper is structured as follows: The next section describes the proposed workflow and the application of frequency mapping technique to JTAG failures. A case study based on a sub 28 nm technology node device will then be presented to demonstrate the benefits in section 3 before the conclusion in the last section.

II. PROPOSED WAFER LEVEL APPROACH TO JTAG FAILURE DEBUG

Fig. 2 depicts the proposed enhanced workflow. On discovery of JTAG failures at sort test, the failure is verified and then characterized using a test procedure that is created to monitor the VDD supply current behavior during JTAG testing. This is termed as dynamic IDD current in this work. Fig. 3 illustrates the dynamic current characteristics on 3 reference dies. Consistent higher current modulations are observed between 450-550 and 1150-1200 test cycles. The signature of all the fail dies are examined and classified. Thereupon, the most efficient and effective FI approach, either static or tester-based, will be determined accordingly, to debug each category of failures. Eventually, all FI hotspots will undergo a circuit analysis, if necessary, to decide the physical failure analysis (PFA) inspection coverage.

For dynamic IDD current on fail dies that show no clear differentiation from reference dies, tester-based FI techniques is required. In this work, we focus on the shift register failures in the JTAG architecture. There are several test data registers in a

JTAG design, namely 1) boundary scan register (BSR) consisting of boundary scan cells (BSCs), 2) bypass register (BR), 3) Device-ID register (for the loading of product information (manufacturer, part number, version number, etc.) and 4) Other user-specified data registers include scan chains, Linear feedback shift register (LFSR) for built-in self-test (BIST). Fig. 4 depicts a typical JTAG state machine schematically. The TMS pin vector determines the state of the TAP. For example, in the beginning of the test sequence, a constant high TMS pin will place the JTAG in a Test-Logic-Reset mode continuously. To prevent race conditions, the state change is executed on the positive or negative edge of TCK after sufficient time for TMS to be stable. With this fundamental test methodology, it is possible to map out the corresponding states to different vector cycles. Fig. 5(a) illustrates the case of exiting from a Run-Test-Idle looping state. *P* refers to a pulse input signal. *1* and *0* refers to a high and low input drive respectively. *X* represents a *Do Not Care* vector and is ignored. It can be observed that a TMS=*0* will lead to a Run-Test-Idle loop regardless of TCK. A TMS=*1* with a pulse on TCK will change the JTAG state to select Scan-DR.

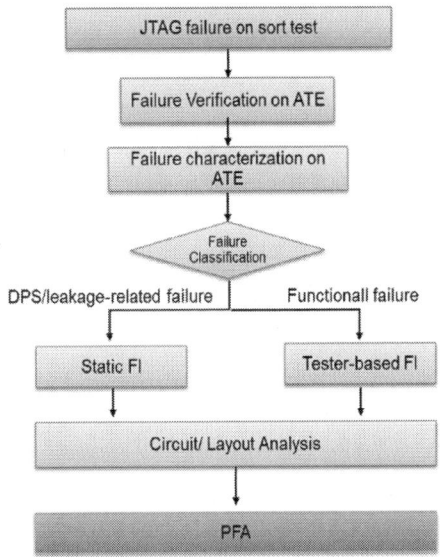

Fig. 2 Proposed wafer-level JTAG fault isolation workflow

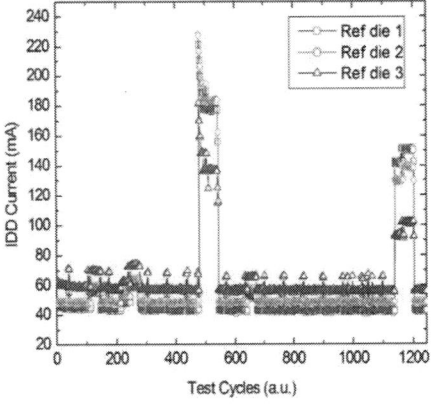

Fig. 3 Dynamic IDD current measured during JTAG testing

978-1-4799-3911-4/14 $31.00 © 2014 IEEE   31

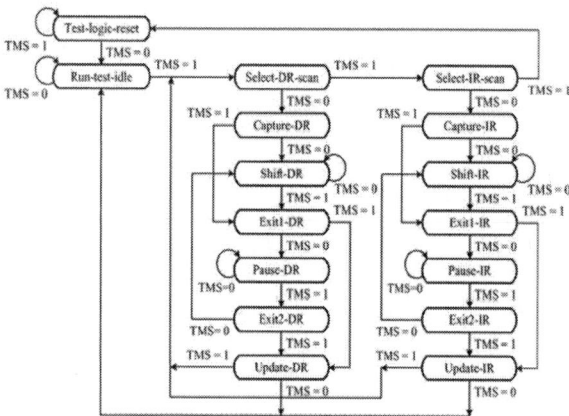

Fig. 4 TAP controller state diagram

Subsequent states can be determined based on the state machine as shown in Fig. 4. In order for frequency mapping to work, the aim is to exercise the JTAG registers using the Shift-DR or Shift-IR test modes in a looping manner. Fig. 5(b) shows the states transition from Run-Test-Idle to Shift-DR on the test pattern. Shift-IR can be executed in the same manner. Once the intended test mode has been attained, a loop can be initiated by maintaining TMS=0. To toggle the registers, it is necessary to modify the vectors on the TDI pin to be periodic on the test cycles of interest. Assume that the locations of the register cells are known, frequency mapping can now be applied in a binary search manner to locate the non-toggling cell. The suspected fail net path is defined as the path between the last togging and the first non-toggling cells.

III. EXPERIMENTAL RESULTS

The ATE and laser diagnostic system used in this work is the ADVANTEST Verigy 93000 and SEMICAPS SOM5000 respectively. The laser wavelength used for frequency mapping is 1300 nm. The case study is based on a sub 28 nm product-like testchip with significant JTAG fallouts. First, all the bad dies have their dynamic IDD current measured for all test cycles and characterized. The bad dies can be classified into 3 groups. Comparing to Fig. 3, dynamic IDD current of bad dies in classification 1 modulates similarly to the good dies as shown in Fig. 7(a) (refer to bad die 1 and 2). In this case, they are likely not related to power domain but logical failures. Tester-based techniques are necessary. In the same figure, it can be observed that the bad dies in classification 2 have dynamic IDD current that stays high at <200 mA throughout the test cycles (refer to bad die 3 and 4). Fig. 7(b) shows the case of bad dies in classification 3 where the dynamic IDD current stays >500 mA (refer to bad die 7 and 8. Bad die 5 and 6 belong to classification 2). Static FI techniques are more appropriate for the latter 2 categories of dies. From the characterization, it is suspected that there are likely at least 2 different JTAG failure mechanisms. Next, the percentage of fail dies in each category is calculated to derive the estimated impact to yield loss. Classification 1, 2 and 3 contributes to 70%, 20% and 10% of yield loss respectively.

Since the impact is the highest for the first category of fail dies, they should be analyzed with priority.

An ATE is hard docked onto a wafer level laser diagnostic tool and a production probecard contacts the wafer. The state machine is modified to loop at the Shift-DR test cycles. 11001100 periodic scan-in data is shifted into the data registers via the TDI pin at a shift frequency of 10 MHz. Fig. 8(a) shows the 10 MHz frequency signals obtained using refractive solid immersion lens (RSIL) on a reference die. The signals are overlaid against the bounding boxes of the register cells.

Test Cycle	TCK	TMS	TDO	Test-Logic Reset	Run-Test Idle	Select DR-Scan	Capture-DR	Shift-DR	Exit1-DR	Pause-DR	Exit2-DR	Update-DR	Select IR-Scan	Capture-IR	Shift-IR	Exit1-IR	Pause-IR	Exit2-IR	Update-IR
23	P	0	X		o														
24	P	0	X		o														
25	P	0	X		o														
26	P	1	X			o													
27	P	1	X										o						
28	P	1	X	o															
29	P	1	X	o															
30	P	1	X	o															

(a)

Test Cycle	TCK	TMS	TDO	Test-Logic Reset	Run-Test Idle	Select DR-Scan	Capture-DR	Shift-DR	Exit1-DR	Pause-DR	Exit2-DR	Update-DR	Select IR-Scan	Capture-IR	Shift-IR	Exit1-IR	Pause-IR	Exit2-IR	Update-IR
46	P	0	X		o														
47	P	1	X			o													
48	P	0	X				o												
49	P	0	X					o											
50	P	0	L					o											
51	P	0	L					o											
52	P	0	L					o											
53	P	0	L					o											

(b)

Fig. 5 Extract of a typical JTAG test pattern illustrating state transition (a) out of Run-Test-Idle and (b) to Shift-DR.

Test Cycle	TCK	TMS	TDI	Test-Logic Reset	Run-Test Idle	Select DR-Scan	Capture-DR	Shift-DR	Exit1-DR	Pause-DR	Exit2-DR	Update-DR	Select IR-Scan	Capture-IR	Shift-IR	Exit1-IR	Pause-IR	Exit2-IR	Update-IR
47	P	1	0			o													
48	P	0	0				o												
49	P	0	0					o											
50	P	0	1					o											
51	P	0	1					o											
52	P	0	1					o											
53	P	0	1					o											
54	P	0	1					o											

Fig. 6 Extract of a typical JTAG test pattern showing TDI pin vectors at shift-DR.

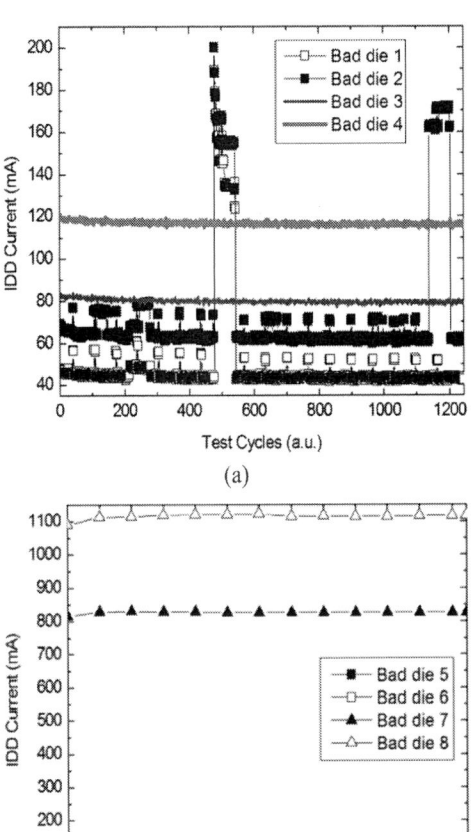

(a)

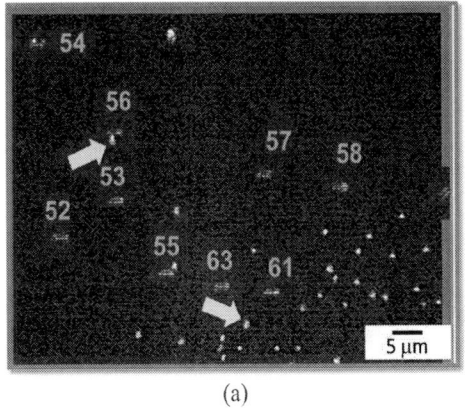

(b)

Fig. 7 Dynamic IDD current during JTAG testing for bad dies (a) under classification 1 where the current modulates similar to reference and classification 2 where current stays high < 200 mA throughout, and (b) under classification 3 where the current stays > 500 mA throughout.

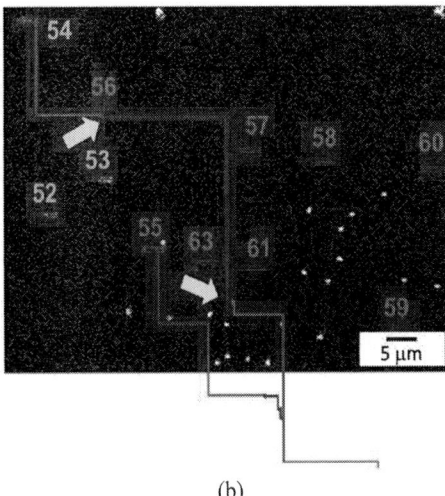

(b)

Fig. 8 Frequency map signals of 10 MHz overlaid with partial Shift-DR registers bounding boxes for (a) good and (b) bad dies under category 1.

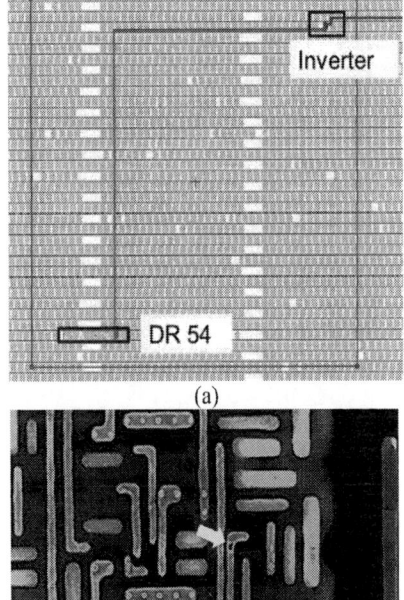

(a)

(b)

Fig. 9 (a) Circuit trace from DR 54 to the first inverter for PFA. (b) Metal voiding discovered at third metal layer.

It is evident that all bounding boxes are accompanied by frequency signals which indicate that they are toggling as expected. Fig. 8(b) presents the results on a bad die. It can be observed that the failure is likely caused by a broken path between cell 54–55 because signals are missing signals from cell 55 onwards. A more detailed analysis on circuits trace

978-1-4799-3911-4/14 $31.00 © 2014 IEEE   33

reveals 2 invertors (indicated in Fig 8(a) and (b)) along this path. The suspect fail net is likely between cell 54 to the first inverter. This path is sent on for PFA. Fig. 9(a) and (b) shows the final trace from data register (DR) 54 to the first inverter and the metal void revealed at the third metallization layer on PFA respectively. This information is feedback for metallization process improvement.

The last 2 groups of bad dies are also analyzed using the same static FI techniques. Fig. 10(a) and (b) present the static FI results on thermal analysis and OBIRCH, respectively, by powering up the VDD of interest. Based on a few dies in category 2 and 3 analysis, it can be observed that the hotspots are consistent in both methods. Further PFA analysis found an inverter PMOS and NMOS short at the polysilicon gate layer in 2 bad dies from each category. This creates a resistive path from power to ground. Leakage exists and hence, induced a resistive

droop affecting the power supply to JTAG. Therefore, there are 2 different failure root causes –one from the frontend of line and the other from the backend of line. Process fixes are identified in both cases.

IV. CONCLUSION

The need for prompt action to address any JTAG failures that occur in the initial phase of product yield ramp is emphasized. In this paper, a comprehensive workflow to manage JTAG failures is proposed. An enhanced failure characterization and classification steps are introduced to 1) predict the number of failure mechanisms involved in the failure and 2) determine the most efficient and effective FI techniques. Fundamentally, static FI is considerably faster than tester-based FI and still valuable, but should be applied appropriately. The success and benefits of the debug flow is demonstrated using a case study based on an advanced technology node device. It is proven that all JTAG fail dies should not be debugged in the same way. Failure characterization on test is crucial to provide additional insights into the number of failure root causes before deciding the FI approach. By incorporating basic knowledge of JTAG state machines and test methodology, frequency map technique has also been demonstrated successfully on JTAG shift registers failures. This proposed workflow is clearly a more systematic and inclusive way, not just for JTAG but general functional failures on test.

It is also worth mentioning that this workflow is best executed on wafer level to benefit from the time saved on packaging to shorten FI turnaround time.

REFERENCES

[1] N Stollon, "On-Chip Instrumentation", Design and debug for system on chio, Springer, pp 31-48, 2011.
[2] V Carlson, "Enhancing In-Circuit Yields", Teradyne Users Groups, 1996.
[3] AW Ley, "A Look at Boundary Scan from a Designer's Perspective," Proceedings of Electronic Design Automation &. Test Asia Conference, 1994.
[4] V Narang et al., "Development of Backside Scanning Capacitance Microscopy Technique for Advanced SOI Microprocessors", Proc Int Symp for Testing and Failure Analysis, pg 94-97, 2006.
[5] P Song et al., "An Advanced Optical Diagnostic Technique of IB< z990 eServer Microprocessor", Proc Int Test Conf, 2005, paper 48.1, pp 1-9.
[6] J Mekkoth et al., "Yield Learning with Layout-aware Advanced Scan Diagnosis," Proc 32nd Int'l Symp for Testing and Failure Analysis, November 2006, pp. 208–418.
[7] SH Goh et al., "Evolution of Wafer Level Tester-Based Diagnostic System: More Than Just a Dynamic Electrical Fault Isolation Tool", Proc Int Symp for Testing and Failure Analysis, pg 587-593, 2013.
[8] SH Goh et al., "Effectiveness of frequency mapping on 28nm device broken scan chain failures", Rev Sci Instrum, Vol 83, Issue 2, 2012, 023702.

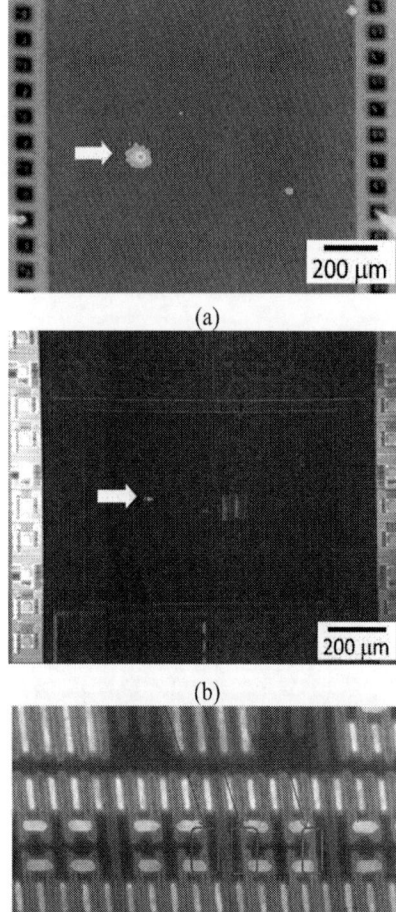

Fig. 10 Static (a) thermal microscopy and (b) OBIRCH signals observed on VDD bias. (c) PFA on hotspot reveals gate tip to tip shorts.

Advanced Package Circuit Modification by µMilling

Christian Hollerith[1], Bernd Krüger[1], Gürcan Gezerci[1], Siegfried Pauthner[1], Gunnar Zimmermann[2]

[1]Infineon Technologies AG, Neubiberg, Germany
[2]Intel Mobile Communication GmbH, Neubiberg, Germany
Phone: +49 8923422785, Email: Christian.hollerith@infineon.com

As packages get smaller and more complex the necessity grows to do electrical modifications in packages to assist design process. Milling machines with accuracy of sub-µm enable a fast and effective approach to do such a kind of modification. In combination with other techniques, cutting of package connections as well as reconnecting of metal lines is possible.

Introduction

Package circuit modification is an important step to assist product development for laminate- or eWLB-packages, especially for optimization of RF-performance. With the opportunity to do such modification it is possible to reduce time and costs for package development, because fixes of design bugs can be tested very fast by producing some reworked parts prior to mass production of new samples. The increasing package complexity, as seen in the last years [1], leads to a growing importance to do circuit modification on package level.

The method of performing circuit modification on die-level using FIB is known for many years [2]. On the other hand FIB circuit modification on package material is difficult because structures to be modified are quite large (> 100 µm) which leads to long processing times with FIB. Additionally FIB-metal connections for rewiring should have quite low resistance in the range of several ohms which is hard to achieve by FIB. FIB is also limited in some cases as it is unable to control accurately the depth of the cut. In this paper some examples will be shown, where a FIB-modification would not be possible due to these restrictions. Therefore it is beneficial to get a fast, reliable and cheap alternative for package modification.

In this article we are proposing µMilling with millers in the range of some 10 µm diameter as an alternative approach to do electrical modifications in the package. With this method it is possible to cut metal lines as well as vias within packages. Also rewiring of connections can be done with the assistance of milling. Therefore it is possible to get the same modification capabilities on package as FIB offers on die level. Also the combination of FIB and µMilling is promising.

Cutting by µMilling

For package modification by µMilling, a KERN-CNC-milling machine is used in combination with millers between 20 µm and 300 µm diameter, see fig. 1. This tool is specified with a precision of less than 1 µm in all 3 axis directions. The rotational frequency is up to 50000 rpm.

Fig. 1: a) KERN CNC-milling machine, b) Sample in milling machine with achieved cut

For aligning the cut position within the sample an optical positioning system is used. This system allows aligning the spindle axis with structures on the sample surface, such as solder balls or visible metal lines. By the use of package layout data or X-ray images and a visible reference point on the surface also structures buried below the surface can be targeted. The z-position, where the miller touches the surface, can also be determined quite easily, because from there on flakes are visible within a live-camera-system that is mounted side by side to the spindle. By slowly lowering the miller in z-direction to the sample surface this point can be determined with an accuracy of about 1 µm. With this arbitrary origin an exact control of the cutting depth is possible.

978-1-4799-3911-4/14 $31.00 © 2014 IEEE

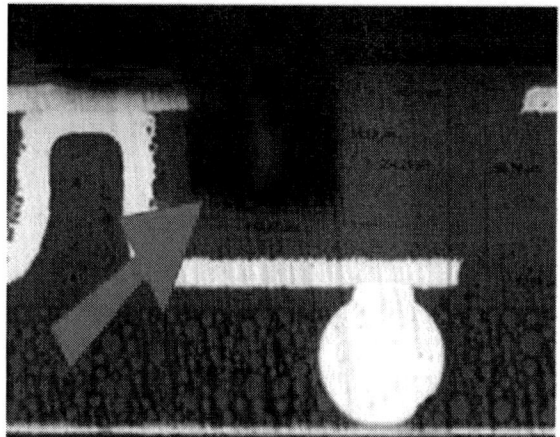

Fig. 2: Cross section view of package with milled hole

At fig. 2, a cross-section of a drilled hole shows the plane surface at the bottom of the hole and the accurate depth of the cut. The quality of the cut can then be examined by an electrical measurement, an optical microscope or in most cases preferred, by X-ray inspection. If the examination reveals that a longer or deeper cut is needed, rework can be done easily, because the accuracy of the assembly in the CNC-mill is extremely high. After the cutting process the milled holes are sealed by epoxy glue to avoid getting shorts between different uncovered metal areas e.g. due to subsequent soldering process.

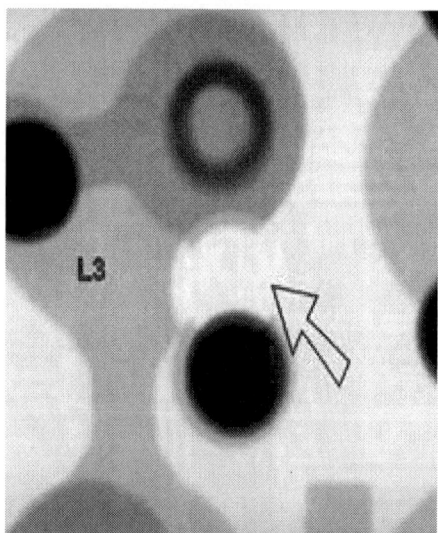

Fig.3: Cut metal line within interposer laminate

Fig. 3 shows the example of a separated metal line in a flip-chip-laminate-package. The modification has been done on the top side of the laminate (side toward die). It is important for this kind of modification to take care that the die is not damaged. This can be difficult because the thickness of the underfiller and therefore the distance between the metal to be

cut and the die surface is much lower than the thickness of the laminate. An additional challenge is that the laminate thickness is not constant but depends on the amount of metal on the laminate sides. These two effects lead to the fact that depth control is quite important. This kind of modification would be very difficult by FIB, not only due to the large processing time but also because of the bad depth control when removing the laminate caused by lots of glass fibers. The challenge would be to stop exactly below the metal line without damaging the die on the whole area of the FIB box.

Fig. 4 shows the cut of a metal via within the laminate, located below a solder ball. In this example the solder ball above the via to be cut should be preserved for soldering. A 200 μm miller was used to drill a hole through the ball until the mid of the via. The metallic circle that can be seen in the middle of the drill hole is the remaining lower part of the via. This shows the accuracy, how structures can be exactly milled that are not visible at the surface. The cut location was aligned with package layout data in this case. Also this example would be very difficult to achieve with FIB, because the solder ball creates a very inhomogeneous surface which would make the control of the cutting depth for the FIB nearly impossible.

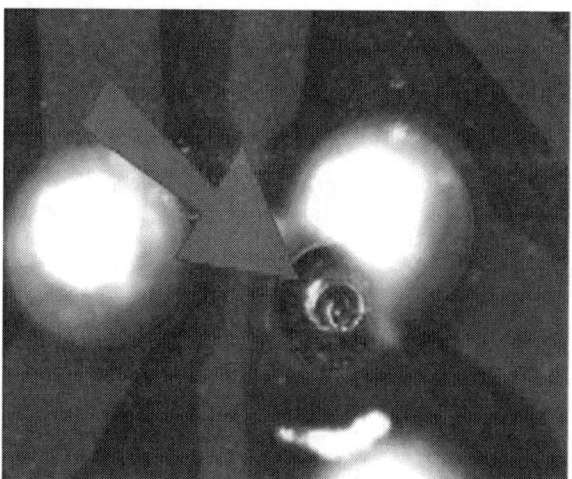

Fig.4: Cut via in laminate-package

Metal cuts are also possible at the RDL of eWLB packages, even in case the RDL runs above the die surface fig. 5. The distance between RDL and die surface is about 6 μm, a separation of RDL needs to be done without damaging the die. Therefore the accuracy of the cutting depth is extremely important. As the accuracy of the milling process is about 1 μm, it is sufficient to reliably cut metal lines in this layer. In the shown example, a metal coil of a shield has been interrupted at two locations to improve the RF properties. The die below this cut has been still fully functional after the modification. This kind of modification can also be done by FIB [3] but cutting is very time consuming. The separation by μMilling helps here to save time and also to decrease the workload for expensive FIB tools.

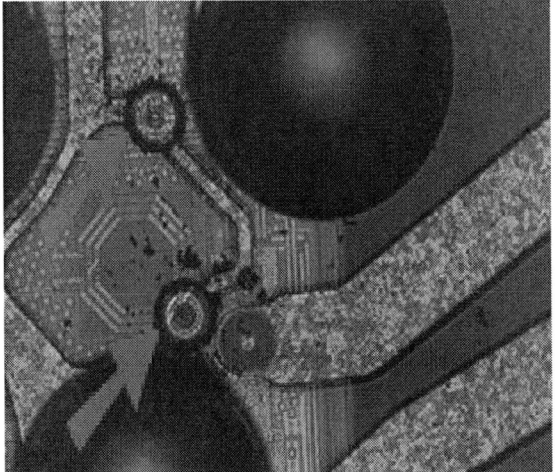

Fig.5: Cut coil in eWLB-package. Die was fully functional

Rewiring by μMilling

Also a rewiring process is available using this method. First step is to remove the solder bumps by desoldering. This is necessary to get a plain surface, which later on assists a polishing process. Afterwards a trench is milled between the metal lines to be connected within package. This trench has to be deep enough to have direct contact to existing Cu- metal lines. Then this trench is filled with silver glue, which enables the new created electrical connection. The ball pitch at BGA-packages can be quite narrow, so it cannot be avoided that excess glue forms shorts to other electrical nets on the package surface. Therefore in the next step, all of the protruding glue is removed by plan-parallel polishing. At the end, only the glue in the trench remains and forms the desired electrical connection. After reballing the sample is fully functional again with additional electrical modification. The electrical resistance of such a modification is in the range of ohms depending on trench size and length of the trench.

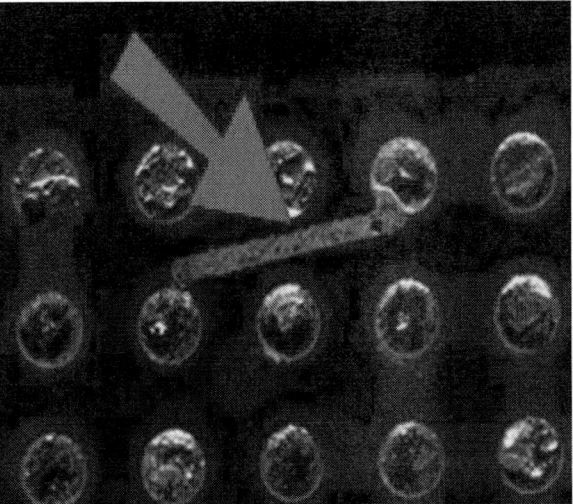

Fig.6: Metal re-connection at Flip Chip package

Fig. 6 shows an example of this method at a flip-chip BGA package. The solder balls have a pitch of 400 μm and two solder balls should be connected by this modification. In this case a trench along a straight line between these two balls could be milled. After the rewiring process the sample could be soldered to a PCB board and tested successfully.

μMilling in combination with FIB circuit edit

μMilling can also be used to assist FIB-circuit edit. In flip-chip laminate packages with advanced CMOS-technologies, most FIB-circuit-edits are done by a trench through the chip backside, because the die backside is easily accessible. For edits at lower metallization levels such as M1 or M2, this works very well, but for upper metallization this often proves to be difficult, as no position can be found, where lower metallization wouldn't be damaged due to the FIB-trench. For these cases it is possible to open a window through the interposer laminate by μMilling to get access to die surface. The position of this window has to be chosen in such a way that no package metal lines are cut. The milling depth of this window is adjusted to stop short before the die surface. Then, the remaining underfiller is removed by plasma etching. Through this created window within the package, the upper die metallization is visible within the FIB and the FIB-modification can be done on the accessible chip frontside.

Fig. 7 shows such a μmilled window through the interposer laminate. The exact position of this window has been controlled by X-ray imaging, which is shown in fig. 8. This check helps to avoid damage at any metal lines within the laminate. It also shows the die metal line, which should be cut, because the line consists of a very thick Cu-metallization that is visible in X-ray too. Therefore an accurate layout overlay between package and die layout was not necessary in this case.

Fig.7: µMilled window for enabling frontside FIB

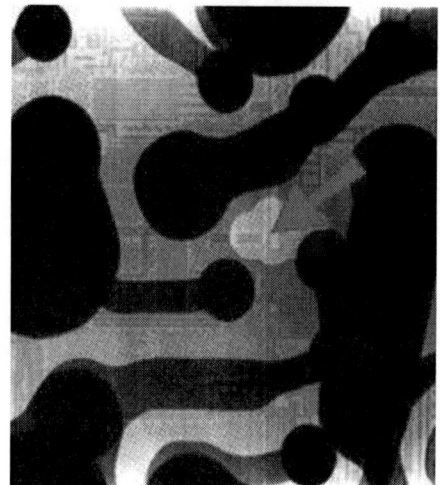

Fig.8: X-ray image of µmilled window; line to be cut by FIB visible in X-ray image (marked by red arrow)

It is also possible to combine FIB circuit edit on die level through backside trench with µMilling on package level. The combination of these methods helps to solve problems that cannot be achieved with one of the methods alone.

Conclusions and outlook

µMilling allows on the one hand the interruption of metal-lines and vias in the package and makes it possible to separate an electrical connection on package level. On the other hand it enables the connection of different nets by rewiring of electrical lines in the package. With these two methods it is now possible to do circuit modification at package level in the same way as on die-level by the use of FIB. In comparison to FIB, µMilling is a very simple and cost efficient method to do package-circuit-edit. But it also offers the advantage of accurate depth control in package cutting. With this advantage applications are possible that would be barely possible by the use of FIB.

The examples shown in this paper had quite simple geometrical shapes and were therefore done with manual control of the milling machine. By the use of the CNC-programming capability also complex routing such as reconnections across larger distances of the package or cutting of inhomogeneously formed metal layers are possible. An example for such inhomogeneous metallization would be double layer eWLB-packages.

References

[1] ITRS, "Assembly and Packaging White Paper on System Level Integration", http://www.itrs.net/papers.html

[2] Herschbein S. et al, "Focused Ion Beam (FIB) for Circuit Edit, Fault isolation and Sample preparation", Proc. 39th ISTFA2013, San Jose, pp. 123-133

[3] Liu S. et al, "Innovative Methodolgies of Circuit Edit on Wafer-level Chip scale Package (WLCSP) Devices, Proc. 36th ISTFA 2010, Addison, pp 359-363

Study and mechanism of static scanning laser fault isolation on embed SRAM function fail

Changqing Chen, huipeng Ng, Ghinboon Ang, J. Lam, Zhihong Mai
GLOBALFOUNDRIES Pte.Ltd
60 Woodlands Industrial Park D Street 2, Singapore 738406
Changqing.CHEN@globalfoundries.com

Abstract

As the technology keep scaling down and IC design becoming more and more complex, failure analysis becomes more and more challenge, especially for static laser analysis. For the foundry FA or process monitoring, SRAM analysis becomes more and more critical. There are two reasons for this: The first one is that SRAM circuit is relative simple which is well known to all, it is also used by fab for monitoring structure. The second reason is the SRAM percentage on-chip keeps increasing. It can occupy more 60% chip area for most logic product. That is also another reason we use the SRAM to monitor our process. SRAM analysis with bit map is relatively easy for FA. But as DFT become more popular, the BIST technical was applied in the SRAM, bit map was provided frequently. The global fault isolation methodology must be employed in the SRAM FA. In this paper, static scanning laser methodology was applied in the SRAM FA which no bit map was provided. Hot spot was observed in the SRAM block edge for some failed units, but some not. Combined with the SRAM schematic and GDS analysis, the defect was successfully found and the failure mechanism was studied, which can successfully link the electrical phenomenon and physical defect. Also, we found the process issue with the FA result.

Background information

Since the SRAM percentage on the chip keeps increasing and it is also the key process development and process monitoring structure. The failure analysis on SRAM is quite critical. The FA on SRAM with bit map is quite common and much more straightforward. But as the technology going forward, DFT and BIST was widely applied in the IC design, the failure analysis becomes more and more challenge on SRAM without bit map. Global fault isolation methodology was must employed in this kind of failure mode. But most of time DC bias doesn't shows any significant difference between good and failed unit, because either the defect location cannot be accessed by DC bias or the defect location only can induce small current change which can be concealed by overall current. Based on our study, the global fault isolation still can be applicable for the second situation. In this paper, an embed SRAM BIST fail was handled. Although the comparable IV was observed, the global fault isolation was still applied on the fault isolation—TIVA analysis.

Experiment and Discussion

Embed SRAM BIST fail was observed in one of our product, 0.18um, no bit map capability was buildup for this product. The SRAM is the normal 6T SRAM. DC IV measurement was performed on the Vdd and Vss. No significant difference was observed. TIVA analysis was performed on several failed units and compared with good unit. Distinct TIVA spot was observed on some failed units as compared with reference one. One more observation is that all of the TIVA spot locate in the SRAM block edge, Figure 1. Based on our experience, this kind of solid spot should be real one. Either defect location or has some kind of relation with the defect. We select one unit for PFA from top down. Nothing abnormal was observed from top metal down to metal 1. PVC also shows no anomaly in the hotspot location at every layer.

Figure1 hot spot of the fault isolation

Further deprocess was performed. The Poly was exposed by BOE. Gross W extrusion was observed in the spot location, Figure 2. Meanwhile, gross W extrusion was also observed within the SRAM block.

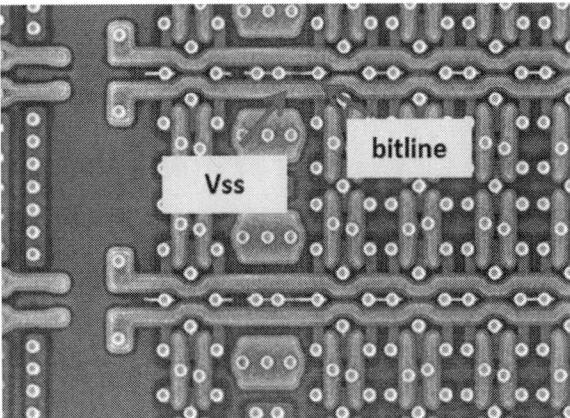

Figure 2 PFA shows W solid short happened in SRAM edge

978-1-4799-3911-4/14 $31.00 © 2014 IEEE

One more PFA observation is that all of the W extrusion locations sit in specific location, bit line contact to bit line contact short. From the process point of view, it is quite easy to understand the root cause of this defect. That means we can easily link the defect to our process. But how can we link the defect with our electrical result: 1. why the hotspot always locates in the SRAM block edge. 2. Why only some of the failed units has the hotspot.

To answer these questions, the in-depth analysis was employed in terms of the circuit and layout of this device. Before that, we select one failed unit without hotspot to do random PFA on the SRAM region. As expected, nothing abnormal was observed in BEOL. But gross W extrusion was also observed in the SRAM block. Then one more question comes out, why this gross W extrusion can induce hotspot in some unit while not in other unit. What's the reason behind.

SEM top down inspection was compared between the sample with hotspot and without hot spot. Some observation was found: for the sample with hot spot, there is a solid W shortness between bit line contact and neighboring contact at SRAM block edge. While for the sample without hot spot, although there is W extrusion, but no solid W shortness happens in the SRAM block edge.

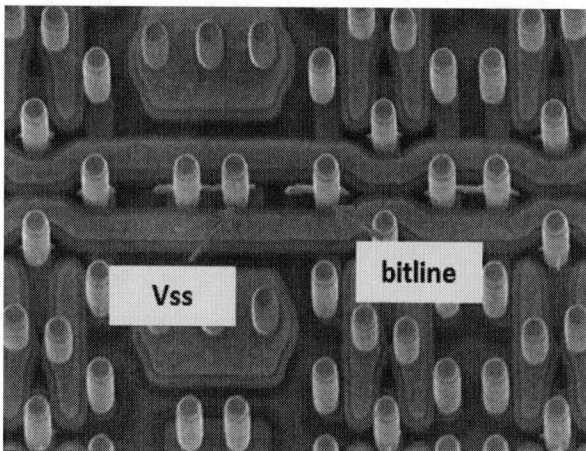

Figure 3 PFA shows no W solid short happened in SRAM edge for the sample without hotspot

In-depth circuit and layout analysis shows that the die edge contact which short with bit line contact is connected with Vss. The Vss in the block edge is highlighted in the GDS layout, figure 4.

In the center of the SRAM block, bit line to bit line short was observed, but this short cannot be accessed by the DC bias, since we can only bias Vdd and Vss. But how about the edge bit line which is short to Vss based on our GDS layout analysis. Can this kind of shortness be accessed by normal DC bias and can it induce hot spot.

In order to answer this question, we have a detailed study of the SRAM peripheral schematic. There is an equalizer PMOS whose source is connected with the Vdd, as highlighted in the Figure 5. Normally, this PMOS is turned off; the bit line cannot be directly accessed by the Vdd. But after the detailed analysis of this circuit and combined with GDS layout, we can confirm this PMOS is turn on.

The reason is as below: under DC bias condition, Vdd and Vss, this PMOS source is connected with Vdd while the gate is floating. But we must bear in mind this PMOS is sitting on the NWELL, while the NWELL is connected with the Vdd. So even the gate is floating or connected with somewhere, the gate is still turn-on. For the Bit line to bit line short, there is no current flow. Because both bit line are short to Vdd for the center cell, there is no current flow, so no spot can be triggered.

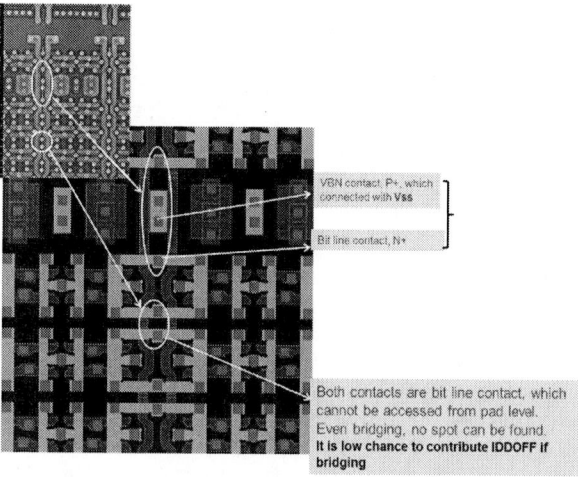

Figure 4 GDS layout of the failed location and analysis

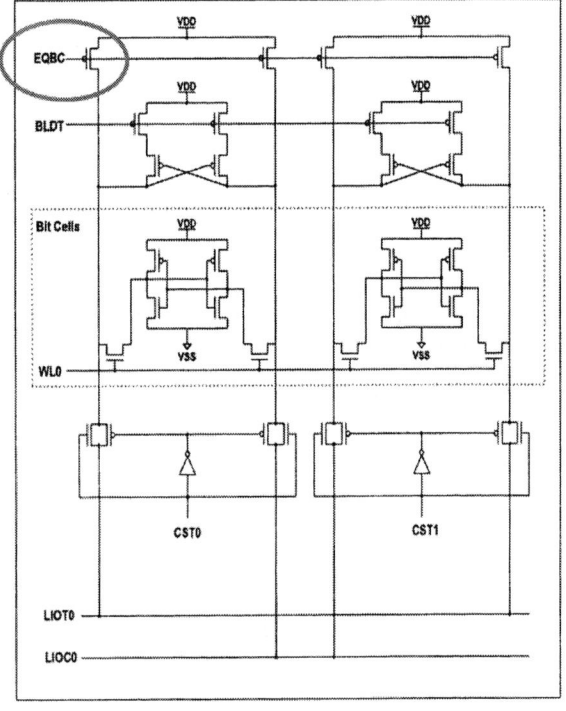

Figure 5 traditional SRAM circuit analysis

But the for the edge bit line, the situation is different, it is bit line short to Vss.

978-1-4799-3911-4/14 $31.00 © 2014 IEEE   40

As shown if the Figur5, there is balance PMOS which is connected with the bit line. If we pull out a single SRAM for

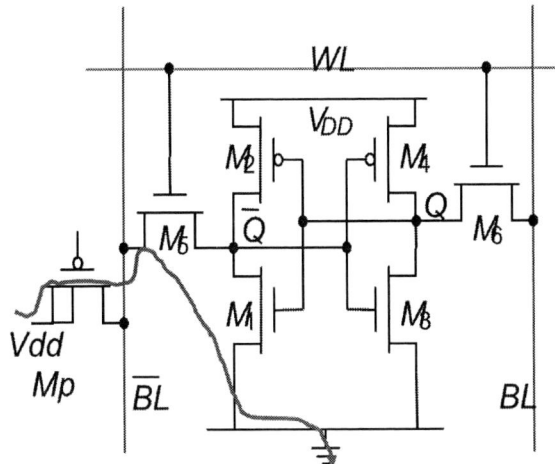

Figure 6 SRAM single bit and analysis

circuit mechanism analysis, it is shown in Figure 6. We can easily find there is a current flow from transistor Mp to Vss when the gate PMOS Mp is floating. That is the reason why the hotspots always locate in the SRAM edge. If there is no solid short happened in the SRAM edge, there is no current flow path. So hot spot cannot be triggered. That is the reason why there is no hot spot in some of the failed die.

Further cross section analysis show W extrusion short

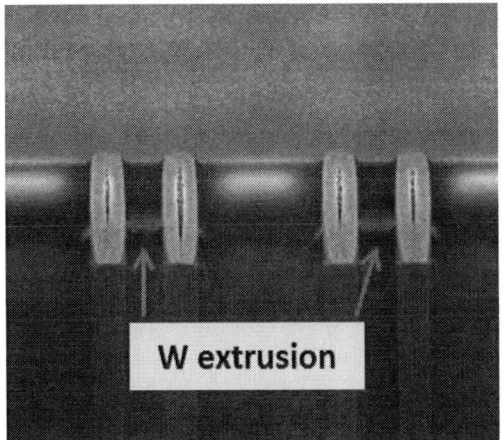

Figure7 cross section result show W short happened

happened in the center location the contact height, Figure 7. Based on our process analysis, we identified that the short is due the film deposition process drift. This drift induces some film interface interaction effect.

Conclusion

Static fault isolation is still capable for some kinds of functional fail, if the defect location can be accessed by the DC bias. Sometimes, even the DC IV curve is comparable; it is still possible to find the defect by static fault isolation, because the DC IV curve is an accumulation result. Some kinds of defect induced change maybe concealed by the overall result. But for these kinds of analysis, the sample

selection is very important. As shown in this paper, not all the sample has hot spot.

For this paper, the BIST/functional fail SRAM was analysis. Since no bit map provided, the normal global fault isolation method was applied in the analysis. The defect and root cause was successfully found. This is a good reference for some kind of functional failure analysis.

References

1. [1] Lawrence C. Wagner, "Failure Analysis of Integrated Circuits: Tools and Techniques," pp. 66-82, 1999. ISBN-10 / ASIN: 0412145618, ISBN-13 / EAN: 9780412145612

[2] JCH Phang. DSH Chanl, M Palaniappanl, JM Chin. " A Review of Laser Induced Techniques for Microelectronic Failure" proceedings of 11th IPFA 2004, Taiwan, 255-261

[3] JCH Phang. DSH Chanl,SL TAN, WB Len, "A review of near infrared photon emission microscopy and spectroscopy", Proceeding of 12th IPFA 2005, Singapore.

Failure analysis for metal bridge defect in logic area of mixed-signal IC

Diwei Fan, Winter Wang,

Product Analysis Engineer of Quality Department in Freescale Semiconductor (China) Limited
No.15 Xinghua Avenue, Xiqing Economic Development Area, Tianjin, China 300385
Phone: 86-22-85684614   Email: B25961@freescale.com

Abstract–In mixed-signal ICs the die surface is divided between the analog circuit and the digital circuit often referred to as the logic area. . Compare with the analog area, the logic area has more complex signals. The metal lines are narrower and closer together. These factors make it very hard to analyze defects such as metal bridges in the logic area. Firstly, the complicated waveform of signals and circuit loops in logic area make the schematic analysis harder. We cannot find the failed signal only through the comparison between reference unit and failed unit. The reason is that in the complicated circuit loop, one signal failure can cause many other signals in the circuit loop to fail. Secondly, if the failed signal is caused by metal bridge defect, since there are many metal lines close to the failed signal metal bridges to several of the metal lines could be the cause of failure. In this paper, we show how many FA techniques such as emission microscopy, microprobe, function, OBIRCH, FIB etc need to be used can be used to find a metal bridge defect causing a failure in the logic area.

Keywords-Logic circuit area; Mixed-signal IC; metal bridge defect; Failure analysis;

I.   INTRODUCTION

The Failure Analysis (FA) process consists of five main steps, namely: failure validation, fault localization, sample preparation and defect tracing, defect characterization and root cause determination [1]. Fault localization is the step in which techniques are used to isolate the defective areas on the die in the failed units. This is the most critical step since it reduces the area required for analysis dramatically [2]. As a component of mixed-signal IC, logic circuit is an important area which defect will occur in. In FA of mixed-signal IC, it is very common that some defects are found in logic circuit area.

Compared with analog circuit, the characterization of logic circuit decides that the FA in logic area is more difficult. From electrical aspect, the outputs of logic devices are often caused by complicated circuit loops and some specific input signals (e.g. clock_osc). So the signals of logic circuit are always some waveform with specific frequency or some pulse with certain period. Due to the electrical feature of logic circuit, we cannot find the failed signal only through comparison between the signal of reference unit and failed unit. Because if one signal is failed, it will make many other signals represent wrong behaviors in this circuit loop. In this situation, we usually need some other FA methods just like schematic analysis and simplify the circuit loop. From physical aspect, the layout of logic circuit area usually put together. Most of the metal lines in logic area are narrower than analog area, and every metal line of logic signal is closed together with other one. It means the layout of logic circuit area is very closely. Due to the physical feature of logic circuit, many of the signals in logic area cannot be probed by microprobe technique, and photon emission technique [3] also hard to get the emission site as analog circuit area. Base on the characterization of logic circuit, if metal bridge defect exist in the logic area, it is more difficult to find out the defect location. Because that the metal lines of signal in logic area is closed with several other metal lines. So for one specific failure mode, if one signal is failed due to connect with other metal line by metal bridge defect, it can be so many possible metal lines induce that. And the location of metal bridge defect can occur at any location among these metal lines. During FA process, we can hardly trace these metal lines one by one in logic area, so we must use some novel method to resolve it.

In this paper, a real case is shared. The whole FA course of this real case is a good solution for the case with metal bridge defect in logic circuit area.

II.   BACKGROUND AND FAILURE ISOLATION

In this case study, the failing device has five high side outputs (HS), every HS can connect with/without load. The DUT is controlled by MCU and the HS can be turned on or off base on the data from MCU (See Fig.1). The behavior of DUT can be reported to MCU by data bus between MCU and DUT. If one of the HS is turned on, the state of this HS will be known by MCU and if no load is connected with this HS, the state of this HS is marked as "OL (open load)" in the software which is controlled by MCU.

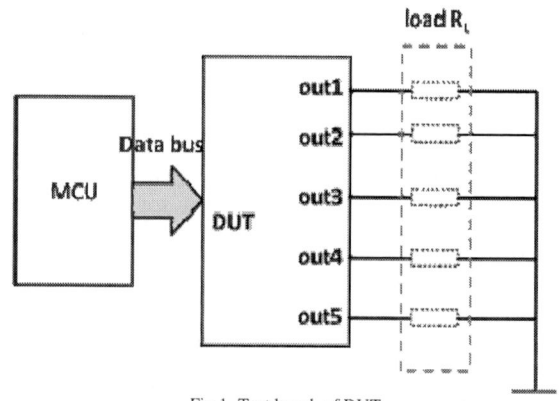

Fig.1. Test bench of DUT

978-1-4799-3911-4/14 $31.00 © 2014 IEEE

The set up condition of this DUT is that HS4 (out4) is turned on and no load is connected. In this situation, the failure mode of the DUT is that no "OL" of HS4 is reported in the software. But for reference unit, it should be report "OL" on HS4 channel. After schematic study, base on the failure mode and our experience previous, we highly suspect the defect is in the logic area. Photon emission is performed according to the setup condition. From the emission result we find that no emission site can be observed in logic area, and the emission result is same as reference unit. The emission result is shown as Fig.2.

Fig.2. Photon Emission result

The emission result cannot provide any useful clue for analysis, so we try to find the key signal about the failure phenomenon. Until now, the schematic analysis is very important, the more analysis on schematic, the less microprobe on signals. Finally we find that the output signal of logic circuit "enable_ol_current4" is abnormal. The waveform of "enable_ol_current4" is shown as Fig.3, the width of pulse is same as reference unit but the voltage level is lower.

The schematic indicate that the signal "enable_ol_current4" is output of logic circuit, and after trace this signal, we find that this signal is generated by two inverters in series (see Fig.4). The input signal of the inverters is n214. In order to isolate the failed signal, we also test the two inverters signals by microprobe technique. We get the waveform of these two inverters signals. The results are also pulse with specific width, but the voltage level can reach the power source of logic circuit (Fig.5). And due to the signal "enable_ol_current4" is a feedback signal which connect to many other devices as input signal in the logic circuit loop, so the signals of two inverters are also different with reference unit, if we want to confirm that the key failed signal is "enable_ol_current4", we'd better do further analysis.

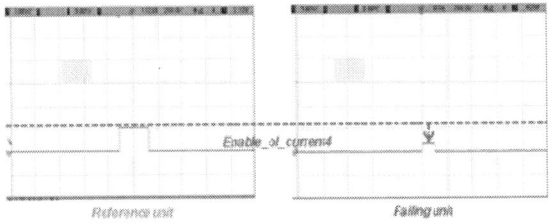

Fig.3. Waveform of "enable_ol_current4"

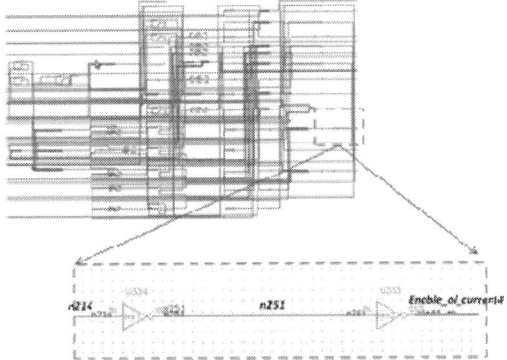

Fig.4. Schematic analysis of signal "enable_ol_current4"

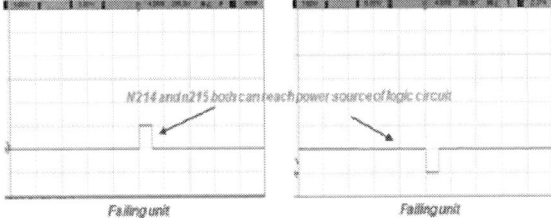

Fig.5. Waveform of two inverters signals of failing unit

In order to further confirm the failed signal, base on the schematic portion of "enable_ol_current4", the n214 net is cut by Focus Ion Beam (FIB), and high level voltage is forced on n214, the voltage of "enable_ol_current4" should be a low voltage (gnd) for reference unit base on the schematic analysis. But after microprobe on "enable_ol_current4" for failing unit, the result is a middle voltage of logic circuit power source. The result can confirm that the signal "enable_ol_current4" is the key failed signal. (See Fig.6). So the next analysis is focused on this signal in logic area.

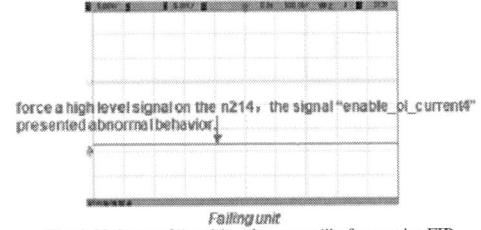

Fig.6. Voltage of "enable_ol_current4" after cut by FIB

III. DEFECT LOCATION BY PHOTON EMISSION AND FUNCTIONAL OBIRCH

In fact, the cutting by FIB is a key point of this case analysis, the reason is that the pulse of logic signal is hard to analyze, after cutting by FIB, we can get a constant voltage instead of pulse. The constant voltage can not only make schematic analysis easier but also can get hot spot easier by photon emission and OBIRCH. For this case, after the cutting on n214 net, the failed signal "enable_ol_current4" is in the middle level voltage when force high level voltage on n214. It means that the signal "enable_ol_current4" is pulled down by a low voltage due to the defect so that "enable_ol_current4" is in middle level voltage. The defect is always in active state when force high level voltage on n214. But before cutting, the defect is only effective when a high level voltage pulse exists. So post-cutting, the abnormal hot spot should be found easier.

The photon emission is performed again base on the new setup condition. Different with previous emission results, two abnormal emission sites are observed in logic area (see Fig.7). After schematic study, base on our previous experience, we find that no one of these two emission sites can be confirmed as defect location. And these two emission sites are all induced by the abnormal middle level voltage of signal "enable_ol_current4" [5]. It means that further analysis should be continued. So next step, in order to perform OBIRCH analysis, we want to find a leakage path in static state. We measure the IV-Curve of "enable_ol_current4" and gnd because that the signal "enable_ol_current4" is pulled down to a lower voltage. But the IV-curve of failing unit is same as reference unit. The normal IV-curve shows that no leakage path is found from "enable_ol_current4" to gnd. On the basic of this finding, the defect occur on a MOSFET is less possible. The failed phenomenon is caused by metal bridge defect in logic area is high possibility. So we try to do some analysis in the next step to confirm that if metal bridge defect exist.

Mag:200x

Fig.7. New photon emission results

Due to the characterization of logic circuit, the metal lines in logic areas are much closed to each other. There are so many possible metal lines can induce the metal bridge defect and produce the failure phenomenon (See Fig.8). But it is difficult to do microprobe to measure IV-Curves on every suspected signal in logic area. So we'd better to try some new method to find hot spot. Because Photon emission can only find two abnormal emission sites which are not defect location. OBIRCH is another good choice to find hot spot. Base on the working principle of OBIRCH, we should find a

leakage path which flow through the metal bridge defect. From the failure phenomenon, we can know that if force n214 a high level voltage, the signal "enable_ol_current4" will be pulled down to a lower voltage when the DUT in turn on state. It means that after the voltage of n214 is forced a high level voltage, there must be always a leakage path form "enable_ol_current4" to gnd through metal defect when DUT in turn on state. The Fig.9 show the leakage path which exist in logic area when DUT in turn on state. In this situation, when the laser of OBIRCH scans the DUT in this set up condition, the defect location should be marked by hot spot.

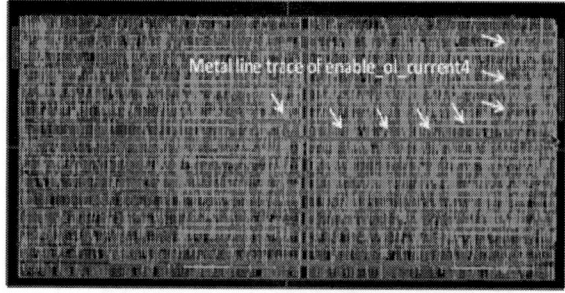

Fig.8. Metal line trace of "enable_ol_current4"

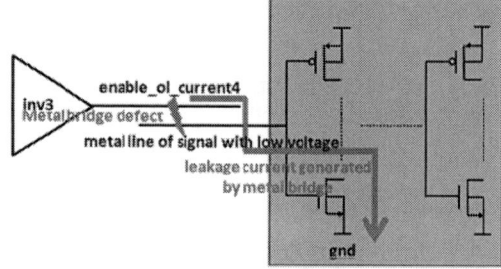

Fig.9. Leakage path of metal bridge defect

Base on the hypothesis previous, we can try to use functional OBIRCH to find the hot spot location. The key point of OBIRCH is that the leakage path can be detected by OBIRCH amplifier. Before cutting by FIB, the signal "enable_ol_current4" represent as a pulse, and the leakage current only exist when the signal at high level voltage. So it is hard to detect the leakage current when the signal is a pulse. After FIB cutting, the signal "enable_ol_current4" is always a middle level voltage when force high level voltage on n214. The leakage current always exist in this condition, So we can use this set up condition to do functional OBIRCH. The set up condition of functional OBIRCH is shown as Fig.10. Firstly we connect net n214 to power source of logic area by FIB so that the n214 can keep high level voltage when DUT is working. Secondly we use OBIRCH amplifier to provide a high level voltage on signal "enable_ol_current4" by microprobe technique. The resistor R is used to clamp the current to avoid over stress on unit. And the bias voltage can be provided by external power source so that the DUT can work in turn on state. When the functional OBIRCH is performed, there will be a leakage current from the power source of OBIRCH amplifier to gnd. And the leakage path must flow through the metal bridge defect.

978-1-4799-3911-4/14 $31.00 © 2014 IEEE   44

As we expect the functional OBIRCH find hot spot successfully, and after layout analysis, the hot spot is just the metal bridge defect location. And we can observe that there is suspected anomaly just at the hot spot location between two metal lines. The functional OBIRCH results and layout image is shown as Fig.11.

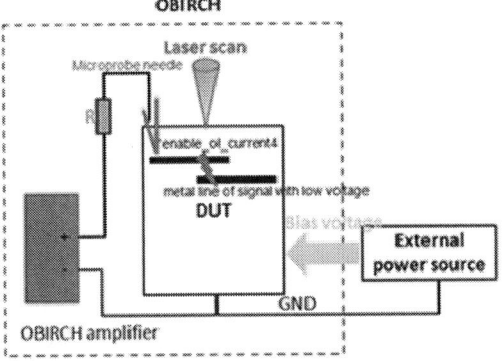

Fig.10. Setup condition of functional OBIRCH

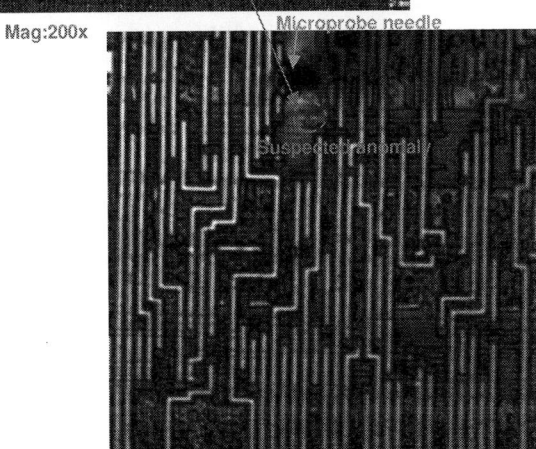

OBIRCH image

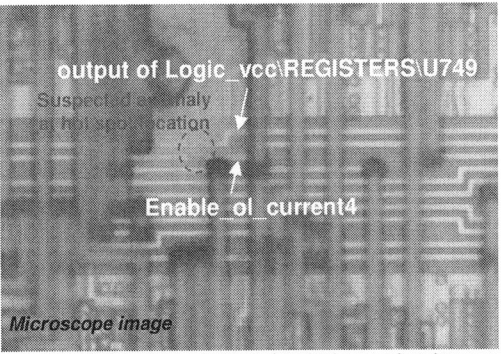

Fig.11. Suspected anomaly is found at hot spot location

FIB cross section was performed at the Metal bridge location. And particle was observed between "enable_ol_current4" and output of register U749. See Fig.12.

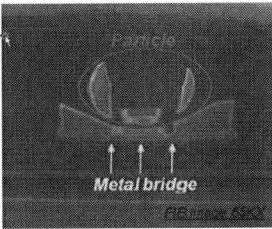

Fig.12. Metal bridge defect was found

IV. SUMMARY

Metal bridge defects in the logic area of a device cause failure by connecting two signals paths together. Generally, the two signal paths are not correlated. Since the two signals are often in different logic blocks, schematic analysis is not helpful in finding the metal bridge defect location. Use of microprobe to find the two abnormal signals are also difficult. In this case study, we use a new method to find the defect. Compared with a pulse, a constant voltage is easier to detect, so we use FIB to edit the circuit to convert a pulse into a constant voltage. Because photon emission cannot find the defect location directly, functional OBIRCH is performed. We try to find a leakage path, and then we use OBIRCH to detect the leakage path. Based on this set up condition, the functional OBIRCH can detect the leakage path and find the hot spot. Finally the physical analysis confirms that the metal bridge defect occurs at the hot spot location identified by OBIRCH.

REFERENCES

[1] Kudva SM, Clark R, et al., "The SEMATECH Failure Analysis Roadmap", Proc In1 Symp Testing & Failure Analysis (ISTFA 1995) 6-10 Nov 95, Santa Clara, California, USA, pg 1-5, 1995

[2] Vallett DP, "Probing the Future of Failure Analysis", Electronic Device Failure Analysis, Vol4, No 4, pg 5-9, 2002

[3] DV Isakov, BWM Tan, "Applications of Scanning Near-field Photon Emission Microscopy", IPFA - 2009, China

[4] DiWei Fan, Li Tian, et al., "Failure analysis of complicated case by functional OBIRCH method", IPFA, pg 87-90, 2013

[5] DiWei Fan, Wang W, et al., "Combined Emission with simulation technique to resolve unstable failure mode sample", ICSE, pg444-447, 2012

Application of Scanning Capacitance Microscopy on SOI Wafer in Die-Level Failure Analysis

Seah Pei Hong[1], Zheng Xin Hua[2], Chng Kheaw Chung[1], Aaron Chin[1],

[1]Systems on Silicon Manufacturing Co. Pte Ltd, 70, Pasir Ris Industrial Drive 1, Singapore 519527
[2]Nanyang Polytechnic, 180 Ang Mo Kio Avenue 8, Singapore 569830
Phone No: (65)-62487000 Fax No: (65)-62487606 Email: pei.hong.seah@ssmc.com

Abstract: **With the presence of Buried Oxide (BOX) layer in Silicon On Insulator (SOI) wafer, local defect isolation by using Conductive Atomic Force Microscopy (C-AFM) in die-level failure analysis is not feasible as electric current is unable to pass through the BOX layer. To overcome this limitation, Scanning Capacitance Microscopy (SCM) is used to perform local defect isolation in die-level failure analysis. Investigation was performed to evaluate the type of SCM probes which gave high signal sensitivity. Case study on sample with leakage through the SOI substrate is demonstrated and presented in this paper.**

I. INTRODUCTION

Silicon on insulator (SOI) technology is a wafer manufacturing solution that is suitable for producing low power and high speed performance devices. SOI wafer is a substrate material stack that is composed of three layers which are a surface silicon layer, over a silicon dioxide insulating layer which is Buried Oxide layer (BOX) and supported by bulk silicon or wafer handler.

With the presence of BOX layer in SOI wafer, local defect isolation by C-AFM, which is using DC bias [1], becomes impossible due to electrical current unable to pass through the BOX layer. C-AFM system setup is shown in Figure 1.

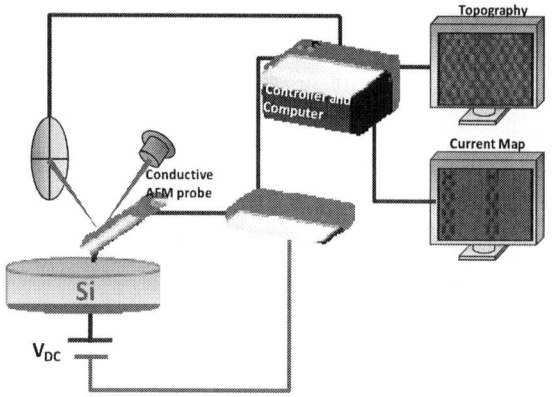

Figure 1: System setup of Conductive Atomic Force Microscopy [2]

Since C-AFM is not feasible, other method is explored to overcome this limitation. SCM technique is commonly used to map the dopant profile with native oxide growth on the

surface. When no oxide is present, a Schottky capacitor is formed. When the probe and surface are in contact, an AC bias is applied and capacitance variations in the sample is generated which can be detected using a capacitance sensor [3](refer to SCM setup in Figure 2).

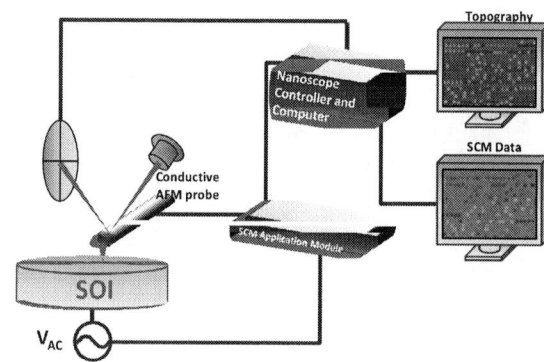

Figure 2: System setup of Scanning Capacitance Microscopy [4]

Experiment was carried out on a SOI sample at Contact level. SCM data revealed that N and P doped diffusion Contact can be distinguished clearly in term of their color contrast. SCM technique was proven to be feasible in isolating local defect in SOI die level failure analysis. A few papers were published in the recent years on SCM technique capability of isolating local defect on SOI wafer [5] & [6].

In this paper, we will discuss the investigation on the SCM probes to improve sensitivity of the SCM technique for better identification of the defective Contact. Besides that, we will also describe the application and effectiveness of SCM technique in die-level failure analysis. A case study presented here was related to localized defect which caused leakage through the substrate. Sample was deprocessed down to Contact level by polishing and SCM analysis was carried out to isolate the source of leakage. Failure mechanism was confirmed with the help of circuit layout analysis. Verification results of physical defect for the issue will also be discussed in details.

II. EXPERIMENTS, ANALYSIS AND DISCUSSION

Conventional platinum (Pt) or platinum-iridium (Pt/Ir) coated silicon probes are used for SCM imaging. However, it

978-1-4799-3911-4/14 $31.00 © 2014 IEEE

was found that the SCM signal to noise (SNR) ratio was not satisfactory when applied to Contact-level SOI devices imaging and the probe tips wore off easily after a few scans.

We have investigated alternative probes for SCM imaging on Contact-level SOI devices. It was found that doped-diamond coated silicon probes, which have been widely used for Contact-level C-AFM imaging, improved the SCM signal to noise (SNR) ratio significantly. Results showed higher dC/dV amplitude and less noisy dC/dV phase (Figure 3 & 4), as compared to Pt/Ir coated probes (Figure 5 & 6). The doped-diamond probes gave higher signal and lower noise levels. As these probes were diamond-coated, they also lasted longer than the Pt/Ir probes.

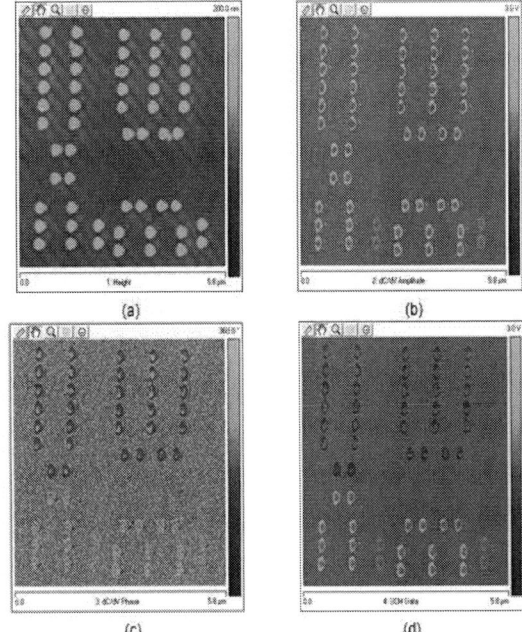

(a) (b)

(c) (d)

Figure 5: Contact-level SCM imaging for SOI devices using conventional Pt/Ir coated probes
(a) Topography (b) dC/dV amplitude (c) dC/dV phase
(d) SCM data

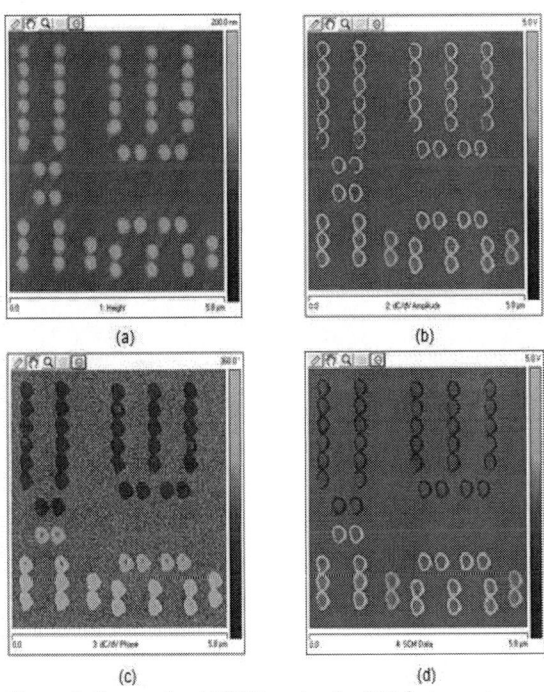

(a) (b)

(c) (d)

Figure 3: Contact-level SCM imaging for SOI devices using doped-diamond coated probes
(a) Topography (b) dC/dV amplitude (c) dC/dV phase
(d) SCM data

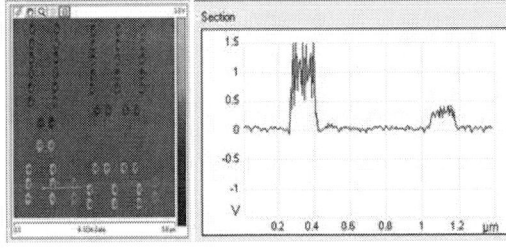

Figure 6: Cross-section profile of SCM data using conventional Pt/Ir coated probes

III. CASE STUDY

Poly Silicon Gate Misalignment Issue

A product suffered 0.6% yield loss due to leakage current with wafer edge cluster failure on wafer e-sort map. From the die level digital block supply and ground I-V curve as shown in Figure 7, additional leakage current of 2mA from 2.7V onwards was measured on the failed devices as compared to good devices. Backside Photon Emission Microscopy (PEM) technique was carried out and systematic emission spots at the specific structure were observed at the decoder of memory area.

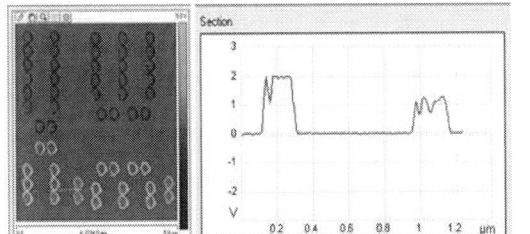

Figure 4: Cross-section profile of SCM data using doped-diamond coated probes

978-1-4799-3911-4/14 $31.00 © 2014 IEEE

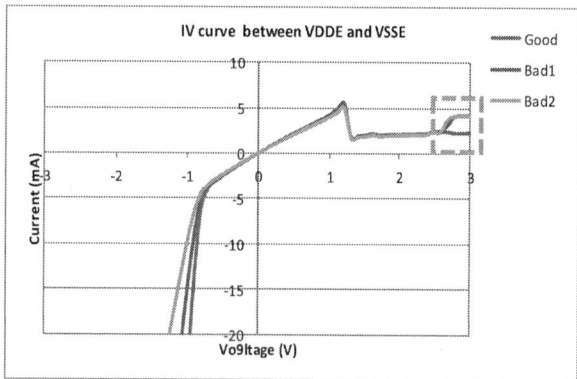

Figure 7: Die level digital block I-V curve comparison between good and failed devices

SCM technique was then performed at the emission site at Contact level and SCM data was obtained for both good device (Figure 8a) and failed device (Figure 8b).

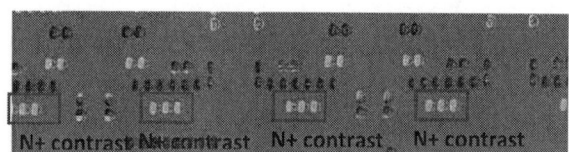

Figure 8(a): SCM data of good device

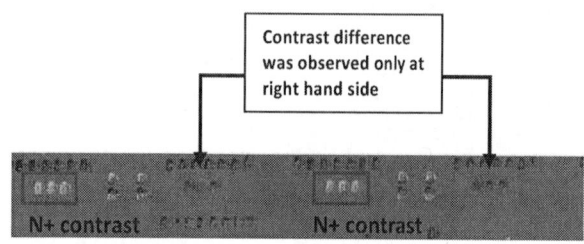

Figure 8(b): SCM data of failed device

After comparing the SCM data between good and failed device, three N+ Contacts at the drain side of the transistor on the failed device were observed to have different capacitance as compared to good device. This phenomenon was occurred systematically on the transistor located at the right side of a pair of mirrored circuit layout structure. Design layout of the abnormal Contact (Figure 9a) was reviewed and suspected leakage current path was established. It was suspected to be due to Poly Gates shifted to the left side and induced Source to Drain leakage at the right side transistor (refer to Figure 9b for the detailed explanation on the circuit design layout).

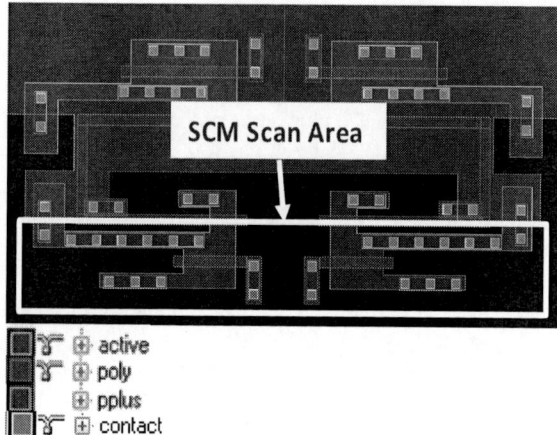

Figure 9a: Design Layout at the SCM scan area

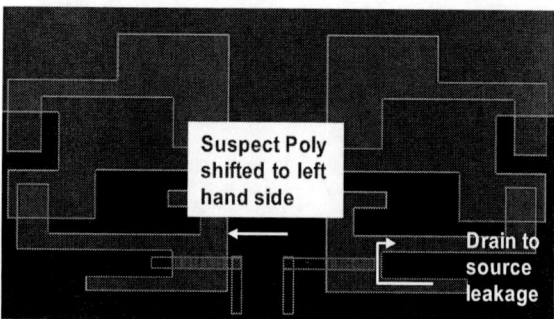

Figure 9b: Suspected leakage current path

This suspected failure mechanism was then verified by performing physical deprocessing to Poly layer. Referring to the Scanning Electron Microscopy (SEM) result in Figure 10, it revealed the Poly Gate at the right side was indeed not fully overlaid across the Active due to photo lithography overlay issue and causing transistor Drain to Source leakage. With the application of SCM technique and circuit design layout analysis, the failure mechanism was identified successfully in a short time.

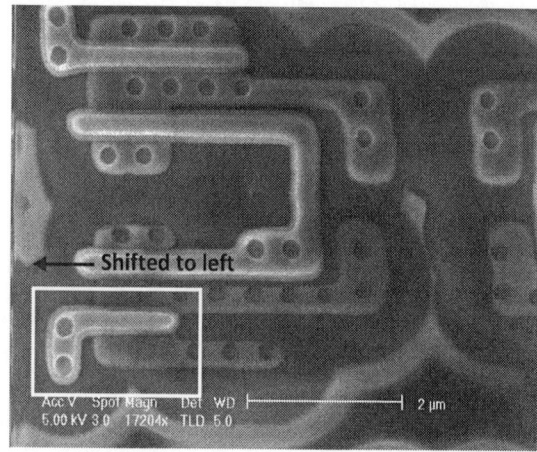

Figure 10: Poly gate at the right side

IV. CONCLUSION

Investigation on selection of SCM probes for contact-level SOI devices was carried out and diamond-coated probes showed much better performance as compared to conventional Pt/Ir probes. Applications of SCM technique at Contact level on SOI wafer have been demonstrated with successful findings. It provides high resolution capacitance map which is able to isolate a single defective Contact and the diffusion layers underneath the Contact. Sample preparation for SCM technique is the same as that of C-AFM technique, and no special treatment such as FIB edit is required. This will greatly increase the success rate of defect isolation on SOI wafer in future.

ACKNOWLEDGMENT

This is a partnership project between Systems on Silicon Manufacturing Company (SSMC) and Nanyang Polytechnic (NYP). The authors would like to thank SSMC Failure Analysis Engineers' effort in supporting sample preparation for SCM analysis and Yu Thi Han of NYP for her support and effort on the tools setup.

REFERENCES

[1] Seah Pei Hong, Zheng Xin Hua, Aaron Chin, Chan Joo Guan, Applications of C-AFM Technique to Identify Localized Implant Related Low Yield Issue, IEEE Proceedings of 19th IPFA, 2012, pp. 1-4

[2] Veeco Application Modules: Dimension and Multimode Manual, Page 11.

[3] http://en.wikipedia.org/wiki/Scanning_capacitance_ microscopy

[4] Veeco Application Modules: Dimension and Multimode Manual, Page 9.

[5] K. Ramanujachar, Scanning Capacitance Microscopy at Transistor Contact Level, *Proceedings of the 30th International Symposium for Testing and Failure Analysis*, 2004, pp. 482-486

[6] Lim Soon Huat, Lwin Hnin-Ei, Vinod Narang, J.M. Chin, Scanning Capacitance Microscopy for Failure Analysis of SOI-Based Advanced Microprocessors, *Proceedings of the 36th International Symposium for Testing and Failure Analysis*, 2010, pp. 309-316

Failure Analysis of Zn-Mo Particle in the Molding Compound Causing Gate-Source Short in Non-Passivated MOSFET Device

C. K. Lau and C. H. Tan

Infineon Technologies (Malaysia) Sdn. Bhd.
Batu Berendam Free Trade Zone, 75350 Melaka, Malaysia
Phone: (606) 2873057   Fax: (606) 2516095   Email: cheekiang.lau@infineon.com

Abstract- **This paper presents the failure analysis steps that localized and revealed the embedded Zn-Mo particle bridging two adjacent metal traces on a non-passivated MOSFET chip by using backside TIVA, frontside parallel polishing and EDX. TIVA emission spot localized between two metal traces indicates a high probability of particle defect.**

I. INTRODUCTION

Fault localization involving non-passivated chip can be very difficult if the electrical failure is caused by a metal particle that shorts two adjacent metal traces. This is due to the problem where the failure recovers after frontside package decapsulation which removes the metal particle all together. With the loss of failure cause, the electrical failure can never be re-induced through whatever methods. Several reoccurrences of failure recoveries prompted the analysts to opt for alternative analysis techniques to confirm and capture the evidence of the metal particle defect. Backside fault localization was selected to preserve the chip surface in its original condition and prevent the failure cause from being modified by the analysis.

A. Thermally Induced Voltage Alteration (TIVA)

TIVA is a fault localization technique that uses laser scanning to detect short circuit on a device. The laser scanning on the chip surface will induce localized heating on the chip surface and causes the power consumption to change. With the device operates in its failing condition and via a constant current source, a laser beam is scanned over the chip surface and changes on the Vcc pin voltage is monitored. Localized heating occurs when the laser hits a short circuit location. The localized heating changes the resistance of the short immediately and leads to the change on the device's power consumption. Changes on the power consumption are processed and overlaid to the optical image of the chip to indicate the location where a power consumption change had occurred. Fig. 1 shows the basic principle of TIVA.

Localization of an early breakdown between the collector-emitter junctions and the local threshold voltage shift in the power cell area of the power transistor can be easily achieved using TIVA [1]. TIVA is able to detect a fault location at a much lower current level, approximately

10% of the current needed by the emission microscope (EMMI) technique. However, TIVA shares the same disadvantage with EMMI that is emission might be blocked by the metallization layer. To overcome this limitation, the backside technique is used since silicon substrate is transparent in infrared wavelengths.

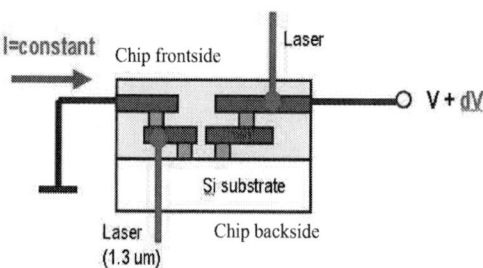

Fig. 1: Basic principle of TIVA.

B. Parallel Polishing and Physical Analysis

Parallel polish was selected as the destructive failure analysis step since decapsulation will cause the failure to recover. Different from the catastrophic destruction which the package decapsulation can cause, parallel polish is safer where the analyst is able to control the destruction speed by using fine polishing material as well as frequent inspection. Most importantly, parallel polishing only grinds off the sample's surface in micrometer range so there is a high chance that the defect will be captured and revealed. Once the defect is revealed, physical analysis occupying the optical microscope, scanning electron microscope (SEM) and energy dispersive X-ray (EDX) will follow to determine the failure mechanism and the failure root cause.

II. EXPERIMENTAL PROCEDURE

The sample is a customer returned unit. The product is a MOSFET which exhibited short circuit failure between gate and source junctions. The sample has a non-passivated chip. To suit the backside TIVA requirements, the sample was subjected to backside grinding to remove the die paddle and to thin down the silicon chip backside. Fig. 2 illustrates the overview of the sample embedded in resin compound and the probing method.

978-1-4799-3911-4/14 $31.00 © 2014 IEEE

The sample was then subjected to backside TIVA analysis using Hamamatsu Phemos-1000 system [2]. The device was powered up in its failing condition i.e. short circuit between gate and source junctions. After localizing the defect location, the sample was parallel polished from the frontside to thin down the molding compound. The parallel polishing started off with grinding the sample on fine silicon carbides until the wedge bonds were visible. The sample was then polished with 3 μm, 1 μm and 0.25 μm polycrystalline diamond compounds on polishing cloth until the chip surface was partially visible under optical microscope inspection. Parallel polishing with the finest polycrystalline diamond compound was continued until the defect was fully visible at the emission spot location. The embedded defect was subsequently inspected using SEM. Elemental analysis using EDX was performed on the defect to determine the failure root cause.

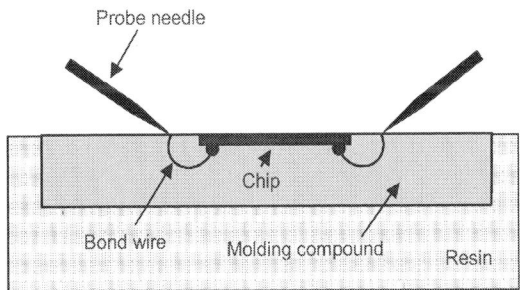

Fig. 2: Overview of the package after parallel backside grinding. Probe needles are used to probe on the bond wires for electrical biasing during backside fault localization.

III. RESULTS AND DISCUSSIONS

A. Fault Localization

Backside TIVA analysis detected a clear emission spot at the edge of the chip, between the gate ring and the source ring as described in Fig. 3, Fig. 4 and Fig. 5. The illuminated image captured during the analysis observed an abnormality at the emission spot location resembling a particle, as shown in Fig. 6. The chip surface structure was clearly seen in the backside TIVA images. Similar fault localization that was performed in the Munich failure analysis laboratory revealed the same observation [3].

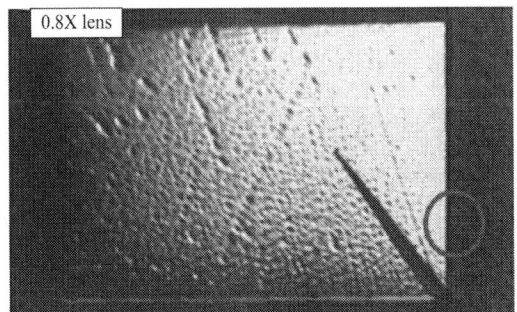

Fig. 3: An anomalous backside emission spot was detected at the edge of the chip.

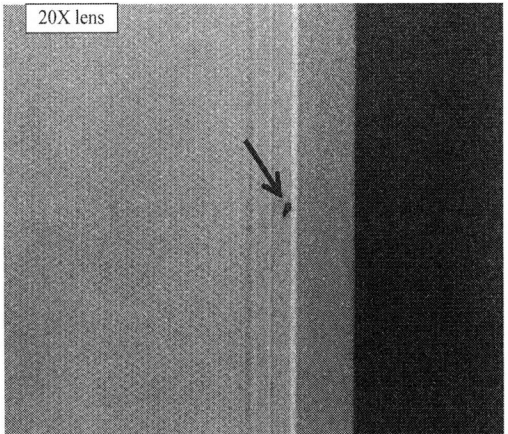

Fig. 4: Backside TIVA emission spot seen under a larger magnification.

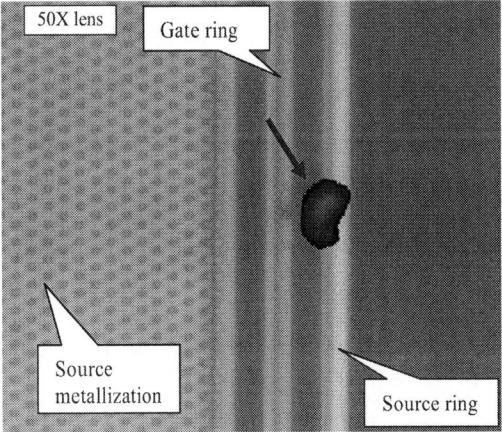

Fig. 5: Backside TIVA utilizing 50X lens detected a very clear emission spot at the edge of the chip between the source and gate rings.

Fig. 6: Illuminated image captured using the backside TIVA observed an abnormality between the source and gate rings.

B. Physical Failure Analysis

Careful parallel polish from the frontside of the sample revealed a particle that touches the gate and source rings, as shown in Fig. 7. The very thin layer of molding compound left on the chip surface was semi-transparent in the dark field optical imaging where the particle, the gate ring and the source ring were partially visible. SEM image of the particle was shown in Fig. 8. To ensure that the SEM was capturing the correct particle, the SEM image was overlaid onto the optical image and the location of the particle was the same, Fig. 9.

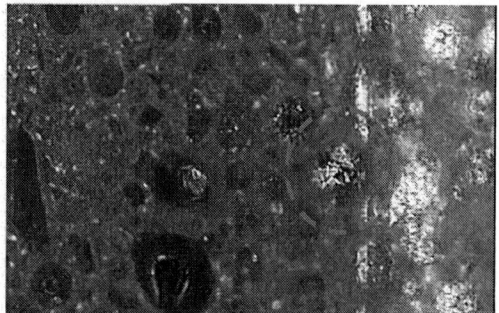

Fig. 7: A particle touching the gate and source rings causing short circuit failure was revealed after parallel polishing.

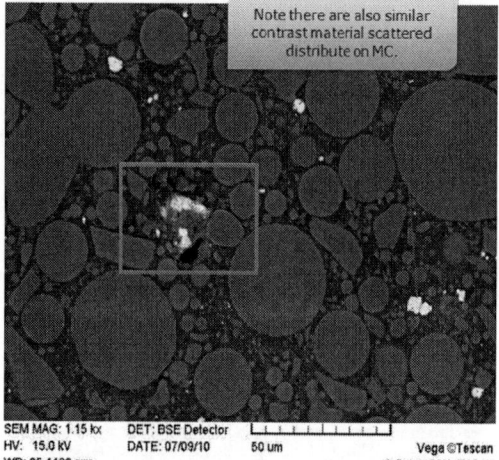

Fig. 8: SEM image of the particle.

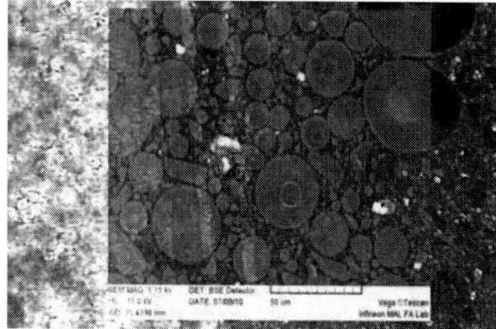

Fig. 9: The location of the particle seen under the SEM was validated by overlaying the SEM image onto the optical image.

In the EDX analysis, the sample was found to contain zinc (Zn) and molybdenum (Mo), as described in Fig. 10. Fig. 11 is the element mapping result which showed concentrated amount of zinc and molybdenum on the particle. The quantitative analysis results showed that the particle consists of about 90% zinc and 10% molybdenum,

which proved that the particle is conductive in nature, and it caused short circuit between two adjacent metal traces. Lastly, the length of the metal particle was measured in SEM imaging to be approximately 20 micrometers as shown in Fig. 12. The size of the particle is larger than the distance between the gate ring and the source ring which is 15 micrometers apart, as shown in Fig. 13. After discussing the results with the assembly engineers, it was learnt that Zn-Mo particles were used in the molding compound as flame retardant material. Discussions with the molding compound supplier were held to get their commitment on controlling the size of the Zn-Mo particles and also to look for non-conductive flame retardant materials. On the other hand, improvements in the wafer fabrication were initiated on possibility to add the passivation layer on the chip surface to prevent recurrence.

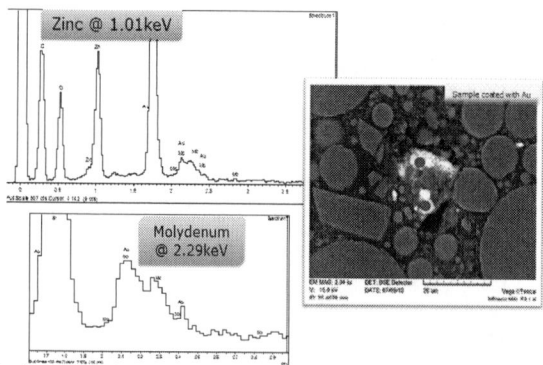

Fig. 10: EDX analysis detected Zinc (Zn) and Molybdenum (Mo) elements in the particle.

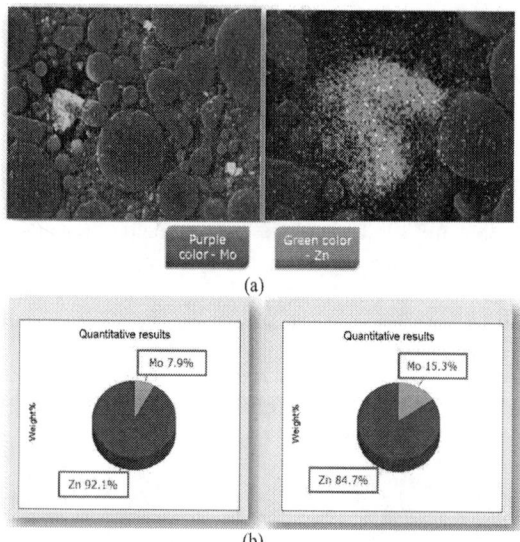

Fig. 11: EDX results. (a) Areas that contain Zn were clearly mapped, Mo's area was not clear due to its small amount and (b) quantitative analysis indicated that the particle was made up of about 90% Zn and 10% Mo.

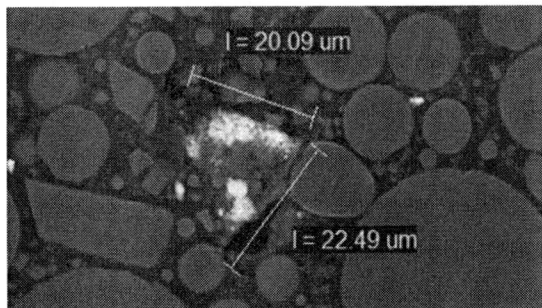

Fig. 12: The Zn-Mo particle's length is approximately 20 micrometers.

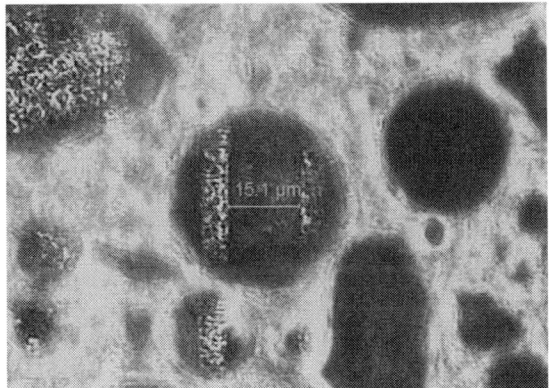

Fig. 13: Distance between the gate and source rings is 15 micrometers.

IV. CONCLUSION

As a conclusion, backside fault localization techniques using TIVA is efficient in localizing a metal particle defect. For non-passivated chip which has leakage or short failure, the analyst needs to anticipate metal particle defect in the sample and pursue backside fault localization followed by careful physical analysis from the frontside.

ACKNOWLEDGMENT

The authors would like to express their gratitude to Infineon Kulim and Melaka colleagues for the constructive advices and appreciation to the management support from Melaka and Munich.

REFERENCES

[1] Reissner, M., "Fault Localization at High Voltage Devices Using Thermally Induced Voltage Alteration (TIVA)". *Proc 18ᵗʰ European Symposium on Reliability of Electron Devices, Failure Physics and Analysis,* Arcachon, France, October. 2007, pp. 1561-1564.

[2] Abd, R. R., Failure Analysis Report # KU10A1-00205. Failure Analysis Lab, Kulim, Kedah, 2010.

[3] Mueller, T. A., Failure Analysis Report # MP10A2-00362. Failure Analysis Lab, Campeon, Munich, 2010.

Thermal-Electric Activated Ions Diffusion on Printed Circuit Board

Lai-Seng Yeoh, Kok-Cheng Chong and Susan Li*

Spansion (Penang) Sdn. Bhd. Phase II, Bayan Lepas Free Industrial Zone, 11900 Penang, MALAYSIA.
*Spansion Inc. 915 DeGuigne Drive, Sunnyvale CA 94088-3453, USA.
Phone: (604)8882201 Email: Lai-Seng.Yeoh@spansion.com

Abstract- **Reliability of PCB is of paramount importance during high temperature application in the field. Good integration among all PCB components is essential to ensure robust PCB performance. Imperfection in the PCB assembly may activate ion diffusion and induce board level contamination, which are detrimental to the device functionality. In this paper, we show that magnesium ions can be thermal-electrically excited from a heat fin. Under the influence of electric field, these ions diffuse through thermal glue and cause fatal failure to the adjacent electronic device. The movement of the ions is governed by various diffusion mechanisms such as interstitial diffusion, grain boundary diffusion, and surface diffusion. The PCB assembly process must be properly controlled and by isolating the thermal glue from the heat fin, the failure risk can be reduced.**

I. INTRODUCTION

A printed circuit board (PCB) is a composite of organic and inorganic materials with external and internal wirings, allowing electronics components to be electrically interconnected and mechanically supported. The PCB supplies power to the components and conducts away heat. Heat management is especially important to maintain the PCB at an acceptable operating temperature. Prolonged exposure to elevated temperature will place the electronics components at the risk of failures related to ionic diffusion, parasitic chemical reactions, and mechanical creep in the bonding materials. Heat fin assisted air cooling systems are usually used for heat convection on the PCB [1]. However, improper mounting of the heat fin and thermal glue can not only cause thermal dissipation failure, but could also trigger thermal-electrically activated ion diffusion, which is the main focus in this paper.

There have been a number of theoretical and experimental studies about ion diffusion on PCB in the literature. For instance, K. Fukunaga et al. investigated ion diffusion in the insulating epoxy layer of a PCB. They observed diffusion of Cu^{2+} and NH^{4+} toward the anode as well as Cl^- and SO_4^{2-} toward the cathode. By using pulsed electroacoustic technique, they discovered that the migrated ions and their composites formed a space charge profile and established conducting paths in the PCB insulation layer [2]. In this paper, we provide a survey of Mg^{2+} ion diffusion on PCB that is not reported in the literature.

Ion diffusion involves transport of ions across solid material. In an ideal solid material, its constituent atoms, molecules, or ions are arranged in an ordered pattern extending in all three spatial dimensions. These constituents are often stacked in a close-packed form and space available for ion diffusion is very limited. At non-zero temperature, deviation from ideal stacking occurs and this results in positional disorder defects which play an important role in governing the ion mobility. There are two types of point defects in the crystal, namely Schottky defect and Frenkel defect. Schottky defect refers to the crystal imperfection where a pair of anion and cation ions disappears and leaves their positions to be vacant. Fenkel defect refers to missing of a single ion from its regular position and wandering in interstitial sites [3].

Both Schottky and Frenkel defects create vacant sites in the crystal and any ion can jump to one of these vacant sites. The previous site of the ion is now vacant and can be occupied by another ion. This process results in transport of ions across the solid material which will give rise to conductivity and this is known as vacancy diffusion (Fig. 1 left). On the other hand, an ion that moves to the interstitial site and induces a Frenkel defect can jump to a neighboring interstitial site and so on. This process leads to a long distance motion of the ion and it is called as interstitial diffusion (Fig. 1 right) [3]. Besides that, there are other types of ion diffusions, such as volume diffusion, grain boundary diffusion, and surface diffusion. In volume diffusion, atoms move through the material from one interstitial site to another. The activation energy is large and the diffusion rate is slow because of the surrounding atoms. In grain boundary diffusion, atoms can diffuse easily along boundaries and interfaces due to poor atom packing. The activation energy is low and the atoms can squeeze their way through the disorder grain boundary. In surface diffusion, atoms diffuse along the material surface and this is an easier process due to less constraint at the surface [4].

There are a number of factors which can influence the ion conductivity during the diffusion process. They are the concentration of charge carriers, the material temperature, the availability of vacant sites, and the ease with which an ion can jump from one site to another. The last factor is controlled by the activation energy, which is the free energy barrier an ion must overcome for making a jump between different sites [3].

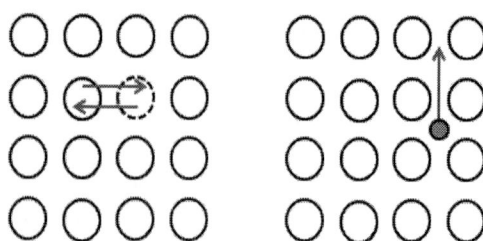

Fig. 1. Vacancy diffusion (left) and interstitial diffusion (right).

II. EXPERIMENTAL PROCEDURE

A customer PCB encountered reliability failure after a certain period of field application. The defective PCB was returned to Spansion together with a good PCB. Failure analysis was requested by the customer in order to determine if the failure is related to the customer PCB or the Spansion flash memory device. The customer reported that the BYTE resistor (which was connected to the BYTE pin of the flash device) was shorted and it caused the entire PCB to malfunction. In the customer's design, the memory device is partially covered by thermal glue so that the device is write-protected (Fig. 2). Failure analysis was carried out at both board level and device level using a variety of tools, including memory tester, curve tracer, multimeter, x-ray, optical microscope, Field Emission Scanning Electron Microscope (FESEM), Energy Dispersive X-ray (EDX), and Fourier Transform Infra-Red (FTIR).

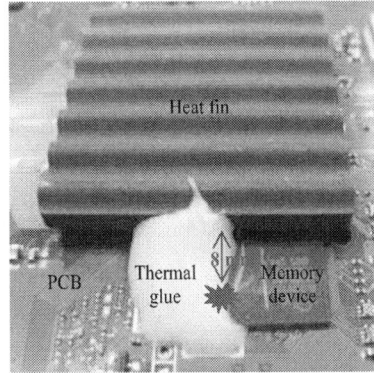

Fig.2. The defective customer-returned PCB containing Spansion flash memory device. The defect location is marked by the explosion symbol.

III. RESULT AND DISCUSSION

A. Analysis Results

First, board level electrical test was performed on the PCB to verify the reported failure. Multimeter measured an abnormal resistance of 8.5 kΩ on the BYTE resistor. A normal resistance should be in the range of MΩ. Curve trace indicated that the BYTE resistor was of possibly shorted to ground (Fig. 3 left). Then, the flash memory device was demounted from the PCB using a hot gun and it was then tested with a memory tester. Testing showed that the isolated device was electrically good, and was able to pass all continuity, parametric, and functional tests. Moreover, the BYTE pin of the isolated device demonstrated normal I-V curve characteristic (Fig. 3 right). This suggested that the short failure was located on the PCB but not on the device. For further confirmation, visual inspection and x-ray examination were performed. Both of these analyses showed that there was no obvious visual anomaly on the device, especially at the area surrounding the BYTE solder ball (Fig. 4).

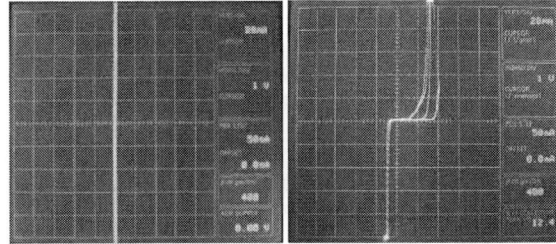

Fig. 3. I-V curves of the BYTE resistor on the PCB (left) and the BYTE pin on the isolated device (right).

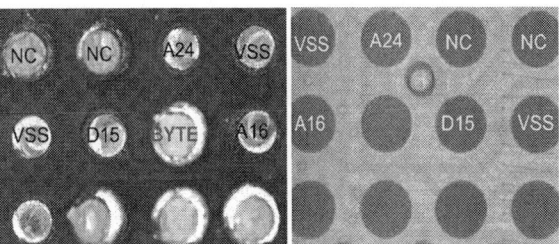

Fig. 4. Optical image (left) and x-ray image (right) of solder balls on the isolated memory device.

Next, board level electrical test was repeated without the memory device on the PCB. Multimeter measurement showed that the BYTE resistor on the PCB was still shorted and its resistance remained at 8.5 kΩ even though the memory device had been removed. This clearly proved that the failure was located on the PCB. Subsequent visual inspection on the PCB revealed residues bridging between the solder pads of BYTE and A24 as well as between A24 and VSS (Fig. 5 & Fig. 6). In other words, BYTE was indirectly shorting to VSS through A24. Such bridging was not seen at other solder pads. The short was further confirmed via electrical probing at these three pads. The resistance between BYTE and VSS was measured to be 8.5 kΩ, which was the same as the resistance on the BYTE resistor. In comparison, resistance at other pins on the board was normal in the range of MΩ. For instance, the resistance between VSS and A16 was 5.6 MΩ.

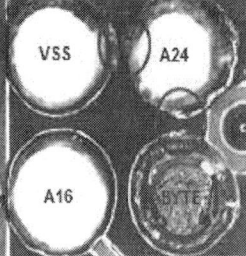

Fig. 5. Residues bridging between solder pads on the PCB.

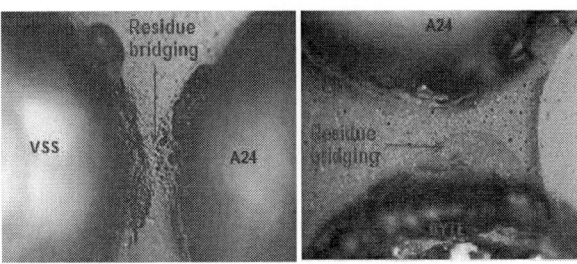

Fig. 6. Close-up views on the residues bridging paths.

The bridging material between BYTE, A24, and VSS was examined using FESEM. It was found to be charging-up and was suspected be a non-conductive polymer. EDX analysis was conducted on the bridging and non-bridging areas. Additional elements of Mg and Br were detected in the bridging area in addition to O, Si and Au (Fig. 7). O and Si were contained in the solder resist compositions, while Au was the sputter-coating material applied prior to FESEM inspection. It is believed that the Mg and Br contamination was likely the cause of the short failure. Since Sn was not detected by EDX, this ruled out the possibility of solder bridging.

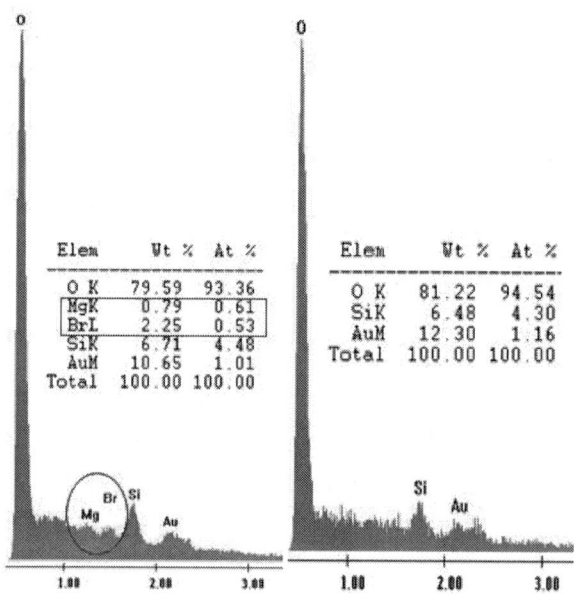

Fig. 7. EDX analysis result of the bridging area (left) and the non-bridging area (right).

Next, FTIR analysis was carried out on the bridging material. The transmission spectrum of the bridging material was found to be a 97%-match to the transmission spectrum of the thermal glue (Fig. 8). The FTIR result suggested that the glue had probably diffused into the PCB area below the device and resulted in a short circuit. One puzzling question is that Mg and Br were not detected in the glue by EDX. So, where did Mg and Br come from? More investigations were carried out and eventually Mg and Br were discovered in the heat fin (Fig. 9), where Mg^{2+} ions were generated.

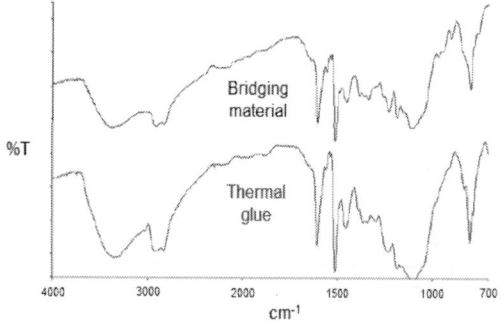

Fig. 8. FTIR spectrums of the bridging material and the thermal glue.

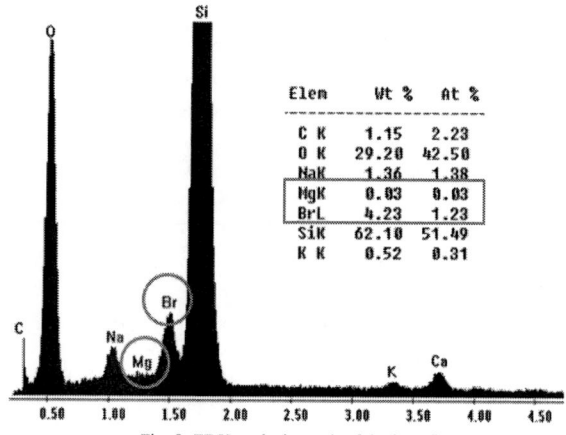

Fig. 9. EDX analysis result of the heat fin.

B. Failure Hypothesis

The above analysis results indicate that the BYTE pin short failure is caused by PCB board contamination. The failure is highly related to the mounting condition of the thermal glue. It is believed that the glue acted as a transit medium for the migration of Mg^{2+} and Br^- ions from the heat fin to the device in the presence of high temperature and electric field. Migration of the ions requires the glue to be fastened on the heat fin in order to establish a continuous charge conduction path from the heat fin to the device. On the other hand, it is observed that the glue is not bonded to the heat fin on the good PCB and therefore the good PCB did not fail the same way in the field (Fig. 10). In order to avoid similar type of failures, thermal glue needs to be carefully applied, and not be fastened or touching on the heat fin.

The solder balls at A16, A24, BYTE, and VSS are located at the package edge (~8 mm from the heat fin) which is partially covered by the thermal glue. When electrical bias is supplied to the device, these solder balls behave as charged spheres and they form four pairs of electric dipoles (A16-BYTE, A16-VSS, A24-BYTE, and A24-VSS). In each dipole, two spheres with opposite charges attract each other. Strong electric field is established in the dipole, where the field lines extend away from the positive sphere to the negative sphere. The electric field has rotational symmetry about an axis passing through both spheres. The vector of the electric field is tangent to the curved field line [5]. When Mg^{2+} and Br^- ions enter the field region, both will experience electrostatic forces and move along the field lines at opposite directions. Mg^{2+} will be attracted toward BYTE or VSS while Br^- will be attracted toward A16 or A24 (Fig. 11).

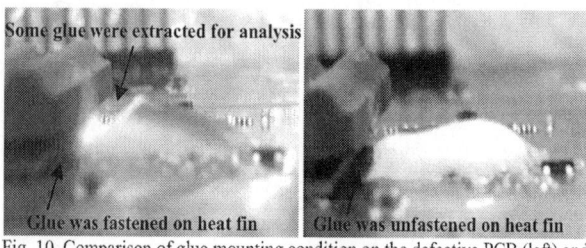

Fig. 10. Comparison of glue mounting condition on the defective PCB (left) and a good PCB (right)

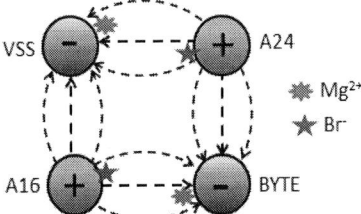

Fig. 11. Four pairs of electric dipole formed among A16, A24, BYTE, and VSS.

Diffusion refers to an observable net flux of atoms, ions, or other species. The extent of diffusion depends on temperature, time, nature and concentration of diffusing species, crystal structure, and composition of the matrix, stoichiometry, and point defects [4]. Mg^{2+} is an anion containing positive electric charge while Br^- is a cation carrying negative electric charge. They are capable to participate in a diffusion process when an electric field is present [6]. Mg^{2+} and Br^- ions are generated in the heat fin and their movement through the heat fin is believed to be governed by interstitial diffusion mechanism (Fig. 1). At the point of contact between the heat fin and the thermal glue, atoms or molecules from the materials diffuse to the contact point in two opposite ways, creating a bridge or a bond between both materials. Mg^{2+} and Br^- ions from the heat fin continue their interstitial diffusion through this bridge or bond and enter the thermal glue. The same principle is applied to the contact point between the thermal glue and the PCB [4].

In the thermal glue, diffusion of Mg^{2+} and Br^- ions can occur by penetrating between the polymer chains rather than moving from one location to another within the chain structure. The sizes of a Mg^{2+} ion and a Br^- ion are relatively smaller than the size of a C-H molecule in the thermal glue. Therefore, the ions can diffuse rapidly. The activation energy required for the ions to squeeze past the surrounding C-H molecules is gained from the electric field attraction at the flash memory device. Besides that, the ions can also move either by volume diffusion or grain boundary diffusion in the thermal glue. Grain boundary diffusion is easier as the atom packing is poor in the grain boundaries and ions can easily squeeze their way through the disordered grain boundary [4]. Perturbation of the ions may provoke chemical reactions with the polymer molecules and cause degradation on the mechanical and physical properties of the glue [6].

On the other hand, movement of Mg^{2+} and Br^- ions along the PCB surface is believed to be governed by surface diffusion where the diffusion barrier is the lowest. As the ions moving at hot temperature, they diffuse together with the amorphous glue. This is evidenced by the detection of both ions and glue on the bridging stuff between the failing solder balls. The amorphous glue is eventually crystallized on the PCB surface [4]. The net movement of the charged Mg^{2+} and Br^- ions will establish an electric current flow. The conductivity of the ions is increased with the PCB operating temperature. The hypothesis of various diffusion mechanisms are illustrated in Fig. 12.

IV. CONCLUSION

Ionic diffusion can create reliability issues on the PCB in the presence of high temperature and electrical bias during the field application. This paper illustrated a real case of board level failure caused by an electrical short due to PCB board contamination and ionic diffusion from the nearby heat fin through the contaminant. To prevent similar type of failures, all PCB assembly processes, including applying glue to cover a portion of the devices on PCB, need to be carefully controlled.

ACKNOWLEDGMENT

The authors would like to render deep appreciation to Gene Daszko, Tony Reyes, and HL Chong for their encouragement on technical paper publication.

REFERENCES

[1] Rao R. Tummala, "Fundamentals of Microsystems Packaging," *McGraw-Hill*, International Ed. 2001.
[2] K. Fukunaga, T.Maeno, and K. Okamoto, "Three-Dimensional Space Charge Observation of Ion Migration in a Metal-Base Printed Circuit Board," *IEEE Transactions on Dielectrics and Electrical Insulation*, Vol. 10, No. 3, June 2003, pp. 458-462.
[3] P. Padma Kumar and S. Yashonath, "Ionic Conduction in the Solid State," *Journal of Chemistry Science*, Vol. 118, No. 1, Jan 2006, pp. 135-154.
[4] Donald R. Askeland and Pradeep P. Phule, "The Science and Engineering of Materials," *Thomson Canada Ltd.* 5th Ed. 2006.
[5] D. Halliday, R. Resnick, and J. Walker, "Fundamentals of Physics," John Wiley & Sons Inc. 5th Ed., 1997.
[6] William D. Callister, "Materials Science and Engineering An Introduction," *John Wiley & Sons Inc.* 6th Ed. 2003.

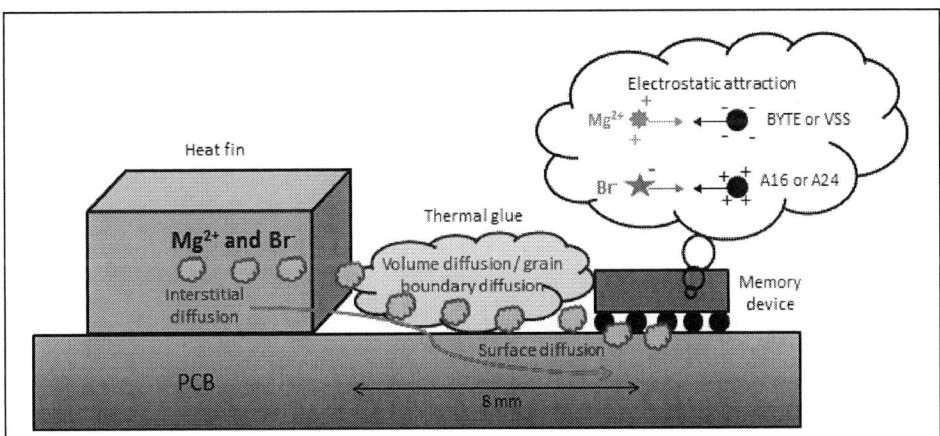

Fig. 12. Various diffusion mechanisms for the migration of Mg^{2+} and Br^- ions from the heat fin to the memory device through the thermal glue.

978-1-4799-3911-4/14 $31.00 © 2014 IEEE

Study on the High Via Resistance by TEM Failure Analysis

Binghai Liu[1], Eddie Er[1], Si Ping Zhao[1], Changqing Chen[2], Ang Ghim Boon[2], Kunihiko Takahashi[3], Chivukula,Subbu[3], Jeffrey Lam[1,2]

1. TEM Failure Analysis, , Product, test and failure analysis Dept. (PTF Singapore), GLOBALFOUNDRIES Singapore Pte Ltd, 60 Woodlands Industrial Park D, Street 2, Singapore 738406
2. Product failure analysis, Product, test and failure analysis Dept. (PTF Singapore), GLOBALFOUNDRIES Singapore Pte Ltd, 60 Woodlands Industrial Park D, Street 2, Singapore 738406
3. Fab2/3/5, Integration & Yield Eng, Process Integration, GLOBALFOUNDRIES Singapore, 60 Woodlands Industrial Park D, Street 2, Singapore 738406
Phone: (+65) 667015555. Fax: (+65) 6362-2938. Email: binghai.liu@globalfoundries.com

Abstract: In this work we reported a case study on ET(electrical testing) failure with via high resistance issue. In order to understand the failure mechanism and root cause behind the high via resistance, detailed TEM (transmission electron microscope) analysis was performed by using various TEM FA (failure analysis) techniques, including EDX, EELS analysis. It was found out that high via resistance arose from the process drift induced Al extrusion and poor barrier metal coverage at via bottom. The correlation between the physical signatures identified by TEM FA and the associated processes were discussed for the root cause understanding.

Keywords: TEM; EFTEM; EELS mapping; failure analysis; ET failure; punch-through via process

I. INTRODUCTION

For Al BEOL (back-end of line) process, in the normal process approach, via etching should stop at top ARC (anti-reflection coating, TiN/Ti) layer with certain gouging into ARC layer by over-etch process, as shown in Fig.1 (a). However, in the old via etching approach, via etching directly punches through ARC layer into Al metal lines, as shown in Fig.1 (b). Compared with normal via etch processes, the punch-through via processes are more challenging in terms of process control, and thus yield and reliability performances are more susceptible to the process drift. It has been reported that the via punch-through process can easily lead to non-uniform interface of W/TiN/Al, resulting in the formation of interfacial void. This will lead to both high via resistance and the long-term device reliability performance. [1]

In this work, we report a typical ET (electrical testing) via2 chain) failure case with punch-through via processes, in which a minor process drift directly led to high via resistance. We will present detailed TEM failure analysis to identify the key physical failure signatures. The detailed discussion was presented on the failure mechanism with correlating the inline process investigations for the root-cause understanding.

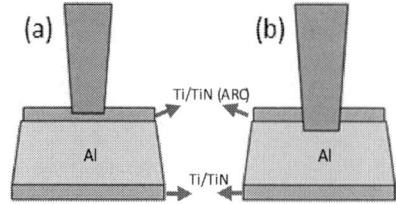

Figure 1. Schematic illustration of (a) normal via etching, and (b) punch-through via etching processes for Al BEOL processes

II. EXPERIMENTAL

Before TEM failure analysis, the detailed electrical fault isolation was performed to identify the defective via with high resistance. Distinct TIVA (thermal induced voltage variation) hotspot was observed.

The TEM samples for these failure analysis were prepared by focused ion beam (FIB, FEI Helios 400S) operated at 30kV. TEM imaging was done by using FEI Titan TEM (80~300kV) at 300kV with Scanning-TEM (STEM) EDX (energy-dispersive x-ray spectroscopy) for elemental analysis. EFTEM (energy-filtered TEM)-EELS (electron energy-loss spectroscopy) mapping was performed by using Gatan GIF (Gatan Imaging Filter, Tridem system at 300kV.

III. RESULTS AND DISCUSSION

In this low yield issue, the ET (via2 chain) full map showed wafer edge pattern with the highest via resistance up to 87 Ω compared with normal resistance of around 2 Ω. In order to identify the defective high-

978-1-4799-3911-4/14 $31.00 © 2014 IEEE

resistant via2, we selected the unit with the highest resistance, and performed electrical fault isolation by using TIVA. Fig.2 showed the distinct TIVA hotspot in the failed via2 chain ET structure.

Figure 2. TIVA hotspot observed at the failed via2 ET chain structure

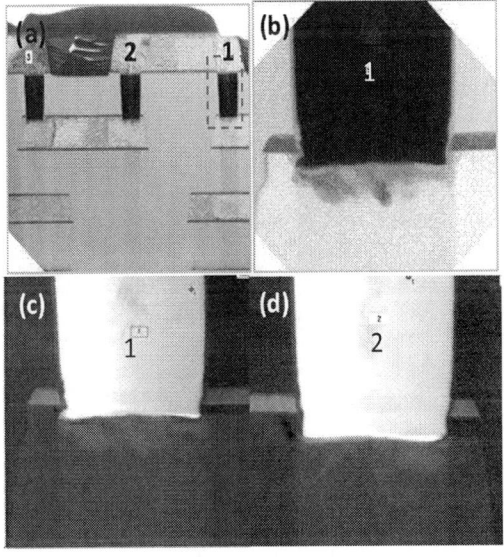

Figure 3. (a)~(b) Bright-field TEM images and (c) STEM image of the defective via2 (the 1st via); (d) reference normal via2

The bright-field TEM images in Fig.3 showed the typical via2 punch-through profile. According to TIVA hotspot position, it was highly suspected that the high-resistance via could be the 1st via2 shown in Fig.3 (a). As seen from Fig.3 (b) and (c), the bottom of the via is relatively uneven when compared with the normal via2 in Fig.3 (d). Under both defective and normal vias, no interfacial void was observed. While voiding was observed at via corners (Fig.3 (c) and (d). The formation of such corner voiding can be ascribed to the intrinsic process weakness of the punch-through via process due to different etch speed of TiN and Al during via etching process. Under STEM Z-contrast,

bright-contrasted layer was observed, indicating some materials diffused to via bottom with higher atomic number than Al, as shown in Fig.3 (c) and (d).

Fig.4 and Fig.5 showed STEM/EDX analysis at the defective via bottom and corner respectively. As showed in Fig.4 (a)~(d), EDX line scan analysis revealed the presence of Ti at the via bottom, indicating the diffusion of Ti into Al. The overlapping of Ti and Al in this region may indicate that Ti-Al compound could be formed with such Ti diffusion. As it is well known, Ti_xAl_y intermetallic compound can be of high resistance phase. The resistivity of Ti_3Al is as high as 210 $\mu\Omega\cdot$cm, compared to 42 $\mu\Omega\cdot$cm of Al and 2.7$\mu\Omega\cdot$cm of Ti.[2] Therefore, we need to understand if such high-resistance Ti_3Al phase could account for the high-resistance ET failure as reported.

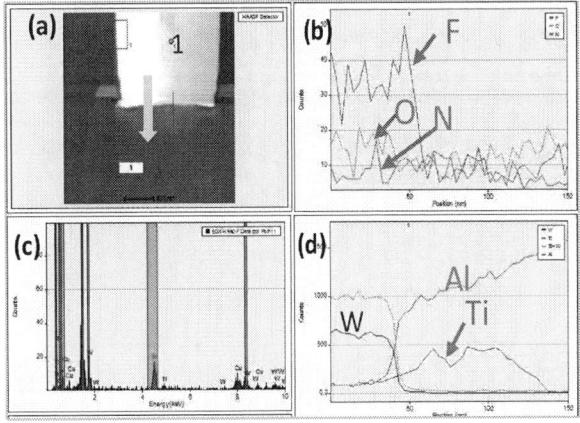

Figure 4 STEM/EDX line scan analysis at the defective via bottom

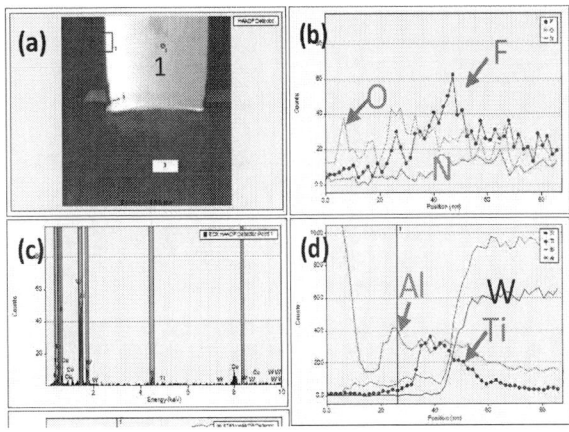

Figure 5 STEM/EDX line scan analysis at the defective via corner

978-1-4799-3911-4/14 $31.00 © 2014 IEEE

Based on TEM results, the Ti-rich layer under the defective via is around 50nm. Assuming there was indeed the formation of high-resistance Ti_3Al phase, with via2 diameter of ~320nm and the resistivity of Ti_3Al phase being 210 $\mu\Omega\cdot cm$, the resistance of the 50nm thick Ti_3Al layer only contribute to the resistance of 1.3Ω. While ET date showed that the resistance of this defective ET is up to 87Ω. In addition, TEM/STEM analysis revealed there was also thick Ti-rich layer under normal vias, which has also been considered to be induced by the intrinsic weakness of the via punch-through process. Therefore, the Ti diffusion into Al and/or the formation Ti_3Al intermetallic compounds cannot explain the ET failure in this low-yield issue.

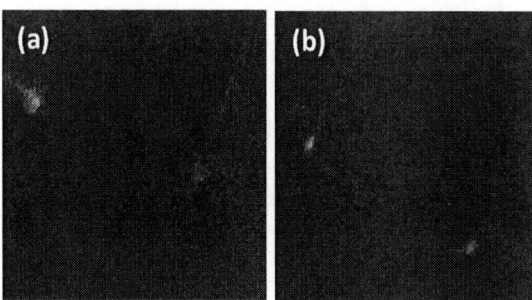

Figure 6 EFTEM-EELS fluorine mapping at (a) defective via; and (b) normal reference via

Another physical signature observed by EDX is the presence of F-rich materials at both via bottom and corners of the high resistance via, as shown in Fig.4 (b) and Fig.5 (b), which is consistent with the results by EELS mapping in Fig.6 (a). However, for the normal vias, both EDX (not shown) and EELS analysis (Fig.6 (b)) also revealed the presence of F at the via bottom and corners. In addition, the counts of F in both EDX and EELS are normally quite low, and the F-rich layer at via bottom is very thin. For punch-through via process, notches at via corners are commonly observed after via etch as mentioned above. With IMP (ionized metal plasma) Ti and CVD (chemical vapor deposition) TiN barrier deposition process, the Ti/TiN barrier normally has poor coverage at these notched via bottom. Therefore, F-rich residue could be easily trapped at notched via corners during W CVD process with WF_X as one of the precursors. As for the F-rich layer under via bottom, it may arise from either F-rich polymer due to incomplete cleaning after via etch or due to F diffusion from those in the notched via corners.

Therefore, both Ti diffusion and the presence of F-rich layer at via corners and bottom arose from the intrinsic process weakness of punch-through via process, and could not explain the high via resistance

abnormality. Hence, further detailed TEM FA was needed in order to understand the root-cause behind.

Fig7 (a)~(d) showed the results of EELS Ti and Al mapping for both normal and defective high-resistance vias. As seen from Fig.7 (a) and (c), compared with normal via, the high-resistance via showed missing or non-continuous Ti barrier coverage, and severe Al out-diffusion was observed at the via bottom. While for normal via, the Ti coverage remains continuous with relatively good conformity at the via bottom although there was also appreciable Ti under diffusion into Al.

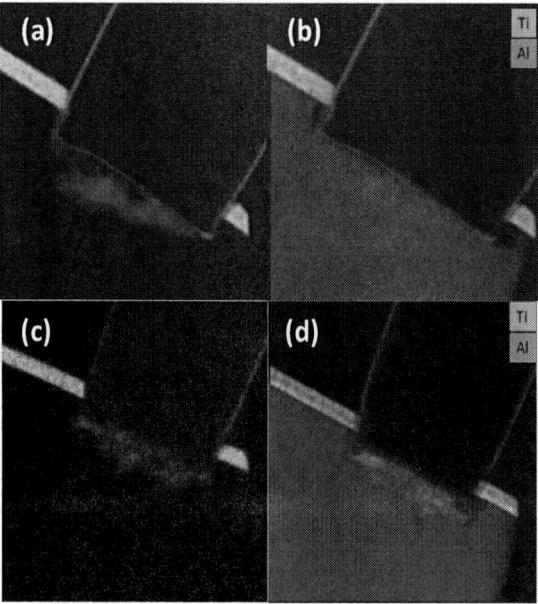

Figure 7 EFTEM-EELS mapping of (a) Ti mapping and (b) Ti/Al mapping of the normal via; (c) Ti mapping and (d) Ti/Al mapping of high-resistance via

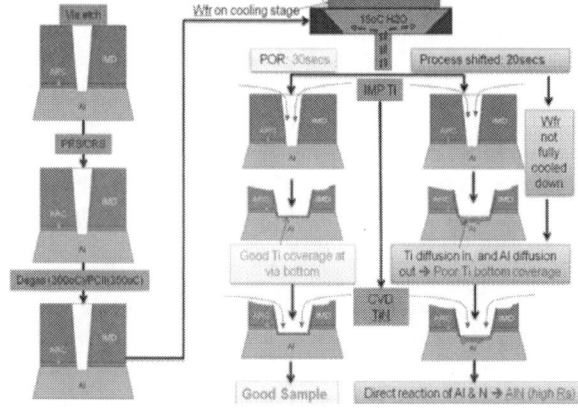

Figure 8 Schematic illustration of the process history/drift and their correlation related to ET failure mechanism

The results indicated that, for high-resistance via, there was more severe Ti diffusion into Al that directly led to the poor Ti barrier metal coverage at the via bottom. We suspected that the high via resistance could be related to the poor Ti barrier coverage at the via bottom, which led to Al out diffusion, and thus Al was directly exposed to the next process step. In this case, Al metal was directly exposed to N plasma during subsequent TiN CVD deposition process by which the high-resistance AlN phase could be formed at the interface with Al directly exposed to N_2 plasma. The formation of AlN dielectric materials at Al and TiN has been ever reported by Okihara M. et al. [3]

Normally the severity of Ti diffusion into Al is thermally-driven process, which is intimately correlated to the temperature effects. In order to verify such hypothesis, we need to understand if there was any process drift which led to the more severe Ti diffusion at the wafer edge as per ET failure map. Therefore, a thorough analysis is needed for the details related to the time frame of affected lots, tool commonality and process history. Based on detailed inline investigations, it was found that there was a process drift related to wafer cooling before IMP (ionized metal plasma) Ti deposition, namely shorter wafer cooling time compared with the POR (process of record) process. It was also matched with the time frame of the effected lots and the tool commonality analysis. As illustrated in Fig.8, before IMP Ti deposition, there was a degas process which is usually performed at a temperature of around 350°C, following which wafers need to be cooled down in a cooling stage. The wafers in the affected lots underwent shorter cooling, leading insufficient wafer cooling before IMP Ti deposition. Therefore, more severe Ti down-diffusion into Al and Al out-diffusion occurred, which resulted in poor Ti barrier coverage at the via bottom. In the subsequent TiN deposition process, with direct Al metal exposure to N_2 plasma, it is highly possible to form AlN high-resistance phase, leading to ET failure. Since cooling water flows from center to edge in the cooling stage, the ET failure at the wafer edge was thus more severe during more insufficient cooling at the wafer edge.

IV. CONCLUSION

In this work, we presented systematic TEM failure analysis for the root cause and failure mechanism understanding of an ET failure issue. With identifying the key physical failure signatures by systematic TEM failure analysis, we correlated physical failure signatures with the inline investigations. It was found that the high via resistance was correlated to the process drift due to insufficient wafer cooling before IMP Ti, which led to severe Ti down-diffusion and Al out-diffusion. Therefore, it was suspected that the high via resistance arise from the formation of AlN high resistance phase with direct Al exposure to N plasma in the subsequent TiN CVD process.

Acknowledgement

The authors would like to thank FIB colleagues in TEM PFA group in GLOBALFOUNDRIES for the sample TEM sample preparation and the colleagues in Electrical Failure Analysis Group for useful discussions.

Reference

[1] C.N. Ho, G. Higelin, C.H. Low, A. See, L. Chan, Plasma Etching Processes for Sub-quarter Micron Devices: Proceedings of the International Symposium, The Electrochemical Society, , pp.269, 2000
[2] Dyos G.T., Farrell T., Electrical Resistivity Handbook, vol 641, pp64, 1992
[3] Okihara M., Hirashita N., Hashimoto, K. ; Onoda, H., Vol.66, No.11, pp 1328, 1995

Failure Analysis of Low-Ohmic Shorts Using Lock-In Thermography

Kannu Wadhwa[1], Rudolf Schlangen[2], Joy Liao[2], Tung Ton[2], Howard Marks[2]

[1] DCG Systems, 3400 W Warren Ave, Fremont, CA 94538

[2] NVidia Corp. 2701 San Thomas Expressway, Santa Clara CA 95050

Phone: (510) 897 6828   Fax: (510 897 6801   Email: kannu_wadhwa@dcgsystems.com

Abstract- **This paper will present the non-destructive Lock-in thermography (LIT) technique and its application in detecting low-ohmic power shorts in 28 nm GPU (Graphics processing units). LIT was successful in detecting power shorts within die and package down to 5 Ohms within seconds, leading to accurate and efficient root cause analysis.**

I. INTRODUCTION

The shrinking technology nodes and increasing interconnect complexity in ICs (Integrated Circuits) today result in some challenges for fault isolation and non-destructive root cause analysis of defects. Electrically active defects are typically localized using standard techniques such as OBIRCH, Emission Microscopy and Liquid Crystal Thermography. With increase in packaged devices, stacked dies increase the need for non-destructive testing, and such methods can no longer be used for functional structures lacking direct access to active areas. For packaged parts, fault analysis is currently done by de-processing the package until a defect is located. However, by doing this we run the risk of accidentally deprocessing the defect out, and/or inducing new defects during the process.

Lock-in thermography is a non-destructive failure analysis technique used to detect electrically active defects (shorts, latch ups, ESD damage, leakage related failures, high resistive opens, GOX breakdowns, etc.) for semiconductor ICs. In measurement for LIT, the device under test is electrically biased with a voltage modulated at an excitation frequency. Modulating the voltage through the device then causes a defect-related heat generation at the short location.

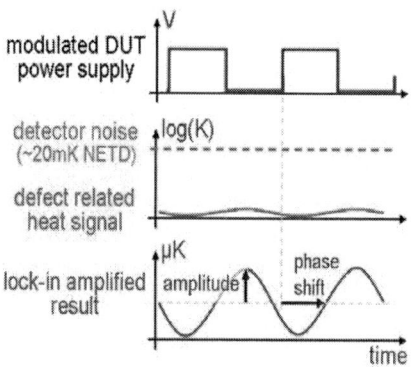

Fig.1. Basic LIT principle

The power dissipated inside the device results in a surface level temperature modulation which is then measured in real time with a highly sensitive IR camera operating between 3-5 μm wavelengths. The result of a standard LIT measurement can be split into two parts; an in-phase part and an out-of-phase part. This further allows the calculation of amplitude and phase. The amplitude result allows us to observe the strength of the signal coming from locally generated heat at defect location, whereas the phase information can be further used to analyze defect depth within the device indicating whether the isolated defect is in a die or package. This clear differentiation between die and package allows LIT being one of the strongest failure analysis techniques. One significant advantage of LIT over steady state thermography is the higher sensitivity allowing for detection down to very low-ohmic shorts buried within a device. As a result, LIT provides hot spot localization with high spatial resolution (~ 1μm on IC level). In this paper, we will present three cases where LIT was successful in localizing low ohmic power shorts where other techniques had failed. The device type analyzed in all the cases is a 28 nm GPU (graphics processing unit). Techniques such as Photon Emission Microscopy, Acoustic Microscopy, Static thermography (IREM-MCT), Static Laser Stimulation (OBIRCH/TIVA) and blind physical failure analysis were unsuccessful in such low ohmic failure cases.

II. CASE 1

The first case reported a low-ohmic power short between VDD and GND in a 28 nm technology GPU. A higher than usual fallout was reported at final test (FT). Initial ATE failure data log indicated power-shorts. During initial curve trace testing, some devices would "recover" and pass subsequent ATE. Because this particular GPU is assembled in two types of packages for different applications, and the power-short phenomenon was only observed on one package type, integrity of package material was suspected to be the cause of failure.

Static thermal emission (TE) was tried on some failed devices with unidentifiable results. In Figure 2 (a), the 1x magnification image with static TE imaging shows an area which is most likely related to a surface level crack, and not related to the short location within the device. After a few unsuccessful attempts, the sample was brought to LIT. Using real time pixel wise Lock-in thermography, we were able to determine the hot spot location at an ESD protection cell within 5 seconds at 1x magnification. Figure 2 (b) and (c) show a wide angle, and 1x magnification results from LIT.

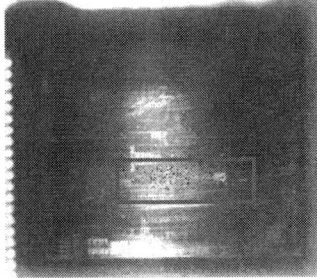

(a) Static 1X TE image showing hot spots over a large area, related to a surface level crack.

(b) LIT 1X image of the failed sample showing hot spot location at an ESD cell.

(c) 10X magnified image with LIT (acquisition time 2 minutes)

Fig.2. 28 nm GPU device Static thermography and LIT results

Optical resolution using the LIT technique is approx. 1.5 micron per pixel. However, further LIT analysis on the failed devices with SIL (Solid immersion lens) allows us to reach sub-micron resolution for optical images. Using the SIL with the lock-in technique, we identified clear areas of fault within the ESD protection cell, as seen in Figure 3(a). Thermal delay information extracted from LIT measurements eliminated the defect location to be in the package. The sample was then sent for physical failure analysis to confirm the location identified by LIT. Since the defect was confirmed not to be in the package, it was removed. An optical inspection on the front-side of the die showed a clear EOS (electrical overstress) damage on the ESD cell on the devices as seen on Figure 3 (b). As failure mechanism was identified, root-cause analysis began. IR inspection using InGaAs and LSM with SIL through the backside of silicon also revealed damage on the ESD protection cell.

Fig 3 (a): LIT image with SIL (Solid Immersion Lens) with ~ 34X (0.5 µm per pixel) magnification showing hot spot within the ESD cell (acquisition time 30seconds)

Fig 3 (b): Optical Image from the front side showing damage on the ESD protection cell.

The ability to examine devices in the package allowed quick inspection on multiple devices, hence speeding up the root-cause analysis. Our first target was the devices that "recovered" during curve trace. LSM inspection found similar EOS damage on the same ESD cell. This indicated that the VDD-GND bridging (resulting from EOS) was fused open. The fact that these "recovered" devices passed subsequent testing also confirmed the ESD cell is properly protecting the internal circuitry of GPU from such overstress events. We then inspected the devices with the package type that did not report any such failure, as well as bare die. No damage was found. Review and comparison of all data from wafer sort to final test revealed that devices in this particular package type were subjected to overstress on ATE at final test. This systematic low-ohmic power-short fallout was resolved with change of test program.

II. CASE 2

Another GPU with power shorts as low as 5 Ohms causing a higher than usual fall out rate was submitted for analysis. Various FA techniques such as static IR emission at 0.5V (500 mA) excitation were used but unsuccessful. Applying high voltages was not an option in this case due to the risk of the failure to fuse open. Lock-in thermography is tried within the safe voltage range of 0.2V. As seen in Figure 4(left), we use a wide angle lens with a large field of view; the hot spot location is identified within 30 seconds. Zooming in on the isolated area, in Figure 4(right) we see the heat generated below the guard ring shining through the areas not covered by metal.

978-1-4799-3911-4/14 $31.00 © 2014 IEEE

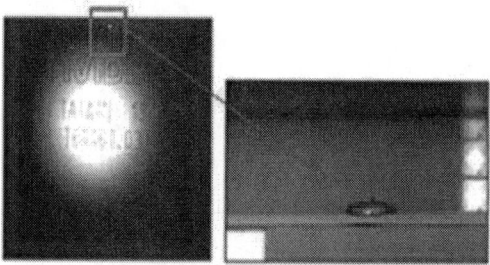

Fig 4 Sample 1: Wide Angle lens LIT image on the left; Zoomed in image at 10X magnification on the right.

A second unit with the same failure was supplied, but the die removed. This second failed unit with similar thermal signature was run with LIT. Die removal via p-lapping left 80% of the underfill material intact on the substrate. In this case, the underfill material is transparent to IR wavelength spectrum, making the metal particle at the defect location visible through the IR lens as seen in Figure 5 (right) image. Cross-sectioning through the location isolated via LIT on a few samples revealed the problem to be a solder mask crack as a result of too much lateral tension from the new underfill material. Figure 6 shows the cross-sectioned result. Change of the underfill material for this design was implemented as corrective action.

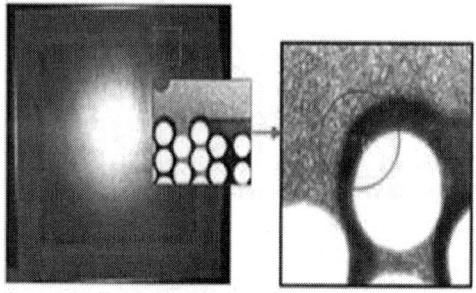

Fig 5 Sample 2: Wide Angle lens LIT image on the left; Zoomed in image at 10X magnification on the right.

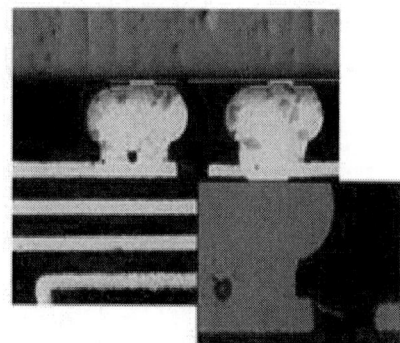

Fig 6: Cross-section results from the location pointed by LIT.

The results from cross-section inspection were then confirmed with the 3D X-ray tool, as seen in Figure 7. Upon analyzing the location from LIT on X-ray, we see a metal stringer clearly visible, shorting to the top metal layer of the substrate.

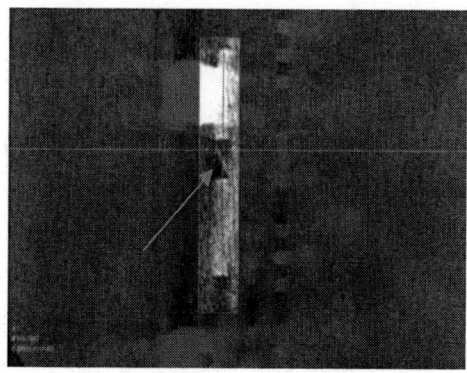

Fig 7: 3D X-ray analysis of the short at the location pointed by LIT shows the solder mask crack clearly.

III. CASE 3

After changing the underfill material on the previous case, we still had one unit that failed from a power short at a different location. Initial tests with a curve tracer (Figure 8 a and b) confirmed the failure mode but CSAM (Confocal scanning acoustic microscopy) that is used to examine the integrity of the package substrate, revealed no anomaly on the device (Figure 9).

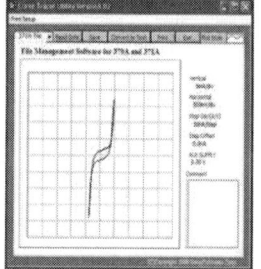

Fig 8 (a): IV Curve for Reference Unit

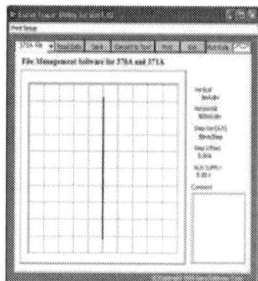

Fig 8 (b): IV Curve for Failed Unit

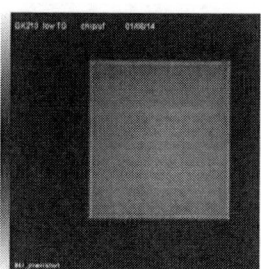

Fig 9: CSAM Image of the whole die showing no anomaly

After a few unsuccessful runs, the sample was sent for a LIT test. With the use of the wide angle lens, and large field of view, we are able to see a clear hot spot location on the top right of the die within 30 seconds (Figure 10 (a)). Thermal delay from the hot spot location shows almost no phase shift, thereby indicating the defect location to be in the die. Since there was no area without metal, and therefore no direct radiation; it is safe to exclude the package. Optical inspection of the front side of the die, after removing the package, showed a clear hot spot location, as seen on Figure 11.

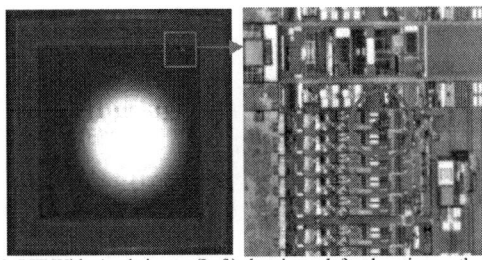

Fig 10: LIT Wide Angle image (Left) showing a defect location on the top right corner of the die; Zoomed in 1X image (Right) with a clear isolation of the short location on the die.

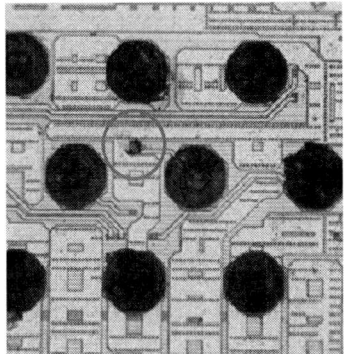

Fig 11: Optical Inspection of the die front side after removing the package shows clear EOS damage.

IV. CONCLUSION

Non-destructive testing is becoming part of the FA workflow for most technology nodes. Differentiating between die and package is increasingly important for accurate and fast physical failure analysis. Lock-in thermography is a promising technique for non-destructive and fast fault isolation proven for up to 28 nm technology nodes with high potential seen for 20 nm nodes. It allows for a higher success rate due to the accuracy of defect isolation within packaged devices. In this paper, we witnessed three case studies where LIT was used to isolate defects in 28 nm GPU packaged devices. Based on the case studies seen in this paper, we have proven that LIT allows for a fast, reliable and successful non-destructive failure analysis. It allows isolation of low-ohmic power shorts in 28 nm technology devices while improving efficiency and accuracy of physical failure analysis. It was able to localize defects in both die (ESD circuit) as well as package (stress due to underfill material). The thermal delay information from LIT measurements further

allows a clear differentiation between die and package level defects, thereby shortening the analysis time. Phase information coming from the device under test is compared with the structural orientation of the device. In case 1, when phase information derives the defect is not in the die, it allows us to remove the die without running the risk of removing the defect. Similarly, if the defect phase information is found to be matching die level analysis, we can remove the package without the risk of accidentally de-capping the defect out. Along with the benefit of accurate isolation of hot spot, LIT also allows a shorter analysis time. Additionally, combining Lock in thermography with phase shift measurements, we can even analyze thick packaged devices with stacked dies up to 16 layers. The differentiation between each die is provided by using a highly sensitive infra-red camera for LIT measurements at various frequencies along with thermal properties of each layer in the stack. Ongoing research and evaluations show promise and potential for using Lockin thermography for future 2.5D and 3D devices using TSV and interposer layers as part of the stack. Additionally, LIT can also be used, in conjunction with a high resolution 3D X-ray to non-destructively isolate shorts in board level FA.

REFERENCES

[1] A.Reverdy: "3-D Defect Localization by Measurement and Modeling of the Dynamics of Heat Transport in Deep Sub-Micron Devices, ISTFA 2007 Desk Reference, Fourth Ed., EDFAS, 739-44 (2002)

[2] Otwin Breitenstein: " Lock-in Thermography – Basic Use for Evaluating Electronic Devices and Materials, 255 pages, Springer, 2010

[3] R. Schlangen et al., "Through Package Defect Localization by Lock-In Thermography", proceeding of IMAPS 2010, p 312-316

[4] Schmidt, C. et al., "Failure analysis of stacked-die devices by combining non-destructive localization and target preparation methods", *Proc. 35th Symposium for Testing and Failure Analysis* (ISTFA), San José 2009

[5] C. Schmidt, "Application of Lock-in Thermography for defect localisation at opened and fully packaged single- and multi-chip devices,"

[6] O. Breitenstein, F. Altmann, T. Riediger, D. Karg, "Lock-in IR microscopy with 1.4. μm resolution by using a solid immersion lens", Proc. 17th EDFAS

A Case Study on The Defective Contact with Schottky Junction Character

Jinglong Li; ChangYan Qi; Yi Che; Quande Zhang; Horse Ma; Jonathon Liu; Motohiko Masuda; Binhai Liu
Freescale Semiconductor (China)Limited
No.15, Xinghua Avenue, Xiqing Economic Development Area, Tianjin, China
Phone: (86 022) 85684345 Email: jinglong.li@freescale.com

Abstract- **Ohmic contacts must be made in any semiconductor device or integrated circuits(IC). Contact failures usually are related to high resistance or open. However, sometimes the defective contact may be Schottky character instead of ohmic contact. In this paper, a case study on such a contact failure is discussed.**

I. Introduction

As is well known, there are two types of metal – semiconductor contacts: non-rectifying ohmic contact and rectifying Schottky barrier junction. The typical I/V characters are shown in Fig.1.

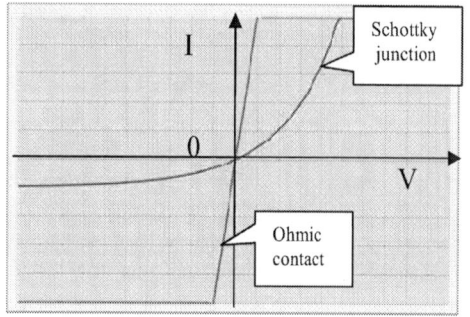

Fig. 1. Typical I/V characters of the Schottky junction and ohmic contact

Metal on lightly doped silicon can make Schottky junction. As a contrast to a p-n junction, it has lower forward voltage drop and faster switching actions.

Metal on heavily doped silicon allows tunneling and make low resistance ohmic contact. It is commonly used as metal to poly silicon (Resistors, Gates) and metal to doped areas (PSD, NSD) interconnections.   An example is shown in Fig.2. The contact is made by aluminum, Ti/TiN and heavily doped silicon.

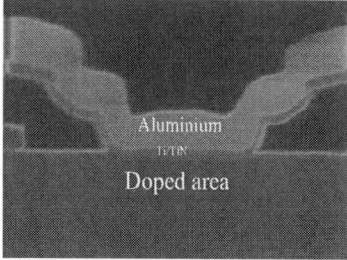

Fig .2. An example of the ohmic contact in SmartMOS5 technology.

Contact failures are commonly high resistance or open. It has been studiedin the past [1-2]. However, in some cases, the defective contact has Schottky junction character instead of a normal ohmic character contact. In this paper, a case study on such a defective contact failure is discussed. After electrical failure characterization, the failing circuit was isolated before de-capsulation. Then infra red optical beam induced resistance change (IR-OBIRCH) inspection and the layout study concluded the defective contact with Schottky junction character has caused due to the implant defect. Finally, physical analysis revealed the defect.

II. Electrical Failure Characterization

A. Background of the failure

An integrated circuit with a customer quality incident (CQI) issue was identified. The CQI unit is a SPI controlled network transceiver in SmartMOS5 technology.

The customer complained that there was an incorrect signal of the software watch dog monitor pin "WDOGB". But the returned CQI unit passed final test, and no functional fault was observed by evaluation board test.

B. Electrical failure characterization

First, I/V curves (WDOGB-VSS and WDOGB-VDD) were traced by semiconductor DC parameter analyzer (Sweeping DC voltage from 0 to 2V, the current clamping at 100μA). The CQI and reference units' curves are shown in Fig.3. Comparing to the reference I/V curve, the CQI unit's WDOGB-VSS and WDOGB-VDD I/V curves are not resistive character, but an early turning on diode character.

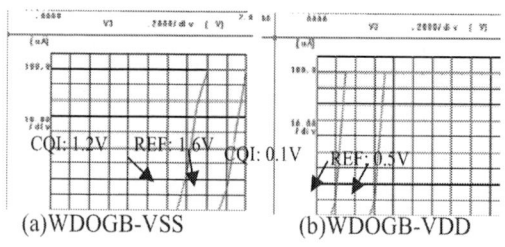

Fig. 3. I/V curves of WDOGB-VSS and WDOGB-VDD.

978-1-4799-3911-4/14 $31.00 © 2014 IEEE

For the reference unit, 0.5V is the turning on voltage. It is a forward biased p-n junction character. But for the CQI unit, the turning on voltage is lower (0.1V). So a forward biased Schottky junction was suspected.

A simple experiment was performed to confirm it. I/V curve of a real Schottky diode 1N5822 was traced (-/+0.1V, -/+10μA), as shown in the Fig. 4(b). The CQI unit (-/+0.1V, -/+1μA) in Fig.4(c) is the similar I/V curve character as 1N5822. Connected 1N5822 to a reference unit, it presented the same failure phenomenon as the CQI unit.

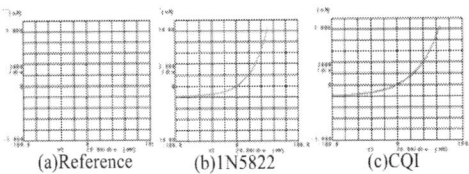

Fig .4. I/V Curves: (a) Reference WDOGB-VSS; (b) IN5822; (c) CQI

So the failure was the Schottky junction character between WDOGB pin and VDD pin. Then schematic and layout were checked. As shown in Fig.5a, WDOGB pin is connected to the output buffer circuit through poly resistors R0 and R1. There are pull up P-channel MOS transistors M1:3 (multi fingers and parallel transistors). M (1:3) body diode is formed by Drain (PSD) and Body (NSD). And it is the p-n junction character observed in the reference I/V curve (WDOGB-VDD).

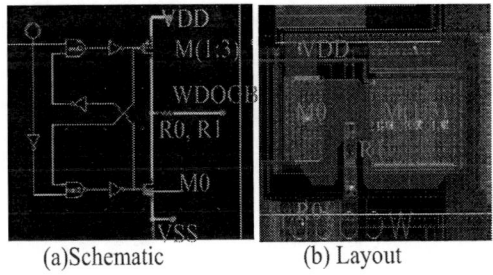

Fig .5. Schematic and layout of the WDOGB output buffer circuit

III. IR-OBIRCH ANALYSIS

The next analysis is to localize the defect. IR-OBIRCH has been proved as an effective method to realize the current path and localize the defect directly [3-5]. In the IR-OBIRCH inspection, metal-semiconductor contacts are important. Because the resistance change of the contacts by laser scanning are the signs of the current path. For example, a metal - poly-metal structure as shown in Fig.6, the current (from metal to poly) is always decreased by laser scanning the contact area. In the other hand, the current (from poly to metal) is increased by laser scanning. According to the result, it is able to judge the

current direction (from green to red).

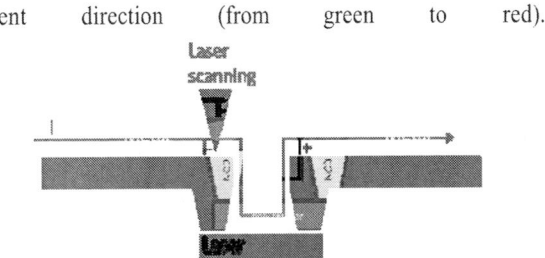

Fig .6. Resistance change of the contacts is the sign of the current path by laser scanning

For this case, to localize the Schottky junction, IR-OBIRCH was performed from backside. Backside grinding was performed to expose silicon. Then IR-OBIRCH was performed. For the CQI unit, WDOGB pin was biased at 0.2V, and VDD was connected to the common GND. In the other hand, a reference unit's WDOGB pin was biased at 0.6V. It is to make sure the junctions are forward biased and the current range is comparable. The IR-OBIRCH inspection results are shown in Fig.7.

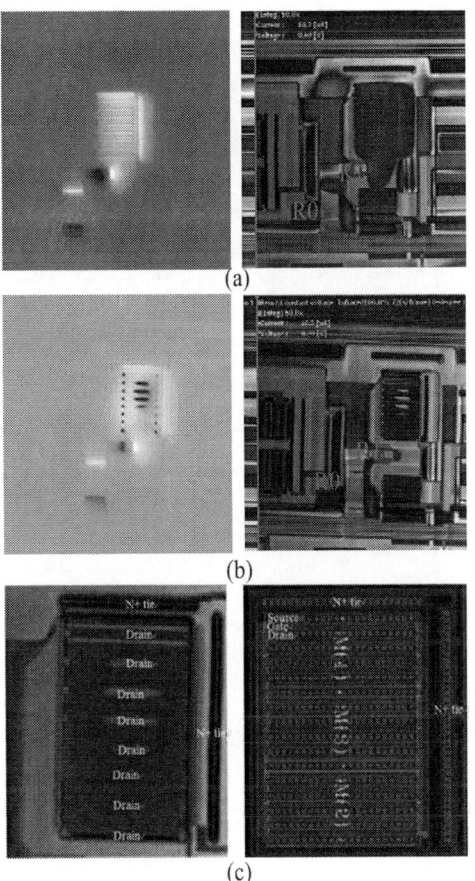

Fig .7. (a) Backside IR-OBIRCH and superimposed images of a reference; (b) Backside IR-OBIRCH and superimposed images of the CQI unit; (c) The CQI units' resistance change area overlaying in layout

In Fig.7a and Fig.7b, the resistance change area of the R0, R1 resistors are observed in the CQI and reference units. Based on the layout, it is in accordance to the metal to poly contacts. So it indicates the current direction from green to red. The different resistance change is observed in M (3:1). For the reference unit, the bright (Red) area is observed in the whole P-MOS area, it is the current through the body diode (p-n junction). The current was increased by laser scanning. Especially the line of contacts in the body (NSD) area presented significant resistance change (much brighter). However, the resistance change in the CQI unit is different. As shown in Fig.7c, the resistance changes are correlated to the contacts in Drain area. And the dark (green) area means current decrease by laser scanning.

According to the cross-section layout of the P-MOS M (1:3) in Fig.8, there are metal-NSD contacts, metal-PSD contacts in this area. The failure should be not at the interface of metal to NSD and PSD, because they do not bypass the bode diode (p-n junction). Only the metal-N-EPI contacting directly is possible to form the Schottky diode and bypass the body diode meanwhile.

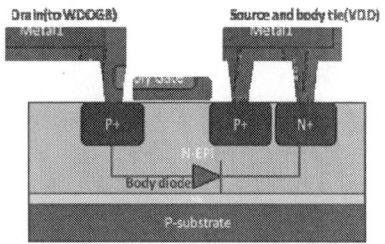

Fig. 8. Cross-section layout of M (1:3)

According to the wafer process flow, the implant dose of PSD area is 4.75E+15 cm-2, NSD area is 6.25E+15 cm-2 and N-EPI is 5.00E+11 cm-2. So PSD and NSD areas are much heavier dose. It is enough to allow tunneling and make low resistance ohmic contact. But N-EPI is lighter dose. As a result, metal to N-EPI contact is the suspect Schottky junction.

IV. PHYSICAL FAILURE ANALYSIS AND FAILURE MECHANISM

The next, physical analysis was performed to reveal the defect. After die surface exposed (with passivation), top down optical inspection in M (1:3) was performed by microscope, but no obvious defect was observed. Then de-process was performed to expose the active area (Fig.9). According to the optical and SEM photos, still no anomaly was observed.

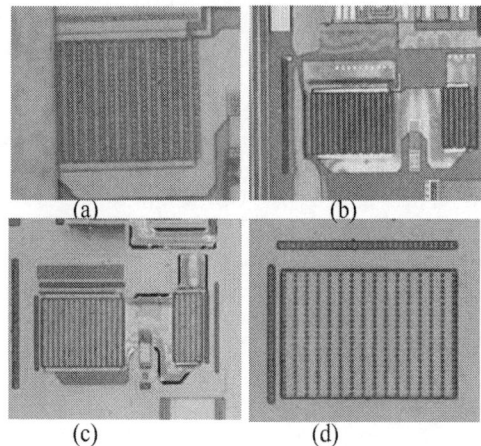

Fig .9. De-process to expose active area. (a): top down optical inspection before de-process; (b): De-layer to metal1 layer; (c): De-layer to poly layer;(d): De-layer to active area.

Then wet stun was performed by N+N acid (mixed acid for implant decoration). Etching for 5seconds, the results of CQI and reference units are shown in Fig.10. Due to the different etching rate, PSD area is deeper than NSD area and it shows dark color by optical microscope inspection.

Fig .10. Posted wet stun, implant defect was revealed. (a): Reference before and after wet stain; (b): CQI unit before and after wet stain.

After wet stun, the anomaly was observed. Parts of the PSD area (top edge, bottom edge and ellipse shape in central area) are not dark color. It means PSD implant missing there.

According to the wafer process, NSD implant step is before PSD implant. Since no anomaly was observed in NSD area (body), NSD implant step was not affected. After NSD implant step done, there are the following PSD photo and PSD implant steps. As shown in Fig.11, photo resist (PR) was deposited first. There was a particle dropped (an ellipse shape) after PR deposition. The next step is PR etching to expose PSD implant area. But PR residue under the particle was not removed. As a result, a part of PSD implant was blocked by the PR residue. After the following wafer process steps, metal to N-EPI Schottky junction character contact was formed finally.

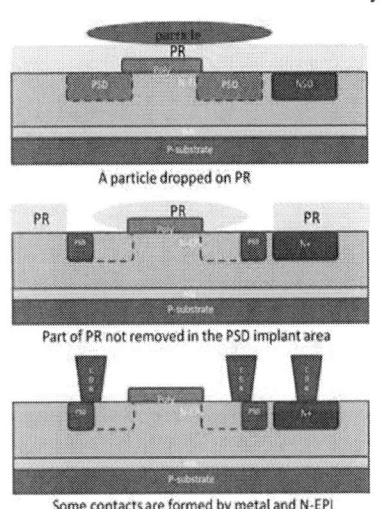

Fig .11. Particle issue in wafer process

V. CONCLUSIONS

Analysis on a Schottky contact failure was performed. Successful electrical failure characterization before de-capsulation is effective to isolate the failing device.

Contacts are very important elements when performing IR-OBIRCH, because the resistance change of contacts area is the reference to realize a current path. In this case, Schottky contact presented the different resistance change. Base on the IR-OBIRCH result, the metal-N-EPI contact failure mechanism was concluded.

After de-processing, no defect was observed directly. However, the failure mechanism has been sure after electrical analysis. So the implant defect was finally revealed by the proper wet stun method.

ACKNOWLEDGMENT

The author would like to thank physical FA team to provide the reference photos and performed de-process and wet stun in this case. Also thanks them to share the information and experience on SMOS5 wafer process.

REFERENCES

[1] I.Österreicher, U. Rossberg, S. Eckl, "Active Voltage Contrast and Seebeck Effect Imaging as Complementary Techniques for Localization of Resistive Interconnections," ISTFA, 2008, pp. 65-69.

[2] J.Y. Dai, S. Ansari, C.L. Tay, S.F. Tee, Eddie Er, S. Redkar, "Failure Mechanism Study for High Resistance Contact in CMOS Devices," IPFA, 2001, pp. 130-133.

[3] Beaudoin, F, Desplats, R, Perdu, P, Boit, C, "Principles of Thermal Laser Stimulation Techniques," Microelectronics Failure Analysis (Materials Park, Ohio: ASM International), 2004, pp. 417–425.

[4] Christian Boit, Clemens Helfmeier, Dmitry Nedospasov, Alexander Fox, "Ultra High Precision Circuit Diagnosis Through Seebeck Generation and Monitoring Charge,"IPFA, 2013, pp.17-21.

[5] Noor Faizah Nordin, "Application of Seebeck Effect Imaging on Failure Analysis of Via Defect," IPFA, 2012, pp.1-4.

Bias Temperature Instability Investigation of Double-gate FinFETs

C.D. Young[1], A. Neugroschel[2], K. Majumdar[3], Z. Wang[1], K. Matthews[3], and C. Hobbs[3]

[1]University of Texas at Dallas
[2]Professor Emeritus at U. Florida – Gainesville, [3]SEMATECH

800 W. Campbell Road, Richardson, TX 75080; chadwin.young@utdallas.edu

Abstract

Double-gate, fin-based Field Effect Transistors (FinFETs) fabricated on silicon-on-insulator (SOI) wafers were subjected to bias temperature instability (BTI) evaluation where focus was placed on the crystallographic sidewall orientation and fin width dependence. For orientation dependence, BTI results at negative stress bias (NBTI) demonstrated that the (110) fin surface degraded more than the (100) surface, because more surface bonds are available in (110) to participate as bond-breaking trap centers during stress. For fin width dependence, positive BTI experienced no dependence on fin width; however, NBTI degradation increased as the fin width narrowed. A plausible cause is a concentration of electrons tunneled from the gate that reside in the SOI fin body. As the fin narrows, the sidewall device channel region moves in closer proximity to these concentrated electrons, which induces more band bending (i.e., increase the surface potential) at the fin/dielectric interface resulting in a higher electric field and hole concentration in this region during stress, leading to more degradation.

Introduction

Multi-gate, Field Effect Transistors (MugFETs) have begun to enter the marketplace [1]. The outstanding attributes of these three-dimensional device structures – which includes excellent immunity to short channel effects, and CMOS compatible processing – are key reasons for their introduction. However, investigating some unique aspects of this type of device structure are critical to ensure their viability to extend to future nodes. These devices typically have transistor device channels on different crystallographic sidewall planes – primarily (110) or (100) sidewall surfaces. Previous results show that the non-planar, fin-like structure and crystal orientation of the sidewalls can have an impact on performance [2, 3] and reliability [4-6]. In addition, FinFETs will need to reduce in size just like planar device technologies have had to scale with each successive technology node. Therefore, understanding the impact of fin width scaling on the key aspects of reliability is important. In this invited work, we will show highlights of our outcomes from investigations of sidewall orientation and fin width dependence on bias temperature instability degradation.

Experiment

Device structures

Silicon-on-insulator (SOI), double-gate FinFETs were processed to form various devices with different fin widths and different sidewall crystallographic orientations. Neutral stress (100) SOI substrates were patterned to produce arrays of fins. Then, a hardmask was placed on top of the fin to decouple the top surface from device operation so that only

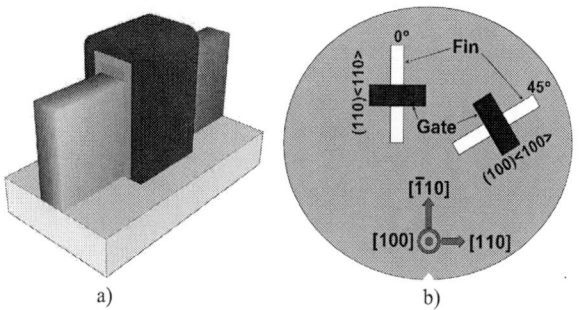

Fig. 1. a) FinFET device structure where the gate wraps around the SOI fin. b) both (110) and (100) orientations are on the same SOI wafer.

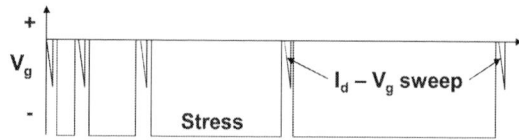

Fig. 2. Example of the stress sequence with interspersed I_d-V_g "sense" measurements for NBTI – polarities are switched for positive BTI.

the device sidewall surfaces are effective under normal biasing conditions (Fig. 1). A typical SiO_2-like interfacial layer and Hafnium-based high-k (~2nm) dielectric stack with a mid-gap metal gate (~10nm) – HK/MG – was capped with polysilicon, giving an EOT of ~1.1 nm. The fin body is slightly p-type ($N_A \cong 2 \times 10^{15}$/cm^3). Typical devices used were comprised of 20 fins in parallel with gate length of 1 μm (mask size). Several different structures were used to obtain robust understanding of the effects of fin width on NBTI. The fin width was 50 nm and 100 nm (mask size), respectively. The actual fin width was reduced approximately 20 nm compare to the mask size in fully fabricated devices.

Electrical Characterization

Since time-to-failure extrapolation is not the intended goal of this work, a self-consistent, conventional NBTI "stress and sense" approach was employed to compare the different aspects under investigation. Figure 2 shows this technique where the stress is interrupted to execute sense I_d-V_g measurements. Threshold voltage (V_t) and subthreshold slope (SS) were extracted from these collected I_d-V_g data (Fig. 3a), and plotted versus stress time (Fig. 3b). The V_t shift demonstrates the overall net effective density of trapped and/or stress generated charge, while the SS can track the increase in interface traps density [7]. The different sidewall orientations or different fin widths of the FinFETs were subjected to BTI stress for 10,000 sec at 125°C unless noted otherwise.

978-1-4799-3911-4/14 $31.00 © 2014 IEEE

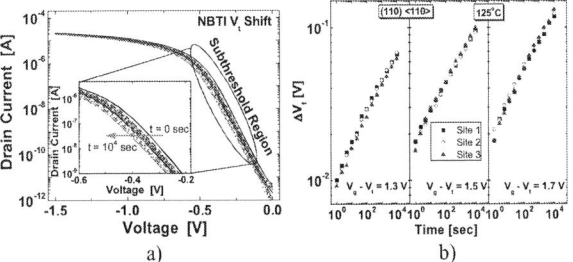

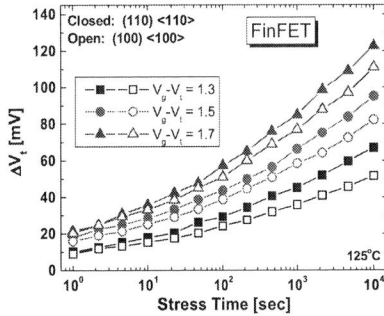

Fig. 3. a) Example data from interspersed I_d-V_g measurements where V_t and subthreshold slope were extracted. b) Example of the extracted V_t shift, ΔV_t, during stress.

Fig. 5. A clear NBTI orientation dependence is seen where the (110) degradation is slightly worse than (100).

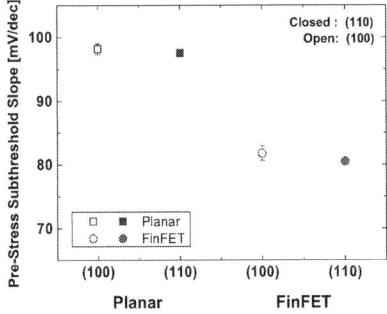

Fig. 4. Pre-stress subthreshold slope for planar and fin-based transistors with different channel surface orientations demonstrating similar SS, respectively, suggesting that the initial state of the interface for different orientations is similar within the given device type.

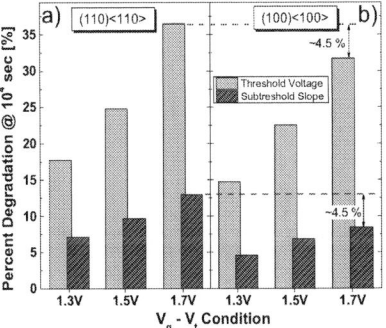

Fig. 6. SS degradation (a sign of interface trap generation) cannot account for all of the V_t shift suggesting that both interface degradation and charge trapping away from the interface are occurring. However, the overall difference between orientations can be attributed to the difference in interface degradation (SS) between the orientations.

Results and Discussion

FinFETs can be fabricated on the (110) or (100) surface sidewalls. Here, the crystal orientation of the fin sidewalls can have an impact on mobility (μ_{eff}) and thereby provides a mobility boost based on orientation (Fig. 1b) [2, 3]. Because of its surface orientation or non-planar fin structure (Fig. 1), the FinFET could be more susceptible to degradation during stress as compared to planar transistors.

Multiple sites and stress bias values were executed during BTI measurements to see overall trends and allow assessment of differences observed when comparing orientation or fin width dependence. An NBTI example is shown if Fig. 3b, where increased stress voltage achieves the expected increased levels of degradation (i.e., threshold voltage shift, ΔV_t). Similar results and further analysis were used to investigate the effects of fin sidewall crystallographic orientation and fin width on BTI characteristics.

Orientation Dependence

This section will discuss the impact of pMOS FinFET sidewall orientation on NBTI degradation. Then, a qualitative comparison will be made to planar pMOS transistors on (100) and (110) substrates that were also subjected to NBTI testing.

Since the impact of crystallographic orientation is being studied, it is important to know the initial condition of the fin sidewalls, and the contribution of interface state generation to the overall V_t shift. The initial subthreshold slope (SS), which can be used to estimate the initial pre-stress interface trap density is similar for both orientations for the particular FinFET and planar MOSFET under investigation, which

demonstrates that neither orientation was initially worse than the other before stress (Fig. 4).

After NBTI stress, a clear orientation dependence is seen (Fig. 5), where the (110) degradation (i.e., increased ΔV_t) is slightly worse than (100). A similar result was obtained for the planar pFETs (not shown). However, the V_t shift is caused by a convolution of charge buildup (trapped charge and/or trap generation) and interface trap generation at the silicon substrate – dielectric interface in HK/MG. Therefore, there is a need to separate the interface trapping from overall charge trapping. To achieve this, the SS was extracted from each measured I_d-V_g measurement during the stress, and therefore, had a corresponding time stamp along with V_t. In Fig. 6, the relative difference of SS degradation from the initial SS (a sign of interface trap generation) cannot account for the entire V_t shift, which suggests that both degradation mechanisms are occurring in HK/MG FinFETs. However, the overall V_t shift of (110) is the same percentage higher as the SS when compared to (100) (e. g., ~4.5% for the 1.7 V condition).

It is also known that silicon dioxide has different growth rates on different starting wafer orientations. Thus, it is important to know if there is a difference in oxide thickness because this would result in different oxide stress fields at the same stress voltage, thereby leading to more degradation for the orientation with the stronger field. Fig. 7 illustrates that the t_{inv} is quite similar since inversion capacitance is the same with no orientation dependence. Thus, this possible model

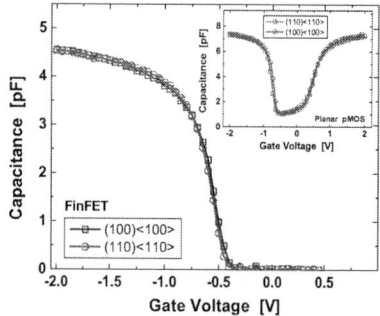

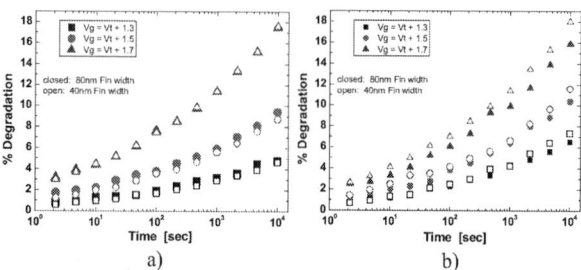

Fig. 9. Comparison of ΔV_t at 10^4 seconds of stress for a) positive BTI and b) negative BTI where a fin width dependence can be seen in this case.

Fig. 7. Irrespective of orientation, the t_{inv} values are similar for finFETs or planar (inset) devices since inversion capacitance is the same, respectively.

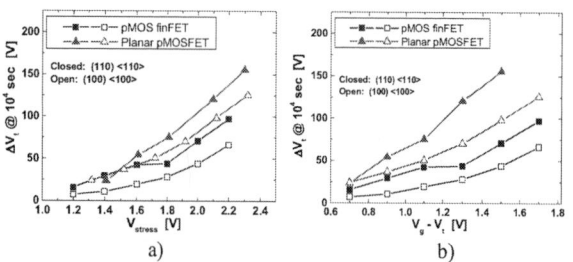

Fig. 8. Comparison of ΔV_t at 10^4 seconds of stress for a) actual stress voltage or b) overdrive (i.e., $V_g - V_t$).

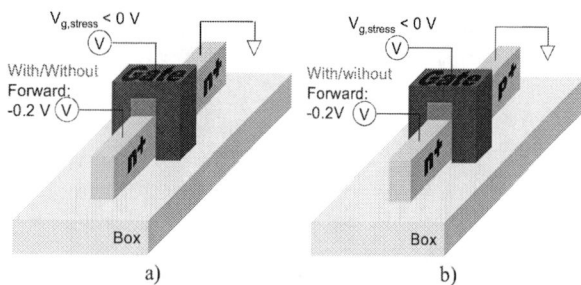

Fig. 10. NBTI stress on a) an nMOS FinFET and b) a gated diode with or without a forward bias to detect the presence (or lack thereof) of electrons residing in the fin body.

of different oxide thickness can be neglected. Therefore, the difference appears to be due to additional interface state generation in (110). The (110) surface orientation is known to have more Si bonds available for bond breakage resulting in more interface trap creation [8]. Therefore, more bonds are available for bond breakage during stress.

Figure 8 demonstrates that both planar and fin devices exhibit larger ΔV_t for (110) when compared to (100). Therefore, the possible mechanism of more bonds available for bond breakage and/or dangling bonds in (110) that result in more degradation compared to (100). In addition, the trends show that pMOS finFETs experience reduced degradation compared to planar pMOS transistors when stressed at similar stress voltages (Fig. 8). Furthermore, the gate dielectric in the planar pMOS devices is actually HfSiO, which have been shown to significantly reduce charge trapping with even as little as 20% Si incorporation, but they demonstrated more overall ΔV_t compared to HfO_2 FinFETs. However, a further analysis is required to determine if finFETs degrade less than planar devices. For instance, detailed investigation of stress field is critical to truly determine if FinFETs are more robust to NBTI than planar pMOS transistors.

Fin Width Dependence

This section will discuss the impact of fin width on BTI for (110) side-wall oriented devices. Fig. 9 illustrates that there is no clear PBTI dependence on fin width [4, 9]; however, fin width dependence is evident for NBTI. Therefore, the focus of the investigation will be on NBTI fin width dependence where several reports in the literature have proposed degradation mechanisms [4, 5, 10]. While the details of our study on these mechanisms can be found in [10], highlights of the results are presented here.

In a model suggested in [4], the (100) surface area on top of the fin would occupy a larger percentage of the overall area for wider fins compared to narrower fins. As the fin width reduces – thereby reducing the top surface contribution to the overall area – the (110) sidewalls begin to dominate the area. Since (110) has been shown to degrade worse than (100), then larger degradation was seen. Since the fin device structures used in this work were double-gate, the model in [4] is not valid in our case since the top of fin was not involved in device operation.

In another model [5], electrons from the valence band of the p^+ polysilicon gate are tunneling across the oxide into the fin body during NBTI where some of the injected electrons are accumulated or stored at the fin center. Therefore, as fin width decreases these stored electrons induce more band bending (i.e., increase the surface potential) at the fin/dielectric interface because the device channel is in closer proximity to the center of the fin where the electrons reside. Thus, the increased band bending results in increased electric field and hole concentration at the interface thereby inducing more damage as the fin narrows. Therefore, the importance of determining if electrons do pile up in the fin body is critical.

This model can be investigated with these HK/MG, double-gate FinFETs. In order to study the effect of electron pile up in the fin body, there needs to be a way to detect if the electrons are actually there. To do this, specific device structures and proper biasing schemes are required to detect the presence (or lack thereof) of electrons in the fin body (Fig. 10) [10]. When applying a small forward bias (FB) to the source or drain of an appropriate fin device structure, the barrier for allowing injected electrons to "leak out" is lowered at the junction during stress and thereby results in a lower concentration of electrons in the fin. If the electrons were allowed to leak out through the forward biased junction during

978-1-4799-3911-4/14 $31.00 © 2014 IEEE

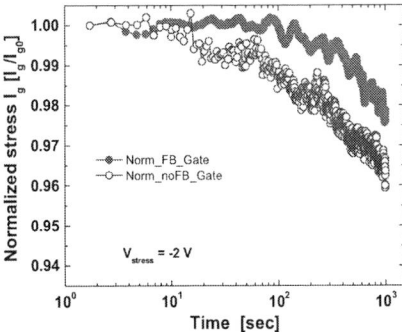

Fig. 11. Stress gate current I_G against stress time for $V_F = 0$ V (noFB) and $V_F = -0.2$V (FB) on a gated diode (n$^-$/p$^-$/p$^+$) structure demonstrating the possibility of electron pile up in the fin body resulting in a greater current reduction with time.

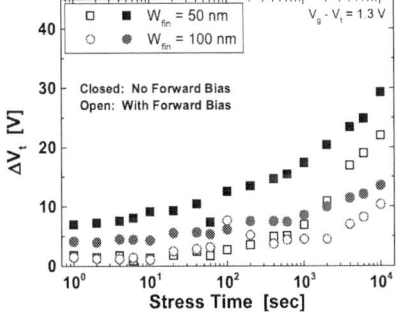

Fig. 12. Effect of the forward bias during NBTI stress (accumulation) on nMOS FinFET where forward bias reduces the amount of degradation due to reduced concentration of electrons in the fin body [10].

stress, then the degradation would be less compared to a device without the forward bias, if the proposed model describes the degradation physics [5, 10].

Compared to the $V_F = 0$V case, forward bias during stress and the reduction in the density of the stored electrons is expected to slightly increase I_G. This is confirmed by the experiment in Fig. 11, where the time dependence of the gate current is measured with and without a forward bias. We speculate that the stored electrons will bias the body slightly more negatively for $V_F=0$V, which can reduce the oxide field and I_G.

Figure 12 demonstrates the effect of the forward bias during stress on NBTI (accumulation stress on nMOS FinFET, Fig. 10a). In addition to the reported fin width dependence shown, Fig. 12 also illustrates a lower overall V_t shift for the stress with FB compared to without FB. This experimental approach helps to demonstrate the plausibility of the effects of electron accumulation (or lack thereof) in the fin body.

Summary

BTI measurements were conducted on SOI, double-gate FinFETs. Sidewall crystallographic orientation effect on NBTI was evaluated. The (110) sidewall was found to be slightly worse than (100) because (110) of higher interfacial bond density available to participate in interface trap generation during stress. In addition, BTI evaluation was performed on devices with different fin widths. Results

demonstrated that, as fin width reduces, NBTI degradation worsens while PBTI demonstrated no fin width dependence. For NBTI, gate-injected electrons that build up in the SOI fin body cause this effect. As the fin width decreased, the sidewall channel moved closer to those electrons. This, in turn, increased band bending, resulting in increased electric field and hole concentration that induced more degradation as the fin narrows.

References

[1] Intel, "Intel Reinvents Transistors Using New 3-D Structure (http://newsroom.intel.com/community/intel_newsroom/blog/2011/05/04/intel-reinvents-transistors-using-new-3-d-structure)," ed, 2011.

[2] V. V. Iyengar, A. Kottantharayil, F. M. Tranjan, M. Jurczak, and K. De Meyer, "Extraction of the Top and Sidewall Mobility in FinFETs and the Impact of Fin-Patterning Processes and Gate Dielectrics on Mobility," *IEEE Transactions on Electron Devices,* vol. 54, pp. 1177-1184, 2007.

[3] C. D. Young, M. O. Baykan, A. Agrawal, H. Madan, K. Akarvardar, C. Hobbs, I. Ok, W. Taylor, C. E. Smith, M. M. Hussain, T. Nishida, S. Thompson, P. Majhi, P. Kirsch, S. Datta, and R. Jammy, "Critical discussion on (100) and (110) orientation dependent transport: nMOS planar and FinFET," in *VLSI Technology (VLSIT), 2011 Symposium on,* 2011, pp. 18-19.

[4] J. J. Kim, M. Cho, L. Pantisano, U. Jung, Y. G. Lee, T. Chiarella, M. Togo, N. Horiguchi, G. Groeseneken, and B. H. Lee, "Process-Dependent N/PBTI Characteristics of TiN Gate FinFETs," *Electron Device Letters, IEEE,* vol. 33, pp. 937-939, 2012.

[5] H. Lee, C.-H. Lee, D. Park, and Y.-K. Choi, "A study of negative-bias temperature instability of SOI and body-tied FinFETs," *Electron Device Letters, IEEE,* vol. 26, pp. 326-328, 2005.

[6] C. Young, K. Akarvardar, M. Baykan, K. Matthews, I. Ok, T. Ngai, K.-W. Ang, J. Pater, C. Smith, and M. Hussain, "(110) and (100) Sidewall-oriented FinFETs: A performance and reliability investigation," *Solid-State Electronics,* vol. 78, pp. 2-10, 2012.

[7] D. K. Schroder, *Semiconductor Material and Device Characterization,* 3rd ed.: John Wiley & Sons, 2006.

[8] S. Maeda, J.-A. Choi, J.-H. Yang, Y.-S. Jin, S.-K. Bae, Y.-W. Kim, and K.-P. Suh, "Negative bias temperature instability in triple gate transistors," in *Reliability Physics Symposium Proceedings, 2004. 42nd Annual. 2004 IEEE International,* 2004, pp. 8-12.

[9] C. Young, A. Neugroschel, K. Matthews, C. Smith, H. Park, M. Hussain, P. Majhi, and G. Bersuker, "Improved interface characterization technique for high-k/metal gated MugFETs utilizing a gated diode structure," in *VLSI technology systems and applications (VLSI-TSA), 2010 International Symposium on,* 2010, pp. 68-69.

[10] C. D. Young, A. Neugroschel, K. Majumdar, K. Matthews, and C. Hobbs, "Comprehensive Investigation of Negative Bias Temperature Instability Dependence on Fin Width of FinFETs," *Submitted to Journal of Applied Physics,* 2014.

978-1-4799-3911-4/14 $31.00 © 2014 IEEE

Short Localization in a Multi Chip BGA Package

Jan Gaudestad,[a] Antonio Orozco,[a] Mark Kimball,[b] Kalon Gopinadhan,[c] and Thirumalai Venkatesan[c]

[a] Neocera LLC, 10000 Virginia Manor Rd. Beltsville, MD 20705, USA
[b] Maxim Integrated, 7250 Evergreen Parkway, Hillsboro, OR 97124, USA
[c] NUSNNI-NanoCore, National University of Singapore, Singapore 117411
Email: Gaudestad@Neocera.com, elekg@nus.edu.sg

Abstract - Magnetic Current Imaging (MCI) has been used for more than a decade to localize shorts and leakages in packages non-destructively. Now that packages are becoming more complex with multiple dies inside the same package, MCI is showing its effectiveness in localizing these complicated shorts non-destructively when the Failure Analysis (FA) engineer does not know from Automated Test Equipment (ATE) if the fault location is in the die or package or which die. We show in this paper that the FA lab can be simplified by the introduction of MCI as a one stop Fault Isolation (FI) tool for all shorts and leakages.

INTRODUCTION

Integration of multiple dies into the same package introduces new challenges to the Failure Analysis (FA) process that requires more advanced non-destructive Electrical Fault Isolation (EFI) techniques to accurately localize the failure to the package or the die; and if localized in the die, then which die and where on the die is the defect located becomes important to address.

Complex devices such as System in Package (SiP), Package on Package (PoP), already commercialized and manufactured for the mobile device industry, and 2.5D Interposers and 3D Through Silicon Vias (TSV) that are expected to be commercialized soon, makes the standard FA work flow time consuming and, in many cases, also inadequate [1]. The first step in the standard FA flow is to categorize the static electrical failures into shorts, leakages or open defects using Automated Test Equipment (ATE). The next step is to isolate the location of the defect using EFI tools; preferably non-destructive EFI tools. Last, if the EFI step has been successful, the FA engineer will do Fault Validation to prove the location of the defect and the root cause of the failure (Tab. 1).

Most commonly deployed EFI techniques in the Semiconductor FA labs are optically based tools. The optical spectra of the signals these tools can detect, or inject, are mainly split into three categories: a) Near Infra Red (NIR, 0.7 - 1.1 μm) using CCD or InGaAs detector, b) Short Wave Infra Red (SWIR, 0.7 - 1.7 μm) using InGaAs or 1.3 μm laser, and c) Mid Wave Infra Red (MWIR, 2-5 μm) using InSb detector. Since all of these wavelengths will be affected by the material used in semiconductor manufacturing assembly in terms of resolution degradation and signal weakening (dependent on both wavelength of the technique and material used by the semiconductor manufacturing company) [2]. In addition, most optical techniques are only designed for shorts and leakage localization, while leaving opens default localization more or

less unresolved. This has lead FA engineers to add more tools to the FA lab with multiple laser wavelengths and multiple optical detectors to cover the entire EFI optical spectrum, making the FA labs more crowded with multiple EFI tools [3].

Tab. 1. Typical Failure Analysis Work Flow. OBIRCH is Optical Beam Induced resistance change and has multiple derivation related to this technique (TIVA and LIVA) [2].

Complex Packages	Electrical Defects	Non Destructive Electrical Fault Isolation	Fault Validation
Failure Analysis Work Flow			
SiP PoP Interposer 3D TSV	Shorts	Magnetic Current Imaging (MCI) Photon Emission (PEM, InGaAs)	Physical FA Laser Ablation Plasma FIB X-Ray 3D X-Ray
	Leakages	Lockin Thermography (LIT, InSb) Laser Scanning Microscopy (LSM, OBIRCH)	
	Opens	Space Domain Reflectometry (SDR) Time Domain Reflectometry (TDR,EOTPR)	

One non-destructive EFI technique that is not affected by the materials typically used in semiconductor manufacturing in terms of resolution degradation and signal weakening is Magnetic Field Imaging (MFI). The only thing affecting the resolution and signal strength is the total distance from the magnetic sensor to the source of the signal; the sample material in between the sensor and the signal source is not affecting the magnetic signal, including magnetic materials used in capacitors [4, 5]. In addition, MFI, with sub techniques Magnetic Current Imaging (MCI) and Space Domain Reflectometry (SDR), can localize all electrical static defects; shorts, leakages and opens, making it the perfect candidate for advanced package EFI as a one stop solution allowing the FA engineer to have fewer EFI tools and techniques in an already crowded FA lab [3].

PRINCIPLE OF MAGNETIC FIELD IMAGING

Magnetic Field Imaging (MFI), as applied to failure analysis (FA) in the semiconductor industry, is based on mapping the magnetic field produced by a current or RF signal injected into the failing structure of the Device Under Test (DUT). The magnetic field image of the DUT is converted into the current density image by using a Fourier Transform inversion technique [6]. In order to determine the fault location, the current image is compared or superimposed on to the circuit diagram or Computer Aided Design (CAD), an optical, infrared or X-ray image, or a current image of a non-failing part.

978-1-4799-3911-4/14 $31.00 © 2014 IEEE

A current carrying conductor generates magnetic field according to the Biot-Savart law [6]. MCI is a sub technique of MFI due to its capability to "look through" any types of materials that are physically covering the signal, thereby allowing for global imaging without physical deprocessing [3, 4]. MCI utilizes two types of sensors: Superconducting Quantum Interference Device (SQUID) sensor for low current and large working distances, including quick overview scans at die level, and a Giant Magneto Resistance (GMR) sensor for sub micron resolution current imaging front side at wafer/die level [5].

A newly developed technique based on MFI, namely Space Domain Reflectometry (SDR), has been successfully used to non-destructively localize opens defects in microchips. SDR injects a continuous wave radio frequency (RF) signal typically at 40-60 MHz into the defective trace and the magnetic field generated by the standing RF wave is imaged by the SQUID sensor and the open defect is where the signal ends and goes to zero [7].

SQUID and GMR sensors differ in both sensitivity and spatial resolution. SQUID is the most sensitive magnetic sensor known [3], and for electronic fault isolation it is typically used to image AC sine currents as low as 500 nA_{p-to-p} (peak-to-peak) at a working distance of several hundred microns between the current path and sensor {Converting sine AC peak-to-peak (p-to-p) current to root-mean-square (rms) current, the p-to-p current needs to be divided by $2 \cdot \sqrt{2}$. Then using that $Power_{rms} = V \cdot I_{rms} = R \cdot I_{rms}^2 = \left(R \cdot I_{p-to-p}^2\right)/\left(2 \cdot \sqrt{2}\right)^2$ and introducing the above minimum current for the SQUID at 500 nA_{p-to-p} we get that minimum power detection using the SQUID is $31.25 \cdot 10^{-15}$ $Watt_{rms}$ (femto watt) at 1Ω resistance}. The SQUID sensor is kept in vacuum at cryogenic temperature, while the DUT is raster-scanned at room temperature and separated from the SQUID sensor enclosure by a thin diamond window. This setup allows localizing currents to within ±3 microns.

The GMR sensor is mostly useful for front side wafer/die applications. Since GMR operates at room temperature and requires no vacuum enclosure, one can bring the sensor to within 1-2 µm from the DUT surface. While GMR has lower magnetic field sensitivity than the SQUID, its smaller size size (<100 nm), and close proximity to the sample, provides for spatial resolution better than 500 nm with AC sine currents as low as 100 μA_{p-to-p} {$7.1 \cdot 10^{-9}$ $Watt_{rms}$ (nano watt) at $1\ \Omega$ resistance}. If the scanning distance is more than 100 µm, the GMR spatial resolution becomes similar to that of the SQUID.

SHORT IN A MULTI CHIP MODULE PACKAGE

A short was found in a high voltage analog switch that was designed into a Multi-Chip Module (MCM) in a Ball Grid Array (BGA) package substrate containing three individual dies. The dies were gold wire bonded to allow for high current, high voltage and high reliability. The sample was pulled out of stock and retested by Quality Assurance (QA)

engineers on ATE, at which time it showed failure for excess supply current on a 200 V supply pin.

The FA engineer first used X-ray in a lower resolution mode to search for the failure in a non destructive way without any visible signs of short failures. Running the X-ray at high resolution in search of defects for the entire sample in a complex MCM is too time consuming. The FA engineer then deprocessed the part using a milling machine removing the top of the mold compound to get direct visible access to the wire bonds. The wire bonds were then severed so that the dies were physically disconnected from the BGA, and manual probing on the solder balls were done. Probing on the BGA then localized the short to the package, as short was still present after cutting the wirebonds, which was the only connection between the solder balls at the substrate level and the dies. The sample was then submitted to MCI for non destructive package EFI, as MCI is the only non-destructive EFI tool available to the FA engineer.

The sample was attached inside the MCI system to an XYZ stage for raster scanning under the SQUID sensor. The SQUID sensor enclosure was brought to within 100 µm of the sample surface (top of the mold compound) to get the highest possible resolution (resolution is dependent on distance only, not the material in between). Thin wires were soldered directly to the failing solder BGAs at the bottom of the sample, while the sensor was scanning from the top side of the MCM (Fig. 2). A current of 1.4 mA was injected into the defect through the shorted solder balls. Since it was unknown where in the sample the short failure location was, a current image was acquired for the entire sample.

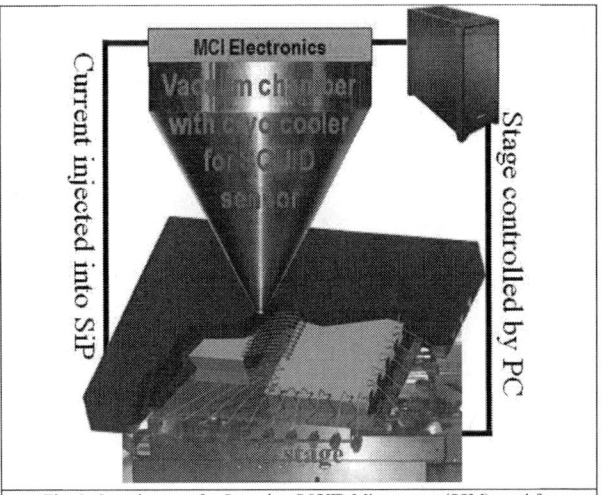

Fig. 2. Sample setup for Scanning SQUID Microscope (SSM) used for package MCI.

MAGNETIC CURRENT IMAGING RESULTS

The first step in using MCI is always to run a quick 15 min overview scan with moderate resolution of the entire sample. The acquired current image was then overlaid the optical image, which also was acquired by the MCI system (Fig. 3a).

In addition, for better understand of the current path, a 2 point alignment technique was used to overlay the current image onto the CAD layout image (Fig. 3b).

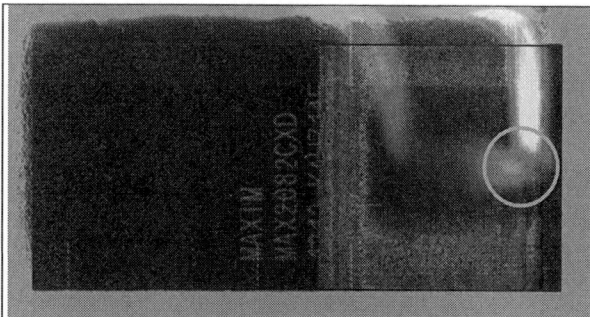

Fig. 3a. Overview scans using the SQUID sensor. One can see that some of the mold compound has been removed for direct wire bond probing.

Fig. 3b. The current image overlaid the CAD layout

After doing the over view scan, the signal generated by the current going through the short location was located to one area of the sample. This smaller area was scanned at a higher resolution for 30 minutes for more accurate fault isolation (Fig. 4).

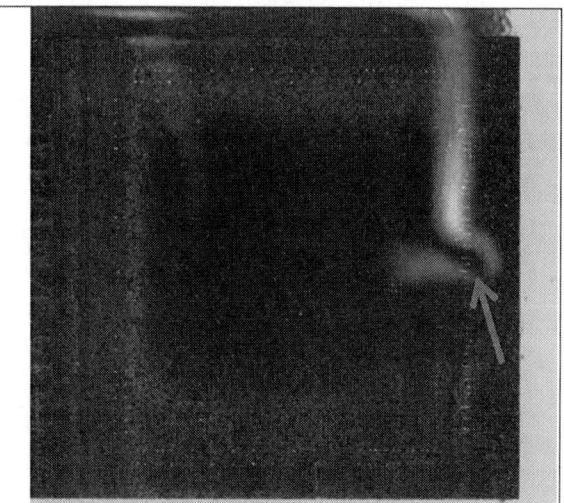

Fig. 4. The short location was found to be in the wire bond. The wirebonds are visible due to the careful milling job done by the FA engineer.

From the high resolution scan where the current image was overlaid the optical image, one could clearly localize the short to a wire bond area (Fig. 4). The next step was to use a high resolution 2D X-ray at the location where MCI indicated the fault location to validate the short location (Fig. 5a).

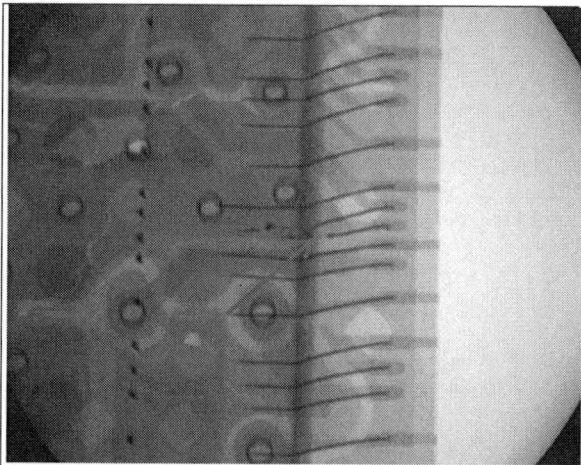

Fig. 5a. X-ray confirming the short location

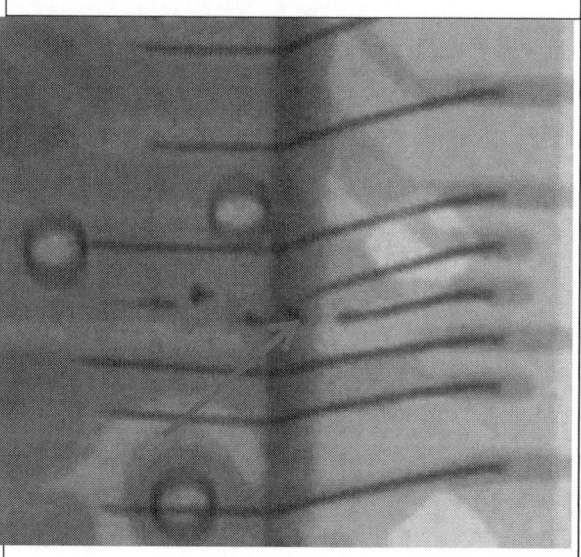

Fig. 5b. X-ray confirming the short locations

A melted wire bond was found to be the root cause for this short (Fig. 5b). Analysis showed that the wire bonds were too close together to handle the large electric fields caused by these high powered devices operating at 200 V. The high electric field strength caused the molding compound dielectric to break down causing the wire bond to melt resulting in a short failure.

It was found that the charred mold compound caused by the 200 V dielectric breakdown, had the same X-ray contrast as undamaged mold compound. Since the conduction paths went through the charred mold compound, X-ray was initially not able to see the failure location. In addition, since the package was a BGA, with multiple circuit boards layers/traces, and solder balls, all these extra features obscure the failure site making X-ray difficult to use for general Fault Isolation.

In a corrective measure, new chip designs included a larger distance between these high powered gold wire bonds.

CONCLUSION

MCI showed that even with complex package and die structures, the signal generated by the current through the short location is easily imaged for then later use high resolution and high contrast X-ray for Fault Validation. This shows that the EFI work flow can be simplified by using MFI for all electrical static failures.

REFERENCES

1. M. Hundt, ST Microelectronics, "Current Trends in Semicondcutor Packaging", iMAPS proceeding January 2004 (http://www.imaps.org/chapters/greatlakes/meetings/GreatLakes_Jan04 webcast.pdf)

2. Falk, R.A, Optometrix, "Advanced LIVA/TIVA Techniques", Proceedings of the 27th International Symposium for Testing and Failure Analysis (Materials Park, Ohio: ASM International): 59–65, ISBN 0-87170-746-2.

3. D. P. Vallett, D. A. Bader, IBM, V. V. Talanov, J. Gaudestad, N. Gagliolo, A. Orozco, Neocera, "Localization of Dead Open in Solder Bump by Space Domain Reflectometry", ISTFA 2012 proceedings p. 17-12.

4. O. Crepel, P. Poirer, P. Descamps, LaMIP Philips Semiconductor, "Magnetic Microscope for ICs Failure Analysis: Comparative Case Studies using SQUID, GMR and MTJ systems", IPFA proceedings 2004

5. J. Gaudestad, N. Gagliolo, and V. V. Talanov, Neocera, R. H. Yeh and C. J. Ma, UMC, "High Resolution Magnetic Current Imaging for Die Level Short Localization", IPFA 2013.

6. J. P. Wikswo, "SQUID sensors: Fundamentals, Fabrication and Applications, Chapter: The Magnetic Inverse Problem for NDE", pp. 629-695, Kluwer Academic Publishers, The Netherlands, 1996.

7. J. Gaudestad, V. Talanov, Neocera LLC, P. C. Huang, TSMC "Space Domain Reflectometry for Opens Detection Location in Micro Bumps", ESREF 2012

An Improved Coffin-Manson Model for Mid-Power LED Wire-bonding Reliability

Bin Zhang, Guoqiao Tao

Philips Lighting Shanghai

No.9, 888 Tianlin Road, Min Hang District Shanghai, Post code: 200233, e-mail:
Bin.Z.Zhang@philips.com ,Guoqiao.Tao@philips.com

Abstract

Thermal shock is usually used for LED wire-bonding accelerated life testing, and the failure is commonly treated as a low-cycle fatigue problem. The lifetime analyses which base on the Coffin-Manson model never consider the modulus saltation of silicone enclosure with the temperature changing. With an extensive DOE, an improved Coffin-Manson model is proposed, which also copes with the glass transition of silicone encapsolent. With this improved model, a more accurate prediction of wire-bonding reliability can be made.

The Category/categories: Novel Devices Reliability

Presentation mode: no preferences

Optimized Thermal Shock Model for Mid-Power LED Wire-bonding Reliability

I. INTRODUCTION

Due to its cost effectiveness, mid-power LEDs have been widely accepted as the energy efficient light source for general lighting. The catastrophic fails of mid-power LED (and COB) are dominated by gold wire breakage. Generally, the reliability is assessed by thermal shock test with Coffin-Manson-Model). This model has been widely used for failure-modes related to materials CTE (temperature expansion coefficient) mismatch. It has been successfully applied to solder-joint reliability [ref1]. However, we are not sure if this approach is still good for the mid-power LED (and COB) gold-wire / silicone system. This is because the silicone material changes its state at around 25 degree C (so called glass temperature). Below this T_glass, the silicone is at glass state, characterized by small CTE and very large modulus. Above, T_glass, the silicone is in Gel state, having large CTE and small modulus. This change of properties will/does have large impact on the results/conclusion of our reliability assessment implying possible wrong prediction of catastrophic fail rates. With an extensive DOE, we propose an improved Coffin-Manson model to cope with the Gel/Glass transition during experiment and applications.

II. II. THE FAILURE MECHANISM OF WIRE-BONDING IN MID-POWER LED DEVICE.

Wire-bonding is the most common method for connecting the chip to electrical pad on LED package. Figure 1 shows the wire-bonding structure in a typical mid-power LED package.

When the LED exposed to high forward currents or high peak transient currents, the gold wire can behave as a fuse. Electrical overstress will cause the gold wire fracture. However, most of the wire-bonding broken failures were not caused by the electrical overstress, but resulted from cumulative fatigue, shown in Fig 2. When LED device powered on/off, the temperature cycle will be generated on wire-bonding. Wire-bonding fatigue takes place when thermo-mechanical stress drives the repetition of thermal expansion and contraction of the expanding materials. It breas when the thermo mechanical stress is higher than the wire-bonding strength. Meanwhile, the level of the CTE and Young's modulus of the encapsulant, as well as the hardness of the die, affects gold wire fatigue. The mismatch of CTE causes the gold wire and chip to generate a significant thermo-mechanical stress in the bonding area, the experienced equation showed as below Equation 1. This results in fatigue broken during thermal cycling (Power on/off in application).

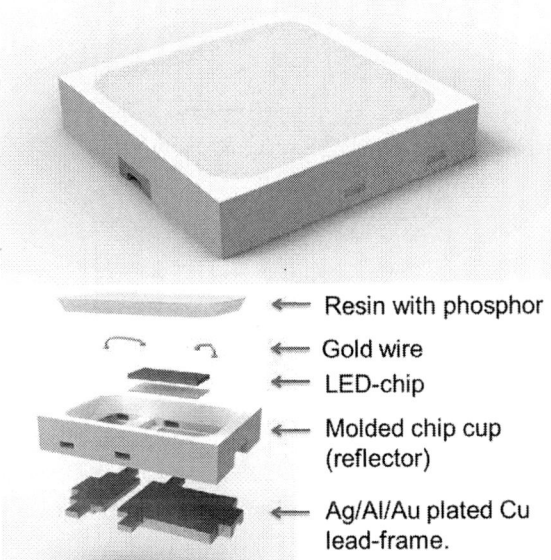

Fig 1. A 3D-drawing of a typical mid-power LED package

- ← Resin with phosphor
- ← Gold wire
- ← LED-chip
- ← Molded chip cup (reflector)
- ← Ag/Al/Au plated Cu lead-frame.

$$\gamma = \frac{[CTE_1 - CTE_2] * L_0 * \Delta T}{h} \qquad (\text{Equation.1})$$

Where, Υ is strain, CTE is Coefficient of Thermal Expansion, L_0 is change length under strain, ΔT is temperature changing, h is thickness.

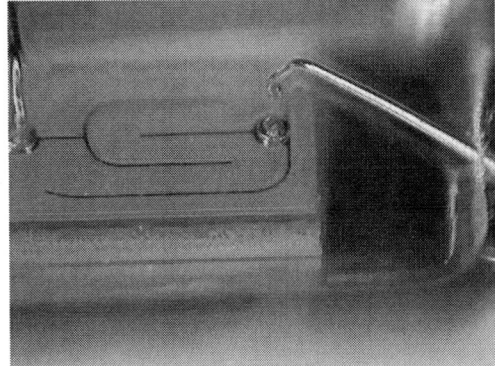

Fig. 2 wire bonding broken

Besides, the wire-bonding process is also critical to gold wire reliability. Fig 3 shows a process issue, a notch found on the neck of the wire-bonding with SEM. The congenital defect of gold wire is the main reason cause the earlier failure.

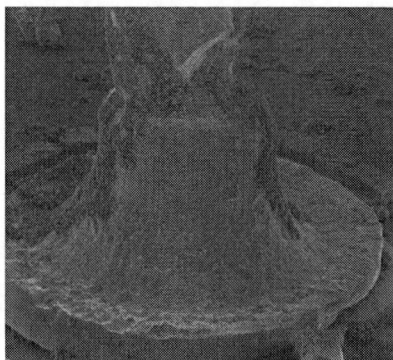

Fig 3 congenital defect on wire-bonding

Number of Failure	Last Inspected	State End Time	Test Condition	ΔT
1	370	456	-40C~150C	190
1	456	552	-40C~150C	190
4	552	688	-40C~150C	190
19	688	826	-40C~150C	190
6	826	940	-40C~150C	190
1	940	1074	-40C~150C	190

Table 3 failure data for Test condition 2

III. ACCELERATED LIFE TEST AND ANALYSIS FOR WIRE-BONDING

Thermal shock is usually used for LED wire-bonding accelerated life testing, and the accelerated factor calculation base on the Coffin-Manson model which is assumed only temperature changing, no ΔCTE, L and h changing (showed in Equation 2).

$$AF = \frac{N_{f1}}{N_{f2}} = (\frac{\varepsilon_{p2}}{\varepsilon_{p1}})^{\alpha} = (\frac{\Delta T_2}{\Delta T_1})^{\alpha} \text{ (Equation.2)}$$

Where, εp is plastic Strain in cycle, α is fatigue factor, ΔT is temperature changing.

According to the wire-bonding accelerated model, we design the experiment for one type of Mid-power LED by thermal shock. See Table1test arrangement for thermal shock.

Test condition	Dwell Time	Sample size(wire)
-40~120C	15min	40
-40C~150C	15min	32

Table 1.Thermal shock experiment

To get the full failure distribution, the test stopped when all the samples failed. Detailed failure point collected as below tables; please refer to Table2 and Table 3.

Number of Failure	Last Inspected	State End Time	Test Condition	ΔT
3	925	1015	-40~120C	160
2	1015	1100	-40~120C	160
4	1100	1152	-40~120C	160
4	1152	1237	-40~120C	160
6	1237	1332	-40~120C	160
6	1332	1468	-40~120C	160
5	1468	1606	-40~120C	160
5	1606	1720	-40~120C	160
5	1720	1854	-40~120C	160

Table 2 failure data for Test condition 1

The failure distributions of two test conditions can be plotted by Weibull++ tool as below picture, showed in Fig 4.

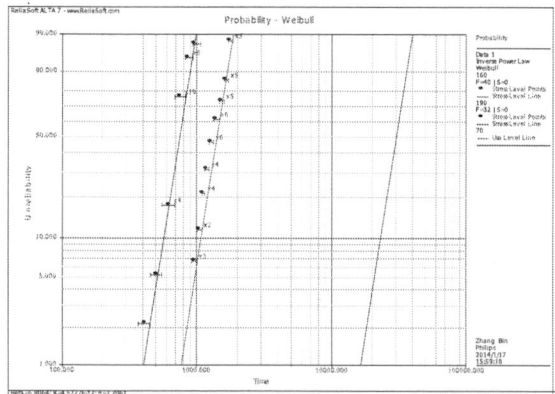

Fig 4. Failure distribution on Weibull

From the analysis result in Weibull++ tool, we can easy know the shape parameter β, scale parameter η and fatigue factor α See table 4.

Model:	Inverse Power Law
Distribution:	Weibull
Analysis:	RRX
Beta:	6.9
η₁	1433
η₂	760
α:	3.7

Table 4 Analysis parameters

It is very convenient to calculate the wire-bonding lifetime in real application with these parameters.

$$AF = (160/85)^{3.7} = 10$$

Where the delta T in real application is 85C (for indoor LED products the Tmax is110, Tmin is 25).

IV. THE ACCELERATED MODEL Optimizing

However, the traditional thermal shock accelerated model never consider the modulus saltation of silicone when temperature changing. Know from silicone datasheet, Phenyl silicone has obviously modulus saltation around 25°C, but Methyl silicon has a stable modulus with temperature changing. Figure5 shows the comparison between Phenyl silicone and Methyl silicon.

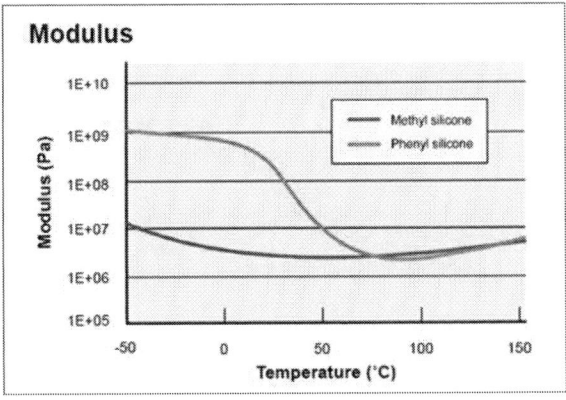

Fig 5.ModulusVS Temperature

Here, the silicone of test samples is Phenyl type due to the better sulphur-resistant performance. Generally, the stress equals to the Modulus*CTE* temperature gradient. From the datasheet, the stress below glass temperature is much higher than the stress above the glass temperature. To simplify the calculation, all the modulus and CTE changing with the temperature changing are normalized to delta Temperature. The ΔT in coffin-mason model was separated as two phases; one is below glass temperature (around 25°C), another is above the glass temperature. See figure 6.

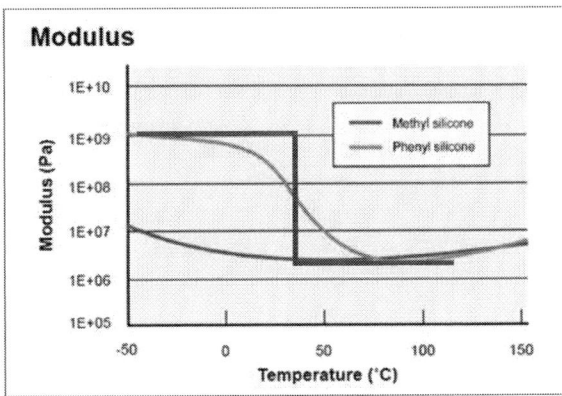

Fig.6 simplify the modulus curve

Hence, the Coffin-Manson model can be modified as below equation, see Equation 3.

$$AF = \frac{N_{f1}}{N_{f2}} = (\frac{\varepsilon_{p2}}{\varepsilon_{p1}})^\alpha = (\frac{n * \Delta T_c + \Delta T_h}{\Delta T_l})^\alpha \text{ (Equation 3)}$$

To build the equation set, third test condition of thermal shock was defined as -60C to 25C, please refer to Equation 4.The failure distribution plotted with Weibull as picture Fig.7.And the scale parameter η can be gained from Weibull++ tool as 480 cycles.

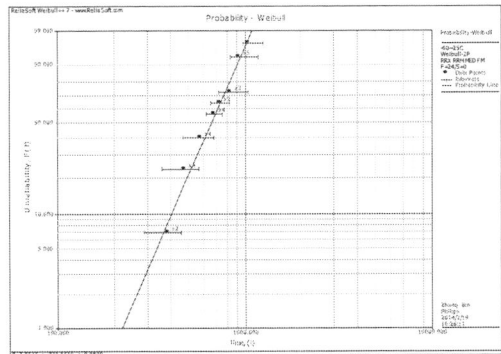

Fig.7 failure distribution of third test condition

$$\begin{cases} AF_1 = (\frac{n * \Delta T_{c1} + \Delta T_{h1}}{n * \Delta T_{c2} + \Delta T_{h2}})^\alpha \\ AF_2 = (\frac{n * \Delta T_{c1} + \Delta T_{h1}}{n * \Delta T_{c3}})^\alpha \end{cases} \text{ (Equation 4)}$$

Substitute into all the parameters to above equations, we can get the parameters of n, α.

$$\begin{cases} n=2.5 \\ \alpha=5.5 \end{cases}$$

Base on the new parameter, the accelerated factor between the accelerated test and real application can be recalculated as number of 440.

To evaluate the result, additional 2 thermal shock tests were executed which had past 3000cyles.There is no any failure in both -10C~150C and 25C~150C.

According to above conclusion, the scale parameter η of test condition (-10C~150C) should be 7750cycles.A predicated failure distribution base on test condition (-40C~120C) as figure 8.

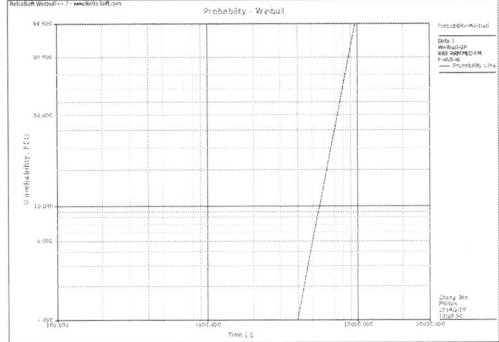

Fig 8 predicated failure distribution.

From this distribution, the B1ponit happened at around 3900cycles, it is very close to the real test result.

V. SUMMARY

Thermal shock is usually used for LED wire-bonding accelerated life testing, and the failure is commonly treated as a low-cycle fatigue problem. The lifetime analyses which base on the Coffin-Manson model never consider the modulus saltation of silicone enclosure with the temperature changing. The

978-1-4799-3911-4/14 $31.00 © 2014 IEEE   81

optimized accelerated model will help us to get more accurate thermal shock result, which will benefit to save the test time and cost. Accelerate to release the products to market.

References

[1]. B. Zhang et al. LED solder joint lifetime evaluation, to be published in China LED Feb 2014.

[2] W.D.van Driel, X.J.Fan, Solid State Lighting Reliability, components to

system，Springer, 2013，ISBN 978-1-4614-3066-7.

[3]Reliasoft, Life Data Analysis Reference, Published by Reliasoft Publishing,

USA.

Quantitative Analysis for Noise Generated from Share Circuitries within DDR3 DRAM

I. Nam[1,2], J. Lim[1], H. Hwang[1], K. Cho[1], and J. Choi[1]

{[1]Samsung Electronics Co./Memory Division, 1-0 Samsungjeonja-ro, Hwaseong-si, Gyeonggi-do 445-330, Republic of Korea}
{[2]Korea University/Department of Electrical Engineering, Anam-dong Seongbuk-Gu, Seoul 136-701, Republic of Korea}
Phone: (82) 31-8096-1791 Fax: (82) 31-8096-1791 Email: allevergreen@hanmail.net

Abstract-Correctable errors, almost single bit errors, can be induced from random data transition in *DRAM*. In the system, most of *CE*s are corrected by error correction code, but there is intermittent system down. Several literatures have reported that noise generated from share circuitries is regarded as a cause of soft failure. Noisy environments are one of the unavoidable factors in the high speed, high density, and low power *DRAM*. For the purpose of finding out noise source, we investigated states of share circuitries with random data transition. *BLSA*, power transistor, power line, and common plate were also researched with *DDR3 DRAM*s. A simple model was proposed with the quantitative analysis. Results show that the soft failure occurs when unexpected combination of noise factors happen at once, because the revealed erratic bits have similar characteristics of normal bits, except for the influence on noise.

I. INTRODUCTION

For the ultra large scale integration (ULSI) or dynamic random access memory (DRAM), interference has been recognized as noise factor which induces the soft errors. Capacitive coupling between adjacent lines is considered an origin of the interference, but the coupling has increased with the shrinkage of technology due to the reduced distance between signal lines. This is accelerated with the constraint that signal lines should be located in the limited area. Moreover, the decrease in current density induces the increase in delay of response time [1-3]. Multi-finger type transistor and share circuitries such as addressing lines, sense amplifiers, and transistors connected to power or ground are also used for high integration. The ultimate goal with these techniques is cost competitiveness although noise is increased at share circuitries.

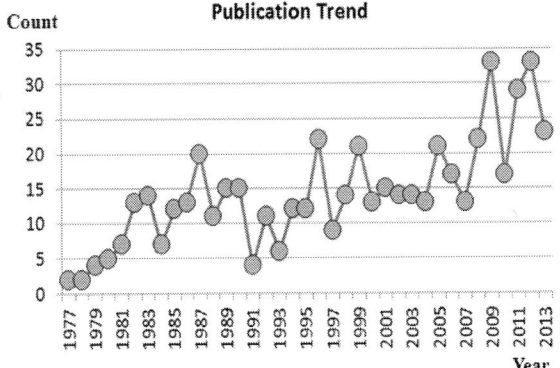

Fig. 1. Publication trend about DRAM errors.

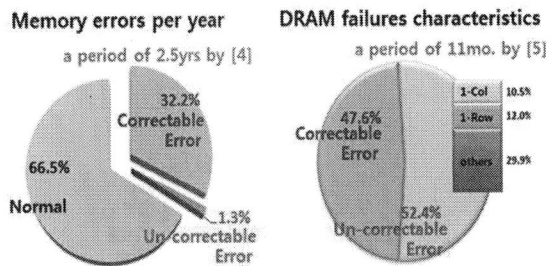

Fig. 2. Attributes of memory errors researched by [4], and [5].

Researches about errors in *DRAM* have gradually increased as shown in Fig. 1. Publications are focused on the soft error and the error correction code (ECC) techniques. Recently, faults in the share circuitries were reported as origin of soft failures [4, 5]. From the systemic perspective, the results showed that a considerable number of correctable error (CE) and un-correctable error (UCE) came from share circuitries in rows, columns and banks within the *DRAM*. Memory errors per year were *CE* = 32.2% and *UCE* = 1.3% per machine [4]. Research about *DRAM* failure characteristics showed *CE* = 47.6% and *UCE* = 52.4% [5], as shown in Fig. 2.

In this paper, we introduce the quantitative analysis of the noise caused in share circuitries by measurement of the delay of read time (tRD) and write time (tWR). Delays were observed and analyzed according to the operating conditions of cross-coupled differential amplifier, including the single-side operation and the finger-type operation. Finger-type operation is similar to the current crowding of multi-finger MOS structure [6]. Effects about array blocks and plate were also experimented with forming a bottleneck of current into local ground line and capacitive coupling. Finally, we proposed the resistor equivalent model about the multi-finger operation of shared transistor for quick noise analysis.

II. EXPERIMENT AND MODELING

For the efficient current density within a restricted area, multi-finger transistors are used as universal approach. However, electrical characteristics are different from existing ones such as normal device or multiple devices in parallel, because multi-finger transistors use shared source and drain regions as shown in Fig. 3. Fig. 4(a) shows the simplified open bitline architecture that cross-coupled differential amplifier, named as bitline sense amplifier (BLSA) in *DRAM*, is located between cell arrays. *BLSA* senses charges in storage capacitor when

978-1-4799-3911-4/14 $31.00 © 2014 IEEE

wordline (WL) is enabled (low to high) and amplifies the charge of bitline (BL) and its complementary bitline (BLB). Therefore, lots of *BLSA* are required to be located in cramped area for the efficiency of chip size. Consequentially, *BLSA* is composed multi-finger transistor with a shared source junction region belongs to both A-th bitline (BL-A) and B-th bitline (BL-B) as shown in Fig. 4.

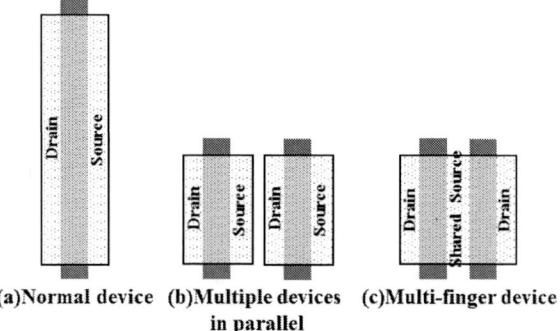

(a)Normal device **(b)Multiple devices in parallel** **(c)Multi-finger device**

Fig. 3. Transistor structures (a) normal (b) multiple (c) multi-finger devices.

When the operating signal is applied to gates of *BLSA* (GATE-A and GATE-B) at the same time, difference in current driving-ability is observed according to the topology between adjacent bitlines (BL-A and BL-B). The n-type transistors in *BLSA* show single-side operation with different phase between *BL-A* (data-0) and *BL-B* (data-1) in Fig. 5(a), while Fig. 5(b) shows finger-type operation with same phase of between *BL-A* and *BL-B* (all data-0). These operations can be explained with a simple resistor equivalent model as shown in Fig. 4(b). The equivalent resistance (R_{Single}) with different topology is

$$R_{DA} + R_{ONA} + R_S + R_{CONTACT}, \quad (1)$$

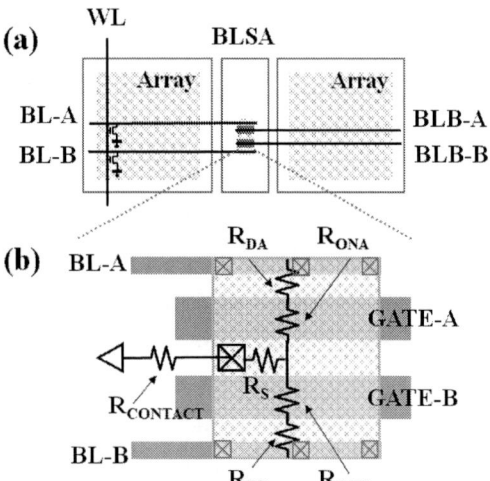

Fig. 4. (a) Simplified open-bitline architecture DRAM (b) equivalent model of multi-finger type BLSA.

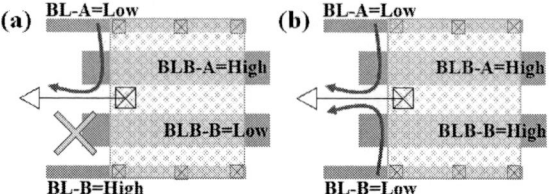

Fig. 5. (a) Single-side operation (b) finger-type operation.

and the equivalent resistance (R_{Finger}) with same topology is

$$\{(R_{CONTACT}+R_S) \times (R_{TA}+R_{TB}) + R_{TA} \times R_{TB}\}/(R_{TA}+R_{TB}), \quad (2)$$
$$R_{TA} = R_{DA} + R_{ONA}, \quad (3)$$
$$R_{TB} = R_{DB} + R_{ONB}, \quad (4)$$

where R_{DA} and R_{DB} are drain resistance, R_{ONA} and R_{ONB} are channel resistance, R_S is common source resistance, and $R_{CONTACT}$ is contact resistance to connect *BLSA* to ground line. Fig. 6 shows the measurement results of *tRD* delay as left axis and degradation rate under the finger-type operation as right axis using a double data rate type three (DDR3) *DRAM*. The average degradation rate of *tRD* under finger-type operation is 16.6%, and it is well correlated with increase in R_{Finger}. R_{Finger} calculated from (2) is 17% larger than R_{Single} calculated from (1).

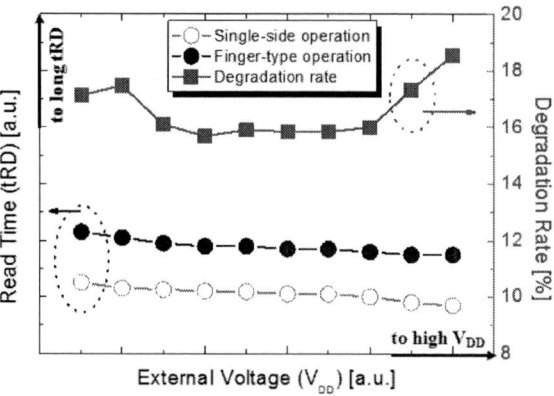

Fig. 6. Measurement results of tRD delay and degradation rate.

To reduce the propagation delay due to the capacitance loading of power drivers, the cell array is organized as a form of lumped array structure, and so *BLSA* region is also divided and located at side of cell array as shown in Fig. 7(a). Each n-type transistor (NT-A to NT-E), as power transistor, is connected between ground line and *BLSA* region, and *BLSAs* discharge the sensing current through these transistors. When operation signal is simultaneously applied from *NT-A* to *NT-E* for supporting page-mode operation in *DRAM*, the local voltage fluctuation (VSS+ΔV) in ground line is observed. This fluctuation affects *tRD* due to the change of operating point of *BLSA*. The *tRD* of data-0 (BL-A) was measured according to the ratio of data-1 to data-0 as shown in Fig. 7(b). Increase in data-1 causes the degradation of *tRD* because bottleneck phenomenon is induced at power transistors (NT-A to NT-E) and common power line. Two types of noise, the noise from adjacent bitlines in a same array and the noise from adjacent arrays, are observed. The noise from adjacent arrays is *tRD* delay of 5 ps/% of data-1 and 100 ps/array. The noise from adjacent bitlines in same array is *tRD* delay of 31 ps/% of data-1.

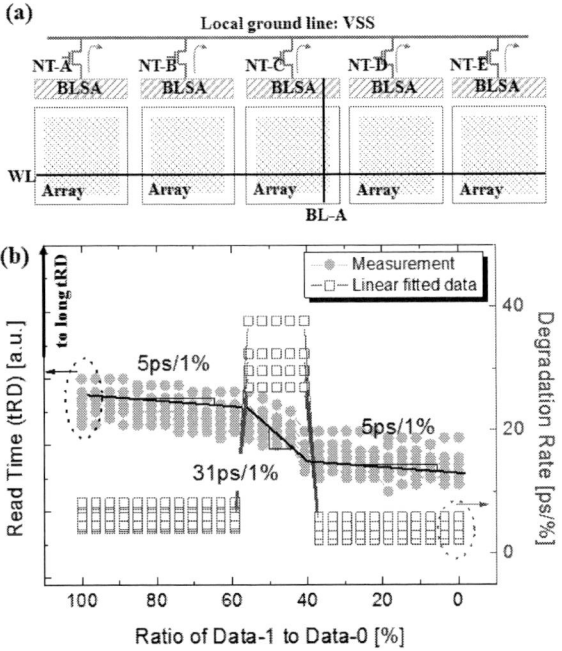

Fig. 7. (a) Test structure for characterization (b) measurement results according to the ratio of data-1 to data-0 and degradation rate of tRD.

When wordline is activated to write the data into storage cell, the capacitive coupling between *BL* to plate or between storage cell to plate affects local plate potential. Although there is a voltage generator for plate (V_P), the slow response time for the local fluctuation degrades *tWR* because the generator is located in a peripheral region away from cell array region. Fig. 8(a) shows a simple architecture. One plate covers over an array, and plates are connected with other plates through plate resistor (R_{PLATE}). The measured *tWR* according to the topology of three plates (Plate-A, Plate-B, and Plate-C) is shown in Fig. 8(b). Herein, the topology of plates is same as topology of arrays. For the case of writing data-0 into storage cell of *BL-A* within *Array-B*, the data topologies of *Plate-A*, *-B* and *-C* were conducted as (010), (011), (111), and (000). The larger *tWR* delay was observed in the case of (010) rather than (011) and (111). The degradation rate of *tWR* is 25% due to *Array-A*, 28% due to *Array-C*, and 85% due to *Array-B*, respectively. These results have consistency with the noise from adjacent bitlines in a same array in Fig. 7(b). Locally decreased V_P due to data topology of *Plate-A* and *Plate-C* affects charges for writing data-0. Therefore, the delay of *tWR* is observed to supply additional charges (ΔQ)

$$\Delta Q = C_{STROGATE} \times V_{FLUCTUATION}. \quad (5)$$

where, $C_{STORAGE}$ is capacitance of storage cell, and $V_{FLUCTUATION}$ is fluctuated plate voltage. Results from finger-type operation, adjacent signal line, shared transistors, and adjacent plate notify

that the induced noise nearby the target cell is dominant, and a combination of many factors causes failures of *DRAM* as shown in Table. I. The (010) case in Fig. 8(b) is well-matched our failure characteristics even though topology of *Array-A*, *-B* and *-C* helps *tRD*.

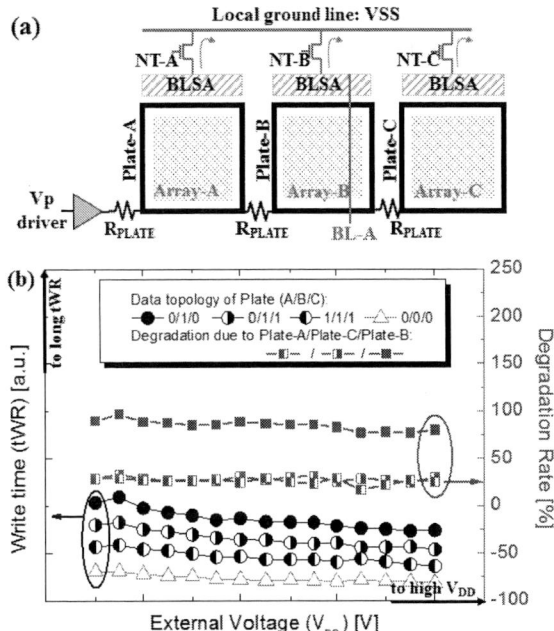

Fig. 8. (a) Test structure for characterization (b) measurement results according to the topology of arrays and degradation rate of tWR.

Table I
CHANGE RATE OF tWR AND tRD

Parameter	Topology of Array-A, -B, and -C	Adjacent Array-A	Self Array-B	Adjacent Array-C
tRD	(111)	+5ps/1%	+31ps/1%	+5ps/1%
tWR	(111)	Ref.	-	-
tWR	(011)	+25%	-	Ref.
tWR	(010)	-	+85%	+28%
tWR	(000)	-	Ref.	-

III. CONCLUSIONS

For minimized chip size without the shrink of process technology, a lot of circuitries are composed common nodes. The noise resulted from the finger-type operation of *BLSA*, the bottleneck phenomenon in power transistor and power lines, and the effect of common plate was analyzed using *tRD* and *tWR* with *DDR3 DRAM*. Moreover, we modeled the noise from adjacent bitlines or arrays quantitatively. The *BLSA* formed as the multi-finger transistor had varying characteristics due to shared junction and contact. The power transistor located in each array was affected the topology of bitlines within an array.

978-1-4799-3911-4/14 $31.00 © 2014 IEEE

The meshed power lines were also affected by locally generated noise because they were located in restricted area with narrow width. Because the shared plate having large capacitance was connected by R_{PLATE}, it had low immunity for locally induced noise. It reveals that the complex noise generated from share circuitries was more effective than it from individual circuitry. Therefore, we presume that the soft failure occurs when unexpected noises are happened with overlapping each other, because the erratic bits which are the origin of soft failure have similar characteristics of normal bits, except for noises.

REFERENCES

[1] K. Itoh, *VLSI Memory chip design*. New York: Springer, 2001.

[2] T. Sekiguchi, K. Itoh, T. Takahashi, M. Sugaya, H. Fujisawa, M. Nakamura, and et al., "A low-impedance open-bitline array for multigigabit DRAM," *Solid-State Circuits, IEEE Journal of*, vol. 37, pp. 487-498, April 2002.

[3] T. Takahashi, T. Sekiguchi, R. Takemura, S. Narui, H. Fujisawa, S. Miyatake, and et al., "A multigigabit DRAM technology with 6F² open-bitline cell, distributed overdriven sensing, and stacked-flash fuse," *Solid-State Circuits, IEEE Journal of*, vol. 36, pp. 1721-1727, November 2001.

[4] B. Schroeder, E. Pinheiro, and W-. D. Weber, "DRAM errors in the wild: A large-scale field study," *ACM SIGMETRICS Performance Evaluation Review*, vol. 37, pp. 193-204, 2009.

[5] V. Sridharan and D. Liberty, "A Study of DRAM Failures in the Field," in *High Performance Computing, Networking, Storage and Analysis (SC), 2012 International Conf. for. IEEE*, 2012, pp. 1-11.

[6] H. Wong, Y. Fu, J. J. Liou, Y. Yue, and H. Iwai, "On the reliability issues of RF CMOS devices," in *Solid-State and Integrated Circuit Technology, 2006. ICSICT'06. 8th International Conf. on. IEEE*, 2006, pp. 1105-1108.

Microstructural Approach to Failure Analysis of Thin Film Transistors

Ju Ho Lee, SungSoon Choi, and Kwan Hun Lee
Reliability Technology Research Center, Korea Electronics Technology Institute (KETI),
68 Yatap-dong, Bundang-gu, Seongnam 463-816, Republic of Korea
Phone: (+82) 31-789-7282 Fax: (+82) 31-789-7299 Email: leejuho@keti.re.kr

Abstract-As size of electronic device miniaturized, microstructural characteristics of materials significantly affect the electrical properties of devices. In this study, microstructural approach to failure analysis of thin film transistors is presented using two examples. Also, we directly demonstrated correlation between electrical properties and microstructural characteristics using transmission electron microscopy.

I. INTRODUCTION

Integration of electronic devices has been a trend over decades and is continuing into the predictable future. According to this tendency, components used in electronic applications were miniaturized at the same time. Among the various electronic components, oxide-based thin film transistors (TFTs) play a role as amplification and switching have attracted a great deal of attention for their applications. As physical size of TFTs miniaturized, microstructural characteristics of materials affect electrical properties of TFTs. Therefore, in this study, we presented microstructural approach to failure analysis of oxide-based TFTs by giving two cases.

II. EXPERIMENTAL DETAILS

About 140 nm thick ZnO thin films were grown on highly conductive *p*-type Si (100) substrates with a thermally grown SiO_2 layer having a thickness of 220 nm by radio-frequency magnetron sputtering. Also, integration of electronic devices has been a trend over decades and is continuing into the predictable future. According to this tendency, components used in electronic applications were miniaturized at the same time. Among the various electronic components, oxide-based thin film transistors (TFTs) play a role as amplification and switching have attracted a great deal of attention for their applications. As physical size of TFTs miniaturized, microstructural characteristics of materials affect electrical properties of TFTs. Therefore, in this study, we presented microstructural approach to failure analysis of oxide-based TFTs by giving two cases.

III. RESULTS AND DISCUSSION

Case I

The schematic diagram of bottom-gated TFT with channel width of 500 μm and channel length of 100 μm which treated in this study is shown in Figure 1.

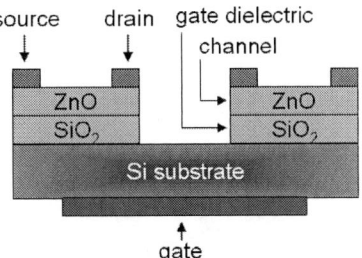

Fig. 1. Schematic diagram of bottom-gated TFT.

Figure 2(a) and 2(b) shows transfer curves for normal and abnormal TFTs, respectively. Noticeably, the drain current-gate voltage [$\log(I_D)$-V_{GS}] curve of the abnormal TFT shows an unusual shape. As shown in Figure 2(b), the $\log(I_D)$-V_{GS} curve has a hump at a V_{GS} of approximately -15 V. In organic TFTs and poly-crystal Si TFTs, it is reported that the hump-shaped transfer characteristics were attributed to their multi-channel operation behavior [1~3].

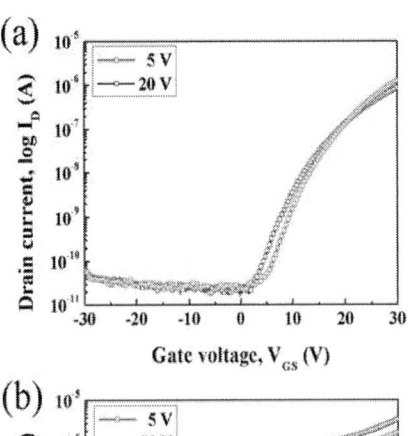

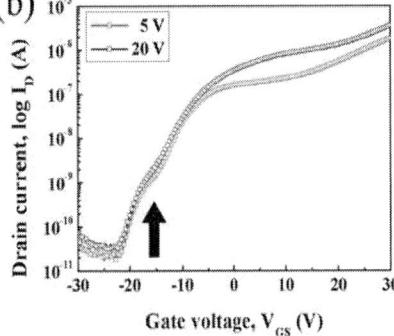

Fig. 2. Transfer curves for (a) normal and (b) abnormal TFTs. The width-to-length ratio of the TFT is 5:1 [Reprinted with permission from *Thin Solid Films* 519, 6801. Copyright 2011 Elsevier].

978-1-4799-3911-4/14 $31.00 © 2014 IEEE

In order to investigate formation reason of hump at V_{GS} of -15 V, microstructural analysis of the synthesized TFTs were conducted using high-resolution transmission electron microscopy (HRTEM). Figure 3 shows the cross-sectional bright-field TEM images of ZnO film/SiO_2 interface of normal and abnormal TFTs. It should be noted that, as shown in inset of Figure 3(b), bright contrasts in TEM images are usually observed when the specimen has electrically insulating characteristics due to the electron charging effect. Therefore, these areas of unusual bright contrast in the bottom regions would be expected to have somewhat different electrical properties. In terms of the electrical performance of the TFTs, the characteristics of the channel layer/gate insulator interface are very important and, consequently, the bright contrast observed near the interface would be expected to critically affect the TFT performance.

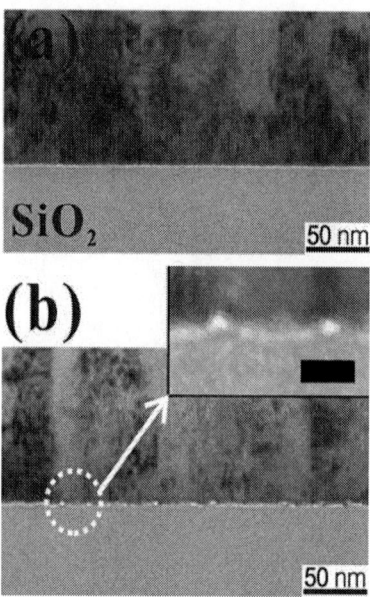

Fig. 3. Cross-sectional bright-field TEM images obtained from ZnO film/SiO_2 interface of (a) normal and (b) abnormal TFTs. The inset of (b) shows the enlarged bright-field TEM image obtained from the circle marked by the dashed line in (b). The scale bar of the inset of (b) indicates 10 nm [Reprinted with permission from *Thin Solid Films* 519, 6801. Copyright 2011 Elsevier].

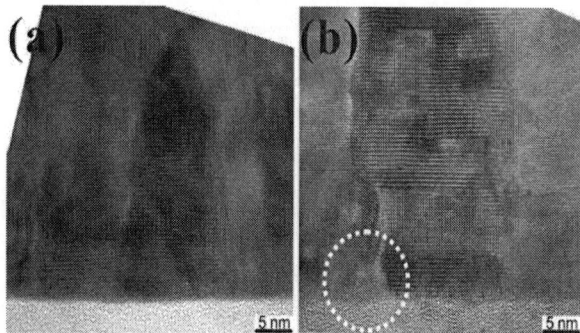

Fig. 4. Cross-sectional bright-field TEM images obtained from ZnO film/SiO2 interface of (a) normal and (b) abnormal TFTs. The dotted circle in (b) indicates a small grain near the ZnO film/SiO_2 interface [Reprinted with permission from *Thin Solid Films* 519, 6801. Copyright 2011 Elsevier].

To examine in greater detail the microstructural characteristics of obtained TFTs, HRTEM observations were conducted. Figure 4 shows the enlarged HRTEM images near the interface between the conducting ZnO film and insulating SiO_2 layer. The bright image (Fig. 3(b)) near the ZnO film/SiO_2 interface of the channel layer correspond not to voids or amorphous phase, but crystalline phases, as shown in the magnified HRTEM image of Figure 4(b).

As can be seen in the Figure 4(b), small grains were only embedded in the junction of the ZnO film/SiO_2 interface of abnormal TFTs. Owing to the formation of small grains, additional grain boundaries were formed in the channel layer. As it is well known, the grain boundary acts as potential barriers to the current flow. The formation of new boundaries induced by the many small grains near the ZnO film/SiO_2 interface is expected to provide an additional potential barrier to electron movement. As a result, the hump near $V_{GS}= -15$ V in the $\log(I_D)$-V_{GS} curve for the abnormal TFT may be due to the high density of trap sites caused by the formation of defects (i.e. small grains) near the channel/gate oxide interface.

Based on the abovementioned failure analysis, we can conclude that the small grains were formed due to Ostwald ripening and growth temperature of channel layer was modified [4].

Case II

Figure 5 shows schematic diagram of alternatively stacked ZnO/Al_2O_3 structures with different thicknesses of insulating layer (i.e. Al_2O_3) grown on glass substrates. The thicknesses of Al_2O_3 layers were varied from 0 to 7 nm, and the stacked numbers of Al_2O_3 layers are the same through all samples.

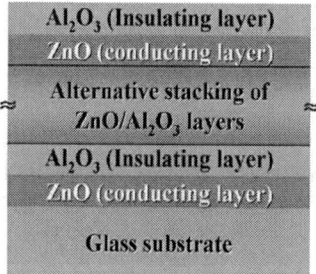

Fig. 5. Schematic diagram of alternatively stacked ZnO/Al_2O_3 structures.

Figure 6 shows cross-sectional bright-field TEM images obtained from the alternatively stacked ZnO/Al_2O_3 structures with different thicknesses of Al_2O_3 grown on glass substrates. From the physical points of views, the thickness difference of each Al_2O_3 layer of samples is barely 2~3 nm.

As shown in Figure 6, even though thickness gap of insulating layer between samples is barely 2~3 nm, but Seebeck properties are completely distinguished corresponding to different thickness of insulating layer. Thus, microstructural characteristics of nanoscaled materials significantly affect their electrical, optical and chemical properties, and this tendency is frequently observed with miniaturization and integration of electronic devices.

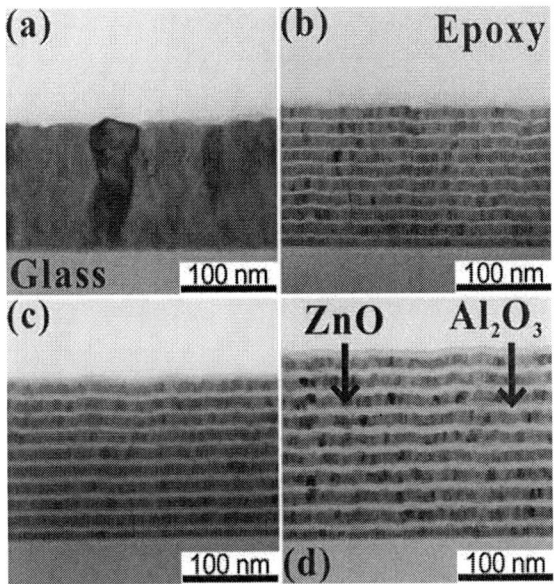

Fig. 6. Cross-sectional bright-field TEM images with different thickness of Al_2O_3: (a) 0, (b) 3, (c) 5, and (d) 7 nm.

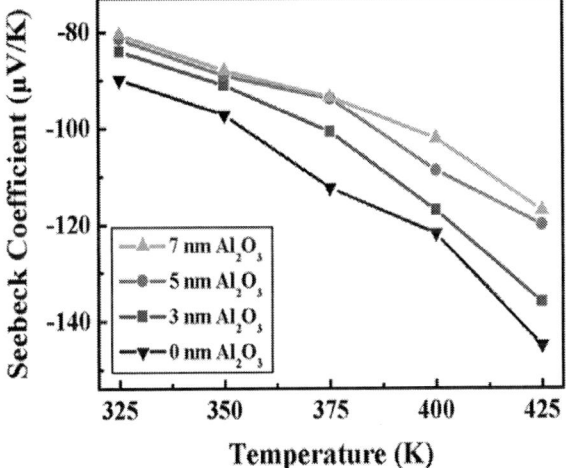

Fig. 7. Seebeck properties of alternatively stacked ZnO/Al_2O_3 structures with different thickness of insulating layer.

IV. CONSLUSIONS

Microstructural characteristics of nanoscaled materials significantly affect their electrical, optical and chemical properties, and this tendency is frequently observed with miniaturization and integration of electronic devices. Therefore, to reveal a failure mechanism in nano era, microstructural approach needs to be performed using high resolution apparatus such as high resolution transmission electron microscope.

ACKNOWLEDGMENT

This work was supported by the Korea Institute for Advancement of Technology (KIAT) funded by Korean Government (Establishment of foundation for Reliability Assessment).

REFERENCES

[1] J. B. Koo, J. H. Lee, C. H. Ku, S. C. Lim, S. H. Kim, J. W. Lim, S. J. Yun, and T. Zyung, "The effect of channel length on turn-on voltage in pentacene-based thin film transistor", *Synthetic Met.*, vol. 156, pp. 633-636, 2006.

[2] C.-M. Yu, H.-C. Lin, T.-Y. Huang, and T.-F. Lei, " H 2 and NH 3 Plasma Passivation on Poly-Si TFTs with Bottom-Sub-Gate Induced Electrical Junctions", *J. Electrochem. Soc.*, vol. 150, pp. G843-G848, 2003.

[3] J. H. Lee, C. H. Ahn, S. Hwang, C. H. Woo, J.-S. Park, H. K. Cho, and J. Y. Lee, "Role of the crystallinity of ZnO films in the electrical properties of bottom-gate thin film transistors", *Thin Solid Films*, vol. 519, pp. 6801-6805, 2011.

[4] L. Ratke, and P. W. Voorhees, *Growth and coarsening: Ripening in material processing*, Springer, 2002, pp. 117.

Characterization Studies of Fluorine-induced Corrosion Crystal Defects on Microchip Al Bondpads Using X-ray Photoelectron Spectroscopy

Hua Younan, Xing Zhen Xiang and Li Xiaomin

WinTech Nano-Technology Services Pte. Ltd.
10 Science Park Road, #03-26 & #03-28, The Alpha
Singapore Science Park II, Singapore 117684
Email:Younan@wintech-nano.com
Tel: 65-92256313

Abstract

In wafer fabrication, Fluorine (F) contamination may cause F-induced corrosion and defects on microchip Al bondpad, resulting in bondpad discoloration or non-stick on pad (NSOP). In the previous paper [1], the authors studied the F-induced corrosion and defects, characterized the composition of the "flower-like" defects and determined the binding energy of Al fluoride $[AlF_6]^{3-}$ using X-ray Photoelectron Spectroscopy (XPS) and Time of Flight Secondary Ion Mass Spectrometry (TOF-SIMS) techniques. In this paper, we further studied F-induced corrosion and defects, and characterized the composition of the "crystal-like" defects using XPS. The experimental results showed that the major component of the "crystal-like" defect was Al fluoride of AlF_3. The percentages of the components of the "crystal-like" defects on the affected bondpad are: Al (22.2%), Al_2O_3 (5.4%), AlF_3 (70.0%) and $[AlF_6]^{3-}$ (2.4%). During high-resolution fitting, the binding energies of Al (72.8eV), Al_2O_3 (74.5eV), AlF_3 (76.3eV) and $[AlF_6]^{3-}$ (78.7eV) were used.

Introduction

In wafer fabrication, F contamination may cause F-induced corrosion and defects on microchip Al bondpad, resulting in bondpad discoloration or non-stick on pad (NSOP). In the past, wafer fab process showed that if F level is higher than the control level (or baseline), which is about 6 at% by Auger analysis [2], it may cause F-induced corrosion and form Al fluoride defects. The Al fluoride can be in the form of AlF_3 or $[AlF_6]^{3-}$, depending on the wafer fab process and wafer storage condition. In the authors' previous paper [3], we studied the F-induced corrosion defects with theoretical analysis and classified F-induced corrosion defects into 4 major defects, which are the "flower-like" defects, "crystal-like" defects, "oxide-like" defects and "cloud-like" defects showed in Figure 1. All these F-induced defects were caused by wafer fab processes and wafer storage.

In our previous paper [1], we characterized the compositions of the "flower-like" defect (refer to the Table 1) and determined the binding energy of Al fluoride $[AlF_6]^{3-}$ using X-ray Photoelectron Spectroscopy (XPS) and Time-of–Flight Secondary Ion Mass Spectroscopy (TOF-SIMS) techniques. Based on the experimental results, besides AlF_3, the major component of the "flower-like" defects was Al fluoride of $[AlF_6]^{3-}$ and its binding energy was determined to be 78.7 eV.

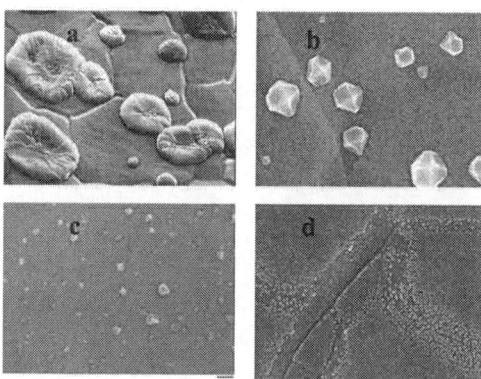

Figure 1. F-induced corrosion defects were classified into 4 major defects, including the "flower-like" defect (a), "crystal-like" defect (b), "oxide-like" defect (c) and "cloud-like" defect (d).

In this paper, we further studied F-induced corrosion and defects, analyzed the failure mechanism and characterized the composition of the "crystal-like" defects using XPS & chemical bonding energies of Al (72.8eV), Al_2O_3 (74.5eV), AlF_3 (76.3eV) and $[AlF_6]^{3-}$ (78.7eV) [1, 5-7].

F-induced Corrosion and Al Fluoride

In wafer fabrication, CF_4 gas is used for bondpad opening process. Thus, on a normal Al bondpad, there is a low level of F contamination, which could not cause F-induced corrosion defects. However, if F contamination is higher than the baseline, reaching certain level on Al bondpads, it may cause F-induced corrosion and chemically react with Al to form Al fluoride, $[AlF_x]^{(x-3)-}$. This is a chemically stable compound and it does not dissolve in water. Therefore, once it is formed, it will be difficult to be cleaned away using a normal cleaning process. In Chemistry, the x in $[AlF_x]^{(x-3)-}$ could be 3 or 6, which depends on the F contamination level & process/wafer storage conditions:

$$X=3 \quad Al + 3F^- \rightarrow AlF_3 + 3e^- \quad (1)$$
$$X=6 \quad Al + 6F^- \rightarrow [AlF_6]^{3-} + 3e^- \quad (2)$$

F-induced corrosion isn't formed immediately, and it needs certain time to chemically react with Al and grow up the defects. For example, when x is 6, the Al fluoride will be $[AlF_6]^{3-}$, observed as the "flower-like" defects on microchip

978-1-4799-3911-4/14 $31.00 © 2014 IEEE

Al bondpads, shown in Figure 1 (a). Al fluoride $[AlF_6]^{3-}$ is a hexno-crystalline complex compound and it is difficult to manually obtain an Al fluoride sample with $[AlF_6]^{3-}$ in the laboratory. In our previous study, the sample was occasionally obtained from a bondpad fluorine contamination case, which was due to wafer shipment packaging foam material [1].

Experimental Results and Discussion

The "Crystal-like" Defects

In this study, the samples were obtained from a real failure analysis case. Discoloration bondpads were found on the affected wafer, which had resulted in NSOP at the assembly house. Scanning electron microscopy (SEM) inspection was performed and the "crystal-like" defects were found on the affected bondpad (Figure 2).

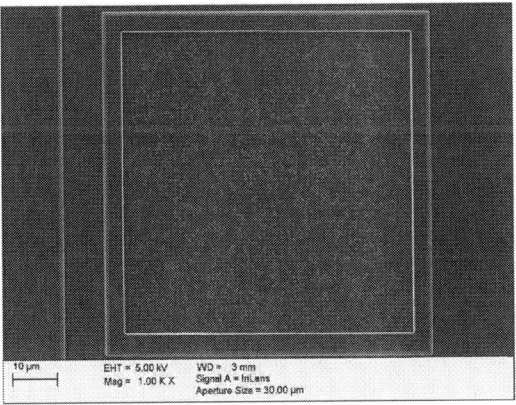

Figure 2. SEM inspection result showed that the "crystal-like" defects were found on the affected bondpads.

The high magnification SEM micrograph showed that the typical sizes of the defect were about 0.2-0.3 um (Figure 3-4), which is similar to the defects, shown in Figure 1 (b).

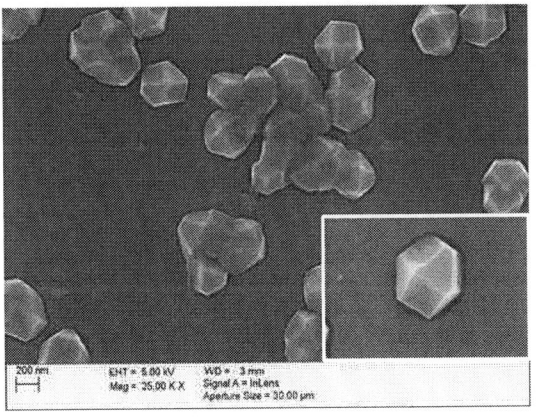

Figure 3. The close-up SEM micrograph showed that the typical sizes of the defects were ~0.2-0.3 um, which is similar to those showed in Figure 1 (b).

However, such defect was not observed on the bondpads of the good wafer (Figure 4).

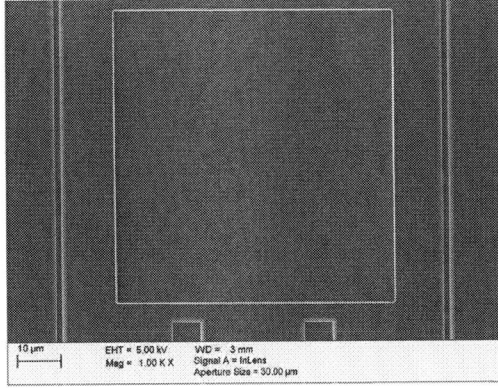

Figure 4. SEM micrograph showed that the "crystal-like" defects were not observed on the bondpad of the good wafer.

EDX elemental analysis revealed the presence of F on the "crystal-like" defects (Figure 5). On the contrary, no fluorine was detected on the bondpads of the good wafer.

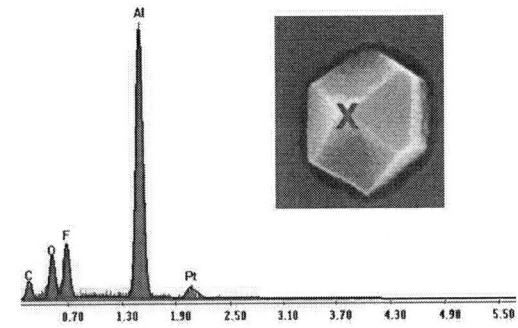

Figure 5. EDX analysis revealed the presence of F on a "crystal-like" defect. On the contrary, no fluorine was detected on the bondpad of the good wafer.

XPS Analysis and Studies of the "Crystal-like" Defects

In this study, XPS/ESCA was used to characterize the composition of the "crystal-like" defect. XPS/ESCA was performed on a PHI Quantera II Scanning XPS Microprobe. Monochromatic Al source was used for high-energy resolution analysis on the bondpads. The area of analysis is about 15 um diameter. An Ar+ sputter gun was operated on a 3 KeV and raster area of 500um x 500um for depth profiling. The sputtering rate was calibrated against SiO_2 standard. Low energy charge compensation electron gun was incorporated in the analysis.

For our XPS tool, the capability of the spot size can reach to the level about 7.5um (actual spot size may be lager). However, it is still much larger than the size of the "crystal-like" defects (about 0.2-0.3um). Therefore, it is difficult to analyze a single such defect using XPS. In this study, we detected many "crystal-like" defects at the XPS spot size area.

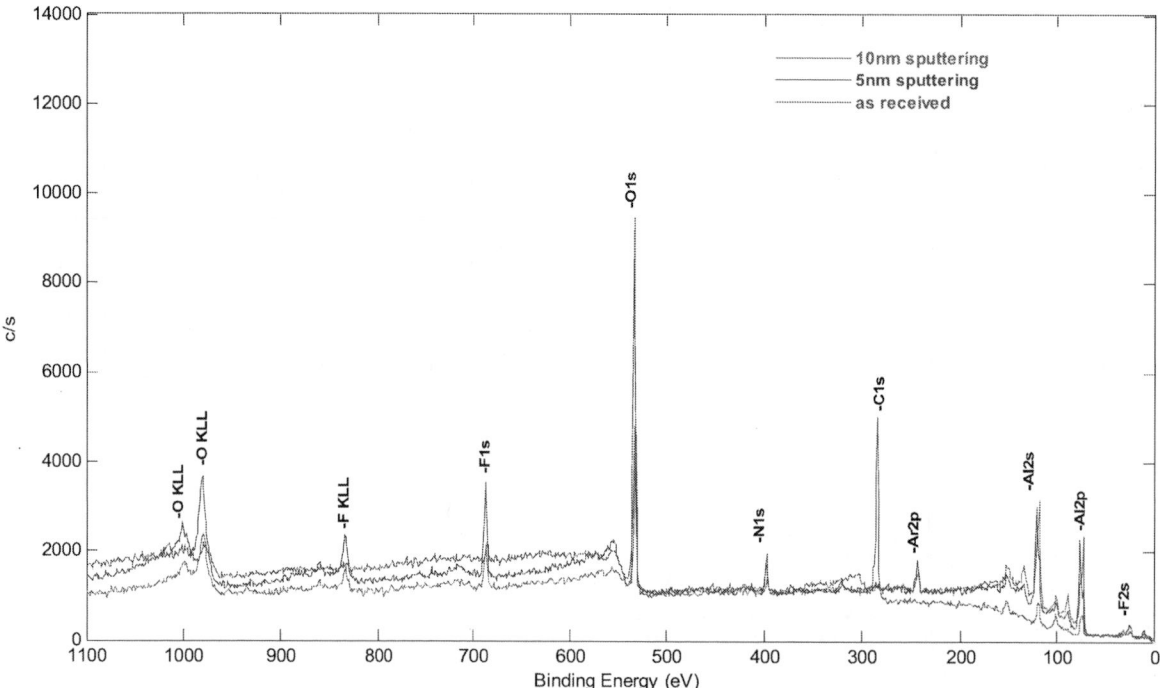

Figure 6. A typical XPS scanning spectrum shows besides Al, the elements of C, N, O & F were detected. High C was detected as received (0A), which was due to sample surface carbon contamination. N peak was due to N2 chemical reaction on the F-induced defects

A typical XPS scanning spectrum is obtained, shown in Figure 6. From the result, one can see that besides Al, the elements of C, N, O & F were detected. For the results as received scanning (0A), high C was detected, which was due to sample surface carbon contamination. Moreover, similar to the previous report [4], N peak was also detected by XPS (Figure 6). Based on our previous studies, we have understood that it was due to N_2 chemical reaction that generated $[NH_4]^+$ ions on the F-induced defects as TOF-SIMS detected $[NH_4]^+$ ions and its fragment ions of NH_3, $[NH_2]^-$ and $[NH]^{2-}$ on the F-induced defects [4]. It is well known that N_2 is chemically stable in the general atmosphere conditions. However, when F-induced corrosion happened on Al bondpads, the chemical energy E^o of Al fluoride AlF_3 or $[AlF_6]^{3-}$ can be much lower than -1.05 V, which can cause the below chemical reaction from left to right and form $[NH_4]^+$ ions [4]:

$$N_2 + 8H_2O + 6e^- \Leftrightarrow 8OH^- + 2[NH_4]^+ \qquad (3)$$
$$(E^o = -1.05 \text{ V})$$

Narrow scan and high-resolution fitting results of XPS Al2p are shown in Figure 8. From the results, one can gain a better understanding of the presence of Al, Al_2O_3, AlF_3 and $[AlF_6]^{3-}$. During high-resolution fitting, the binding energies of Al (72.8eV), Al_2O_3 (74.5eV), AlF_3 (76.3eV) and $[AlF_6]^{3-}$ (78.7eV) [1, 5-7] were used. According to the fitting results of Al2P electron peak, the percentages of the components of the "crystal-like" defects on the affected bondpad are: Al element (22.2%), Al_2O_3 (5.4%), AlF_3 (70.0%) and $[AlF_6]^{3-}$ (2.4%).

Interestingly, unlike previous case of the "flower-like" defect showing that $[AlF_6]^{3-}$ was the major component shown in Figure 7, AlF_3 was found to be the major component in the "crystal-like" defect shown in the Figure 6. The results of the "flower-like" defect and "crystal-like" defect are showed in Table 1. Present study is important for further understanding of the types of F-induced defects and the formation mechanisms of AlF_3 that lead to Al bondpads NSOP issue.

Conclusions

In this study, we further studied the F-induced corrosion and defects on microchip Al bondpads and characterized the composition of the "crystal-like" defects using XPS. During high-resolution XPS Al2p fitting, the binding energies of Al (72.8eV), Al_2O_3 (74.5eV), AlF_3 (76.3eV) and $[AlF_6]^{3-}$ (78.7eV) were used. According to the high-resolution fitting results of Al2P electron peak, the percentages of the components of the "crystal-like" defects on the affected bondpad are: Al element (22.2%), Al_2O_3 (5.4%), AlF_3 (70.0%) and $[AlF_6]^{3-}$ (2.4%). It is concluded that the fluoride of $[AlF_6]^{3-}$ was the major component of the "flower-like" defects, while the fluoride of AlF_3 was the major component in the "crystal-like" defects.

Acknowledgments

The authors would like to thank Dr Gui Dong, Dr Lee Hwang Sheng and Miss Sun Fangda for their technical support and contributions to this technical paper.

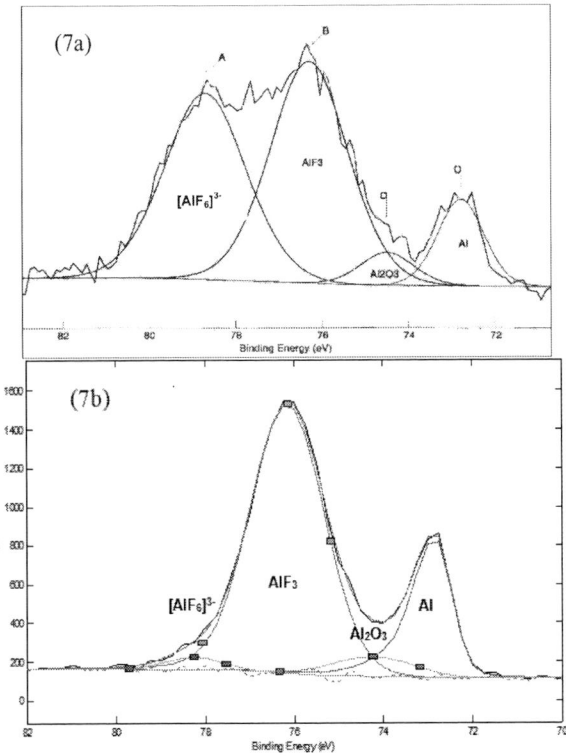

Figure 7. High-resolution XPS Al2p fitting result for (a) the "flower-like" defects [1] and (b) the "crystal-like" defects (in this study). The best fitting binding energy of Al Hexafluoride $[AlF_6]^{3-}$ was at 78.7eV for the "flower-like" defects when the binding energies of Al (72.8eV), Al_2O_3 (74.5eV) and AlF_3 (76.3eV) were used. Similar Al, Al_2O_3, AlF_3 and $[AlF_6]^{3-}$ binding energies were also applied for determining the components of the "crystal-like" defects. According to the high-resolution fitting results of XPS Al2P, the percentages of the components of the "crystal-like" defects are: Al (22.2%), Al_2O_3 (5.4%), AlF_3 (70.0%) and $[AlF_6]^{3-}$ (2.4%).

Table 1. Comparison of the XPS fitting Results of Al2p for the "Flower-like" Defects [3] and "Crystal-like" Defects (this study)

Element/Compound	Binding Energy (eV)	The "Flower-like" Defects (%)	The "Crystal-like" Defects (%)
Al	72.8[a]	10.4	22.2
Al_2O_3	74.5[a]	4.6	5.4
AlF_3	76.3[a]	45.6	70.0
$[AlF_6]^{3-}$	78.7[b]	39.4	2.4

Note: The Binding energy data for "a" are from [5-7] and for "b" is from [1].

References

1. Hua Younan, Characterization of Binding Energy of Al Hexafluoride $[AlF_6]^{3-}$ in X-Ray Photoelectron Spectroscopy. The Proceedings of the 13th International Physics and Failure Analysis, 03-07 July, Meritus Mandarin Hotel, Singapore, pp.214-216 (2006).

2. Hua Younan et al, Simulation Studies on Fluorine Spec Limit for Process Monitoring of Microchip Al Bondpads in Wafer Fabrication. The Proceedings of the 39th International Symposium for Testing and Failure Analysis (ISTFA 2013), November 3-7, 2013. San Jose, California, USA (2013).

3. Hua Younan et al, Studies of Fluorine-induced Corrosion Defects on Microchip Al Bondpads and Elimination Solutions. The Proceedings of the 34th International Symposium for Testing and Failure Analysis (ISTFA 2008), Nov 2-6, Portland, Oregon, USA, p285-290 (2008).

4. Hua Younan et al, A Study on Fluorine-Induced Corrosion on Microchip Aluminium Bondpads. The proceedings from the 29th International Symposium for Testing & Failure Analysis (ISTFA'2003), 2-6 Nov. 2003, Santa Clara, California, USA, p249-255 (2003).

5. NIST X-ray Photoelectron Spectroscopy Database, Version 2, 1997, National Institute of Standards and Technology, USA.

6. Handbook of X-ray Photoelectron Spectroscopy, Perkin-Elmer Corporation, USA (1992).

7. John F. Moulder et al, "Handbook of X-ray Photoelectron Spectroscopy". Edited by Jill Chastain, Perkin-Elmer Corporation, USA p. 213-214 (1992).

Characterization of Wet Chemical Etching for Effective Backside Sample Preparation on Devices with Exposed Pads

Andrew C. Sabate and Rowin V. Galarce

Product Analysis Laboratory

ON Semiconductor Malaysia Sbn Bhd

Lot 122, Senawang Industrial Estate, 70450 Seremban, Negeri Sembilan, W. Malaysia

Andrew.Sabate@onsemi.com and Rowin.Galarce@onsemi.com

Abstract

With the growing complexity of Integrated Circuit (IC) design having multiple metallization layers and copper wire bonded devices, most of the time backside fault isolation is a better approach. This paper evaluated wet chemical backside sample preparation as an alternative method for the traditional milling/polishing backside sample preparation.

I. Introduction

Integrated Circuit (IC) today has become more complex with some wafer fab process involve multilayer metals with top metal layers covering large portion of the die. Some die fabrication uses copper power metal which can be easily peeled off by the usual decapsulation process. Assembly process also has somewhat shifted to copper wire bonds and it also gives problem on the decapsulation of copper wire devices. These are the arising hindrances for the front side analysis which can be resolved through backside fault isolation.

Backside sample preparation allows you to observe the die substrate level directly through the use of infrared (IR) scopes. Die backside analysis avoids thick top metal layer which gives good results for fault isolation technique such as Laser Signal Injection Microscopy (LSIM) and Photon Emission Microscopy (PEM). Among the traditional backside sample preparation techniques used are backside polishing and mechanical milling.

Backside polishing is the common backside sample preparation technique used in failure analysis. Polishing can either be done using a handheld tripod polishing tool or by using a precision polisher. The process is to thin down the device until the die backside is expose. Another traditional backside sample preparation technique is the mechanical milling. This technique is done by a precision milling tool which uses precision drill bit. Only a specific opening will be milled out as outlined through the machine console/machine PC.

With the existence of the traditional backside sample preparation technique, this paper aims to discuss fast and cost effective alternative method for the backside sample preparation on devices with exposed pad/flag using wet chemical etching. This paper will also discuss the effect of different chemicals,

temperatures and condition of the unit on the evaluation. The experimentation done utilized the existing resources inside the Failure Analysis (FA) Lab.

Figure 1. Image showing the results of mechanical polishing. Electrical connection can only be accessed through exposed wire bonds using micro-probing.

II. Backside sample preparation using wet chemical etching

The materials used for the evaluation were as follows: 69%-70% non-fuming Nitric Acid (HNO_3), Aqua Regia (HNO_3+3 HCl), 98% fuming Nitric Acid (HNO_3) and 5:1 combination of fuming nitric acid and sulfuric acid were used for exposed pad etching experimentation. Sulfuric acid (H_2SO_4), Solder etch **(99% methanesulfonic; 70%Nitric Acid; DiH$_2$0)**, Hydrochloric acid (**HCl**) and Hydrogen peroxide (**H$_2$0$_2$**) were used for die-attach removal. High temperature tape and hot plate were also used for experimentation. All of the chemicals for evaluation are readily available inside the Lab.

Evaluation was done in consideration of the parameters: chemical and hot plate temperature. Etch time and the condition of the unit after etching will be the basis of the result of evaluation.

Evaluation started by covering the device package with the high temperature tape. The high temperature tape will serve as the masking of the unit from the effect of chemicals. Opening on the high temperature tape was made specifically at the exposed pad. The cut should be aligned to the position of the die. The alignment can be check by either using the wire bonding diagram specification or doing the X-ray examination.

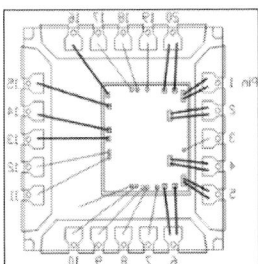

Figure 2. Wire bonding diagram of the device at bottom view perspective.

Figure 3. Actual unit covered by high temperature tape with an opening aligned to position of the die.

A beaker filled with chemical is set to a temperature based on the DOE. After the chemical temperature stabilizes, the unit was soaked to the chemical and the progress of expose pad/flag etching was checked time to time. After the expose pad/flag was etched out, the unit was cleaned through running water and subsequently cleaned using acetone.

After the removal of the exposed pad/flag, remnants of die-attach epoxy was still visible. The die-attach epoxy residue was then etched using different chemicals under evaluation. The unit was heated up and a drop of chemical was used to soften the die-attach epoxy. A cotton swab dipped to acetone was used to clean out the residual epoxy.

Figure 4. Photo showing sample unit dipped at 69%-70% non-fuming Nitric Acid at 150C.

The procedure together with the DOE was evaluated until a reasonable result was taken.

Exposed pad wet chemical etching evaluation results

UNIT TYPE	TEMPERATURE	ETCH TIME	RESULT
Chemical: AQUA REGIA			
FRESH UNIT	80°C	40 minutes	Acid attacks the flag/paddle but not totally the die attach portion
	150°C	17 minutes	Acid attacks the flag/paddle but not totally the die attach portion
USED UNIT	80°C	32 minutes	Acid attacks the flag/paddle but not totally the die attach portion
	150°C	15 minutes	Acid attacks the flag/paddle but not totally the die attach portion

(A) Results for AQUA REGIA

UNIT TYPE	TEMPERATURE	ETCH TIME	RESULT
Chemical: 69%-70% Nitric Acid			
FRESH UNIT	80°C	1 minute and 30 seconds	Acid attacks the flag/paddle and some portion of the die attach
	150°C	1 minute	Acid attacks the flag/paddle and some portion of the die attach
USED UNIT	80°C	1 minute and 15 seconds	Acid attacks the flag/paddle and some portion of the die attach
	150°C	less than 1 minute	Acid attacks the flag/paddle and some portion of the die attach

(B) Results for non-fuming 69%-70% Nitric Acid

UNIT TYPE	TEMPERATURE	ETCH TIME	RESULT
Chemical: 5:1 (Fuming Nitric Acid:Sulfuric Acid)			
FRESH UNIT	80°C	-	Acid attacks the high temp tape and melted the tape
	150°C	-	-
USED UNIT	80°C	-	Acid attacks the high temp tape and melted the tape
	150°C	-	-

(C) Results for 5:1 (Fuming Nitric Acid: Sulfuric Acid)

UNIT TYPE	TEMPERATURE	ETCH TIME	RESULT
Chemical: 98% Fuming Nitric Acid			
FRESH UNIT	80°C	-	Acid attacks the high temp tape and melted the tape together with the device
	150°C	-	-
USED UNIT	80°C	-	Acid attacks the high temp tape and melted the tape together with the device
	150°C	-	-

(D) Results for 98% Fuming Nitric Acid

Table 1. Tables showing the evaluation results of exposed pad wet chemical etching using (A) Aqua Regia (HNO_3+3 HCl) (B) 69%-70% non-fuming Nitric Acid (HNO_3) (C) 5:1 combination of fuming nitric acid and sulfuric acid (D) 98% fuming Nitric Acid (HNO_3). The evaluation on non-fuming 69%-70% Nitric Acid yielded the best results.

Evaluation results showed that AQUA REGIA (HNO_3+3 HCl) manage to etch the exposed pad, however it took 15-30 minutes to completely etch out the exposed pad. Both 5:1 solution and 98% Fuming Nitric Acid melted the tape and then eventually attack the molding compound. The strong acid solution was found not suited for the application. The non-fuming 69%-70% Nitric Acid yielded the fastest and good results. Etching the exposed pad within 1 -2 mins with minimal package abnormality.

After the exposed pad was removed, remnants of die-attach material was observed obstructing the direct access to die backside. Evaluation with different chemicals to remove the die-attach was done. See table 2.

Die attach remnant etching evaluation results

Die Attach Remnants Removal			
Chemical Used	TEMPERATURE	ETCH TIME	RESULT
Sulfuric Acid	100°C	15 seconds	Die attach remnants were totally etched when drops of sulfuric acid was applied on the back portion of the paddle and susequently cleaned with cotton buds dipped
Solder Etch (15ml 99% Methanesulfonic; 4ml 70%Nitric acid; 25ml DiH2O)	150°C	2 minutes	Die attach remnants were totally etched when unit was dipped on solder etch and subsequently cleaned with cotton buds dipped at acetone.
Metal Etch (HCl)	150°C	-	Acid did not etch the die attach remnants
H2O2	150°C	-	Acid did not etch the die attach remnants

Table 2. Tables showing the results of die-attach remnants removal.

Hydrochloric acid (**HCl**) and Hydrogen peroxide (**H$_2$O$_2$**) have no effect on the die-attach material. While, Solder etch (**99% methanesulfonic; 70%Nitric Acid; DiH$_2$0**) was successful with an etch time of 2 minutes. However, Sulfuric acid (**H$_2$SO$_4$**) reacts faster on the die attach material with an etch rate of 15 seconds.

Figure 5. Optical photo showing the result of evaluation on 69%-70% nitric acid at 80C with some residual die-attach.

Figure 6. Optical photo showing the evaluation results of die-attach remnant removal using Sulfuric Acid.

III. Wet chemical etching versus mechanical backside sample preparation technique

Mechanical polishing of integrated circuit with plastic package uses either handheld tripod polishing tool or by using a precision polisher. The sample needs to be attached on a fixture using wax or glue. The sample should alos be flat on top of the fixture to avoid un-even polishing. If precision polisher is use, the equipment should be calibrated prior using it. The backside sample preparation technique using mechanical polishing will take almost an hour to complete the process. Backside fault isolation after mechanical polishing is only possible using 0.7/0.5um needle thru micro-probing.

Mechanical milling using precision drilling tool is another method of mechanical backside sample preparation technique. It uses precision drilling machine where you can indicate the dimension of opening on the sample. The device is positioned in sample holder to keep the unit in place during the drilling operation. Accurate dimension is needed specially the depth to be milled. Inaccurate depth will result to mechanically damaged sample. The backside sample preparation technique using mechanical milling will take almost an hour to complete the process. Backside fault isolation after mechanical milling is possible using 1.0/0.7/0.5um needle thru micro-probing at pins. Since the pins are still intact, electrical verification thru bench testing is still possible.

Wet chemical etching is an alternative method for backside sample preparation technique. It uses high temperature tape to create masking effect from the 69%-70% non-fuming Nitric Acid which is heated at any temperature in between 80^{0}C to 150^{0}C. The acid will only attack the opening created on the tape specifically aligned to the position of the die. This technique has a very low risk of inducing damage on the die. The backside sample preparation technique using wet chemical etching will take less than 15 minutes to complete the process. Backside fault isolation after wet chemical etching is possible using 1.0/0.7/0.5um needle thru micro-probing at pins. Since the pins are still intact, electrical verification thru bench testing is still possible.

Factors	Backside sample preparation technique		
	Wet chemical etching	Mechanical Milling	Mechanical Polishing
Time	Less than 15 minutes	Almost 1 hour	Almost 1 hour
Electrical Access	- Leads are still intact and analyst can easily probe. - Bench testing is also applicable for further electrical verification. - can use 1/0.7/0.5 um needle for the pin microprobing	- Leads are still intact and analyst can easily probe. - Bench testing is also applicable for further electrical verification. - can use 1/0.7/0.5 um needle for the pin microprobing	- Difficulty on probing the pads since it was polished. - Difficulty on bench testing - can use 0.7/0.5 um needle for microprobing - electrical connection is only accessible through the expose wire bond.
Cost	Less Expensive; High temperature tape, hot plate, chemicals used (69%-70% nitric acid, sulfuric acid and acetone for cleaning)	Expensive; Precision milling tool	Expensive; Precision polisher machine
Risk	Over etching of molding compound; less risk on the die.	Over milled. Mechanical damage induced on the die.	Un-even polishing; over polished die.

Table 3. Comparison of mechanical and wet chemical etching backside sample preparation technique.

IV. Failure analysis using wet chemical etching for backside sample preparation technique

The effectiveness of the backside sample preparation using wet chemical etching was tested by applying the technique to several cases. Case number S131128-034[1] and ONIT#:440614[2] with Dual Flat No-leads (DFN) package with exposed pad were taken. The devices for both case numbers are having multiple metal layers. Both cases are failing continuity test parameter, shorted.

The units for both cases have undergone backside sample preparation technique using wet chemical etching. The die backside was successfully exposed leaving the pins still intact. Fault isolation using Optical Beam Induced Resistance Change (OBIRCH) analysis was employed to locate the fault site. The unit was biased using micro-probing with 1um probe needle accessing the device through the pins. All units were successfully fault isolated.

The units die backside was easily inspected using the IR microscope. Backside IR inspection showed substrate level damage. The cases were successfully resolved with minimal time requirements for the backside sample preparation.

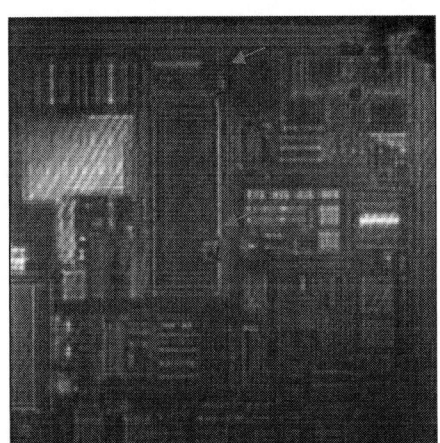

Figure 7. Backside OBIRCH fault isolation result on case number S131128-034 utilizing wet chemical backside preparation.[1]

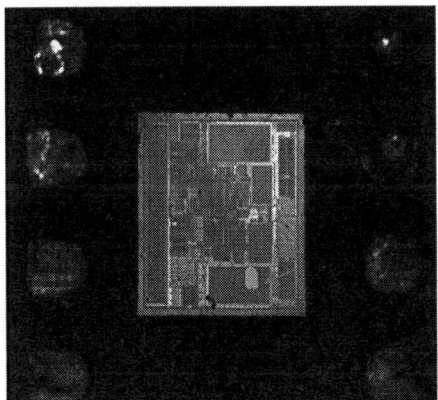

Figure 8. Backside OBIRCH fault isolation result on case ONIT#:440614 utilizing the wet chemical backside preparation.[2]

V. Conclusion

In this paper, we presented a methodology for backside sample preparation on devices with exposed pad/flag using wet chemical etching and compared it with mechanical backside polishing. Using the non-fuming nitric acid with 69%-70% concentration and any temperature between 80C to 150C, a backside sample preparation can be achieved in less than 15 minutes. With the device package temperature set to 100C, a drop of Sulfuric Acid can easily clean out the epoxy die-attach remnant.

Fast, low cost, less risk and effective backside sample preparation can be achieved using wet chemical etching on the packages with exposed pads without cashing out for machine purchase.

ACKNOWLEDGMENT

We would like to thank Nik Yusof for driving us to complete the paper. Thanks to team: Huey Shan Foo, Anton Van Alferez, Yusri and Amirul for doing the evaluation.

REFERENCES

[1] Devarajan Subramaniam, "Failure analysis report, S131128-034" On-semiconductor, Senewang, Negeri Sembilan Malaysia, December 2013.

[2] Francis Nikolai Lupena, "Failure analysis report, ONIT#:440614" On-semiconductor, Senewang, Negeri Sembilan Malaysia, February 2014.

Library Setup For Epitaxial Layer Dopant Profile Using Spreading Resistance Profiling Analysis

Lim Saw Sing, Lim Chan Way

Infineon Technologies (Kulim) Sdn Bhd

Lot 10&11, Jalan Hi-Tech 7, Industrial Phase II,

Kulim Hi-Tech Park,

09000 Kulim, Kedah.

SawSing.Lim@infineon.com, ChanWay.Lim@infineon.com

Abstract -The paper describes an approach to establish library for epitaxial layer monitoring using spreading resistance profiling (SRP) technique. This library can be used as complementary technique for conventional epitaxial monitoring such as inline four-point probe (FPP) or surface charge profiler (SCP).

I. Introduction

The control of epitaxial doping level is commonly monitored by using four-point probe (FPP) or surface charge profiler (SCP). The FFP technique is destructive method. This technique consists of two outer contact probes for current sourcing and two inner probes are used for measuring voltage drop across surface of sample [Fig. 1]. The voltage and current ratio obtained from these probes provide the sheet resistance (Rs) of the samples. This technique is fast but it is constraint by epitaxial layer conductivity must be opposite type compared to substrate. Secondly, the sheet resistance is from average resistance of epitaxial layering which unable to resolve in-depth of epitaxial layer [1].

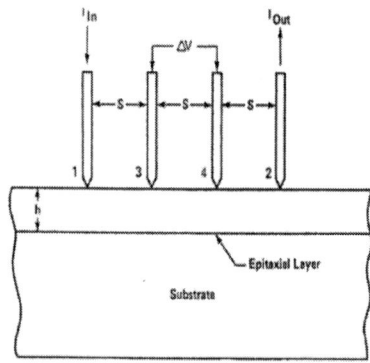

Fig. 1: Typical four points probes method [1].

On the other hand, SCP is contactless technique. The principle is using illumination to create electron-hole pair phenomenon. The electron hole pair is then separated by surface electrical field form depletion layer. The excess of minority carriers causes surface potential barrier of illumination changed. The surface photo voltage changes are detected by electrode and this signal is identified as doping conductivity and concentration [Fig.2]. This technique is non-destructive but it only provides surface dose density measurement and it is dependent on sample surface condition [2].

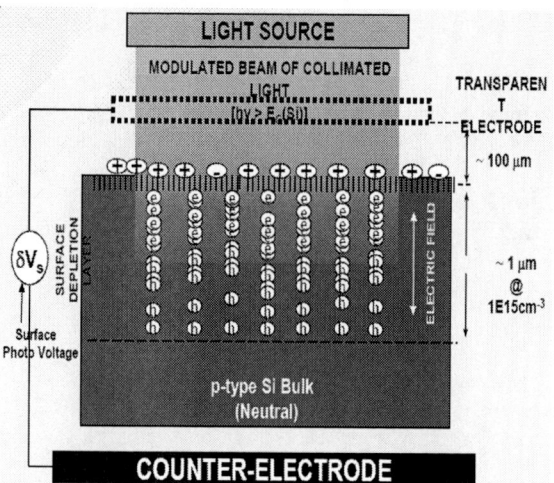

Fig.2: Surface charge profiler method

SRP is two point probes technique. The two probes' surfaces consist of stable and sufficient micro contacts. When small voltage (5mV) supply to two probes, current allow to flow between the probes, spreading resistance of sample is measured [Fig 3]. The measured resistance can be converted to resistivity parameter using known resistivity versus measurement resistance calibration curve. Once SRP resistivity is calculated, it is straightforward to determine Rs value of sample, which is reciprocal of integrated resistivity depth profile (1).

$$Rs = 1/ [\int n(x)\ dx] \qquad (1)$$

Besides, the resistivity data can be converted to dopant concentration information by using mobility formula (2).

$$n\ (x) = 1/\ (q\ \rho\ \mu) \qquad (2)$$

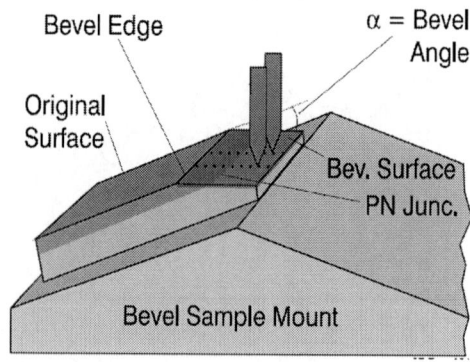

Fig.3:Spreading resistance profiling method.

II. PROBLEM STATEMENT

Average Rs measurement from inline FPP and surface dopant concentration from SCP may not sufficient to confirm deviation of epitaxial layer. Besides, stand-alone results from FPP or SCP techniques sometimes are unable to decide for wafer disposition due to no commonality found from epitaxial tool or chamber. Therefore, SRP analysis can be complementary alternative verification option. There were some studies discussed on quantification of spreading resistance probe operations [3]. However, there are limited literatures to show correlation of FPP, SCP results versus SRP technique. In order to clear the doubts of uncertainty epitaxial Rs or dose obtain from different techniques, setup epitaxial library work is demonstrated. This work will facilitate SRP analyst analyzed epitaxial layer from different type of samples versus FPP or SCP especially in case of marginal drifted.

III. EXPERIMENT SETUP

Three groups of epitaxial layer had been selected for the library setup. Group determination is based on epitaxial thickness, resistivity and different epitaxial process block. Each group consists of three samples included target resistivity, out of upper limit (USL) and out of lower limit (LSL) samples, respectively. Accuracy of quantification technique was verified using round robin sample which is control within 10%. Relative comparison can be made for target sample and out of limit sample to understand sensitivity and detectability of SRP tool.

Samples were cleaved into 2mm-3mm width and about 5mm length. Bevel angle was selected based on epitaxy thickness. Deeper bevel angle is required for thicker epitaxial layer sample. Bevel angle selection can be determined using below formula (1).

$$\Delta z = \Delta x \sin \alpha \qquad (1)$$

The samples were beveled using 0.1diamond compound with oil on 5um conditioned glass plate. The surface roughness of samples is crucial. Rougher surface normally will yield lower resistance because of poor contact being applied during measurement. To obtain reproducible surface roughness, samples were prepared using the same of slurry and keep consistent speed of glass plate rotation.

Measurement wise, reproducible electrical contact after many times of measurement is crucial to avoid any drift resistance due to poor probe condition. Therefore, probes quality check and calibration were performed prior to measurement. To gain better relative comparison and minimize the measurement variation due to different condition of probes, the samples from same group were measured in the same batch of run.

On data assessment part, the noise level was reduced using the same smoothing scheme and factor. Also, data

algorithms correction factor was standardize to ensure data extracted correctly.

III. RESULTS AND DISCUSSION

The first group (G1) is thin n-epitaxial layer on p-substrate. G1 is selected to represent the sensitivity of SRP technique on resistivity range 0.4 ohm-cm to 0.6 ohm-cm, 6 μm of epitaxial thickness. SRP result showed consistent profile of epitaxial resistivity from 0 μm up to 5 μm [Fig4]. The resistivity profile for USL and LSL samples showed shifted higher and lower respectively compared to target sample.

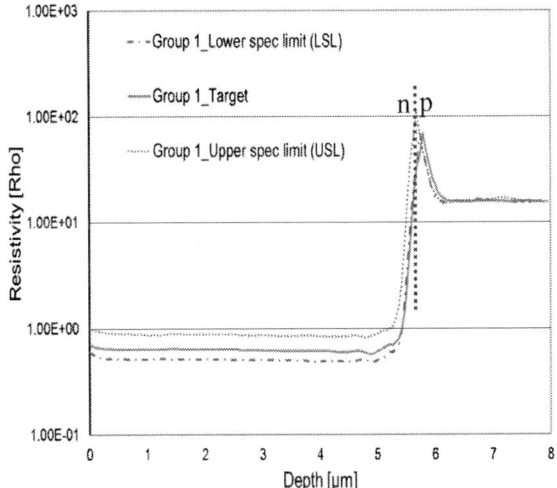

Fig. 4: SRP resistivity depth profiling for G1 samples.

Further quantify SRP Rs value for epitaxial layer from 0 μm up to PN junction (about 6 μm) showed USL-Target-LSL correlation as similar to inline FPP Rs [Fig 5]. Comparing SRP Rs to FPP Rs values, the results showed the highest delta was about 14%.

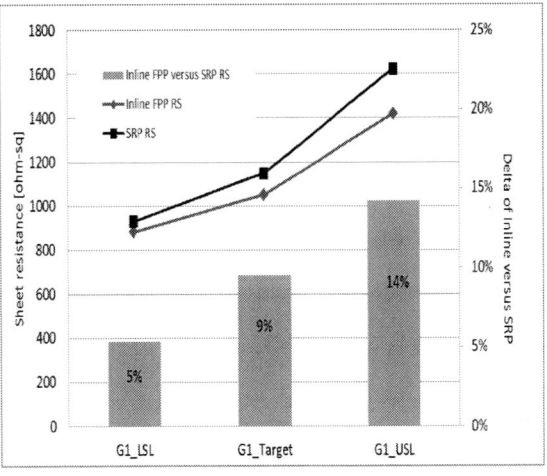

Fig.5: Correlation of FPP Rs - SRP Rs value for G1.

Furthermore, SRP Rs quantification is sensitive to resolve deviation of Epi. For example figure 6, the percentage delta of Rs value between USL-target is 16% from inline FPP technique whereas SRP Rs comparison showed 19%.

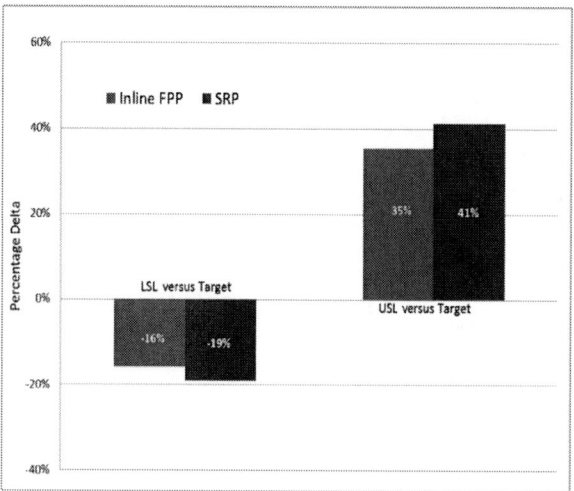

Fig.6: Relative comparison of delta for LSL-Target and USL-Target for G1 sample.

Second group (G2) is selected to represent lower resistivity range 0.09 ohm-cm to 0.2 ohm-cm of epitaxial resistivity compared to G1. The epitaxial layer is ~ 8.5 μm. SRP results again showed small gap of epitaxial resistivity profile between USL- target and LSL - target samples [Fig 7].

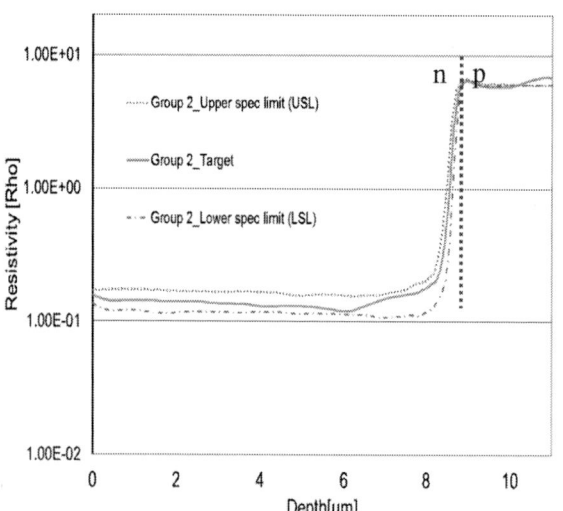

Fig. 7: SRP resistivity depth profiling for G2 samples.

The Rs quantification results from 0 μm up to PN junction (about 8 μm) also showed upward trend from LSL-Target-USL sample. Comparing the SRP Rs- FPP Rs values, results are linear correlation with highest delta was about 22% [Fig 8]. Delta of USL-target and LSL–target showed

significant difference, which were 12% and 25%, respectively which indicate SRP technique can also resolve in case if out of specification [Fig 9].

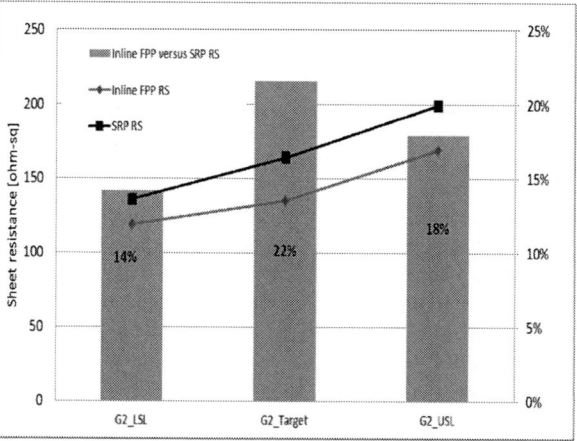

Fig.8: Correlation of FPP Rs - SRP Rs value for G2.

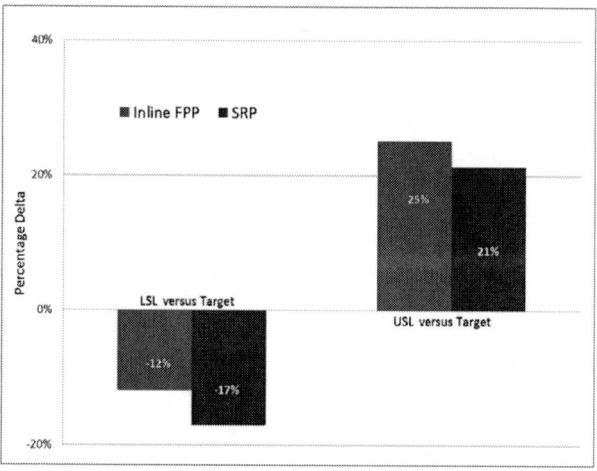

Fig.9: Relative comparison of delta for LSL-Target and USL-Target for G2 sample.

The third group (G3) is selected because their inline monitoring was quantified in term of dose density instead of Rs value. The epitaxial layer is only ~ 2 μm and underneath is buried layer. The gap of epitaxial resistivity profile between USL and LSL versus target samples can be resolved by SRP technique [Fig.9].

978-1-4799-3911-4/14 $31.00 © 2014 IEEE   100

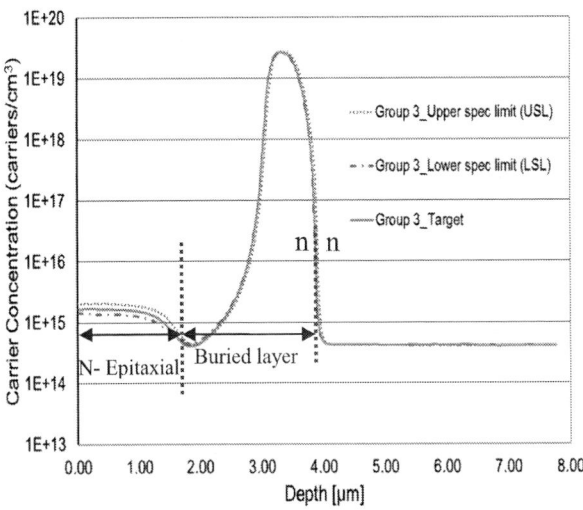

Fig.9: SRP carrier concentration profiling for G3 samples.

Further carrier density quantification from 0 μm to 2 μm showed linear correlation from samples LSL, target, and USL samples. Comparing SCP and SRP results, no consistent factor correlation was observed. However, delta of SCP-SRP technique was less 10% [Fig10]. Whereas USL-target and LSL-target SRP comparison delta were showed the difference of 17% and 23%, respectively which were much better resolution compared to SCP results [Fig.11].

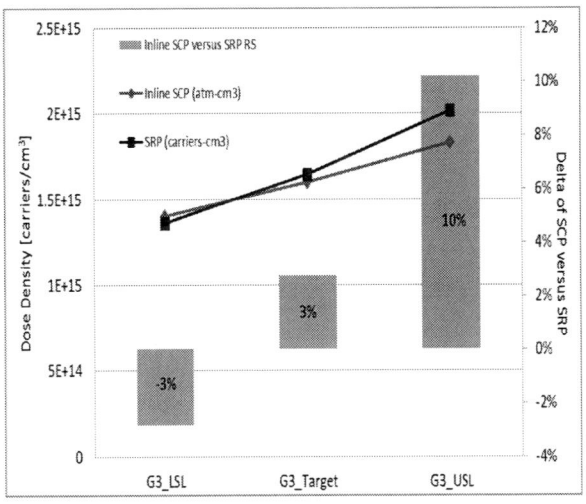

Fig.10: Correlation of SCP dopant density - SRP carrier density for G3 samples.

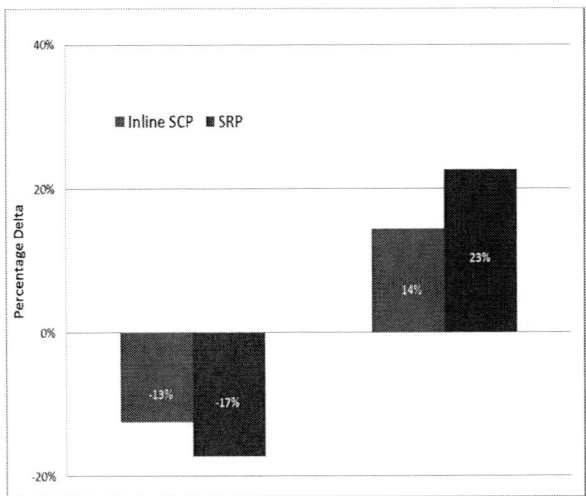

Fig.11: Relative comparison of delta for LSL-Target and USL-Target for G3 sample.

Conclusions

The workflow of setup epitaxial layer library is demonstrated. Correlation of FPP-SRP techniques showed linear correlation regardless of resistivity range and epitaxial thickness. Whereas SCP versus SRP techniques showed no direct factor of correlation but smaller delta compared to FPP-SRP relation. SRP technique is proven sensitive to resolve in case of out of specification. The epitaxial library setup work is beneficial for SRP analyst to provide conclusive results for case of out of specification and enable fast feedback for lot disposition decision.

IV. ACKNOWLEDGEMENTS

The author would like to acknowledge the fruitful discussion and guidance by Ms. Lim Siew Ping, and Mr. Yong Foo Khong. The author would also like to express his gratitude to Mr. Andreev Andrea and Mr. Tobias Bianga, Mr. Zolts Durko for his knowledge sharing.

V. REFERENCE

1. Roland Levy, Roland Albert Levy "Microelectronic Materials and Processes" Kluwer Academic Publishers (1989) [A book reference].

2. Thomas Jaehrling, "Advanced Non-Contact Metrology for Semiconductor and PV Material and Process Control" [A reference from product presentation from Semilab].

3. T.Clarysse, W.Vandervorst. " Quantification of spreading resistance probe operations (8 July 1999) [An article reference].

Decapsulation of Silver-Alloy Wire-Bonded Devices

Francois KERISIT[1,2], Matthew J. Lefevre[1],Bernadette DOMENGES[2],Wilfrid PRELLIER[3], and Michael OBEIN[1]

[1]Digit Concept, Chemin de St Sulpice, 14740 Secqueville en Bessin, France

[2]LaMIPS, CRISMAT-NXP semiconductors-Presto Engineering joint laboratory,
CNRS-UMR 6508, ENSICAEN, UCBN, Presto-Engineering Europe, 2 rue de la Girafe, 14000 Caen, France

[3]CRISMAT Laboratory, CNRS-UMR 6508, ENSICAEN, UCBN, 6 Bd du Maréchal Juin, 14050 Caen, France

f.kerisit@digit-concept.com, +33 2 31 354 354

Abstract

In order to reduce costs and improve the bonding process, silver has been recently introduced as an alternative to common bonding wire metals (gold, aluminum, copper), leading to new failure analysis issues. This study compares the efficiency of wet and dry chemistries for decapsulation on three Ag-based alloy wires.

Introduction

New developments of silver alloy bonding wire have emphasized specific problems due to silver alloy properties. Whereas this new type of wiring materials seemed to fulfill most challenges, like physical propertiesand reliability [1,2],it was suggested that epoxy molding compound (EMC) should be adapted in order to ease decapsulation[3]. Indeed, afterthe packaging industry moved from gold to copper wires, the failure analysis community had to come up with new decapsulation techniques [4,5]. Again, a new type of bonding will raise new problems of decapsulation. Furthermore, people facing failure analysis cases do not always have all required information on the type of EMC and the true composition of the bonding wires.

Twomajor techniques of decapsulation regarding wire-bonded devices are known: wet (acid) and dry etching (plasma). LASER ablation or milling are used for pre-opening, This study will compare the capabilities of these techniques on three different types of Ag-based wiring integrated circuits(IC).

Experimental

Three types of ICs were chosen because of their different silver alloys wiring (Table 1). All samples were pre-opened using an infra-red pulsed laser system (SesameLASER) to reduce the process time for both wet and dry chemistry (Figure 1). Wet chemistry was performed with automated equipment running on nitric and sulfuric acid (SesameACIDECu). Dry chemistry was performed with an RF plasma chamber running on three gases at 13.56 MHz (SesamePLASMADcap).

The decapsulated samples were controlled with first optical microscopy and second scanning electron microscopy (SEM ensures much better resolution), looking for any damage on the wires and/or chips.

IC #	Dia. (µm)	Ag (%)	Au (%)	Pd (%)
1	20	87-90	n/a	n/a
2	18	95	4.5	0.5
3	15	90	10	-

Table 1: Wire diameter and composition for each IC type (supplier data).

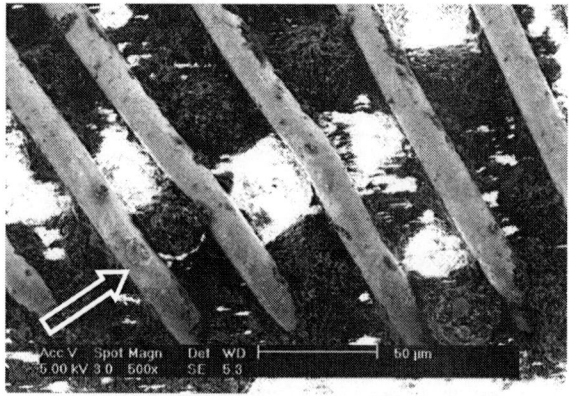

Figure 1: SEM picture of undamaged wires after laser ablation, exhibiting only one crater (arrow).

1. Wet Mixed Chemistry

The main method for plastic package decapsulation is wet chemistry, mixing nitric and fuming (f) or non-fuming (nf) sulfuric acids. This removes EMC efficiently while preserving Au wires. But, like copper, silver is attacked by nitric and sulfuric acids, see reactions (1) and (2).

$$3\,Ag + 4\,HNO_3 \rightarrow 3\,AgNO_3 + 2\,H_2O + NO \quad (1)$$
$$2\,Ag + H_2SO_4 \rightarrow Ag_2SO_4 + H_2 \quad (2)$$

Based on our experience of copper wired ICs, we first tried three different recipes for opening type #1 device (Table 2), varying the nitric (N) to sulfuric (S) acid ratio, and temperature.

Recipe	HNO$_3$	H$_2$SO$_4$	Ratio N:S	Temp.
1	100% f	20% f	3:1	10 °C
2	100% f	95% nf	1:4	100 °C
3	-	95% nf	0:1	180 °C

Table 2: Wet mixed chemistry recipes for IC #1

In all cases, most of the wires were either severely thinned or completely broken through by the time the die was fully exposed. The idea of the 10 °C opening was of course, as in the case of Cu wires, that the lower temperature would reduce the activity of the acid attack more on the wires than on the EMC, thus allowing us to etch down to the die before the wires would be significantly damaged [4]. It appears, unfortunately, that the Ag wires are even more sensitive to acid attack than the Cu ones. We then tried the alternate Cu recipe, which is the 1:4 mixture at 100 °C, using normal

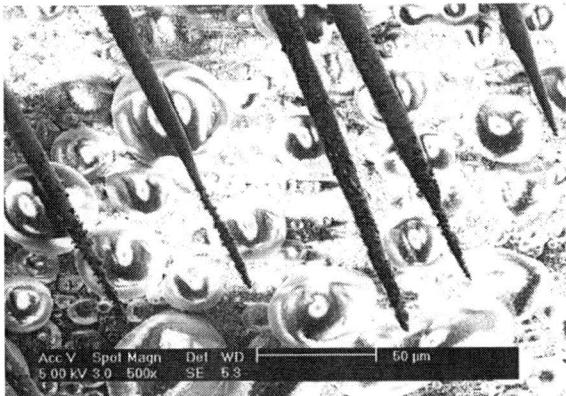

Figure 2: Thinned and broken wires after acid decapsulation, type #2 IC

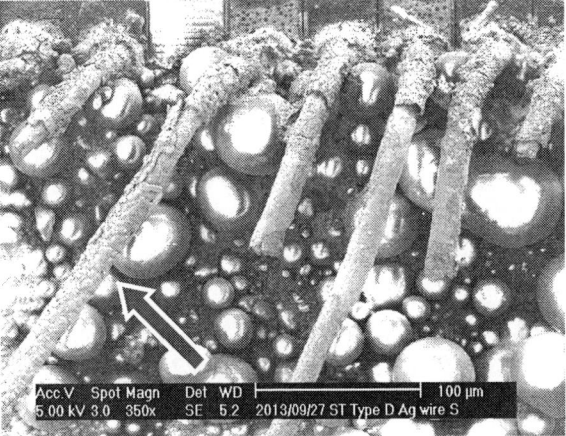

Figure 3: SEM image after plasma decapsulation at 100 W, type #2 IC; broken wires and some wires are still embedded in resin (arrow)

sulfuric acid as the fuming variety is too reactive at higher temperatures. This recipe resulted in all the wires either being completely eaten through or severely attacked and discolored. Finally the pure sulfuric recipe was tried as silver jewelry is often cleaned using a dilute sulfuric acid solution (2%). Such a dilute solution is not reactive enough to remove the EMC however, so normal H2SO4 95% was used at 180 °C (sulfuric acid becomes active enough to etch efficiently EMC only around 160 °C). Unfortunately, most of the wires were broken and discolored even before reaching the die.

To ensure EMC removal, sulfuric acid activity can be enhanced with increasing temperature or nitric acid proportion, keeping in mind that silver is more sensitive than copper. For type #2 IC, containing more Ag, we only tried a 2:1 ratio at 10 °C for 120 s test sample. Likewise, wires are attacked, severely thinned and broken (Figure 2).

Therefore, we went on processing the samples with plasma decapsulation.

2. Dry etching

Samples from type #2 and #3 IC have been processed in the plasma chamber. The first trial conditions used for plasma etching were those recommended by the plasma manufacturer, that is: 40% of CF4 - 60% of O2 gas mixture, 100W power, temperature of 80 °C, 4000 mTorr pressure and 10 minutes cycles with a CO2 blast between each cycle.

On type #2 ICs, this plasma appeared not selective enough as a lot of resin can still be observed at the surface of the wires but also too reactive as some wires were completely broken (Figure 3). At first sight, images might as well suggest that wires are strongly attacked by the plasma and the uneven aspect of their surface is due to reaction product between Ag and the plasma. This was invalidated by the measurement of the wire diameters and Energy Dispersive X-ray Spectrometry analyses.

Therefore, the power was reduced down to 50 W. Only one wire was cut but much more resin is left. (Figure 4).

For type #3 IC, the sample was processed at 50 W also. All wires are attacked, yet none are broken. But we observe cracks on the balls and passivation damages when trying to fully expose the pads (Figure 5).

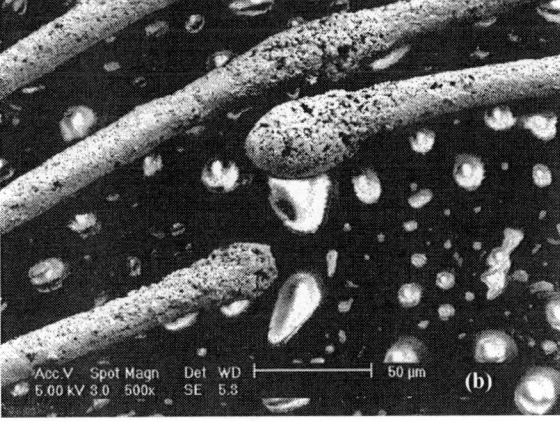

Figure 4: SEM images after plasma decapsulation at 50 W and 80 °C, type #2 IC. a) wires are still embedded in resin, b) only one single broken wire

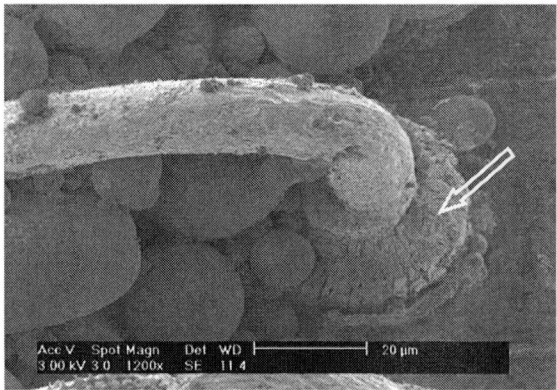

Figure 5: SEM image showing cracks on the ball, type #3 IC

In order to increase selectivity between wire and resin, we tried to use the equipment out of the specifications recommended by the manufacturer. Therefore we lowered the temperature of the chiller down to 25°C. Moreover this should supposedly lower the temperature at die level. This decrease in temperature will make the plasma also less reactive and then reduce the resin etch rate. A total etch time of 200 minutes was necessary to completely expose the die. On one hand, it is a very long duration for a single sample. On the other hand, this allows a step by step procedure with a precise control over the etching. Table 3 summarizes etch times for acid and plasma etching, showing that a lower reactivity induces a longer etch time but better results.

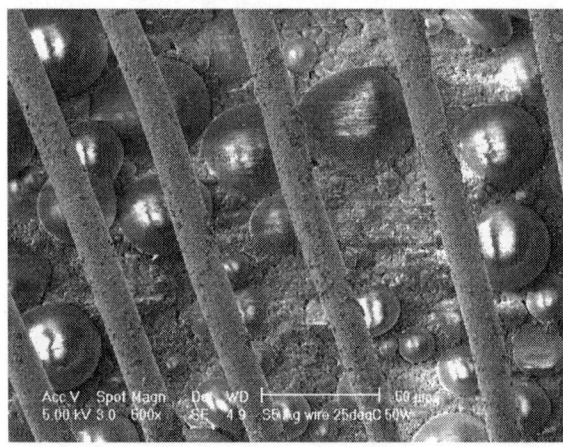

Figure 6: SEM image of slightly attacked wires after plasma decapsulation at 50 W and 25 °C, type #2 IC.

SEM observation shows that none of the wire is broken (Figure 6). The surface appears to be slightly modified, but the overall result is good. Small spheres remain on the wires after etching (Figure 7). Investigation with EDXS-SEM showed it was mostly silicon, therefore remaining silica fillers. Wire diameters have been measured to show to that when EMC has been removed, and then the wire is etched, as the diameter decreases (Table 3). A reference value of wire diameter was obtained after laser ablation, while wires are partially exposed.

Figure 7: SEM close-up image of wires with remaining silica fillers after decapsulation at 50 W and 25 °C, type #2 IC.

Technique	Mean Dia. (µm)	Diameter Range	Etch time (min)
Laser*	16.0	15.5 – 17.0	0.5
Acid	n/a	n/a	2
P. 80°C 100W	17.2	15.8 – 20.0	20
P. 80°C, 50W	19.2	16.2 – 21.4	60
P. 25°C, 50W	14.8	14.1 – 15.3	200

Table 3: Measured wire diameter, type #2 IC
* laser ablation pre-opening on all samples

Conclusion

This study shows that acid decapsulation is not successful for all types of our Ag-based wiring devices and might also depend on the EMC. Indeed the EMC will define the acid recipe used, and therefore the etch time and activity of the mix on the wires. Plasma decapsulation seems to offer more latitude regarding EMC etching, yet still damaging the Ag wires.

We showed that wires can be exposed using proper plasma parameters. This solution might damage sensitive die passivation, but it already provides a way to inspect the package and to do mechanical tests (*i.e.* shear and pull tests).

References

1. Kai, L.J. et al, "Silver alloy wire bonding," *62nd Electronic Components and Technology Conference*, 2012 IEEE 62nd, San Diego, CA, May 2012, pp.1163-1168.
2. Chen, Q.J. *et al*, "Cu wire and beyond - Ag wire an alternative to Cu?," *12th Electronics Packaging Technology Conference*, Singapore, Dec. 2010, pp. 591-596
3. Lan, A. *et al*, "Interconnection Challenge of wire bonding – Ag alloy wire," *Proc.15th Electronics Packaging Technology Conference*, Singapore, Dec. 2013.
4. Lefevre, M. *et al*, "New method for decapsulation of copper wire devices using LASER and sub-ambient temperature chemical etch," *Proc. 37thInternational*

Symposium for Testing and Failure Analysis, San Jose, CA, Nov. 2011, pp. 248–255).

5. Murali, S.; Srikanth, N., "Acid Decapsulation of Epoxy Molded IC Packages With Copper Wire Bonds," Electronics Packaging Manufacturing, IEEE Transactions on Electronics Packaging Manufacturing, vol.29, no.3, pp.179,183, July 2006

High Energy Electron Degradation of the Bonding Connections

P. R. Laskowski[a], M. Olszacki[a], M. Al Bahri[b], P. Pons[b], M. Jasiorski[c]

a) National Centre for Nuclear Research, 05-400 Otwock, POLAND
b) Université de Toulouse ; UPS, INSA, INP, ISAE ; LAAS ; F-31077 Toulouse, FRANCE
c) Wroclaw University of Technology, 50-370 Wroclaw, POLAND
Phone: +48604069141, Fax: +48 22 77 93 481, Email: piotr.laskowski@ncbj.gov.pl

Abstract

The influence of the high energy electrons on the wire bonding connections has been investigated. To degrade the devices 6MeV electron source has been employed. The irradiation resulted in both: bonding rupture force and ohmic resistance decrease. Device degradation level was additionally evaluated by means of the electron microscopy inspection.

I. INTRODUCTION

Reliability of the devices operating in the high radiation environments is still being an influential obstacle in the design process. Space applications and some of the medical ones, demand high endurance of the electronic equipment to destructive irradiation effects. Most of the literature is focusing on the side of the semiconductor elements of the device; however, for the complete functionality, the wire bonding aspect cannot be neglected. There were some attempts do standarize the FA tests under irradiation like Sr-90/Y-90 electron irradiation of epoxy encapsulated devices in order to induce the ionization in the thin oxide layers, therefore the main impact was focused on the chip and not on the wire bonding [1,2]. Another attempt which was more focused on the bonding aspect was the irradiation of the solder bumps in order to check the mechanical stability of the bumps and their adhesion to the substrate [3]. However, it has to be emphasized that none of these works was focused on the wire bonding. In this paper we would like to present a case study of the devices irradiated by a high energy electron flux. To characterize the influence of the radiation two parameters have been measured: the resistance and the rupture force of the bonding wires. For both of the factors downward trend has been revealed. For the root cause explanation of the phenomena that have occurred in the DUTs the additional electron microcopy inspection has been conducted. Finally, in the last chapter of this paper all the results were summed up and concluded.

II. MEASUREMENT SETUP AND SAMPLES

Irradiation has been performed in the test bench consisting of a linear accelerator. The tool has been designed and produced in National Centre for Nuclear Research in Poland.

Prior to measurements, the test bench has been calibrated with the Farmer Ionization Chamber method (Fig. 1.).

Fig. 1. Linear accelerator during calibration process.

During measurements samples were placed inside a special chuck-holder to ensure repeatability of irradiations. Samples have been irradiated by high energy electron (6 MeV) beam with 7mm spot size. To verify the destructive influence of the radiation two parameters have been measured. For the resistance measurements probe station and a digital multimeter have been employed. The bonding rupture force inspection was performed by means of the micro tensile testing machine (Fig. 2.).

978-1-4799-3911-4/14 $31.00 © 2014 IEEE

Fig. 2. Device under test (DUT) during the bonding rupture process.

Each of the wires was caught by micro hook and pulled up till the moment of rupture, with a displayed force value.

The DUTs (Fig. 3.) were the pressure sensors fabricated using standard MEMS process, where the PYREX 7740 glass was micromachined, using deep HF etching. Then the Al electrode was deposited at the bottom of the cavity and the external contact was formed. As next, the silicon wafer was KOH etched in order to form 45 um thick membranes and then, anodically bonded to the glass. At the end the cavity was sealed using parylene coating. The sensors were glued to the TO3 case using standard epoxy based glue. Both the case pins and the wires were made of aluminum. For the connections, the wedge bonding technique has been applied and the diameter of the wires was around 25μm each. However, for our investigation, these were only employed as the bonding examples.

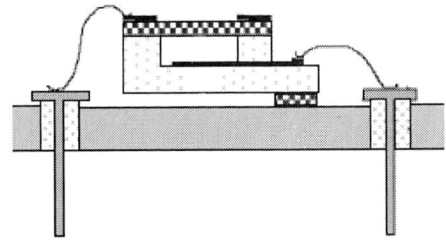

Fig. 3. Cross-section schematic of the DUT.

III. EXPERIMENTAL AND DISCUSSION

Since the tests of this paper were destructive, all the DUTs were divided in two groups. First group of the devices was assigned to irradiation process and in oposite, the second saved for the repture tests without any additional stress influence. The only parameter that could be measured before and after radiation stress exactly for the same sample was the ohmic resistance of the wires. After resistance meaurements all the DUTs assigned to the first group were placed one by one into electron accelerator chamber. The flux was calibrated to cover all the wires by a 7mm diameter spot size. The radiation dose delivered to the devices was equal to 100kGy. After the irradiation process electrical resistance measurements were repeated. As a final stage of the research the rupture tests were performed. DUTs were placed on the microscope stage and by means of the micro hook each of the wires were caught possibly closest to the centre and pulled up till the rupture occurred. The forces of each rupture were measured and displayed by the testing tool.

For the lucid visualization of the obtained data, these were divided into two groups assigned as: dev1 and dev2, where dev1 is equal to non-irradiated devices and dev2 is equal to the devices to which 100kGy dose has been applied. The measurement data is presented in Fig. 4. To expedite the data analysis both the resistance and the rupture force were plotted in one graph. Analyzing the values one can observe that for the irradiated devices both parameters represent downward trend.

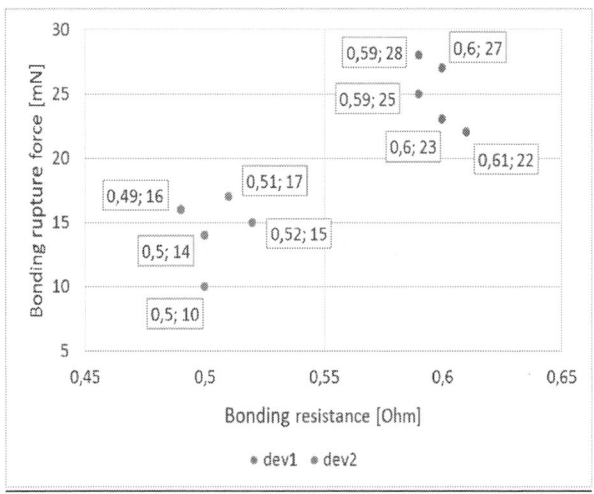

Fig. 4. Bonding rupture force vs. bonding resistance for degraded and non-degraded devices

Delivering 6MeV energy to a single device results in a number of physical phenomena. Firstly the ionization process occurs. In case of some materials this phenomenon has a significant influence on their physical properties and is able to modify permanently the crystal lattice structure; however, in case of the metals and especially aluminum the influence of this process on its permanent properties change has no influence and could be negligible for this research. Second phenomenon that should be taken into account is additional current generation. In case of the tested devices and analyzing the 6MeV energy and 100kGy dose mean currents that can be generated, is in the 100μA range; therefore similarly to the ionization process could be neglected. The third phenomenon that typically is recognized as the consequence of the irradiation is the crystal lattice thermal excitement and the mentioned process was crucial for the research of this paper. Electrons are the smallest among the atomic particles; however since their energy was equal to 6MeV, these were able to permanently modify the crystal lattice. The electrons while hitting the Al crystal, stimulate the lattice vibrations. Kinetic energy of the electrons is transferred into these vibrations and finally into a local heat generation. Since the temperature gradient as a time function is relatively high, it results in grain boundary melting. In consequence one can notice grain size growth (Fig. 5.), which results in conductivity increase and as a final consequence, resistance decrease [4,5].

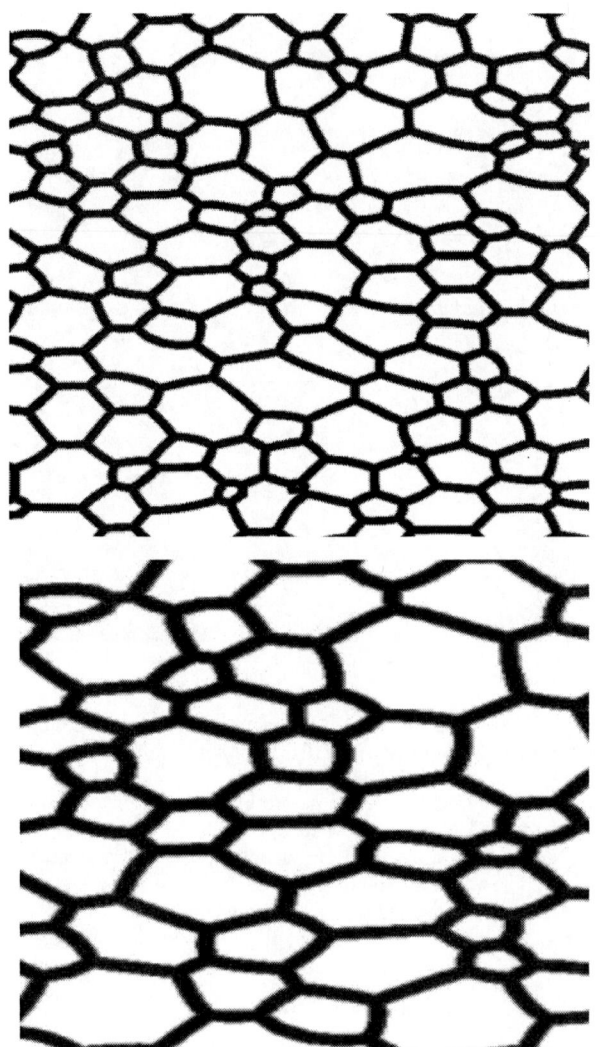

Fig. 5. Grain size growth caused by the irradiation process.

On top of that, the higher grain size, the stiffer the crystal lattice and finally brittleness increases. This phenomenon was related to the decrease of the bonding rupture force. To support this theory additional scanning electron microscopy inspection has been conducted. In Fig. 6 one can notice two images related to dev1 – non-irradiated sample.

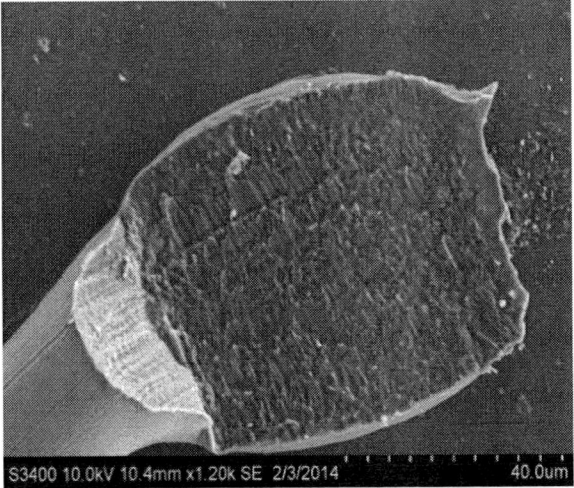

Fig. 6. Electron microscopy image for non-irradiated bonding.

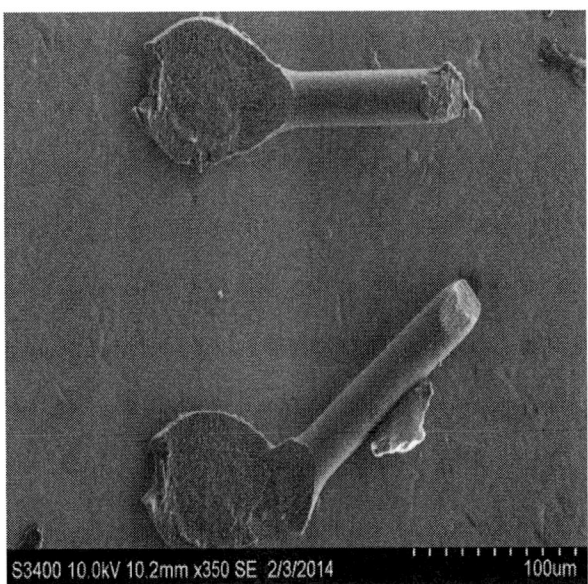

Fig. 7. Electron microscopy image for irradiated bonding.

Additionally electron microscopy inspection has been conducted. Finally, the root causes of such material behavior have been highlighted and discussed. Performed research confirmed that high energy electron irradiation process of the bonding wires resulted not only in their ohmic resistance reduction but also in decrease of their resistance for the rupture stress testing.

ACKNOWLEDGEMENTS

We would like to thank our colleague Dr. Slawek Wronka and his team members for their support during the accelerating process and Mr. Teodorczyk from ITME for his help with the testing. Special thanks go to Lukasz Potrzebka for his remarkable tips.

And in Fig.7. a picture of the irradiated device (dev2) is exposed. In the properly bonded device the strength of the solder is around 25mN and the strength of the wire around 100mN, therefore in case of the non-irradiated devices the ruptures always take place on the junction point of the soldering. In case of the irradiated devices the situation looks thoroughly different. Since the wire brittleness significantly increased the solders remain almost untouched while the bonding rupture took place on the wire side.

Presented work was partially financed by the ERA-NET MNT DOSIMEMS project.

REFERENCES

[1] G. Thuesen, P. Buch Guldager, J. Leif Jørgensen, "Application Specific Radiation Tests for Cots EEE Components", *4th International Symposium of the International Academy of Astronautics*, pp. 353-356, 2003.

[2] ASTM F 1263-94. Standard Guide for Analysis of Overtest Data in Radiation Testing of Electronic Parts.

[3] S. Kwan, J.A. Appel, G. Chiodini, D. Christian, S. Cihangir and F. Reygadas et al., "A Study of Thermal Cycling and Radiation Effects on Indium and Solder Bump Bonds", *FERMILAB-Conf-01/377-E*, December 2001.

[4] M.E. Day, M. Delfino, J.A. Fair, W. Tsai, "Correlation of electrical resistivity and grain size in sputtered titanium films", *Thin Solid Films Volume 254*, Issues 1–2, pp. 285-290, 1 January 1995.

[5] M. Hosseini, B. Yasaei, "Effect of Grain Size and Microstructures on Resistivity of Mn^Co^N. Thermistor", *Ceramics International 24* (1998) 543±545

IV. SUMMARY AND CONCLUSION

The influence of the high energy electrons on the aluminum bonding has been investigated. The radiation related degradation resulted not only in the ohmic resistance but also bonding rupture force decrease.

An unique sample pretreatment method for circuit edit on WLCSP devices

Sheng-Yu Chen, Chino Wang, and Puma Wu
Integrated service technology
No.19, Pu-ding Rd., Hsinchu 30072, Taiwan
sy_chen@istgroup.com

Abstract

In the present work, the thick organic passivation layer of WLCSP products was uniformly removed before the focused ion beam (FIB) circuit edit (CE) process. By this approach, the cycle time of the whole CE process can be remarkably decreased and the successful rate can be effectively increased. In addition, the suggested flows for successful CE under solder bumps and RDLs have been proposed.

I. CHALLENGE OF FIB CE ON WLCSP PRODUCTS

FIB is a widely used technique for CE and IC debug [1]. For ICs of conventional package types, FIB CE can be performed after a simple decap procedure. However, for wafer-level chip-scale-package (WLCSP) products, FIB CE would encounter difficulties due to the presence of thick organic passivation layers at package-level. A previous investigation has proposed a method for FIB CE on WLCSP devices [2]. In their study, the organic passivation layer was directly milled by FIB, to expose the target metal for edit. However, it was time-consuming and may impact yield.

Figure 1 shows a cross-sectional FIB micrograph of a WLCSP sample. The thickness of the package-level organic passivation layer is as high as 25 μm, which is unfavorable for FIB CE on chip-level interconnects. The high aspect-ratio contact window formed by direct FIB mill would degrade the image quality at the bottom of contact and cause discontinuity in metal deposition. This can be seen in the inset of Fig.1, which is a 30°-tilt FIB micrograph of a FIB milled contact window. Both the image quality and metal deposition issues would impact the successful rate of CE. Furthermore, the whole CE process is time-consuming.

Beside the issue of the thick organic passivation, the presence of solder bumps and Cu RDLs in WLCSP products would block the target metal for FIB CE, as illustrated in Fig. 2. In the present work, we uniformly removed the thick organic layer before the FIB CE process. As a result, the cycle time of whole CE process can be remarkably decreased and the successful rate can be effectively increased. In addition, the suggested flows for successfully CE under solder bumps and RDLs have been proposed.

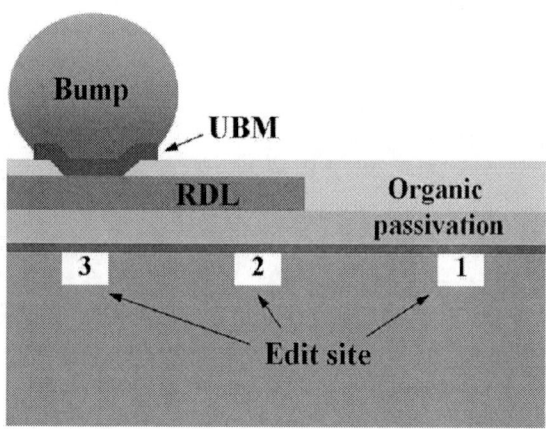

Fig. 2. A cross-sectional schematic illustration of a WLCSP product. The edit site-1 corresponds to the case of the target metal blocked by organic passivation layers only. The edit site-2 corresponds to the case of the target metal blocked by RDLs and organic passivation layers. The edit site-3 corresponds to the case of the target metal blocked by a solder bump, RDLs, and organic passivation layers.

II. EDIT SITES BLOCKED BY THICK ORGANIC PASSIVATION LAYERS

There are two main problems in removing organic passivation layers by directly FIB mill. First, the yield of CE would suffer impact due to the high aspect ratio contact window. Second, the whole procedure is time-consuming. The problems can be solved if the thick organic passivation layers can be uniformly removed before FIB CE procedure without yield impact. There are many investigations in removing organic passivation layers [3-5]. In general, the organic passivation can be uniformly removed by chemical wet etching process or plasma dry etching process. Figure 3 shows the top-view FIB micrograph of the WLCSP sample after uniformly removal of whole organic passivation layers. All

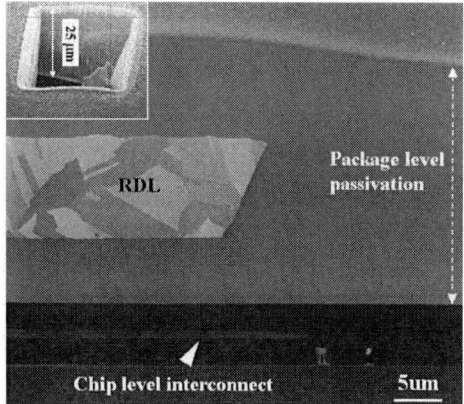

Fig. 1. A cross-sectional FIB micrograph of the WLCSP sample.

978-1-4799-3911-4/14 $31.00 © 2014 IEEE

the solder bumps and RDLs didn't suffer impact during the passivation removing process. There's also no impact on the IC's electrical performance. After the organic passivation layers were uniformly removed, the following FIB CE process would be more reliable and time-saving.

Figure 4 shows a FIB connecting process on the passivation-removed sample. Figure 4(a) shows a FIB micrograph of two FIB contact windows formation. The target metals at chip-level have been exposed inside the contact windows. The thickness of the remained package-level passivation layer and the underlying chip-level passivation layer are about 2 μm and 1 μm, respectively. The contact window with small aspect ratio (about 0.3) ensures reliable metal deposition and good image quality inside it. Comparing to the direct FIB mill method, there are three remarkable advantages: First, the cycle time of each contact window formation can be shorten from 1.5 hr to 0.5 hr. Second, the cycle time of each contact deposition can be shorten from 30 min to 5 min. Third, the continuity of the metal interconnect can be improved, as shown in Fig. 3 (b).

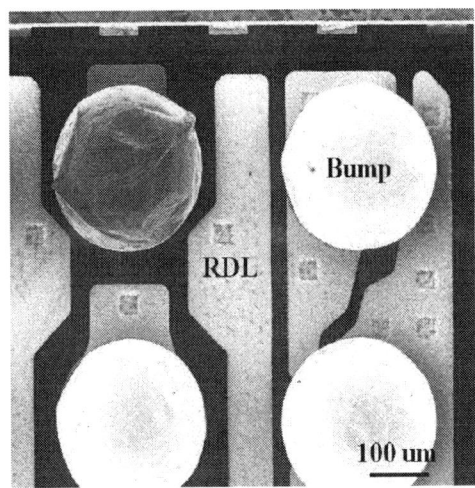

Fig. 3. A top-view FIB micrograph of the WLCSP sample after the organic passivation layers were uniformly removed.

(a)

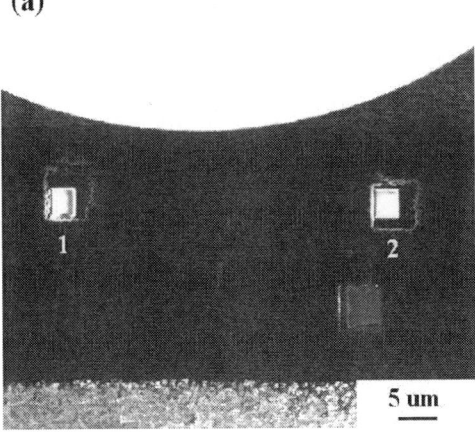

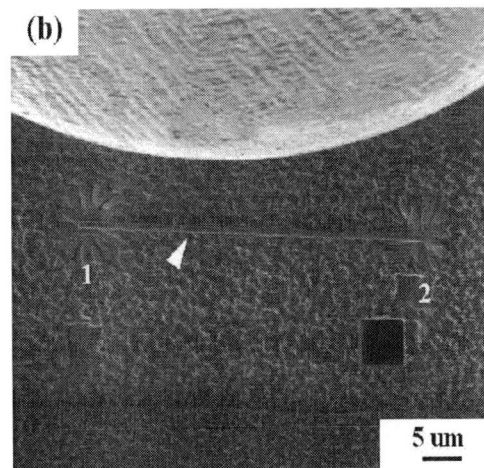

Fig. 4. Top-view micrographs of a FIB connecting process on the organic passivation-removed sample: (a) contact windows formation and target metal exposed, and (b) contact deposition and interconnect formation.

III. EDIT SITES BLOCKED BY RDLS

When the edit site is blocked by RDLs, as illustrated in Fig. 2 (edit site-2), the whole circuit edit procedure would become complicated and time-consuming. It is always better to avoid designate edit sites under RDLs. However, sometimes it would be inevitable to perform CE under RDLs to meet some specific debug or FA requirements.

In the present work, a suggested flow for CE under RDLs can be divided into several steps: First, the 1st organic passivation layer was uniformly etched away. Second, the Cu RDLs was partially etched away by FIB, to form a rectangular contact window. The H_2O-gas assisted etching (GAE) was utilized in this step to enhance the etching rate of Cu. Third, the 2nd organic passivation layer was partially etched away in the same contact window by FIB. The GAE gas utilized in this step is also H_2O. Before the fourth step, all the blocking layers in package-level were removed. Then the chip-level passivation can be etched away by the XeF_2 GAE process and the target metal for edit can be exposed inside the contact window. Figure 5 is a schematic illustration of the suggested flow. It's worth mentioning that the area of the contact window should not be too small since the depth of the contact window could be up to several tens μ m. The image quality at the bottom of the contact would degrade to impact yield if the contact aspect ratio is too high. The aspect ratio of contact windows in such case is suggested to be less than 2.

Figure 6(a) and 6(b) show the target metal to be exposed and cut, respectively. To form the 30 μ m x 20 μ m contact window, the cycle time of RDL removal, 2nd organic passivation removal, and chip-level passivation removal are 60 min, 30min, and 10min, respectively. The most time-consuming step is RDL removal. Reducing contact area can significantly shorten the whole cycle time. However, it would be a tradeoff between cycle time and yield in determining the contact area.

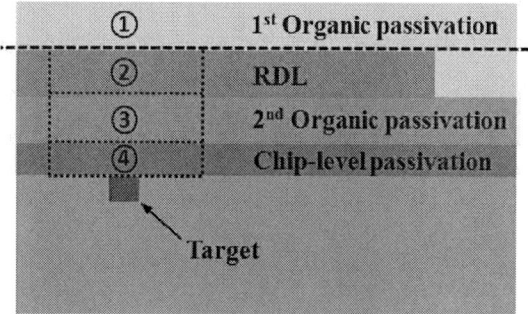

Fig. 5. A schematic illustration of the suggested flow for the case of target metals blocked by RDLs. The first step represents 1st organic passivation removal. The second step represents RDL removal. The third step represents 2nd organic passivation removal. The fourth step represents chip-level passivation removal.

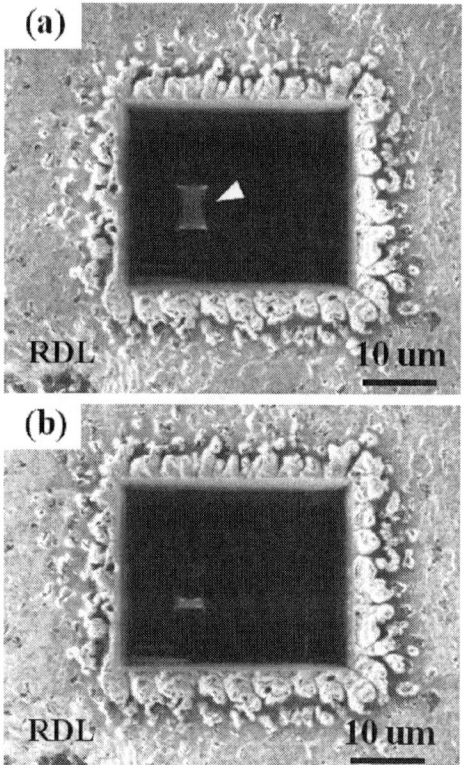

Fig. 6. (a) The target metal (as labeled) is exposed, and (b) after cut, inside the contact window.

IV. EDIT SITES BLOCKED BY SOLDER BUMPS

The most complicated case for FIB CE on WLCSP products is when the target metal is blocked by solder bumps, as illustrated in Fig. 2 (edit site-3). In the present study, we have overcame such problem and successfully performed FIB CE under a solder bump, as shown in Fig. 7. A suggested flow for CE under solder bumps was proposed. The steps of solder bump removal and re-attachment are key in the flow.

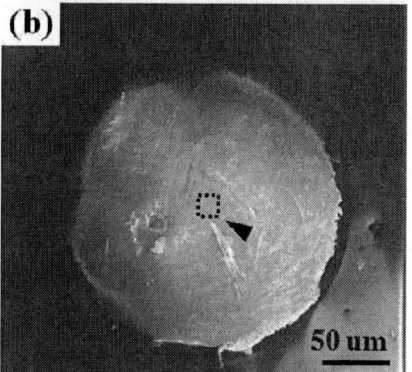

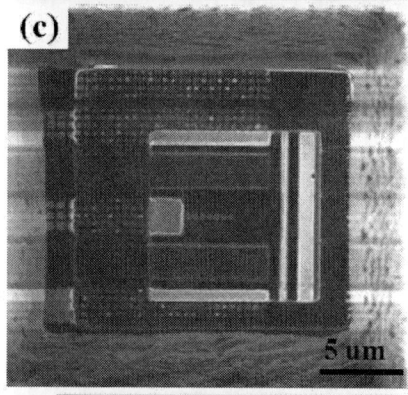

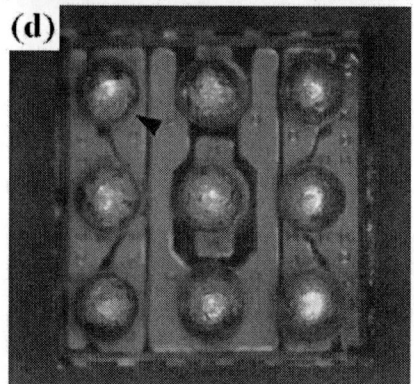

Fig. 7. (a) An OM image shows the solder bump was removed (as labeled). (b) An room-in FIB micrograph show the surface morphology after the bump was removed. The rectangle of dotted line defines the edit site. (c) An room-in FIB micrograph of the edit site. The target metal is exposed. (d) An OM image taken after another solder bump was attached.

The procedure of FIB CE under a solder bump can be divided into several steps. First, the solder bump which blocks the target metal was removed by a physical cutting process, as shown in Fig. 7(a) and 7(b). Second, the contact window can be then formed through the multiple layers composed of UBM, RDL, organic passivation, and chip-level passivation, to expose the target metal, as shown in Fig. 7(c). After the CE was done, the SiO_2-based insulator layer deposited by FIB was used to cap all exposed metals inside the contact. The final step is to attach another solder bump onto the UBM by local heating, as shown in Fig. 7(d). The processed IC can pass the final test (FT) and meet the designer's requirement.

V. CONCLUSIONS

In the present work, the thick organic passivation layer of WLCSP products was uniformly removed before the FIB CE process. By this approach, the cycle time of whole CE process can be remarkably decreased and the successful rate can be increased. In addition, the suggested flows for successful CE under solder bumps and RDLs were proposed. As a result, there will be no limitation for FIB CE on WLCSP products.

REFERENCES

[1] D.M. Donnet and H. Roberts, "FIB applications for semiconductor device failure analysis," *Microscopy of Semiconducting Materials*, vol. 107, pp. 403-408, 1995.

[2] T. C. Liu, C. Chen, S. T. Liu, M. L. Chang, and Jandel Lin, "Innovative methodologies of circuit edit by focused ion beam (FIB) on wafer-level chip-scale-package (WLCSP) devices," *J. Mater. Sci: Mater. Electron* , vol. 22, pp. 1536–1541, 2011.

[3] B. Mimoun, H. M. Pham, V. Henneken, and R. Dekker, "Residue-free plasma etching of polyimide coatings for small pitch vias with improved step coverage," *J. Vac. Sci. Technol. B*, vol. 31, pp. 21201, 2013.

[4] W. H. Juan and S. W. Pang, "High aspect ratio polyimide etching using an oxygen plasma generated by electron cyclotron resonance source," *J. Vac. Sci. Technol. B*, vol. 12, pp. 422, 1994.

[5] G. Turban and M. Rapeaux, "Dry etching of polyimide in O_2-CF_4 and O_2-SF_6 plasmas," *J. Electrochem. Soc*, vol. 130, pp. 2231-2236, 1983.

ICP-RIE Platinum (Pt) Sputter Etching

R. G. Mendaros, M. T. Marcelo

Worldwide Product Analysis

Analog Devices General Trias (ADGT)

Gateway Business Park, Brgy. Javalera, Gen. Trias, Cavite, Philippines 4107

Phone: +63 2 867703, Email: raymond.mendaros@analog.com, Mary-Jane.Marcelo@analog.com

Abstract-One of the coating materials that is used to reduced electron charging effect during Scanning Electron Microscope (SEM) imaging is Platinum (Pt). Removing Pt coating for parts requiring further electrical testing or deprocessing has been a challenge in failure analysis. This paper discusses the established methodology in removing Pt coating using the ICP-RIE sputter etching technique.

I. INTRODUCTION

In semiconductor failure analysis, several coating materials are used to reduce electron charging effects during Scanning Electron Microscope (SEM) imaging. The most common coating materials are gold (Au), gold/palladium (Au/Pd), platinum (Pt), silver (Ag), chromium (Cr) and iridium (Ir) [1]. Electron charging is the buildup of negative charges on a specimen irradiated with an electron beam. Charging may occur in a SEM when there is poor electrical conductivity of the specimen [2]. Sputter coating is a deposition process to cover a specimen with a thin layer of conducting material, typically a metal, such as a gold/palladium (Au/Pd) alloy. A conductive coating is needed to prevent charging of a specimen with an electron beam in conventional SEM mode (high vacuum, high voltage). While metal coatings are also useful for increasing signal to noise ratio (heavy metals are good secondary electron emitters), they are of inferior quality when X-ray spectroscopy is employed [3].

There are four plasma etching processes. These are: sputtering, pure chemical etching, ion energy-driven etching and ion-enhanced inhibitor etching [4]. In this work, the focus would be on plasma etching through the sputtering technique. Sputtering is the removal of atoms caused by high energetic ions hitting the surface. This process is highly anisotropic. If the ions have sufficiently high energy, they can knock atoms out of the material to be etched without chemical reaction. Removing atoms by sputtering with an inert gas is called "ion milling" or "ion etching" [5].

In this study, the coating material of interest is Pt. There are various reasons, in the context of failure analysis applications, why the coating material has to be removed from the surface of the IC being analyzed, as follows:

1. Further electrical testing or characterization is needed on the IC. Testing required may be whole-chip, a specific circuitry or at the component level. Pt coating covers the surface of the chip and therefore causes electrical shorting on the exposed metallizations, bond pads and bonding wires.

2. There is a requirement for further fault isolation analysis such as photon emission microscopy (PEM), laser scanning injection microscopy (LSIM), circuit probing analysis or other applicable techniques.

3. There is a need to perform further layer-by-layer chip deprocessing. To effectively deprocess samples using the commonly-used deprocessing etchants, the Pt coating has to be removed first.

II. EXPERIMENTAL RESULTS, ANALYSIS AND DISCUSSION

Electrically good samples were collected. Curve Tracer (CT) analysis and bench testing have been conducted to document their initial response. Samples are decapsulated to expose their die surfaces. Curve trace and bench testing were performed to confirm that the parts' electrical responses did not change after the decapsulation process. Samples are coated with Pt with a platinum coater. Curve trace and bench testing were done to document the parts' electrical response after the coating process. Resistive shorts now appear during curve tracing. The switch functionality of the IC is also failing during bench testing. Energy Dispersive X-ray (EDX) was performed to document the presence of coating on top of the die. The Pt removal has been evaluated using and ICP-RIE system to remove the coating material. The plasma recipe, a sputter etch parameter, that was successful in removing the Pt coating are as follows:

- ➤ Low pressure
- ➤ High ICP power
- ➤ Low RIE power
- ➤ Argon Gas
- ➤ Ten minutes process etch

Optical Photo Documentation of the die surface

The samples were inspected using a metallurgical microscope that can magnify the specimen from 100X to 2000X. Using bright field illumination, photos of the samples were taken. A noticeable colour difference on the coated samples were noted. See figures 1a-1c.

978-1-4799-3911-4/14 $31.00 © 2014 IEEE

Figure 1a. Optical photo showing the initial condition of the sample after chemical decapsulation.

Figure 1b. Optical photo after the Pt coating. A noticeable colour difference is observed indicative of the Pt coating.

Figure 1c. Optical photo after the Pt etching. The bondpad surfaces became shiny and reflective.

Curve Trace Analysis Documentation

Curve Trace Analysis is an FA technique performed to characterize and analyze the current vs. voltage behavior of the external pins using a curve tracer tool. A curve trace tool is an instrument that is used to stimulate, measure and display the I-V pin curve characteristics of an IC. Voltage is continuously swept

across the path being traced. The resulting current across the path is measured.

Curve Trace analysis on the evaluated unit confirms shifts in its electrical response. The unit's I/V curve response changed to resistive curve after it has been coated with Pt. The part recovered after the Pt removal. Refer to Figures 2a-2c for the I/V curve plot documentations.

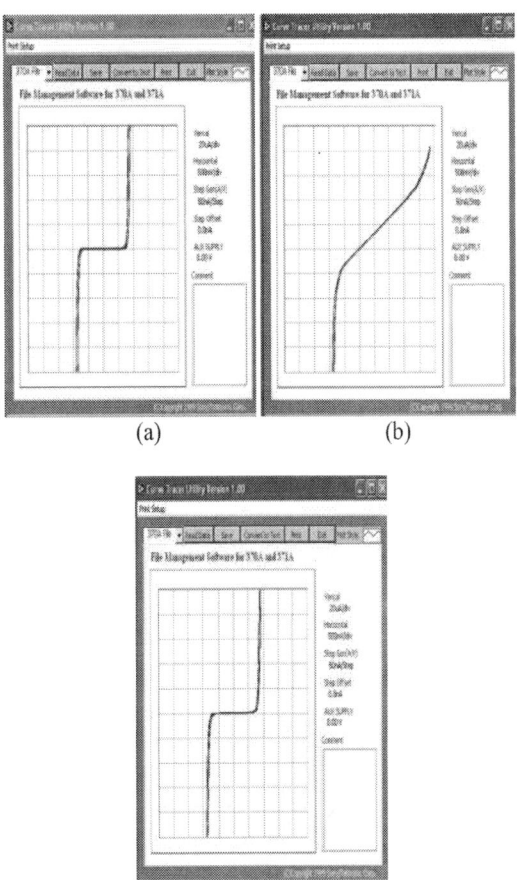

Figure 2: (a) I/V curve response of the sample after chemical decapsulation. (b) Curve Tracer analysis result after Pt coating. (c) CT result post Pt etching.

Bench Testing Analysis Documentation:

The evaluated part underwent bench testing for functionality checking. Using the switch parameter of an Analog Devices Incorporated (ADI) switch device, the output is monitored using an oscilloscope. Ideally, the waveform at the input should be identical to the processed output of the switch device. There was an observable failure after the part has been subjected to Pt coating. The part however recovered from failure when the Pt coating has been removed using the ICP-RIE system. See Figures 3a-3c for the oscilloscope plots.

978-1-4799-3911-4/14 $31.00 © 2014 IEEE

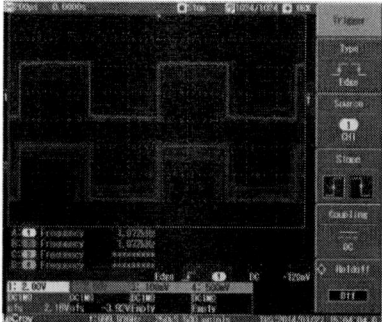

Figure 3a. Initial bench response of the sample post chemical decapsulation.

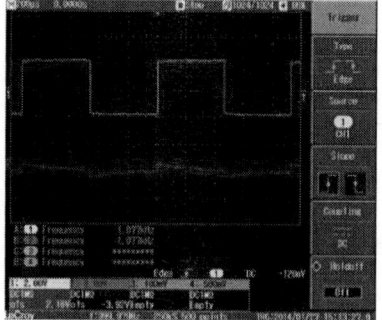

Figure 3b. Bench response post Pt coating.

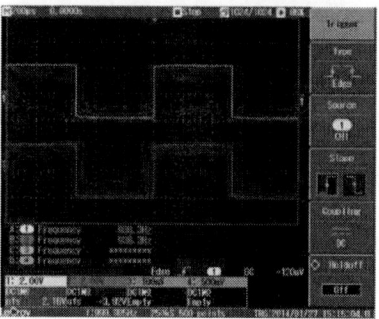

Figure 3c. Bench response post Pt etching. The part recovered from failure.

Scanning Electron Microscopy (SEM) / Energy Dispersive X-ray (EDX) Documentation:

SEM is a technique used to inspect topographies of a specimen at very high magnifications offering a better depth of field while EDX is a technique used to identify the elemental composition of a specimen or an area.

SEM/ EDX analysis results of the evaluated sample are shown in Figures 4a-4c.

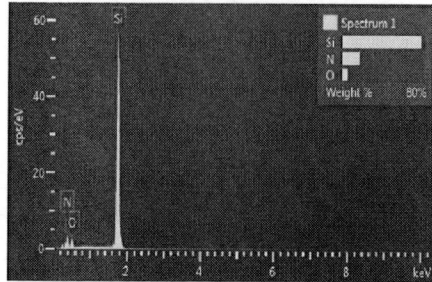

Figure 4a. EDX analysis post chemical decapsulation. The elements detected, nitrogen (N), oxygen (O) and silicon (Si) are coming from the passivation material (SiN and SiO$_2$)

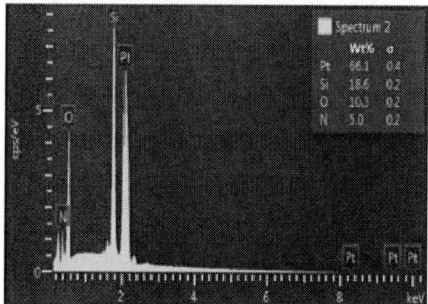

Figure 4b. EDX analysis post Pt Coating. Pt has been detected and displayed in the spectrum by the system.

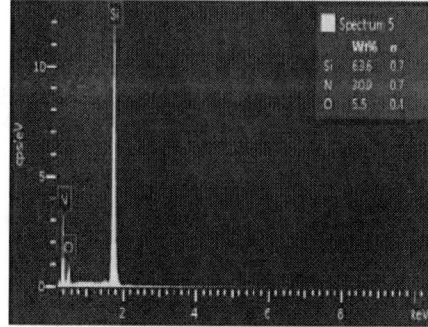

Figure 4c. EDX analysis post Pt Etching. There are no more traces of Pt coating.

CONCLUSION

The evaluated ICP/ RIE recipe has been successful in removing the Pt coating on top of passivation and aluminium metallizations. It was confirmed through optical inspections, curve tracing analysis, bench testing and EDX analysis that the Pt coatings has been removed from the surface of the chip.

ACKNOWLEDGMENT

The authors wish to acknowledge the Worldwide Product Analysis Management (WWPA) of Analog Devices Incorporated (ADI) for allowing the external publication of this technical paper.

REFERENCES

[1] Hoflinger G. (2013, May 28), Brief Introduction to Coating Technology for Electron Microscopy. *Science Lab*. Retrieved April 28, 2014, from http://www.leica-microsystems.com/science-lab/brief-introduction-to-coating-technology-for-electron-microscopy/

[2] A. Halfpenny, *Eliminating charging in the SEM*. Journal of the Virtual Explorer, http://virtualexplorer.com.au/article/2011/272/ebsd-data-from-rocks-and-minerals/charge.html

[3] Sputter Coating. *Wikipedia*. Retrieved April 28, 2014, from http://en.wikipedia.org/wiki/Sputter_coating

[4] P. Magnusson, *Plasma Etching; Ion sputtering in a RIE*. KTH Royal Institute of Technology, http://www.kth.se/polopoly_fs/1.254043!/Menu/general/column-content/attachment/report22.pdf

[5] Wikipedia, *Sputtering*. Retrieved April 28, 2014, from http://en.wikipedia.org/wiki/Sputtering

Spatial correction in dynamic photon emission by affine transformation matrix estimation

S. Chef[1,2], S. Jacquir[2], P. Perdu[1], K. Sanchez[1] and S. Binczak[2]

[1]CNES, DCT/AQ/LE, Bpi 1414, 18 Avenue Edouard Belin, 31401 Toulouse, France
[2]Le2i UMR CNRS 6306, University of Burgundy, 9 Avenue Alain Savary, 21000 Dijon, France
Phone: +33380399035   Email: samuel.chef@u-bourgogne.fr

Abstract- **Photon emission microscopy and Time Resolved Imaging have proved their efficiency for defect localization on VLSI. A common process to find defect candidate locations is to draw a comparison between acquisitions on a normally working device and a faulty one. In order to be accurate and meaningful, this method requires that the acquisition scene remains the same between the two parts. In practice, it can be difficult to set. In this paper, a method to correct position by affine matrix transformation is suggested. It is based on image features detection, description and matching and affine transformation estimation.**

I. INTRODUCTION

Defect localization often is the critical part of the failure analysis process. Optical techniques are tools of choice as invasiveness is minimized and few machining of the sample is required. Among these techniques, photon emission microscopy has been used for almost thirty years [1]. Around a decade and a half ago, a better understanding of the physical phenomena implied in the process and an improvement of the detection and acquisition hardware had made possible the recording of the positions and time of emitted photons during transistors' switching [2]. This is known as dynamic photon emission or Time Resolved Imaging (TRI).

Regardless of the acquisition mode (static or dynamic), the common process used to highlight differences between databases acquired on a good and on a bad device relies on spatial comparison. In static mode, acquired photons are displayed in the two dimensions (x,y). A subtraction between the images is used to isolate potential faulty nodes. In dynamic mode, the photons characterized in the three dimensions (x,y,t) are projected in the (x,y) plan and the same operation is applied to find candidates.

Between two samples, it can be difficult to ensure the exact same scene. Indeed, it requires that the two samples have the same orientation and the camera is located on the exact same region of the circuit. There also can be some changes along the z-axis. Camera should be set slightly closer or further from the component in order to get focus. A potential cause of this is the sample preparation that cannot be exactly the same for every part (substrate thickness, etc) or the mounting of the device on test board. As a result, the field of view can be a bit different. All of this leads to a lot of remaining pixels after the subtraction and finding the spot related to the defect can be challenging.

One solution to overcome the limitations of the spatial projection is to directly use a 3D process, like in [3]. Unfortunately, if the defect does not exhibit an emission with singular statistical properties, a comparison between databases shall also be considered. Once again, a correct positioning of photons is required so that databases are comparable.

The difference between the two acquisition scenes can be interpreted as a geometrical transformation. Indeed, the difference of orientation between the two samples is a geometric rotation, camera position in the (x,y) plan is a translation. The change in the field of view is a scale modification. All of these transformations are linear and can be modeled by an affine transformation matrix. In this paper, a process to estimate this matrix using photon emission data is reported. Once it has been estimated, it becomes possible to make some corrections and have comparable data to find candidate locations for the defect. As long as the algorithm has been designed for 2D signals (i.e. images), it can be applied on both static and dynamic emission.

The remaining of this paper is organized as follows: in the second section, the method is introduced in details. Some results are given in the third section. In the fourth section, the application boundaries are discussed. Finally, a conclusion finishes this paper.

II. METHOD DESCRIPTION

A. General Overview

The natural process to find how an object has moved between two moments consists in finding some points of interest and see how they have changed in the same reference system. As photon emission is a stochastic phenomenon, it seems meaningless to try to compare the coordinates of single photons in both databases. On the other hand, emission should occur from the same nodes and for a long enough acquisition time, in the (x,y) projection, a node belonging to both databases should have similar properties in size and shape. As a consequence, images of photon emission have some specific interest points which might be detectable by some image processing algorithms.

Once a specific point has been found for one image, we need to find its match in the second one to estimate the move parameters. It can be done by a characterization using its neighborhood. A vector of *n* dimensions named descriptor is built. Two points having similar vectors in the descriptor space are more likely to be the same point of interest in the two images. After matching, the motion can be estimated and expressed as an affine transformation matrix. If this matrix is named T, its expression is the following :

$$T = \begin{bmatrix} A\cos r & A\sin r & 0 \\ -A\sin r & A\cos r & 0 \\ T_X & T_Y & 1 \end{bmatrix}, \qquad (1)$$

where A is the scale factor, r is the rotation angle and T_X and T_Y are translation vector coordinates.

Finally, the inverse transformation is applied to the image in order to restore the image. To sum-up here are the key steps of the process:

1) Features detection.
2) Descriptors building.
3) Features matching.
4) Motion estimation.
5) Inverse transformation

B. Interest points detection

The identification of matching points between images of a moving scene has drawn the attention of the image processing community for a long time. For instance Harris suggested an algorithm to detect and built descriptor of corners in [4] in 1988. In 2004, Lowe reported an algorithm to find scale space maxima and named Scale-Invariant Feature Transform (SIFT) [5]. One of the drawbacks of the method is its computational cost. In order to speed things up, a new algorithm called Speeded-Up Robust Features (SURF) has been introduced by Bay *et al.* in 2008 in [6].

A common approach to detect interest points is to build a scale space representation of an image and search for local maxima. Usually, this representation requires two steps: first, the image is convolved with a Gaussian filter of parameter σ and then, in a second time, is sub-sampled. The larger the σ, the coarser the result. Instead of doing so, the SURF algorithm computes the scale space transform using several tricks to accelerate the process.

The Gaussian filter responses are first approximated with box-filters. The aim is to get results closed to the ones with Gaussians but at a shorter computation time because of the sparseness of the filter response. Another point is that the convolution operation is a costly one. The use of integral images helps to compute it faster. An integral image is defined as $I_\Sigma(x,y) = \sum_{i=0}^{x}\sum_{j=0}^{y} I(i,j)$, where $I(i,j)$ denotes a pixel of the actual image. Considering these two tricks, for each pixel $\mathbf{X} = (x,y)$, a Hessian matrix H at scale s is computed :

$$H(X,\sigma) = \begin{bmatrix} L_{xx}(X,s) & L_{xy}(X,s) \\ L_{xy}(X,s) & L_{yy}(X,s) \end{bmatrix}, \qquad (2)$$

where L_{ij} denotes the result of the convolution of the image in $\mathbf{X}$ with the filter in the direction ij. Finally, a blob map at each scale s is build using the determinant of the matrix H:

$$\det(H) = L_{xx}.L_{yy} - (w.L_{xy})^2, \qquad (3)$$

where w is a weight used to balance the determinant equation. Points of interest are found by searching for the blob map maxima at the different scales.

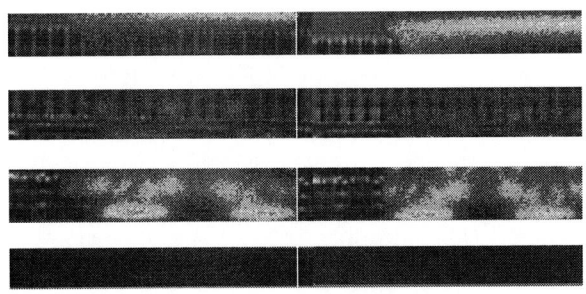

Fig. 1. Superposition over pattern of the two acquisitions.

C. Descriptor building and features matching

For each key point identified, a descriptor of 64 or 128 components, depending on user's choice, is build from a neighborhood of 20 pixels. A smaller descriptor makes computation faster but brings a loss of precision. It is based on a 2D wavelet transform using Haar wavelet. Once again, the wavelet is chosen so that it allows a faster process thanks to integral images. Matching between images' points is achieved by searching for nearest neighbor in the descriptor space. Indexing structures like kd-tree can be considered to accelerate the process.

D. Affine transformation : Motion Estimation

Once matching has been done, a matrix of transformation (translation + rotation + scale factor) is found by using robust estimator such as Random Sample and Consensus (RANSAC) [7], M-Estimator Sample Based Consensus (MSAC) [8]. The second algorithm is an upgrade of the first one. RANSAC is based on the maximization of inliers (points that fit to the model). A set of points is randomly chosen and a first estimation of the parameters is performed. Then the model is applied to the data. Those which fit to the model are labeled as inliers. The process is repeated on a new random set including freshly labeled inliers and the model is updated. It continues until the number of inliers is maximized.

The threshold for the minimum number of inliers can be difficult to set. Especially, if the potential number of inliers is low because there are only too few matching points, the estimation can be erroneous. Several algorithms tried to overcome the parameter selection issue. Among them, MSAC choose to penalize outlier according to one model and give a score to inliers according to how well they fit. There is a finer quantification of a model quality.

At the end of the process, the estimated matrix has the following shape :

$$T = \begin{bmatrix} I & L & O \\ J & M & P \\ K & N & Q \end{bmatrix},$$

where the matrix coefficients are real values. The parameters of the transformation can be estimated as follow:

$$\hat{r} = \frac{\arctan\left(\frac{L}{M}\right) + \arctan\left(-\frac{J}{I}\right)}{2}, \begin{bmatrix} \hat{T}_X \\ \hat{T}_Y \end{bmatrix} = \begin{bmatrix} K \\ N \end{bmatrix},$$

$$\hat{A} = \frac{I + M}{2 \cos \hat{r}}.$$

The hat in the expression above indicates an estimated value.

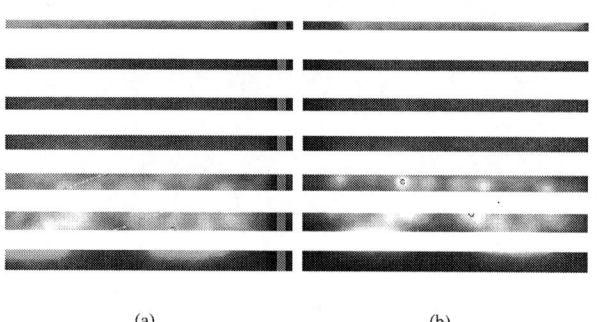

(a)   (b)

Fig. 2. Superposition of the emission images of the two acquisitions before (left) and after correction(right). Circles and crosses indicates the location of interest points

III. APPLICATION AND RESULTS

A. First Example

The process is applied on TRI images. This kind of images can be complicated to deal with because of their low resolution.

The images have been acquired using Hamamatsu TriPHEMOS on two microcontrollers built in 0,15 μm technology. The superposition of the two emissions over pattern images are available in Fig. 1. Hereafter, left image will be referred as component A and the right one as component B. It can be seen that there is a small rotation of the device between the two acquisitions, in addition to a small translation. At first sight, a change of scale (parameter A in the matrix T expression in (1)) does not seem to occur so we can expect it to be close to one and be neglected.

In order to highlight the differences, the two emission images have been superimposed in Fig. 2 (a). A preprocessing step composed of a median filtering was required. Without it, the results were absurd. The detected interest points have been marked with red circles for component A and green crosses for component B. Thirteen correspondences have been kept after SURF detection and matching. More than the half of them seems to indicate different transformation tensors. Superposition after inverse transform is available in Fig. 2 (b). Despite the high value of outliers, MSAC algorithm estimates a translation vector $\begin{bmatrix} Tx \\ Ty \end{bmatrix} = \begin{bmatrix} -22,2 \\ -36,7 \end{bmatrix}$; a rotation angle $r = 1,06$ °; a scale factor $A \sim 1$.

When there is overlapping of pixels of the same intensity, it appears in white in Fig. 2 (b). As it can be seen, the geometric transformation matrix found by the MSAC algorithm is correct as there is overlapping for most of the acquired area. Left and right sides are in single color (red on the left and cyan on the right), meaning that these parts of the picture have only been acquired for one of the devices. As a consequence, no advance study can be carried out for these locations. No comparison is possible as no data exist for both databases but only for one of them.

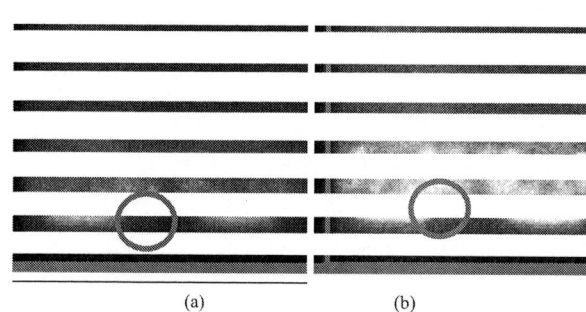

(a)   (b)

Fig. 1. Superposition of images in the case of suspicious spots (circled in red) without (a) and with (b) spatial correction.

In order to quantify the correction quality, the mean square error (MSE) between the two images is computed. It is done before and after spatial correction. The MSE is defined as :

$$\text{MSE} = \sum_{i=0}^{x} \sum_{j=0}^{y} \left(A(i,j) - B(i,j)\right)^2, \quad (4)$$

where x and y are the dimensions of the images and A and B are the two images to compare. In this case, before spatial correction and with intensity normalization (meaning that I(x,y) is in [0;1]) and spatial correction, $\text{MSE}_B = 1,23 \cdot 10^{-2}$ a.u. whereas after transformation recovery, $\text{MSE}_R = 9,0 \cdot 10^{-3}$ a.u.

B. Second Example

The databases have acquired on the same devices, with the same setup but at another location. As the center of rotation of the device was closer to this area, the spatial difference between the two databases was smaller. Superposition is available in Fig. 3 (a).

This case is a bit more particular than the previous one as there is one spot that exist in one image and not on the other. It is circled in red in Fig. (a) and (b). Even if the position of the two samples seems to be quite the same, because of a small misalignment, some differences exist at other locations. To discard any other potential difference, the process for spatial correction is applied. This time, good results have been achieved without any filtering. Twenty seven matching features have been found.

The transformation matrix estimation gives the following parameters: $r = -0,038°$; $\begin{bmatrix} Tx \\ Ty \end{bmatrix} = \begin{bmatrix} -3,7 \\ -4,9 \end{bmatrix}$; $A = 1,0038 \sim 1$.

Once again, the values of the translation vector are given in pixels and the scale factor does not have any unit. The images superposition after spatial recovery is reported in Fig. 3 (b). The suspicious spot can better be seen. The mean square error computation gives the following results: $\text{MSE}_B = 4,4 \cdot 10^{-3}$ a.u.

978-1-4799-3911-4/14 $31.00 © 2014 IEEE

and $MSE_R = 2,9.10^{-3}$ a.u. Both scores have been calculated after images intensity normalization.

Many differences between two acquisitions can exist due to the random nature of the photon emission process and the small changes of samples orientation. As a conclusion of this second application, even if a spot outcomes from the unprocessed images superposition, a spatial correction can help to discard the small meaningless differences and focus on the true defect related candidate spot.

IV. DISCUSSION ON PREPROCESSING

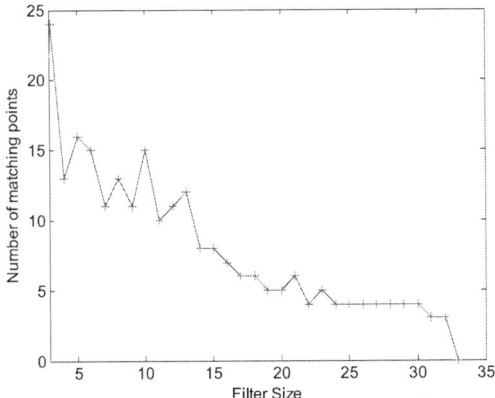

Fig. 2. Number of matching points as a function of the filter size for the images from Fig. 1 and 2.

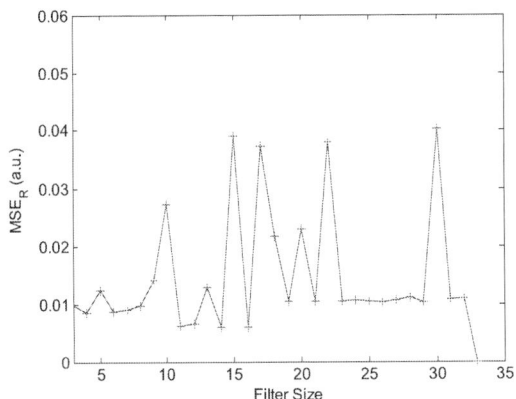

Figure 3 : MSE after inverse transformation as a function of the filter size.

As it was the case for the first application example, some prior processing like filtering is sometimes required to get good results. One of the reason why is there are too many wrong matching between points. Filter like the median one acts like a low pass filter and the resulting image resolution is lowered. For the images used in Fig. 1, a study of the number of matching points found as a function of the filter size has been carried out. Results are available in Fig. 4. The size of the neighborhood used for the filtering is varying from 3 pixels to 33 pixels. The

resolution of the effective area of the TRI image in quite small (around 200x200 pixels) so the largest filter uses almost 25 % of the image, which is huge. As the figure shows, the larger the filter size, the smaller the number of matching points. Because the estimation has been performed with fewer points, recovery should be less reliable.

A study to find an optimal filter size according to the mean square error has been carried out and results is reported in Fig. 5. Before any processing, the MSE value was of $1,23.10^{-2}$ a.u. A minimum of $6,1.10^{-3}$ a.u. is found for filter size of 16 pixels but the image after recovery looked like there was a shear transform, which is different from the simple rotation and translation considered in this paper. On the other hand, the recovered images corresponding to filter sizes with high MSE in Fig. 5 was effectively poorly recovered. As a conclusion, MSE can possibly be a metric to quantify how bad is the recovery but is not enough to quantify its goodness.

V. CONCLUSION

Spatial comparison is a key procedure to analyze data from photon emission and finding candidate spots for defect localization. Although it requires a good matching between the two scenes. Because it can be difficult to ensure when the acquisitions are not performed on the same sample, a method for spatial recovery based on interest points detection and characterization has been studied. Thanks to these features, an affine transformation matrix is estimated and applied to correct the images. It has been shown it can help to emphasize a suspicious spot, even if no correction was required at first sight. In order to grant success of the recovery, a preprocessing can be useful but the parameters remain empirically defined. In addition, the small number of detected features challenges the estimation.

ACKNOWLEDGMENT

Authors would like to thank Renesas Electronics (databases), Hamamatsu Hpks for their technical support (TriPhemos) and the Regional Council of Burgundy for its financial support.

REFERENCES

[1] N. Khurana, C.L. Chiang, "Analysis of product hot electron problems by gated emission microscopy," *Proceedings of the 24th Annual IEEE Reliability Physics Symposium,* pp. 189-194, 1986.

[2] J.C. Tsang, J.A. Kash, "Picosecond hot electron light emission from submicron complementary metal-oxide-semiconductor circuits," *Applied Physics Letters,* Vol. 70, No. 7, pp. 889-891, 1997.

[3] S. Chef, P. Perdu, G. Bascoul, S. Jacquir, K. Sanchez, S. Binczak "News statistical post-processing approach for precise fault and defect localization in TRI database acquired on complex VLSI," *Proceedings of the 20th IEEE International Symposium on the Physical and Failure Analysis of Integrated Circuits,* pp. 136-141, 2013.

[4] C. Harris, M. Stephens, "A combined corner and edge detector," *Proceedings of the 1988 Alvey Vision Conference,* pp. 23.1-23.6, 1988.

[5] D.G. Lowe, "Distinctive image features from scale invariant keypoints," *International Journal of Computer Vision,* Vol. 60, No. 2, pp. 91-110, 2004.

[6] H. Bay, T. Tuytelaars, L. Van Gool, "Speeded-up robust features (SURF)," *Computer vision and image understanding,* Vol. 110, No. 3, pp. 346-359, 1997.

[7] M.A. Fischler, R.C. Bolles, "Random sample consensus: a paradigm for model fitting with applications to image analysis and automated cartography," *Communications of the ACM,* Vol. 24, No. 6, pp. 381-395, 1981.

[8] P.H. Torr, A. Zisseman, "Robust computation and parametization of multiple view relations," *Proceedings of the 6th International Conference on Computer Vision (ICCV),* pp. 727-732, 1998.

XPS and TEM Studies of Oxidation States on Sn Solder Ball

Shen Yiqiang, Chen Yixin, Lee Hwang Sheng, Chow Shue Yin, Xing Zhen Xiang,
Hua Younan and Li Xiaomin

WinTech Nano-Technology Services Pte. Ltd.
10 Science Park Road, #03-26 & #03-28, The Alpha
Singapore Science Park II, Singapore 117684
Email: *Yiqiang@wintech-nano.com*

Abstract: **A Sn oxide layer on the surface of Sn solder balls plays an important role in the semiconductor packaging industry. This paper shows a comprehensive analysis of the Sn oxide layer by XPS depth profiles. The distribution of Sn with different oxidation states can be derived from curves fitting Sn3d5/2 peaks. Moreover, the oxide layer thicknesses obtained from XPS demonstrate a linear correlation with the values from TEM measurements.**

I. INTRODUCTION

Solder balls play a crucial role in the semiconductor industries because they are not only act as electric connections, but also as mechanical support. The high temperature reflow and bonding process make the formation of surface oxide on solder balls unavoidable. Therefore, the properties of the surface oxide can affect the bonding performance in both the electric and mechanical aspects significantly. Research has shown that the surface oxide formed on a solder ball would degrade the solderability and affect the reliability of the whole unit [1][2]. Most of the previous studies on the solder ball utilized TEM+EDX, by using its ultra-high lateral resolution to determine the oxide layer thickness and chemical composition. However, besides these information, the chemical state of the Sn oxide is also important since Sn has two types of oxides, SnO and SnO_2, which show different properties of electric conductivity. SnO_2 is an N-type material with a wide band gap, while SnO is a P-type semiconductor that is highly conductive [3-6]. Different Sn oxidation states may result in different electric properties of the solder joint. Since it is difficult to determine the oxidation state from the EDX results obtained from TEM, in this study, we used XPS to determine the Sn oxidation state. XPS depth profiles, which sputtered through the oxide layer, were performed and different Sn oxidation states were separated by curve fitting the Sn3d5/2 XPS peak. Then, the Sn oxide layer thicknesses derived from XPS depth profiling are compared to the ones obtained from TEM. A good agreement was shown between the oxide thicknesses obtained from both techniques when the thickness exceeds 10nm. However, when the oxide layer is less than 10nm, the thickness obtained from XPS was underestimated. Hence, another way to derive the oxide layer thickness should be adopted.

II. EXPERIMENT

Sn-Pb solder balls were used in this study. Four groups of solder balls with different oxide layer thickness were prepared. The nominal thickness for group #1, #2, #3 and #4 is 30nm, 20nm, 10nm and 3nm respectively. These nominal thicknesses are based on the TEM (transmission electron microscope) measurement, achieved by FEI Tecnai G2 with 200kV accelerate voltage. These TEM samples were prepared by FEI Helios 600i FIB (focused ion beam). In order to prevent the further growth of oxide layer on the solder balls, the samples were immediately inserted into the TEM vacuum chamber after preparation. In order to obtain information on the oxidation states and correlate it to the TEM thickness measurement results, XPS (X-ray photoelectron spectroscope, Ulvac-Phi Quantera II with Al Kα target) depth profiles were performed on the neighbouring solder balls that had gone through the FIB-TEM process. During the depth profile, certain cycles of Argon sputter gun were used to remove layers of the materials. The duration of each sputter cycle was fixed and after each sputter cycle, XPS spectra of O1s and Sn3d5/2 were collected. The depth profile can then be constructed with the atomic percentage against the sputter time. By correlating the sputter time with the thickness measurements obtained from TEM, the sputter rate (nm/min) can be determined and the sputter time can be converted into depth. The depth profiles were used to determine the oxide layer thickness, while the Sn3d5/2 spectra were mainly used for oxidation state analysis. The test was conducted on the top area of each solder ball. In order to minimize the influence of the circular arc shaped solder ball on the photoelectron take-off angle and the sputter gun incident angle, the detection area was set to 40um x 40um. Both the ion and electron neutralizers were switched on to minimize the possible peak shift in the XPS spectra. Then, the energy scale of XPS spectra was calibrated to the standards of Au, Ag and Cu before the measurements. Three repetitive tests of XPS were done for each group to ensure reliability.

978-1-4799-3911-4/14 $31.00 © 2014 IEEE

III. RESULTS AND DISCUSSION

The peak positions for Sn, SnO and SnO$_2$ are fixed at 483.02eV, 484.59eV and 485.32eV during the peak deconvolution [7] and the FWHM (full width half maximum) for Sn, SnO and SnO$_2$ peaks are kept the same. Since SnO$_2$ and SnO are an insulator and a semi-insulator respectively, the shapes of their XPS peaks are symmetric and fit the Gaussian-Lorentz function. On the other hand, metallic Sn should have an asymmetric XPS peak shape. The Sn 3d5/2 peak in the pure Sn region, after the sputtering of the oxide layer, is used to determine the tail length of this asymmetric peak. For example, Fig 1 shows the XPS deconvolution of the different chemical states of group #3 Sn3d5/2. The Sn3d5/2 XPS peak can be well deconvoluted by the method mentioned above, at all sputter times. As shown in Fig.1, when the sputter time t=0 min, the outermost surface consists of high content of SnO$_2$ and SnO. When the sputter time t=0.1 min, the heavily oxidized SnO$_2$ layer is partially removed and the SnO and metal Sn that lie underneath become dominant. Finally, when the sputter time t=0.2 min, both oxides are mostly removed and the main composition becomes metal Sn. After the deconvolution of Sn3d5/2 peak at each sputter time, the changes of the percentages among the different chemical states can be plotted against the sputter time, as shown in Fig. 2. The composition of SnO$_2$ keeps decreasing with the sputter time, while that of SnO shows a plateau region in all groups. Such interesting information cannot be obtained from TEM+EDX measurement and more detailed studies about such oxide distributions are undergoing.

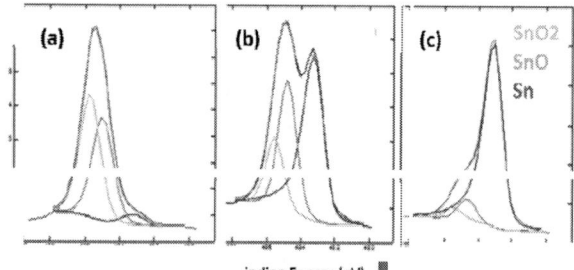

Fig. 1 Sn 3d5/2 XPS peak of group #3 at different sputter times. (a) sputter time=0 min, (b) sputter time=0.1min, (c) sputter time=0.2min. After the peak decovolution, Sn, SnO and SnO$_2$ peaks are shown in blue, green and orange respectively.

The oxide layer thickness was derived from the profile of O1s and Sn3d5/2 versus sputter time. The point with half of the maximum oxygen intensity was used as the interface between oxide and Sn metal [8]. The different sputter times needed to reach the oxide/metal interface are 2.4min, 1.6min, 0.9min and 0.06 min for group #1, #2, #3 and #4 respectively. On the other hand, the oxide layer thickness measurement results obtained from TEM are 32nm (± 10.0nm), 21.1nm (± 0.7nm), 11.6nm (± 4.1nm) and 3.02nm (± 1.9nm) for group #1, #2, #3 and #4 respectively (Fig. 3). As shown in Fig. 4, the sputter time and thickness of the group #1, #2 and #3 show a good linear correlation. The sputter rate can be determined from Fig.4 as the gradient of the curve, which is 13nm/min. By using this sputter

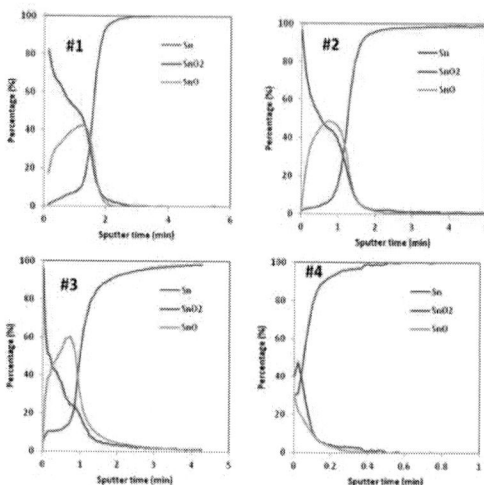

Fig. 2 Depth profiles of different Sn oxidation states in four groups. The composition of SnO$_2$ keeps decreasing with the sputter time while that of SnO shows a plateau region in all the groups.

rate, the sputter time can then be converted into depth. Although the other three groups show good linear correlations, group #1 shows a large deviation from the linearity. This is resulted from errors of both TEM and XPS. Under the high magnification of TEM, it is difficult to determine the interface between the thin oxide layer and the metal Sn. Moreover, the rough surface results in the coverage of the thin oxide layer to be non-uniform, which also leads to errors in the TEM thickness measurements. On the other hand, in the case of ultra-thin oxide layer, the usage of half of the maximum oxygen intensity as the interface between oxide and Sn metal is no longer applicable as the oxide layer thickness is comparable or even less than the information depth of XPS. Indeed, the Sn3d5/2 peak of group #1 when the sputter time = 0min (Fig.5) shows a high content of metal Sn, which shows that the oxide layer is too thin to block the metal Sn signal coming from underneath. In order to eliminate the effects of the surface roughness and take the XPS information depth into account, another method of utilizing XPS spectra is developed to determine the tin oxide layer thickness. According to the Beer-Lambert law, the photoelectron intensity will be attenuated by a thin layer (thickness t) above it:

$$I = I_0 e^{(-t/\lambda\cos\theta)} \qquad \text{Equation (1)}$$

Where I is the attenuated electron intensity, I_0 is the un-attenuated electron intensity at depth t, λ is the attenuation length, θ is the take-off angle of electron with respect to the norm of the surface.

When considering a thin layer material (dt) at depth t, the total electron intensity of this layer can be obtained through the following integration, from depth t to t=0:

$$\int_t^0 Idt = \int_t^0 I_0 e^{(-t/\lambda\cos\theta)}dt \qquad \text{Equation (2)}$$

In our case, the oxide layer thickness is t_{ox} can be derived by:

$$t_{ox} = \lambda\cos\theta \ln(I_{total}/I_{Sn}) \qquad \text{Equation (3)}$$

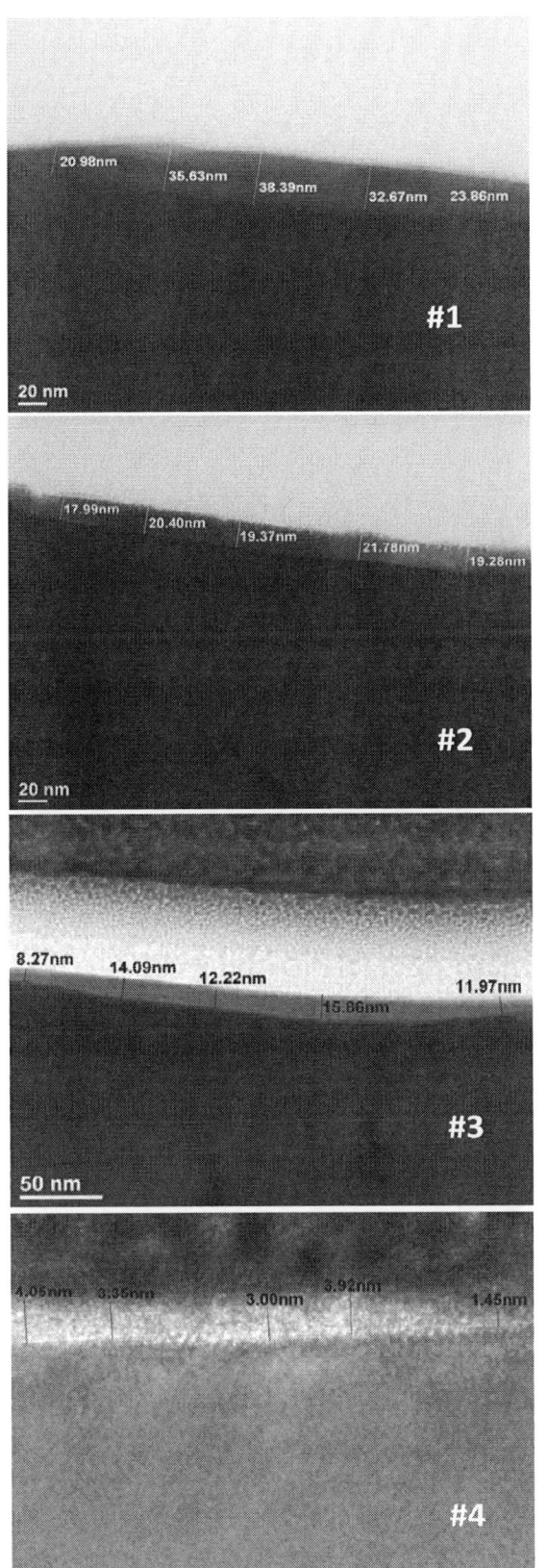

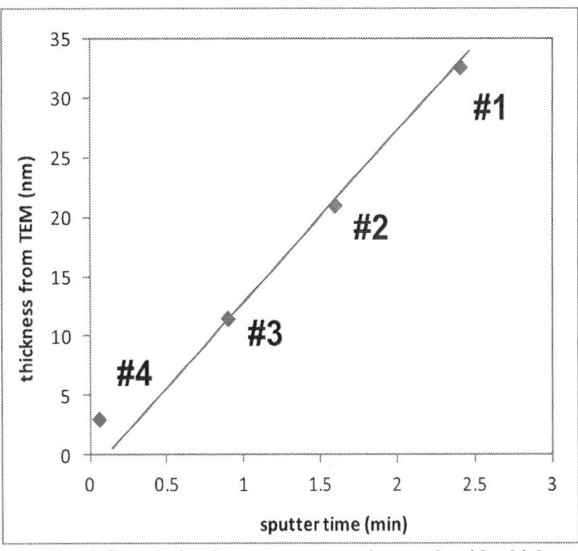

Fig. 4 Correlation between sputter time and oxide thickness measured by TEM. The sputter time and thickness of the group #1, #2 and #3 show good linear correlation. These three points are linearly fitted and the gradient of the fitting line is 13nm/min, which is the sputter rate.

where $I_{total} = I_{Sn} + I_{ox}$, I_{Sn} and I_{ox} are the XPS electron intensities of metal Sn and Sn oxides respectively, obtainable from the XPS spectra of Sn3d5/2 after the peak deconvolution. Based on the peak deconvolution results, shown in Fig. 5, since $I_{Sn} = 23395$ counts/sec, $I_{SnO} = 20138$ counts/sec and $I_{SnO2} = 32632$ counts/sec, $I_{ox} = 52770$ counts/sec. The photoelectron take-off angle with respect to the norm of the surface is 45°. In order to simply the calculation, the attenuation length λ can be estimated with the average IMFP (inelastic mean free path) of SnO and SnO$_2$, which are 1.96nm and 1.80nm respectively. Therefore, λ is estimated to be 1.88nm and according to Equation (3), t_{ox} can be determined to be 1.93nm. Since the factors degrading the TEM thickness measurement, such as surface roughness and uniformity of the oxide layer, are reflected in the XPS peak intensities, this method is more appropriate to estimate the oxide layer thickness over a large area.

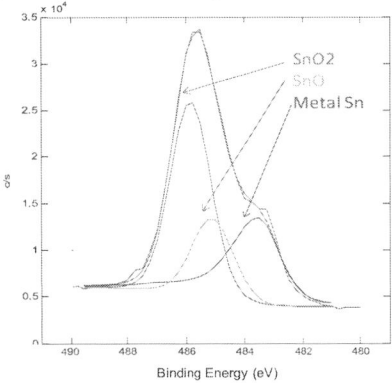

Fig. 5 Sn3d5/2 peak of group #1 when sputter time = 0min. Large content of metal Sn is observed from the surface, which indicates the information depth of XPS is larger than the oxide layer thickness.

Fig. 3 TEM images for all four groups. The thickness measurements of the oxide layers are indicated.

IV. CONCLUSIONS

XPS depth profiles demonstrate the capability to determine the Sn chemical states distribution along the depth. The distribution of different Sn chemical states can be obtained by deconvoluting the Sn3d5/2 peaks. Additionally, the Sn oxide layer thickness also can be obtained from XPS depth profile. Excellent linearity is shown between the thickness values from XPS depth profile and TEM when the oxide layer is thicker than 10nm. When the oxide layer thickness is less than or comparable to the information depth of XPS, another method is proposed to estimate the oxide layer thickness from the XPS spectra.

REFERENCES

[1]. Hetschel, T., Wolter, K. J., "Wettability effects of immersion tin final finishes with lead free solder", Electronics System-Integration Technology Conference, Page561 – 566, 2008.

[2] Bradley, E., Banerji, K., "Effect of PCB finish on the reliability and wettability of ball grid array packages" Components, Packaging, and Manufacturing Technology, Part B: Advanced Packaging, Volume:19 , Issue: 2, Page320 – 330, 1996.

[3]Hwang S., Kim Y., Lee J., "Irregular Electrical Conduction Types in Tin Oxide Thin Films Induced by Nanoscale Phase Separation" Journal of the American Ceramic Society Volume 95, Issue 1, Page 324-327, 2012.

[4]Swanepoel, R. "Determination of surface roughness and optical constants of inhomogeneous amorphous silicon films" Journal of Physics E, Volume 17, Issue 10, Page 896, 1984.

[5] Guo, W., Fu, L., Zhang, Y. et al "Microstructure, optical, and electrical properties of p-type SnO thin films" Applied Physics Letters, Vol.96, Issue4, Page 042113, 2009.

[6] Sanon, G., Rup, R., and Mansingh, A., "Band-gap narrowing and band structure in degenerate tin oxide (SnO2) films" Physical Review B, Volume 44, Page 5672 –5674, 1991.

[7] Themlin,J., Mohammed Chtaib, M., Henrard, L. et al. "Characterization of tin oxides by x-ray photoemission spectroscopy", Physical Review B, Volume 46, Page 2460-2466.

[8] Hofmann, S., "Quantitative depth profiling in surface analysis: A review, surface and interface analysis", Volume 2, Issue4, Page148-160, 1980.

Inverted Scan Transducer Mount Technique: A Cost Effective Acoustic Scanning of IGBT Modules for Failure Analysis

Em Julius De La Cruz[1], Sheenel Karl De La Rea[1], Stephen McDonough[2], SF Chai[3]

ON Semiconductor Malaysia Sdn. Bhd. [1], OKOS Solutions LCC USA[2], QES (Kuala Lumpur) Sdn. Bhd.[3]

Lot 122, Senawang Industrial Estate, 70450 Seremban, Negeri Sembilan, W. Malaysia

Abstract:

Acoustic Scanning for IGBT modules is a critical process to find anomalies that could lead to field failures. However, the cost to build this capability for failure analysis use is relatively expensive. This paper aims at evaluating a cost effective acoustic scanning technique for IGBT modules suitable for failure.

I. INTRODUCTION

IGBT (Insulated-Gate Bipolar Transistor) modules are gaining popularity in the market because they provide high efficiency and fast switching in aircraft, electric cars, trains and other critical environments. Large IGBT modules typically consist of many devices in parallel and can have very high current handling capabilities in the order of hundreds of amperes with blocking voltages of up to 6000V, equating to hundreds of kilowatts.

Because they are high-voltage and high-power switches, IGBT modules generate a great deal of heat that must be dissipated at a rate sufficient to avoid over-heating.

The most common causes of overheating in IGBT modules are the following:

1. Voids, delaminations or other gaps within or adjacent to a solder thermal interface material. Even if they are very thin, gaps are efficient insulators.
2. Warping or tilting of ceramic element that may cause differential thickness of solder.

Figure 1 shows an example of an IGBT module. An IGBT module is typically composed of multiple layers. Figure 2 illustrates the different layers of an IGBT module. The base plate is normally a Nickel-plated Copper; where a Die-Bonded Copper (DBC) assembly is mounted by a solder material, followed by another layer of solder die attach to mount the IGBT die. The importance of this solder die attachment layer is one of the most uncontrolled processes in today's power module fabrication which is practiced by the IGBT modules manufacturers. Typically a significant amount of solder die attached was dispensed on the area to be attached, that brought two surfaces together then reflowed on an oven. The paste is manually dispensed, which results a random thickness layer from one surface to another surface. As a result, the level of voiding on the paste is extremely significant.

Figure 1. An example of IGBT module.

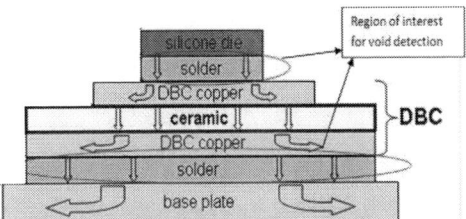

Figure 2. Typical IGBT layers. Solder layers are the region of interest to be checked for voids.

Voids, delaminations and other gaps at the module's internal interfaces can form even during assembly. Any type of gap in IGBT modules may grow larger as a result of repeated thermal cycling. At some point the gap becomes large enough to overheat the die, and the module fails electrically. Hence, effective screening methods to detect defects during product development stage and during production are needed to screen out possible defects.

X-Ray imaging and acoustic micrography imaging are the most common tools for flaw detection. Usually, X-Ray imaging is not suitable to all types of materials. X-Ray imaging is based on the variation of X-Ray attenuation by a solid body. However, the sensitivity of this technique is dependent on the thickness of a material, which the beam needs to penetrate. Acoustic image scanning uses reflected ultrasonic signals to generate images representative of internal structures of semiconductor devices, metals, plastics and composites. As semiconductor devices become more complex and multi-layered, extraction of different variants of information from reflected ultrasonic signal, to help end-users interpret the data, becomes critical [1].

978-1-4799-3911-4/14 $31.00 © 2014 IEEE

However, manufacturers of IGBTs have little interest in conventional scanning method where water comes in contact with the die at the top of the module. For IGBT modules, the usual method used in protecting the critical components from direct water exposure while scanning is to flip the sample so the critical side of the sample is facing down, away from the flow of water but this is critical and can cause contamination from water splash back.

To solve this problem, there is an available tool in the market utilizing inverted transducer that uses a water plume. Water plume uses inverted waterfall/squirted transducer from underneath and the plume of water constantly makes contact while the system scans the sample, the water level stays below the top. This non-immersion technology keeps the top of the module dry and gives good acoustic access to the internal interfaces.

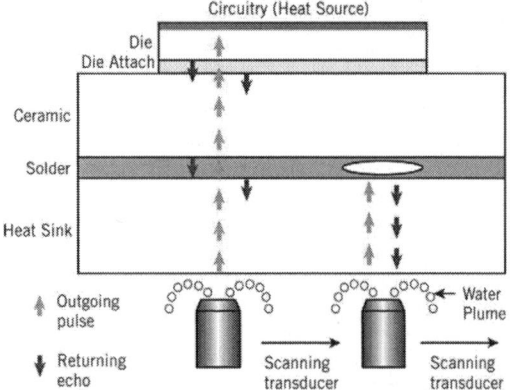

Figure 3. The water plume technique scans IGBT modules from the bottom side, leaving the topside circuitry dry. The right side of Figure 3 demonstrates what happens when the pulse encounters a gap, even a gap as thin as 1 micron. Because solder and air have profoundly different acoustic properties, the solder-to-air interface at the bottom of the void or other gap reflects virtually all of the pulse back to the transducer.

Figure 3 illustrates water plume technique mechanism. The transducer scans the bottom of the heat sink, pulsing ultrasound into its surface. Ultrasound is propagated upward through the heat sink, and sends back echoes from both the bottom and top of the solder layer. At each of these interfaces, a portion of the ultrasound is reflected back to the transducer, and another portion travels on. The pulse then reaches the die attach and sends back echoes from the top and bottom interfaces of the die attach. Figures 4 and 5 are acoustic images taken from water plume technique. Ultrasound is "interface-sensitive" because it is reflected only from the interfaces between both solid materials and gaps but not from the bulk of homogeneous materials. The ultrasonic frequencies used for imaging IGBTs are typically from 30 MHz to 50 MHz [2].

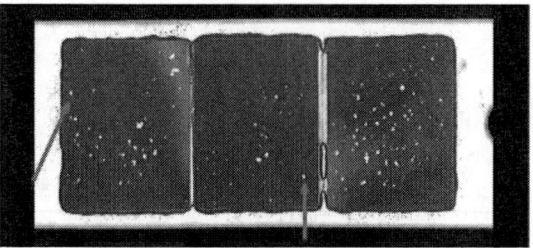

Figure 4. Acoustic image of Base to Solder interface using water plume technique. The image was taken at 50MHz frequency. White areas indicate void/gap.

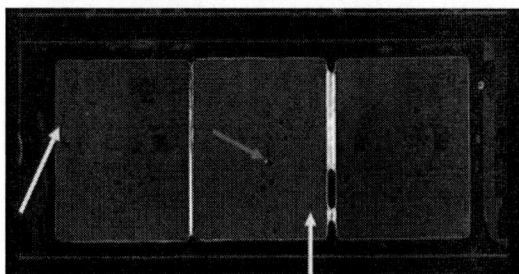

Figure 5. Acoustic image of solder to ceramic interface using water plume technique at 50MhZ frequency.

The system can do voids calculation by providing binary images into good and bad pixels and gives percentage of the defect area. The user can also define the accept and reject criteria.

The same water plume system is also capable of thickness measurement. The technique utilizes Time Difference Mode which evaluates the thickness of the solder at every x-y coordinate into which the scanning transducer pulses ultrasound. At each x-y location, the scanning transducer receives two return echo signals, one from the heat sink-to-solder interface and one from the solder-top late interface. The difference between the arrival times of the two echoes is measured and converted into the thickness of the solder at that x-y location. Scanning the entire area of the ceramic plate yields a map of solder thickness [3].

The transducer pulses ultrasound into thousands of coordinates per second. The resulting acoustic image is a detailed map of solder thickness across the IGBT module. Figure 6 is the Time Difference Mode acoustic image of one portion of a large multi-die IGBT module, imaged from the heat sink side. The ceramic pieces and the die are thus below the heat sink; it is the near side of the ceramic pieces that forms the rectangles seen in the image. This image displays solder thickness rather than the acoustic reflectivity of each feature. The thickest areas solder are pink, while the thinnest areas are orange. Most of the numerous irregularly shaped voids in the solder are red, not because of their high reflectivity but because they are very close to the top of the solder and the Time Difference software measures the distance from the top of the solder to the top of the void. Some smaller voids deeper in the solder are blue [4].

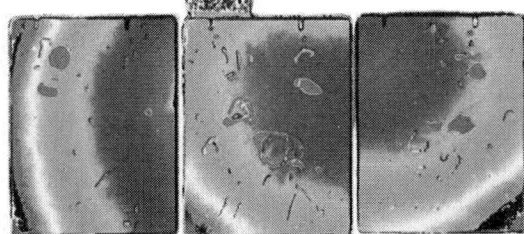

Figure 6. Time Difference Mode image showing thickness variations. The solder in the unit at the far left is thicker toward its right edge (pink) and much thinner toward the left edge (orange). The cause of the variation in the thickness of the solder may be a tilted ceramic piece beneath the solder.

The water plume technique is a breakthrough solution to perform non-destructive imaging and evaluation of IGBT modules. It is also a good screening tool at production line to check the quality of the modules. However, capital to own this capability is relatively expensive, more so if this capability is to be brought in failure analysis. This motivated the authors to evaluate a cost effective acoustic scanning technique that will produce similar imaging quality and capability as the water plume technique.

The Scan Transducer Mount Technique was evaluated as a potential cost-effective method of acoustic imaging for IGBT modules. The objective is to evaluate the image quality of scan transducer mount technique in detecting and calculating voids on IGBT modules and to evaluate the capability to perform contour scanning, measurement and warp compensation.

II. EVALUATION RESULT AND DATA ANALYSIS

Scan transducer mount technique utilizes the concept of bottom scan technique where a frequency transducer is mounted from the bottom of the tank such that scanning can be performed from the bottom, thus keeps the top of the module dry. This is done by modifying the existing SAT machine tank to be deeper and building a U-shaped arm to hold the transducer (refer to Figure 7 and 8). The transducer arm can hold a high frequency transducer.

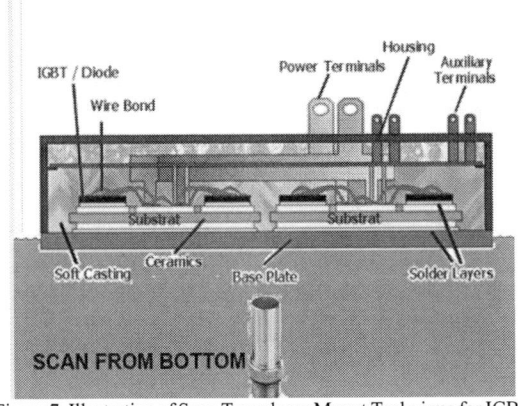

Figure 7. Illustration of Scan Transducer Mount Technique for IGBT module acoustic scanning.

Figure 8. Actual set-up of Scan Transducer Mount Technique.

The Scanning Acoustic Machine used for this study utilizes high frequency digital pulse receiver and 12-bit high speed digitizer. The transducer is immersed in water at the bottom of the tank, in which the water was used as a medium for ultrasound signals. The transducer scans the base plate of the module, pulsing ultrasound into its surface. The ultrasound is propagated upward through the base plate and reflected back to the transducer and is turned into a pulsed electronic signal that is used to display the returning signals in a time and amplitude raster scan to produce the image. Ultrasound pulsed signal into the base plate will create an image from the ceramic layer above the base plate up to the solder die attachments on the top module. Refer to Figure 9.

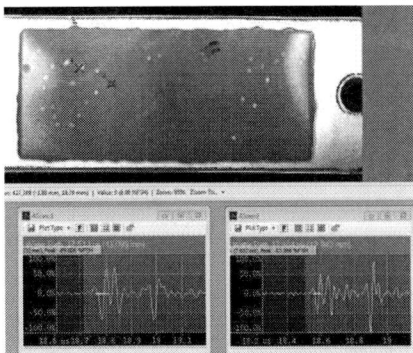

Figure 9. Acoustic scan image of the Base to Solder interface. White areas indicate void/gap.

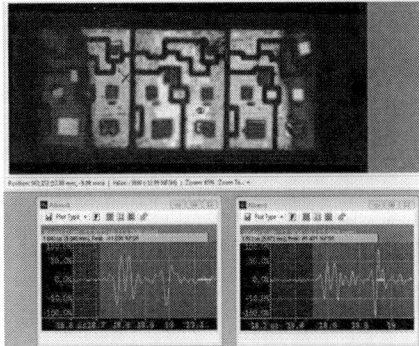

Figure 10. Acoustic image of solder to ceramic interface using at

50MhZ frequency.

It is necessary to get adequate signal penetration and layer separation. Based on the evaluation, a 50 MHz non delay line type produced the best images. Scanning images on sample IGBT module showed capability to scan through layers and detect solder voids. Void calculation can be performed using the software built-in to the machine. The machine's cluster analysis feature provides extensive reporting of individual voids and percentage voids of entire package or specific regions of interest as shown in Figure 10.

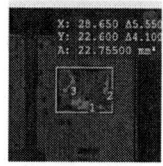

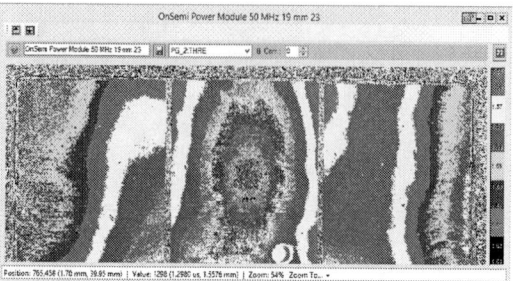

Figure 12. Thickness measurement using Time of Flight (TOF).

#	Count (px)	Size (px)	Area	Bounding Area	Bounds (mm)	Bounds (in)
0	1770	19.45% Total area	4.43 mm², 0.01 in²	—	—	—
1	1101	12.10%	2.75 mm², 0.00 in²	6.04 mm², 0.01 in²	3.45 x 1.75	0.14 x 0.07
2	358	3.93%	0.90 mm², 0.00 in²	2.07 mm², 0.00 in²	0.70 x 2.95	0.03 x 0.12
3	311	3.42%	0.78 mm², 0.00 in²	2.23 mm², 0.00 in²	1.35 x 1.65	0.05 x 0.06

Figure 10. Void calculation can be performed using built-in software in the machine and can be processed using excel file or other similar application.

Thickness measurement was performed using the Time of Flight (TOF) concept. The TOF imaging will show the minimum and maximum TOF results. As shown in Figures 11, to do a thickness measurement a surface follower gate and a data gate is needed. The Follower gate is placed on the first interface to be measured and the data gate will be placed on the second or next interface and the live thickness measurement can be obtained .

III. CONCLUSION

Based on evaluation results, the Scan Transducer Technique was found to be a feasible method for acoustic scanning of IGBT modules. The cost of ownership is relatively cheaper. A complete set of water plume technology machine would cost roughly USD250K. A complete set of Scan Transducer Mount Technique costs around USD150K. The cost will be even lower if there is already an existing system that can be modified by adding a deeper water tank and transducer arm.

The scanning technique provides capability to inspect and aid in failure analysis of IGBT modules, specifically in checking for solder voids. The technique was found to be reliable and repeatable in defect detection and integrity assessment of IGBT modules.

REFERENCES

[1] Hari Polu & Steve McDonough, "Scanning Acoustic Microscopy (SAM) Imaging Tecniques" OKOS Solutions, LLC.

[2] Tom Adams, "Inverted Acoustic System Cuts IGBT Failures," Power Electronics Technology Exclusive Insight; 09/02/2011, p1.

[3] Tom Adams, "Size Up Component Defects Non-Destructively," U.S. Tech; August 2013; p2.

[4] Tom Adams, "Acoustically Mapping IGBT Module Solder Thickness," Power Electronics Technology Exclusive Insight; 11/5/2012, p1.

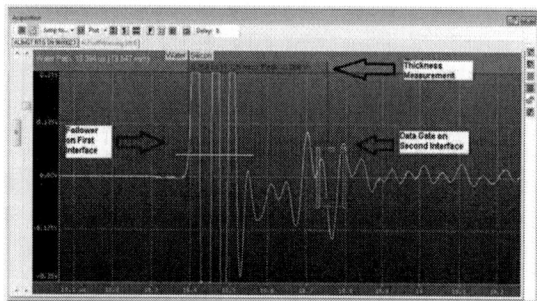

Figure 11. Thickness measurement using Digital Scope.

Based on the result, thickness measurement is able to identify warpage or thickness variations of the interface. Contour Scan was performed as shown in Figure 12 using Surface Mapping. As warping is a mechanical problem, the machine maker has designed this feature to compensate by moving Z axis during scan axis movement.

ATR-FTIR, DUAL BEAM FIB-SEM, TEM and TOF-SIMS Studies on High Temperature and Moisture Induced "White Haze" Following the Pattern of Electrodes in Touch Panels

Chen Yixin, Hao Meng, Shao Jingjing, Lee Esther, Khoo Bing Sheng, Chooi Meailing, Li Kai, Xin Qiuju, Kon Cambridge, Lee Hwang Sheng, Shen Yiqiang, Song Lu, Xing Zhenxiang, Zhou Yongkai, Feng Yang, Fu Chao, Hua Younan and Li Xiaomin

WinTech Nano-Technology Services Pte. Ltd.
10 Science Park Road, #03-26 & #03-28, The Alpha
Singapore Science Park II, Singapore 117684
Phone: (65) 92256313. Fax: (65) 67772462. Email: yixin@wintech-nano.com

Abstract - **White haze or the so called mura effect has been recognized as a common defect in touch panels. Nevertheless, the underlying mechanism has not been fully understood and clearly investigated. In this study, a comprehensive characterization study using the ATR-FTIR, DUAL BEAM FIB-SEM, TEM and TOF-SIMS on the high temperature and moisture induced white haze, which follows the pattern of electrodes in touch panels, is first reported. It is suspected that the white haze is a moisture induced reflection alteration phenomenon of the OCA (optically clear adhesive), while the electrodes related pattern is highly dependent on the local variation in hygroscopic swelling.**

I. INTRODUCTION

Researchers have observed that the OCAs can turn hazy after being subjected to the high temperature and humidity accelerated aging test. Usually, the haze would disappear and the transparency is able to reappear over time upon storage. However, in the touch panel industry, no one has reported that the OCA related white haze may follow the electrodes related pattern and carried out an in-depth analysis on the effect. In this paper, the electrodes related white haze effect has been comprehensively studied by various advanced characterization tools including the ATR-FTIR, DUAL BEAM FIB-SEM, TEM and TOF-SIMS. Clegg and Collyer reported that the humidity affects the performance of the polymer, as the polymer absorbs moisture and holds water molecules rather firmly by hydrogen bonding, which causes a slow variation of properties dimensions [1]. The OCA studied in this work is one type of polymers and should follow the above rule. We first provided the evidence of moisture residence at the interface of the OCA and cover glass after being subjected to the high temperature and humidity accelerated aging test using ATR-FTIR analysis, further demonstrating the electrodes related pattern of the white haze is highly dependent on the location variations in the hygroscopic swelling of the OCA by using the methodology of combining the FIB-SEM-TEM structural analysis and the TOF-SIMS ion intensity analysis.

II. . EXPERIMENTAL RESULTS

A. High temperature and humidity accelerated aging test

Several touch panels were subjected to the high temperature and humidity accelerated aging test, under the condition at temperature of 90°C and relative humidity of 80%HR. The white haze effect appears in some of the testing panels, while others remained intact. We denote the former ones being "bad panel" and the later ones being "good panel". Moreover, the coverage area of the white haze is related to the storage duration. Fig. 1a shows that the edge of the bad panel turned hazy after being subjected to a 12-hour aging test, while Fig. 1b shows that the patterned white haze appears over the whole area after being subjected to a 24-hour aging test. The above phenomenon indicates that the white haze started from the edge and extended to the whole area of the panel with the aging time Increasing. Upon two-day storage, 80% of the bad panels returned transparent. All bad panels returned transparent upon one- week storage.

B. Sample preparation

Immediately after a 24-hour aging test, both the bad panel and the good panel were mechanically peeled off and divided into two parts as illustrated in Fig. 2. The resulting two surfaces of the glass and the OCA were then exposed. The surface of the OCA is ready to be analyzed using ATR-FTIR, and the surface of the glass is ready to be characterized and investigated using the DUAL BEAM FIB-SEM, TEM and TOF-SIMS. At the surface of the glass, the pattern with alternating dark and bright areas, following the electrode pattern can be retained. Whereas, at the OCA surface, no such pattern is observable. After one-week storage, one bad panel after recovering from the white haze effect was subjected to the same sample preparation procedure as explained above and the OCA surface will be used for the ATR-FTIR analysis.

978-1-4799-3911-4/14 $31.00 © 2014 IEEE

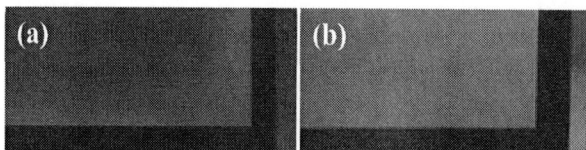

Fig. 1. Images showing a) the edge of the touch panel laminates turned hazy after being subjected to the 12-hour of aging test b) a patterned white haze appears over the whole touch panel laminates after being subjected to the 24-hour of aging test. At the surface of the glass, the pattern with alternating dark and bright areas, following the electrode pattern can be observed, whereas, at the OCA surface, no such pattern is observable.

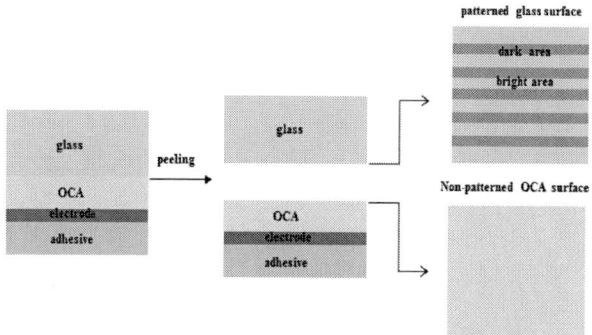

Fig. 2. Schematic illustration of the sample preparation procedure: the touch panel laminates were mechanically peeled off being divided into two parts. The resulting two surfaces of the glass and the OCA were then exposed and ready to be analyzed for different purposes.

C. ATR-FTIR analysis

FTIR analysis on the resulting OCA surfaces of the good panel, bad panel when the white haze just appeared and the bad panel after recovering from the white haze effect upon one-week storage were analyzed using the ATR mode of FTIR. Overlaying the three spectra (Fig. 3), the most intensive O-H stretching and H-O-H bending peaks were observed in the bad panel when the white haze just appears, whereas, the weakest O-H stretching and H-O-H bending peaks were detected in the good panel, indicating the least contained moisture. The similarities between the bad panel after recovering from the "white haze" and the good panel indicates that the "white haze" effect can be recovered due to the disappearance of most of the moisture.

Furthermore, the fact that other main peaks including C-H and C=O stretching peaks are nearly the same indicates the structure of OCA remained intact after the aging test and no other materials formed.

D. Dual Beam FIB-SEM and TEM characterization

The FIB cross sectional milling procedure and the in-situ lift out method in the DUAL BEAM FIB-SEM system were used to prepare the site specific cross sectional SEM and TEM lamella samples on the resulting glass surfaces of the good, the dark and bright areas of the bad panel. SEM top view and cross sectional images and the TEM cross sectional images were obtained to analyze the three surfaces. The SEM is operated at 5kev for the top view imaging and 2kev for the cross sectional imaging, while the TEM is operated at 200kev for the imaging.

three surfaces indicate different morphologies. Dispersed nano scaled particles can be observed in the bad panel, whereas a relatively smooth surface were obtained from the good panel. For the bad panel, compared to the dark area, larger and more condensed particles in the bright area were observed. SEM and TEM cross sectional images (Fig. 4a, 4b, 4c and Fig. 5a, 5b, 5c) confirmed that a nano scaled layer exists in all three surfaces. However, the layer on top of the good sample is thinnest and flattest, whereas the layer on top of the bright area of the bad panel is thickest and most rough. It is supposed that the nano scaled particles could be the OCA's residues, whose sizes and densities can cause difference in the reflection indexes. This result indicates that the electrodes related pattern in the original panel could depend on the local variations in hygroscopic swelling of the OCA. Nevertheless, this white haze pattern remained on the glass surface is not the same as the real white haze appeared on the original panel and it is not reversible.

E. TOF-SIMS analysis

The TOF-SIMS surface study using a Bi^+ gun under the condition of 25Kev and 1.0pA has been carried out on the OCA, the resulting glass surfaces of the good and bad panel. From the comparison among three mass spectra for both the positive and negative ions mode (Fig. 6), it is observed that there is no obvious variation among main peaks in both the peak positions and intensities, indicating the nano-scaled top layer and particles at all surfaces are residues from the OCA. Then the TOF-SIMS ion intensity analysis was carried out on the resulting glass surfaces of the dark/bright areas of the bad panel and the good panel. As shown in Fig. 7, although the concentrations of Na^+ and Mg^+ ions are below 0.1%, compared to the good panel, the concentrations of Na^+ and Mg^+ in the bad panel are much higher, while within the bad panel, the concentrations of Na^+ and Mg^+ in the bright area are slightly higher than those in the dark area, indicating more moisture contained in the bright area. This result further confirmed that the electrodes related pattern depends on the local variations in the hygroscopic swelling of the OCA.

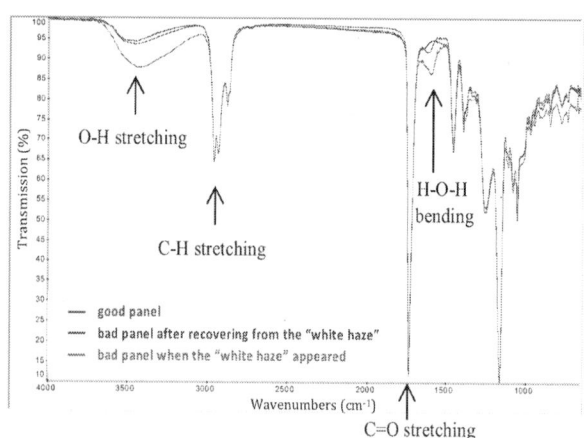

Fig. 3. ATR-FTIR spectra comparison of the resulting OCA surfaces of the good panel, bad panel when the white haze just appeared and the bad panel

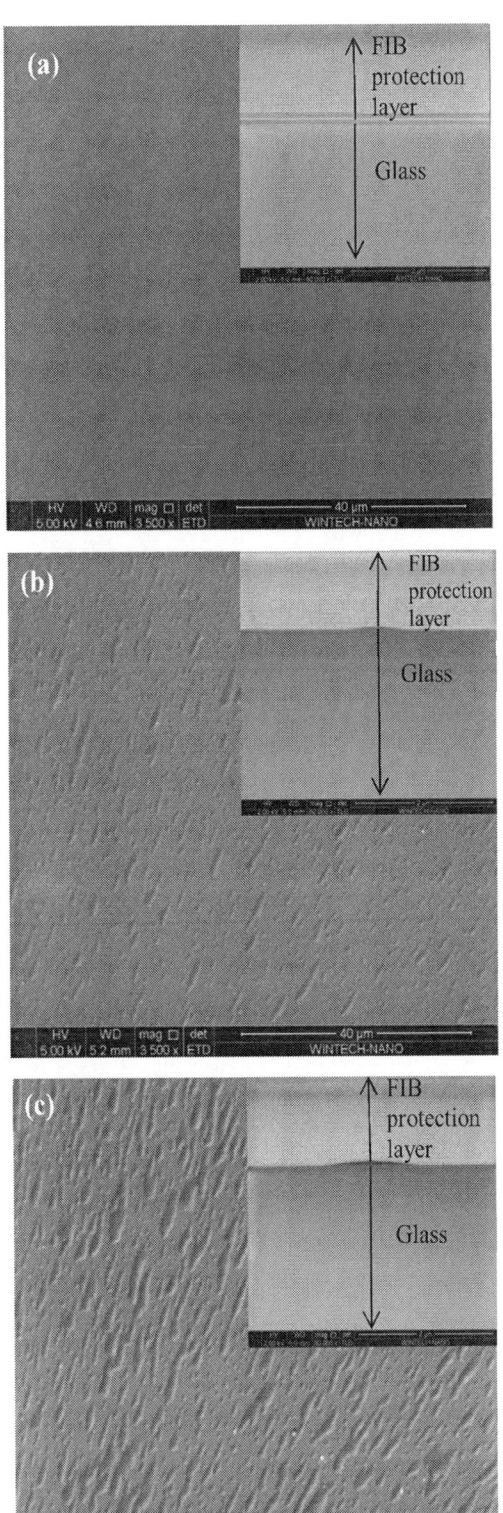

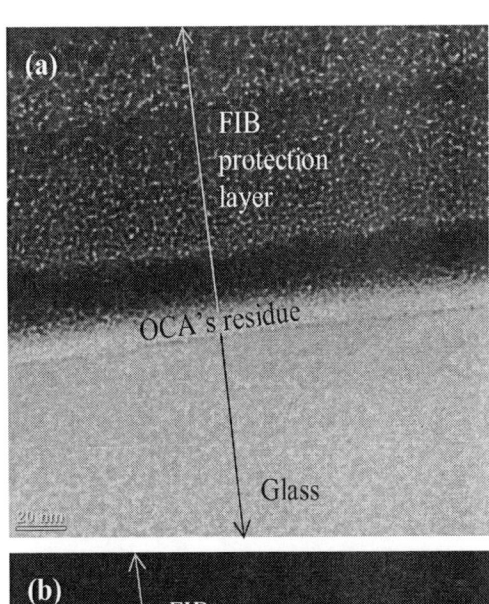

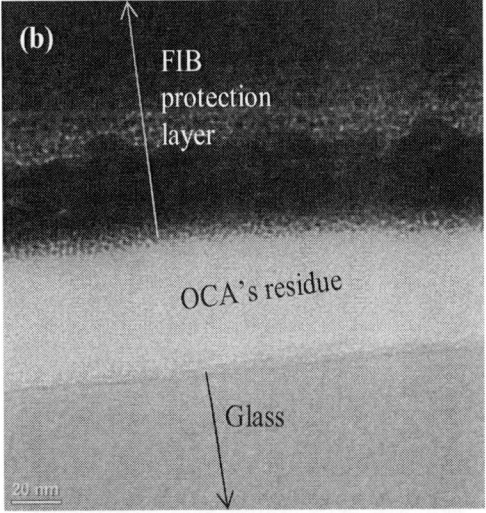

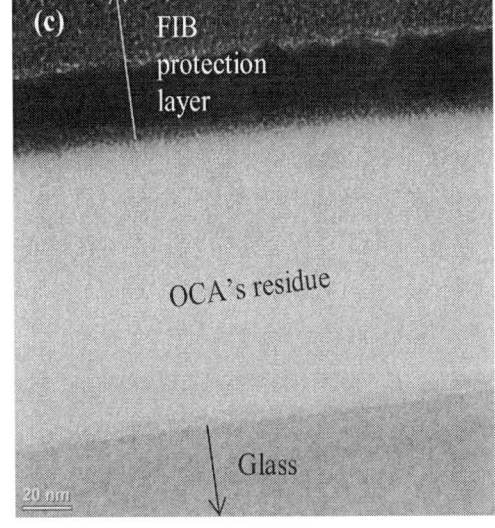

Fig. 4. SEM top view and cross sectional (top right corner) images on the surface of the a) good panel; b) dark area in the bad panel; c) bright area in the bad panel.

Fig. 5. TEM Images of the site specific lamellae, which were prepared using the DUAL BEAM FIB-SEM in-situ lift out method, for the surfaces of the a)good panel; and the surface particles indicated in Fig. 4 of the b) dark area in the bad panel; c) bright area in the bad panel.

978-1-4799-3911-4/14 $31.00 © 2014 IEEE

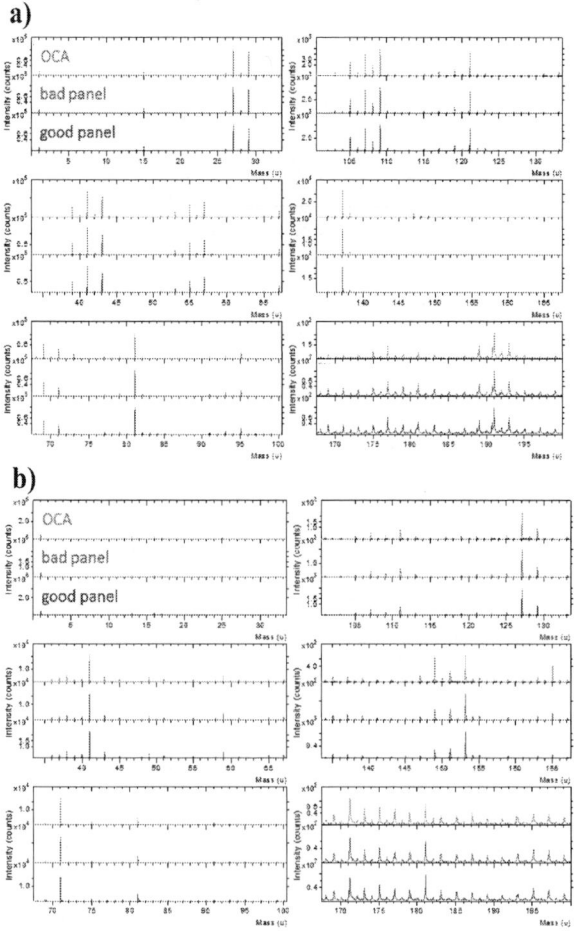

Fig. 6. TOF-SIMS surface mass spectra comparison among the OCA, the surface of the good panel and the bad panel using the a) positive ion mode and b) negative ion mode.

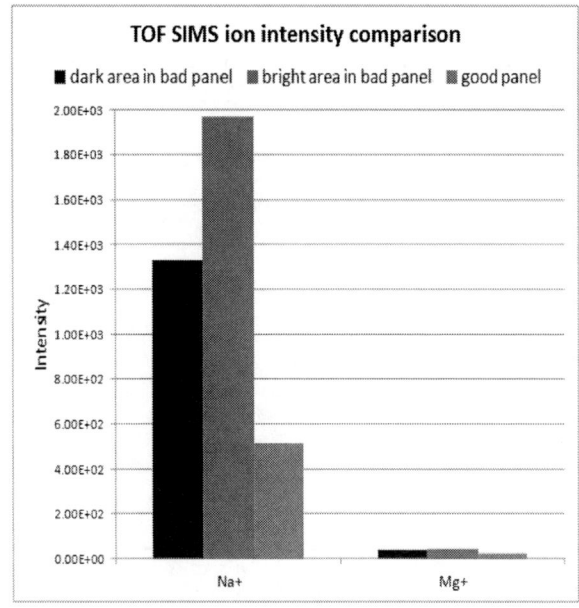

Fig. 7. TOF-SIMS surface Na$^+$ and Mg$^+$ intensity comparison among the

III. DISCUSSIONS

The white haze effect in touch panels after being subjected to a high temperature and humidity accelerated aging test has been extensively studied.

The ATR-FTIR result proves that the existence of the moisture between the OCA and the glass after the aging test and the white haze appearing in the bad panel is suggested to be a moisture induced reflection alteration phenomenon of the OCA. Both the results of ATR-FTIR and the TOF-SIMS surface analysis indicate that the structure of the OCA remained intact after the aging test and no other materials formed. Furthermore, the results of the DUAL BEAM FIB-SEM and TEM characterization and the TOF-SIMS surface ion intensity analysis imply that the electrodes related pattern of the white haze is highly dependent on the location variations in the hygroscopic swelling of the OCA.

IV. CONCLUSION

The underlying mechanism of the high temperature and moisture induced white haze, which follows the pattern of electrodes in touch panels has been comprehensively investigated by ATR-FTIR, DUAL BEAM FIB-SEM, TEM and TOF-SIMS. In summary, based on the analytical results of this study, the white haze effect is suspected to be a moisture induced reflection alteration phenomenon of the OCA, while the electrodes related pattern is highly dependent on the local variations in the hygroscopic swelling of the OCA. It is proved that the white haze effect is reversible and can be recovered after losing moisture.

ACKNOWLEDGMENTS

The authors would like to thank all colleagues who contributed to this work and the IPFA 2014 committee members for their great support.

REFERENCES

1. Clegg, D. W and A. A. Collyer, The Structure & Properties of Polymeric Materials, Prentice-Hall: The Institute of Materials, 1993.

On-Chip Device and Circuit Diagnostics on Advanced Technology Nodes by Nanoprobing

M. K. Dawood*, T. H. Ng, P. K. Tan, H. Tan, S. James, P. S. Limin, H. H. Yap, J. Lam and Z. H. Mai

GLOBALFOUNDRIES Singapore Pte. Ltd.
60 Woodlands Industrial Park D Street 2 Singapore 738406
Email: *mohammedkhalidbin.dawood@globalfoundries.com*

Abstract- **It is becoming increasingly challenging for conventional failure analysis methods to identify the failure mechanism at circuit level in an integrated chip. This paper demonstrates the utilization of nanoprobing for on-chip device and circuit debugging for defect localization at circuit level. FIB circuit edit was first performed to isolate the intended circuit. Next nanoprobing was performed on higher metal layer to identify the cause of failure. Nanoprobing was then performed at the contact level to verify the source of failure.**

I. INTRODUCTION

Advancements in transistor dimensional scaling and fabrication result in greater difficulty and complexity in identifying and understanding the failure mechanism at circuit level as well as at the transistor level in an integrated circuit (IC) chip.

Nanoprobing is becoming an increasingly critical tool for identifying non-visual failures via electrical characterization in current electrical FA metrology for fault isolation [1,2]. A nanoprobing system comprises of a scanning electron microscope (SEM) integrated with either 4, 6 or 8 nano-manipulators with tungsten (W) nanotips which are connected to a parametric analyzer for device characterization. Unlike conventional current-voltage characterization, whereby a wafer is mounted onto a probe station and probes are landed on bond pads to characterize a device or structure, nanoprobing involves controlling the nano-manipulators through an SEM computer monitor and landing tungsten nanotips at a particular area of interest at contact or a metal layer for direct transistor, device or circuit characterization. A system with 8 nano-probes allows for greater circuit analysis such as studying the static noise margins (SNM) in SRAM cells [3]. In addition, the ability to electrically characterize a particular failing location at circuit level opens up a new avenue for on-chip circuit and device diagnostics. This work demonstrates the capability of nanoprobing for such on-chip, device and circuit analysis by presenting FA case studies on advanced technology nodes to demonstrate circuit level diagnostics.

II. RESULTS AND DISCUSSION

A. Experimental setup

FA samples were decapped from a packaged die before undergoing conventional PFA to the desired layer of interest. Samples were delayered with a polisher with slurry containing silica particles of 0.05 μm and 3 μm in size. The samples were mounted onto an SEM stud with copper tape before being loaded into DCG's nProber system in a Zeiss' Supra 55 SEM. To minimize e-beam induced damages to the transistors, nanoprobing was carried out at a low acceleration voltage of 500eV [4]. FIB was performed in a Helios 450s from FEI.

B. FIB circuit edit and nanoprobing for circuit level debugging

A failed 28nm technology node IO circuit was found to be incapable of performing pull down operation. The failing location and its corresponding connections were provided by the customer. The suspected failing location was localized to a circuit made up of two cascaded PMOS transistors named PG PMOS and VREFP PMOS, as shown schematically in Fig. 1. The cascaded PMOS circuit is made up of 42 pairs of dual finger transistors. It would be time consuming and tedious to probe each of the 42 pairs of transistors at the contact layer, to determine the failed device/s. To mitigate this, the circuit was nanoprobed at a higher metal layer, metal 2 (M2), where all the transistors were connected to each other. However at M2 layer, the region of interest is connected to the rest of the IO circuit. In order to isolate the intended region, FIB cut was performed to isolate the drain, source and gate terminals of the cascaded PMOS structure. Fig. 1 also shows an SEM image of the circuit at M2 layer after FIB circuit edit. Locations of FIB cut are circled.

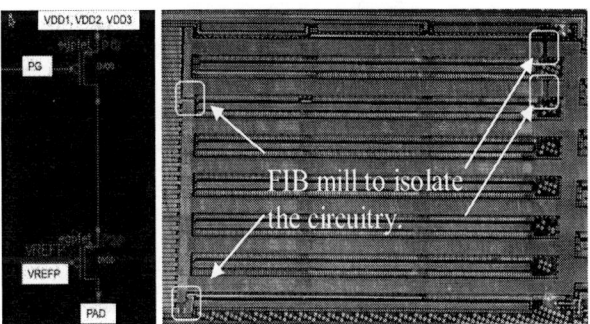

Fig.1. Schematic diagram of two cascaded PMOS devices and an SEM image of the PMOS circuit after FIB circuit edit. Location of FIB cuts are circled in white.

Next nanoprobing was performed on the sample at M2 layer. The biasing conditions are tabulated in Table I. Test 1 measures the Id-Vd characteristics of both PMOS at different gate voltage. Test 2 measures the Id-Vg of PG PMOS at different VREFP gate voltages and Test 3 measures the Id-Vg of VREFP PMOS at different PG gate voltages. The three tests are designed to study the individual transistor performance with respect to each gate terminal. Current limit of 2mA was set for all terminals.

Fig. 2 shows the I-V curves of the 3 tests in Table I. Fig. 2a, 2b and 2c shows the I-V curves for Tests 1, 2 and 3 respectively. The respective y-axis in Fig. 2 shows the current out the respective terminals. PMOS characteristics were observed for the cascaded PMOS device. However, an abnormally high gate current, on the order of milliamps (mA) through terminal VREFP (which is the gate terminal of PMOS VREFP) is observed. High gate current leakage to drain terminal (VREFP to PAD) was observed, which explains why the IO circuit was unable to perform a pull down operation.

	VDD (V)	PAD (V)	PG (V)	VREFP (V)	Test Summary
Test1	3.3	Sweep Voltage: (-3.3) to 3.3 Step: 0.05	P1=P2= 0, 3.3, 2.5, 2.6, 2.7, 2.8, 2.9		Id-Vd of both PMOS transistors with different Vg steps
Test2	3.3	0	Sweep Voltage: 0 to 3.3 Step: 0.05	3.3, 1.58, 0	Id-Vg of PG PMOS at different VREFP gate voltage.
Test3	3.3	0	3.3, 0	Sweep Voltage: 0 to 3.3 Step: 0.05	Id-Vg of VREFP PMOS transistor at different PG gate voltage.

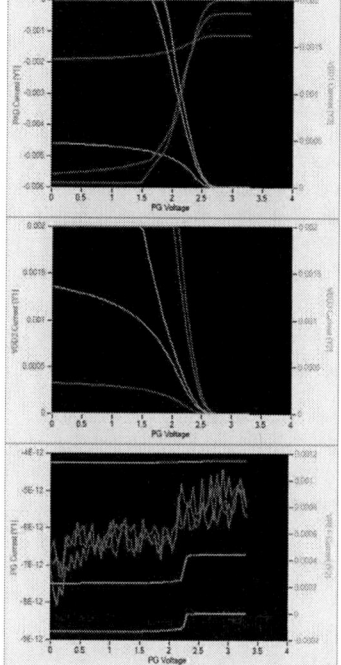

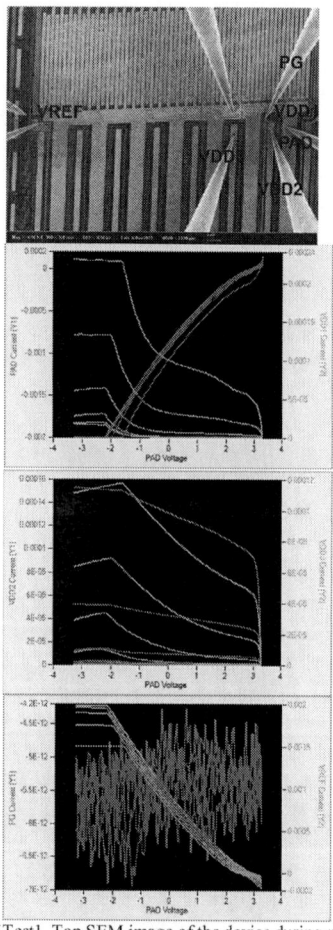

Fig. 2b. Id-Vg of Test 2. The first I-V curve plots PAD and VDD1 current with respect to PG voltage. The second I-V curve plots VDD2 and VDD3 current with respect to PG voltage and the third plots PG and VREFP current with PG Voltage. An abnormally large VREFP current is observed showing high gate leakage in the PMOS device.

Table II
BIASING CONDITION FOR CASCADED PMOS TRANSISTORS AT CA

	Drain (V)	Source (V)	Bulk (V)	G1 (V)	G2 (V)	Test Summary
Id-Vd G1	Sweep Voltage: 0 to -3.3	0	0	-3.3 to 0V Step 1.1V	-3.3	Id-Vd of G1 PMOS with G2 turned on
Id-Vg G1	-3.3	0	0	Sweep Voltage: 0 to -3.3 Step: 0.05	Step -3.3 and 0	Id-Vg of G1 PMOS with G2 turned on and off
Leakage Test	0	0	0	Sweep Voltage: (-3.3) to 3.3	0	

Fig. 2a. Id-Vd of Test1. Top SEM image of the device during nanoprobing with the respective terminals labeled. The first I-V curve plots PAD and VDD1 current with respect to PAD voltage. The second I-V curve plots VDD2 and VDD3 current with respect to PAD voltage and the third plots PG and VREFP current with PAD Voltage. An abnormally large VREFP current is observed showing high gate leakage in the PMOS device.

Table I
BIASING CONDITION FOR CASCADED PMOS TRANSISTORS

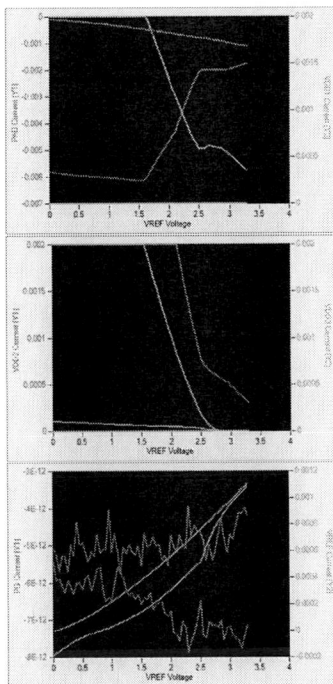

Fig. 2c. Id-Vg of Test 3. The first I-V curve plots PAD and VDD1 current with respect to VREFP voltage. The second I-V curve plots VDD2 and VDD3 current with respect to VREFP voltage and the third plots PG and VREFP current with VREFP Voltage. Abnormal Id-Vg curves are observed due to the large leakage VREFP current.

To further isolate the failing transistor/s, the sample was delayered from M2 layer to tungsten contact (CA) layer. SEM inspection revealed an abnormally bright passive voltage contrast (PVC) on the VREFP terminal (gate G1 terminal) of one PMOS finger as shown in Fig. 3. This device was further nanoprobed based on the biasing conditions in Table II which are the Id-Vd and Id-Vg tests for the PMOS device.

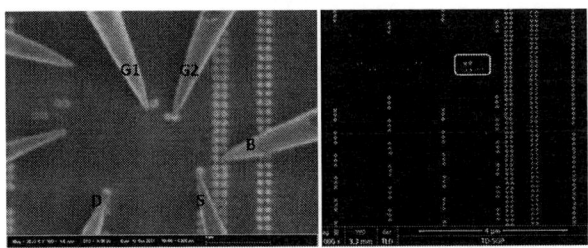

Fig. 3. Left image shows a PVC SEM image of an abnormally bright gate contact. While the right SEM shows the faulty PMOS device during nanoprobing. The corresponding terminals are labelled accordingly.

Fig. 4 shows the I-V curves listed in Table 2 of the faulty PMOS device with bright gate passive voltage contrast (PVC). Figure 4 studies the Id-Vd and Id-Vg of the dual gate PMOS by varying G1 and turning on G2. Terminals G1, G2, drain and source corresponds to terminals VREFP, PG, Pad and VDD at M2 layer. The I-V curves show high gate (G1) to drain leakage as observed during nanoprobing at M2. Fig. 5 shows similar I-V

curves of a neighboring PMOS device which shows normal transistor behavior as a reference.

Table III
BIASING CONDITION FOR CASCADED NMOS TRANSISTORS AT CA

	Drain (V)	Source (V)	Bulk (V)	G1 (V)	G2 (V)	Test Summary
Id-Vd G1	Sweep Voltage: 0 to 3.3 And sweep -3.3 to 3.3	0	0	3.3 to 0V Step 1.1V	3.3	Id-Vd of G1 NMOS with G2 turned on
Id-Vg G1	3.3	0	0	Sweep Voltage: 0 to 3.3 Step: 0.05	Step 3.3 and 0	Id-Vg of G1 NMOS with G2 turned on and off
Leakage Test	0	0	0	Sweep Voltage: (-3.3) to 3.3	0	

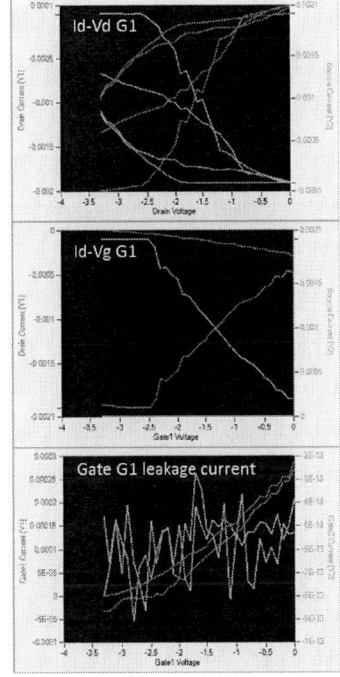

Fig. 4. Id-Vd, Id-Vg and gate leakage with respect to G1.

978-1-4799-3911-4/14 $31.00 © 2014 IEEE

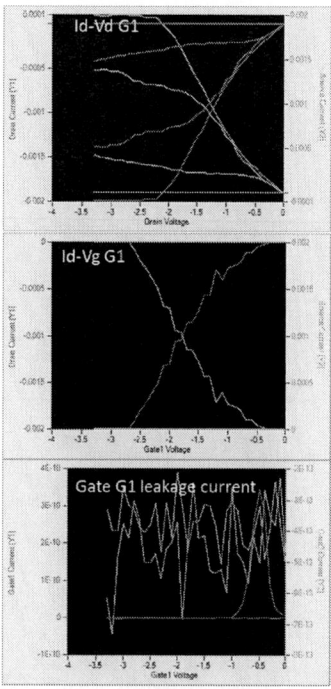

Figure 5: Id-Vd, Id-Vg and gate leakage with respect to G1 of a good reference PMOS.

After nanoprobing at CA, the sample was exposed to a Buffered-oxide etch (BOE) to remove the dielectric in order to expose the poly silicon (Poly). SEM investigation as shown in Fig. 6 revealed no abnormality in the poly, silicide and tungsten contacts. Cross section FIB performed across the poly revealed no abnormalities. The failure is attributed to a failed gate dielectric along the poly.

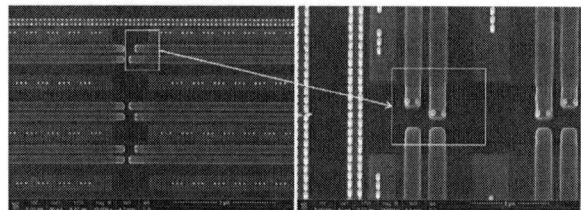

Fig. 6. SEM inspection on abnormally bright PVC on contact after BOE.

Another failed die with a similar failure at the NMOS region of the IO circuit was deprocessed down to CA layer. Similar to the case above, no abnormality was observed at back-end-of-line (BEOL) metal layers while bright PVC was observed on a gate contact as shown in Fig. 7. The sample was nanoprobed according to the biasing conditions shown in Table III. Fig. 8 shows the I-V plots of the failed NMOS device.

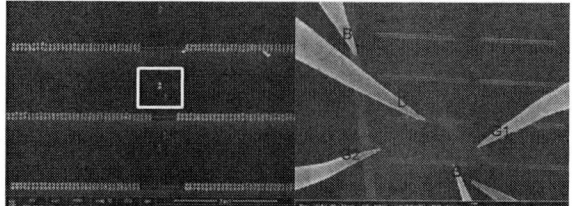

Fig. 7. Left SEM image showing bright PVC and right SEM image of the failed NMOS during nanoprobing with the corresponding terminals labelled accordingly.

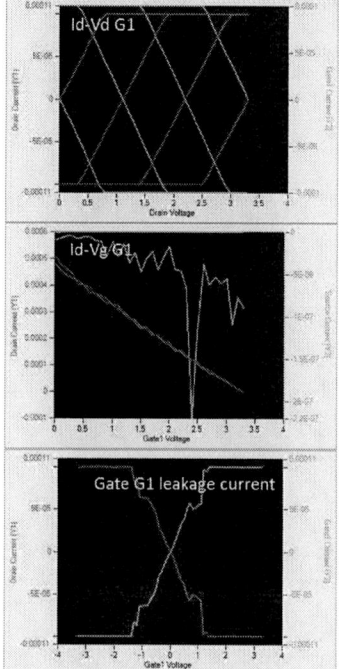

Fig. 8. Id-Vd, Id-Vg and gate leakage with respect to G1 of the failed NMOS device.

Fig. 8 shows no transistor behavior at the failed device. Instead a severe gate to drain short was observed as shown from the Id-Vd curve. Fig. 9 shows electron beam induced current (EBIC) composite image performed on a contact of terminal G1. A very strong EBIC signal is observed extending out to the drain junction and not to the source junction. This result helps to isolate that the leakage path is to the drain instead of source.

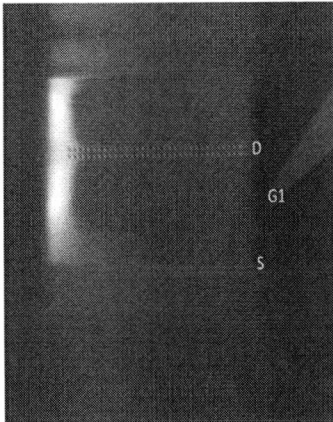

Fig. 9. EBIC composite image showing the strong EBIC signal under G1 to drain.

The sample then was subjected to BOE to expose the Poly gate. SEM inspection revealed an abnormality at a particular location between the poly and the silicon near the drain terminal, as shown in Fig. 10. TEM images shown in Fig. 11 (across the poly) revealed the occurrence of poly breakdown. In addition, the O-map shows gate oxide breakdown occurred at this location which matches the gate-drain short observed during nanoprobing.

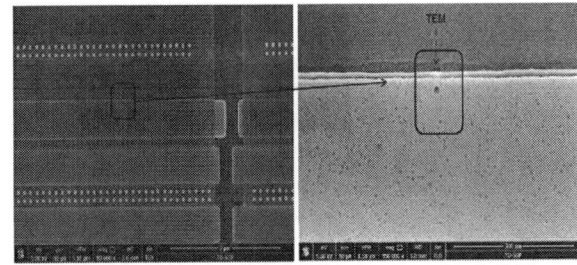

Fig. 10. SEM images of the failed NMOS device after etching in BOE was performed to expose the poly. An abnormality was observed between the gate and the silicon underneath it.

III. CONCLUSION

In this work we demonstrate using nanoprobing to characterize a failed circuit on chip. We demonstrated the application of FIB circuit edit followed by nanoprobing to isolate a failed circuit and to identify the cause of failure. FIB circuit edit allowed the isolation of a portion of the IO circuit from the rest of the circuit at metal layers. Nanoprobing on this edited circuit allowed us to verify and identify the cause of failure. Further delayering down to contact layer revealed the likely cause of the failure. Nanoprobing at the contact layer further verified the root cause of failure. The ability to characterize and verify electrical faults of circuit on-chip by nanoprobing is essential for rapid fault isolation for yield improvement.

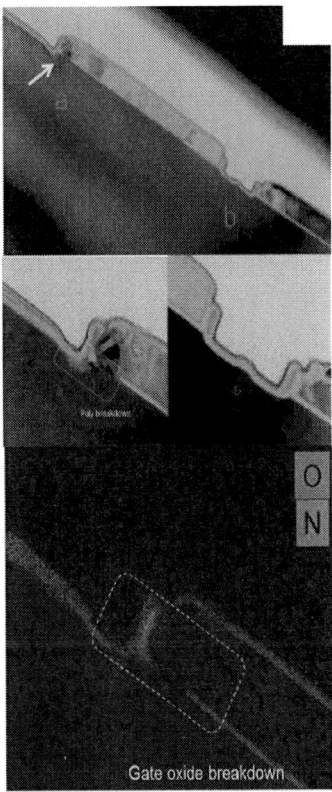

Figure 11: TEM images showing poly breakdown observed at the abnormality. (Bottom) O and N element mapping result shows the presence of oxide breakdown.

ACKNOWLEDGMENT

We would like to thank the Failure Analysis Department (Singapore) of GLOBALFOUNDRIES for their kind support in TEM and EDX works.

REFERENCES

[1] C-M Shen, T-C Chuang, S-C Lin, L-F Wen, C-M Huang, "Combining the Nano-probing Technique with Mathematics to model and identify Non-Visual Failures" Proc of 33rd ISTFA 2007, pp214-218.

[2] S. L. Toh, Z. H. Mai, P. K. Tan, E. Hendarto, H. Tan, Q. F. Wang, J. L. Cai, Q. Deng, T. H. Ng, Y. W. Goh, J. Lam, L. C. Hsia, "Use of nanoprobing as the diagnostic tool for nanoscaled devices," Proc of 14th IPFA 2007, pp. 53-58.

[3] M. K. Dawood, T. H. Ng, P. K. Tan, H. Tan, S. James, P. S. Limin, Y. Huang, J. Lam, Z. H. Mai, "Study of Static Noise Margin and Circuit Analysis on Advanced Technology Node SRAM Devices by Nanoprobing." Proc of 39th ISTFA 2013, pp 505-510.

4] T. H. Ng, M. K. Dawood, P. K. Tan, H. Tan, C. K. Oh, J. C. Lam, Z. H. Mai, "Study of Static Noise Margin, Cell Stability and Influence of Electron Beam on sub-30nm SRAM Using SEM-based Nanoprobing with 8 Nanoprobes" Proc of 38th ISTFA 2012, pp 112-117.

Simple, Novel and Low Cost Numerical Aperture Increasing Lens System for High Resolution Infrared Image in Backside Failure Analysis

Li Tian

Product Analysis Laboratory of Quality Department in Freescale Semiconductor (China) Limited

Phone: 86-22-85684614. Email: B19628@freescale.com

Abstract— **As is known to all, we could capture clearer infrared (IR) image from backside as Si substrate was thinner. But if we needed higher resolution image with conventional optical objective lens, we must introduce numerical aperture increasing lens(NAIL) technology or shorter wavelength light to improve numerical aperture (NA) in objective space. Now some vendors can provide NanoLens with NAIL but it is very expensive. In this paper, we proposed one simple and novel system of NAIL. Firstly, we fabricated two NAILs (one R≈3mm, the other R≈5mm) with glass material, and captured higher resolution IR image with NAIL help. Secondly, we found clearer image from smaller size NAIL by comparing IR images. Then, we studied how to moving NAIL on backside surface of die. Two moving methods were designed and we discussed their advantage and disadvantage, one of them was used in FA experiment. Although there were some limitation and disadvantage for this system, we believed this simple, novel and low cost NAIL system was beneficial to our FA from backside.**

Keywords-Fialure analysis; NAIL system; T-4-10 probe needle; IR image;

I. INTRODUCTION

As the development of VLSI and scaling down & multi-metal-layer of semiconductor devices, there was the obstacle for failure analysis (FA) from front side of device. Meanwhile, smaller metal line size and multi-metal-layer of semiconductor device brought the obstacle for FA from front side. So, FA from backside was developed in microelectronics yield in recent years. As is known to all, we could capture clearer infrared (IR) image from backside as Si substrate was thinner. But if we needed higher resolution image with conventional optical objective lens, we must develop new technology. To improve resolution (resolution= λ/2NA and NA=n*sin$u$/2) of backside IR image, we needed to enlarge aperture angle with optical structure or increase refractive index of objective space to improve numerical aperture(NA). Recently some researchers reported and studied some methods, for example Solid Immersion Lens(SIL), NAIL, shorter wavelength light in objective space and so on[1-4]. Now some vendors can provide NanoLens for improve resolution for backside IR image, but it is very expensive.

In many FA labs there is OBIRCH or similar equipment and in usual OBIRCH (Optical Beam Induced Resistor Change) equipment (for example HAMAMATSU PHEMOS1000 in our lab) there was a laser system with 1340nm and this wavelength was limited by energy band of Si. So, wavelength of laser couldn't be shorter. On the other hand,

NA depended on medium refractive index n and aperture angle u. With limited tools recently we only improved aperture angle to developed NAIL or SIL by inducing optical structure.

So, in this paper we proposed one simple, novel and low cost NAIL system. Firstly, we introduced principle numerical aperture increasing technology and fabricated two NAIL samples with glass material (one: R≈3mm; the other: R≈5mm). Then, higher resolution IR image was captured with NAIL help and we found clearer image from smaller size NAIL by comparing IR images. Secondly, we studied how to moving NAIL on backside surface of die. Two moving methods were designed and we discussed their advantage and disadvantage, one of them was used in FA experiment. Finally, we summarized characteristic of this system. Although there were limitations and disadvantages on this system, we believed this simple, novel and low cost NAIL system was beneficial to our FA from backside.

II. PRICIPLE OF NAIL

As is kwon to all resolution=λ/2NA and NA= n*sin$u$/2, so reducing the wavelength or increasing the aperture angle can improve NA and the resolution. Here, n also can be adjusted to lager, but it is limited to n_{Si} due to total reflection phenomenon (total reflection can decrease the amount of transmission light-beam). NAIL subsurface microscope technique is a significant method for capturing IR image and collecting emission light from backside in FA field. The NAIL is placed on the backside surface of a sample as illustrated in figure1 (a). Ideally the NAIL is made of the same material as the sample, and for FA Si is best material. The convex surface of the NAIL is spherical with a radius of curvature of R. Light passes through the sample and NAIL from an objective space in the sample at a vertical depth of X, see light path in figure1 (b). To increase the NA without introducing an additional aberration, the vertical thickness of the lens is selected to be $D=R(1+1/n)$-X. The addition of NAIL increases the conventional optical microscope's light-gathering and resolving power from that object. Increasing NA, which is ultimately limited to n, improves the resolution, which is limited to ≈ $\lambda_0/2n$ laterally and ≈ λ_0/n longitudinally.

Addition of the NAIL to a standard microscope increases the NA by a factor of n^2, up to NA = n. In Si at λ_0 =1μm, the NA is increased by a factor of 13, up to NA=3.6, corresponding to a lateral spatial resolution limit of 0.14μm.

The NA increase also allows for the use of a smaller NA microscope objective lens without sacrificing the spatial resolution of the NAIL microscope. A microscope objective lens with NA=0.3 is sufficient to achieve the highest spatial resolution for a Si NAIL microscope.

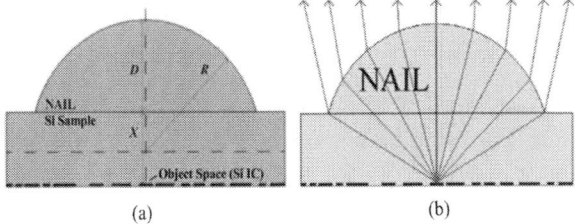

(a)   (b)

Fig.1 A NAIL placed on backside surface of sample (a) and emission light path form circuit with NAIL (b)

III.   HIGHER RESOLUTION IR IMAGE

According to above principle of NAIL, there are centric and aplanatic structure for NAIL, aplanatic structure is applied in our study. Convex surface of the NAIL was important optical structure for improving NA. Following we fabricated two NAIL samples with glass material and their radius of curvature are about 3mm and 5mm respectively. The reason of choosing 3mm & 5mm size: (1) NAIL must be limited it size due to our small size of die and package; (2) Polishing of NAIL is more difficult for smaller size. Then, higher resolution IR image was captured with 1340nm laser system and by comparing we found there was higher resolution with smaller size NAIL.

A.   fabrication of NAIL smaple

For NAIL there is a convex surface and a planar and the planar contacts the surface of a sample closely. Ideally the NAIL is made of the same material as the sample, both polished to allow intimate contact to avoid reflections at the planar interface. In our study, we only used glass material (refractive index is about 1.5) initially to polish NAIL with precise numerical control method, because we didn't find Si with enough thickness. In figure2 design diagram of our NAIL was showed, our NAIL should be hemispheric. But its height D is smaller than radius R, and it must be satisfy the rule (R-d)<D<R. After polishing our NAIL pictures are showed in figure3.

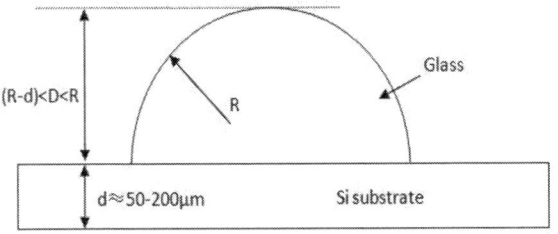

Fig.2 design diagram of our NAIL

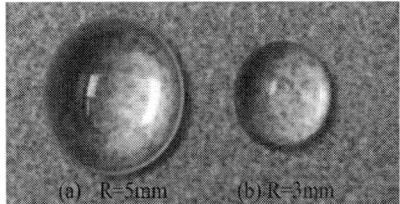

Top view

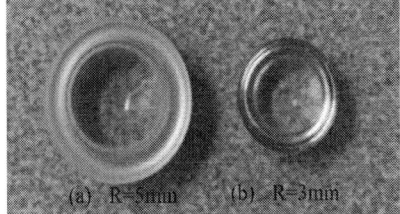

Bottom view

Fig.3 Top and Bottom view of two NAIL samples

B.   Capturing IR iamge

Thanks to high transmission index for 1340nm IR laser in Si substrate (curve of transmission index of Si substrate is showed in figure4[3-4]) and glass NAIL, incidence IR laser can be propagated to circuit layer from backside through NAIL and Si substrate and then is reflected back to objective space. The reflected light is detected by laser system to image the circuit IR pattern in software. The NAIL should be placed on Si backside; meanwhile the planar contact with backside surface of IC closely. Then IR image of circuit layer is observed through NAIL and Si substrate. The configuration of light path system is illustrated in Figure5.

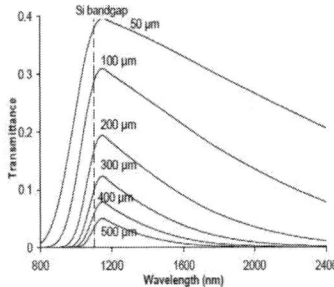

Fig.4 Photon transmittance of p-Si doped at $10^{19} cm^{-3}$ for different backside thicknesses

Some issue caused our result of experiment don't match to theoretical analysis perfectly, for example effect of chromatic aberration, scattering and so on. The microscopes are designed to minimize the effects of chromatic aberration over the range of mid-infrared wavelengths they collect. And refraction at the spherical surface of a NAIL imparts a chromatic aberration. Although we did our best to minimize the gap between planar of NAIL and Si backside surface, a finite space still remained (planar don't match to backside surface perfectly in practice). Chromatic aberration were related to a radius of curvature of R and there was different NA for different value of R, so we chose proper value of R in polishing NAIL (one is about 3mm; the other is about 5mm). Additive oil between planar and

978-1-4799-3911-4/14 $31.00 © 2014 IEEE   141

backside surface could reduce gap between them, which could increase reflected light beam input to objective lens and make IR image clearer.

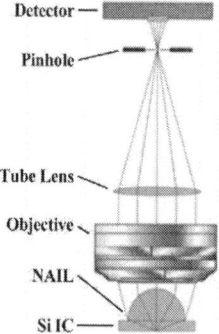

Figure5. Configurations of laser scan system, NAIL and Si IC

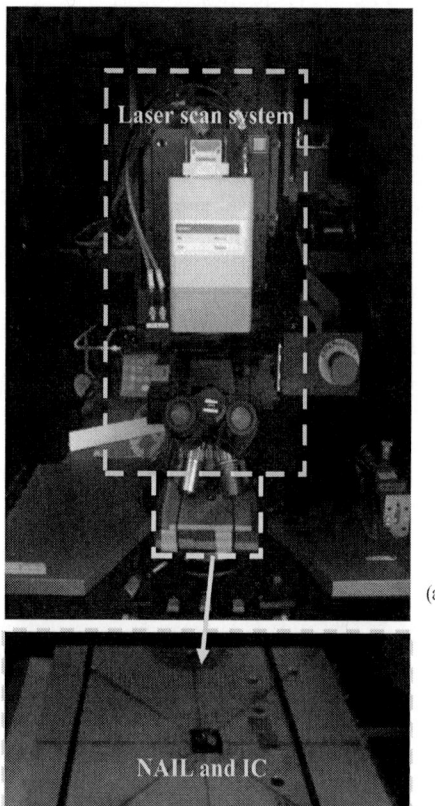

Fig.6 Our laser scan system (a), NAIL on Si IC (b)

For our setup, we employed 1340nm infrared (IR) laser scan system in HAMAMATSU PHEMOS1000. Our glass NAIL was placed on backside surface of device with oil between them. Setup was showed in figure6. Following we captured IR image from backside with two NAIL under different magnification, see pictures in figure7&8.

C. Discussion & Conclusion

To observe resolution of IR images clearer, we captured IR pattern from backside without NAIL, and IR images with 3mm&5mm NAIL under same magnification was placed in one drawing canvas. In figure7&8 IR images were cpatured under 20X & 50X objective lens and 10X eye lens. Process of IC device applied in experiment was SmartMOS8 0.25um techongly. We polished backside down to about 150um. On observed location there were many parallel metal lines, the separation distance of them was about 0.5um, and matel line width was also about 0.5um. By comparsion in figure7 we could make out single metal line with 3mm NAIL, but don't observe metal lines clearly with 5mm NAIL and without NAIL. In figure8, we could distinguish adjacent metal lines with 3mm & 5mm NAIL and image from 3mm NAIL was clearer than 5mm NAIL, while single metal lines wasn't still observed without NAIL obviously. in addition, we compared image IR with 5mm NAIL and without NAIL under 100X objective lens and 10X eye lens. As a result, clearer image was captured with 5mm NAIL as illustrated in figure9. The reason of no image with 3mm NAIL in figure9 was that unsymmetrical profile in polishing NAIL caused astigmatism of image and other factor was alignment issue for center axis of optical lens.

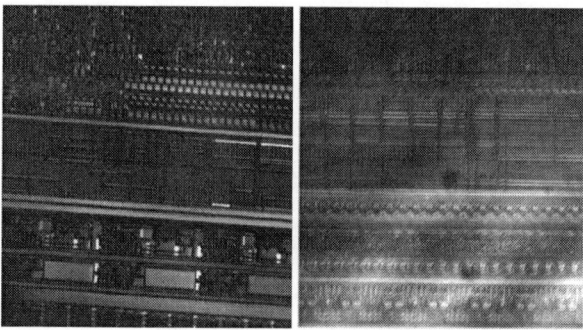

With R≈5mm NAIL   With R≈3mm NAIL

Without NAIL

Fig.7 IR image (optical lens magnification:200X)

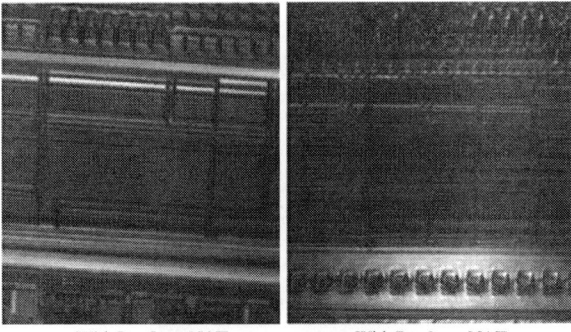

With R≈5mm NAIL   With R≈3mm NAIL

Without NAIL

Fig.8 IR image (optical lens magnification:500X)

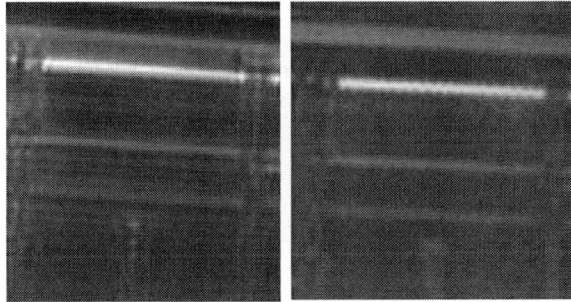

With R≈5mm NAIL   Without NAIL

Fig.9 IR image (optical lens magnification:1000X)

We also compared IR image with different thickness Si substrate without NAIL. With the help of thickness measurement method by using IR laser Si substrate thickness was obtained and clearer image with thinner Si substrate was observed [7]. See pictures in figure10.

According to above experiment and discussion, we could conclude: (1) there was higher resolution and clearer IR image for thinner Si substrate; (2) IR image with higher resolution could capture with smaller size NAIL.

I. NAIL MOVING METHOD

Our NAIL wasn't fixed on lens together, so it wasn't moved as lens moved synchronously. When capturing IR image with NAIL we occurred alignment issue for center axis of optical lens. Here we must design NAIL moving method to help analyst to capture better IR image. In this study, we proposed two NAIL moving designs and summarized their advantage and disadvantage respectively.

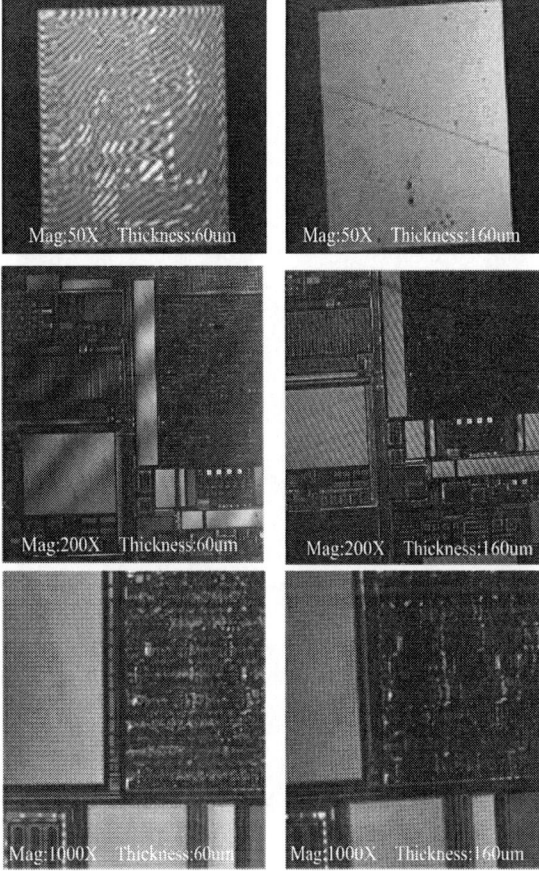

Fig.10 IR image with different thickness Si substrate without NAIL

A. Design of NAIL moving probe needle

Main idea of the first design was using probe needle to move NAIL with probe head. In our lab, we performed daily microprobing job with Karl Suss PH120 or PH150 probehead, which could let NAIL moving along X &Y axis. And then we needed to find one kind of probe needle which connected NAIL and probehead. And then we designed new probe needle as illustrated in figure11. It had two structures: ring and stick, they were connected on one point. When we moved NAIL, NAIL probe needle was stuck in arm of probehead and ring structure of NAIL probe needle covered on NAIL closely. We showed design of NAIL moving with probehead in figure12.

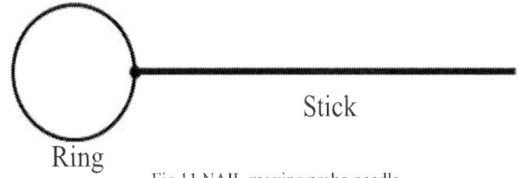

Fig.11 NAIL moving probe needle

978-1-4799-3911-4/14 $31.00 © 2014 IEEE   143

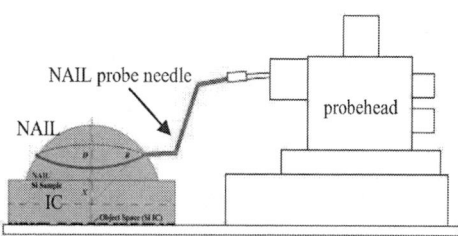

Fig.12 NAIL moving with probehead

With above design we proposed one fabrication method of NAIL moving probe needle, which was finished with old T-4-10 microprobing needle. After preparing a cylinder with proper radius of curvature of R, we wound one end of T-4-10 needle on the stick and other end was kept straight line. The cylinders maybe were pencil, steel stick and so on, meanwhile we highlighted R of the cylinder should be a little smaller than R of NAIL, but R of the cylinder was too small to make it move NAIL difficultly. The design and real fabrication of NAIL moving probe needle were showed in figure13 and figure14 respectively.

We performed practical operation for moving NAIL with our new probe needle, Karl Suss PH150 probehead and Karl Suss PM8 probestation. As a result, we found NAIL could be moved along X & Y axis perfectly. The NAIL moving setup photo was showed in figure15.

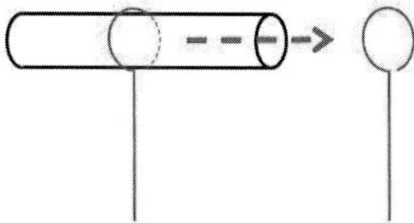

Fig.13 Design diagram of NAIL moving probe needle

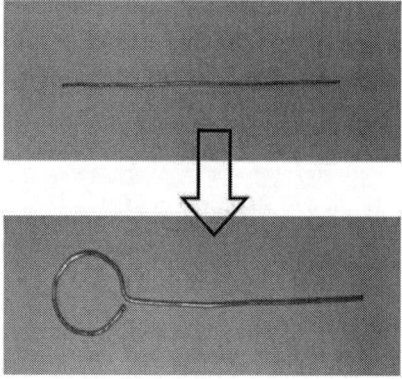

Fig.14 Fabrication of NAIL moving probe needle with T-4-10 mciroprobing needle

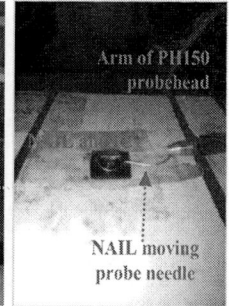

Fig.15 Practical operation of moving NAIL with new probe needle

B. Alignment method of NAIL and lens

The second method for NAIL moving was proposing one installation, which connected NAIL and objective lens together to align their center axis. Design diagram was showed in figure16. Here, NAIL structure for this alignment installation should be different with conventional NAIL. In our design there were some characteristics: (1) NAIL structure must have one ring groove around NAIL to fix NAIL on backside surface stably; (2) the spring was designed in strut structure to provide proper pressure on NAIL and avoid to separate alignment installation from ring groove of NAIL; (3) elastic or gluey material should be fixed between objective lens and alignment installation to ensure proper friction force or no any slippage between them; (4) hard material should be applied for alignment installation, for example stainless steel and so on.

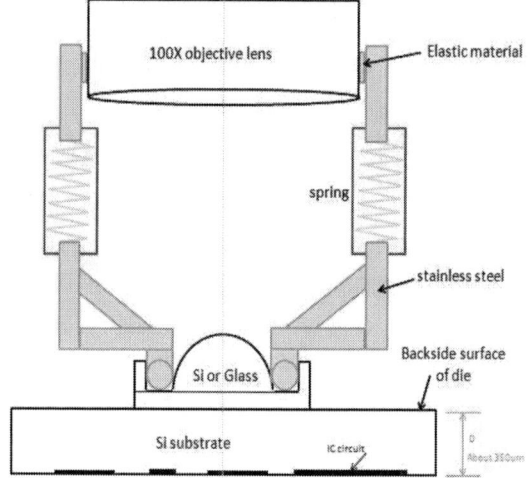

Fig.16 Diagram of alignment method of NAIL and lens

C. Discussion for two methods

For above two methods, they had respective advantage and disadvantage. NAIL moving probe needle method was simple and low cost, but it couldn't finish alignment task very well. With our alignment installation we aligned center axis of NAIL and objective lens perfectly, but its disadvantage were complicated fabrication. In practical operation we could move lens to align NAIL with first method, and the center axis of NAIL and lens should be on same straight line when IR image was clearest. If we needed NAIL for FA, best choice should be NAIL moving probe needle, because it was simple, low cost and could satisfy our requirement in FA.

II. SUMMARY

To improve resolution of IR image from backside we applied NAIL to captured image. But only NAIL on backside surface couldn't be moving to the location where we wanted to observe. So, we proposed one simple, novel and low cost NAIL system in this paper. Firstly, we introduced principle numerical aperture increasing technology and fabricated two NAIL samples with glass material (one: $R \approx 3mm$; the other: $R \approx 5mm$). Then, higher resolution IR image was captured with NAIL help and we found clearer image from smaller size NAIL by comparing IR images. Secondly, we studied how to moving NAIL on backside surface of die. Two moving methods were designed and we discussed their advantage and disadvantage, one of them was used in FA experiment perfectly. Although there were limitations and disadvantages on this system, we believed this simple, novel and low cost NAIL system was beneficial to our FA from backside.

ACKNOWLEDGMENT

For this work and study, I would like to thank for cooperation with Gaojie Wen, Miao Wu, Chunlei Wu, Diwei Fan, and Dong Wang. Meanwhile I sincerely appreciate my colleagues in Tianjin product analysis laboratory of Freescale semiconductor.

REFERENCES

[1] S. B. Ippolito, B. B. Goldberg, and M. S. Ünlü, "Theoretical analysis of numerical aperture increasing lens microscopy", Journal of Applied Physics, Volume97, issue 5, 2005.

[2] S. B. Ippolito, B. B. Goldberg, and M. S. Ünlü, "High spatial resolution subsurface microscopy", Journal of Applied Physics, Volume 78, Number 26, 2001.

[3] S. B. Ippolito, S. A. Thorne, M. G. Eraslan, B. B. Goldberg, and M. S. Ünlü, "High spatial resolution subsurface thermal emission microscopy", Journal of Applied Physics, Volume 84, Number 22, 2004.

[4] Li Tian, Kuibo Lan, Gaojie Wen, Miao Wu, Chunlei Wu, Diwei Fan, Dong Wang, "A Novel Optical Structure of Numerical Aperture Increasing Lens (NAIL) for Resolution Improvement in Backside Failure Analysis", Proc. of 20th Interna-tional Symposium on the Physical and Failure Analysis of Integrated Circuit, SuZhou, 2013, pp.509-511.

[5] JCH Phang, DSH Chan, M Palaniappan, JM Chin, B Davis, M Bruce, G Gilfeather, CM Chua, LS Koh, HY Ng, and SH Tan, "A Review of Laser Induced Techniques for Microelectronic Failure Analysis", Proc. of 11th Interna-tional Symposium on the Physical and Failure Analysis of Integrated Circuit, Taiwan, 2004, pp.255-261.

[6] JCH Phang, DSH Chan, SL Tan, WB Len, KH Yim, LS Koh, CM Chua, and LJ Balk, "A Review of Near Infrared Photon Emission Microscopy and Spectroscopy", Proc. of 12th International Symposium on the Physical and Failure Analysis of Integrated Circuit, Singapore, 2005, pp.275-281.

[7] Li Tian, Miao Wu, Gaojie Wen, Chunlei Wu, Diwei Fan, Dong Wang, "Thickness Measurement of Si Substrate with Infrared Laser of Optical Beam Induced Resistor Change (OBIRCH) in Failure Analysis", Proc. of 20th Interna-tional Symposium on the Physical and Failure Analysis of Integrated Circuit, SuZhou, 2013, pp.32-34.

978-1-4799-3911-4/14 $31.00 © 2014 IEEE

Localized FIB Delayering on Advanced Process Technologies

David Donnet, Oleg Sidorov, Pete Carleson, Chad Rue, Roger Alvis and Surendra Madala

FEI Company

5350 NE Dawson Creek Drive Hillsboro Or 97124, USA

Email: david.donnet@fei.com

Abstract-Good control over beam and chemistry conditions are required to enable uniform delayering of advanced process technologies in the FIB. The introduction of newer, thinner and more beam sensitive materials have made delayering more complicated. We shall introduce a new chemistry for device delayering and present results from both Ga and Xe ion beams showing its improvement over existing chemistries.

INTRODUCTION

Delayering of IC devices is an important tool for semiconductor failure analysis; reverse engineering and circuit edit activities [1]. Once a defect has been localized, it is necessary to isolate, inspect and perform failure analysis. One method to enable this is to remove layer after layer until the defect is exposed. Traditionally, this was performed by mechanical polishing, followed by observation in either an optical or electron microscope. While this method can provide reproducible results over large areas (which is important for reverse engineering), it is neither site-specific nor controllable, and the functionality of the chip is lost due to the destructive nature of the polishing.

The focused ion beam (FIB) offers the means to retain chip functionality whilst removing only localized regions during delayering [2]. This is a great advantage, allowing further analysis (voltage contrast, defect analysis, electrical probing etc.) or circuit edit to be performed once the layer of interest has been reached. However, it is no easy task to delayer modern IC devices in the FIB due to the large number of different layers (with vastly differing milling rates) which are present with varying thicknesses and mechanical properties. The complexity is compounded by the fact that multiple materials can exist together in a single layer, thus requiring what is called mixed field delayering [3]. Fig. 1 clearly demonstrates the difficulties that arise when trying to delayer in mixed field regions with the bare FIB beam only and the necessity to use gas assisted etching (GAE) to achieve planar delayering.

CONVENTIONAL GAE FOR DELAYERING

For the most commonly available Ga FIB tools, alternating XeF_2 assisted etch (for dielectric removal) and H_2O assisted etch (for mixed field removal) is a robust and reproducible method that produces planar delayering under the correct beam and chemistry conditions. Excellent success has been realized

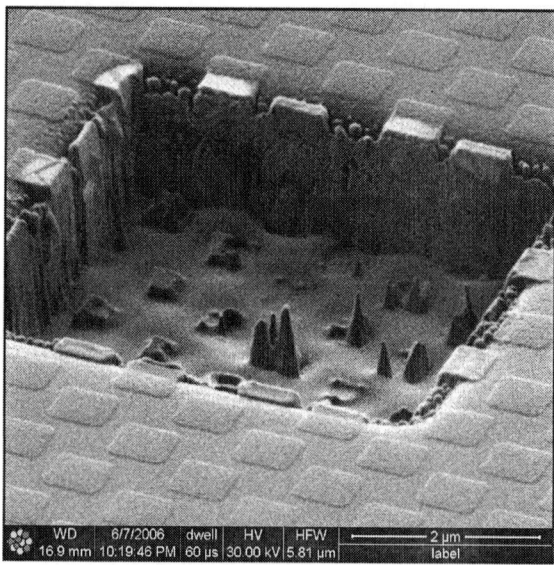

Fig. 1. Illustration of problems encountered when trying to delayer mixed field regions using bare beam milling.

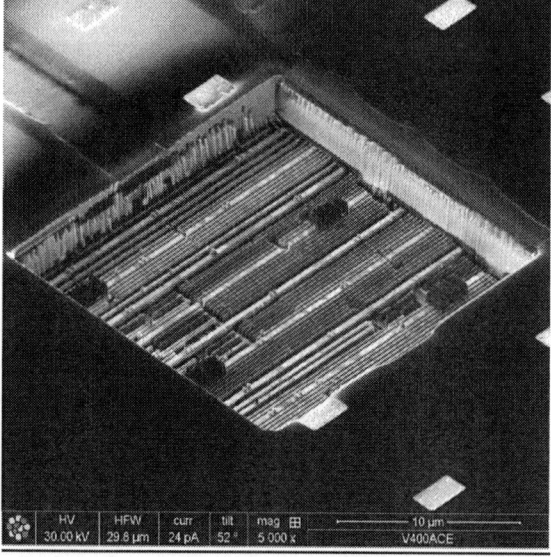

Fig. 2. Conductive redeposition is a consequence of delayering using conventional beam chemistry

in copper devices down to the 32 nm node. One disadvantage however of this technique is that the sidewalls of the trench are bright, which could indicate that some conductive material has been redeposited during the process which could affect device functionality. This is illustrated in Fig. 2. Decreased dimensions and the introduction of new materials make delayering at aggressive technology nodes ever more challenging. In particular, the switch from Si_3N_4 to SiC etch stop layers plays a significant role, due to the different interactions that layer has with currently used gas chemistries, whereby volatile compounds are formed which inhibit planar delayering and lead to uncontrollable dielectric etch. An example of this can be found in Fig. 3 where up to 3 different Cu metal layers can be observed simultaneously when attempting to delayer a 28 nm device using the XeF_2/H_2O method.

One method to work around this introduction of unwanted topography during the delayering process has recently been published [4]. This approach involves iteratively back filling the regions of topography with oxide followed by GAE to retain planarity.

In the next sections, a simpler method to retain delayering planarity will be introduced making use of a newly available beam chemistry which is able to planarise mixed field regions with excellent results from both Ga and Xe ion beams.

D_x BEAM CHEMISTRY FOR DELAYERING

A new beam chemistry D_x has been developed to deliver planar delayering even at the most aggressive technology nodes. In addition to better end results, the process is simpler since it is continuous without the need to stop the milling for changing gas chemistries when different layers are encountered. In addition, controllability is improved which is an important consideration when dealing with sub 100 nm thick layers. A 28 nm device with 9 metal layers in the low-k dielectric region was delayered over an area of 20 x 20 μm using an ion beam current of 1 nA throughout. This is a significantly larger area than was possible using the older techniques and yet excellent planarity is retained, all the way down to the poly-silicon level, as illustrated in Fig. 4.

In comparison to conventional delayering, another significant benefit of using the D_x beam chemistry is that the sidewalls of the milled trench remain free from conductive redeposition. An example of this is presented in Fig. 5 where cleanly milled sidewalls can be clearly observed.

The chemistry in the presence of Ga or Xe Plasma beam is able to volatilize the Copper Metallization along with SiO2 and low-k dielectrics thus reducing the redeposition. Any resulting compounds formed are non-conductive as the Cu metallization is well isolated.

The absence of the formation of redeposition when using D_x could expand its possibilities in advanced circuit edit. First tests on copper lines have shown that no shorting occurs when adjacent lines are cut, making it an excellent candidate to replace H_2O for copper line cutting. An example of copper line cutting is presented in Fig. 6. In this work, the lower line in

Fig. 3. Uncontrollable dielectric etching occurs when the traditional XeF_2/H_2O method is attempted on devices with SiC etch stop layers.

the image was cut with only the bare beam and it is clear that redeposition prohibits correct cutting of the line. When the correct beam parameters and D_x assisted milling is performed then voltage contrast shows the line is cut and the dark sidewalls of the metal show no presence of redeposition. The uppermost line in the image was cut with different beam conditions during D_x assisted milling and this time redeposition is present. Work on this particular application is at an early phase and will be reported upon fully at a later date.

Fig. 4. Poly-silicon exposed uniformly over an area of 20 x 20 μm following a 9 metal stack delayering with the D_x beam chemistry

978-1-4799-3911-4/14 $31.00 © 2014 IEEE

Fig. 5. Tilted FIB view of a device locally delayered using D_x. The absence of redeposition on the sidewalls is clearly observed.

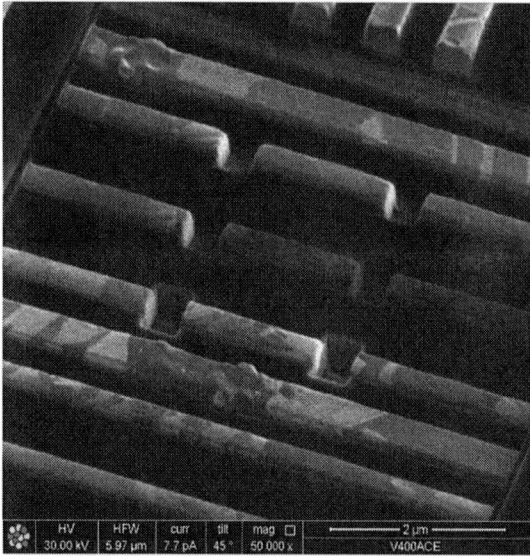

Fig. 6. FIB voltage contrast following the cutting of adjacent copper lines with and without D_x chemistry. The absence of redeposition is important for the correct cutting of minimum spacing lines.

Whilst excellent planarity has been achieved using D_x with a conventional Ga FIB, the obvious drawback is the area which can be delayered within a reasonable timeframe. The results presented here were achieved in less than 2hrs and delayering larger areas would become extremely time consuming.

The introduction of plasma FIB (PFIB) systems using a Xe source [5] where beam currents in the order of 1 μA are available can result in a 20X increase material removal speed compared to Ga FIB systems (where beam currents are limited to 50 – 100nA). This has opened up new applications related to 3D integration such as wide field of view cross-sectioning and large volume material removal as well as package level editing on length scales simply not feasible on a Ga FIB [6].

With the introduction of D_x, the opportunity to perform large area (at least 100 x 100 μm) delayering in the PFIB can now be realized.

Delayering in the FIB can be performed from either the frontside or backside of the device dependent on packaging constraints and which layers need to be removed. In this PFIB example, backside delayering is presented. Large scale XeF_2 assisted silicon trenching is extremely efficient in the PFIB. As shown in Fig. 7, this etching creates voids where the diffusion blocks were. When only the bare beam is used this topography propagates and planarity cannot be recovered.

In the same backside trench, D_x assisted milling was also performed with significantly different results. The milling box size was 50 x 50 μm and a beam current of 15 nA was used. Time required to remove each layer was approximately 4 mins. Fig. 8 shows the results of D_x assisted milling for different layers. Despite the added problem of holes being created during via milling, the planarising properties of D_x are particularly well suited to overcome these difficulties.

As a final comparison of the advantage that D_x assisted milling gives, cross-sections in a dualbeam were made of the process following delayering down to M3 to illuminate the difference in planarity.

Fig. 7. Trench floor following XeF_2 assisted removal of silicon (a). Deep voids are present where diffusion was present. Subsequent bare beam milling cannot remove this topography, resulting in uneven exposure of metal layer below (b).

When D_x is applied, an enormous improvement in planarity can be observed, as shown in Fig. 9.

Most beam chemistries are used to enhance the selectivity of etching one material over another (e.g. metal over dielectric). This is actually contrary to the requirements for planar delayering, but it is also clear that bare beam milling through mixed field regions cannot achieve the uniformity required for further analysis once the layer of interest has been reached.

Whilst the mechanism that D_x uses to attain planarity is not fully understood and currently the focus of investigation, it is clear that D_x depresses any selectivity that exists between the metal and dielectric layers. It is this that makes it particularly well suited to the planar removal of mixed fields regions, especially when ultra-low-k dielectrics and SiC capping layers are present.

Subtle changes in beam parameters can reduce this "selectivity" further and whilst this would certainly slow the process more, it could prove very useful for future even more advanced technologies where even more control is required to endpoint successfully on the correct layer.

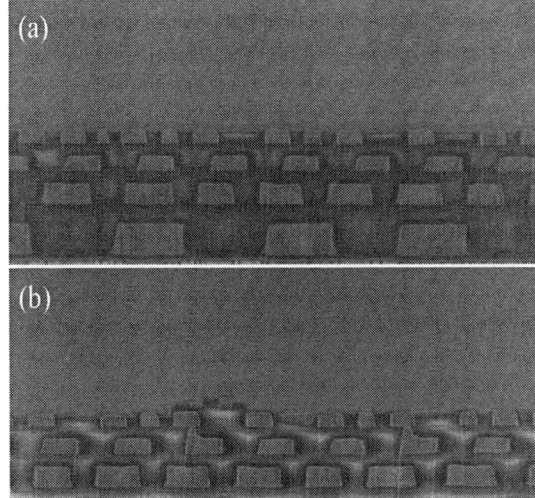

Fig. 9. Comparative cross-sectional images following delayering from backside down to M3 with (a) and without (b) D_x assisted milling. Bare beam milling results in non-uniformities at different length scales. Depositions are applied to the surface to aid the cross-sectioning.

CONCLUSIONS

The new D_x beam chemistry enables excellent planarity during delayering of mixed field regions, especially for technology nodes of 28 nm and below where traditional methods are no longer viable. Excellent results have been achieved using both Ga and Xe FIB beams for frontside and backside delayering. An additional benefit to switching to D_x, for delayering is that the sidewalls of the milled trench are no longer conductive, thus device functionality can be retained where necessary.

A further potential interesting application for D_x is that of copper line cutting. Provisional results are very promising but more work is required and this will be presented at a later date.

The ability to control the behavior of D_x through changes in beam parameters makes it a very powerful and versatile chemistry with potential for multiple applications.

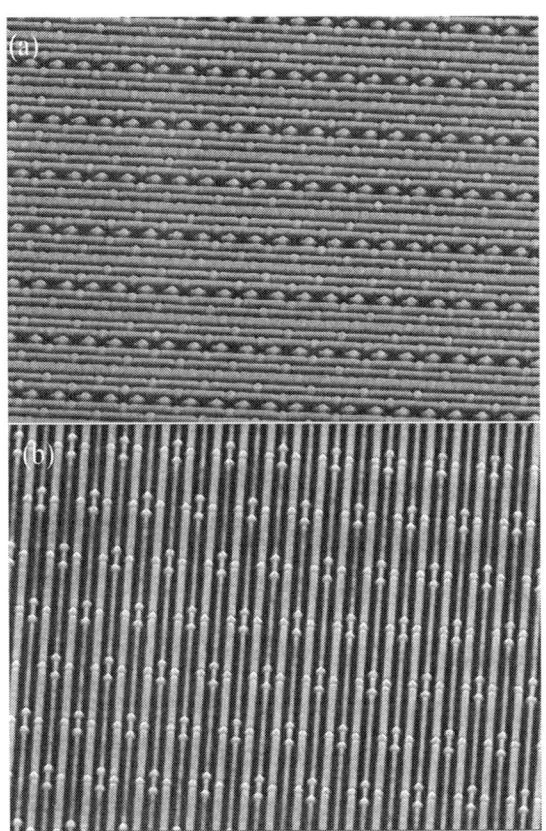

Fig. 8. Images following Dx assisted milling to M1 (a) and M2 (b) revealing excellent planarity of large areas.

REFERENCES

1. K. S. Wills and S.Perungulam "Delayering Techniques" Microelectronics Failure Analysis Desk Reference 5th Edition; p440.
2. D. Casey et al; "Advanced sub 0.13 µm Cu devices- Failure Analysis and Circuit Edit with Improved FIB Chemical Processes and Beam Characteristics" Proc. 28th ISTFA 2002 p553.
3. C. Rue; "Mixed Field Deprocessing" Presentation at EFUG 2006, Wuppertal, Germany
4. D. Serre, L Ma and P. Gounet, "Deprocessing of latest copper technologies for circuit edit and failure analysis" Presentation at EFUG 2013, Arcachon France
5. R.J. Young, "Site-Specific Analysis of Advanced Packaging Enabled by Focused Ion Beams" Electronic Device Failure Analysis, Vol. 13(1), 12-19, 2011
6. M. Gonzales et al "Developing Functional Prototypes by package modification using Plasma FIB technology" IPFA 2012

An Innovative Method to Overcome Signal Instability during TDR measurement of Power MOSFET

S.Y.Tan, K.K.Ng, S.Y.Gan, and C.K.Sin

Infineon Technologies (Malaysia) Sdn. Bhd.
Batu Berendam Free Trade Zone, 75350 Melaka, Malaysia
Phone: (606) 2873636 Fax: (606) 2515069 Email: SzeYee.Tan@infineon.com
Phone: (606) 2873574 Fax: (606) 2515069 Email: KiongKay.Ng@infineon.com
Phone: (606) 2873206 Fax: (606) 2515069 Email: Sue-Yin.Gan@infineon.com
Phone: (606) 2873695 Fax: (606) 2515069 Email: Victor.Sin@infineon.com

Abstract

Hand probing is the most common technique being applied in Time Domain Reflectometry (TDR) measurement. It is a simple and easy method, but it produced instability on overall signal. Therefore, a new test fixture is designed to maximize its reproducibility. Repeatability test will be used to show its effectiveness on Power MOSFET. In addition, with the test fixture, the standard deviation for impedance results was $\sigma=0.12$ based on the same part for eight times measurement, compare to $\sigma=3.26$ with hand probing. Therefore a significant improvement on the repeatability test was clearly demonstrated.

I. INTRODUCTION

TDR is a well-known measurement technique for characterizing and localizing the differences of impedance (impedance mismatch) in a transmission path [1]. Nowadays, it has been widely used in semiconductor industry as a non-destructive test on electrical failure device. Failures such as open, short and resistive contact can be detected quickly and easily.

In general, TDR measurement is made by launching a series of fast pulses into a close or open circuit and capturing the resultant impedance from the reflected signal which caused by the impedance mismatch of the circuit (or device) [2] as showed in Figure 1.

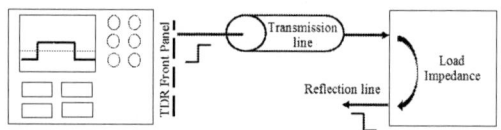

Figure 1: Diagram of TDR

One important TDR measurements regarding impedance mismatch is the reflection coefficient, ρ (rho). The coefficient ρ is the ratio of the reflected pulse amplitude to the incident pulse amplitude:

$$\rho = \frac{V_{reflected}}{V_{incident}}$$

(1)

The coefficient ρ can also rewrite in relation to characteristic impedance and load impedance [3]:

$$\rho = \frac{Z_L - Z_o}{Z_L + Z_o}$$

(2)

The Z_L and Z_o are the load impedance and transmission characteristic impedance respectively.

Assume that a uniform impedance of a transmission line is $Z_o=50$ ohm. If load impedance same as Z_o (which is matched impedance), there is no reflection observed and the reflection coefficient is zero. If any mismatch in the signal path, the reflection coefficient will be either in positive value or negative value. For example, when Z_L is infinity, the reflection signal is equal to incident signal and it's having the same polarity ($\rho = 1$). On the other hand, when the reading of Z_L is 0, the reflection signal is equal to incident signal but opposite in polarity ($\rho = -1$). Figure 2 is illustrating the resultant for these two conditions.

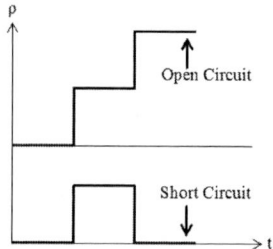

Figure 2: The open circuit waveform noted as p = 1 whereas short circuit waveform noted as p = -1.

As mention early, TDR measures the reflection impedance that resulted from impedance mismatch between the load impedance (DUT) and internal impedance of transmission line. The discrepancies between the measured and expected characteristics of the electrical path, helps to identify the type of failure and fault location [3]. In view of this, it is crucial to have stable and repeatable TDR measurement for reliable result interpretation.

Therefore, the interconnect point between the DUT and the test instrument play an important role in the study as the most accurate result will represent the actual root cause of the device under test (DUT). Thus, an appropriate measuring method is under consideration.

Hand probing

Hand probe is one of the most common methods use to contact the internal transmission line to DUT due to its convenience and off-the-shelf availability as shown in Figure 3. [3]. The TDR probe tip acts as a direct electrical contact between TDR instrument and DUT. [4]

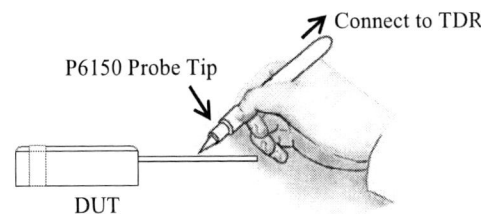

Figure 3: TDR probe hold by hand to make a contact between TDR instrument and DUT.

However, hand probe produces unstable impedance measurement result, due to the inconsistency on the positioning of the contact point, contact angle, and the force applied on the probe during the measurement. Subsequently, this will affect the integrity of the impedance measurement on TDR. More detail of each respective factor will be outline in later section.

As such, a good quality impedance measurement requires a good quality interconnect. In order to compensate the signal distortion caused by hand probing, an innovative approach is applied to improve the signal integrity. This is achieved by a careful design of a test fixture that can minimize these inconsistency factors. Apart from that, we also have to consider the impedance matching of the test fixture as mismatches and variations of the test fixture impedance can cause reflections that would indirectly decrease the signal quality. Therefore conventional test fixture used in Curve Tracer 370A is not suitable to use in TDR system. With this in mind, it is designed with a total impedance of 50 Ω in order to match the impedance of internal impedance.

After fabricating the test fixture, confirmations are carried out to validate how well the fixture is in improving the signal stability and quality. Our assessment shows that the test fixture is able to provide stable signal with good repeatability on impedance measurement on three-leaded Power MOSFET packages.

II. EXPERIMENTAL PROCEDURE

First and foremost, a golden device was selected for the entire experiment. The same device was used for every single measurement so that there would not point out the wrong information due to "varies in static characteristic". The experiment began with 8 repeatability test on DUT with hand probing as shown in Figure 4 and each maximum impedance result was recorded respectively. Standard deviation was calculated based on the measurements. Next, we screened out the three factors which cause the inconsistence of TDR signal. To figure out each of the factors, only one factor was allowed to vary in one time, the others must be remained constant. Repeatability test was applied for each factor and results were capture by TDR.

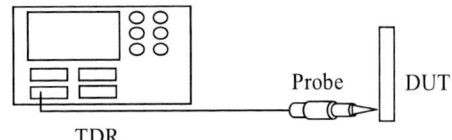

Figure 4: TDR probe makes a connection between DUT and TDR.

Secondly, we captured the signal between two types of connection to the TDR: hand probing and conventional test fixture (used for three-leaded Power MOSFET in Curve Tracer 370A). Signals were captured with two conditions, where the first condition was signal pin shorted to GND and the other was with DUT. Total of four results were captured by TDR with these setup as shown in Figure 5.

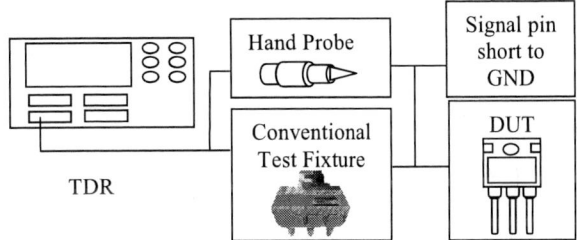

Figure 5: Block diagram of two types of connection (hand probe and conventional test fixture) to TDR with two conditions (signal pin shorted to GND and with DUT)

Thirdly, a new test fixture was designed based on the IPC standard [5] and standard PCB layout [6,7] (for impedance matching). The design of new test fixture would be elaborated more in next section. Repeatability test was done once again for 8 times and measurement was recorded down for comparison purposed.

Finally, a real case study was carried out. This study showed the TDR was used with the aid of new test fixture to encounter an open contact issue in a Power MOSFET.

Fabricating a new Test fixture

A poorly designed substrate with inappropriately selected materials can degrade the electrical performance of signal transmission increase the emission and crosstalk. Also, this will cause it more susceptible to external noise [6].

First of all, selecting an appropriate number of the layer of the new test fixture is extremely important. It will direct influence the refection signal of TDR especially impedance mismatch. Generally, there is no fundamental information about how many layers should be used. [7]. It depends on how the PCB was designed to match the impedance needed.

Therefore, we had chosen a four layers (stack-up) PCB for the test fixture instead of two layers PCB. This is because four layers (stack-up) PCB provides easy routing for ground and power planes and relaxes routing considerations compared to a two layers PCB. [8]. By the way, it also allows for distributed RF decoupling of a DC power plane sandwiched

978-1-4799-3911-4/14 $31.00 © 2014 IEEE   151

between two layers of predominantly ground plane [9]. While for two-layer (stack-up) PCB, it is quite challenge to implement 50 ohm impedance due to many factors like signal routing and component placement. With additional two planes in the middle of the PCB board, it was able to control the 50 Ohm impedance. Furthermore, an enhancement on the capacitive decoupling and reduction of the electromagnetic interference (EMI) could be achieved [6].

The four layers PCB cross-sectional view is summarized in Figure 6.

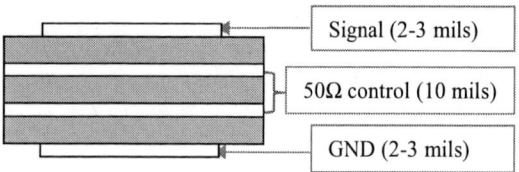

Figure 6: This figure shows four layers (stack-up) PCB cut-out view. The green colour is representing the core.

Apart from that, the selection of dielectric material (FR4), spacing between two layers and trace dimension are important too due to it can significantly influence the board impedance and crosstalk. [7,10]. It is strongly recommended that the impedance matching for this fixture must meet the 50 ohm so that the signal integrity would not be affected or being distorted. Therefore, the designed test fixture must meet the standard (IPC specification [5]).

Figure 7 illustrates the designed test fixture used for three-leaded Power MOSFET and the new configuration of the TDR system is shown in Figure 8. The new analysing structure consists of a coaxial cable with one end connected to the test fixture and the other end linked to TDR system.

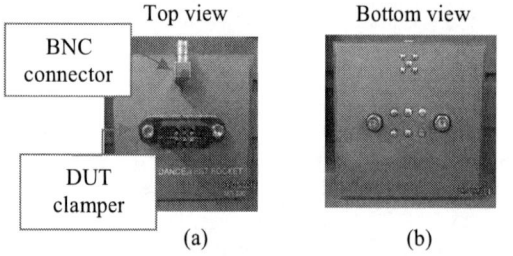

Figure 7: A typical top and bottom view (a) and (b) of TDR test fixture for three-leaded Power MOSFET.

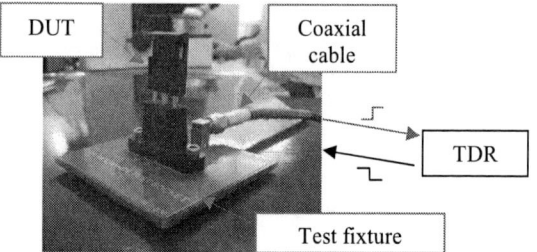

Figure 8: A new hardware setup for TDR measurement. TDR signal will transmit into the device through test fixture and coaxial cable, and then the reflection signal will captured and displayed on screen.

III. EXPERIMENTAL RESULTS AND DISCUSSION

A. Hand Probing

During the application of hand held TDR probe tip to do the TDR measurement, we always having difficulty to maintain the signal integrity. Figure 9 illustrates a real scenario/situation where the DUT was subjected to TDR measurement and repeated for 8 times at the same interface by hand probing. Closed up view on the waveform showed the inconsistent signal (especially the peak of the impedance) was captured for 8 times.

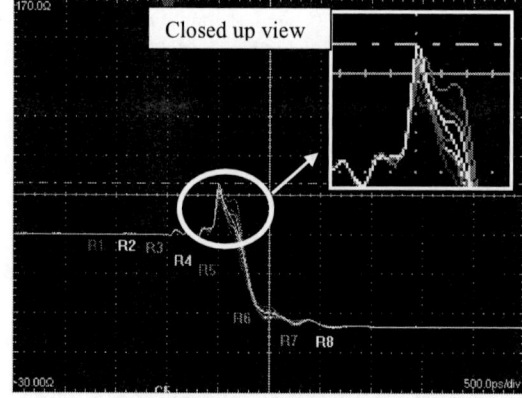

Figure 9: Repeatability test on the same device was performed. The results were displayed from R1 to R8.

Maximum impedance measurement for each result was listed out in Table 1. It is clearly seen that hand probing did not provide an accurate result on impedance measurement and the standard deviation based on the results was σ=3.26.

Table 1: Result for the maximum impedance of each test with hand probe.

Number of test	Value (Ohm)
R1	65.90
R2	68.78
R3	71.48
R4	72.54
R5	70.24
R6	74.10
R7	73.51
R8	76.21

It cannot be denied that the instability of TDR signal came from many factors. Among these factors, we have successfully screen out three crucial factors. The three factors were found out due to the human error. They are inconsistence interconnection point, force insertion, phase angle.

I. Positioning of the contact point at the interface

The most obvious unstable TDR waveform encountered was positioning of the contact location, (in Figure 10) where every contact location reflect a unique TDR waveform respectively.

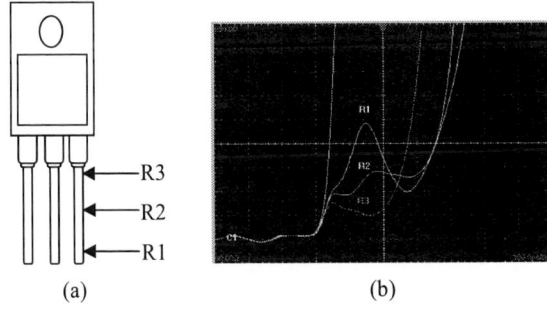

(a)   (b)

Figure 10: Three difference positioning of contact locations (a) were applied – R1, R2 and R3. The waveform of R1, R2 and R3 (b) were displayed the difference contact position of the failed device respectively by referring the three position in part (a).

II. Contact angle at the interface

The second challenge was the inconsistency of the probe contact angle at the same interconnect location (refer in Figure 11) affect the impedance measurement.

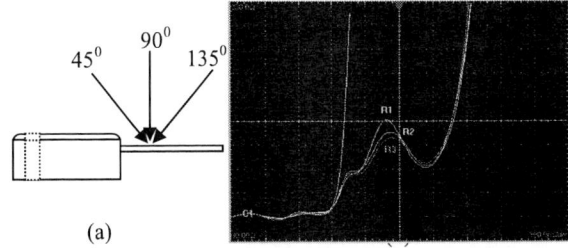

(a)

Figure 11: Three difference probe angles (45^0, 90^0 and 135^0) were applied to the approximate same position on the external lead of DUT (a). The waveform of R1, R2 and R3 were illustrating angle of 45^0, 90^0 and 135^0 respectively (b).

III. Force applied on the interface

The third challenge that affects the quality of the signal in the TDR system was the force applied from the TDR probe to the DUT. The impedance increase with higher force applied to the DUT was captured by TDR system and vice versa (in Figure 12).

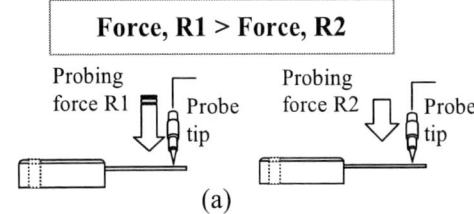

(a)

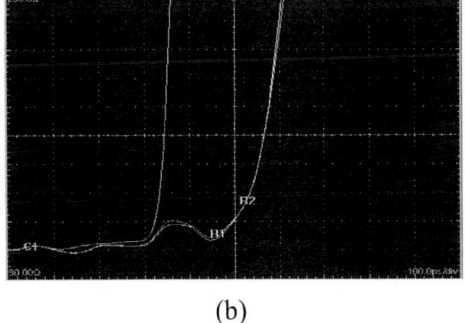

(b)

Figure 12: The force applied for R1 is higher than R2 (a). The waveform captured on difference force insertion (b).

B. Test Fixture

Test fixture is a better solution to eliminate these factors. However, as the impedance matching is vital, conventional test fixture is not suitable for TDR measurement.

I. Conventional Test Fixture

Figure 13 and 14 indicated that conventional test fixture where signal pin shorted to GND as well as with DUT reveals high impedance as compare to hand probe. These tell us that conventional test fixture use in Curve Tracer 370 does not meet the 50 ohm impedance which is crucial for TDR system.

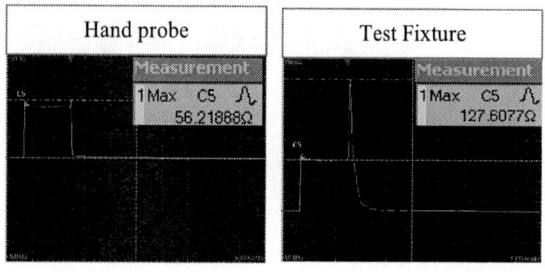

Figure 13: Waveform captured by hand probe (a) and conventional test fixture (b) with signal pin shorted to GND.

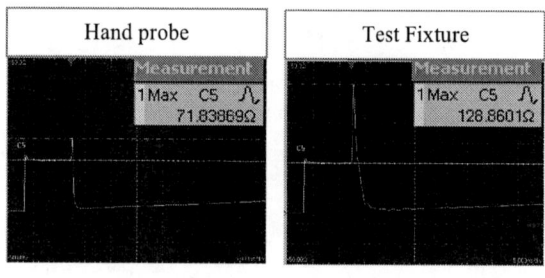

Figure 14: Waveform captured by hand probe (a) and conventional test fixture (b) with DUT.

II. New Test Fixture with impedance matching to the TDR

With the new test fixture, TDR impedance characteristic for repeatability test was performed to determine how consistent the result is with the aid of a test fixture. Based on the TDR setup in Figure 11, the DUT was plugged-in and plugged-out from the test fixture for 8 times repeatedly (R1 to R8). The reflection signals were then captured by the TDR as shown in Figure 15 and maximum impedance for each test was recorded in Table 2. Eventually, the end result showed that the signal integrity did not degrade and fluctuate at all (refer to the closed up view) with respect to the measurement made at different intervals. The standard deviation was σ=0.12 in the overall which showed significant improvement in term of consistency. In other words, usage of test fixture is able to provide a stable and consistent waveform.

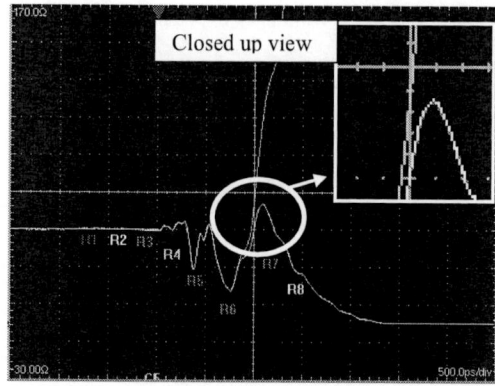

Figure 15: This figure showed waveform captured for 8 (R1-R8) repeatability tests with the new test fixture.

Table 2: Result for the maximum impedance of each test with new test fixture.

Number of test	Value (Ohm)
R1	64.18
R2	64.06
R3	63.86
R4	64.04
R5	64.25
R6	63.96
R7	64.13
R8	64.05

Furthermore, there is an evaluation of the test fixture's end-point in the TDR signal. This is crucial for the analyst to understand the initial interface signal between the test fixture and DUT in order to prevent misinterpretation correspond to the TDR result. As showed in Figure 16, a clear, precise and stable TDR result was obtained and the initial interface signal of the DUT can be easily distinguish from the test fixture signal (refer to the circle with yellow colour).

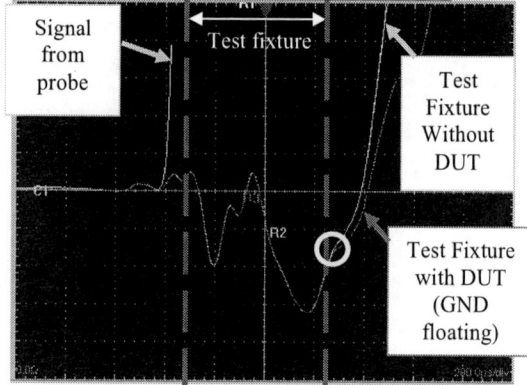

Figure 16: This figure illustrates that the signal recognition for test fixture and DUT.

Finally, a real Failure Analysis with the open (Gate) failure has been confirmed by using TDR. The result was illustrated in Figure 17. A red dotted line was drawn vertically according to the yellow arrow which stated in Figure 16. As this failure analysis was having the open failure in the device, therefore, TDR has been used to measure and observed the behaviour of the open signature of the device (refer to the failed unit waveform). A good unit, from the waveform is behaving like a short circuit signature. On the other words, TDR signal pulse able to transmit over the good unit (refer to the good unit waveform). From the TDR curve, it was clearly shown that the open contact was very near to the lead location. After decapsulation of the failure unit, a broken wedge near the lead was confirmed under Scanning Electron Microscope inspection as showed in Figure 18.

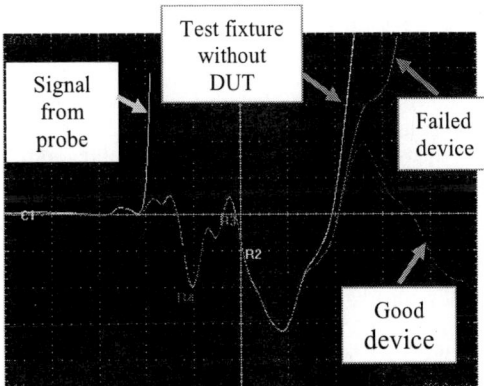

Figure 17: TDR impedance result on gate open failure device.

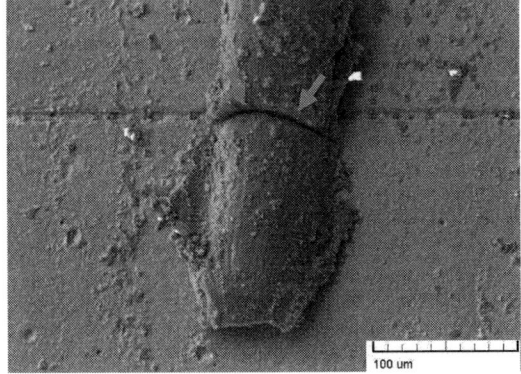

Figure 18: Broken gate heel (red arrows) was captured under Scanning Electron Microscopy (SEM).

IV. CONCLUSION

In summary, during TDR measurement, inconsistence of probing angle, positioning in contact point and force applied to the prober were encountered. Therefore, a four layer stack up test fixture was designed with some specification which stated in IPC standard. The result was successfully proved that with the aid of TDR test fixture, the signal obtained is a constant and reliable one, where the error is minimizes down

to $\sigma=0.12$ from $\sigma=3.26$ as compare to conventional hand probe method. Eventually, the analyst is able to conduct his or her task using TDR equipment with confident in order to provide reliable results in Failure Analysis.

ACKNOWLEDGMENT

The authors will like to express their gratitude to Miss Lo Chea Wee (IFMY QM FA O 3) who provide us the IPFA sample to perform TDR analysing. Not to forget, the authors will also like to express their very great appreciation to Dr. Goerlich Siegfried and Mr. Chew TT from Infineon for his valuable and constructive comments to this technical paper.

REFERENCES

[1] "Electronic Package Fault Isolation Using TDR," Published in EDFAS Microelectronic Failure Analysis Desk Reference, 5th edition, 2004.
[2] Bernard Hyland, "Propagation Delay Measurement Using TDR (Time Domain Reflectometry)," Application Note 4395 in Maxim Integrated, vol. 68, 2010.
[3] TDR Impedance Measurements: A Foundation for Signal Integrity," Application Note, Nov 2007.
[4] "Probe Fundamental," October, 2008
[5] "IPC-TM-650 Test Methods Manual," March 2004
[6] Barry Olney, "Multiple PCB Stackup Planning. In-Circuit Design," 2011.
[7] "High-Speed Layout Guidelines," Application Reports SCAA082, Nov 2006
[8] Suyash Jain, "Layout Review Techniques for Low Power RF Designs," Application Note AN098 SWRA367A, Aug 2012
[9] "RF Design guidelines: PCB Layout and Circuit Optimization," Application Note AN 1200.04,Nov 2007
[10] Bill Hargin, "PCB Signal-Layer Selection – 1. In-Circuit Design," Aug 2012

In-depth description for the FA case with Gate-to-Source or Drain short by Nanoprobing analysis

LiLung Lai, Oscar Zhang, Ling Zhu, Feng Qian, and Mason Sun
Semiconductor Manufacturing International (Shanghai) Corp.

E-mail: LiLung_Lai@smics.com

Abstract–Nanoprobing analysis has become standard analytical technique in the modern semiconductor FA lab. In this paper, we describe the use of nanoprobing to investigate cases of Gate-to-Source or Gate-to-Drain shorts and follow up the data generated by nanoprobing with physical analysis. The paper provide discussion of the electrical details and the physical mechanisms.

I. INTRODUCTION

During ramp-up of the yield of a 90nm process in a 200mm Fab, we encountered several failure modes in a test vehicle SRAM. Single bit (SB) failures were the major yield killer. However, at the early stage, Single Bit-Line (SBL) fails, see Fig.1, also caused around 15% of yield loss. In this paper, we describe a case study of Single Bit-Line (SBL) failure analysis (FA) and emphasize the value of Nanoprobing analysis.

The failure analysis is challenged for the methodology due to no global fault isolation along the long Bit-Line. The first BVC isolated the possible failure location at the somewhere of periphery and proceeded in Nanoprobing analysis. The Gate-to-Source or Drain short was found but no Source-to-Drain short in the PMOS. The deduction to possible root cause deduction and following PFA to target transistors were executed in the FA procedure. The final root cause is related to Gox weak quality. The corrective action is the replacement of new raw wafer, which successfully eliminated the SBL failure and enhanced the yield.

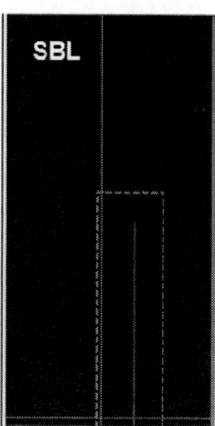

Fig. 1 SBL fail pattern from the engineering test

II. Case Background and Summary of the FA process

A. Background

At the beginning of 90nm yield ramp up at 200mm Fab, we encountered the severe issue of Single Bit-Line (SBL) failure. From Electrical Failure Analysis (EFA), we know the SBL is timing-relate soft fail. Passed window can be found in fast testing (shorter testing time). The debugging process was challenged for initial FA and very urgent priority for the problem solving. We planned the OBIRCH analysis at the first step of FA by static probing but no hot spots were found. Afterward, FA team went layer-by-layer SEM inspection along the BL region and observed Bright-PVC, see Fig.2, at the periphery of cell array, which is correlated to Bit-line operation. After layout analysis, we know it is transmission function of BL operation. To further investigate the failure mechanism, the Nanorprobing measurement to device characteristic is necessary for next step of analysis. We already publish the basic postulates of Nanoprobing analysis [1] and hope to extend the capability to identify the location of failure.

B. Nanoprobing measurement and analysis

Then, we implemented Nanoprobing analysis for the transistors. The failure signature in Nanoporbing in Fig.3 is the gate-to-source or gate-to-drain short in PMOS but lack of source-to-drain direct short. According to the electrical data and theoretical deduction, the weak Gox induced the gate-to-source or drain leakage becomes the most possible root cause.

The Fig.4 illustrates the possible mechanism with boron penetration through Gox.

978-1-4799-3911-4/14 $31.00 © 2014 IEEE

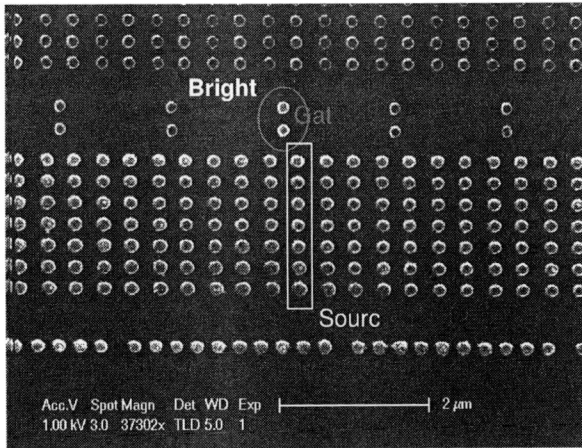

Fig. 2 The bright VC in SEM was found at the region of transmission gate

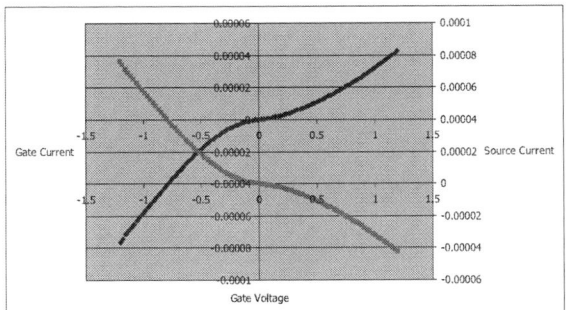

Fig. 3 An example of Nanoprobing showing Gate-to-Source short.

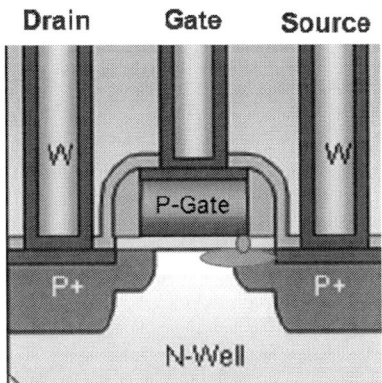

Fig. 4 The possible model of G-S/D short in PMOS

III. Further Investigation for Possible Root Cause

The bright VC was observed at the periphery along the region of fail bit-line, including transmission and pre-charge gates. The common signature is abnormal PMOS behaviour in Nanoprobing data for all measurements. There are different signatures for the different samples. This attached two examples in the submitted files; see Fig.5 and Fig.6 for Nanoprobing analysis.

With several Nanoprobing data analysis, we have conclusion of Gox weakness by induction. Owing to the electrical connection of gate-to-channel, the direct broken of Gox or tunnelling through Gox in both Gate and Channel could be the mechanism. However, only BVC is observed in gate but not in Drain or Source Contact, see Fig.7. Suppose that the direct electrical connection in gate and Silicon should perform the same VC. Using Nanoprobing analysis in Fig.8, we can verify no short in between gate-to-channel via bulk biasing. Therefore, the tunnelling instead of short for gate-to-channel is our speculation. Thus, local thinning and lack of Nitrogen (N) in the Gox becomes the highly possible model of the mechanism. The Boron (B) penetration through Gox into channel short Gate with either Drain or Source (P+) region cause the electrical failure. Fig.9 illustrates our model for the abnormal characteristic in the PMOS of fail SBL.

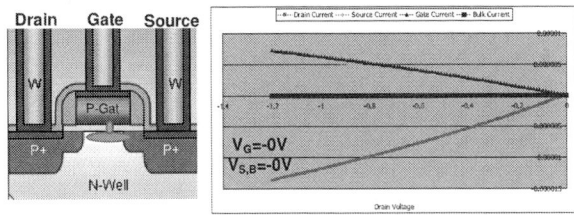

Fig. 5 Nanoprobing data in one of fail PMOS show Drain leak current to Gate and Source at the same condition

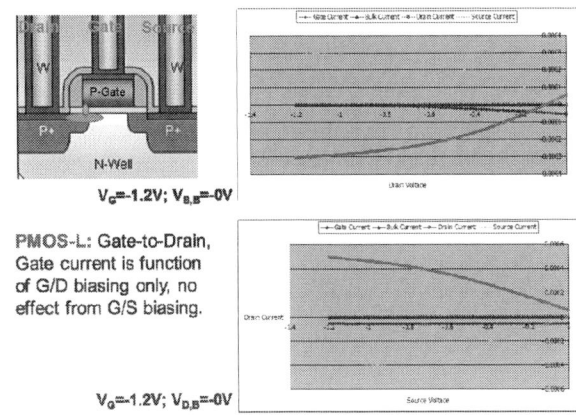

Fig. 6 Nanoprobing data of the one fail PMOS show Gate-to-Drain short but no leakage to Source

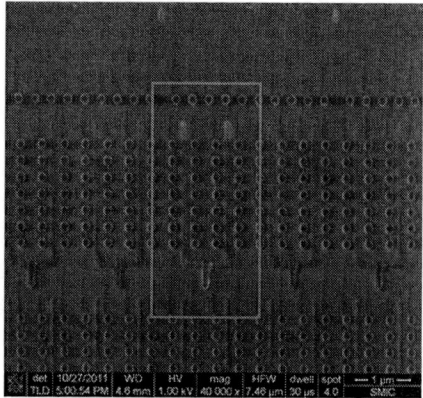

Fig. 7 The Bright VC in SEM only exhibits on gate but not in S/D AA Contact.

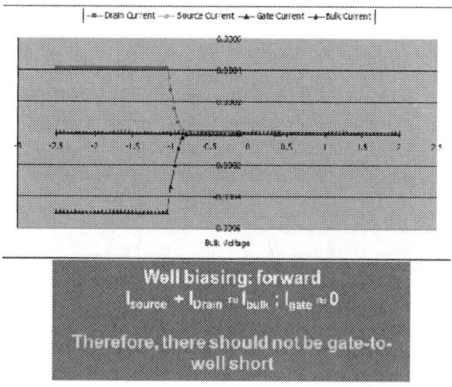

Fig. 8 The Nanoprobing data via well biasing observe no gate current and confirm no direct short through gate oxide.

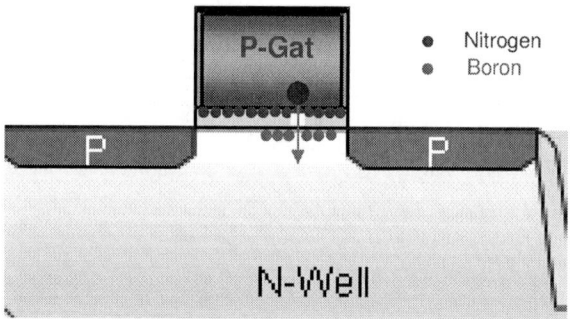

Fig. 9 Proposed model showing Local thin SiO$_2$ (Dopant penetration) and resulting tunnelling current from gate to Silicon channel

IV. Stress experiment and following PFA

A. Stress

To confirm the possible issue of Gox, we designed the stress

effect by Nanoprobing to measure the breakdown voltage and leakage current after stress. The Fig.10 describes the difference of performance in between fail and passed location. The data confirm the Gox weakness in fail sample and no direct short in between gate-to-Si channel.

B. PFA

After the stress test, we implement the following PFA including Wet/SEM and X-TEM ways for defect inspection. Using sample preparation of uncovering Gox and following wet stain, SEM imaging reveals the abnormality underneath the gate, see Fig.11. The next PFA from X-TEM once through the channel shows the abnormal defect in the Gox, see Fig.12. Thus, we have physical evidence for the failure analysis and conclude the defect in the gate oxide.

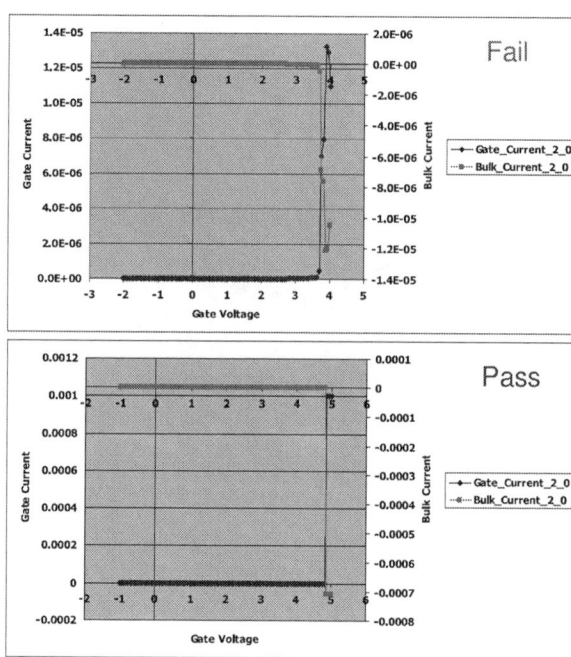

Fig.10 Fail address breakdown at ~3.7V with current ~10uA and Pass address breakdown at ~4.7V with current >1mA

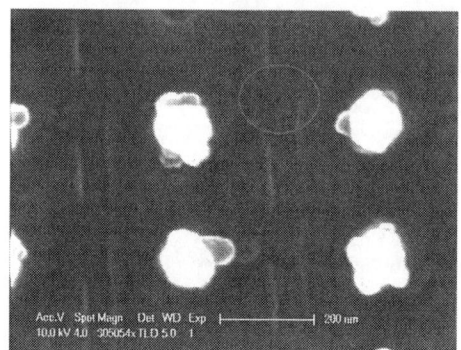

978-1-4799-3911-4/14 $31.00 © 2014 IEEE   158

Fig. 11 The defect observed at the Silicon underneath the gate.

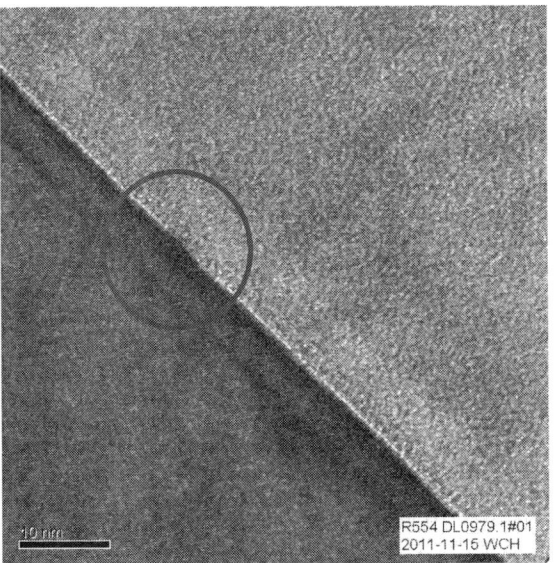

Fig.12 X-TEM reveals an abnormal Gate-to-Silicon (Active Area) bridge, which meets the electrical failure model.

III. Discussion and Summary

A. Discussion for Possible Mechanism

Circuit analysis is able to explain the soft failure of Single Bit-Line due to gate leakage to Source or Drain in the transmission or pre-charge region The gradual leakage from the gate would reduce the voltage of the gate and finally destroy the function of the gate and cause BL/BLB comparison failure under normal test conditions. At the region of pre-charge, see Fig.13, it would cause loss of the BL/BLB signal gradually. At the area of transmission, it will disturb the BL/BLB gradually during read process. If we adjust the testing time to be faster, then it may pass. Thus, the timing-relate soft SBL failure comes from the abnormal performance of gate leakage.

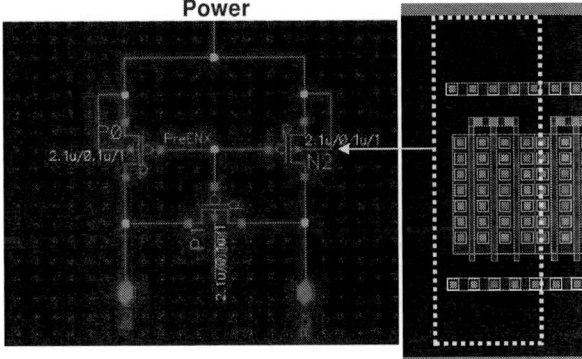

Fig.13 The circuit and Layout of Pre-charge region help to interpret the mechanism of failure from gate leakage in PMOS.

B. Summary

The SBL FA is very challenging for all kind of Memory type products. At the first step, we tried to localize the failure using PEM or OBIRCH but even with proper set-up, we didn't successfully find the hot spot. The challenge of fault isolation by the way of hot spot for Bit-line failure could be a hot-topic of future development. Layout tracing and Passive Voltage Contrast (PVC) proved to be the key and most effective method to isolate the failure at the periphery for SBL fails. Afterwards, Nanoprobing analysis, stress-up and direct PFA provide more information and clue of root cause. According to the FA finding, the Fab team identified probable source of yield issue and quickly delivered a solution to correct the problem. This case offers a good lesson for the yield debugging process.

Finally, we changed the raw wafer material and resolved the issue which increased the yield significantly.

REFERENCES

[1] LiLung. Lai, "The Unique and Completed Characteristics of Device Behaviors in the Nanoprobing Analysis and Application for Missing LDD", ISTFA 2013, 222 -227.

A Study of Isolation Test on FullPAK Device

S.Y. Gan, Lokman Alias and W.Y. Ng

Infineon Technologies (M) Sdn. Bhd.

Batu Berendam Free Trade Zone,

75350 Melaka.

Phone: (06) 287 3206 No. Fax: (06) 2323279

Email: Sue-Yin.Gan@infineon.com, Lokman.Alias@infineon.com & WanYee.Ng@infineon.com

Abstract

The isolation test condition and the root cause of isolation failures are always the main concerns of the manufacturers and customers. Several of studies have been carried out to evaluate the optimize settings so that the device meets the required quality standard. The company is benefited from the isolation test activity which is also known as Dielectric Withstand Voltage test or Isolation Test, to enable the screening of the mold compound rejects electrically. The test is used to verify the insulation of the mold compound whether it is sufficient enough or not to protect the user from electric shock by measuring and checking the mold compound compactness so that any reject which lead to mold voids can be filtered out. 'Mold voids' is actually referring to the air pockets trap within the mold compound itself which is generated from the incompactness of mold process. This is where failure analysis comes in to reveal the underlying root cause failure of the package insulation. Different kinds of method have been used here, for example, X-ray, SAM (Scanning Acoustic Microscopy), SEM (Scanning Electron Microscopy), electrical test verifications, cross section and etcetera.

Introduction

Dielectric strength of an insulating material is the maximum electric field strength that it can withstand intrinsically without breaking down [1]. Isolation test is used to stress the insulation of a product far beyond what it would encounter during normal operation. High voltage is applied from the mains-input lines to the chassis of the product for a specified length of time to check the integrity of the insulation by monitoring resulting leakage current [2].The product is safe to be used in nominal operating conditions, if breakdown (no excessive amount of leakage current flowing through) does not occur. Insulation materials often contain cavities or voids which are filled with gas or liquid medium of lower breakdown strength than solid. Discharged induced breakdown occurs when electrical discharges occur on the surface or in voids of electrical insulation.

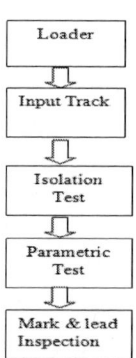

Figure 1: Isolation Test Flow

Isolation test is conducted at ambient temperature by applying ykVAC at 1 second by fully integrated test handlers. All the devices must go through the isolation test before any electrical test to save manufacturing cost and time. This is because assembly defects at mold compounds, i.e. mold voids or piping holes can be sorted out at first place using isolation test hardware before units are proceed to parametric test. The details of test sequence as shown in Fig 1.

Test Methodology

Isolation test can be performed either in AC or DC voltage. For AC voltage, terminology of Voltage Root Mean Square (RMS) is applied. The RMS value is the effective value of a varying voltage. Refer to Fig. 2, peak voltage (Vpeak) is the highest voltage magnitude of a sinusoidal wave whereas RMS values (Vrms) is lower than Vpeak and can be obtained using (1)[3][4]. Vrms is the equivalent to a steady DC (constant) value which gives the same effect in terms of power. To test the devices regarding isolation the AC Vpeak voltage must be equivalent to a DC voltage (Vdc) (2) [3] [5]. For example, to test the device using 3kVrms@1s, the Vdc equivalent will be 4.242kVdc@1s. For Infineon, TO-220 FullPAK isolation testing is done using AC voltage, which is Vrms.

$$Vrms = Vpeak \, / \, \sqrt{2}$$

(1)

$$Vdc = Vpeak = Vrms \times \sqrt{2}$$

(2)

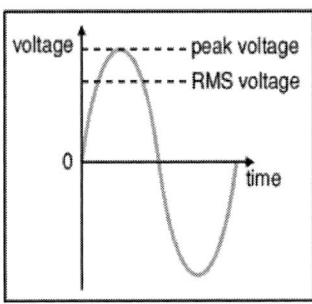

Figure 2: Difference between Vpeak and Vrms (AC) [4]

Based on UL 1557 & EN60950 standard, it is stated that rated isolation RMS voltage for 60 sec or 120% of the rated isolation RMS voltage for 1s needs to be applied. Due to testing is carried out in production environment, short test time is vital important. Therefore, the test condition is defined as below:

Rated Isolation RMS Voltage = 2.5kVrms * 120%
(Refer to data sheet)   = 3kVrms @ 1s

(3)

The 3kVrms@1s stated in (3) is the minimum requirement. To ensure the products achieve higher quality guaranteed in datasheet, Infineon perform testing with more stringent conditions, which is ykVrms@1s. Evaluations have been carried out to ensure the testing is done correctly without weakening the devices with higher voltages. The isolation contact test is designed in such a way where top and bottom of the package are directly contact with a metal block. DUT's leads connected to the high voltage and the metal block connects to ground. An alternating current is applied, so the leads and heat sink (metal block) can be either connected to positive or negative voltages.

Experiments and Results

CASE STUDY 1
Isolation Voltage Vs Package Robustness
To perform this evaluation, sample size of 30 pieces of tested good units were used. 10 pieces each for different voltages range from xkVrms@1s, ykVrms@1s and zkVrms@1s. Tested good unit means units tested and pass in isolation test and electrical test. From Fig. 3, xkVrms@1s and ykVrms@1s were having much lower leakages and far from the upper side limit (USL) with small deviations as compared to zkVrms@1s. At zkVrms@1s, leakage increase significantly. 3 out of 10 pieces of units fallout from the upper side limit. Therefore, zkVrms@1s is not recommended to be used because it overstresses the package insulation and causing degradation. The tested good parts start to have isolation failure once the Vrms is set to zkVrms. Burnt mold observed as illustrated in Fig. 9. Therefore, ykVrms measurement is recommended for productive screening of mold voids defect in TO-220 FullPAK device.

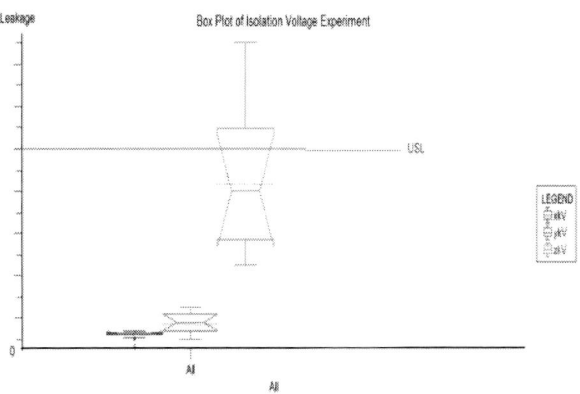

Figure 3: High leakage fall-out at test condition zkVrms at 1s

CASE STUDY 2
Repetitive Isolation Test vs Package Robustness
Another experiment has also been carried out whereby repetitive high voltage applied onto the DUTs to find out the optimum test condition and to monitor the package robustness. Again, another 30 pieces of tested good units were used and 10 pieces each for different voltages. From Fig 4, it is again observed that good parts tested with zkVrms start to have higher leakage. The deviations at zkVrms@1s is more than 20% and go up to 180%, exiting the upper side limit as compared to ykVrms@1s where all the good parts are still far below the upper side limit where deviations is about 20% only. The drift percentage analysis is shown in Fig 5. Therefore, it is proven that ykVrms@1s test condition is the optimum settings for Infineon TO-220 FullPAK devices.

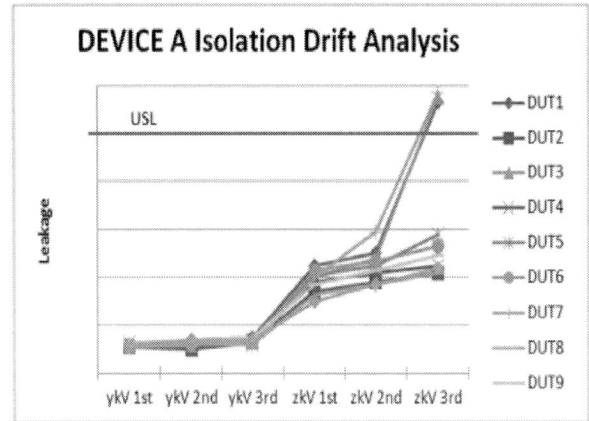

Figure 4: 3 times repetitive test at zkVrms @1s overstress and DUTs get weakened.

978-1-4799-3911-4/14 $31.00 © 2014 IEEE   161

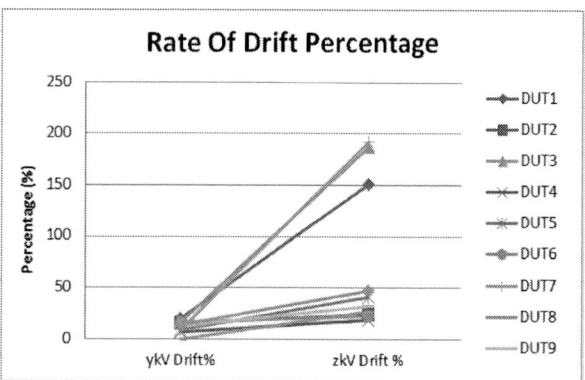

Figure 5: Drift percentage after 3 times repetitive isolation test at ykVrms@1s & zkVrms@1s. At ykVrms@1s, all devices pass and has about 20% drift. At zkVrms @1s, 3/10 pieces of devices failed with drift more than 20% and up to 180%.

CASE STUDY 3
Mold Voids vs Package Robustness

After the isolation test, one piece of good and reject units are respectively subject to mold voids depth measurements. From the comparative analysis as refer to Fig. 6 and Fig. 7, conclusion can be made that good part has relatively smaller lateral X-Y region and shallow depth trench which is about 75um (Z Height) as compared to reject part which has wider lateral X-Y region and deeper trench about which is about 216 um (Z Height). Deeper trench means thinner mold compound to the heat sink. Mold compound's dielectric strength is inverse proportional with the thickness. This explains why thinner mold compound breakdowns at lower voltage.

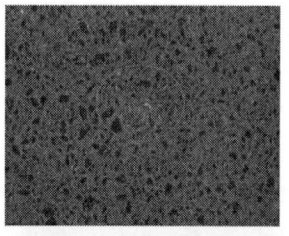 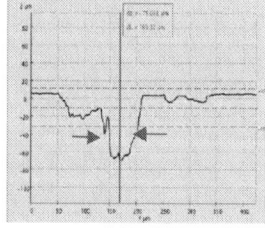

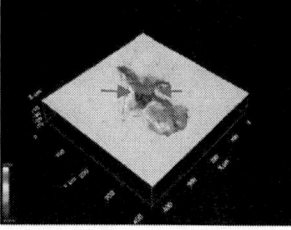

Figure 6: Good part which has shallow mold void area, refer to blue color code in 3D diagram.

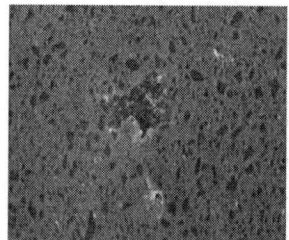 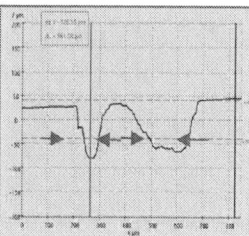

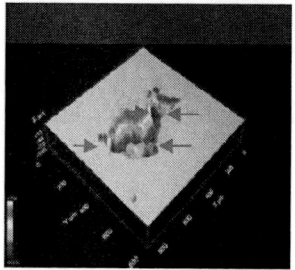

Figure 7: Reject part which has wider and deeper mold voids refer to blue color code in 3D diagram.

Root Cause Findings

Various FA approach has been carried out to rule out the root cause of internal voids towards package robustness. Fig. 8 illustrates the standard FA flow in order to carry out the investigation. Different kinds of method have been used here, for example, X-ray, SAM (Scanning Acoustic Microscopy), SEM (Scanning Electron Microscopy), electrical test verifications, cross sectioning and etcetera. The FA results are shown from Fig. 9 to Fig. 15.

```
Optical Inspection, LPS/ HPS
          ↓
     X-ray Inspection
          ↓
 Electrical Verification, EV
          ↓
Scanning Acoustic Microscopy, SAM
          ↓
       Depth Profile
          ↓
     Cross-sectioning
          ↓
Scanning Electron Microscopy, SEM
```

Figure 8: Standard Failure Analysis Flow

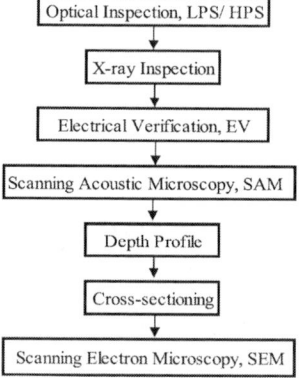

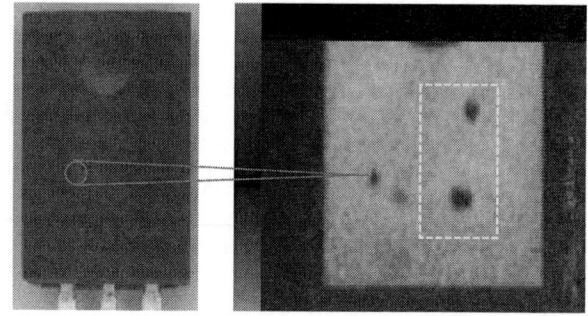

Figure 9: Optical inspection (photo at left side) detects one void but SAM (photo at right side) found more than one internal voids, highlighted in square box.

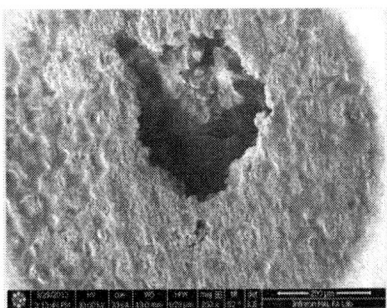

Figure 10: Mold voids and burnt compound under SEM inspection.

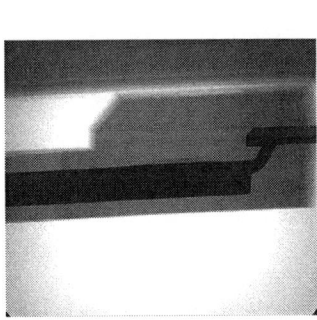

Figure 11: No visual defect found under X-ray inspection due to low density on mold compound.

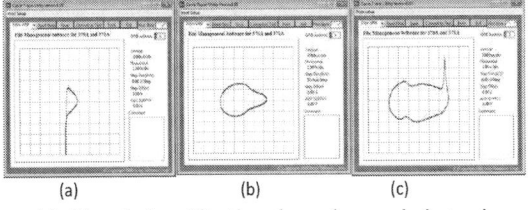

| (a) | (b) | (c) |

Figure 12: Electrical verification showed normal pin to pin characteristic on MOSFET device. (a) graph showed Drain-Source Lissajous curve with floating gate pin; (b) graph showed Gate-Drain Lissajous curve with floating source pin; and (c) graph showed Gate-Source Lissajous curve with floating drain pin.

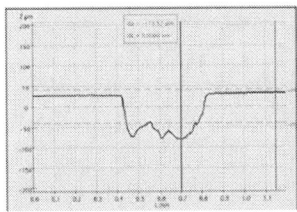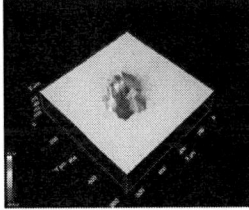

Figure 13: Depth profile measurement (Reject device) revealed internal voids depth/width respectively.

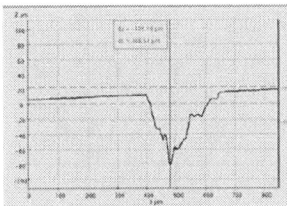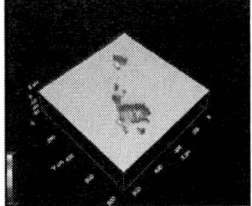

Figure 14: Depth profile measurement (Good device) revealed internal voids depth/width for comparison.

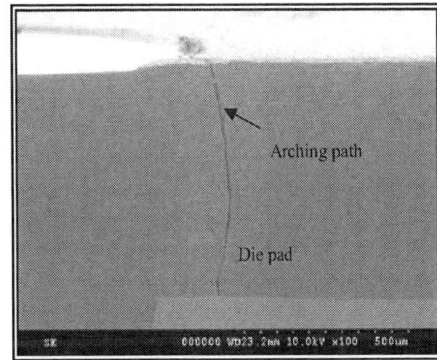

Figure 15: Cross-section photo show an arching path from voids (at bottom package) penetrates towards die pad.

Some reject parts exhibit more than one internal void that being screened out by SAM at bottom layer, which is between mold compound and lead frame interface layers. X-ray inspection has limitation to detect any internal voids or arching path underneath the lead frame as observed in Fig.11 due to some voids is under minimal size. On the other hand, electrical verification shows the part pass the pin to pin characterization tests as shown in Fig. 12, for standard MOSFET characterization test such as Drain-Source, Gate-Drain and Gate-Source [6]. Depth profiler revealed the depth is nearly similar for all samples with around 70-80um in the width. Therefore, the size and volume of the void is much bigger for the fail device. Among the FA technique, the most effective method to detect the defect is by cross-sectioning as shown in Fig. 15. An arching path was revealed from the bottom package penetrates towards lead frame area which causing isolation test failure.

Figure 16: Formation of air trap or void(s) inside the cavity before final compaction.

Based on the FA findings, the failure mechanism was being rule out due to internal voids in which the mold compound's top and bottom meeting point is located on the bottom cavity which traps the air on bottom package causing internal voids after final compaction as shown in Fig. 16. An arching would occurred during high voltage test because the package insulation is weak. The internal void occur due to unbalance

978-1-4799-3911-4/14 $31.00 © 2014 IEEE

compound flow inside the cavity. Internal voids occurred due to the top cavity fills faster compare to bottom cavity causing blockage of air vents. As a result, the compounds starts to turn into low viscosity and do not allow the trapped air to outgas.

. Therefore, optimization at End of Line (EOL) process is vital. Mold parameter to flush out the trapped air inside the cavity before final compaction needs to be improved and further optimized. Evaluations are ongoing for optimizations and continuous improvement.

Conclusions

Based on case study 1 and 2, optimized test conditions have been defined. Early failure defects, such as mold voids can be sorted out effectively. This ensures the tested units are having good quality in terms of package integrity. This would save a lot of time and cost in terms of disposition of the production lot. From case study 3, failure analysis prove that both internal and external voids or piping holes can cause the isolation failures. Deeper trench (Z) or thinner mold compound to the heat sink is more critical and it is the culprit that can cause isolation failures. The failure analysis carried out show isolation failure rejects actually passing electrical test which means test setup used is capable to do package insulation screening without affecting the internal chip. The arching path penetrate from the package to the die pad also shows that the failure is genuine.

Acknowledgments

The authors would like to express their gratitude to Infineon MAL for the good team work and also the continuous support provided by Dr. Goerlich Siegfried and Mr. Chew TT from Infineon for their valuable and constructive comments to this technical paper.

References

[1] http://en.wikipedia.org/wiki/Dielectric_strength

[2] http://www.asresearch.com/support/faqs/dielectric.shtml

[3]http://www.cirris.com/testing/guidelines/ac_hipot_testing.html

[4]http://www.raeng.org.uk/education/diploma/maths/pdf/exemplars_engineering/8_RMS.pdf

[5] http://electrical-engineering-portal.com/what-is-hipot-testing-dielectric-strength-test

[6] Hubert Beermann, "Interpretation of Power DMOS Transistor Characteristics Measured With Curve Tracer," *Microelectronics Failure Analysis Desk Reference Sixth Edition,2005.*

Study on Sensitive Character of Unexpected High Impedance Circuit in VLSI Failure analysis

Gaojie Wen

Freescale Semiconductor (China) Limited, Tianjin
No.15, Xinghua Avenue, Xiqing Economic Development area, Tianjin, China
Phone: (+86) No.22 85684232/13682170769 Email: b16245@freescale.com

Abstract- **Microprobe analysis plays an important role in failure analysis as it could reveal the failed signal directly and help to isolate the final failed device. But when met unexpected high impedance circuit, the real signal couldn't be measured as high impedance site was sensitive to probe needle and strong light. One real Case and experiment was studied in this paper to show how high impedance circuit was sensitive and how to find the root cause of this unstable failure efficiently.**

I. INTRODUCTION

Due to the decreasing size on complicated mixed signal integrated circuit, special functional failures which were sensitive to time, temperature or light were difficult to handle. It would take long time to perform failure analysis and may not surely to find the root cause. Meanwhile, unstable failed sample also had high risk to recover in failure analysis procedure. Known the typical sensitive structure was very helpful in failure analysis.

When handling a function failure case, photo EMMI was helpful in indicating the further analysis direction, but if EMMI failed to reveal abnormal emitted spot, microprobe is needed to find the root failed site step by step. [1] Focused on the abnormal leakage which got by passive microprobe, OBIRCH (Optical Beam Induced Resistance Change) can be used to locate the final failed device. For most of internal circuit, passive and active microprobe could reveal true failed control signal while not affect the original failure. But when met special internal circuit such as unexpected high impedance circuit, microprobe couldn't get real signal on failed sample due to its sensitive character. The root cause may be lost If the failed signal couldn't be got.

In internal CMOS logic control circuit, every site had a pull up CMOS and pull down CMOS. When the pull up CMOS and pull down CMOS didn't work at the same time, the middle site of circuit would in an uncertain state. It was sensitive to factors which could contact with it in many ways, for instance: contact directly by probe needle, strong light contact by photo electron effect and thermal motion of molecules by high temperature. When well known of the property of high impedance circuit, it wouldn't fall into confusion when met one single signal sensitive and much analysis cycle time could be saved.

One time delay related functional failure case was presented in this paper to show high impedance circuit property and how to find the root cause efficiently. The experiment was also performed by changing conditions to study more properties about high impedance circuit.

II. REAL CASE ANALYSIS

One mixed signal IC was returned by customer for CAN (Controller Area Network) communication failure after sent SPI (Serial Peripheral Interface) command about 30 seconds (Fig.1). As RX signal was controlled by CANH and CANL, the RX failure was only a result. The CAN failure was also found had relationship with temperature which would fail sooner at high temperature. Post decap, this failed sample had the same phenomenon with before decap and didn't show any sensitive tendency to normal visible light. As the sample would fail automatically after working some time, the soft defect location method would difficult to use focused on the temperature sensitivity.

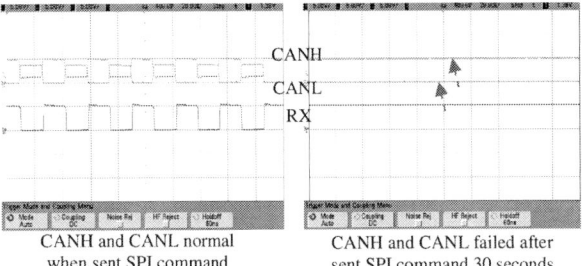

CANH and CANL normal when sent SPI command	CANH and CANL failed after sent SPI command 30 seconds

Figure 1. CANH and CANL signals on failed sample

EMMI analysis didn't reveal abnormal emitted spot on failed sample when in failed condition. Since EMMI analysis didn't provide the direct direction for further analysis, schematic analysis need to perform to determine which kind of circuit is most suspected to cause the CAN failure. To confirm the main failed site, microprobe analysis was needed to employ based on the CAN communication failure. Nearly all the modern smart IC had the self-protection capability by monitoring its temperature, low voltage and high current to prevent the fatal damage. As the CAN failure also had relationship with time, the THERMAL_SHUT module which was used to monitor the IC temperature was highly suspected according to the function. [2]

When performing microprobe analysis, there were two kinds of needle which were usually used in microprobe analysis: passive needle and active needle (12C and 18B). The passive needle is used to get stable digital signal which connected with digit multi-meter; Active needle was used to get wave signal which connected with oscilloscope. Active needle contained low input capacitance and high input impedance, it could reduce affection to the circuit while improve the signal measurement precision (Fig 2).

978-1-4799-3911-4/14 $31.00 © 2014 IEEE

(1). Microprobe needle comparisons

parameter Type	Input resistance	Input capacitance (1fF=10⁻⁵pF)	Voltage range
12-C Active Probe	10^6 ohms	100 fF	-10 to +20V
18-B Active Probe	10^{14} ohms	20 fF	0 to +15V
Passive Probe	10^{12} ohms	150000 fF	-15 to +15V

Figure 2. Property comparison between passive probe needle and active probe needle

Active microprobe (12C) was performed focused on the THERMAL_SHUT module of CAN block and monitored the CAN signal at the same time. CANHIGHTEMP signal was found to be failed and wrongly shut the CAN communication after sent SPI command about 30 seconds. Microprobe in internal circuit of THERMAL_SHUT module revealed sensitive site on TP8. When put active probe needle on TP8 site, the CAN signal wouldn't fail and failed sample recovered. When took away the needle from test pad, the CAN failure happened again. The signal on TP8 should be a high level signal which could be deduced by latter failed signals. As the front pulls up control signals were normal when compared with reference, the failure was isolated on TP8 site. But how to find the root cause was a tough job as the real high level signal on TP8 couldn't be got. When probe needle was put on the high impedance site, it would involve another circuit which contained input resistance to GND. Since the abnormal high level signal on high impedance site was not stable, it could be pulled down by the probe needle easily and became low. Meanwhile, the low level signal was a normal one and the failed sample recovered.

Suspected IV curves were measured focused on several signals which can pull up the TP8 signal, but no anomaly was found [3]. When failure was isolated on one single signal and no main control signal failure was found, metal bridge or particle was highly suspected based on the failed signal wiring in layout. As the TP8 signal had a long wire in layout, further layout analysis and microprobe would take a long time (Fig 3).

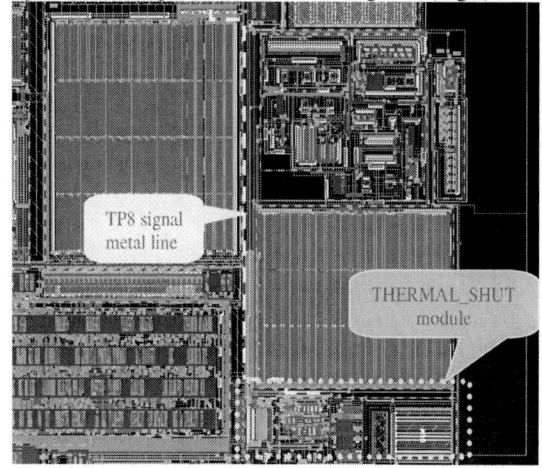

Figure 3. Failed TP8 signal had a long wire in layout

The schematic structure was also studied focused on the sensitive TP8 signal. The pull up signal and main pull down

signal was normal. But if the bias pull down signal was also abnormal, the Q55, M1 and M2 didn't work at the same time; therefore, the TP8 site would in a high impedance state. This kind of high impedance circuit was forbidden in internal control circuit as it was in an uncertain state. Further probe confirmed the VBIAS was abnormal 0V and caused M2 didn't allow bias current flow through it. TP8 site was in high impedance state (Fig.4).

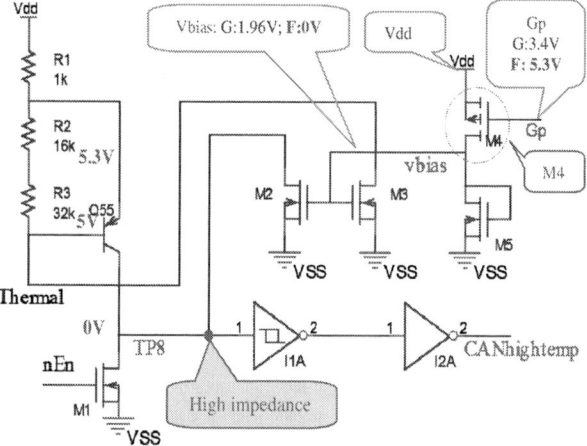

Figure 4. Simple schematic of unexpected high impedance circuit in THERMAL_SHUT module

Finally, high leakage was found between VDD (Source of M4) and Gp (Gate of M4) which indicated the main failure was on M4 (Fig.5). OBIRCH was set up according to the leakage and bias voltage was forced between VDD and Gp, abnormal resistance change spot was found on M4. Deprocess focused on abnormal resistance change spot on M4 confirmed the gate oxide rupture was the root cause (Fig.6).

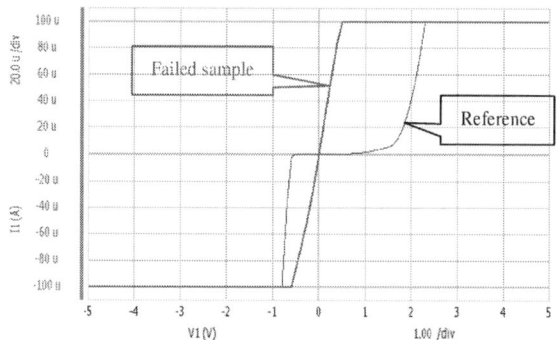

Figure 5. IV curve between Source and Gate of M4 revealed high leakage on failed sample

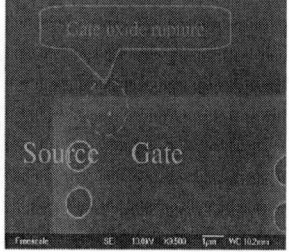

Figure 6. OBIRCH got resistance change spot on M4 and gate oxide rupture was confirmed by chemical deprocess

Schematic structure analysis combined with microprobe data was essential in failure analysis. But the awareness of special circuit property was efficient in finding the root cause. High impedance circuit was sensitive to probe needle and strong light. When met this kind of sensitive condition in failure analysis, the quick understanding of its possibility can shorten much analysis cycle time.

III. EXPERIMENT

The high impedance circuit was sensitive to probe needle when measured the real signal on TP8 site using oscilloscope. What's more, the failed sample also showed character of sensitive to strong light which was from the optical microscope. Study was performed on real failed sample to know how the probe needle and strong light affected the high impedance circuit performance. To double confirm the high impedance circuit sensitive character, one reference sample was used to simulate high impedance circuit by connecting the VBIAS signal to GND. Then EVB verification, microprobe analysis and optical inspection using strong light of microscope all revealed the same character with the real failed sample. [4] Experiment was performed focused on two kinds of phenomenon on high impedance circuit: probe needle sensitive and light sensitive.

1.) Probe needle sensitive:

Load effect when measuring the signal. When measured an internal signal using microprobe, it was connected with multi-meter or oscilloscope. The multi-meter or oscilloscope may play as a load role in probe analysis and this may affect some current precise circuit. To confirm if the load effect lead to the probe sensitive character, the probe head was disconnected with oscilloscope or multi-meter and kept the probe needle floating. Based on this kind of setup condition, when put probe needle on TP8 test pad, the failed signal disappeared and failed sample also recovered. When took away the floating needle from TP8 test pad, failure appeared again. It meant that the load effect of oscilloscope or multi-meter couldn't lead to sensitive character on high impedance circuit.

Probe head and probe station parasitic effect. The probe head was adsorbed on probe station through compressed air, the probe head and probe station were stainless steel and contacted closely. To confirm if there was parasitic capacitance between the probe head and probe station that affect the high impedance circuit sensitive to probe needle, the probe head was isolated from the station through a smooth insulating plastic sheet. What's more, the probe head was in floating state. Then put probe needle on the sensitive test pad in high impedance circuit. At this condition, when put probe needle on TP8, the sample would also recover. It indicated the probe head and probe station won't involve parasitic capacitance and affect the high impedance circuit.

Probe needle parameter effect. There were three types of probe needles were usually used when performing microprobe analysis: Passive needle, 12C active needle and18B active needle. 18B active needle is the most precise needle in measurement as it had highest input impedance and lowest input capacitance which can assure lowest affection to the real signal in circuit. 18B active probe needle was tried to if it can avoid the sensitive character on this kind of special circuit. 18B active was put on the insulating and floating probe head. When put probe needle on TP8 site, the failure would disappear too. The high impedance circuit was also sensitive to the most precise 18B active needle. Besides that, the passive needle was also tried and the result was the same.

To conclusion, the high impedance circuit was a complete probe needle contact sensitive circuit regardless what kind of probe needle and whatever it connected (Fig.7).

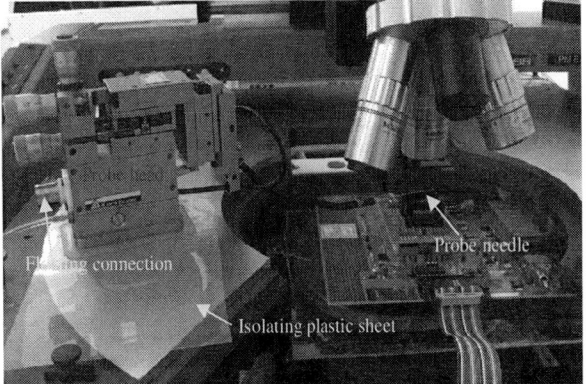

Figure 7. Experiment revealed high impedance circuit was sensitive to insulating and floating probe needle contact

2.) Light sensitive:

When performing microprobe analysis under microscope, the strong light of optical lens would be shut while measuring the signal to prevent potential affection. For normal decaped IC, strong light wouldn't affect the function. But for high impedance circuit, when strong light from high magnification lens shined on high impedance circuit area, failed sample would recover; when strong light was moved away, sample failed again. It indicated the high impedance circuit was also sensitive to strong light, but not sensitive to normal visible light. When strong light was shined on high impedance site, the particle of light would cause pressure or photoelectron effect and induce sub-current to make the TP8 site to be a low level signal stable state. This effect would also cause the failed sample recovered. Using the strong light of high magnification optical microscope to scan the die surface or suspected circuit area can detect the failed circuit. [5]

On reference simulated sample, the high impedance circuit was made by Pt connection using FIB. When scan the die surface using high magnification lens, it was the high impedance circuit that sensitive to light and not the Pt connection location. This may indicate that when got light sensitive area, there may be high impedance circuit existed, but not assured the defect located there. This method could provide another efficient and low cost approach to isolate the high impedance circuit failure (Fig.8).

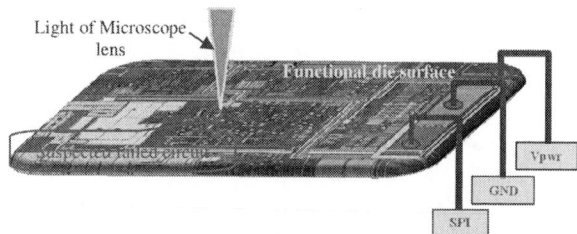

Figure 8. Strong light of microscope lens scanning the die surface which was in functional mode can isolate the high impedance circuit area

Experiment on high impedance circuit confirmed that it had sensitive character to many kinds of material which could contact with it directly. For instance, the probe needle which can contact it by probing on test pad; strong light can contact it with photon. When well known the property of high impedance circuit in failure analysis, much complex layout analysis and microprobe work can be reduced; therefore much cycle time can be saved.

V. SUMMARY

Failure analysis on decreasing size IC is facing more and more challenges and many circuit characters are important for finding the root cause. Unexpected high impedance circuit is in an unstable state and it is sensitive to many other factors which could contact with it in different ways. This may bring unexpected obstacle to find the defect efficiently. When well known of property of high impedance circuit, much analysis cycle time on microprobe and layout analysis could be saved, what's more, the success rate was also promoted when met complex functional cases.

ACKNOWLEDGMENT

Thanks physical failure analysis team of Tian Jin failure analysis lab for their great efforts in cases analysis. Also thanks to other colleagues of electronic failure analysis team of Tian Jin FA lab who gave much assistance in case analyzing.

REFERENCE

[1] Alberto Tosi, Franco Stellari, Andrea Pigozzi. "HOT-CARRIER PHOTOEMISSION IN SCALED CMOS TECHNOLOGIES: A CHALLENGE FOR EMISSION BASED TESTING AND DIAGNOSTICS" IPRS2006, 595-601.
[2] Michael Rubin, "LOCALIZATION AND ANALYSIS OF FUNCTIONAL FAILURES IN DEEP SUBMICRON ADVANCED ASIC PRODUCTS" IRPS2003, 588-589.
[3]DiWei Fan, Miao Wu, Joe Yu. "Novel Failure Isolation Techniques for Circuits Sensitive to Microprobe", ICDEA 2012, 401-404.
[4] Gaojie Wen, Diwei Fan, Li Tian, "Study on Light Sensitive Functional Failures in VLSI Failure Analysis", IPFA 2013, 186-189.
[5] Teh Swee Thian and Teoh Wan Yen, "Illumination-sensitive failure mechanism-A case study on transient Icc failure", IPFA1999, 73-76.

978-1-4799-3911-4/14 $31.00 © 2014 IEEE

Case Study of Embedded Memory Failure Analysis For Dislocation Issue

Cheng Wei Tang, Shin Chia Lin, Yi Chen Lin, Mei Ying Hsiao, Yau Shan Wu, Chi Lin
Reliability Testing & Failure Analysis Department, Powerchip Technology Corp.
No. 12 Li-Hsin Rd. 1, Hsin Chu Science Park, Hsin Chu, Taiwan, R.O.C
Phone: 886-3-5795000 EXT.3031 Fax: 886-3-5792142 Email: stanleyt@powerchip .com

Abstract- **Embedded memory is an integrated on-chip memory that supports the logic core to accomplish intended functions. High-performance embedded memory is a key component in VLSI, because of its high-speed and wide bus-width capability, which eliminates inter-chip communication.**
In this paper, embedded memory device and CMOS logic is integrated on-chip and it is a more complex of process technology compared with stand-alone memory. For the process engineering, we are always confronted with many problems, especially, the dislocation that causes some failures.

I. INTRODUCTION

Embedded memory device and CMOS Logic is integrated on-chip and it is the complexity of process technology with larger die size compared with stand-alone memory. People are always confronted with many problems due to difference of process and material during the development. Therefore, the failure analysis is becoming more difficult. In this paper, a developed new product which integrated logic IC, SRAM and NOR flash on one chip is presented. In the new product manufacturing process, there exists some failure mode such as program failure, power short and Isb high failure. In these issue, the Isb current higher failure is the most serious one. The aim of this study is focused on the failure analysis of the dislocation which causes the stand-by current failure.

II. FAILURE ANALYSES

In general, the failure analysis challenges are failure site isolation and physical analysis. We usually use some method and tools for failure analysis application. According to the stand-by current failure investigation, we performed analysis in electrical, chemical and physical failure analyses.

First, we checked the I-V curve of the stand-by failure sample to identify the leakage current abnormal or high resistance issue. IR-OBIRCH (Infrared Optical Beam Induced Resistance Change) and EMMI (Emission Microscope) are used to detect the failure site. Both of IR-OBIRCH and EMMI are considered powerful electrical fault isolation tools.

The nano probing technique is more important as devices become more complex. Sometime, we are confronted with substrate defect coming from anneal or implant process. Nano probing provided I-V curves measurement of electrical characterization for the device. Nano probing analysis is to measure the I-V curves of failed devices and confirm abnormal transistor characterization.

Transmission electron microscopy (TEM) provided topographical, morphological and crystalline information. This information is useful in the study of crystals and metals, but also has industrial applications. In order to observe the small failure, TEM was used to observe the small defect which was not observed with focused ion beam (FIB).

(a) Electrical Failure Analysis

Based on the wafer test result, we selected the stand-by higher current failure sample for IR-OBIRCH (Infrared Optical Beam Induced Resistance Change) and EMMI (Emission Microscope) analysis. First, we selected the stand-by current failure samples for IR-OBIRCH analysis. Based on the IR-OBIRCH result, we found some abnormal hot spot (Fig.1 a & b). And we used EMMI to double confirm the abnormal hot spot. Based on the IR-OBIRCH and EMMI analysis, the IR-OBIRCH and EMMI detection result showed some abnormal hot spots in stand-by leakage current failure (Fig.1-2).

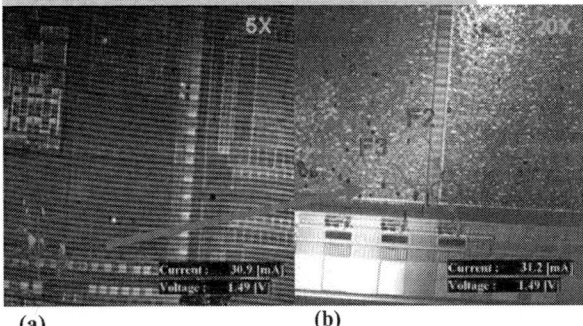

(a)   **(b)**

Fig.1 (a) (b) IR-OBIRCH detection result showed some abnormal hot spots in stand-by leakage current failure.

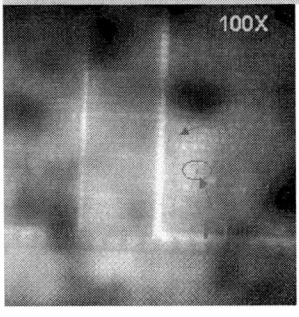

(c)

Fig.1 (c) EMMI detection result showed some abnormal hot spots in stand-by leakage current failure.

(b) Nano Probing Electrical measure

After abnormal hot spot layout checking, the abnormal area is localized to the X area. (Fig2). In order to identify which MOS abnormal, nano probe measure is used to measure the I-V curves of the devices G1, G2 and G3. (Fig3). First, we forced a voltage from -1V to 2V and checked the leakage current of G1、 G2 and G3 by nano probing measure. Fig.4 shows the nano-probing measurement result. The IV curve was normal for G1, Source-Drain punch was found for G2 and VT shift for G3. Based on the nano-probing result, we found that G2 and G3 have some leakage current at the same AA region and identified NMOS circuit has leakage issue.

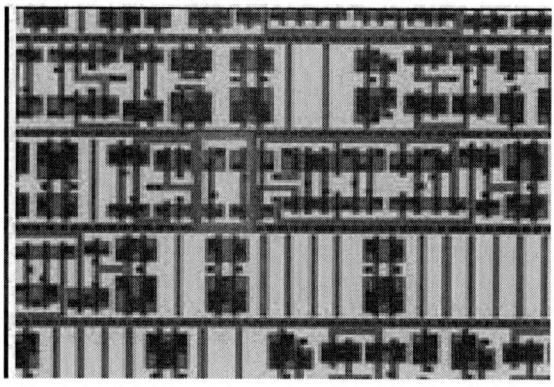

Fig2: hot spot area by layout checking

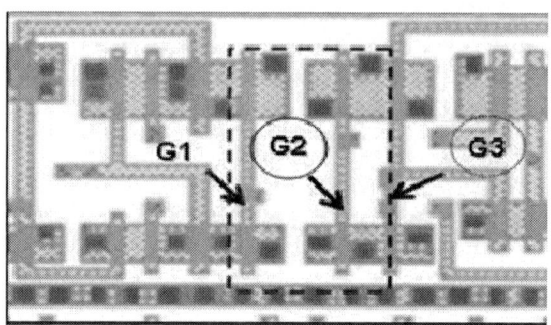

Fig.3 Nano-probing measurement for G1, G2 and G3.

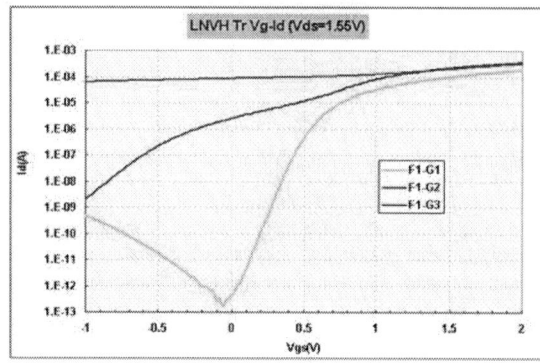

Fig.4 Nano-Probing measurement result

(c) Physical Failure Analysis & Plane-View TEM Analysis

Through investigation, nano-probing measurement displayed the abnormal AA. Based on above result, the failure site was narrow down to the same AA transistors. We select the failure location for plane-view TEM sample preparation and TEM analysis. We found the abnormality after TEM analysis. Fig.5 shows the plane-view TEM result. The line type deformation was found in the fail transistor. Based on the Plane-View TEM analysis result, we found dislocation and source-drain punch issue. The dislocation will induce leak when dislocation cross source and drain region. The TEM result matches the nano-probe measurement.

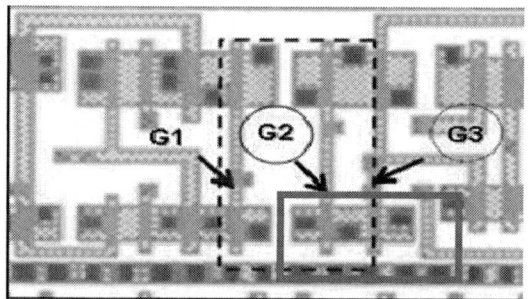

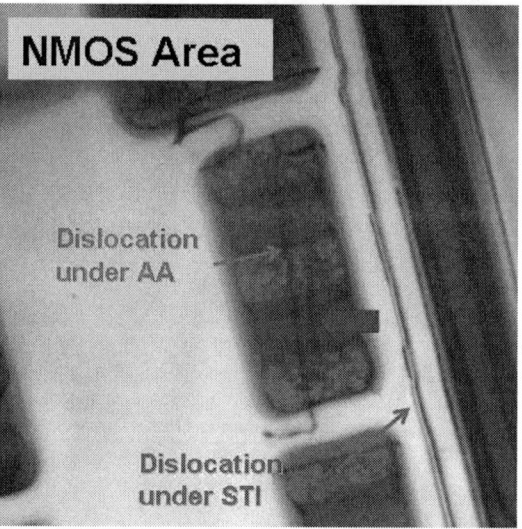

Fig.5 Plane-View TEM result shows Dislocation cross Source-Drain area.

(d) Cross-Section TEM Analysis

The plane-view TEM result displayed the dislocation issue. We prepared the cross-section TEM sample for cross-section profile check and crystal analysis from plane-view TEM sample. Cross-section TEM sample preparation is very difficult. Therefore, we used Dual-Beam FIB to prepare the high resolution transmission electron microscopy (HRTEM) sample for cross-section profile check (Ion-Beam condition: 5kV, 50pA). Fig. 6 shows cross-section TEM result. We found that the dislocation was located in STI corner of implantation region. Based on the HRTEM result, the dislocation was the reason of localized weak point.

978-1-4799-3911-4/14 $31.00 © 2014 IEEE

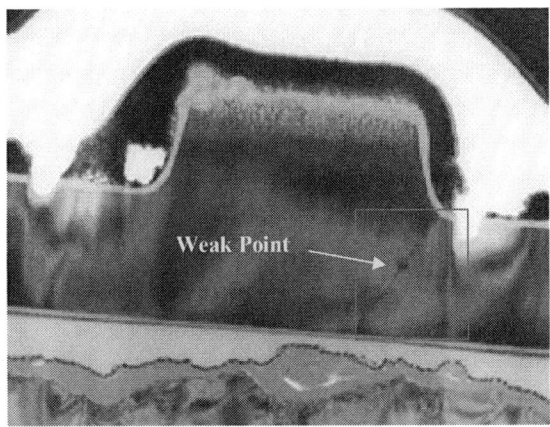

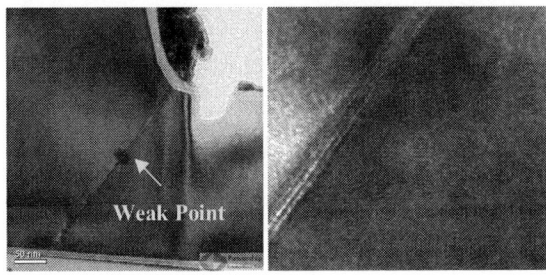

Fig.5 Cross-section TEM Image

III. DISLOCATION MECHANISM

The dislocation forms by two steps, the first step is nucleation; the second step is growth and propagation. In the nucleation process, oxidation induced high stress and volume expansion of SiO2 from Si.

Based on the cross-section TEM result, we found the dislocation happened near the STI corner and the weak point located the well implant region. In this case, we suspected the weak point defect comes from high energy well implant. Then, D2 and D3 form by compressive stress induced dislocation growth and propagation. The compressive stress (at D2 & D3) suspected it comes from thermal expansion mismatch. This defect of dislocation was not be well treated by anneal process.

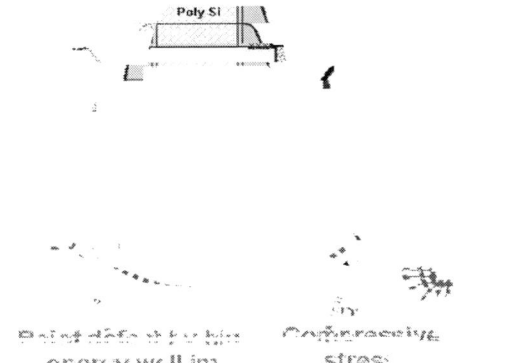

Fig. 6 Dislocation formed

IV. DISLOCATION IMPOVEMENT

Based on the dislocation mechanism, the process integration engineers provided two steps for improving the dislocation issue. They reduced the temperature of the furnaces original recipe and anneal time in anneal process. Based on the PFA result, we improved the compressive stress issue. But, this solution is effective in the wafer center region than wafer edge region. Next step, Rapid thermal Anneal process was implemented to repair the defect of crystal. The result shows the dislocation improved in wafer edge region successfully. The two solutions solved the dislocation mechanism. Table.1 showed that the dislocation has been improved in wafer center and wafer edge region.

Table.1 Improved result for dislocation

V. CONCLUSION

Embedded memory device and CMOS Logic integrated on-chip and it is a more complex of process technology than stand-alone memory, because too heavy thermal process from thermal budget review. Thus, we always confronted dislocation problem in some failure modes.

For this dislocation investigation, we considered this failure to come from STI corner and anneal process intrinsic defect. However, we identified dislocation issue by TEM analysis due to IR-OBIRCH/EMMI detected the abnormal hot spot and nano probing measured the I-V curve. We found out the abnormal AA and defect and identified the root cause. Through investigation result displayed the dislocation mechanism. Thus, we implement the actions to improve the dislocation issue during the development of new product.

REFERENCES

[1] Lawrence C. Wagner "Failure Analysis Challenges". Texas Instruments Incorporated, 12500 TI Boulevard, Dallas, TX USA

[2] Sang In Kim, Hyung Mo Yang, Hee Seong Yang, Ju Hyeon Ahn, Seok Sik Kim, Yu Gyun Shin*, Ki Hyun Hwang*, Ji Woon Rim*, Woon Kyung Lee**, Han Soo Kim**, Sun Kyu Whang**, Ji Woong Sue*** and Han Ku Cho "Failure analysis on the standby current due to dislocation in STI structure of flash memory". IEEE, Jul 2, 2012

Electrical and Physical Analysis of a 28nm FPGA Programmable Delay Circuit Single Tap Delay Failure

Chow Yew Meng, Bai Haonan, Grace Tan, Peter F Salinas, Johney Ou Yang
Xilinx Asia Pacific Pte Ltd
5 Changi Business Park Vista, Singapore 486040
Tel: (+65) 6407-3000 Email: yewmeng@xilinx.com

Abstract

This failure analysis is based on a 28nm FPGA IDelay logic block which features an all programmable, 32-tap delay line. Each tap delay is carefully calibrated to provide an absolute delay value of 78ps independent of process voltage, and temperature variations. To locate the failing IDelay site, scan chain methodology was utilized. Combinations of delay tests were created to localize the defect within the IDelay block and the failure was isolated to a single tap delay circuit. Photon emission analysis validated the electrical analysis with an emission successfully detected at the suspect area. Physical failure analysis utilizing a combination of AFP current contrast imaging and nano-probing analysis at the contact layer further isolated the area of interest to a specific transistor. Die delayering and SEM high beam inspection did not show any anomalies, but subsequent TEM analysis revealed diffusion bridging at the failure location.

I. INTRODUCTION

This case study is of a 28nm field-programmable gate array (FPGA) device. The device had exhibited basic functional failure during ATE testing. An input delay (IDELAY) pattern failure was identified based on the ATE result.

IDelay logic block features an all programmable, 32-tap delay line (Fig. 1). It can be applied to the combinatorial input path, registered input path, or both. It can also be accessed directly from the FPGA logic. IDELAY allows incoming signals to be delayed on an individual input pin basis. The tap delay resolution is contiguously calibrated by the use of an IDELAYCTRL reference clock. The delay tap can be pre-defined, read/write from FPGA logic, or increased/decreased from any current value. There are various ways of testing each delay function. The failure complexity laid in the fact that any of the 32 tap failure would cause the IDELAY block to fail.

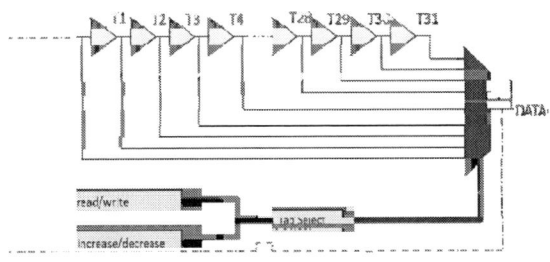

Fig 1.IDELAY circuit

Most of the ATE tests are today switched to Multiple Input Signature Register (MISR) testing; it is designed for compression of the response data which would save testing time. MISRs collect response streams into a signature and the signatures would provide a Pass/Fail indication. This comes with a disadvantage for debugging in that a defect cannot generally be located using the failing signature. In this investigation, the pattern that can capture the failure is known from ATE test, however the location of the failing IDELAY site is impossible to be isolated based on the failing vector as this pattern is using (MISR) test. The only information that can be obtained from ATE test is there is one IDELAY failure in the device.

In the next section, electrical tests focusing on IDELAY functions will be introduced. In subsequent sections, this paper will demonstrate how such a combination of electrical fault isolation as well as an array of failure analysis tools successfully determined the cause of failure.

II. FAULT ISOLATION

To proceed with isolation, a new Non-MISR pattern was created to isolate the failure using scan chain methodology. How the pattern works is that each IDELAY output is connect to a flip-flop (FF) and shifted out to another IDELAY. Basically all IDELAYs are cascaded into a scan chain. Data will be shifted into the first delay unit and passed on one by one until the last one (Fig. 2).

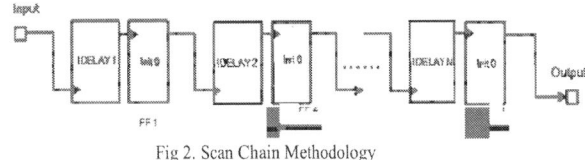

Fig 2. Scan Chain Methodology

To cover all the output stuck-at-low and stuck-at-high scenarios, the scan chain was tested through four steps:

1. FFs default initial value 0, Walk a 1 in sea of 0s.
2. Fill up all FFs with 1s.
3. Walk a 0 in sea of 1s.
4. Fill up all FFs with 0s again.

There are 32 delay taps for each IDELAY; to fully test all the taps, tap increase/decrease testing was used. IDELAY was

978-1-4799-3911-4/14 $31.00 © 2014 IEEE

set to tap 0 at the start, all the four steps of scan chain tests were performed through all the IDELAY blocks. After finishing testing tap 0, the delay tap was automatically increased to 1 and the four steps of scan chain tests repeated for delay tap 1. This process was repeated up to delay tap 31. This way all the delay taps were covered, together with all the other functional resources used to assist the tap testing in IDELAY.

Execution of this test detected a failure when testing delay tap 19 (Fig. 3). At the first step when a 1 was walked through the chain of 0s at vector 46404, after shifting through all the FFs, the 1 should be shifted out to vector 47001. However there was a failure here indicated with "/" (Fig. 3), which means the output received a 0 instead of the expected value 1. Theoretically by using the following 3 steps, the IDELAY block could be isolated by figuring which FF was the first to fail. The first FF in the chain failed indicates the corresponding failing IDELAY. But considering that this was an output stuck at low failure and all FFs are initialized with 0s, the second step of filling all FFs with 1s would not be successfully implemented. Whatever value the FFs are the output will shift out a 0. It was not practical to isolate the IDELAY block as all FFs shifted out data are 0s. So at this stage the failure could only be isolated to be due to one of the IDELAY blocks when the delay tap was set to 19.

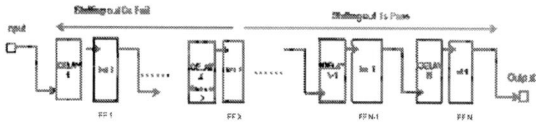

Fig 3. Result 1

In this case, how did we isolate this stuck-at-low failure? Since all FFs are actually defect free, the only thing that could cause this device to fail would be if one of the IDELAY outputs was stuck at low. So if all FFs initial values are changed to 1s, the IDELAY stuck-at-low failure would be passed to the FF. The output register will keep shifting out the passing FFs 1s until the stuck at low FF. Just by knowing the number of the passing FFs, the failing FF would be identified, and so the IDELAY block (Fig. 4).

Fig 4. Initialize 1 Scan Chain

Since the failure was only at tap 19, a new test was pre-defined at delay tap 19. With all FFs initial value set to 1s, a 0 was walked through the chain of 1s. The failing IDELAY was isolated at vector 46663 indicated with a "/" (Fig. 5). The passing FFs were 258 (46663 -46405). There were a total of 596 IDELAYs tested in the chain. The 338th IDELAY in the chain was the one that was stuck-at-low. This was site IDELAY_X1Y220.

Fig 5. Result 2

At this point the failure had been isolated to a single delay tap 19 of a single IDELAY block X1Y220. From the schematics, a failure at one of the delay tap circuit, would normally cause all the delay taps downstream to fail as well. For a single tap failure, the hypothesis was that the failure was at the end of the output route where there was no connection to any other delay tap circuits. There were two branches of output for each IDELAY tap; both combine into one final delay output. Both of the output paths could be defective. The area of interest was identified on the die layout (Fig. 6):

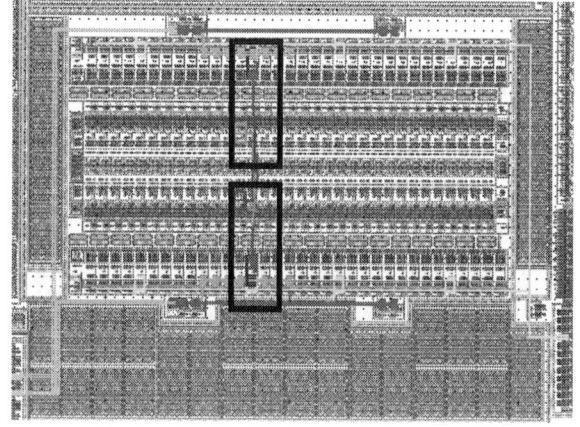

Fig 6. Region of interest (boxed in black) identified by fault isolation

As the circuit area of the IDELAY was relatively large, even for the single tap 19 output, to validate the failure hypothesis and further minimize the area of interest, photon emission microscopy was employed.

III. FAULT ISOLATION HYPOTHESIS VALIDATION

With the failing circuitry and corresponding region of interest identified by fault isolation, a further step was taken to validate the failure hypothesis. Photon emission analysis was performed with the failing input delay pattern continually looping through the device. An emission was detected with macro-lens (Fig. 7); the same emission was consistently detected when the analysis was repeated at sequentially higher magnifications up to 100x.

Fig 7. Photon Emission detected at the area of interest.

The emission location was mapped onto the die layout and confirmed to be within the region of interest identified by fault isolation. This validated the hypothesis that the device failure was at the circuitry identified by fault isolation.

IV. PHYSICAL FAILURE ANALYSIS OF DEPROCESSING AND SEM INSPECTION

The metal paths associated with the failing circuitry were traced and highlighted on the die layout drawing.

The die was extracted from the package for deprocessing. The region of interest was demarcated on the die via laser marks. Since the failing circuitry was from metal 4 to diffusion, the die was parallel polished from the passivation layer to inter-metal dielectric 4 (IMD4), at which inspection of the region of interest commenced. A combination of reactive ion etch, diamond film and diamond suspensions were utilized for die lapping, in order to minimize polishing artefacts while maintaining sample surface planarity, particularly over the region of interest.

After Metal 5 removal, IMD4 was thinned down and Metal 4 of the region of interest was inspected by SEM (scanning electrons microscopy). Since the metal paths associated with the failing circuitry had already been highlighted on the die layout drawing, the metal 4 features corresponding to the failing circuitry could be precisely located and inspected on SEM. However, no Metal 4 anomalies were observed. Metal 4 was subsequently polished away and IMD3 thinned down while ensuring some thickness of via 3 remained after polishing. Voltage contrast (VC) analysis was performed at via 3 level but no abnormal contrasts were detected over the region of interest. SEM inspection of metal 3 similarly showed no anomalies.

This iterative process of parallel lapping, VC analysis and SEM inspection was applied from metal 4 to metal 1 but all with no anomalies found. Metal 1 was then removed and passive voltage contrast (PVC) was performed with the region of interest at the contact level. This time, a leaky poly contact was detected within the region of interest and specifically at one part of the failing circuitry (Fig. 8). The leaky poly contact appeared bright in the PVC image; however since poly gates are by definition floating, normal poly contacts would build up negative charge during PVC analysis and appear dark. A bright poly contact would suggest some form of leakage path that prevents charge build up at the contact/poly, thus giving a bright appearance in the PVC image.

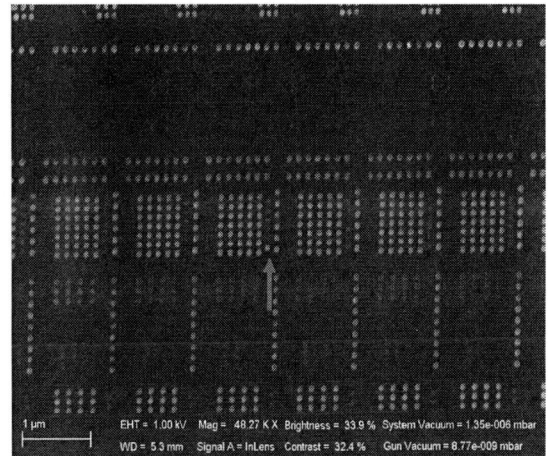

Fig 8. Abnormal PVC contrast detected at failing circuitry.

V. AFP NANOPROBING AND TEM ANALYSIS

Pico-current imaging was performed over the region of interest using an Atomic Force Probing (AFP) system. A current image was generated by rastering a biased probe needle over the region of interest: leakage was detected at the same poly contact detected with prior PVC analysis (Fig. 9; top).

Nano probing was then performed with the failing poly contact probed with respect to the neighbouring source/drain diffusion contacts. The failing poly was found to be shorted to the diffusion on both sides of the poly (Fig. 9; bottom). However, at this stage high KeV SEM inspection of the area did not reveal any physical anomalies.

978-1-4799-3911-4/14 $31.00 © 2014 IEEE   175

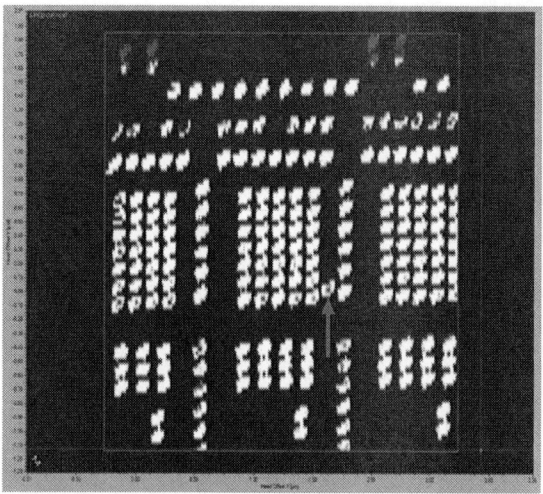

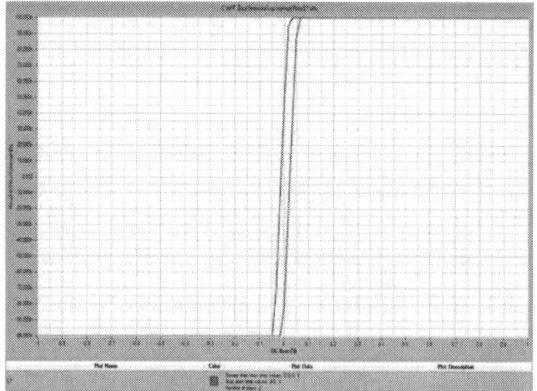

Fig 9. AFP current imaging confirmed the leaky poly contact (top); nanoprobing showed the leaky poly was shorted to both the source & drain diffusions on either side of the poly (bottom).

Planar Transmission Electron Microscope (P-TEM) analysis of the failing poly area showed some form of bridging material at one end of the transistor area with the failing poly (red box indicated in Fig. 10). The material appeared to run beneath the poly, which bridged the source and drain diffusions on either side.

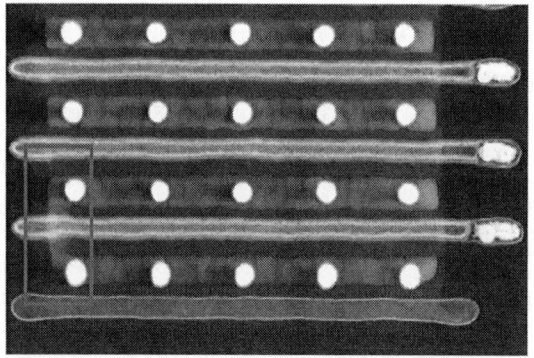

Fig 10. Planar TEM (top) showing source-drain diffusion bridging across the failing poly.

Subsequent cross-sectional TEM (X-TEM) analysis of the bridging area confirmed a diffusion-level bridging between the source and drain (red circle indicated in Fig 11).

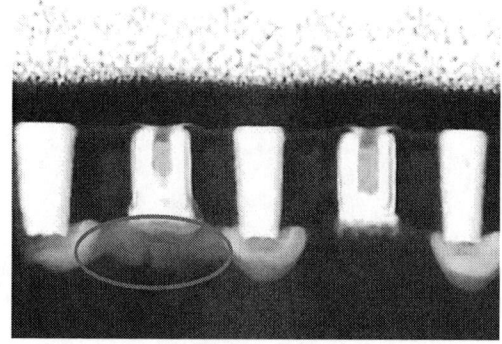

Fig 11. Cross-sectional TEM (below) showing that the bridging was beneath the poly, at the channel/diffusion level.

Energy Dispersive Spectrometry (EDS) analysis of the material showed high nickel content (Fig. 12), indicating that the bridging material could be related to nickel silicide from the source/drain areas.

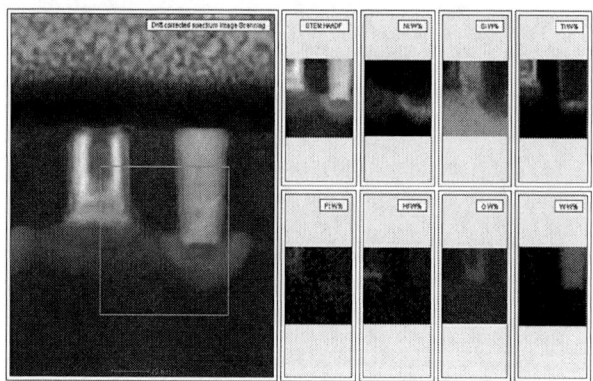

Fig 12. EDX mapping showing Ni in the bridging material

Conclusions

The successful failure analysis of a complex failure in a 28nm device has been presented in this paper. Electrical fault isolation using scan chain methodology and combinations of delay tests identified the failing circuitry. This hypothesis was further validated using photon emission analysis with the device in the failing state. Subsequent physical analysis employed a wide array of tools, including polisher, SEM, voltage contrast, AFP current imagin, nano probing, TEM and EDX mapping analysis and so on. Through this combination of multiple electrical/physical analysis tools, the cause of the input delay failure was successfully established to be due to nickel silicide bridging between the source and drain of a transistor within the failing circuitry.

References

1. Fayez Elguibaly and M. Watheq El-Kharashi, Multiple-Input Signature Registers: An Improved Design, IEEE Pacific Rim Conference, vol.2, 1997 pp. 519 -522.

2. Xilinx production and support document "7-series Generation FPGA User Guide" vol.1.7 pp. 114-116, July 2012.

3. Xilinx production and support document "7-series Generation FPGA User Guide" vol.1.7 pp. 120, July 2012.

4. Swaminathan Subramaniam *et al*, Transmission Electron Microscopy for Failure Analysis of Integrated Circuits, Microelectronic Failure Analysis Desk Reference, 5th Ed, 2004 pp. 595-614.

5. J. Colvin *et al*, Atomic Force Microscopy: Modes and Analytical Techniques with Scanning Pobe Microscopy, Microelectronic Failure Analysis Desk Reference, 5th Ed, 2004 pp. 615-627.

6. Yang Jing *et al*, Electrical Diagnosis of Temperature-Dependent Global Clock Failures using Probeless Isolation and Pattern Commonality Analysis, IPFA2012.

Finding a new type of in-line failure mechanism "Floating Antenna Effect" and its solution

Yutian ZHANG, Junzhi SANG PhD., Yun XU, Zhimin ZENG.

Shanghai Huahong Grace Semiconductor Manufacturing Corporation

1188 Chuanqiao Road, Pudong New Area, Shanghai, P.R.China, 201206 Email: zhangyt@263.net

Abstract- **On a new embedded flash platform we have got high failure rate on charge pump test. It is impossible to directly force external current into failure signal path to do FA but we did it under dynamic condition. Afterwards we have captured metal short locations by emission scope which is rarely seen. Then we used in-line KLA SEM VC scan to define the real failure mechanism – a new type of antenna effect, and finally solved it with combined process condition changes. We named the new type of antenna effect "Floating Antenna Effect".**

I. INTRODUCTION

The low yield of a new product did not meet mass production conditions. Base on some basic analysis we found most of failures were charge pump voltage (HV output) failure. By solving this ~15% failure, we can improve the yield to target, and pull the whole platform a big step forward by eliminating the reliability concern for those "newly passed" chips after stress. Using EMMI scope with dynamic test condition, we located a metal short. Such short has rarely been captured by EMMI because of its emitting wavelength. Then PFA was performed to check the short location. But it was still far from the solution. After in-line inspection and careful layout check, the root cause was discovered. When a piece of floating metal pattern is big enough, its collected charge during ILD deposition could possibly damage an adjacent metal which is well grounded. Based on this finding, we modified the layout and process to avoid it happen on whole process platform.

II. ANALYSIS STEPS

A series of analysis steps were taken by the following order, to lead us from basic failure phenomenon to root cause, then to solution.

1. Stability test: Retested all failure chips and found out that some failing chips became passing chip after each round of retest (Fig.1). Suspect there were weak bridging defect which can be burnt out by applying stress.

2. Characterization test: Select some chips that always failed after several retest, and define them as hard failure chips. Use ATE to set test conditions at failure mode, then force external voltage source into test pin to collect I-V curve. The I-V curve showed the failure to be ohmic (Chart 1). Suspect there was direct-short bridging, and should be related to a hard defect.

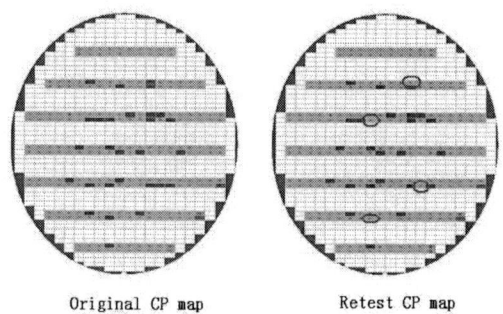

Fig.1, CP bin map comparison: We can see some failed chips become passed ones during retest.

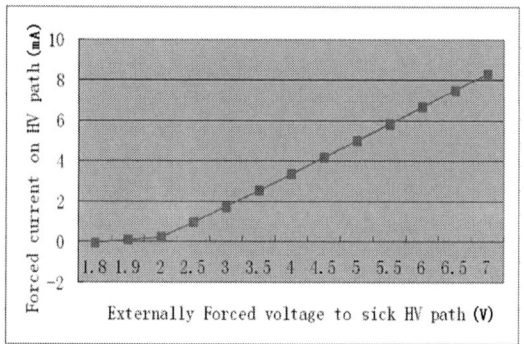

Chart 1, I-V character of the suspected leakage path. The current was forced by probe card from a test point which can be switched to the faulty net.

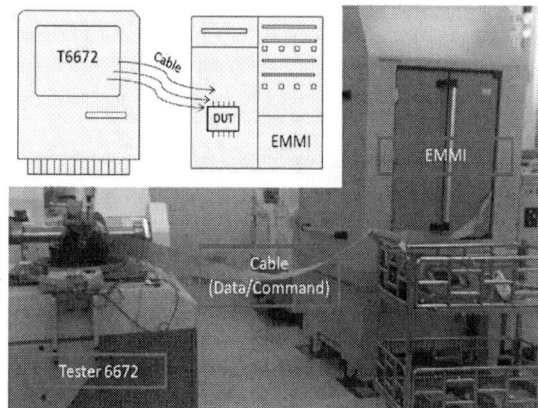

Fig. 2, Dynamic EMMI environment and connections: ATE , Cable and photon emission scope

3. Failure localization: Next we have to locate the bridging point. That required us to force dynamic signal and driving current into failure HV path. However, there were too many pins

defined in datasheet to drive the chip to the failure mode, we had to probe the chip while biased by ATE and set Hi-Z state to those possibly unnecessary signals one-by-one, then successfully entered the failure mode with less than 40 pins. That allowed us possible to prepare a sample with DIP 40 wire bonds for EMMI sample holder. See Fig.2 for dynamic EMMI connections.

We captured a clear hot spot by this dynamic EMMI analysis.(Fig. 3) with cooled–CCD Photon Emission Microscopy (PHEMOS-1000 the product of HAMAMATSU Japan)

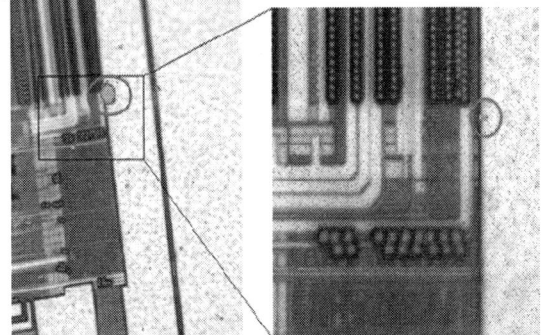

Fig. 3, hotspot captured by EMMI scope.

Then we checked the layout (Fig. 4) for possible failing transistors, but no active transistor was found near this hot spot area.

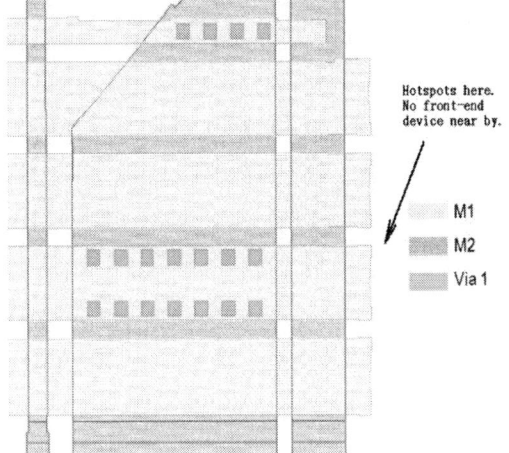

Fig. 4, Layout check: multiple metal line-end were found at the hotspot location. As a result of OPC, line ends were a little expanded to correct optical shrink. We then suspected there is process weakness to leave a direct short here (but it did not happen.)

That was an unusual situation because most of hot spots captured by EMMI was related to front-end light emission, maybe by carrier combination or electron transition between the band gap .etc. [1][2] Metal bridge may usually emit IR light with wavelength longer than 1000nm and could be captured by InGaAs or MerCad detectors only.

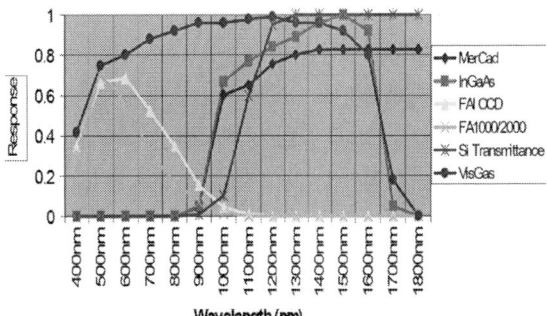

Chart 2, Sensitivity character comparison among multiple detectors. [3] EMMI microscope used the cool CCD with sensitive wavelength (400nm~1000nm) similar to FAI CCD or FA1000/2000 CCD.

4. PFA: Physical analysis was then performed with focus on the hot spot area. After anomaly was found at the hot spot location (attach a top view photo). Using FIB we simply found the melt-bridging or worn-out bubble defect on both 2 types of failed samples: dice failed on first test but passed on retest (sample 1 on Fig. 5); dice failed on first and all test. (sample 2 on Fig. 5)

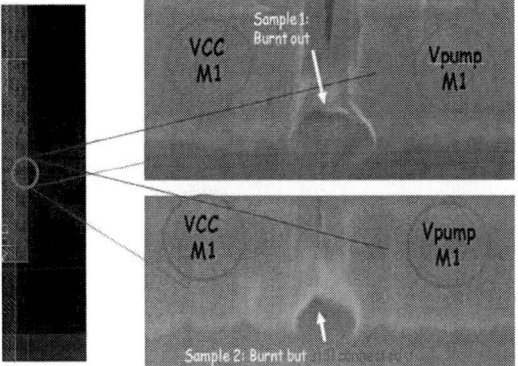

Fig.5 PFA: FIB check failure location. The burnt location may be a mess, the melt location were short, the burnt-out location were OK during CP test.

And sample 2 was sent for TEM analysis (Fig. 6) .

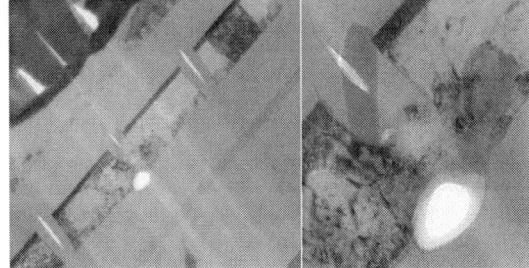

Fig. 6, PFA: TEM check bridging location.

The original metal bottom profile was no longer kept because of the burn out. Thus we can take adjacent metal spaces for original profile reference. (Fig. 7)

978-1-4799-3911-4/14 $31.00 © 2014 IEEE   179

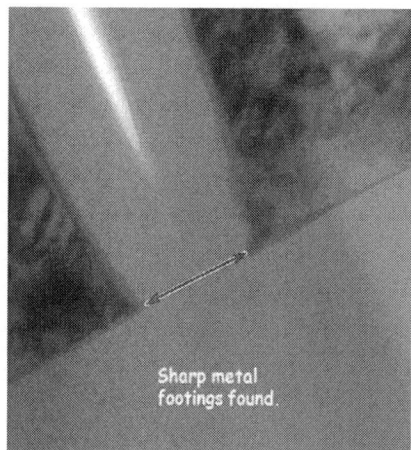

Fig. 7, Some sharp metal footings were found at metal bottom. That may bring a high electron field or even a zap between metals.

III. FAILURE MECHANISM DISCUSSION

The bridge may exist after wafer process as a worse case of metal footing but we cannot see it. Some bridges should be weak and some should be solid. Weak bridges cannot withstand the HV charge pump output then could be broken down and passed the test with reliability risk. Strong bridges were also burnt due to high current but not broken open. Then the melt residue stuck in the area causing the chip failed at specific failure mode so we could have captured the hotspot on it by proper means. Once we knew the problem, then we moved on for finding the solution to fix it.

There are still some questions that need to be answered: When was the metal burnt? Was the bridge intrinsic as an etch residue? The layout of bridging point has passed DRC as well during design rule verification. That means both the photo and etch process should have appropriate process capability to build a clear and safe space between the 2 metal lines. At the same time, why did the adjacent similar structure with the very same spaces do not get short or burnt?

IV. DIGGING FOR ROOT CAUSE

1. DRC study: such metal width/ space/ line-end combination can be found everywhere in the layout. OPC also defined a reasonable space between multiple line-ends where we found the burnt point. No violation was found.

2. KLA defect scan on in-line wafers: Originally we intended to let KLA check the ADI (after development inspection) step, the AEI (after etch inspection) step and ASI (inspection after PR dry removal) steps to find the burnt step, but no metal were burnt after these steps. Fortunately we found in KLA image (Fig. 8) that the faulty metal line (charge pump voltage path) was dark during KLA SEM VC scan after PR removal. That means the metal was safe before that and will be floating during future ILD deposition step. So there's no electrical path to discharge this metal line during ILD deposition process with mass plasma/charge accumulation. As the size of that piece of metal is really large, we highly suspect the discharge took place at metal line end with smallest space made by OPC enlargement. It is

also because we found no defect or residue using this KLA SEM scan.

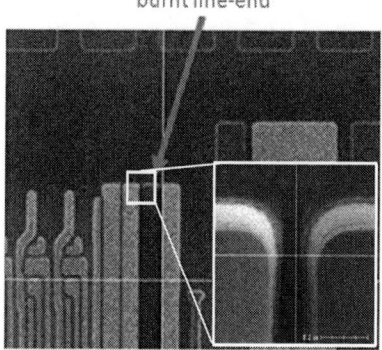

Fig. 8, KLA scan result: The metal space were clean with long footings. The HV path was a piece of floating metal after etch. That means the plasma charge in future steps such as PR ashing, ILD deposition cannot be well discharged. A zap may take place here because it is the smallest space between the charged voltage and a discharge path.

As both of the dry etch and PR ashing were finished before KLA scan but no defects were found then, the only suspect has been narrowed down to ILD deposition step. A new type of failure mechanism named "Floating Antenna Effect" was defined. We need to find the solutions.

V. PROCESS / DESIGN EXPERIMENT AND SOLUTIONS

In next steps we tried to study all factors may lead to this floating antenna failure.

1. Metal film stack experiment: we changed the Ti/TiN/Al film thickness combination and anneal condition .etc. to tune the alloy at metal bottom. (chart 3 & table 1)

2. Metal etch experiment: we changed the metal dry etch recipe to eliminate metal footings. (chart 3 & table 1)

3. ILD deposition experiment: We also changed ILD deposition recipe to reduce plasma charge.(chart 3 & table 1)

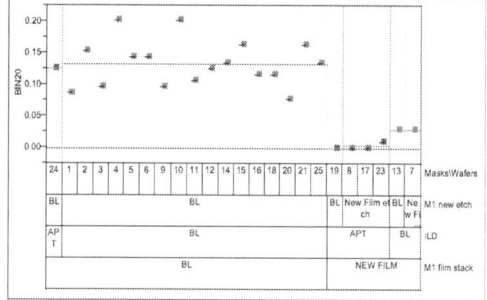

Chart 3, Yield loss of target failure "bin20" VS. each split conditions.

Recipe change	Result
M1 etch	Ineffective
M1 film stack	Effective
ILD deposition	Obviously effective

Table 1, DOE summary. The yield result in the DOE lot showed most effective solutions for Floating Antenna Effect is to combine "new metal film stack + new ILD deposition" recipe. At the same time we can find the metal etching new recipe doesn't work.

4. We also tried to modify the product layout (Fig. 9) to verify the failure model and that have given us direct effective response. The failure can be completely eliminated.

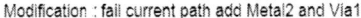

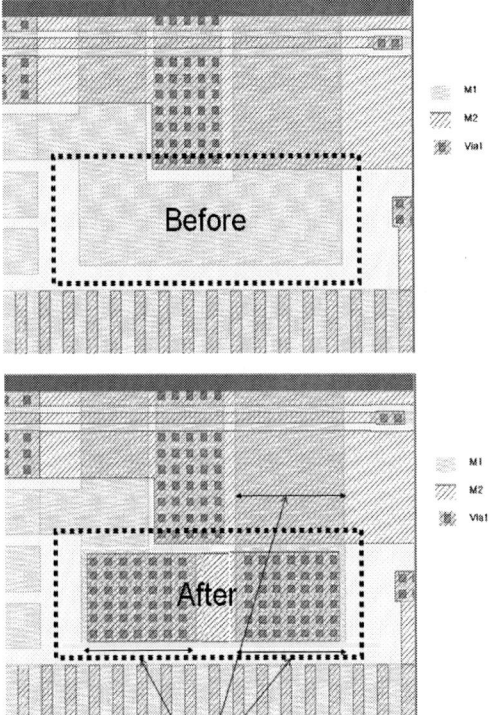

Fig. 9, The layout was modified to add a bridge to cut the large floating metal into 2 parts. Then most of the charge cannot be sent to the narrow line-end spaces to bring risk of zap.

CONCLUSIONS

A new in-line failure mode was found and named "Floating Antenna Effect". That happens when following conditions were met (Fig. 10):

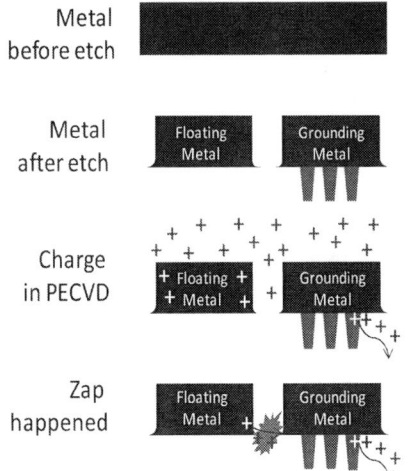

Fig. 10, Failure mechanism: Plasma charge induced zap between metal footings

1. A large piece of metal was designed to be floating before future layers built on, and another piece of well grounded metal was laid close enough to the previous floating metal.

2. The metal etching + metal film stack process could leave a tiny and sharp footing at metal bottom.

3. An ILD deposition step with strong power provided sufficient plasma charge.

To solve this problem, we can lower down the plasma power, change the metal dep./etch process to leave no footing, or add a design rule to avoid large floating metal polygons.

ACKNOWLEDGMENT

Thanks to all managers for release so many resources to us and all fellow from within the company and out of it. That made a whole foundry a big FA lab to solve a complicated problem.

Great thanks for the guidance of Dr. Junzhi Sang, Yun Xu and Susan Li. All their input was so helpful and meaningful to this paper.

REFERENCES

[1] Xuanlong Chen, Xianjun Kuang and Guangning Xu, "Application of EMMI Contrast Method in Failure Analysis," *Reliability Research and Analysis Center, China CEPREI Laboratory, China,* Guangzhou 510610, E-mail: chenxl@ceprei.com

[2] Jim Colvin "A Tutorial on FA Methods and Failure," *FA Instruments, Inc. www.fainstruments.com,* 2381 Zanker Rd. Suite 150 San Jose, pp. 16

[3] Richard J. Ross, "Fundamentals of photon emission (PEM) in silicon", *Microelectronics failure analysis desk reference, sixth edition,* pp.279, October 2011.

[4] Alexandre Acovic and Joerg Dreybrodt, "Study of process influences on the break down limit of High Voltage transistors in an embedded EEPROM CMOS technology by using EMMI and Nanoprobing," *EM Microelectronic,* Marin SA, Rue des Sors 3, 2074 Marin, Switzerland, E-Mail: jdreybrodt@emmicroelectronic.com

[5] Yuansheng Wang[1,2], Xiao Hong[2], Chang Zeng[2], Ping Lai[2] and Yun Huang[2], "Reliability Evaluation and Failure Analysis of AlGaN/GaN High Electron Mobility Transistor by Photo Emission Microscope" *1 Guangdong University of Technology, 2 Science and Technology on Reliability Physics and Application of Electronic Component Laboratory, The 5th Electronics Research Institute of the Ministry of Industry and Information Technology,* E-mail: foxroz2003@hotmail.com

[6] GUAN SIONG LEE, Aaron Chin, KOK KENG PEE and FONG LING CHOW, "Method to Increase Defect Localization Success Rate on Open Failure by Combining Circuit Layout Analysis with Photon Emission," in *Magnetism,* vol. III, G.T. Rado and H. Suhl, Eds. New York: Academic, 1963, pp. 271-350

ABOUT THE AUTHORS

Yutian ZHANG, BA. E.E., graduated from Xi'an Jiaotong University in '2003, Section Manager & Senior Principal Engineer. Product Dept. in Shanghai Huahong Grace Semiconductor Manufacturing Corporation. Have focused on integrated circuit failure analysis / debug / yield enhancement/ product engineering for 10+ years. e-mail: zhangyt@263.net

Junzhi SANG, PhD., graduated from Mie University, Japan, 2000. Now Senior Director of Shanghai Huahong Grace Semiconductor Manufacturing Corporation. His research interests are in DFT (Design for Testability), solution for mass production testing technology and FA (Failure Analysis) for VLSI.

Failure Analysis Methodology for the Localization of Thin and Ultra-Thin Metal Barrier Residue

A.C.T. Quah, N. Dayanand, S.P. Neo, G.B. Ang, M. GUNAWARDANA, H.H. Ma, Z.H. Mai, J. Lam
Product, Test & Failure Analysis, GLOBALFOUNDRIES Singapore Pte. Ltd
60 Woodlands, Industrial Park D Street 2, Singapore 738406

Phone No:(65) 6670-1517, Email: Alfred.QUAH@globalfoundries.com

Abstract

This paper describes several case studies which used a combination of laser induced techniques, photon emission microscopy and layout analysis, together with the identification of common failure signatures that are associated with CMP under-polish, for the effective localization of thin and ultra-thin Ta barrier residue in the backend of line Cu metallization stack.

I. INTRODUCTION

Failure analysis (FA) is an integral step for the development and manufacturing of semiconductor integrated circuits (IC) and fault localization is the most crucial step in the entire FA cycle. In the wafer foundry industry, a marginal process drift could result in severe yield loss and excursion which leads to wafer scrap and loss in revenue. Thus, early inline detection and short FA cycle time are critical to maintain line quality and profitability. In today's ultra-large-scale advanced IC technology, Cu is used as the interconnect material in the backend-of-line process due to its low electrical resistivity and resistance to electromigration. Chemical Mechanical Polishing (CMP) is used to achieve planarization in the Cu single and dual damascene process. With increasing backend metal stacks beyond 10 layers and the use of low-k and ultra-low inter-metal dielectric layers, achieving global uniform planarization is challenging. Slight process drift could result in CMP under-polish, leaving behind thin or ultra-thin Ta barrier residue resulting in resistive short or over-polish impacting product speed performance. Both scenarios would result in unfavourable yield loss.

Today's die-level static fault localization relies mainly on photon emission microscopy [1] and laser induced techniques, like Thermal Induced Voltage Alteration (TIVA) [2]. It is well-understood that PEM is effective to localize front-end of line defects resulting in junction leakages, gate oxide breakdown and bridging nodes and laser induced techniques are effective for localizing a whole host of resistive opens and shorts resulting in static leakages and yield loss. In this paper, 3 case studies are described with different severity of barrier residue from CMP under-polish resulting in resistive short for on different technology nodes. While it is uncommon to use PEM to localize backend-of-line bridging defects, the case studies would illustrate that as barrier residue becomes thinner, laser induced techniques becomes less effective. The use of PEM, layout analysis and the correlation with common failure signatures that are associated with CMP under-polish becomes critical to effectively localize thin and ultra-thin Ta barrier residue. It is also highlighted that for barrier residue defect, cross-section FIB and TEM is more effective for defect tracing than top down physical failure analysis (PFA) approach.

II. RESULTS AND DISCUSSION

Case Study 1: Thick Cu/Ta Residue

The first case study describes a low yield issue due to CMP under-polish on a 90nm product resulting in extreme edge ring pattern failure. In this case, the under-polish resulted in barrier residue which was effectively localized by TIVA analysis. Fig 1(a) shows the wafer sort map to illustrate the extreme edge power short ring pattern failure with failing die highlighted in red. TIVA analysis on several units within the same failure region observed exclusive TIVA spots at the same location as shown in Fig 1(b). Top down delayering observed no anomaly from top metal to metal 3. However, after metal 3 removal, optical inspection observed greenish yellow discoloration at metal 2 as shown in Fig 1(c) in the circled region. Further progressive FIB observed metal 2 barrier bridge as shown in Fig 1(d) at the discolouration site. The cross-section TEM results in Fig 2 show that the bridge is made up of full Ta metal barrier with also a thin layer of Cu.

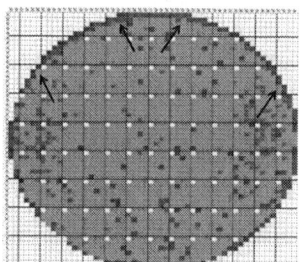

(a) Extreme edge ring pattern failure

(b) Systematic TIVA spots

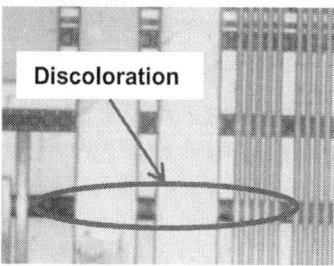

(c) Discolouration from optical Inspection

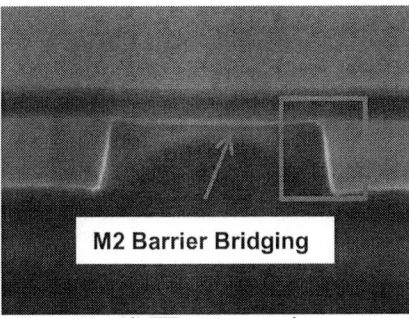

(d) FIB cross-section

Fig 1 Failure analysis of extreme edge power short ring pattern failure

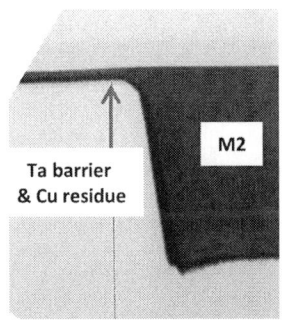

Fig 2. Cross-section TEM showing barrier and Cu residue

In this case study, a conventional top down physical delayering physical failure analysis approach was adopted. However, the success to the PFA lies in the careful observation of discoloration during the optical inspection step. This was possible for this issue as the residue is thick (Ta + Cu) and the bridge location is of large metal pitch resulting in observable abnormal colour contrast. For thin and ultra-thin residue, the colour contrast would be too faint to be observable and top down PFA approach would not be successful as the use of reactive ion etching (RIE) to expose metal for subsequent SEM inspection would etch away the residue. Progressive cross-sectional viewing with FIB followed by TEM analysis is thus a more effective approach for barrier defect viewing. Therefore, a good sensitivity to failure signatures and symptoms barrier bridge is critical in the use of the right PFA approach to achieve first time success and shorter FA cycle time.

Case Study 2: Thin Ta Residue

The 2nd case study describes a 65nm low yield issue with partial ring shape failing pattern highlighted in pink as shown in Fig 3(a). The partial ring shape failing pattern with no passing dies within the patch strongly correlates to possible CMP issue and thus barrier residue. Fab investigation revealed strong correlation with metal 6 CMP process. Although leakages were observed on the failing units, TIVA analysis was not effective in localizing the defect location. Instead PEM analysis observed similar exclusive 2 emissions spots, as shown in Fig 3(b) on several dies. As there was no metal 6 running on top of these emission spots locations, the hotspots were suspected to be

induced emissions due suspected barrier bridge elsewhere. With these signatures which indicate strongly to barrier residue, further path tracing was then engaged. From the 2 emission spot locations, two long paths which stretches > 1000 um were identified. The 2 paths were running close to each other and at the box region shown in Fig. 3(c), it was observed that metal 6 were routed side by side each other at critical design metal pitch. Progressive cross-section FIB at the boxed suspected failure region, as shown in Fig 3(d), observed metal 6 barrier bridge. Subsequent cross-section TEM analysis verified thin barrier residue with no Cu as shown in Fig. 3(e).

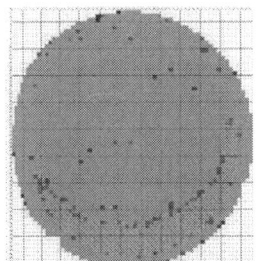

(a) Partial ring failure pattern

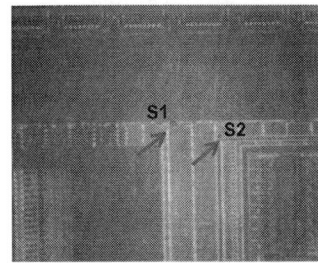

(b) Exclusive PEM spots

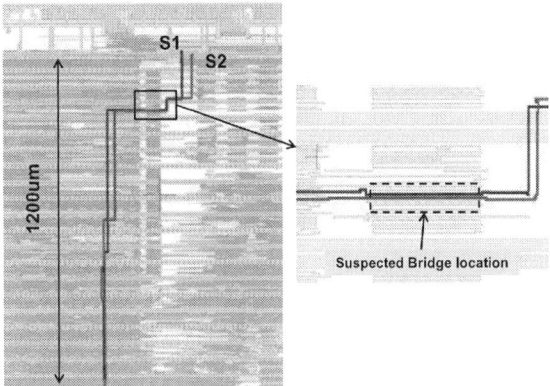

(c) Layout trace and analysis derived suspected failure location

978-1-4799-3911-4/14 $31.00 © 2014 IEEE

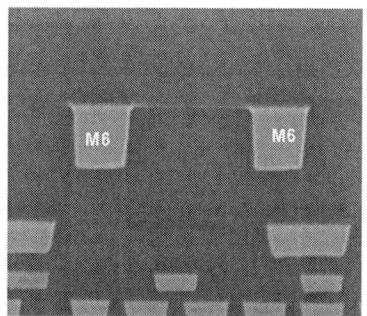

(d) Cross-section FIB on suspected fail region

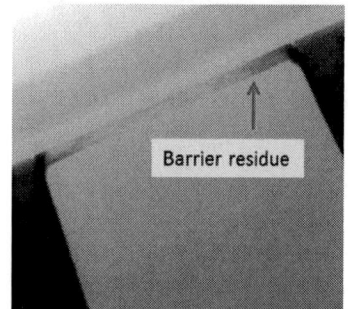

(e) TEM cross-section on Ta residue only

Fig 3 Failure analysis of extreme edge power short ring pattern failure

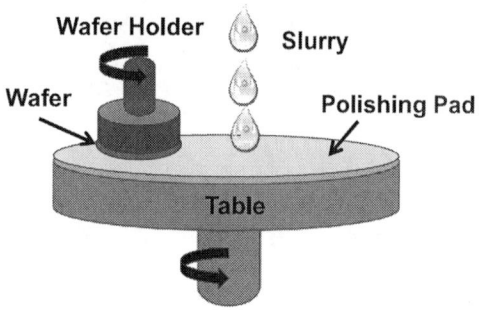

Fig 4 Chemical mechanical polishing process

In wafer foundry, wafer sort failure pattern plays an important role in providing tell-tale signs on which process step has possibly drifted. From there, possible defect types being metal bridge, broken, via open, front-end or back-end of line related issues can be predicted. Suitable fault localization and physical failure approach can then be used to shorten the FA cycle time. As CMP involved rotational movement of the wafer holder on a rotating polishing pad as shown in Fig. 4, a drift in the polishing rate, be it due to the wear out or insufficient seasoning of the polishing pad, would result in a ring or partial ring shape failure pattern as illustrated in case study 1 and 2. Although, there are other processes for example, etch or photoresist coat, clean or even wafer movement in load lock which also involve circular motion [3-4], failure due to CMP under-polish is typically a patch effect where within the continuous or discontinuous partial ring patch as shown in Fig. 1(a) and Fig. 3(a) respectively, there are no passing dies within. In fact, it is the observation of this characteristic CMP

under-polish wafer map signature in case study 2 that directed a FIB cross-section PFA approach instead of top down PFA approach achieving first unit success.

Case Study 3: Ultra-thin Ta Residue

Fig. 5 describes a case of ultra-thin barrier residue on a 40nm product resulting in a reticle pattern SCAN failure as shown in Fig 5(a). From Fab investigation, backend of line was suspected. Marginal leakage was measured from several units. As reticle pattern failure typically correlates with lithography issue which would result in metal bridging, TIVA analysis was first employed for fault localization. However, no distinctive TIVA spots were observed. Several PFA checks on potential lithography weak points observed smaller metal 5 spacing but no metal bridging was found. Interestingly further fault localization with PEM observed strong exclusive emission spots on several failure dies. From layout analysis, it was discovered that the emission sites were located in the boundary region between analogue circuitry and logic blocks, as shown in Fig 5(b) where there are no front-end transistors structures. Only floating dummy poly and active structures were present. These emissions were thus concluded to be thermal emissions from the back-end of line defects. Progressive FIB observed barrier anomaly in Fig 5(c) and subsequent cross-section TEM with EDX linescan verified an ultra-thin layer of Ta barrier residue shorting the two metal lines. As PEM is generally used for localizing front-end of line defects and junction leakages, this is an interesting case to highlight the use of PEM to localize ultra-thin conductive residue in the back-end of line.

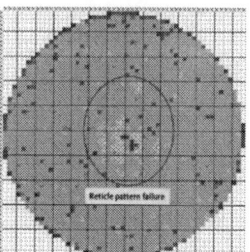

(a) Center Reticle Pattern

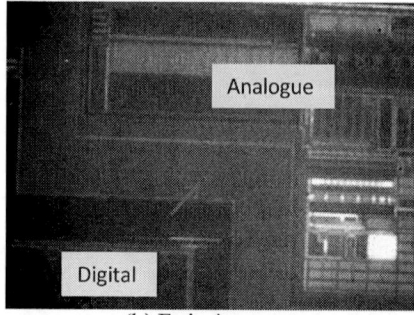

(b) Emission spot

978-1-4799-3911-4/14 $31.00 © 2014 IEEE

(c) FIB cross-section

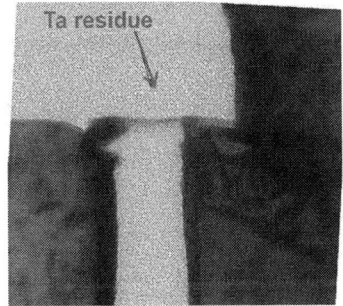

(d) TEM cross-section

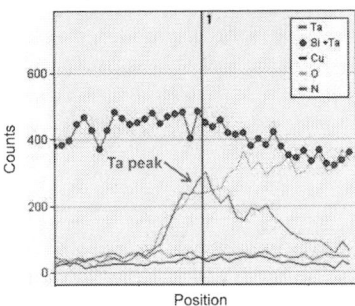

(e) TEM EDX linescan analysis

Fig 5 Failure analysis of extreme edge power short ring pattern failure

In this case study, as the failure is due to process drift in both the lithography and CMP steps, with dominant effect from lithography, the failure pattern differs from the typical partial ring patch failures that is observed in case studies 1 and 2. In addition, it can be observed from these case studies that the failure locations from CMP under-polish were similar among the failure dies in the same portion of the region. The hotspot locations for all 3 case studies were identical with neighbouring failure dies. These systematic failure locations indicate local non-uniformity in polishing rate due to difference in pattern density on the existing (case study 1) or the underlying metal layers (case study 2 & 3).

III CONCLUSIONS

In conclusion, we have described 3 low yield case studies on different severity of barrier residue defect and different technology nodes. It is evident that on advanced technology

node, barrier defects could be much more subtle resulting in marginal or no observation leakages in the power domains. As the killer residue gets thinner, conventional fault localization approach with laser induced techniques on backend-of-line metal bridging defects is insufficient. Photon emission microscopy together with layout analysis should be used in complimentary to localize ultra-thin barrier residue or to deduce the failure locations from localizing the symptoms of the failure. These case studies have also illustrated that barrier residue from CMP under-polish would result in ring shape failure pattern with systematic failure locations due to pattern density differences resulting in local non-uniformity in polishing rate. It was also highlighted to use cross-sectional FIB or TEM as better PFA approach for defect tracing.

ACKNOWLEDGMENT

The authors would like to thank the EFA, XTEM and for the data and Wang SK from AVS team for the engagement in the discussion.

REFERENCES

[1] J.C.H. Phang et al , "A Review of Near Infrared Photon Emission Microscopy and Spectroscopy", Proc. Intl. Symp. Physical & Failure Analysis of Integrated Circuits (IPFA), pp. 275-281, 2005.

[2] J.C.H. Phang et al, "A Review of Laser Induced Techniques for Microelectronic Failure Analysis", Proc. Intl. Symp. Physical & Failure Analysis of Integrated Circuits (IPFA), pp. 255-26, 2004.

[3] G.B. Ang et al, "Failure Analysis Methodology on Systematic defect in ADC_PLL Ring Pattern due to Plasma De-chuck Process", Proc Int Symp Testing & Failure Analysis (ISTFA), pp58-61, 2010

[4] A.C.T Quah et al, "Failure Analysis Methodology on Unique 68mm Single Ring Pattern due to Load Lock Burr", Proc Int Symp Testing & Failure Analysis (ISTFA), pp 349-353, 2011

Idss Failure Investigated by SIMS Profiling and TCAD Simulation

Lei Zhu, M. B. Bai, X. P. Wang, Y. H. Huang, Kenny Ong, A.B.S. Sumarlina, W. G. Park, Z. Q. Mo, Peck Y. Zheng,
S. P. Zhao, Jeffrey Lam

GLOBALFOUNDRIES Singapore Pte Ltd

60 Woodlands Industrial Park D, Street 2, Singapore 738406

Tel: (+65) 66701537, Fax: (+65) 63604370, E-mail: lei.zhu@globalfoundries.com

ABSTRACT

A Power MOSFET Idss failure case was studied by SIMS profiling that showed a deeper junction depth between the body/source. TCAD simulation was used to understand the mechanism of the failure.

INTRODUCTION

Power Metal Oxide Semiconductor Field Effect Transistors (MOSFETs) are commonly used power devices due to their low gate drive power and fast switching speed. When no bias is applied to the Gate, the Power MOSFET is capable of supporting a high Drain voltage through the reverse-biased P-body and N- Epi junction. In high voltage devices, most of the applied voltage is supported by the lightly doped Epi layer. A thicker and more lightly doped Epi supports higher breakdown voltage but with increased on-resistance. In lower voltage devices, the P-body doping becomes comparable to the N- Epi layer and supports part of the applied voltage. The breakdown voltage (BVDSS) is usually defined as the drain to source voltage when leakage current reaches a specific limit (for example, 250uA). The leakage current flowing between source and drain is denoted by Idss. The failure happens if the leakage value exceeds high specification limit and Idss is the most famous yield killer for Power MOSFET device during die sort. Base on device characterization understanding, Idss failure may come from low junction avalanche breakdown voltage, soft leakage from channel/edge termination area, or incomplete PN junction structure.

In this paper, a case study of Idss failure for a particular type of power MOSFET was reported. SIMS was used to study the Idss in relation to the junction depth formed by the body/source.

THE FAILURE CASE ANALYSIS AND DISCUSSION

The case happened on one particular proto type device. After initial physical failure analysis, it was understood that the failure is not due to junction avalanche breakdown voltage since breakdown voltage value is normal. It was neither due to aluminum spiking since no abnormal pin was found at leakage spot after delayering, nor abnormal edge termination since leakage spot distributed at active area randomly. Hence source/body doping at channel area was suspected. Fig.1 shows the Idss as a function of different wafers measured which may shoot up to 1E-4 level for the failed lots.

SIMS analysis was performed on fully processed chips to understand the doping difference on good/fail device. The source is doped with As dose of 6.3E15 atoms/cm2 and the body is doped with B dose of 2.6E13 atoms/cm2. Thus the As and B profile were obtained by using Cs+ and O+ gun of the dynamic SIMS (Cameca WF) respectively. The impact energy is at 9KeV. The SIMS profiles of the dopants are shown in

Fig.2. The junction depth for the good device is about 120nm while that for the failed is about 160nm. Thus the junction depth formed with failed device is 40nm deeper than that of the good one. The result gives a strong evidence that source doping (As) was more deeper compared to normal device.

After checking the drive-in recipes, it was found that a long source drive-in process was used for the proto lot that might be the major killer for this Idss fail. For example, the time that the failed device went through the source drive-in is about 155min whereas the good one at 40min both with the same temperature.

SIMULATION

In order to understand the impact of the longer drive-in time on the device performance, a TCAD simulation was carried out. Fig.3 shows the result which indicates deeper Source junction due to longer drive-in time, which is consistent with SIMS depth profiling results. For the failed sample, it is observed that shorting between source to drain occurs likely due to larger diffusion of boron away from the oxide which increases with longer oxidation process during source drive-in as shown in highlighted circle in Fig 3. An enlarged figure of this region seen in Fig 4 shows the positive net doping at the region near the channel, and this can be proven from the doping profile taken from the lateral cut along this region shown in Fig 5. There is lesser contribution of channel doping from boron and hence larger contribution from the n-type dopants in this case phosphorus from the epi, which increases with longer source drive-in time. Hence, this contributes to the source-drain punchthrough leakage which is responsible for the larger leakage current, which can be observed from further electrical simulation with high drain bias (Vdrain=60V) and other terminals grounded (Fig 6 and 7). Fig.6 shows electron current flows from source to drain (at the bottom) through the channel, for the failed sample, which undergoes source drive-in 155min, whereas for the normal sample which undergoes source drive-in 40min, no leakage current flows from source to drain at the labeled current level. In addition, impact ionization values shown in Fig 7 are significantly higher for failed sample compared to normal sample, further explaining the larger leakage observed.

CONCLUSION

SIMS profiling of the failed Idss wafers shows a deeper junction depth of N+ source. This result is consistent with the TCAD simulation that the deeper source junction due to longer process drive-in time may result in significant leakage from Drain to Source. However, the reason for the large Idss is likely due to the increased boron diffusion away from the oxide which increases with longer source drive-in time, resulting in larger contribution from the phosphorus near the channel. This results in shorting between source-drain and hence significantly larger leakage.

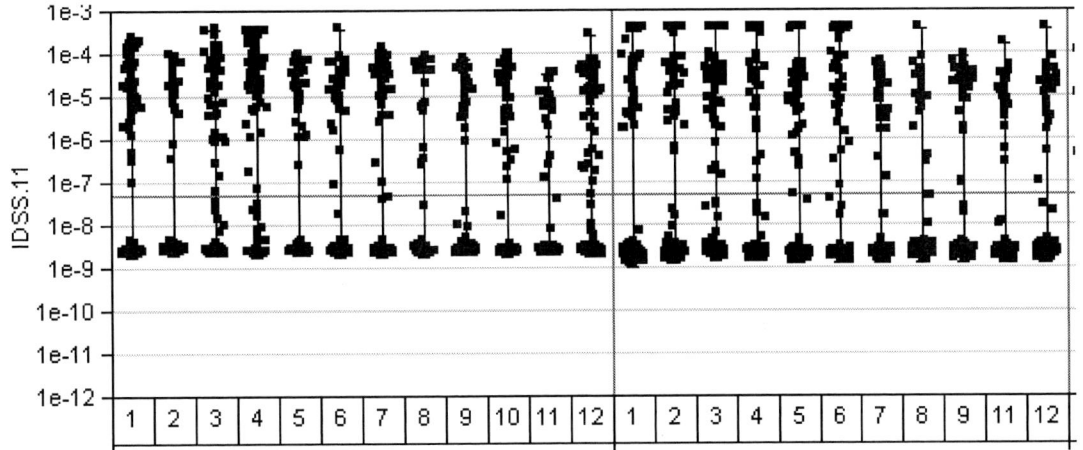

Fig.1 Idss variation measured for the a couple of different wafers

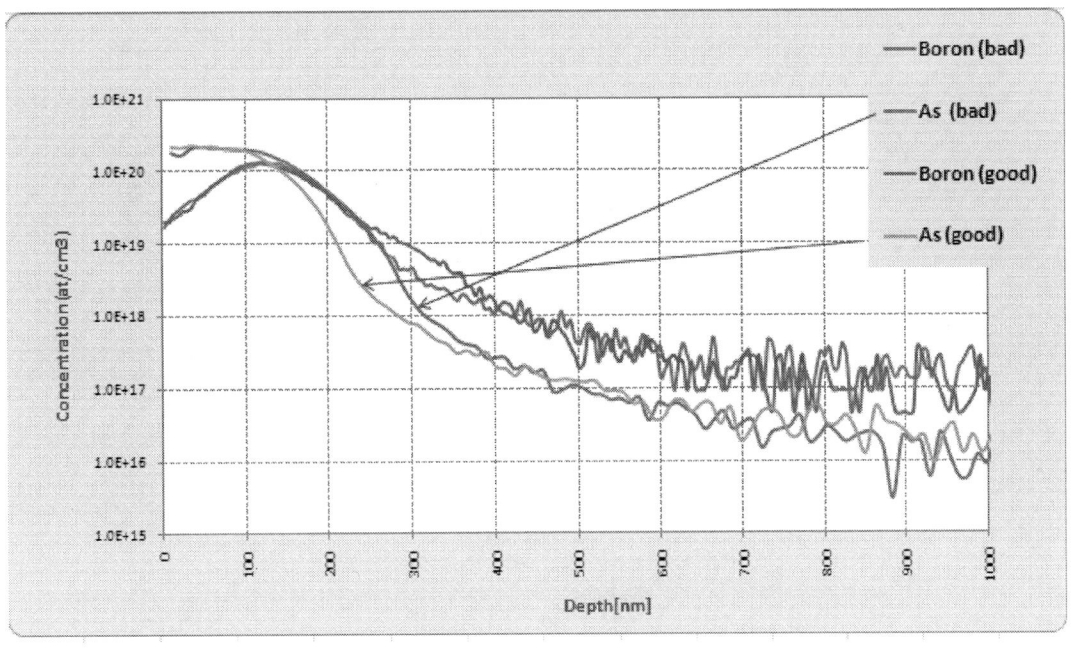

Fig.2 SIMS depth profile of the dopant (failed vs normal)

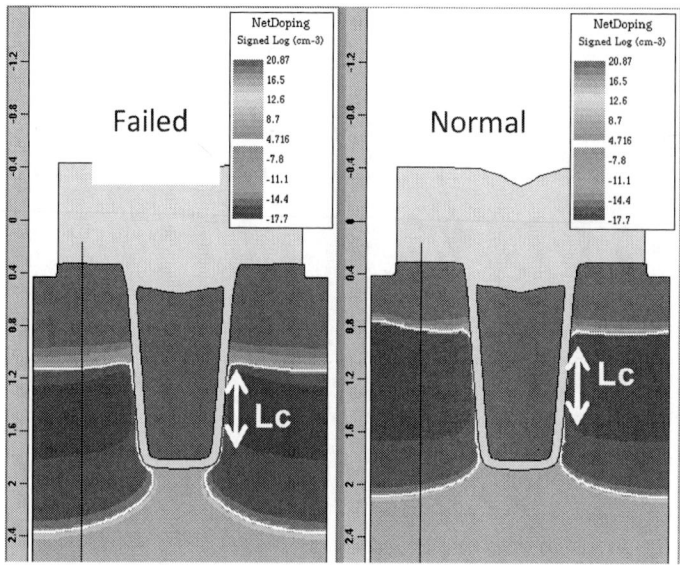

Fig.3 TCAD simulation showing a deeper

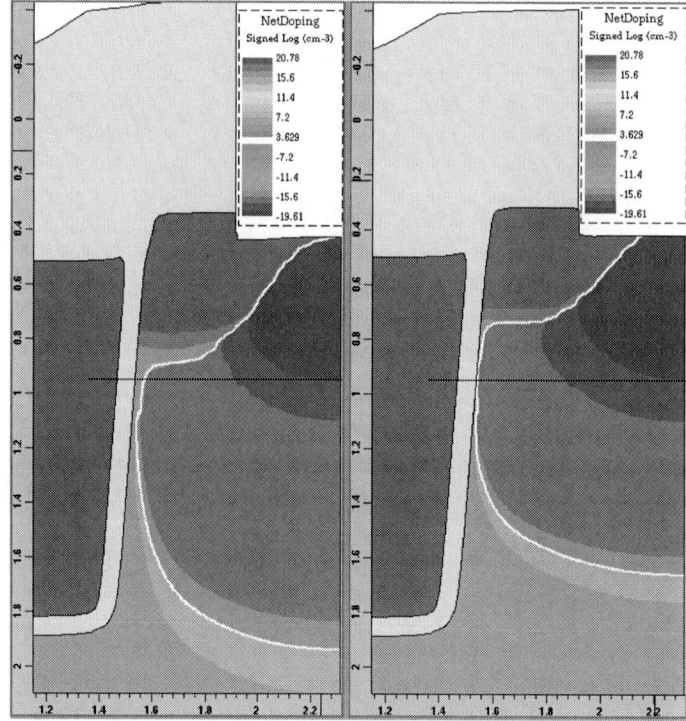

Fig 4. TCAD simulation showing the 2D net doping profiles taken from the encircled green region in Fig 3. Note the region highlighted in green shows the failed sample having more contribution from the n-type dopants (phosphorus from the epi) than p-type dopants (boron) near the channel likely due to increased boron diffusion away from the oxide

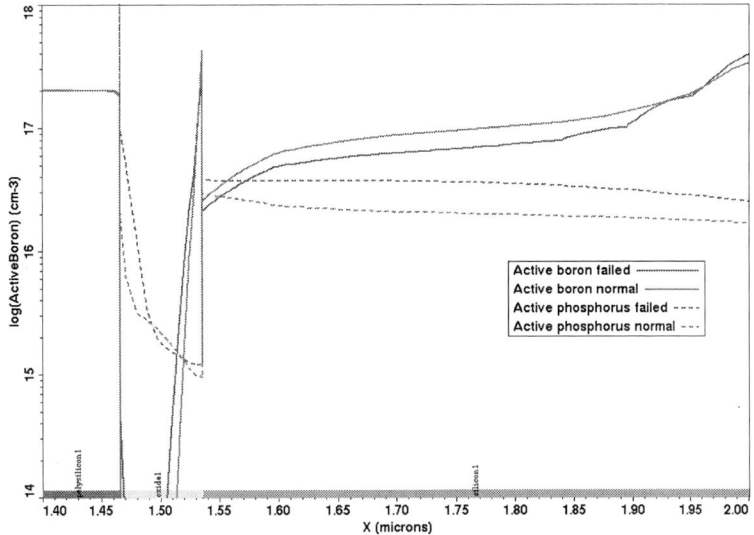

Fig 5. TCAD simulation of the cut line in the horizontal direction seen in Fig 4. Lesser boron and phosphorus concentration exists near the channel for the failed sample, causing larger contribution from the phosphorus to the channel doping concentration

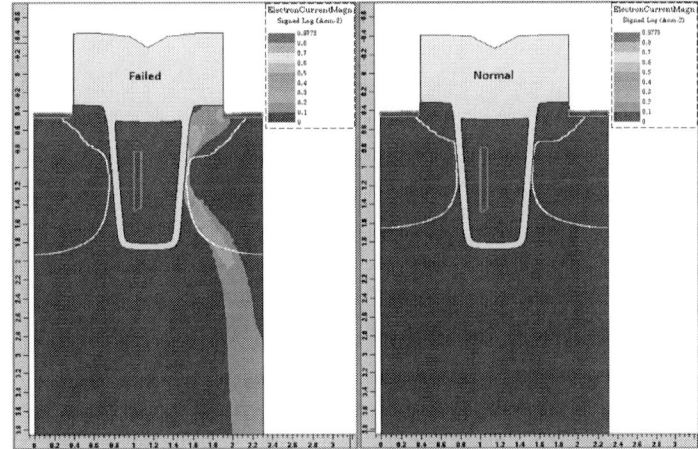

Fig.6 TCAD simulation showing leakage current flow from source to drain at high drain bias of Vd=60V for failed sample

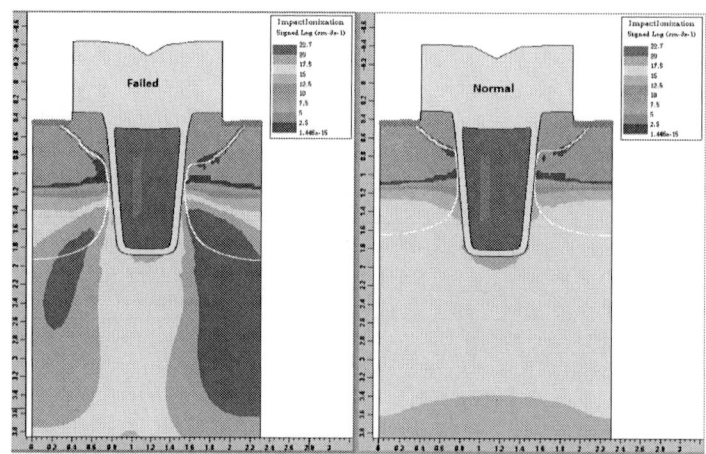

Fig 7. TCAD simulation showing higher impact ionization at high drain bias of Vd=60V for failed sample

Fault Isolation by Conquering Obstruction Effect of Resistor in Complex Cases Analysis

Gaojie Wen

Freescale Semiconductor (China) Limited, Tianjin
No.15, Xinghua Avenue, Xiqing Economic Development area, Tianjin, China
Phone: (+86) No.22 85684232/13682170769   Email: b16245@freescale.com

Abstract- **Resistor plays a key role in circuit designer's view as it was indispensable for the whole circuit function. But for integrated circuit failure analysis, the effect of resistor contained both guidance and as well as blocks to find the root cause. The obstruction effect of resistor was seldom studied in failure analysis. This paper presented how resistor to be block for failure analysis. And how to conquer the obstruction effect was also studied which will be helpful in analyzing complicated failure cases.**

I. INTRODUCTION

Impactful analysis method plays a significant effect in failure analysis as it could reduce much cycle time based on the condition of smaller metal line size and more metal layers. Many analysis strategies had been studied and employed in analyzing the capacitor and MOSFET related failures. But the resistor related failures have seldom been studied. Resistor was a passive device in integrated circuit and it seldom failed except the serious EOS damage. When resistor was a component of failed device, it would bring obstacle to find the root cause. Well known of the resistor's property would be efficient to determine the defect. [1]

Resistor plays a key role in circuit designer's view as it was indispensable for the whole circuit function. It was mainly used not only to divide voltage and limit current, but also compose the oscillator and feedback circuit. But for IC failure analysis, the resistor was in an uncertain situation which could provide both guidance and blocks for finding the root cause. The guidance effect of resistor was mainly on the OBIRCH (Optical Beam Induced Resistance Change) result which could provide current path for further schematic analysis. But the obstruction effect contained many aspects including big resistor would cover real defect when performing OBIRCH analysis and cause wrong action on transistor by inducing abnormal bias current data. The resistor blocked failure analysis not only by its circuit property under OBIRCH analysis, but also by failed directly.

OBIRCH and FIB (Focused Ion Beam) were useful analysis method for resistor related failure analysis, much precious cycle time can be saved and root cause can be found efficiently when combined with schematic and layout analysis. Two real cases were presented in this paper to show how resistor blocked us in failure analysis procedure and how to conquer the obstruction.

II. CASE1

One 250nm high voltage mixed signal ASIC was analyzed due to leakage between PORT0_SENSE and PORT1_SENSE pin

(Fig.1). The Auto-Test Equipment and evaluation board revealed current consumption failure on PORT0_SENSE pin and PORT1_SENSE pin. OBIRCH analysis should be considered firstly due to the pin to pin leakage.

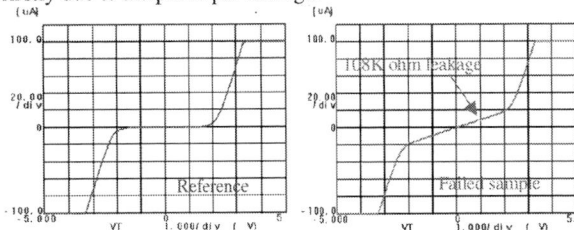

Figure 1. IV curve revealed leakage on failed sample between PORT0_SENSE pin and PORT1_SENSE pin

Since the sample had four metal layers and double poly layers, front side analysis may be difficult to get abnormal resistance change spot. Backside milling was employed to expose the die from bottom side. OBIRCH analysis was performed when biased proper voltage on two failed pins and abnormal resistance change spots were found on resistors (Fig.2). But they were not the defect and only indicated there was current flow through it. It seemed OBIRCH can't detect the defect location directly by this leakage failure. Backside EMMI analysis was also performed when in failed mode but no useful direction was provided.

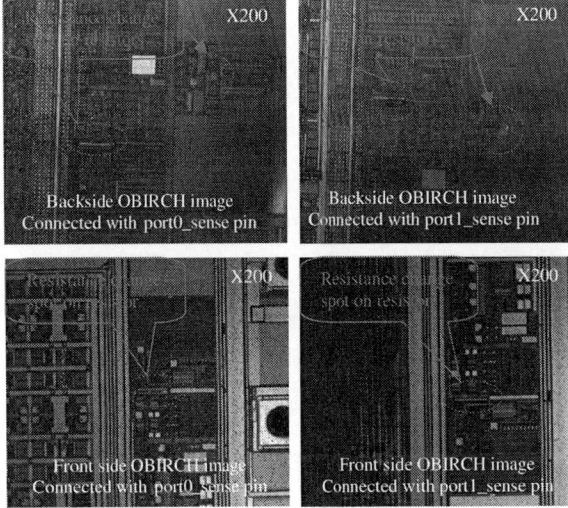

Figure 2. Pin to pin Backside and front side OBIRCH image revealed resistance change spot on same resistors.

As backside analysis didn't give much help in finding the root cause except the resistance change spot on resistors, it was remolded from backside and decapped from front side to check if front side analysis can be efficient. Front side OBIRCH analysis also revealed resistance change spots on resistors which were consistent with backside OBIRCH result. [2]

Schematic analysis based on the resistance change spots on resistors revealed a special circuit between PORT0_SENSE and PORT1_SENSE pin. It was a current sense circuit which contained big resistors to monitor the output current and provide feedback to internal control logic. Four poly resistors in schematic indicated there was total 105k ohm between PORT0_SENSE and PORT1_SENSE pins. But the leakage IV curve between PORT0_SENSE and PORT1_SENSE pins was about 108k ohm. That meant there was about 3k ohm leakage in internal circuit. Four poly resistors which had heavy diffusion were in series connection in the suspected failed leakage circuit. When performing OBIRCH analysis, laser would heat the die surface one pixel by one pixel, when met big resistor, on metal to poly interface, the heating would cause resistance increase which in turn reduces the current. In OBIRCH image, this is displayed as a green spot after super impose. On the poly to metal interface was just the reverse and the OBIRCH image displayed a red spot after super impose. In this close circuit, little change on big resistor would affect the whole circuit current much, that displayed a much brighter spot; meanwhile, when met small resistor or resistance leakage defect, current change which caused by laser heating have little influence on whole circuit current and this may not displayed on screen. OBIRCH would not get the resistance change spot on lower resistance defect if it connected with big resistors.

There was not much probe work allowed to perform on this kind of four metal layers sample. What's more, there was long wire in layout on both PORT0_SENSE and PORT1_SENSE internal signal metal lines. The emphasis of analyzing this kind of failure should also be put on the OBIRCH strategy and avoid other functional analysis direction. The efficient way of conquering big resistor obstruction was to remove the big resistors in leakage circuit using FIB (Fig.3). Navigation was used to assist FIB in removing the big resistor in the pin to pin leakage path. Test pads were made on PORT0_SENSE and PORT1_SENSE internal signal metal 2 lines. IV curve was measured again by microprobe after bypassed the big resistor and a 3k ohm leakage was found (Fig.4).

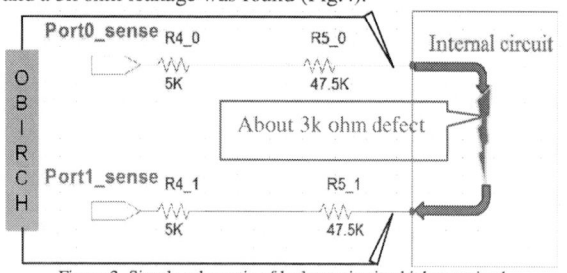

Figure 3. Simple schematic of leakage circuit which contained big resistors and OBIRCH setup solution

This finding confirmed the suspicion which was proposed before. New setup condition OBIRCH analysis was performed

by providing bias voltage to the high leakage between PORT0_SENSE and PORT1_SENSE internal signal metal line directly. As these two signals had long metal lines in layout, the abnormal resistance change spot was finally located between these two signals' internal metal 3 lines when tracing the current path (Fig.5). Deprocess analysis revealed metal defect was the root   cause   (Fig.6).

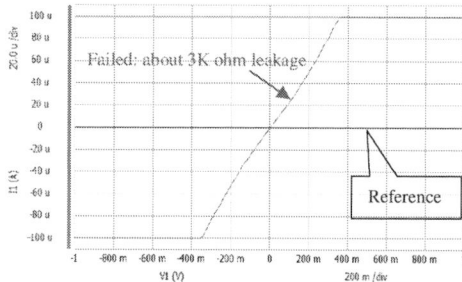

Figure 4. A 3K ohm resistance leakage was got on PORT0_SENSE and PORT1_SENSE internal signal metal lines after removing the big resistor

Figure 5. Front side OBIRCH result revealed abnormal resistance change spot on metal lines after removed obstruction of big resistors.

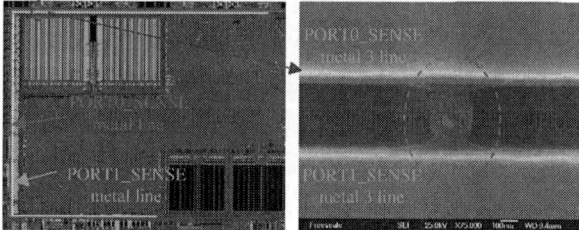

Figure 6. PORT0_SENSE and PORT1_SENSE internal long metal lines in layout and Metal defect was confirmed to be the root cause.

The real defect couldn't be located directly if big resistor was in the leakage circuit. FIB assist bypass big resistor strategy can successfully remove the big resistor obstruction effect in OBIRCH analysis and the defect can be determined directly.

III. CASE 2

Another mixed signal sample failed on wrong high temperature warning after powered on about 2 minutes (Fig.7). It passed ATE due to the delayed time failure longer than ATE test time. The EVB (Electronic Verification Board) can detect the wrong high temperature warning failure at room temperature. As this was a room temperature failure and the delayed time not stable, soft defect location strategy may not be efficient in finding the root cause.

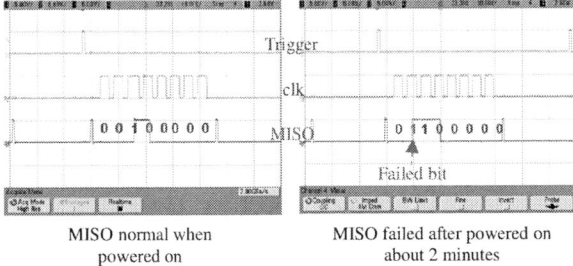

MISO normal when powered on

MISO failed after powered on about 2 minutes

Figure 7. MISO signal failed after powered on about 2 minutes

EMMI analysis was performed post decapsulation, but it didn't provide useful direction for further analysis. After studied schematic and layout focused on the wrong high temperature warning signal, temperature sense module was highly suspected to be failed. Since there were seven high temperature warning modules on this sample, logic calculation microprobe was performed based on the failure. The abnormal warning signal was found on COLTEMP which was generated in HS3 thermal module. But no abnormal IV curve was determined focused on the failed signal---COLTEMP. Detailed schematic was analyzed focused on this thermal module. The temperature sense transistor was located close to the power MOS to monitor the temperature and generate warning signal. PN junction conduction threshold would decrease when temperature became higher. When the voltage difference between emitter junction and Base junction lower than the bias voltage which gained by divided voltage resistors, this circuit would generate a warning signal on COLTEMP. [3] It couldn't be determined whether the failed COLTEMP signal was generated by transistor or bridged with other signal metal lines as the base junction voltage was nearly same with reference on oscilloscope.

The failed COLTEMP signal had a long wire on layout and it was difficult to determine where the defect location was. Further microprobe analysis would take long time and was not an efficient analysis method based on this situation. Failure isolation can be deepened by cutting the failed signal into two sides and trace it in two directions. This strategy can help failure analysis dig the root cause deeply. FIB was used to cut the COLTEMP signal to deep isolate the failure (Fig.8). [4]

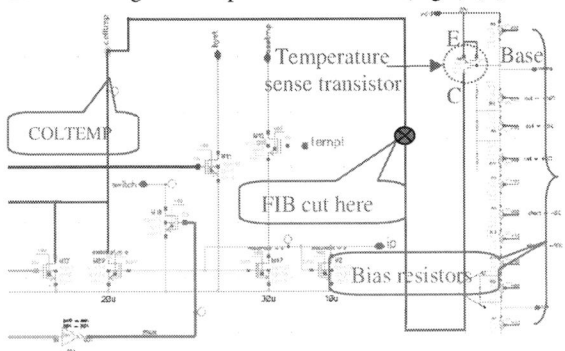

Figure 8. Schematic of thermal module which contained bias resistors and FIB cut location

Then COLTEMP signal which connected with temperature sense transistor was confirmed still to be failed. The resistors were focused on as it provided bias voltage and current which controlled the collector junction node. If resistance became low, the bias current would become larger and pull up the voltage on collector junction node and output a warning signal. The resistors were measured on failed sample and abnormal about 52K ohm resistance was found between base junction and GND while the reference was about 97K ohm (Fig.9). The low resistance on bias voltage circuit can cause wrong bias current and affect the bias voltage. The temperature sense transistor will wrongly work when the bias voltage on base junction became lower; finally output the wrong warning signal after power on 2 minutes.

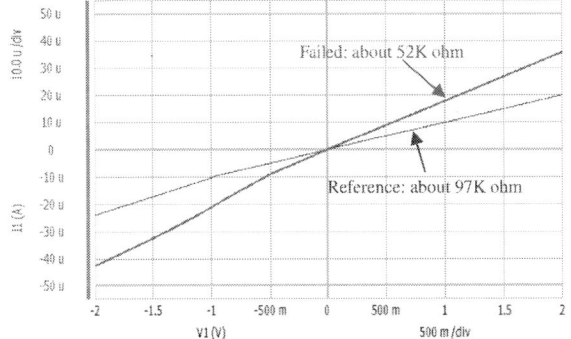

Figure 9. IV curve between base junction and GND revealed lower bias resistance on failed sample

High magnification optical inspection on the bias resistors revealed abnormal image on R3. These bias voltage resistors were fabricated by active area and in series connection. Cross section was performed using FIB and active area damage was found (Fig.10). This damage can lead to the R3 short to gnd which can cause lower resistance phenomenon. The resistor failed directly and lead to the time delay failure.

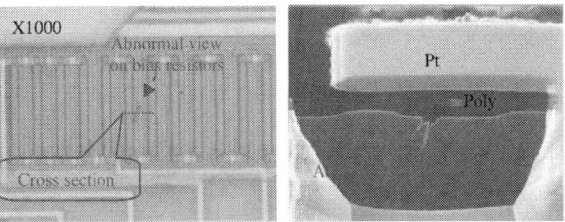

Figure 10. Optical image and cross section of abnormal view location in thermal module revealed active area resistor damaged.

The failed resistors on bias current circuit was difficult to detect as it may only induce the bias current failure which may not easy to get in internal circuit. But the failed resistors can be found by FIB isolation and resistance measurement directly. Resistor was seldom failed but didn't mean never failed. The obstruction effect of resistor can be conquered by don't neglect the possibility of resistor failed itself.

978-1-4799-3911-4/14 $31.00 © 2014 IEEE   193

V. Summary

Resistors could provide guidance and as well as obstruction to failure analysis. On the one hand, it could provide current path indication in OBIRCH analysis and help to find defect; on the other hand, it could cover the real defect or lead to wrong bias current and mislead the analysis direction. Knowing the guidance effect of resistor can help the failure analysis. However, it would also promote failure analysis capacity when well known of the resistor's obstruction effect and how to conquer it. By conquering the obstruction effect, no much probe work needed in decreasing metal line size and more metal layers integrate circuit chip, ultimately, much cycle time can be saved.

Acknowledgment

Thanks physical failure analysis team of Tian Jin failure analysis lab for their great efforts in cases analysis. Also thanks to other colleagues of electronic failure analysis team of Tian Jin FA lab who gave much assistance in case analyzing.

Reference

[1] Clement Huang, Mingte Lin, James W. Liang, "Degradation and Failure Analysis of Polysilicon Resistor connecting with Tungsten contact and Copper line", IRPS 2011, 731-733.

[2] Jinglong Li; Gaojie Wen; Joe Yu; Grace Song, "Die-Level Leakage Current Path Analysis Based on IR-OBIRCH Technology", IPFA2012, 1-4.

[3] Sheng-Huang Lee ; Dept. of Electr. & Comput. Eng. "Multi-threshold transistors cell for Low Voltage temperature sensing applications", Circuits and Systems (MWSCAS), 2011. 1-4.

[4] Gaojie Wen, Binghai Liu, Winter Wang, "Failure Isolation Using FIB Assist Photon Emission Microscopy Analysis and Microprobe Analysis", IPFA 2011, 140-143.

Analysis of insertion force of electric connector based on FEM

Ying Li[1], Fulong Zhu[1*], Yanming Chen[2], Ke Duan[1], Kai Tang[1], Sheng Liu[1]

1. Institute of Microsystems, School of Mechanical Science and Engineering, HuaZhong University of Science and Technology, 1037 Luoyu Road, Wuhan, Hubei province, 430074, P. R. China
2. Hubei Institute of Measurement and Testing Technology, No.2 Maodianshan middle road, Automotive Electronics Industry Park, East Lake High-Tech development Zone, Wuhan, Hubei province,430223, P. R. China,
Phone: (0086-027)87557830-801. Fax: (0086-27) 87557074. Email: zhufulong@hust.edu.cn

Abstract-The paper investigates the insertion force and contact reliability of N electric connector. A finite element model (FEM) of the contacts was created and simulation of the contact force was completed by ANSYS. Impact of the friction coefficient, shrink range, length of socket, and groove width on the insertion force was analyzed by changing the structural parameters. Variation curves of the insertion force for different structural parameters were studied. Simulation results showed that the shrink range and length of socket have a great influence on the insertion force. In contrast, the friction coefficient and groove width have a much smaller effect on the contact resistance of N electric connector. The optimized connector structure reduces the contact resistance and improves the contact reliability.

I. INTRODUCTION

Electric connector is a device for transmitting electrical signal in circuits by inserting and separating the contacts. With the development of science and technology, electronic devices have got wide application in electronic, aerospace, and other industries fields. Meanwhile, the structure and category of electronic devices are more and more complex. Polchow et al. introduced the application of electric connector in public transportation [1]. The contact performance of electric connector is directly related to whether the structure is reasonable and reliable or not. Therefore, the performance and reliability of electric connector have great effect on the quality and reliability of electronic devices. Generally, the contact reliability depends on the design, technology, manufacturing, and management of contacts. Fig. 1 shows some typical contacts of the slotted cylindrical electric connector.

Usually, researchers do not pay sufficient attention to the structure of electric connector. But so far, equipment cannot work well without reliable electric connectors. In addition, the failure of contact pair of an electric connector has a negative effect on equipment performance [2, 3]. The structure of contact pair plays the major role in improving the reliability of electric connector. Scholars have made a lot of efforts in this field. Some researchers have discussed the influences of contact geometry on the reliability of connection [4-8]. Leidner presented a novel method to characterize the mechanical quality of electrical contacts [9]. Angadi et al. proposed a method of considering rough surface contact [10].

The smaller contact resistance induces less electrical signal lose. Contact resistance mainly depends on the contact force. Insertion force is directly related to the contact force to some extent. Insertion force is one of the important mechanical properties of contacts and, at the same time, it is convenient to estimate and measure. Although the increase of insertion force will decrease the contact resistance in certain range, the contacts are close enough when the insertion force reached a certain value; after this the contact resistance will not decrease obviously. In addition, an excessive insertion force will make wearing of the contacts faster and decrease life of electric connector. Quantitative analysis of contact resistance of electric connector can be carried out by studying the insertion force. Finally, we can find out the impact factors and improve the electrical properties of electric connectors.

II. MODEL AND ANALYSIS METHOD

A. Design Variables and Constants

In this paper, a typical N electric connector was selected as the object. The contacts of N electric connector are made up of the pin and socket. The socket consists of four grooves. A simple model of the N electric connector is shown in Fig. 2.

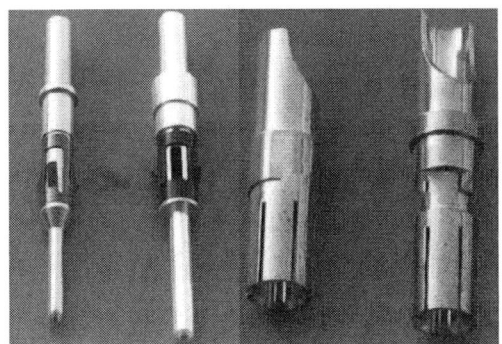

Fig. 1. Contacts of N electric connector

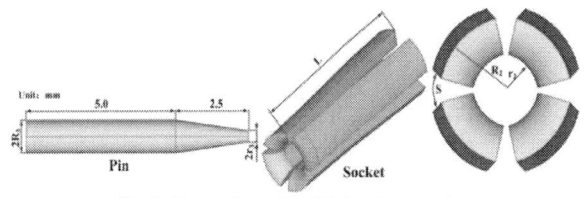

Fig. 2. Geometric model of N electric connector

978-1-4799-3911-4/14 $31.00 © 2014 IEEE

The main structural parameters are L (length of socket) and S (groove width). The value of L and S can be adjusted easily changing the size parameter. L is 5mm and S is 0.4mm. $2R_2$=2.0mm is the external diameter of socket, $2r_2$=1.1mm is the internal diameter of socket, $2R_3$=1.0mm is the external diameter of pin, and $2r_3$=0.3mm is the internal diameter of pin. The external diameter of pin (1.0mm) is smaller than the internal diameter of socket (1.1mm) to avoid the fitting issues. For a slot socket contact, the shell nosing treatment of contact piece is realized by mechanical means. Shell nosing decreases the internal radius of socket (r_2) and the initial value of shell nosing (μ) is 0.15mm. In this condition, an interference fit is obtained and an insertion force that satisfies the national standards can be achieved between the socket contact and pin contact. Fig. 3 illustrates the changes of socket after shell nosing treatment. The shrink range between the socket and pin is $\sigma = R_3 - r_2 + \mu$. Different shrink ranges can be achieved by changing μ. The friction coefficient can be adjusted by altering the settings of finite element analysis.

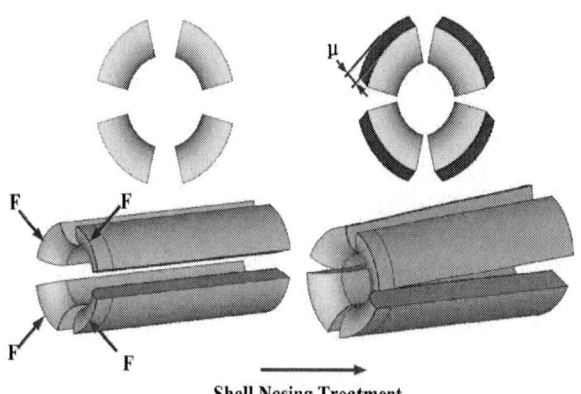

Shell Nosing Treatment

Fig. 3. Changes of socket after shell nosing treatment

B. Finite Element Models

Three-Dimension Motion of electric connector was simulated with the finite element contact analysis method. FEM is an effective method in simulating the real physics system with the method of mathematical approximation. With the development of computer technologies and computational methods, FEM has been widely applied in engineering design and scientific fields. In this paper, a FEM model of electric connector was constructed by ANSYS [11]. Only a quarter of the FEM model was meshed because of the symmetry of the geometrical structure. The mechanical boundary and contact conditions were applied. Fig. 4 illustrates that all DOFs for bottom of socket were constrained, and the UX and UY DOFs of pin were constrained. In order to simulate the process of inserting, the UZ was set at 1.5mm. In this simulation, the applied displacement loading UZ was divided into a serious of load increments. For nonlinear computation, the solution is reached when the calculated force convergence norm is within the defined force criterion.

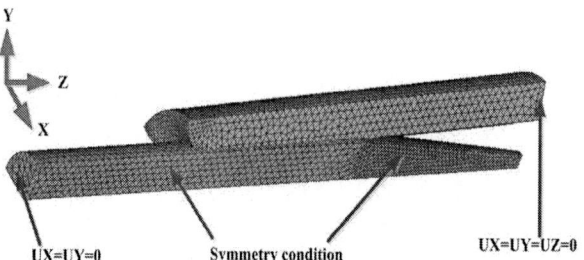

Fig. 4. Boundary conditions used for N electric connector

The two contact samples are made of tin bronze alloys whose properties are: the Modulus of Elasticity is 1.1E11Pa, 0.33 is the Poisson's Ratio. The FEM model was meshed with 3-D 8-Node Structural Solid185 for Solid185 element is suitable for modeling general 3-D solid structures. The degrees of freedom are displacements UX, UY and UZ. Symmetric boundary conditions were applied on the symmetric surfaces of FEM model because of the symmetric structure of the FEM model. The contact surfaces of pin were defined as contact surfaces, the contact surfaces of socket were target surfaces. Contact surfaces were meshed with Contact174 elements and target surfaces were meshed with Target170 elements. When contact areas are not clear and a non-negligible sliding is expected, this kind of surface-to-surface model is used [12, 13]. In order to simulate the real status of the contact, a friction coefficient between the contact surface and target surface (f=0.2) was defined.

The variation curves of insertion force were obtained. One of friction coefficient, shrink range, length of socket and groove width was changed while other parameters were kept the same, and relationships between structural parameters and insertion force were obtained. The effects of friction coefficient and shrink range, length of socket, groove width on insertion force were studied by comparative method. The mechanical analysis of contacts was accomplished and the theoretical value of insertion force was calculated [14]. Comparing the theoretical values with the simulation values, we find the results of the two calculation methods basically identical. Finally, the sensitivity of contact insertion force on different contact parameters was estimated.

III. RESULTS AND DISCUSSION

The distribution of Von Mises stress of N electric connector is shown in Fig. 5. Computational result of stress distribution showed that stress concentration existed at the root of socket. The maximum stress was 368 MPa which has exceeded the yield limit. However, the stress at root of the socket varied rapidly: only a small part of the socket had excessive stress, the stress in most area of socket is in the specified limits. Therefore, the N electric connector can work safely and steadily.

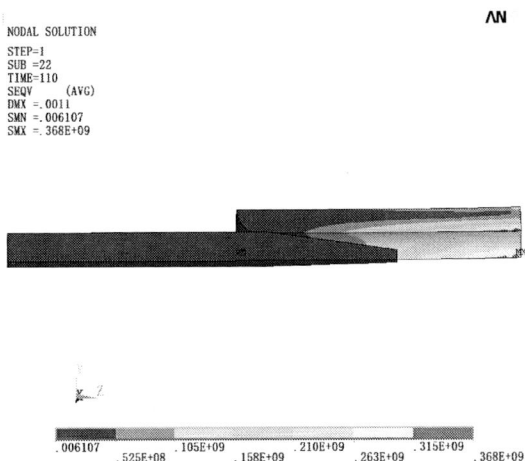

```
NODAL SOLUTION
STEP=1
SUB =22
TIME=110
SEQV    (AVG)
DMX =.0011
SMN =.006107
SMX =.368E+09
```

.006107   .525E+08   .105E+09   .158E+09   .210E+09   .263E+09   .315E+09   .368E+09

Fig. 5. The distribution of Von Mises stress of N electric connector

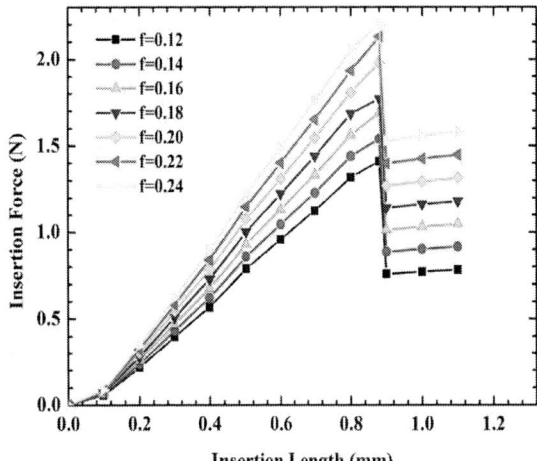

Fig. 7. Insertion force versus Insertion length under different friction coefficient

Fig. 6 clearly showed the effect of the shrink range on the insertion force. Change tendency of the insertion force for different shrink ranges was almost equal. At the initial stage of contact, there was a good linear relationship between shrink range and insertion force. When the shrink range varied from 0.1mm to 0.15mm, the insertion force in steady state varied from 1.30N to 2.05N. The peak of the insertion length was delayed at a larger shrink range because of the larger deflection of socket. Simulation results have shown that the shrink range has a significant impact on the reliability of electric connector contacts. Designers should choose an appropriate shrink range for their needs.

Fig. 7 presented the relationship between insertion force under different friction coefficient and insertion length. At the beginning of inserting, the insertion force increased sharply and reached the maximum value of 2 N. There existed certain linear relations between the insertion force and the insertion length. When the contact came to a steady state, the insertion force decreased rapidly and remained constant because of the change of friction coefficient and insertion angle of connector. The sensitivity to the friction coefficient is less than that to the shrink range.

To obtain the influence of length of socket on the insertion force, we selected different lengths of socket: 5.0mm, 5.25mm, 5.5mm, 5.75mm, and 6.0mm for the transfer curves in Fig. 8. These curves share the same variation trend with the curves in Fig. 7. When the contact is stable, the curves indicated that with the length of socket increasing from 5mm to 6mm, the insertion force decreases from 1.2N to 0.7N. The sensitivity of insertion force to the length of socket was remarkable.

Fig.9 illustrated the change curves of insertion force under different groove width. Like the curves above, the curves in Fig. 9 consisted of linear exponential increase process, steep decrease process and stable process. But these curves in Fig. 9 not only shared the same variation trend, but also had the almost same position. When the contact came to a steady state, although the groove width changed from 0.3mm to 0.4mm, the value of insertion force varied a bit, moving from 1.48N to 1.27N. The groove width has the smallest influence among the four structural parameters. The friction coefficient, shrink range, and length of socket should be preferentially considered in design phase.

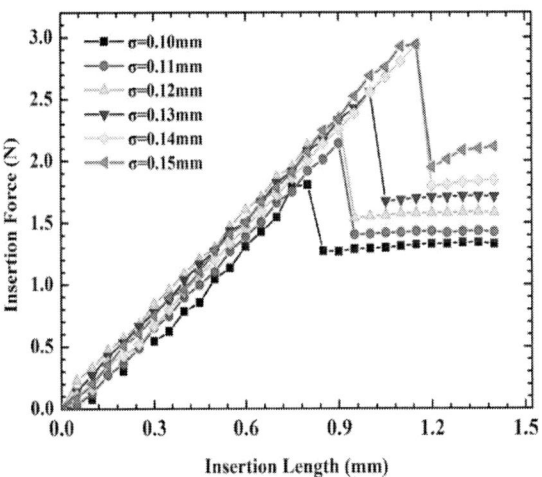

Fig. 6. Insertion force versus Insertion length under different shrink range

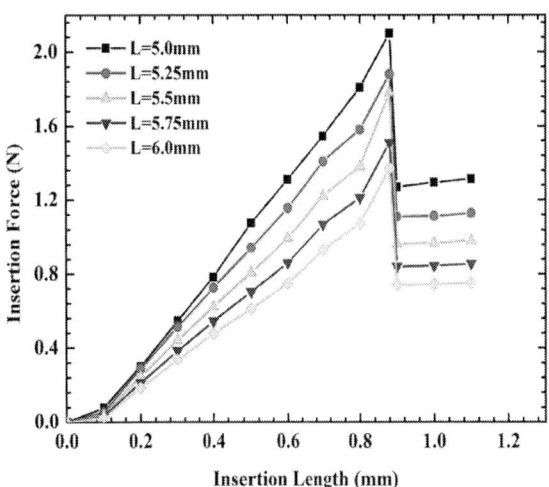

Fig. 8. Insertion force versus Insertion length under different length of socket

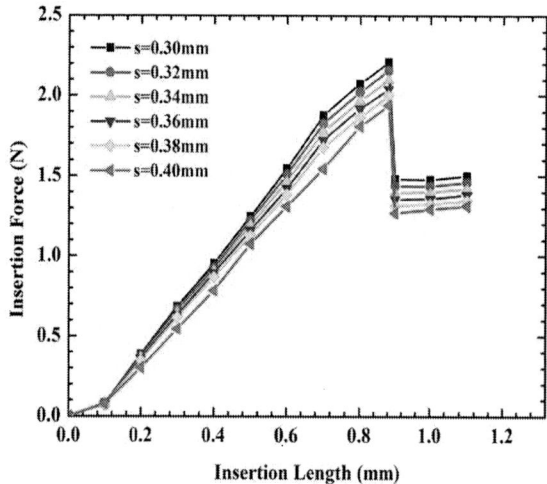

Fig. 9. Insertion force versus Insertion length under different groove width

IV. CONCLUSIONS

The purpose of this paper was to research the effects of different structural parameters on N electric connector in order to improve the reliability of N electric connector. Mechanical analysis of the insertion force and contacts has been carried out using ANSYS. Sensitivity of influence on contact performance caused by structural parameters of contacts was obtained. Comparison had been made between the results of the model with different structural parameters. The key factors that influence the contact performance were determined. The shrink range and length of socket have a significant effect on the insertion force. In contrast, friction coefficient and groove width have a much smaller influence on contact performance of electric connector. Increasing the shrink range or decreasing the length of socket decreases the contact resistance and improves the reliability of contact to some extent. The numerical simulation based on FEM made it possible to obtain the optimal structural design.

Consequently, the optimal design of contacts has been chosen in order to minimize the contact resistance. It can be achieved by varying the following parameters: shrink range and length of socket.

ACKNOWLEDGMENT

This work is supported by the Aeronautical Science Foundation of China for the Central Universities, HUST: 2013TS023. The authors would like to express their sincere gratitude to the supervisor, Mr. Wei Zhang, for his instructive advices and useful suggestions.

REFERENCES

[1] J.R. Polchow. *et al*, "A multi-physics finite element analysis of round pin high power connectors," in *Electrical Contacts, 2010 Proceedings of the 56th IEEE Holm Conference on*, Oct. 2008, pp. 1-9.

[2] G.P. Luo, J.G. Lu, and J.G. Zhang, "Failure analysis on bolt-type power connector's application," in *Electrical Contacts, 1999 Proceedings of the 45th IEEE Holm Conference on*, Oct. 1999, pp. 77-86.

[3] Y.L. Zhou, Z. Ning, and L.J. Xu, "Failure analysis of a kind of low power connector," *Journal of Zhejiang University SCIENCE A*, Vol. 8, No. 3, pp. 384-392, 2007.

[4] Y.-L. Hsu, Y.-C. Hsu, and M.-S. Hsu, "Shape optimal design of contact springs of electronic connectors," *Journal of Electronic Packaging*, Vol. 124, No. 3, pp. 178-183, 2002.

[5] I. Sawchyn, S. Edward, "Optimizing force and geometry parameters in design of reduced insertion force connectors," *IEEE Trans. CHMT*, Vol. 15, No. 6, pp. 1025-1033, 1992.

[6] H. Jochen, and E. Bernhard, "Shape optimization of connector contacts for reduced wear and reduced insertion force," *AMP J. Technol*, Vol. 2, pp. 42-46, 1992.

[7] B. Amine, "Influence of shapes, contact forces and high copper alloys on the contact resistance and temperature," in *Engineering Mechanics, Structures and Engineering Geology, 2009 Proceedings of the 2th WSEAS Conference on*, July. 2009, pp. 139-144.

[8] K. Mashimo, *et al*, "Computational Modeling and Analysis of a Contact Pair for the Prediction of Fretting Dependent Electrical Contact Resistance," in *Electrical Contacts, 2011 IEEE 57th Holm Conference on*, Sept. 2011, pp. 1-6.

[9] M. Leidner, M. Myers, H. Schmidt and H.F. Schlaak, "A new simulation approach to characterazing the mechanical and electrical qualities of a connector contact," *Eur. Phys. J*, Vol. 49, No. 2, pp. 165-170, 2010.

[10] S.V. Angadi, W.E. Wilson, R. L. Jackson, G.T. Flowers and B.I Ricket, "A multi-physics finite element model of an electrical connector considering rough surface contact," in *Electrical Contacts, 2008 Proceedings of the 54th IEEE Holm Conference on*, Oct. 2008, pp. 168-177.

[11] Madenci, E, Guven, I, The Finite Element Method and Application in Engineering Using ANSYS, Springer-Verlag, pp. 118-121, 2007.

[12] A. Beloufa, "Numerical and experimental optimization of mechanical stress, contact temperature and electrical contact resistance of power automotive connector," *International Journal of Mechanics*, Vol. 4, No. 4, pp. 94-104, 2010.

[13] X. Fei. *et al*, "Analysis and Prediction of Vibration-Induced Fretting Motion in a Blade Receptacle Connector Pair," *IEEE Trans. CPT*, Vol. 32, No. 3, pp. 585-592, 2009.

[14] H.Y. Liu, "The characteristic research of contact insertion and separation force in connector," in *Electrical Contacts, 1990 Proceedings of the 76th IEEE Holm Conference on*, Aug. 1990, pp. 619-624.

Observation of Long Term Potentiation in Papain-Based Memory Devices

A. Bag, M. K. Hota[†*], S. Mallik, C. K. Maiti

VLSI Engineering Laboratory, Department of Electronics and ECE, Indian Institute of Technology, Kharagpur 721302, India
[†]*currently with* Materials Science and Engineering, King Abdullah University of Science & Technology (KAUST), Thuwal
23955-6900, Saudi Arabia

Phone: +91-3222-281475   Fax: +91-32222-55303   [*]Email: mksan21@gmail.com

Abstract

Biological synaptic behavior in terms of long term potentiation has been observed in papain-based (plant protein) memory devices (memristors) for the first time. Improvement in long term potentiation depends on pulse amplitude and width (duration). Continuous/repetitive dc voltage sweep leads to an increase in memristor conductivity leading to a long term memory in the 'learning' processes.

Introduction

The idea that changes in the effectiveness of synapses within miscellaneous neural circuits could arbitrate the storage of information acquired during learning has a long history [1-3]. The discovery of long term potentiation (LTP) boasts to this concept in an area of the brain, named as the hippocampal formation, which had been implicated in memory from clinical observations of amnesia [4]. Bliss *et al.* has pointed out that the time scale of LTP is long enough to be potentially useful for information storage or in other words the LTP may be described using an analogy to electronic memory devices [5]. In the biological nervous system, a synapse is a nano-gap structure that permits nerve signals (memory information) from one neuron (memory cell) to another neuron via axon potential exchange method. In contrast, recent non-volatile memory research has focused on neuromorphic engineering to design artificial neural systems (synapse), the physical architecture and design principles of which are based on the knowledge from biology, physics, mathematics, computer science and engineering [6-7]. Several solid state devices such as, phase change, conducting bridge, ferroelectric memory, and FET based devices *etc.* reported in the past behave as artificial synapse [8-11]. Among them memristors are considered as most promising due to their ability to mimic the electronic and chemical actions of a biological synapses and neurons which are responsible for registering stimuli (touch, sound, and light), controlling muscles and forming memories in biological systems [12].

We selected papain, a protein that is obtained from natural green papaya, to understand the synaptic behavior in terms of long term potentiation in biomaterial-based devices. There are few reasons that stimulated our choice of papain as the memristive media. Firstly, it is water soluble and bio-degradable material. Secondly, papain can form uniform films with medium compactness and it is suitable for memory device applications. Moreover, papain is a readily and abundantly available material; it can be obtained from the natural green papaya by scoring the neck of the fruit, hence its

cost is minimal. In this study, papain-based memristor devices are used for the first time as synaptic device for the demonstration of analogous biological memory behavior in terms of long term potentiation.

Experimental

Metal-Insulator-Metal (MIM) structure is used as the memristor memory cell or the synapse. The aqueous solution of plant protein, papain, extracted from the natural green papaya used as the memristive media. After proper cleaning of ITO (~100 nm) coated glass substrate, papain was spin-coated on ITO directly at room temperature and kept in an oven for 30 min at constant temperature of 40°C to dry and make the film hard. Thickness of the papain film was measured by surface profilometer (Veeco Dektak 150) and was found to be around ~400 nm. Finally, the top electrode (Al) was deposited through a metal shadow mask having a circular area of 1.96×10^{-3} cm^2. High speed pulse current–voltage (*I–V*) characteristics of the devices were measured using Agilent B1500A semiconductor device analyzer system. The bottom electrode, ITO was grounded and the external 'stimuli' was applied on the top electrode with proper probing system.

Results and Discussions

The surface morphology of the spin coated protein films were studied using atomic force microscopy (AFM) imaging in height profile mode with a scan area of 25×25 and 5×5 μm, as shown in Figs. 1 (a) and 1(b), respectively. It can be observed that the papain films consist of large islands or grains which may act as a polycrystalline surface which however, leads to the large hysteresis behavior of the devices. From AFM morphology, the root mean square (RMS) roughness (R_q) was calculated using by the formula given bellow;

$$R_q = \sqrt{\frac{1}{n_x n_y} \sum_{i=1}^{n_x} \sum_{j=1}^{n_y} \left| Z(i,j) - Z_{avg} \right|^2} \qquad (1)$$

where, $Z(i,j)$ denotes the topography data for the surface after specimen tilt-correction, Z_{avg} is the average surface height, and i and j correspond to pixels in x- and y-directions. The maximum numbers of pixels in two directions are given by n_x and n_y [13]. For calculating RMS value of surface roughness, several scans were performed with several test samples of papain films. The average surface roughness was estimated in the range of ~75-78 nm with a scan area of 5×5 μm. The field emission scanning electron microscope (FESEM) images of

978-1-4799-3911-4/14 $31.00 © 2014 IEEE

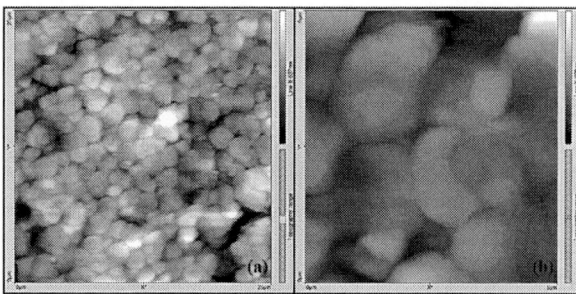

Fig 1: AFM surface morphology of papain on ITO coated glass substrates.

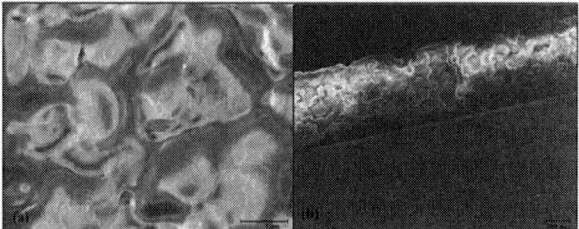

Fig 2: FESEM image of the (a) top view and (b) cross sectional view of the papain thin film.

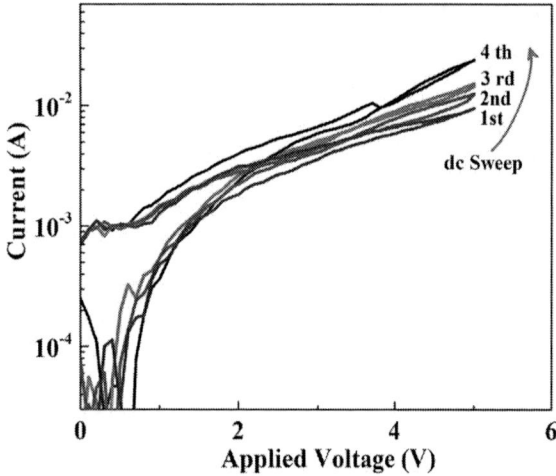

Fig 3: Current-Voltage characteristics of the papain-based memristor device. Continuous positive voltage sweep increases the memristor conductivity state continuously.

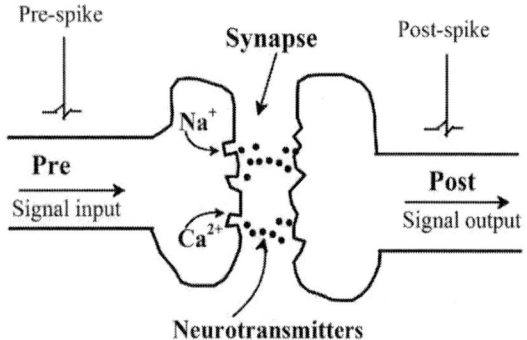

Fig. 4: Schematic presentation of a biological synapse system.

the papain thin film in top and cross-sectional view are shown in Figs. 2 (a) and 2 (b), respectively. Presence of similar kind (like Fig. 1) of non-spherical granules with varying sizes of grains in the films were also observed with FESEM characteristics of papain films, as shown in Fig. 2 (a).The thickness of the film obtained from the cross section analysis is about 410 nm, which is close with the thickness measured by the surface profilometer.

The memristive behavior observed in the papain-based memristor structures is shown in Fig. 3 with positive polarity of bias applied on the top electrode. It can be observed that continuous sweeping of dc bias leads to the increase of the conductivity level of the device, which can be explained as follows. In the first dc bias voltage sweeping from 0 V to 5 V, first memory window was formed, which is corresponding to the formation of the initial conducting path or filaments in the protein film. However, during the RESET operation (reverse voltage sweeping) most of the filaments ruptured, only few may exists. After the formation of first memory window, a further gradual increase in the current level was observed with subsequent bias voltage sweepings, as shown in Fig. 3. The increase in current level in the memory device corresponds to the increase of the number of residual conducting filaments in the film during subsequent bias voltage sweepings [14-15]. However, after a certain numbers of dc sweeping, the increases of current level saturates, because no further new filaments formation possible with the same bias or in other word, the device reached its memory limit. The origin of the memory performance in papain based devices is not yet clearly understood. However, it can be explained using the filament formation hypothesis [16]. It is assumed that a redox process in the bulk of the papain film plays a critical role. During the dc bias application on the top electrode, an electrochemical change in form of the formation of closely bonded positive and negative domains occurs, this leads to the formation of the electric dipoles via a local redox process. When, the applied dc bias is sufficient enough, the dipoles align along the electric field direction and form a conducting path or filament. However, during reverse dc sweeping, the alignment of the dipoles get 'disturbed' and ultimately, the rupture of the filament occurs. This back and forth effects of the alignment and misalignment of the 'dipoles' leads to the memory window in the papain based memory devices. The history of the switching events can be stored in the papain based memory devices as the concentration of 'residual conducting filaments' in their operations. Neuroplastic behavior, which refers to changes that occur in the organization of the human brain as a result of 'experience', can be demonstrated by using this characteristic, as follows.

In neurobiological systems, synaptic weights and strength of synaptic connections are the measure of a memory. A simple schematic diagram of a synapse between a pre- and a post-synaptic neuron is shown in Fig. 4. The strength of a synaptic weight depends on the concentrations of ionic species (*e.g.*, Ca^{2+}, Na^+, K^+ etc.) that 'fire' a neuro-transmittance. The neuro-transmittance is analogous to the above mentioned dipoles is assumed to form in the papain based memory systems. It is reported recently that the retention loss in the memristor systems bears remarkable similarities to memory

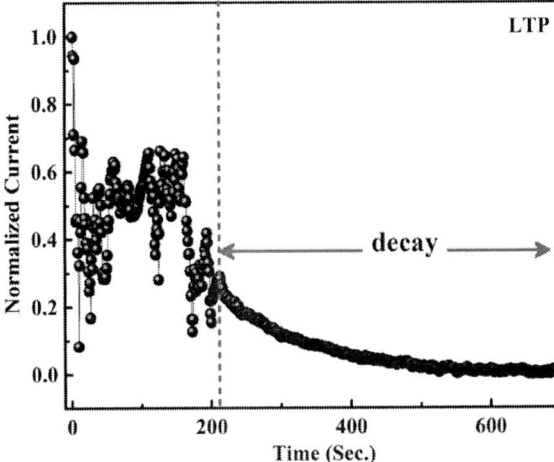

Fig. 5: Retention characteristics of the papain memristor, analogous to the memory loss of a biological synapse with 1.0 V stimuli.

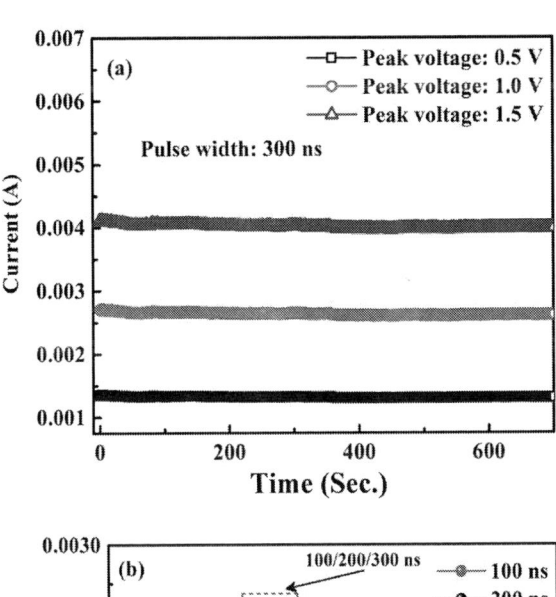

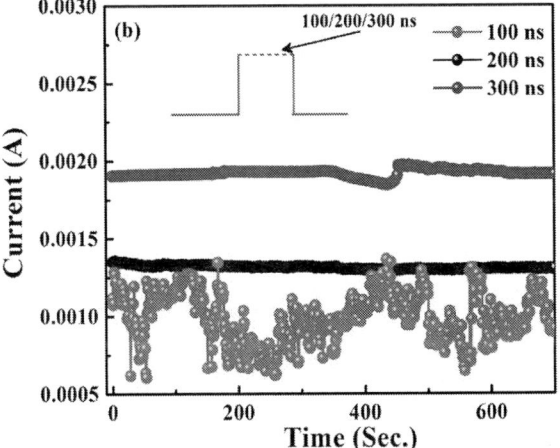

Fig. 6: The conductivity level variation under the application of different (a) pulse amplitude and (b) pulse width (+1 V as read voltage).

loss in biological systems [12]. In our study, the competing effect of memory loss and memory strengthening upon the

application of external stimuli, retention characteristics of the papain based memristor devices were investigated.

Figure 5 shows the retention behavior of the memristor device after the application of a short pulse width of 300 ns with +1 V as peak amplitude. After application of the stimulation pulse, the device conductance in form of retention test was investigated. It is suggested that two memory regimes exist in the papain memristor devices, consisting of a regime with short relaxation time and another regime with much longer relaxation time. During first ~200 sec no significant memory loss was observed. It is due to the existence of large numbers of induced charge due to the pulsing effect and also the 'residual filaments'. However, after ~200 sec, the conductivity of the device starts decreasing continuously in an exponential form which lasts up to 700 sec, which can be considered as the LTP of the synapse. The current fading process after a sequence of input pulses in the papain memory devices was systemically investigated. The effect of different stimulation strength *i.e.,* different peak amplitude on the conductivity of the memory device is shown in Fig. 6 (a). On the other hand, conductivity by means of increasing current can also be observed with increasing pulse width of the external stimuli, as shown in Fig. 6 (b). The current level was observed to increase gradually as the strength (amplitude and width) of applied voltage pulses was increased. Hence, multiple resistance states were observed by applying voltage pulses with different pulse strength. However, the high current achieved for positive pulses automatically faded over time but did not return to the original state even during the measurement time after the application of the pulses, which indicates a partial memorization.

This external stimulus dependency is analogous to a biological system, which can be explained as follows. A stimulus (action potential) at the pre-synaptic neuron forms a channel for ionic species caused to release of neuro-transmittance and finally enhancing the synaptic transmission. Once, the stimulation is terminated, it requires a finite time for the residual ionic species to decay to its equilibrium level. If the stimulus is strong enough due to its amplitude and width, then it takes long time to reach to its equilibrium level, which is the so called LTP. Also repetition of the stimuli in short interval of times leads to strong LTP. This repetition of the external stimuli or practice of the 'practical experience' to an event causes improvement of the memory functionality of a human brain, which is analogous to the continuous sweeping of the dc bias which leads to the increase of the conductivity level of the memristor (see Fig. 3). In other words, this phenomenon is similar to the so called 'learning' process. However, there is a maximum limit of the biological memory to store the information and then it start fading after a certain period of time.

Conclusions

In summary, we have demonstrated the long term potentiation in papain-based memristor devices under different external stimuli observed. We show experimentally that the retention loss in the device bears striking resemblance to memory loss in biological systems. The memristor conductance or memory state also improved upon increasing of the peak amplitude and the pulse width. Considering memory as a fundamental building block in learning and decision making in biological systems, the demonstration of such functionalities in a biomaterial based memory devices, may lead to the realization of the neuromorphic and artificial neural networks.

References

1. J. Konorski, "Conditioned reflexes and neuron organization," Cambridge University Press (Cambridge, UK, 1948).

2. R. G. M Morris, "D. O. Hebb: The Organization of Behavior Wiley: New York; 1949," *Brain Research Bulletin*, Vol. 50, No. 5–6 (1999), pp. 437.

3. E. G. Jones, "Santiago Ramón y Cajal and the Croonian Lecture, March 1894," *Trends Neurosci*, Vol. 17, No. 5 (1994), pp. 190–192.

4. W. B. Scoville, and B. Milner, "Loss of recent memory after bilateral hippocampal lesions," *J. Neurol. Neurosurg. Psychiatry*, Vol. 20, No. 1 (1957), pp. 11–21.

5. T. V. P. Bliss, and T. Lomo, "Long-lasting potentiation of synaptic transmission in the dentate area of the anaesthetized rabbit following stimulation of the perforant path," *J Physiol.*, Vol. 232, No. 2 (1973), pp. 313-356.

6. H. Markram, J. Lubke, M. Frotscher, and B. Sakmann, "Regulation of Synaptic Efficacy by Coincidence of Postsynaptic APs and EPSPs," *Science*, Vol. 275, No. 5297 (1997), pp. 213-215.

7. G. Bi, and M. Poo, "Synaptic modification by correlated activity: Hebb's postulate revisited," *Ann. Rev. Neurosci.*, Vol. 24 (2001), pp. 139-166.

8. S. Lai, "Current status of the phase change memory and its future," in *Proc. IEEE International Electron Devices Meeting 2003, IEDM Technical Digest*, Dec. 8-10, 2003, pp. 10.1.1–4.

9. S. H. Jo, T. Chang, I. Ebong, B. B. Bhadviya, P. Mazumder, and W. Lu, "Nanoscale Memristor Device as Synapse in Neuromorphic Systems," *Nano Lett.*, Vol. 10, No. 4 (2010), pp 1297–1301.

10. A. Chanthbouala *et al.*, "A ferroelectric memristor," *Nat. Mater.*, Vol. 11, No. 10 (2012), pp. 860–864.

11. G. Agnus, W. Zhao, V. Derycke, A. Filoramo, Y. Lhuillier, S. Lenfant, D. Vuillaume, C. Gamrat, and J.-P. Bourgoin, "Two-Terminal Carbon Nanotube Programmable Devices for Adaptive Architectures," *Adv. Mater.*, Vol. 22, No. 6 (2010), pp. 702–706.

12. T. Chang, S.-H. Jo, and W. Lu, "Short-Term Memory to Long-Term Memory Transition in a Nanoscale Memristor," *ACS Nano*, Vol. 5, No. 9 (2011), pp 7669–7676.

13. B. B. Mandal, S. Das, K. Choudhury, and S. C. Kundu, "Implications of silk film RGD availability and surface roughness on cytoskeletal organization and proliferation of primary rat bone marrow cells," Tissue Engineering A, Vol. 16, No. 7 (2010), pp. 2391-2403.

14. M. K. Hota, J. A. Caraveo-Frescas, M. A. McLachlan, and H. N. Alshareef, "Electroforming-free resistive switching memory effect in transparent p-type tin monoxide," *Appl. Phys. Lett.*, Vol. 104, No. 15 (2014), pp. 152104-1-4.

15. M. K. Hota, C. Mukherjee, T. Das, and C. K. Maiti, "Bipolar Resistive Switching in $Al/HfO_2/In_{0.53}Ga_{0.47}As$ MIS Structures," *ECS J. Solid State Sci. Technol.*, Vol. 1, No. 6 (2012), pp. N149-N152.

16. M. K. Hota, M. K. Bera, and C. K. Maiti, "Switching Mechanism in Au Nanodot Embedded Nb_2O_5 Memristors," *J. Nanoscience and Nanotech.*, Vol. 14, No. 5 (2014), pp. 3538-3544.

978-1-4799-3911-4/14 $31.00 © 2014 IEEE

Bipolar Resistive Switching in Different Plant and Animal Proteins

A. Bag[*], M. K. Hota[†], S. Mallik, C. K. Maiti

VLSI Engineering Laboratory, Department of Electronics and ECE, Indian Institute of Technology, Kharagpur 721302, India
[†]*currently with* Materials Science and Engineering, King Abdullah University of Science & Technology (KAUST),
Thuwal 23955-6900, Saudi Arabia
Phone: +91-3222-281475;   Fax: +91-32222-55303   [*]Email: abag@iitkgp.ac.in

Abstract

We report bipolar resistive switching phenomena observed in different types of plant and animal proteins. Using protein as the switching medium, resistive switching devices have been fabricated with conducting indium tin oxide (ITO) and Al as bottom and top electrodes, respectively. A clockwise bipolar resistive switching phenomenon is observed in all proteins. It is shown that the resistive switching phenomena originate from the local redox process in the protein and the ion exchange from the top electrode/protein interface.

Introduction

Digital-type resistive switching showing transition between two resistance states with substantial resistance change in a resistive switching media (normally, metal-oxide layer) has been investigated for applications such as, next generation digital nonvolatile memory and electronic switches. These types of non-volatile resistive switching memories have attracted serious attention due to several advantages of nondestructive readout, low power consumption, simple device architecture and high density of data storage [1-3]. As a result intensive investigations on the switching mechanisms in different resistive media, such as solid electrolytes, binary metals and perovskite oxides and even organic materials are in progress [4-7].

It is well known that proteins perform specific functions in all physiological or biochemical processes such as support, catalysis, signaling, storage, charge transfer and so forth. The diversity of these natural molecules is vast, tailoring their structures to fit the variable and complex requirements of both the biological and non-biological world is achievable by leveraging on the rapidly developing bioengineering field [8, 9]. It is expected that bioengineering may provide an alternative approach to tune the structural and electronic properties of functional molecules leading to further developments in the field of molecular electronics. Recently, a few attempts have been made to fabricate resistive switching memory devices with bio-materials such as proteins [10, 11]. However, reports on the studies of bio-materials as resistive switching media are rather scarce as compared to its inorganic counterpart. Among different bio-materials, proteins are most promising bio-electronic material as they are eco-friendly.

In this work, we report the use of both (plant and animal) types of proteins (papain, rubber latex, bovine serum, and wasp silk) as resistive switching media for bipolar switching applications.

Experimental

Indium tin oxide (~100 nm) coated glass substrates were cleaned sequentially with acetone, ethanol and deionized (DI) water, respectively, in an ultrasonic water bath. Different types of proteins were then spin-coated onto the cleaned substrates separately at room temperature and then kept in an oven for 30 min at constant temperature of 40°C. Thickness of the protein films were measured by surface profilometer (Veeco Dektak 150) and were found to be around ~400 nm for all films. The bovine serum (animal protein) and papain (plant protein) solutions were purchased from Sigma-Aldrich. The wasp silk (animal protein) aqueous solution was prepared from the wasp hives. Natural rubbed latex (plant protein) was collected from the Pará rubber tree, Hevea brasiliensis. Finally, the top electrode, Al was deposited through a metal shadow mask having a circular area of 1.96×10^{-3} cm^2 using vacuum evaporation system. The surface morphology of the protein films were investigated using atomic force microscopy (AFM) micrographs (Model NanoSurf Easyscan2). The chemical properties, such as, structural conformations of the protein films are elucidated by studying the chemical bonding states using Fourier transform infrared attenuated total reflectance (FTIR-ATR) spectra using model Agilent Cary 630. The FTIR-ATR spectrum was acquired over the range of 1800-1400 cm^{-1} with a spectral resolution of 2 cm^{-1}. The current–voltage (*I-V*) characteristics of the devices were measured using Agilent B1500A semiconductor device parameter analyzer system. The schematic device structure with experimental setup for the electrical measurements is shown in Fig. 1. The bottom electrode, ITO was grounded and the external bias was applied on top electrode using a probing system.

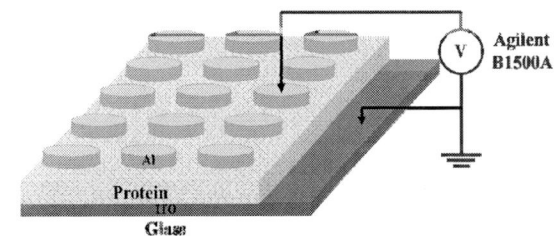

Fig. 1: Schematic of ITO/Protein/Al device with experimental setup.

Results and Discussions

The surface morphology of the spin coated protein films were studied using AFM in height profile mode with a raster-scan type, as shown in Figs. 2 (a) - (d). All the scans were performed in 10 x 10 µm area. It may be observed that all the protein films consist of islands or grains. However, the grains in papain film are rather large. This may act as a polycrystalline surface of the film, which ultimately leads to the large hysteresis behavior of the papain device. The root mean square (rms) roughness (R_q) was calculated using the relation;

$$R_q = \sqrt{\frac{1}{n_x n_y} \sum_{i=1}^{n_x} \sum_{j=1}^{n_y} \left| Z(i,j) - Z_{avg} \right|^2}$$

where, $Z(i, j)$ denotes the topography data for the surface after specimen tilt-correction, Z_{avg} is the average surface height, and i and j correspond to pixels in x and y direction. The maximum numbers of pixels in two directions are given by n_x and n_y [12]. It is observed that the surfaces of all the protein films contain peaks and valleys. For calculating RMS value of surface roughness, several scans were performed with different samples of protein films. The surface roughness of different protein films is given below in Table I.

Table I
ROUGHNESS VALUES OF DIFFERENT PROTEIN FILMS

Proteins →	Bovine serum	Wasp silk	Papain	Rubber latex
Roughness (nm) →	21.32±9	39±7	78.88±20	21.32±9

The average surface roughness was estimated for all the samples in the range of ~21 to ~78 nm. The chemical properties of the papain film are elucidated by studying the chemical bonding states using Fourier transform infrared attenuated total reflectance (FTIR-ATR), spectra using model Agilent Cary 630 FTIR system.

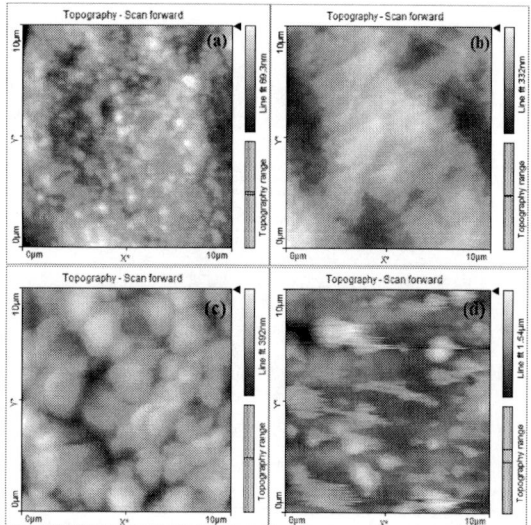

Fig. 2: AFM surface morphology of (a) bovine serum, (b) wasp silk, (c) papain and (d) rubber latex on ITO coated glass substrates.

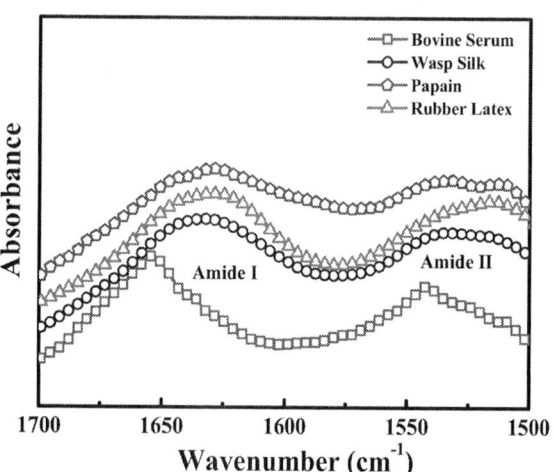

Fig. 3: FTIR spectra of different proteins.

The FTIR-ATR spectrum was acquired over the range of 1800-1400 cm^{-1} with a spectral resolution of 2 cm^{-1}. Figure 3 presents the delocalized vibrations of secondary structures of the protein films in the range of amide-I and amide-II. It is clearly visible that both the amide-I and amide-II conformations are present in all protein films.

The current–voltage (I-V) characteristics of protein based resistive switching devices are shown in Figs. 4(a)-(d). Initially, all the devices are in high resistance (virgin) state (HRS). However, an interesting phenomenon occurs in the conductivity level of the devices, when an external bias voltage is swept across the device using a special programing method ($0 \to V_{Max}(-ve) \to 0 \to V_{Max}(+ve) \to 0$), which leads to the resistive memory activation in the devices. During the application of a negative sweep bias starting from 0 V on the top electrode, the current of the devices increases rapidly, when it reached a threshold voltage (V_{SET}) and the device achieved its low resistance state (LRS) or ON state. It holds ON state until a RESET voltage (in this case it is a +ve bias) was applied on the top electrode. The device RESET back to its initial like HRS state under application of a sufficiently high positive bias (V_{RESET}) on the top electrode. In contrast, it may be noted that initial electroforming process (by means of high dc voltage or pulse voltage stressing) is necessary to activate a resistive switching in the devices. In addition, these kinds of electroforming process may lead to partially damage of the memory cells and as well as lowering the yields. However, in our study all the devices have shown resistive switching behavior without the application of any external electroforming process, i.e., self-activation process occurs during the voltage sweeping as mentioned above. This may be an important advantage of using a bio-material as resistive switching layer [13-15]. Among all the proteins, presented in this study, rubber latex based memory devices show relatively low bias resistive switching behavior, which may be useful for low power applications such as, in artificial synaptic devices. On the other hand, papain based devices shows more stability in its LRS and HRS state with relatively high ON/OFF ratio. This is probably due to the presence of relatively more grains

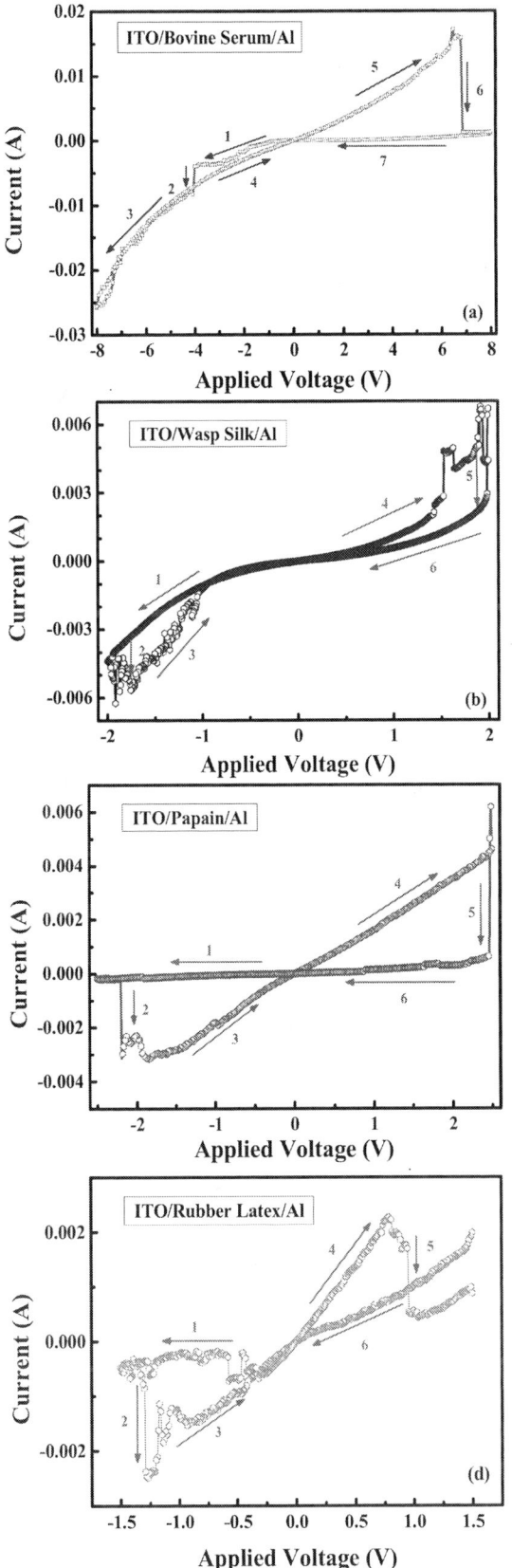

Fig. 4: Pinched hysteretic current-voltage characteristics of memory devices.

in the surface of the protein film. It is assumed that the enhancement of the electric field at the 'tips' of the grains can help in controlled filament formation/rupture which leads to the uniformity in the switching parameters. Similar results have been reported by other research groups recently [16]. On the other hand, relatively smooth surface of other proteins films leads to low switching performance. Moreover, all the proteins show resistive switching behavior and may be utilized for future memory applications. The origin of the resistive memory behavior in protein-based devices is not yet fully understood. However, it is assumed that a redox process in the bulk of the film and as well as in the Al/protein interface plays a critical role. During the application of a bipolar dc voltage sweep, electrochemical changes in the protein films via local redox process is possible. This leads to the formation of closely associate positive and negative regions in the film and forms electrical dipole. These dipoles tends to align when an effective bias was applied on top electrode and make a conducting path (filament) when applied voltage is close to V_{SET}. However, this path gets destroyed or become 'normal' under V_{RESET}. In this condition, the film become less conducting, i.e., HRS state is achieved. On the other hand, the interface between the top electrode and the protein films changes under the bias condition on the top electrode. Since, Al is used as top electrode for all the devices and Al has large negative Gibb's free energy for formation of AlO_x, it is assumed that during positive bias on the top electrode, it gets partially oxidized and forms AlO_x as interfacial layer which causes lowering the current level in positive bias condition and hence the HRS state is achieved. However, this interfacial layer gets dissolved via a reduction process under negative bias on top electrode and current increases in LRS condition as may be observed in the current-voltage characteristics (Fig. 4). Finally, it is observed that the back and forth effects of the formation and the dissolution of the conducting path and the interfacial layer leads to the bipolar resistive switching in protein-based devices.

Conclusions

In summary, plant and animal proteins have been used as resistive switching media for the fabrication of memory devices. Surface roughness of the films is found very critical for the resistive switching performance of the respective devices. Low bias bipolar resistive switching phenomena have been observed in all proteins with self-electroforming process. Interestingly, papain based devices show better memory effect as compared to other devices. It may be expected that protein-based memory devices may find applications in future bio-inspired devices/circuit applications.

References

1. R. Waser, and M. Aono, "Nanoionics-based resistive switching memories," *Nat. Mater.*, Vol. 6, No. 11 (2007), pp. 833-840.
2. B. Cho, S. Song, Y. Ji, T.-W. Kim, and T. Lee "Organic Resistive Memory Devices: Performance Enhancement, Integration, and Advanced Architectures," *Adv. Funct. Mater.*, Vol. 21, No. 15 (2011), pp. 2806- 2829.

3. M.D. Ventra, and Y.V. Pershin, "Memory materials: a unifying description," *Mater. Today*, Vol. 14, No. 12 (2011), pp. 584–591.

4. M. N. Kozicki, C. Gopalan, M. Balakrishnan, M. Park, and M. Mitkova, "Nonvolatile memory based on solid electrolytes," in *Proc. Non-Volatile Memory Technology Symposium 2004*, Nov. 15-17, 2004, pp. 10-17.

5. I. G. Baek, M. S. Lee, S. Seo, M. J. Lee, D. H. Seo, D.-S. Suh, J. C. Park, S. O. Park, H. S. Kim, I. K. Yoo, U.-In. Chung, and J. T. Moon, "Highly scalable nonvolatile resistive memory using simple binary oxide driven by asymmetric unipolar voltage pulses," in *Proc. IEEE International Electron Devices Meeting 2004, IEDM Technical Digest*, Dec. 13-15, 2004, pp. 587-590.

6. Y. Watanabe, J. G. Bednorz, A. Bietsch, Ch. Gerber, D. Widmer, and A. Beck, "Current-driven insulator–conductor transition and nonvolatile memory in chromium-doped $SrTiO_3$ single crystals," *Appl. Phys. Lett.*, Vol. 78, No. 23 (2001), pp. 3738-3740.

7. W. L. Kwan, R. J. Tseng , W. Wu , Q. Pei, and Y. Yang, "Stackable Resistive Memory Device Using Photo Cross-linkable Copolymer," in *Proc. IEEE International Electron Devices Meeting 2007, IEDM Technical Digest*, Dec. 10-12, 2007, pp. 237-240.

8. S. C Lee, M. A. Ruegsegger, M. Ferrari, "Biological Molecules in Nanodevices," in The Encyclopedia of Nanoscience and Nanotechnology, American Scientific Publishers. (California, 2004), pp. 309-327.

9. K. D. Bhalerao , E. Eteshola , M. Keener , and S. C. Lee , "Nanodevice design through the functional abstraction of biological macromolecules," *Appl. Phys. Lett.*, Vol. 87, No.14 (2005), pp. 143902-1-3.

10. M. K. Hota, M. K. Bera, B. Kundu, S. C. Kundu, and C. K. Maiti, "A Natural Silk Fibroin Protein-Based Transparent Bio-Memristor," *Adv. Funct. Mater.*, Vol. 22, No. 21 (2012), pp. 4493- 4499.

11. C. Mukherjee, M. K. Hota, D. Naskar, S. C. Kundu, and C. K. Maiti, "Resistive switching in natural silk fibroin protein-based bio-memristors," *Phys. Status Solidi A*, Vol. 210, No. 9 (2013), pp. 1797-1805.

12. B. B. Mandal, S. Das, K. Choudhury, and S. C. Kundu, "Implications of silk film RGD availability and surface roughness on cytoskeletal organization and proliferation of primary rat bone marrow cells," *Tissue Engineering A*, Vol. 16, No. 7 (2010), pp. 2391-2403.

13. C.-Y. Lin, C.-Y. Wu, C.-Y. Wu, C. Hu, and T.-Y. Tseng, "Bistable Resistive Switching in Al_2O_3 Memory Thin Films," *J. Electrochem. Soc.*, Vol. 154, No. 9 (2007), pp. G189-G192.

14. Y. H. Do, J. S. Kwak, J. P. Hong, K. H. Jung, and H. S. Kim, "Al electrode dependent transition to bipolar resistive switching characteristics in pure TiO_2 films," *J. Appl. Phys.*, Vol. 104, No. 11 (2008), pp. 114512-1-4.

15. Y.-T. Tsai, T.-C. Chang, C.-C. Lin, L.-S. Chiang, S.-C. Chen, S. M. Sze, et al., "Effect of Top Electrode Material on Resistive Switching Characteristics in MnO_2 Nonvolatile Memory Devices," *ECS Transactions*, Vol. 41, No. 3 (2011), pp. 475-482.

16. Y.- C. Huang, W.- L. Tsai, C.- H. Chou, C.- Y. Wan, C. Hsiao, and H.- C. Cheng, "High-performance programmable metallization cell memory with the pyramid-structured electrode," *IEEE Elecron Device Lett.*, Vol. 34, No. 10 (2013), pp. 1244-1246.

Reliability prediction and real world for LED lamps

G.Mura and M.Vanzi

{ DIEE - Department of Electrical and Electronic Engineering - University of Cagliari }

{ Piazza D'Armi – 09134 Cagliari - Italy }

Phone: (+39) 0706755775   Fax: (+39) 0706755900   Email: gmura@diee.unica.it

Abstract- **The paper focuses on Reliability of Reliability Predictions by comparison with available Reliability Data Sheet and Accelerated Stress Test results on commercially available devices. The striking difference in the predicted MTTFs (Mean Time To Failure) is discussed.**

I. INTRODUCTION

In today's competitive electronic market, in order to obtain high product reliability, consideration of reliability issues should be integrated from the beginning of the design phase. This leads to the concept of Reliability Prediction passing through the Design for Reliability.

The term "Reliability Predictions" describes the method used to estimate the constant failure rate during the useful life of a product. It has historically been used to indicate the process of applying mathematical models and data for the purpose of estimating field-reliability of a system before empirical data are available for the system. For this reason it provides the quantitative baseline needed to assess progress in reliability engineering.

MIL-STD-217F, is a widely used standard that defines two prediction methods that vary in the degree of information required to be provided. Part Stress Analysis Prediction that is applicable during later design phases and Parts Count Reliability Prediction that is applicable during early design phase or proposal phases of a project. [1]

A wide range of other Reliability Prediction methods has been used as Reliability engineering tool for decades [2] and at the same time there has been debates regarding the appropriate approaches to be used in predicting the reliability of electronic systems [3, 4, 5, 6].

On the other hands, accelerated lifetime tests estimate the failure rate of a component during use. These test are focused on the specific stresses of the environment (such as ambient temperature, humidity, voltage and current) in which the component is used and observations are made of the failures using the conditions of these stresses as parameters [7, 8,9,10].

During accelerated life-tests electronic devices are run under severe conditions and fail sooner than in standard operation. The life time distribution under usual conditions is then estimated by extrapolation. This is faster and also cheaper than testing at usual condition which is usually inconvenient because lifetime is generally long.

The paper starts from the job of a group of students during the courses of Reliability held at the University of Cagliari, because of the striking evidence and clearness of their results in pointing out "Reliability of Reliability". In particular, students tested some commercial LED lamps under four constant current levels, identifying the distribution function for the cumulative failures, checking the consistency of the results at all stress level, finding a power-law ruling the acceleration and finally extrapolating their data to the nominal operating conditions of the devices.

The results of such academic exercise have then been compared with the Reliability Data Sheet, available on the web, issued by the Manufacturer. The MTBF (Mean Time Between Failure) was as different as 1.2 years for the Life Test prediction and 2500 years from the official datasheet. Anyway, once stated the calculation constraints and the initial assumptions, both results are correct and honest towards their respective experimental datasets. The discussion raised among the students about the relative meaning of some reliability concepts is worth of a wider audience.

This is the aim of this paper.

II. THE DEVICES AND THEIR FAILURES

Commercial HLMP orange LED lamps have been selected for the test. They are supplied in standard T-1 ¾ package and their datasheets include Reliability tests and Reliability predictions. In particular, an operating condition of 30 mA at 55°C is stated to lead to a MTBF exceeding 20,000,000 hours, that is in the range of more than 2000 years. Even considering an utilization factor of 0.25, that means the use of the devices 168 hours per week, this means some 500 years of continuous operation.

Failures in LEDs were classified as catastrophic and degradation.

Catastrophic failures are sudden and drastic changes in the operating characteristics resulting in the complete loss of useful performance.

Degradation failures occur when the power luminosity decays is below a predefined power luminosity value. ASSIST (Alliance for Solid-State Illumination Systems and Technologies) guidelines specify a lumen maintenance of 70%, corresponding to a 30% reduction in initial light output, as the end of useful life for general lighting.[11]

III. EXPERIMENTALS

36 LEDs lamps have been tested at room temperature under constant current conditions

The nominal operating current is indicated by the manufacturer at 30 mA, 55°C, that is more severe under the thermal point of view than the proposed lifetest, that is performed at room temperature. After a proper step stress, the

---

This work is partially supported by Sardinia Regional Government (P.O.R. Sardegna F.S.E. Operational Programme of the Autonomous Region of Sardinia, European Social Fund 2007–2013 – Axis IV Human Resources, Objective 1.3, Line of Activity 1.3.1 ''Avviso di chiamata per il finanziamento di Assegni di Ricerca'').

978-1-4799-3911-4/14 $31.00 © 2014 IEEE

stress levels have then been chosen at 70 mA, 80 mA, 90 mA, 100 mA, and applied to groups of 9 devices each.

The data were acquired at time intervals (2, 4, 8, 16, 32, 64, 128 hours). The results are summarized in fig. 1.

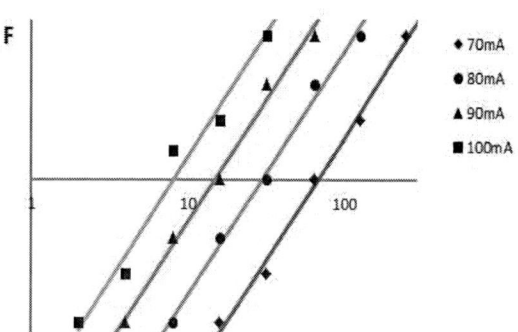

Fig.1 Life tests result

They state some evident cardinal points:

1) The experimental points align along straight lines on a log-normal plot, which identifies the real distribution function of failures.

2) The straight lines are parallel. This does not identify the failure mechanism, but tells that it is the same for all cases and roughly anticipates that ln(t) changes proportionally to the applied stress.

3) The spacing of the lines is related to the stress level by a simple Eyring Inverse Power Rule, as demonstrated in the following by the perfect alignment of the points in the Arrhenius chart (fig.2).

4) This allows to predict the cumulative failure function at operating conditions, that is a straight line, parallel to the experimental ones in fig. 1, intersecting to 50% level at a MTTF (utilization factor = 1, that means continuous operation) of 1.2 years.

In order to further explain the points, above, fig.3 reports the same data of fig.1, complemented with the predictions for the same operating conditions given by the lifetest and the zero-failure reliability demonstration.

The latter is merely based on the plot of the cumulative failure function for an exponential distribution with a nominal MTTF of about 22,000,000 hours, as reported by the available Reliability Data Sheet.

For the Life Test prediction at 30 mA an inverse power law has been first supposed and then checked.

The inverse power law model states that the life (t50%) is inversely proportional to the stress (I) with the power of an exponent (α) that depends on the individual product and/or product material.

$$t_{50\%} = \frac{A}{I^{\alpha}}$$

Both A and α are constants to be calculated from experimental data. If plotting t50% as a function of $1/I\alpha$ (Arrhenius plot, even

if the Arrhenius Law is not called into play) for three or more data pairs one obtains a straight line, the power rule is assumed to be confirmed.

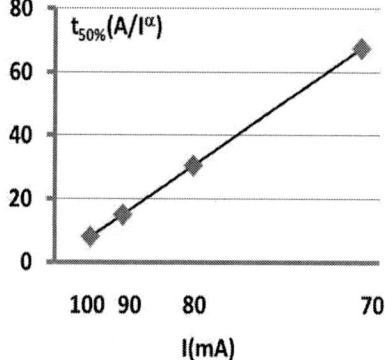

Fig.2 Arrhenius plot for the Power rule

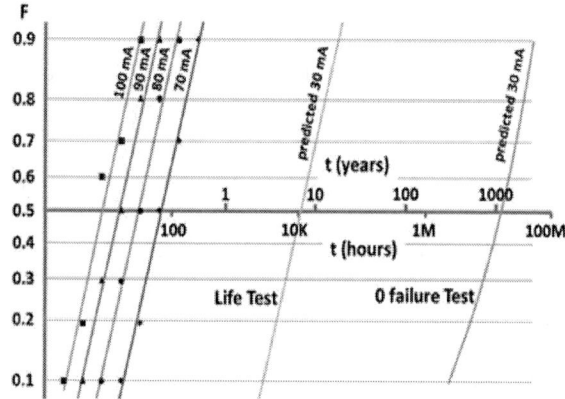

Fig.3 Prediction data from Life Test and Reliability Demonstration

IV. DISCUSSION

Data from the Life Test (LT) and the Reliability Data Sheets (RDS) should be compared with some caution. The rough comparison of the LT prediction of 1.2 years and the 2500 years for the MTTF seem to neglect many different situations.

First, the RDS data are referred to an utilization time of 0.25, and then the expected life should be divided by 4 before comparing with the LT.

Second, the room temperature of the LT and the 55°C of the zero-failure tests reported by RDS seem to make the two results not comparable.

Both points can be solved by recalculating the RDS data for continuous life at room temperature. The document, indeed, duly reports predictions for different temperatures, at the two confidence levels of 60% and 90%. The discussion about confidence levels will be given in the following. From RDS it results clear that the estimated MTTF at 55°C has been calculated at 60% confidence level, and that it is the ambient temperature. This is an important point, because a set of calculated junction temperatures is also provided for each situation, but it does not enter the Reliability calculations.

The temperature dependence of the MTTF is calculated assuming a thermal acceleration factor given by a pure Arrhenius Law for an activation energy of 0.43 eV, referring this value to the MIL-STD -217 document [1].

It follows that the predicted MTTF, for a temperature comparable with the LT (300°K), rises to more than 66,000,000 hours, corresponding to 7500 years. Rescaling this value to continuous operation, one gets some 1900 years of expected median life. This is the value to be compared with the 1.2 years prediction of the Life Test.

Fig.4 reports the expected lifetimes for the Life Test, assuming a lognormal distribution, for the nominal MTTF of the RDS at 0.25 utilization factor and for the corresponding continuous life.

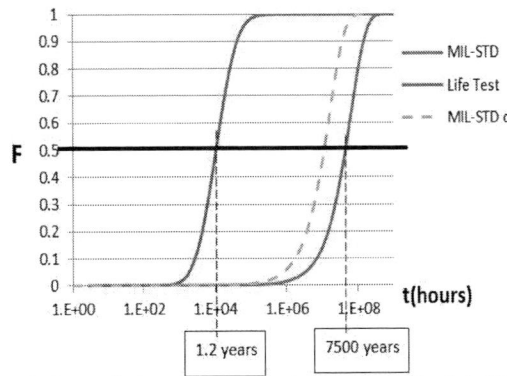

Fig 4 Lifetime prediction and lifetime obtained by lifetest. The dashed line represents the expected prediction at continuous operation, which should be compared with the Life Test data

Before proceeding with the discussion, a marginal point can be also commented: the RDS indicates MTBF instead of MTTF. The same Manufacturer in other data sheets uses MTTF for similar data and even, in at least one case, interchanges the two in the same document. It is the Author's opinion that MTTF should be the correct term. In any case, any textbook recalls that the two terms differ for the MTTR (Main Time To Repair) that in any case implies the removal of the failed LED. It should be clear that with times to failure in the order of thousands of years, the MTTR would be negligible with respect to the MTTF, leading it to coincide with the MTBF.

Facing now the kernel of the problem, one must say that nothing wrong has been done by the manufacturer. It is the demonstration method that should be investigated. The following points should be considered.

The standard prediction methods imply exponential distribution. This is a one-parameter function that predicts that the probability R of survival of n devices after t hours, if a given failure rate λ is given, is $R = \exp(-n\lambda t)$

Conversely, if n devices all survive along a time t, the failure rate is said to be demonstrated with a *confidence level* and is usually assumed at 60%.

An intriguing point, in teaching Reliability, was the discussion on the relationship between R and the confidence level, because of the evident equivalence R=1-CL that comes when the *Chi-square* method is applied for the zero-failure case, leading to the well-known formula:

$$\lambda_{CL} = \frac{\chi^2_{2;(1-CL)}}{2nt} \underset{0\,failures}{=} -\frac{\ln(1-CL)}{nt}$$

The ambiguity comes for the good-sense interpretation of the words "confidence level". If, indeed, it is logically clear that high failure rate (low MTTF) corresponds to low Reliability R, it sounds less intuitive that a high confidence level implies a worse (higher) failure rate. It is a matter of definitions, and probably the term "confidence" has been historically assigned in a way that for many people (starting with students) is anti-intuitive. But it is the state of things.

One more substantial point is that acceleration effects are only assumed, in Reliability Demonstration as in the studied RDS, to be thermal, and then different levels of current injection are first converted into a thermal stress by considering the dissipated power and then applying a tabulated (not measured) general acceleration parameter (the activation energy) that the Reliability handbooks indicate for LEDs at 0.43 eV.

It should be clear that on one side the use of uncontrolled numerical values in exponential functions (as the Arrhenius law is) can introduce extremely large errors, and that converting a current stress into a thermal one neglects all current-specific degradation mechanisms. But what is more important is that the exponential law is known to be valid for random failures and not for aging. This is the key point that the accelerated stress test puts into evidence: degradation takes place because of internal, not casual mechanisms.

On the other hand, the accelerated Life Test predicts the expected MTTF at operating conditions by extrapolating a numerical value for the logarithm of the MTTF, based on experimental logarithmic data that are few in number and usually quite far from it, which is surely a source of error. It then happens that the prediction of MTTF=1.2 years also tells, for the measured lognormal distributions, that along the 2000 hours at 30 mA at least 100 devices out of 2000 should have failed, which was not. But, if one allows a simple 20% increase of the extrapolated log(MTTF) one would predict no failures.

Anyway, a ratio of two orders of magnitude remains between the two predictions. Good sense and experience tell that also LEDs fail during a human life, which supports the Life Test method for any real estimate of Reliability. The role of Reliability Demonstration remains that of a relative tool for monitoring reliability improvements and possibly comparing different sources for a same kind of device. This point, duly explained for students, not always is equally clear at user level, where some excess confidence is attributed to millenary MTTFs.

CONCLUSIONS

The paper focuses on Reliability of Reliability Predictions, based on available documents and educational experiments on commercially available devices.

As reported in MIL-STD-217F "A Reliability prediction should never be assumed to represent the expected field reliability as measured by the user- This does not negate its value as reliability engineering tool; note that none of applications discussed above requires the predicted reliability to match the field measurement".

The impressive difference in predictions between Life Test and Reliability Demonstration procedures, that has been discussed before, points out that the basic limit of Reliability Prediction is its dependence on correct application by the end user.

Those who correctly apply the models and use the information on a conscientious reliability program will find the prediction a useful tool. Those who view the prediction only as a number which must exceed a specified value can usually find a way to achieve their goal without any impact on the system [1].

On the other hand zero-failure reliability testing is a test conducted to substantiate that a given design is better than a requirement or a previous design.

The final consideration is once again a warning against the many risks introduced for more and more small and medium end users of microelectronic devices by the lack of reliability culture.

ACKNOWLEDGMENT

The authors would like to thank Fabrizio Marcia for his contribution to the experimental setup.

REFERENCES

[1] US MIL-STD-217F Reliability Prediction of electronic equipment
[2] Denson, W, "The History of Reliability Prediction," *IEEE Trans. On Reliab.,* vol. 47, No. 3, pp. 321-328,1998
[3] Walker, E, *The Design Analysis Handbook: A Practical Guide to Design Validation*, Elsevier, New York, 1998
[4] Foucher, B. et al, "A review of reliability prediction methods for electronic devices "*Microel. Reliab.*,vol.42, I.8, pp. 1155.1162, 2002
[5] Pecht, M. et al "Predicting the reliability of electronic equipment" *Proc. of the IEEE*, vol. 82, I.7, pp.992-1004, 1994
[6] Cassanelli, G. et al. "Reliability predictions in electronic industrial applications" *Microel. Reliab.*, vol. 45, I.9–11, pp.1321–1326, 2005
[7] Leonard, C. et al , "Failure Prediction Methodology Calculations Can Mislead: Use Them Wisely, Not Blindly" *Proc. of the IEEE NAECON*, pp. 1248-1253,1989
[8] Renesas Electronics, Reliability Testing and Reliability Prediction – Reliability handbook, 2010
[9] US MIL- STD-883 Test Method Standard Microcircuits
[10] JIS- C-7021 Japan Industrial Standard
[11] ASSIST recommends: LED Life for General Lighting, 2007

Hot Carrier Injection on Back Biasing Double-Gate FinFET with 10 and 25-nm Fin Width

Wen-Teng Chang*[1], Li-Gong Cin[1], Wen-Kuan Yeh[1], Po-Ying Chen[2]

[1]Department of Electrical Engineering, National University of Kaohsiung, 811 Taiwan
[2]Department of Information Engineering, I-Shou University, Kaohsiung, Taiwan
Phone: +886-7-5919434 Fax: +886-7-5919374 Email: wtchang@nuk.edu.tw

Abstract- The effects of hot carrier injection on double-gate FinFETs with fin widths (W_{fin}) of 10 and 25 nm with positive and negative back biases are compared in this study. The FinFETs with a positive bias and narrow W_{fin} exhibit a large current tuning range but experiences high degradation after stress. By contrast, a negative back bias alleviates degradation but also eliminates the drive current.

I. INTRODUCTION

FinFETs are one of the mainstream device structures for digital/logic and memory technologies. They are commonly utilized to resolve the short-channel effect (SCE) compared with conventional planar transistors [1], which employ single-gate MOSFETs. FinFETs utilize gates to surround a channel; thus, fin width (W_{fin}) is an important parameter. A thin W_{fin} improves the subthreshold leakage and SCE of FinFETs [2]. Even with the benefit of thin W_{fin}, fin scaling is accompanied by large reliability degradation under hot carrier degradation [3]. Furthermore, ballistic transport becomes prominent when the channel length of FinFETs is reduced to 15 nm [4]. The volume inversion of double-gate FinFETs has been reported as significant for W_{fin} below 9 nm; on the contrary, inversion carriers are confined near the Si/SiO₂ interface at a thickness of 32 nm, similar to the behavior of classical single-gate MOSFETs.

Meanwhile, their ultra-thin body structure allows FinFETs to modulate threshold voltage (V_T) with a back-gate scheme [5]. Negative back bias on FinFETs provides the circuit with the capability to reduce current leakage and power consumption but with the disadvantage of speed. On the contrary, positive back bias improves the speed but increases power consumption. A proper circuit design that employs both positive and negative back biases has been reported to improve leakage reduction and retain circuit delay [6]. However, the reliability of the back-gate scheme under such critical dimensions is seldom addressed. This study explores the effect of hot carrier degradation on W_{fin} and back bias with hot carrier injection (HCI).

II. EXPERIMENTS

FinFETs were formed by a shallow trench isolation (STI) process. The undoped n-channel bulk FinFETs employ TiN/HfO₂ high-k/metal gate stacks on two sides of a silicon body with an equivalent oxide thickness of 1.0 nm to 1.6 nm by atomic layer deposition. SiN was employed on top of the body.

The channel orientation is <110>, and the fin sidewall is (110). The fin height is 30 nm, and the channel length is 45 nm. The W_{fin} values of 10 and 25 nm were compared with W_{fin} = 10 nm within the width quantization region. The current–voltage curves were measured based on HP-4156B semiconductor parameters. The drain voltage (V_D) was set to 0.05 V to operate the device at the linear region. The tunneling current (I_G–V_G) was equal to the ground, drain, and source voltages, that is, V_D = V_S = 0. The back bias applies positive and negative voltages of 0.5 V compared with the unbiased voltage with and without HCI. HCI applies equal potentials ranging from 1.4 V to 1.7 V at intervals of 0.1 V on the gate and drain. The electrical properties of the devices were measured every 20 min to 100 min.

III. RESULTS AND DISCUSSION

A. Back Biases with 10 and 25 nm Fin Widths

Fig. 1(a) shows the effect of back bias (V_B) on drain current (I_D) with V_B of -0.5, 0 (unbiased), and 0.5 V. The figure indicates that the FinFET with W_{fin} = 10 nm has comparatively larger tuning current range (2.4%), with V_B ranging from -0.5 V to 0.5 V, compared with the FinFET with W_{fin} = 25 nm (0.6%). The I_D for positive V_B (0.5 V) is induced because of the increase in V_T, whereas the I_D with a negative V_B (-0.5 V) is suppressed. This shift was also observed from I_D as a function of V_G in Fig. 1(b). Off-state I_D leakage is associated with V_B. A positive V_B results in a forward-biased p–n junction between the p-type substrate and n-type drain. This condition significantly increases gate-induced drain leakage (GIDL) and subthreshold leakage. Additionally, the off-state leakage current is more significant for the device with W_{fin} = 10 nm than for that with W_{fin} = 25 nm. Regardless of W_{fin}, the tunneling current is visually unaffected at positive or negative V_B but shifts with back biases [Fig. 1(c)].

B. Hot Carrier Stress with Positive and Negative Back Biases

Fig. 2(a) presents the I_D plot pre-HCI and post-HCI. I_D is degraded because of the shift in V_T. However, a negative V_B should move holes away from the interface. I_D degradation is thus alleviated with negative V_B compared with positive and neutral biases. The I_D–V_G plots pre-HCI and post-HCI for different V_B values also indicate superior subthreshold leakage with negative bias. However, GIDL significantly deteriorates with positive V_B (0.5 V) after HCI [Fig. 2(b)]. This phenomenon resulted from the creation of interfacial charges during HCI, which in turn resulted in substantial tunneling leakage because of the forward-biased p–n junction. Fig. 3 provides a summary

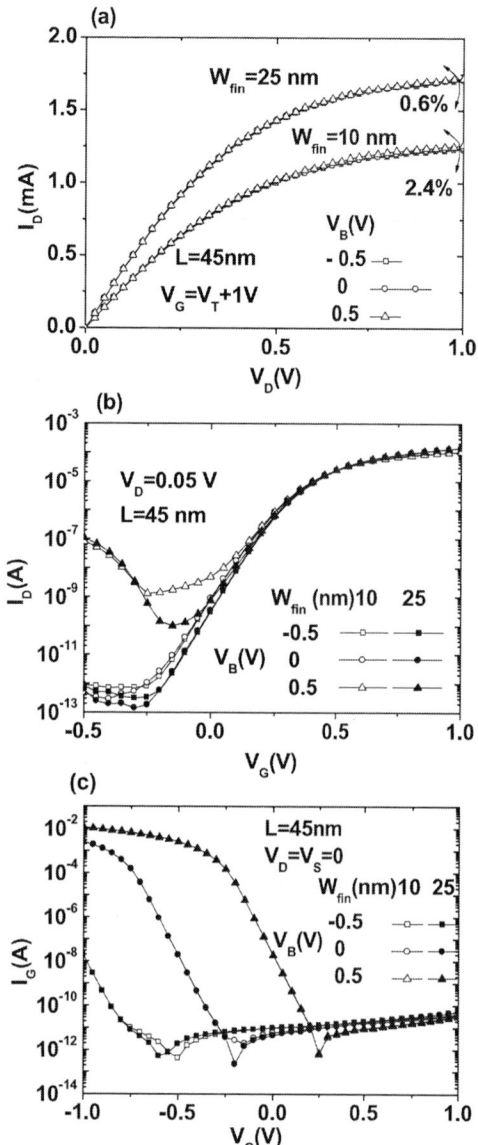

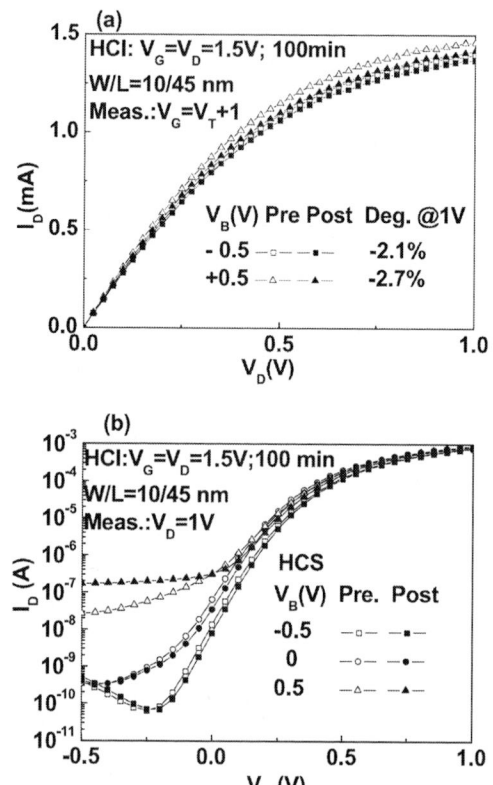

Fig. 2. (a) I_D as a function of V_D for pre-HCI and post-HCI up to 100 min with V_B of -0.5 and 0.5 V; (b) I_D as a function of V_G for pre-HCI and post-HCI.

resulted in high degradation. This result agrees with the degradation monitored from the SRAM circuit, which shows that HCI causes serious degradation in forward biasing [7].

C. Hot Carrier Stress with FinFET Width of 10 and 25 nm

Fig. 4(a) presents ID as a function of VG for pre-HCI and post-HCI in the FinFETs with Wfin = 10 and 25 nm under a stress voltage of 1.5 V for 100 min. ID degradation with Wfin = 10 nm is higher than that with Wfin = 25 nm because of higher interfacial state. However, HCI generates an insignificant effect on gate leakage variation because the interface–state is distributed near the drain [7]. An increase in HCI stress degrades ID [Fig. 5(a)] and VT variation [Fig. 5(b)] more significantly for the FinFET with small W_{fin} than that with large W_{fin}. A previous study has shown that degradation with small W_{fin} is governed by bias temperature instability owing to corner effect, whereas that with large W_{fin} is governed by hot carrier injection [8]. In other words, HCI-induced degradation increases with W_{fin}. The increase in HCI degrades the FinFET and results in a high degree of interfacial traps and threshold voltage. This result agrees with that in another study, indicating that a FinFET with

Fig. 1. (a) I_D as a function of V_D, (b) I_D as a function of V_G (W_{fin} = 10 and 25 nm), and (c) I_G as a function of V_G (W_{fin} = 10 and 25 nm) at back biases of -0.5, 0, and 0.5 V.

of gate current degradation with V_B of -0.5, 0 (unbiased), and 0.5 V after HCI for 100 min. Although the drain current exhibits high driving capability at the positive bias [Fig. 1(a)], it suffers from high degradation. By contrast, the drain current at the negative bias is low but exhibits low current degradation. Although HCI degradation is mainly caused by channel hot electron, the enhanced drive current at forward biasing (0.5 V)

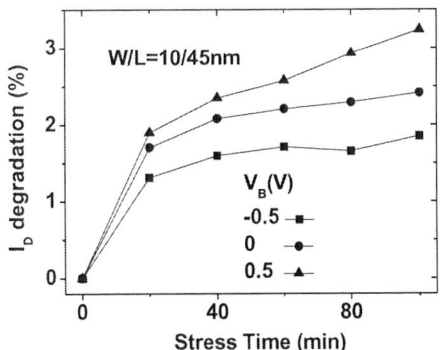

Fig. 3. I_D degradation for HCI from 0 min to 100 min at an interval of 20 min.

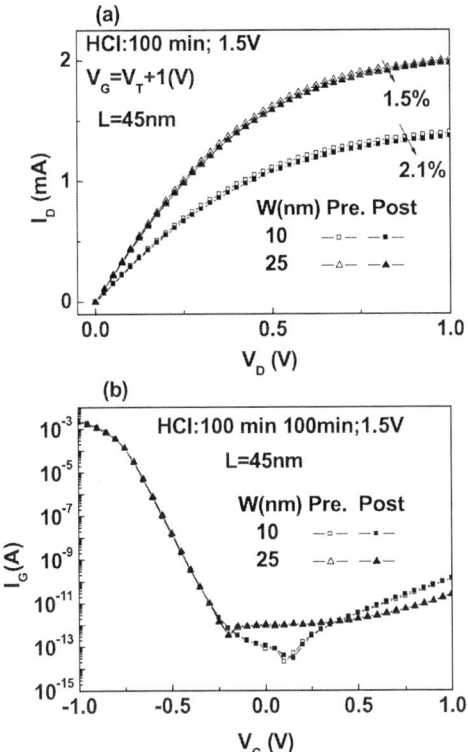

Fig. 4. (a) I_D as a function of V_D and (b) gate leakage current before and after HCI of 100 min for W_{fin} of 10 and 25 nm.

small W_{fin} results in high degradation because of the small side-channel V_T although the FinFET utilized in this particular experiment had W_{fin} above 40 nm [9].

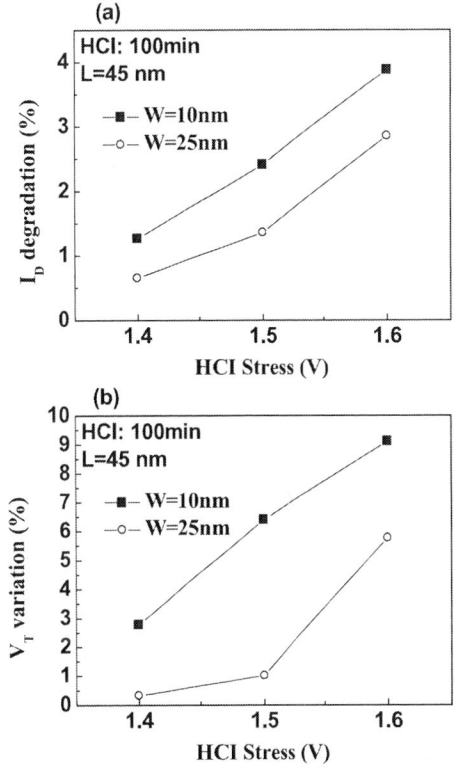

Fig. 5. (a) I_D and (b) V_T variation as a function of hot carrier stress voltage for W = 10 and 25 nm.

IV. CONCLUSIONS

The effects of the W_{fin} of double-gate FinFETs and that of devices with HCI and back biases were compared. A narrow W_{fin} is close to the quantum width. Double-gate FinFETs with narrow W_{fin} (W_{fin} = 10 nm) exhibit low current drive but high current degradation and threshold voltage variation during HCI. These devices also exhibit a large current tuning range but present high hot carrier degradation and threshold voltage variation at positive back bias. However, this trend depends on the doping concentration of Si. A negative back bias causes hot hole injection during HCI and thus alleviates degradation compared with unbiased and positive-biased devices.

ACKNOWLEDGMENT

The authors thank the National Science Council for their financial support under contract number 101-2221-E-390-001-MY2 and the staff of the United Microelectronics Corporation for the information that they have provided.

REFERENCES

[1] The International Technology Roadmap for Semiconductor (ITRS), 2012.

[2] G. Pei, J. Kedzierski, P. Oldiges, L. Meikei, E.C.-C. Kan, "FinFET design considerations based on 3-D simulation and analytical modeling," *IEEE Trans. Electron Devices*, vol. 49, no. 8, pp. 1411–1419, Aug. 2002.

[3] S. Chabukswar, D. Maji, C.R. Manoj, K.G. Anil, V. Ramgopal Rao, F. Crupi, P. Magnone, G. Giusi, C. Pace, N. Collaert, "Implications of fin width scaling on variability and reliability of high-k metal gate FinFETs," *Microelectronic Engineering*, vol. 87, no. 10, pp. 1963–1967, Oct. 2010.

[4] J. S. Martin, A. Bournel, P. Dollfus, "On the ballistic transport in nanometer-scaled DG MOSFETs," *IEEE Trans. Electron Devices*, vol. 51, no. 7, pp. 1148–1155, Jul. 2004.

[5] C.-P. Lin, B.-Y. Tsui, "Impact of back gate bias on hot-carrier effects of n-channel tri-gate FETs (TGFETs)," in *VLSI-TSA Tech. Dig.*, 2006, pp. 1–2.

[6] D. Baccarin, D. Esseni, M. Alioto, "Mixed FBB/RBB: A novel low-leakage technique for Finfet forced stacks," *IEEE Trans. VLSI System*, vol. 20, no. 8, pp. 1467–1472, Aug. 2012.

[7] C. Ma, L. Zhang, C. Zhang, X. Zhang, J. He, X. Zhang, "A physical based model to predict performance degradation of FinFET accounting for interface state distribution effect due to hot carrier injection," *Microelectronic Reliabilities,* vol. 51, no. 2, pp. 337–341, Feb. 2011.

[8] D. H. Lee, S. M. Lee, C. G. Yu, J. T. Park, "A guideline for the optimum fin width considering hot-carrier and NBTI degradation in MuGFETs," *IEEE Electron Device Letters,* vol. 32, no. 9, pp. 1176–1178, Sep. 2011.

[9] S.-Y. Kim, J.-H. Lee, "Hot carrier-induced degradation in bulk FinFETs,' *IEEE Electron Device Letters*, vol. 26, no. 8, pp. 566–568, Aug. 2005.

Junction induced Variation and Reliability for Ultra-Thin-Body and Bulk Oxide MOSFETs

Wen-Kuan Yeh[1], Wen-Teng Chang[1], Po-Ying Chen[2] and Cheng-Li Lin[3]

1. Department of Electrical Engineering, National University of Kaohsiung, 2. Department of Information Engineering, I-Shou University, 3. Department of Electronic Engineering, Feng Chia University
No. 700 Kaohsiung University Rd., Nan-Tzu Dist., Kaohsiung, Taiwan
Tel: 886-7-5919372 Fax: 886-7-5919374 E-mail: wkyeh@nuk.edu.tw

Abstract- In this work, we investigate the impact of junction dose distribution (LDD/halo) on device characteristic variation and symmetry for ultra-thin body and bulk oxide silicon on insulator (UTBB SOI) nMOSFET. The device performance and hot carrier induced degradations have also been examined. High junction doping profile will enhances the device's driving capability and sub-threshold swing, but makes the transistor forward and reverse characteristics unsymmetrical. Compared to high dose junction profile UTBB-SOI device, low dose junction profile device is less sensitive to substrate bias effect. After hot carrier stressing, low junction dose device with lower impact ionization exhibits better device reliability than high junction dose one.
Keywords—UTBB-SOI, VARIANCE, STRESSING

I. INTRODUCTION

As aggressive MOSFET scaled toward 20nm node, planar bulk MOSFET is approaching to the imension limitation for well controlling in operation due mainly to serious short channel effects will degrading device subthreshold slope. Although SCE can be suppressed and by increasing channel dopant concentration, but it will also increases the variability in the V_{TH}. Thus, this is a one of the reason why the supply voltage V_{DD} cannot be decreased as the scaling law. [1] Fully depleted silicon on insulator (FD-SOI) transistors [2] and Ultrathin-body and Bulk Oxide (UTBB) SOI MOSFET [3] has been attracted a lot of interest in recently year due to they can suppress device's SCE and improve device's subthreshold slope effectively for reducing the off-state leakage. With this ultra-thin bulk oxide layer, the lateral electrostatic coupling between drain and channel through buried bulk oxide can be suppressed effectively [4]. However higher channel dopant concentration still used to control device's V_{TH} and SCE with the risk of device's characteristic variation due to totally substrate dose in ultra thin Si is difficult to control. [5] And the influence of random doping fluctuation is still a problem [6] to make this UTBB SOI MOSFET can be easily adopted for commercial use. It is well known that quantum mechanical effects due to the variation of doping concentration greatly influence carrier transport in UTBB-SOI device [7]. But the impacts of doping concentration in UTBB-SOI performance are still lack of well investigation especially in device variance and the effect in device reliability. In this work, we investigate the impact of substrate doping concentration on device characteristic variability and the sensitivity to substrate bias. Related device performance and hot carrier induced device degradation of UTBB-SOI MOSFET were also inspected.

II. EXPERIMENT

MOSFET was formed on undoped substrate with 10nm thick SOI wafer and 20nm SiO2 BOX using 28nm CMOS technology node process. Different substrate doping concentrations for S/D extension and halo implantation were implemented with range of dose concentration from 1013 to 1015. For gate formation, Hf-based oxide was deposited by atomic layer deposition (ALD) method followed by a TiN gate metal film deposition. After the gate stack formation, an appropriate post laser spiking annealing was used to activate source/drain junction. Finally, an advanced metallization process was performed to finish this 28nm high k/ metal gate UTBB SOI CMOSFET. Figure 1(a) shows the schematic illustration and (b) TEM of the Hf-based high-k/metal gate UTBB SOI MOSFET structure. After DC electrical measurements, the samples were then stressed by constant voltage stress for hot-carrier stressing with $|V_D| = |V_G| = 0 - 3V$ under room temperature for 100 minutes.

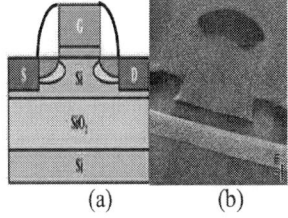

(a)   (b)

Figure 1 (a) The schematic illustration and (b) TEM of the Hf-based high-k/metal gate Ultra Thin Body and Bulk Oxide (UTBB) MOSFET structure with different substrate dose concentration substrate.

III. RESULTS AND DISCUSSION

Figure 2 compare device's driving capability and source to drain resistance with different substrate dose. It was found that device with higher substrate dose shows lower total source-to-drain total channel resistance (Fig. 2(a)), and thus higher device's driving capability (Fig. 2(b)). We found that the source-to-drain resistance is very sensitive to substrate dose concentration and increases apparently as substrate dose increases especially for p-channel UTBB SOI MOSFET. Fig. 3 shows net doping profile for both UTBB SOI nMOSFETs and proved that higher net channel dose concentration can be obtained in device with higher LDD/halo junction doping device.

978-1-4799-3911-4/14 $31.00 © 2014 IEEE

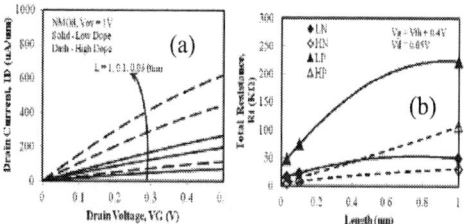

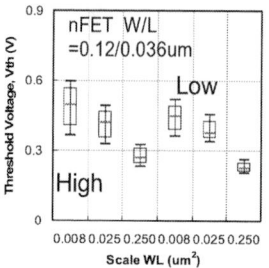

Figure 2 Higher substrate dose device shows (a) higher driving capability with (b) lower total channel resistance.

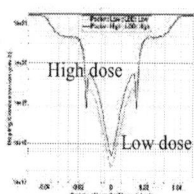

Figure 5 Lower substrate dose device shows shows lower device's threshold voltage sensitive to device channel effect.

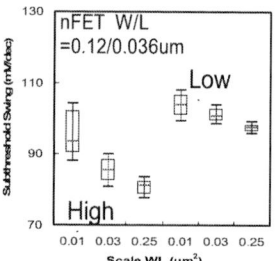

Figure 3 The simulation of doping profile UTBB SOI nMOSFET with high and low junction doping.

Figure 4 investigates the device symmetrical behavior from forward (solid lines with $V_D=V_{DS}$ and $V_S=0$) to reverse (dash lines with $V_S=V_{DS}$ and $V_D=0$) characteristics for different junction doses. It is clear in Fig. 4 that low junction dose FDSOI nMOSFET shows better symmetrical characteristics (less deviation measured from forward to reverse and only shown the median measured data point) in threshold voltage variability and lower sensitivity to substrate bias (less deviation measured from $V_B = -0.5V$ to 0.5V). Owing to higher junction barrier on the source/substrate junction and the drain/substrate junction, high junction dose FDSOI nMOSFET shows higher device sensitivity to substrate bias. High junction dose FDSOI nMOSFET also shows larger threshold voltage variability, especially for devices with smaller W×L area than those of low junction dose nMOSFETs, as shown in Fig. 5.

Figure 6 Lower substrate dose device shows lower device's subthreshold swing sensitive to device channel effect.

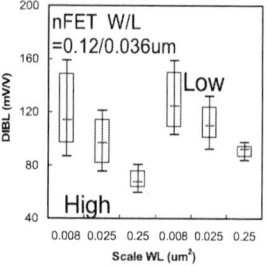

Figure 7 Lower substrate dose device shows (a) lower device's DIBL sensitive to device channel effect.

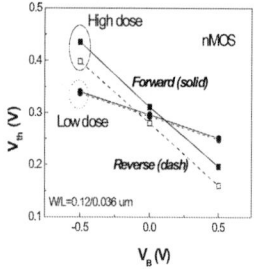

Figure 4 Lower substrate dose device shows better symmetrical characteristic (forward to reverse measurement) as well as lower sensitive to substrate bias.

According to [8], the dominant factors on the threshold voltage variability of the SOI MOSFET are random doping fluctuation (RDF) and SOI thickness variability. Thus for this FDSOI MOSFET, the device symmetry and V_{TH} variability to device dimension could be due presumably to junction doping distribution resulting in device characteristic variability. Compared to low junction dose FDSOI nMOSFET, high junction dose nMOSFET possess lower device's sub-threshold swing (SS) (see Fig. 6), and lower drain induced barrier lowering (DIBL) (Fig. 7).

In order to inspect the totally device's V_{TH} variability in these SOI MOSFET with different doping substrate, a Pelgrom coefficient A_{vt} [10] is often used as an index of the device's V_{TH} variation, which plotted was checked with various dimension of device within a die for comparison, as shown Fig. 8. It is apparently that higher substrate dose UTBB SOI nMOSFET shows larger threshold voltage variance totally, which very fit to our measurement.

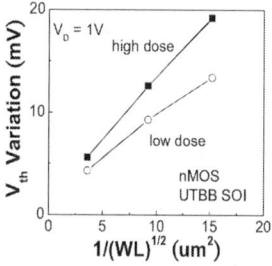

Figure 8 The comparison of the Pelgram coefficient Avt plotted for UTBB SOI MOSFET structure, higher dose nMOSFET shows larger Vth variance within a die

We stress these devices using constant drain and gate voltage ($V_{GS} = V_{DS} = 0$ to 3 volts) to inspect the hot carrier induced device degradation. Compared to low substrate dose device, higher substrate dose nMOSFET shows more serious hot carrier induced drain current degradation (Fig 9), and hot carrier induced threshold voltage degradation (Fig 10);

both degradation become more serious especially with positive substrate bias ($V_B > 0$) due to forward junction biasing ($V_{BS} > 0$) will reduce the barrier height in source/substrate; thus, enhancing electron impact ionization rate and further resulting in more serious hot carrier induced I_D and V_{TH} degradation. But rather than vise versa because negative substrate biased can increase the barrier height in source/substrate, thus suppressing carrier impact ionization and suppressing hot carrier stressing I_D and V_{TH} degradation. In order to compare the I_D degradation for both samples, the fresh I_D-V_D curves were normalized together. And we found that the hot carrier induced I_D and V_{TH} degradation become serious as stressing voltage (V_{stress}) increase especially for high substrate dose SOI nMOSFET, as shown in Fig. 11.

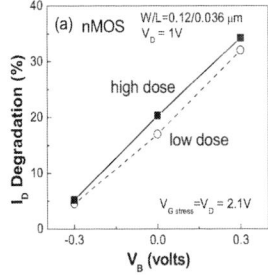

Figure 9 Higher substrate dose device shows higher hot carrier induced drain current degradation, become more serious especially with forward substrate bias.

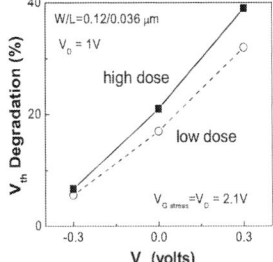

Figure 10 Higher substrate dose device shows higher hot carrier induced device's V_{TH} degradation.

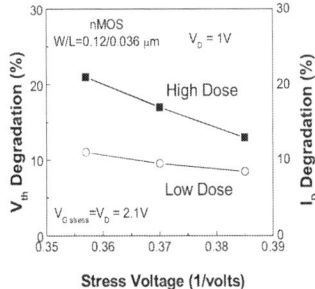

Figure 11 Higher substrate dose device shows higher hot carrier induced device's threshold voltage and driving capability degradation especially at higher stressing voltage.

If we inspect both substrate bias (V_B) and V_{stress} impact on hot carriers stressing device degradation, and we can found that hot carrier stressing I_D and V_{TH} degradation become more serious as V_{stress} especially with forward V_B and slow down as V_{stress} with negative V_B. Thus we can understand that higher impact ionization was happened on higher substrate dose device which will cause higher hot carrier induced device's I_D and V_{TH} degradation, and become more serious as V_{stress} increased; rather than vise versa. Detail hot carrier induced I_D-V_G characteristic degradation including before and after 100 minutes constant voltage stress with different substrate doping concentration was showing in Fig. 12 for device with forward substrate bias, and showing in Fig. 13 for with reverse substrate bias. It is apparently that higher device's threshold voltage and subthreshold swing were affected after hot carrier stressing. After 100 minutes hot carrier stressing, we found that device with forward substrate bias will cause more serious subthreshold swing slope degradation with lower V_{TH} and off state leakage, which is more serious than the device with reverse substrate bias does, and this result is consisting with device stressed by various constant voltage.

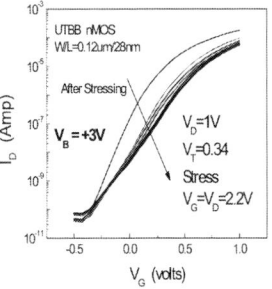

Figure 12 The hot carrier induced device's degradation in I_D -V_G for UTBB SOI nMOSFET with +0.3V substrate bias.

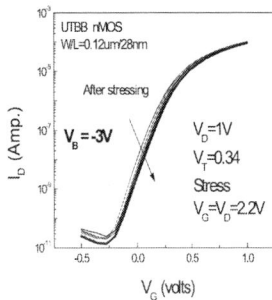

Figure 13 The hot carrier induced device's degradation in I_D -V_G for UTBB SOI nMOSFET with -0.3V substrate bias, gate leakage become more serious after 100 minute stressing.

After 100 minutes hot carrier stressing, we found that UTBB SOI nMOSFET with forward substrate bias (Fig. 14) shows more serious I_G degradation than the device with reverse substrate bias does (Fig. 15). The device's reliability can be predicted for 10-years device's characteristic of 10% ID degradation, as shown in Fig. 16. It shows that if the UTBB SOI device wants to achieve 10-years operation with 10% I_D degradation, lower dose UTBB device can sustain to 10 years easier than higher dose device does at same operation voltage

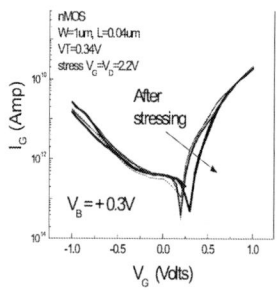

Figure 14 The hot carrier induced device's degradation in I_G-V_G for UTBB SOI nMOSFET with +0.3V substrate bias, gate leakage become more serious after 100 minute stressing.

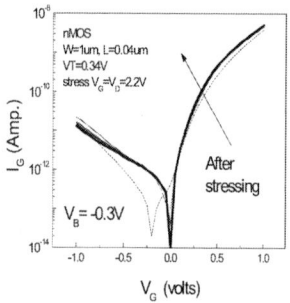

Figure 15 The hot carrier induced device's degradation in I_G-V_G for UTBB SOI nMOSFET with -0.3V substrate bias, gate leakage become more serious after 100 minute stressing.

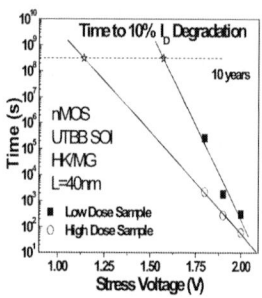

Figure 16 The operation voltage for 10% hot carrier induced I_D degradation in 10-years prediction, the operation voltage is around 0.5V in low dose nFET which larger than in high dose device.

IV. CONCLUSIONS

The impact of junction dose concentration on device characteristic variance and sensitivity to substrate bias has been studied. The higher junction dose UTBB SOI nMOSFET shows larger effective channel effect which improving device's subthreshold swing and driving capability, but suffers more devices characteristic variations and hot carrier induced device degradation. After constant-voltage stressing and hot carrier injection, the low junction dose device shows better reliability performance in hot carrier induced I_D and V_{TH} degradations than those of the high junction dose device. In addition, the low junction dose UTBB-SOI nMOSFET shows less device's characteristic variance and is also less sensitive to substrate back bias. Because of the substrate biasing induced junction barrier

lowering, UTBB SOI nMOSFET with a forward substrate bias shows more significant hot carrier stressing induced I_G degradation than the device with a reverse substrate bias.

ACKNOWLEDGMENT

This work was supported by the National Science Council under Contract NSC 102-2221-E-390 -023 -MY2, NSC102-2221-E-035-081 and the authors would like to thank UMC staff for their supporting

REFERENCES

[1] T. Sakurai *et. al.* , ISQED, p. 417, 2000.
[2] V. Barral *et. al.* , IEDM. p. 61, 2010.
[3] L. Clavelier *et. al.* , IEDM Tech. Dig. p.42, 2010.
[4] O. Weber *et. al.* , IEDM p. 1, 2008
[5] Y. Morita *et. al.* , Symposium on VLSI Technology, p. 166, 2008.
[6] T. Ohtou *et. al.* , IEEE EDL, Vol. 28, No. 8, , p. 740, 2007.
[7] K. Uchida *et. al.* , IEDM. p. 47, 2002.
[8] T. Ohtou *et. al.* , *IEEE EDL*, Vol. 28, No. 8, , pp. 740-742, 2007.

Palladium-Copper Inter-diffusion during Copper Activation for Electroless Nickel Plating Process on Copper Power Metal

Poo Khai Yee[1], Wan Tatt Wai[2], Yong Foo Khong[3]

Infineon Technologies (Kulim) Sdn Bhd

Lot 10 & 11, Jalan Hi-Tech 7, Industrial Zone Phase II,

Kulim Hi-Tech Park, 09000 Kulim, Kedah Darul Aman, Malaysia

Phone: +6044278741[1], +6044278079[2]

Email: Khaiyee.Poo@infineon.com[1], Tattwai.Wan@infineon.com[2], FooKhong.Yong@infineon.com[3]

Abstract - **Electroless Nickel deposition on Copper metallization required a thin Palladium layer (< 5nm) as catalyst for Copper activation purpose. By using several physical failure analysis approaches, Palladium-Copper inter-diffusion behavior was proven affected by flow rate variation of chemical solution during Copper activation process.**

I. INTRODUCTION

Copper has been replacing gold material in wire bonding of integrated circuits, due to lower manufacturing cost in microelectronic industries. However, there are still several technical challenges to be resolved. Copper wire bonding on aluminum bond pad has much lower shear strength. As copper has great hardness properties, higher ultrasonic energy during wire bonding is required. This aggravates the problem of aluminum pad deformation (Aluminum splash) and has high potential to damage the fragile structure underneath the bond pad. As such copper bond pad is nowadays preferred than aluminum bond pad due to its advantages on bond pad damage resistance for devices bonded with copper wire [1].

Copper has lower resistivity from electrical point of view compared to aluminum and is becoming the main material to be used as interconnect when the microelectronics devices are shrinking. As resistance is inversely proportional to the cross-sectional area, shrinking of interconnect dimension will increase the resistance and results in higher RC time delay in the device. Thus, the resistivity of interconnect material has to be decreased to achieve optimum time delay in the device. The bulk resistivity of aluminum ($2.65 \mu\Omega$cm) is higher as compared to copper ($1.68\ \mu\Omega$cm). The utilization of copper interconnect is expected to reduce the time delay for about 40% [2].

Copper enables higher current density application by minimizing heat generation and power consumption. When a metal conductor carries large current density, it is subjected to a failure mechanism well known as electromigration where diffusion phenomenon happens. Aluminum is very susceptible to electromigation damage due to its low melting temperature. As such diffusion is relatively easier in aluminum compared to copper [2,3]. Nevertheless, copper also offer significant reduction in manufacturing cost through electroplating process which involves electrolyte.

Nickel offers an attractive alternative to copper, since it is relatively harder than copper. Therefore, nickel is suitable to apply as an encasement on copper and follow by electroless palladium (Figure 1). Nickel and palladium metallization on copper not only able to provide additional mechanical protection for underlying structure, but it also able to provide great protection from corrosive environments. Nickel has very low porosity properties and under the optimal process condition, an uniform nickel layer can be produced [4]. Diffusion of nickel into copper is also very slow which provides highly stable metal interface which is not easily affected by high current density [1,3]. Palladium is typically selected as a thin protection layer for nickel due to its hard, nobility, corrosion resistance and strong mechanical properties. Besides, palladium thickness also needs to be sufficient to prevent inter-diffusion between nickel and wire bond.

Nickel is deposited on copper using a cost effective electroless method [5] which is also named as autocatalytic plating. It involves presence of a chemical reducing agent in solution to reduce metallic ions to metal state when two metals have different electrochemical potential. Electroless nickel plating is a high selectivity process and is able to intricate small dimension area with high degree uniformity without the need of photoresist mask [2,4].

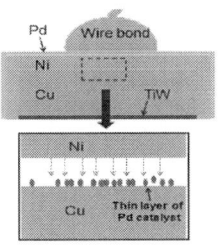

Fig 1. Schematic diagram of Cross Section view of Cu pad with Ni (top) and Cu activation process prior electroless Nickel deposition (bottom).

However, electroless nickel deposition on copper requires a thin palladium layer (< 5nm) as catalyst for activation purpose because copper is more electropositive or more noble than nickel. The activation is undergoing chemical oxidation-reduction reaction when copper surface is immersed into an acidic chemical solution which consists of palladium (Pd) ions. When copper is immersed in a palladium chemical solution, the copper atoms (less noble) are dissolved and spontaneously replaced by palladium atoms from solution. The two autocatalytic reactions can be represented as follow:

$$\text{Oxidation: } Cu^0 \rightarrow Cu^{2+} + 2e^-, E^0 = 0.34V$$
$$\underline{\text{Reduction: } Pd^{2+} + 2e^- \rightarrow Pd, E^0 = 0.99V}$$
$$\text{Overall reaction: } Cu^0 + Pd^{2+} \rightarrow Cu^{2+} + Pd, E_{all} = 1.33V$$

In generally, electroless nickel plating requires the following steps: cleaning, catalyzing and activation with various chemical solutions (Figure 2). Rinsing is generally required between the steps. The surface preparation is very crucial as the contamination may create inactive surface and will not initiate electroless deposition. Many factors may impact the quality of electroless nickel plating such as surface cleaning, temperature, pH and concentration of chemical bath and tool hardware applied [5]. Small variations during electroless plating process may have big impacts on the following step and negatively influence the bonding performance. In this paper, the behavior of palladium catalyst is found to be very crucial for the electroless nickel deposition process. The correlation between chemical solution flow dynamic during copper activation process and the palladium-copper inter-diffusion anomaly is investigated using several physical failure analysis approaches.

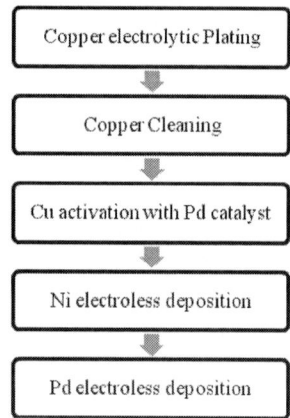

Fig 2. General process flow of copper electrolytic plating and subsequent electroless deposition of Ni and Pd.

II. EXPERIMENTAL SETUP

Experiment was first carried out using a manual beaker method in room temperature condition. Two copper plated blanket wafers were immersed into chemical solution containing Pd ions with two different facing directions, i.e. one wafer facing upward and another wafer facing downward into the chemical solution. Wafer was supported by two parallel rods in the chemical solution. A magnetic stirrer was placed in every beaker of chemical solution in order to get a different flow rate condition during copper activation process. Stronger flow dynamic (high flow rate) is expected on copper wafer which is facing down as compared to copper wafer which is facing up. This is because the flow rate is weakened by transfer the circulation force from the bottom to the top side of the beaker (Figure 3). Wafers were taken out for inspection after 3 minutes of reaction time. Discoloration with dark stain was observed on copper wafer surface under high flow rate environment and no abnormality (stainless) was found on copper wafer surface under low flow rate environment as shown in Figure 4. This observation had indicated the solution flow dynamic did affect the copper activation process. These two blanket wafers were subsequently analyzed using Auger Electron Spectroscopy (AES) technique; however the result revealed no significant elemental differences on wafer surface.

Similar experiment undergoing same copper activation and flow dynamic were set up using copper structured wafers with Titanium Tungsten (TiW) as barrier structure underneath. On few sample wafers, an additional wet acidic chemical etching was applied after copper activation to both wafers experiencing different rate to partially remove copper and expose boundary of Titanium Tungsten underneath and subsequently, proceed with electroless Nickel deposition. Wafers upon wet etching and after electroless Nickel deposition were inspected using several physical material characterization methods.

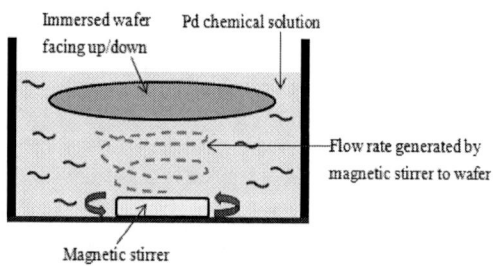

Fig 3. Manual beaker experiments set up using magnetic stirrer with Cu wafer facing downward or upwards.

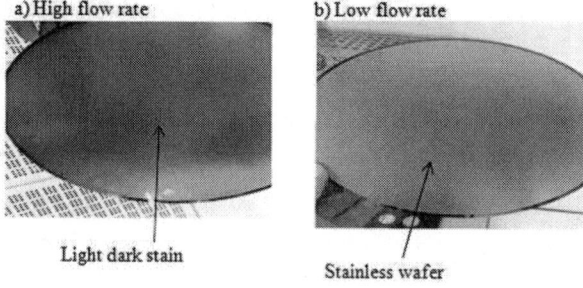

Fig 4. Cu wafer (a) with high flow rate observed discoloration with dark stain after Cu activation process but Cu wafer with slow rate was stainless.

III. RESULT AND DISCUSSION

Copper structured wafers sample without nickel layer were inspected using Scanning Electron Microscopy (SEM) technique upon partial removal of copper layer. Flakes type residue partially adhering to the boundary of TiW barrier and laying on copper surface was found on wafer from high flow rate environment whereas relatively clean surface without residue was observed on wafer from low flow rate environment (Figure 5). Subsequent Energy Dispersive Spectroscopy (EDS) analysis confirmed the flake residue to be Cu-Pd interlayer residue (Figure 6). Apparently this interlayer residue was induced during Cu activation process under high flow rate environment. The residue had turned into Cu-Pd metallic compound and was not able to be removed by copper wet etchants (Figure 7).

Structured wafers going through the electroless nickel deposition process was investigated with Scanning Transmission Electron Microscopy (STEM) technique. The sample was prepared using Focus Ion Beam (FIB) to produce a thin cross section of lamella. STEM dark field imaging observed massive voids at the interface of Cu and Ni layer on wafer sample from high flow rate environment. EDS elemental line profile carried out across the massive voids area revealed presence of ~100nm Cu-Pd(catalyst)-Ni layer inter-metallic diffusion region. This diffusion was very minimum (< 5nm) on wafer sample from low flow rate environment (Figure 8& 9).

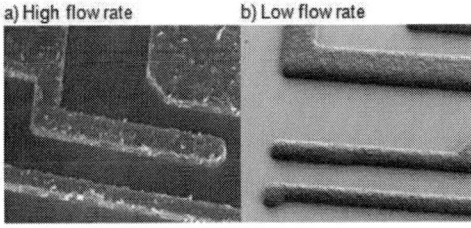

Fig 5. SEM inspections in tilted view on Cu structure wafer which went through high flow rate and low flow rate Cu activation after additional copper wet etching process.

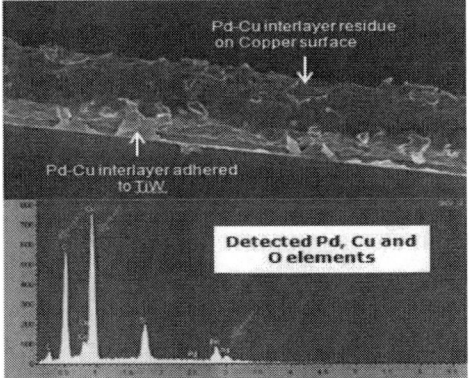

Fig 6. Result of SEM and EDS on high flow rate sample after applied additional copper wet etching.

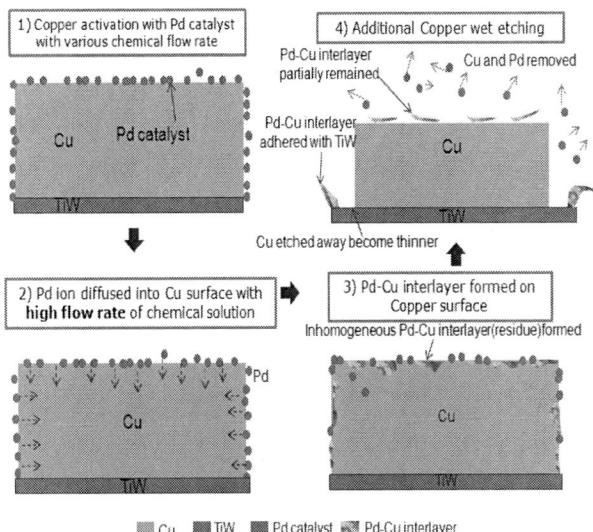

Fig 7. Mechanism of Pd-Cu interlayer formation on wafer from high flow rate environment prior to electroless nickel deposition.

Based on the two analysis result outlined in previous section, it's predicted that Pd-Cu inter diffusion had occurred and at the same time resulted in near surface void on wafer from high flow rate environment prior to electroless nickel deposition. This created an area which was prone to nickel diffusion upon electroless deposition process. These voids would have been enlarged under high current application; it also produces an easy path for crack propagation as ductility of material has decreased. In summary high flow rate during copper activation process should be avoided to prevent reliability issue.

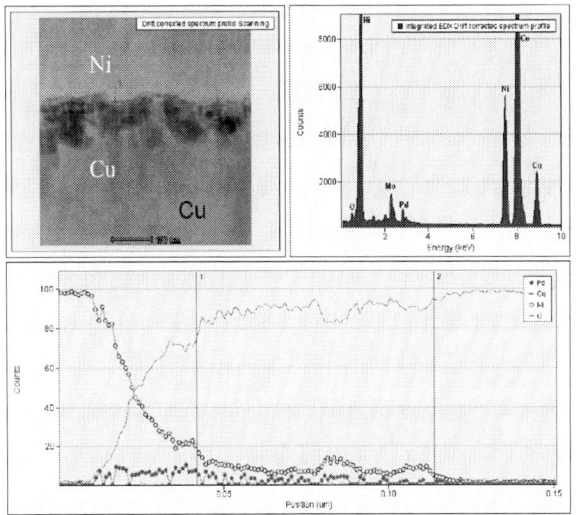

Fig 8. STEM-EDS result revealed the metallic inter-diffusion occurred between Ni, Pd and Cu near to metal void area on high flow rate sample (top).

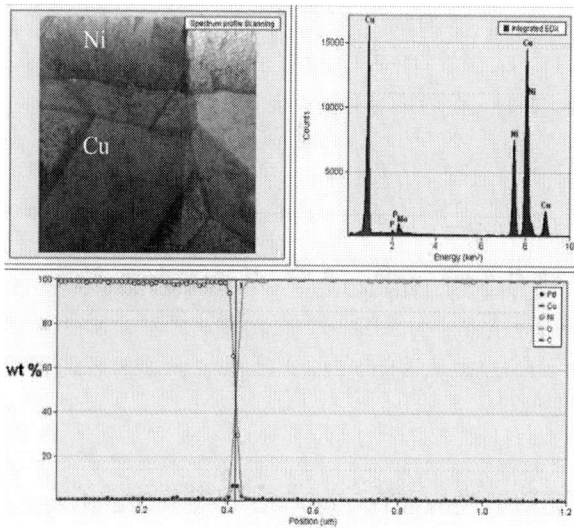

Fig 9. STEM-EDS result revealed Pd-Cu interlayer was negligible on low flow rate sample (bottom).

IV. CONCLUSION

The advantages of using copper as interconnect metallization was discussed with respect to excellent shear strength; reduced bond damaged its reliability impact. Copper activation and its subsequent process condition were very crucial and could potentially cause reliability drawbacks as observed in the analysis result. Several physical failure analysis approaches had successfully proven that palladium catalyst ion diffusion was caused by high flow rate during copper activation process.

ACKNOWLEDGMENT

The author would like to thank the Unit Process Development Staff Engineer Dr. Wan Tatt Wai, Yong Foo Khong and FA lab manager Kenny Gan Chye Siong for their support and willingness to share valuable technical discussion.

REFERENCES

[1] Clauberg. H et al, "Nickel-palladium bond pads for copper wire bonding", Microelectronics Reliability 51, 2011, pp75-80.

[2] Uziel. L, "Copper Metallization of Semiconductor Interconnects, Issues and Prospects," Chemical Engineering Department and The Yeager Center for Electrochemical Sceiences Case Western Reserved Univeristy (Cleveland), 2000.

[3] Lloyd. JR, "Reliability of Copper Metallization", Lloyd Technology Associates,Inc, 1998.

[4] Parkinson. R, "Properties and application of electroless nickel" Nicke Development Institute, 1997.

[5] M. Schlesinger et al, Electroless Depostion of Nickel, Modern Electroplating, 5th Edition (John Wiley &Sons, Inc, 2010), Chapter 18, pp 447-458.

SRAM Failure Analysis Evolution Driven by Technology Scaling

Zhigang Song

IBM Systems and Technology, 2070 Route 52, Hopewell Junction, NY, USA
Phone: (01)8458921691 Fax: (01)8458926094 Email: zsong@us.ibm.com

Abstract - **Demand for high speed and more function microelectronic devices has driven semiconductor industry to continue developing technologies with ever-shrinking geometry. During technology development, Static Random Access Memory (SRAM) is often chosen as the process qualification and yield learning vehicle. Thus SRAM failure analysis is the major activity in any microelectronic device failure analysis lab. Conventional physical failure analysis in old technology nodes has achieved high success rate since the SRAM bitcell failures can be precisely localized by functional test and the defect causing such failures is within the failing bitcells. However, As SRAM feature size decreases with technology scaling down, the size of the defect causing SRAM failure also scales down. Some of the defects are so tiny that they are invisible in ultra-high resolution SEM. On the other hand, the SRAM bitcell number greatly increases, and thus the SRAM design, especially address decoder scheme becomes more complex. More and complicated SRAM logic type failures arise. Therefore, the conventional physical failure analysis has faced increasing challenges and encountered low success rate. This paper will talk about how SRAM failure analysis evolves to maintain high success rate.**

I. INTRODUCTION

Demand for high speed and more function microelectronic devices has driven ever-shrinking semiconductor technology. Static Random Access Memory (SRAM), which serves as caches in microprocessor or System on Chip (SoC) device adopts even more aggressive design rules than logic circuitry in the same technology generation. Thus, SRAM is more prone to manufacturing defects and more sensitive to process variation. For this sake, SRAM is often chosen to be the process qualification vehicle during technology development or yield learning vehicle during product manufacturing, and consequently failure analysis of SRAM is the main feedback for process improvement and yield learning [1, 2].

SRAM is organized as array structure of single bit storage cells. Each cell can be selected and access through address decoder logic circuitry and other circuitry, like sense amplifier, multiplexer, driver and control logic. A SRAM operates in such a way that the row decoder selects a row, also known as WL, by setting the WL at logic high for all the cells along the selected row. The contents of these cells become available to be transferred to their respective bitline and bitline-bar. The column decoder is then used to select the particular column, also known as BL, to write or read the content of the cell at the

intersection point of the selected row and column. The data in the bitcell are ultimately routed to drive an output pin. Originating from the array structure, SRAM failures appear in different patterns, the simplest ones are bitcell failures, such as single bit failure, pair bit failure or quad bit failure etc. For those cases, the defect causing the failure is located exactly within the failing cell or at the shared stacked interconnects for pair bit failure or quad bit failure. Besides the simplest bitcell failures, there are other failures with various regular spatial patterns, like BL type, WL type and entire block failure. These failures are called SRAM logic type failures because they are usually caused by a defect within the SRAM address decoder logic circuitry. With the advent of hierarchy decoder scheme in modern SRAM designs, the SRAM logic type failures are becoming even more complex.

As technology keeps shrinking, SRAM failure analysis is increasingly challenged. For addressing ever-decreasing success rate, not only physical failure analysis, but also the defect localization methodology, especially for SRAM logic type failure, need to evolve.

II. EVOLUTION IN PHYSICAL FAILURE ANALYSIS

As semiconductor technology scales down, interlayer dielectric (ILD) thickness decreases and new materials are introduced, which have resulted in evolution in physical failure analysis. At quarter micron and above technology nodes, the interconnect metallization is aluminum with aluminum or tungsten via. At that time, the main sample de-processing methodology is etch, namely dry etch with plasma or Reactive Ion Etch (RIE) to remove ILD and wet chemical etch to remove metallization. When the metallization was exposed by dry etch, Scanning Electron Microscope (SEM) secondary electron (SE) imaging was employed to inspect the sample and image the defect. This de-processing methodology with pure etch easily made the sample dirty, often causing artificial defect, which would be carried on in the subsequent layers by micro-mask mechanism of dry etch. When copper metallization replaced the aluminum metallization in 0.13um technology, the commonly used de-processing methodology was half etch and half polish, namely parallel polishing to remove metallization and RIE or plasma etch to remove ILD. Although any particle introduced during polish step caused an artificial defect after dry etch, it would not likely be carried on in the next layer. The SEM inspection and imaging still employed secondary electrons since the inspected metallization was exposed. As technology further scaled down into nanometer region and low k materials, like SiCOH, was introduced as the ILD insulator films, the de-processing methodology with pure parallel polish was adopted and the SEM inspection and imaging more often used backscattered electrons (BSE) than secondary electrons. This pure polish methodology greatly reduced the likelihood of causing artificial defect and also improved the efficiency of sample de-processing. Why BSE imaging is superior to the SE imaging for nanometer technology nodes with the pure polish de-processing methodology? It is because of the different characteristic of BSE and SE, and also the sample morphology after polish, namely the metallization is still embedded in ILD with a smooth surface. SEs generate from inelastic electron scattering event caused by the interaction between the sample's electrons and the incident electrons and thus SE usually has less than 50eV of energy, so that only the SEs within few nanometers depth below the sample surface can escape from the sample and reach the detector to form an image [3]. As a result, the SE imaging has high resolution for sample surface, but the information depth is shallow (only a few nanometers). While the BSEs generate from wide angle Rutherford scattering events resulting from the interaction between the incident electrons and the sample atomic nuclei [4], and thus most BSEs have energies not too far below the incident electron energy and can escape from a depth of up to about half the penetration depth below the sample surface [5]. It enables the BSE imaging to have deeper information depth than SE imaging, and image the features from significant depth within the sample [6]. Furthermore, the contrast of BSE image is mainly atomic number contrast, namely Z contrast [7]. These features make BSE imaging more advantageous than SE imaging to image a subtle defects embedded in dielectric with smooth surface. For advanced semiconductor integrated circuits, there are a lot of small features with high Z materials, such as Cu with Ta/TaN, W, or NiSi on poly and active region immersed in a low Z matrix, such as SiCOH, or SiO2, which thickness is well within the BSE information depth. For example, the contact dielectric thickness is approximately 200 nm for 32nm technology node. When the sample is prepared with parallel polish without any etch, the possible defect is often embedded in dielectrics with smooth surface. For imaging such defect, long information depth and atomic number (Z) contrast make BSE imaging be the better choice [8]. Fig. 1 and Fig. 2 show the SE and BSE images at contact level for a singe cell failure, respectively. Fig. 1 does not show any abnormality, while, Fig. 2 clearly shows a poly stringer, which causes the local poly and Vdd contact short, leading to a single bit failure. The poly stringer, which was likely silicided, had higher mean atomic number than the surrounding materials so that it showed bright contrast in the BSE image. Localizing and imaging them before exposing them will avoid the risk of destroying the subtle defect by dry or chemical etch, and also keep the defect intact and well wrapped for follow up analysis like Cross-sectional Transmission Electron Microscopy (XTEM) analysis.

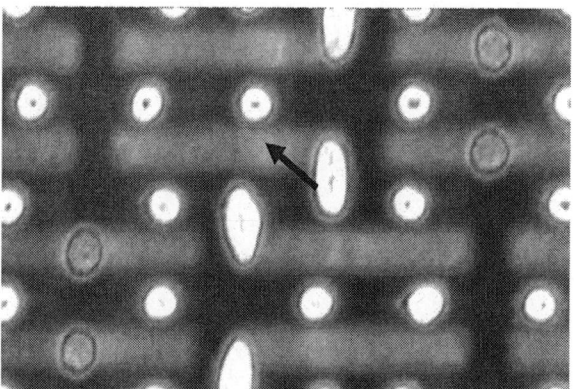

Fig. 1: SE image at contact level of a SCF showing normal

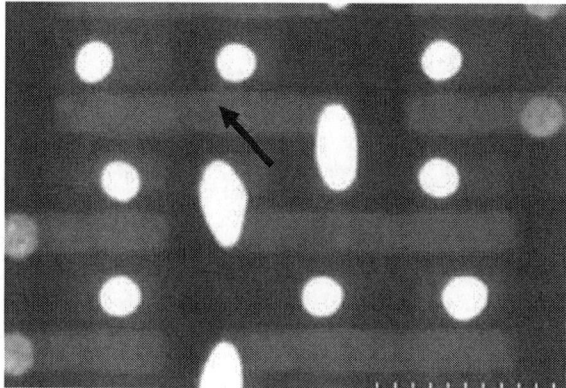

Fig. 2: BSE image at contact level of the same SCF showing a poly stringer

III. EVOLUTION IN ELECTRICAL FAILURE ANALYSIS

SRAM bitcell failures are the simplest and also the most analyzed ones. They can be precisely localized by functional test and the defect causing such failures is within the failing bitcells. In old technology nodes, such as submicron technology, conventional physical failure analysis usually employed Passive Voltage Contrast (PVC) technique [9], SEM inspection, Focus Ion Beam (FIB) cross-section and TEM analysis to locate and image the defect and has achieved quite high success rate. However, with technology scaling down, SRAM feature size and power supply voltage decrease, resulting in increased transistor mismatch and reduced functional marginality. Thus, the size of the defect causing SRAM failure also decreases. Some of these defects are so tiny that they are invisible in ultra high resolution SEM, or even TEM if the exact defective location is not known. Relying only on conventional physical failure analysis resulted in reduced failure analysis success rate. In order to increase the success rate, more sensitive and accurate fault isolation is required before TEM analysis. Electrical nano probing, either SEM-chamber based probe [10, 11] or Atomic Force Probe [12, 13] can characterize individual transistor in a SRAM failing bitcell and in a reference bitcell. Comparing the transistor characteristics in failing bitcell and reference bitcell, it is possible to pinpoint the defect to a single source, drain or gate stack terminal. It greatly increases the likeliness of finding a

978-1-4799-3911-4/14 $31.00 © 2014 IEEE

visible defect in subsequent TEM analysis, thus improving the failure analysis success rate. Furthermore, AFP has other two useful applications often used for initial fault isolation before electrical probing. Pico-current imaging [14, 15] can detect leakage or high resistance issue with much more sensitive than passive voltage contrast and Nanoprobe Capacitance Voltage Spectroscope (NCVS) [16, 17] imaging can pick up any abnormal response of capacitance change with voltage. Nowadays, electrical failure analysis is becoming an important step for SRAM failure analysis and its weight will further increase with the foreseeable smaller bitcell. It suggests that the SRAM failure analysis team not only has skill set in physics, chemistry and materials background for characterizing defects, but now must also include microelectronic and semiconductor device background skills for device analysis.

A. Single Cell Failure

During 28nm technology development, several lots showed Single Cell Failure (SCF) preferentially at end of BL. Conventional physical failure analysis did not fully identify the root cause. One SCF was de-processed to contact level and submitted to AFP probing. AFP probing found a defective pull-down transistor after comparing the Id and Is versus Vg semi-log plots of the defective pull-down transistor and a reference pull-down transistor in a good bitcell at linear mode (Vd at 0.05V) and saturation mode (Vd at 1.0V) respectively, as shown in Fig. 3. These plots indicated that both Ioff and Ion for the defective pull-down transistor were lower than those for a reference pull-down transistor at both linear and saturation modes. Comparing the measurement results with source drain swap, namely ground was set as source or as drain, had found that the defective pull-down transistor was asymmetric at saturation mode. It suggested that either the ground side or the internal node side of the defective pull-down transistor had high resistance. To further identify which side had high resistance, the Id and Is versus Vg of the defective pull-down transistor and a reference pull-down transistor at saturation mode was plotted linearly, as shown in Fig. 4. These plots pointed to a high resistance at the ground side. When the ground side was set to source during AFP probing (see Fig. 4a), the voltage dropping on the R (representative of high resistance) brought the Vs potential to IsR, which reduced the Vgs to Vg-IsR, and thus it required higher Vg, appearing as higher Vt, to turn on the transistor. When the ground side was set to drain during AFP probing (see Fig. 4b), in this configuration, the R was at drain side, the R limited the Ion increasing even when the transistor was fully turn-on. The R was estimated to be around 0.1 Mohm. After the side with high resistance was identified, the sample was submitted to XTEM analysis. XTEM image, as shown in Fig. 5 clearly showed poor NiSi formation under the contact at the high resistance side (see Fig. 5a), comparing to good NiSi formation at a reference site (see Fig. 5b).

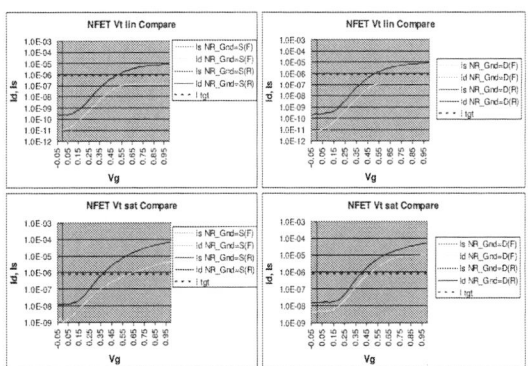

Fig. 3: The Id and Is versus Vg semi-log plots of the defective pull-down transistor and a reference pull-down transistor in a good bitcell at linear mode (Vd at 0.05V) and saturation mode (Vd at 1.0V), respectively

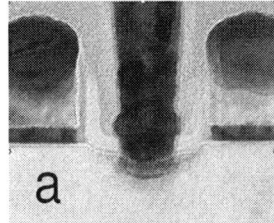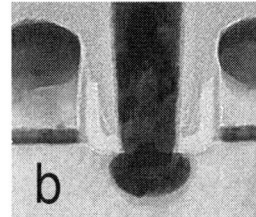

Fig. 4: The Id and Is versus Vg linear plots of the defective pull-down transistor and a reference pull-down transistor at saturation mode

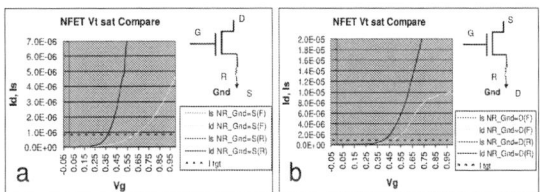

Fig. 5: XTEM images showing poor NiSi formation at the high resistance site (a) and good NiSi formation at a reference site (b)

B. Soft Horizontal Pair Bit Failure

Several 32nm wafers suffered from low yield due to soft Horizontal Pair Bit Failure (HPF). Conventional physical failure analysis did not find any defect. One sample with the soft HPF was submitted to AFP probing at contact level. First, NCVS analysis was tried and it was found that the contrast of the pass-gate contact of the HPF was slightly brighter than other pass-gate contacts for good bitcells in the NCVS image, as shown in Fig. 6. Then both pass-gate transistors were probed and it was found that both of them had around 100mV higher Vt than a pass-gate transistor in a good bitcell. Fig. 7 is the Id and Is versus Vg semilog plots for a failing pass-gate transistor and a reference one at saturation mode. It was different from previous case, since the failing pass-gate transistor was symmetric and no low Ioff were observed. It was pure Vt shift issue, implying that the issue was related to the gate stack (well/GOX/poly). However, in this case of HPF, both pass-gate transistors showed the same electrical signature, the follow-up XTEM analysis was focused on the poly and contact stack. XTEM image, as shown in Fig. 8, did not show any physical defect. However, EDX spectrum of line scan crossing the contact-poly stack showed that As was not

978-1-4799-3911-4/14 $31.00 © 2014 IEEE 225

diffused properly in the failing pass-gate transistor (see EDX spectrum at right side of Fig. 8). The area with less implant concentration could form a depletion layer when voltage was applied on the poly gate, leading to higher Vt.

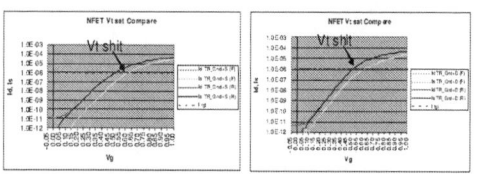

Fig. 6: AFP NCVS image showing slightly brighter contrast with the contact for the HPF

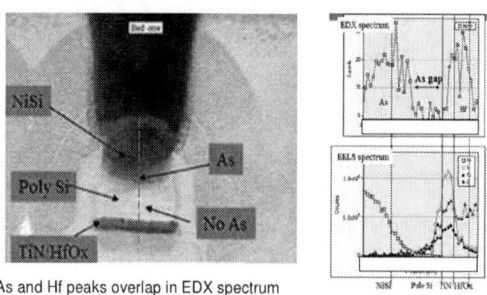

Fig. 7: The Id and Is versus Vg semilog plots for a failing pass-gate transistor and a reference one at saturation mode

As and Hf peaks overlap in EDX spectrum
Left is As signal and right is Hf in EDX spectrum

Fig. 8: XTEM image of the contact-poly stack and EDX line scan of this stack for the HPF

IV. EVOLUTION IN LOGIC TYPE FAILURE ANALYSIS

Besides bitcell failures, there are some SRAM failures with various regular spatial patterns, called SRAM logic type failures, for instance, regular repeated bitline (BL) failure, regular repeated word line (WL) failure or entire block failure. For such failures, the defects causing the failures are often not at the failure location, thus, failure analysis for them is more difficult. In order to cope with the challenge, layout tracing [18], Photoemission Microscope (PEM) analysis [19] and laser stimulation techniques such as OBIRCH, TIVA [20, 21], CPA [22], LVP [23] or LVI [24, 25] etc techniques or tools, which are traditionally employed in logic product failure analysis, are now used prior to SRAM physical failure analysis. It is also worthwhile to mention that as modern SRAM design includes Jtag and scan chain concepts, failure analysis on SRAM logic type failures is becoming indispensable for SRAM non-zero yield breakthrough at the beginning stage of technology

development. With this evolution, logic diagnostic toolset are used in addition to physical failure analysis toolset and SRAM bitmapping system.

A. WL Type Failure

A 48M SRAM chip suffered from a WL type failure within sub-block 192BLx1024WL. Fig. 9 shows a chip level failure bitmap and the WL type failure pattern to be analyzed, as shown at right side. It was noticed that there are 16 solid WLs with regular interval. Considering an one fault only assumption, the defect should be within the WL decoder circuit, which controls these 16 WLs, leading to hard failure for the 16 WLs. The other partially failing WLs were likely secondary effect due to interference among WLs. So, the failure analysis focused on the hard failure of the 16 WLs. After looking at the 16 WLs more carefully, it was found that each failed WL in 64 WL group was 20 WLs from the boundary and with mirror image in every two 64 WL groups. The WL type failure pattern is graphically showed in Fig. 10. Based on the failure pattern, layout tracing was performed starting from every failing WL and tracing up to WL decoder. Layout tracing from all 16 failing WLs ended up at the same WL decoder. The traced layout was highlighted and is shown in Fig. 11. It was found that the WL decoder was located at the middle of the sub-block 192BLx1024WL and two signal lines from the WL decoder were routed to top and bottom sub-blocks 192BLx512WL respectively. According to one fault only assumption, the defective area causing all 16 WL failure in the sub-block 192BLx1024WL was highly suspected in the WL decoder area, which is showed at the right side of Fig. 11. After the suspected defective area was isolated, the chip was de-processed layer by layer and SEM inspection was focused on the WL decoder area. Finally, distorted and bridging poly lines were observed, as shown in Fig. 12. To further understand the root cause, the defect was subjected to FIB cross-section. FIB cross-sectional image in Fig. 13 showed that the defective poly was abnormally higher than normal poly and its bottom was shorted to contact. It was hypothesized that a big poly silicon particle dropped here during poly deposition, which then affected the subsequent poly patterning.

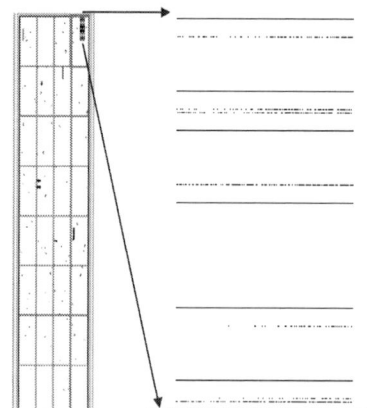

Fig. 9: A chip level failure bitmap with zoom-in image of a WL type failure within sub-block 192BLx1024WL at right side

978-1-4799-3911-4/14 $31.00 © 2014 IEEE
226

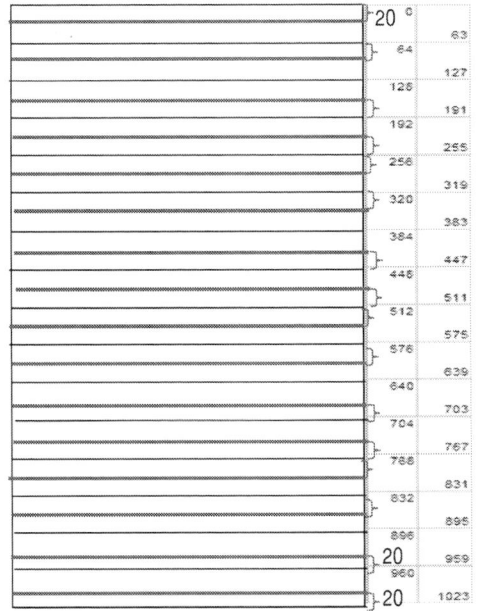

Fig. 10: The 16 WL hard failure pattern in a sub-block 192BLx1024WL

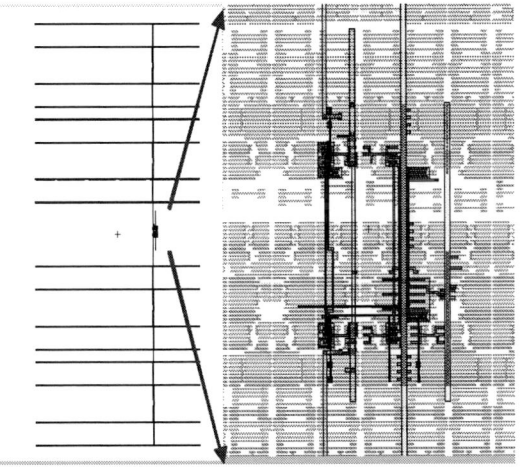

Fig. 11: The layout tracing results with the suspected defective WL decoder shown at right side

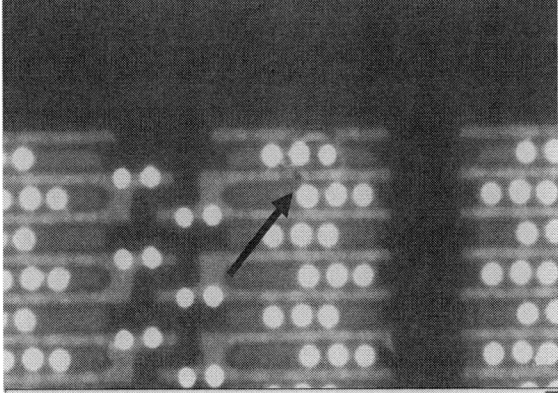

Fig. 12: SEM image showing distorted and bridging poly pattern

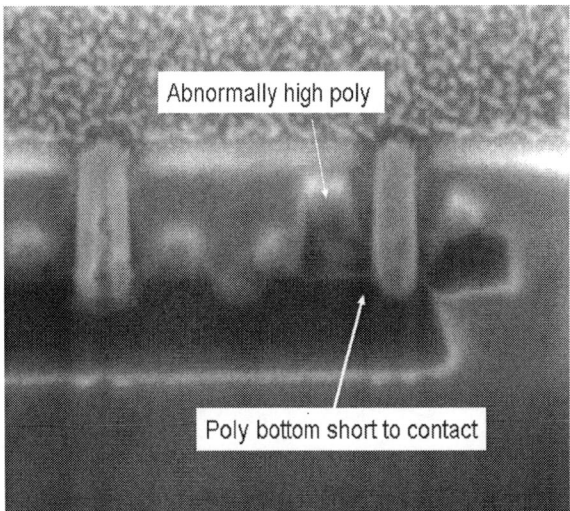

Fig. 13: FIB cross-sectional image showing abnormally high poly and poly bottom short to contact

B. BL Type Failure

Another 48M SRAM chip suffered from BL type failure within sub-block 192BLx512WL. Its failure bitmap pattern is shown in Fig. 14. Analysis of the failure pattern found that there were two solid BLs (BLs 3 and 4) failing at middle of every 8 BL group, as shown at right side of the Fig. 14. Similarly, the other partially failing BLs were secondary effect due to interference among BLs. Based on the failure pattern, layout tracing started from these solid BLs and traced up to a BL decoder. Fig. 15 shows the layout tracing result, which suggested that there were three possibilities causing the reported failure: (1) the signal line shorting to adjacent lines, like Vdd and Vss, (2) the signal line having open interconnect from the BL decoder, (3) a BL decoder malfunction. Bearing these suspected defects in mind, the subsequent physical failure analysis found an undersized contact within the highlighted BL decoder by layout tracing, as shown in Fig. 16 SEM image. FIB cross-section on the undersized contact showed that the contact was actually under-etched (See Fig. 17).

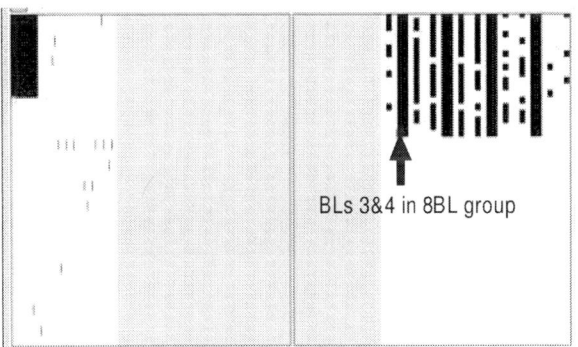

Fig. 14: The BL type failure pattern with BLs 3 and 4 hard failure in every 8BL group in a sub-block 192BLx512WL

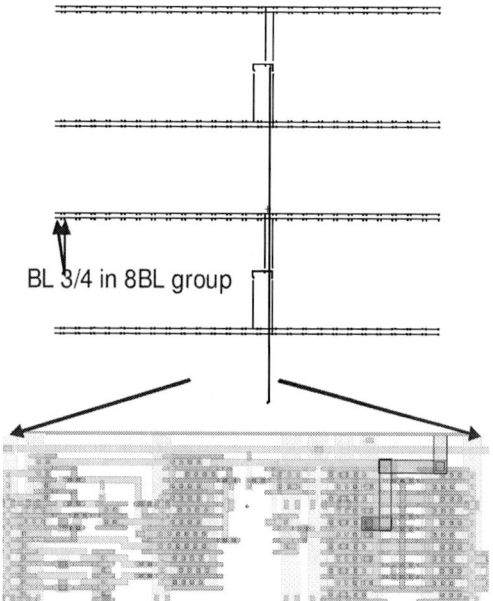

Fig. 15: The layout tracing results with the highly suspected defective BL decoder shown at bottom

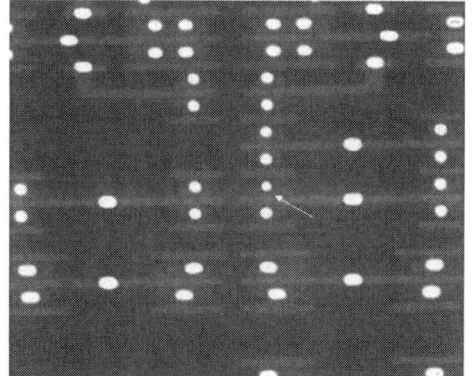

Fig. 16: SEM backscattering electron image showing an undersized contact

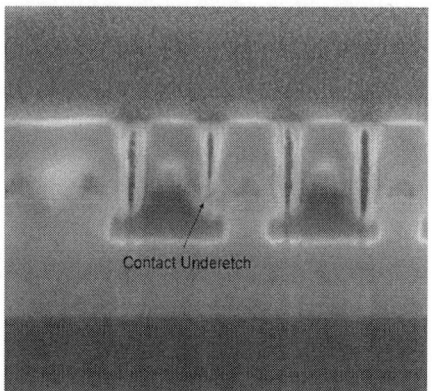

Fig. 17: FIB cross-sectional image of the undersized contact showing contact under-etch

C. Soft BL Type Failure

Another case of BL type failure showed a different failure pattern within sub-block 192BLx512WL, as shown in Fig. 18.

For this case, no solid BLs were observed and they failed at low voltage only, namely it was a soft failure. Although layout tracing could be used to identify the responsible BL decoder, Critical Parameter Analysis (CPA) is a fault isolation method that can be advantageously used to precisely localize soft failures. CPA (Similar to SDL, Soft Defect Localization [26, 27]) is a technique which scans a laser beam across the chip when it is under repeated test at the pass/fail boundary for a given critical parameter. CPA can be labeled with "positive" or "negative" depending on the passing or failing condition change of the device under test (DUT) from the laser stimulation. If the DUT is operating in a failing condition, and the laser stimulation causes it to pass, the CPA is labeled with "positive". A positive CPA spot is referred to the site where laser illuminates on to make the DUT pass from failing condition. For this case of soft BL type failure, positive CPA analysis found a CPA spot in a BL decoder (see Fig. 19). Follow-up AFP probing on several NFETs and PFETs near the CPA spot found a leaky PFET, as labeled with P6 (see Fig. 20). Fig. 21 is the Id and Is versus Vg semilog plots of the leaky PFET and a reference PFET at linear and saturation modes with different source drain setting respectively. These plots suggest that the leaky PFET had leakage path between source and drain and the leakage current increased with drain voltage. Why a leaky P6 caused a positive CPA spot at N6? It was because N6 and P6 actually formed a voltage divider, as shown in the schematic shown in Fig. 22. When laser illuminated N6, its Ion was higher than normal and compensated the leakage in P6, making the circuit to pass from its initial failing condition. The subsequent XTEM analysis on the leaky PFET showed Si dislocation at the interface of channel SiGe and Si (see XTEM image in Fig. 23), responsible for the source-drain leakage.

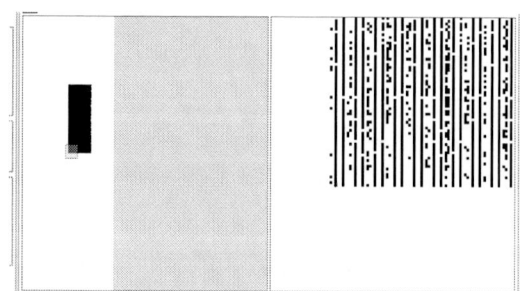

Fig. 18: Another BL type failure pattern without any BL hard failure in a sub-block 192BLx512WL

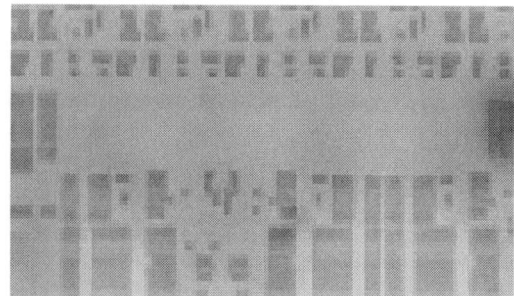

Fig. 19: CPA analysis showing a CPA spot in a BL decoder

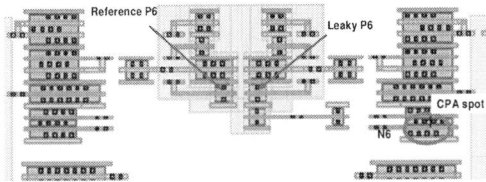

Fig. 20: Layout view with the leaky PFET and CPA spot labeled

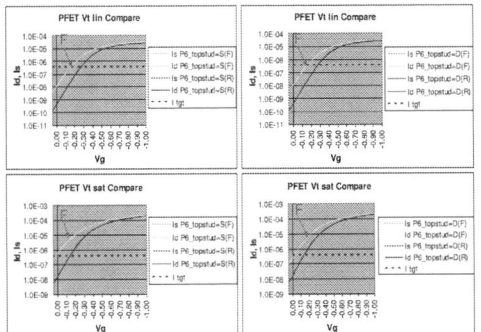

Fig. 21: The Id and Is versus Vg semilog plots of the leaky PFET and a reference PFET at linear and saturation modes with different source drain setting respectively

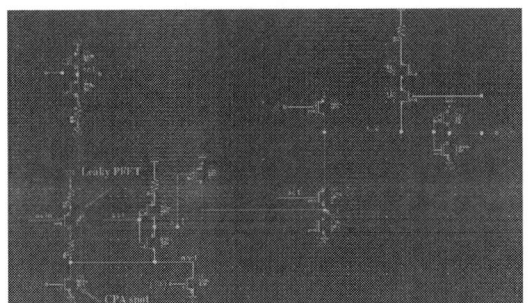

Fig. 22: The BL decoder schematic with leaky PFET and CPA spot (NFET) labeled

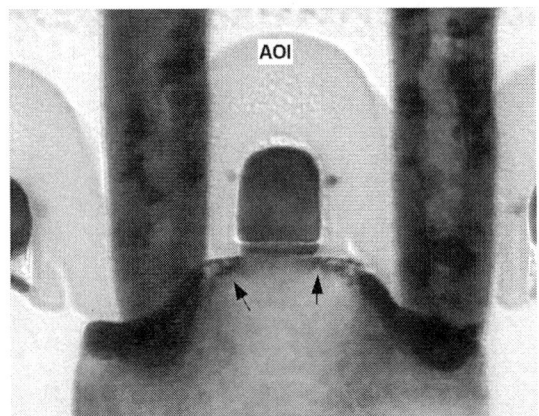

Fig. 23: XTEM image of the leaky PFET showing Si dislocation at the interface of the channel SiGe and Si

D. Entire Block Failure

Several 8M SRAM wafers suffered from zero yield with major issue of entire block failure at the beginning stage of technology development. Fig. 24 shows a typical wafer map with failure bitmap for a chip at right side. For this case, no

WL or BL pattern was able to be identified. Thus, without the assistance of the SRAM designer, layout tracing was impossible. In order to narrow down the suspected defective area for physical failure analysis, three failing chips were subjected to PEM analysis. PEM analysis on three failing chips showed emission at a repeated particular circuit (See Fig. 25). PEM was also done on a reference chip and no emission was detected at the same given circuit. The 100X PEM image showing the emission at the particular circuit is displayed in Fig. 26. It implied that a systematic defect in this circuit caused the entire block failure. Based on the PEM analysis, subsequent physical analysis found poly broken at all the emission sites. Fig. 27 is the SEM image of an example of the poly broken.

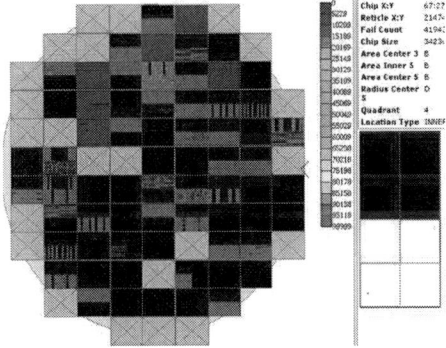

Fig. 24: A wafer map showing lots of entire block failures with one chip-level failure bitmap at right side

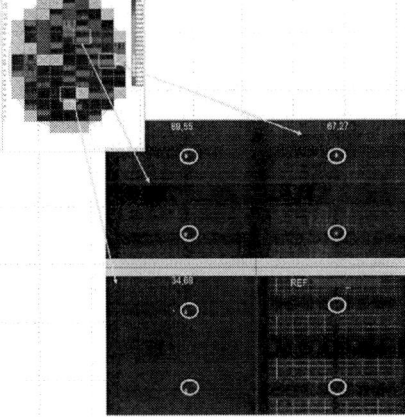

Fig. 25: PEM image showing emission sites at a repeated particular circuit for three failing chips and no emission at the same circuit for a reference chip

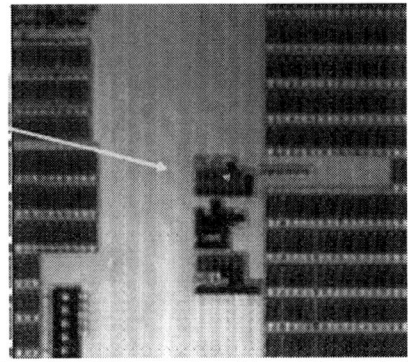

Fig. 26: A 100X PEM image of one of the emission sites in Fig. 25

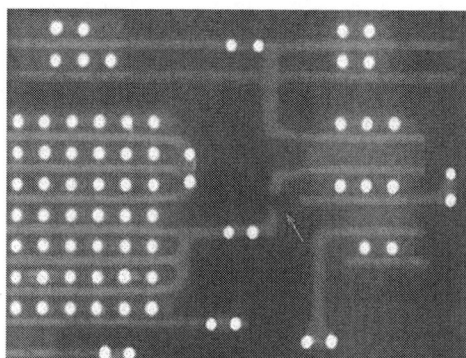

Fig. 27: SEM image showing an example of poly broken

V. CONCLUSIONS

As semiconductor technology keeps scaling down, SRAM failure analysis has experienced evolution to maintain the high success rate achieved with conventional physical failure analysis methodology in old technology nodes. This evolution occurred in physical failure analysis methodology, electrical failure analysis and SRAM logic type failure analysis. Keeping abreast of this evolution has required expanding the toolset of SRAM failure analysis with some tools traditionally employed in logic product failure analysis. Finally, in-depth electrical and device knowledge is now required to perform successful SRAM failure analysis.

ACKNOWLEDGMENTS

This work was performed at Semiconductor Research & Development Center, IBM Microelectronics Division, Hopewell Junction, New York 12533. The author would like to thank Andrew Dalton, Felix Beaudoin, Stephen Lucarini, John Sylvestri, Laura Safran, Terry Kane, Sweta Pendyala, David Albert, Michael Tenney, Manuel Villallobos and Richard Oldrey, Yunyu Wang, Jinghong Li, Thuy Su and Aaron Shore for their help in one way or another.

REFERENCES

[1] S. Naik, F. Agricola, W. Maly, "Failure Analysis of High Density CMOS SRAMs", IEEE Des Test Comput, 10(2), P. 13 (1993).

[2] F. Beaudoin, S. Lucarini, F. Towler, S. Wu, Z. Song, D. Albert, L. Safran, J. Sylvestri, G. Karve, X. Yu, E. Kachir and N. Jungmann, "Challenges and Benefits of Product-Like SRAM in Technology Development", ISTFA' 2011 Proceeding, P. 362 (2011).

[3] T.E. Everhart, O.C. Wells and C. W. Oatley, "Factors Affecting Contrast and Resolution in the Scanning Electron Microscope", J. Electron Control, 7 (1959), pp. 97.

[4] Niedrig, "Scanning Electron Microscopy", SEM Inc., Chicago, IL, (1978), pp. 841.

[5] L. Reimer, "Scanning Electron Microscopy: Physics of Image Formation and Microanalysis", Springer Seriers Optical Sciences, Vol. 45 (Springer Berlin, 1985), pp. 131.

[6] L. M. Gignac, M. Kawasaki, S. H. Boettcher and O.C. Wells, "Imaging and Analysis of Substrate Cu Interconnects by Detecting Backscattered Electrons in the Scanning Electron Microscope", J. Appl. Physics, 97 (2005), pp. 114506.

[7] J. L. Abraham and P.B. DeNee, " Scanning Electron Microscope Histochemistry sing Backscatter Electrons and Metal Stains", Lancet, 19 (1973), pp. 1125.

[8] Z. G. Song, H. S. Song, J. Yu and T. Su, "Backscattered Electron Imaging for Embedded Subtle Defects in 32nm Processes", ISTFA 2010 Proceedings, P. 108 (2010).

[9] Z. G. Song, J. Y. Dai, S. Ansari, C. K. Oh and S. Redkar, "Front-end Processing Defect Localization by Contact-level Passive Voltage Contrast Technique and Root Cause Analysis", IPFA' 2002 Proceedings, P. 97 (2002).

[10] C-M Shen, T-C Chuang, S-C Lin, L-F Wen, C-M Huang, "Combining the Nano-probing Technique with athematics to model and identify Non-Visual Failures", ISTFA' 2007 Processing, P. 214 (2007).

[11] J. C. Lin, W. S. Wu, "Using Nano-Probing Technique to Clarify Nickel Silicide beyond Process Window Causing Device Failure", ISTFA 2010 Proceedings, P. 236 (2010).

[12] Jon C. Lee, J. H. Chuang, "Fault Localization in Contact Level by Using Conductive Atomic Force Microscopy," ISTFA' 2003 Processing, P. 413 (2003).

[13] R. Mulder, S. Subramanian, T. Chrastecky, "Low Voltage, Low Current AFP Characterization of Non-Visible Soft Transistor Defects", ISTFA 2008 Proceedings, P. 428 (2008).

[14] T. X. Tong, A. N. Erickson, "Current Image Atomic Force Microscopy (CI-AFM) combined with Atomic Force probing (AFP) for location and characterization of advanced technology node", ISTFA 2004 Proceedings, P. 42 (2004).

[15] H. B. Zhang, W. Lee, R. D. Lin, W. T. Leong, "Applications of C-AFM Analysis Techniques at Advanced IC on SRAM Soft Failure", IPFA 2010 Proceedings, (2010).

[16] T. Kane, M. Tenney, A. Erickson, S. Phan, "Challenges of Atomic Force Probe Characterization of Logic Based Embedded DRAM for on-Processor Applications", ISTFA 2007 Proceedings, P. 46 (2007).

[17] T. Kane and M. P. Tenney, A. Erickson and P. Harris, "Calibration of Nanoprobe Capacitance-Voltage Spectroscopy (NCVS)", ISTFA 2008 Proceedings, P. 204 (2008).

[18] Z. G. Song, F. Beaudoin, S. Lucarini, J. Sylvestri, L. Safran, M. Villallobos and R. Oldrey, "Failure Analysis for SRAM Logic Type Failures", ISTFA 2013 Proceedings, P. 105 (2013).

[19] N. Khurana and C. L. Chiang, "Analysis of Product Hot Electron Problem by Gated Emission Microscopy", Proceedings, IRPS, P. 189 (1986).

[20] K. Nikawa and S. Tozaki, "Novel OBIC Observation Method for Detecting Defects in Al Stripes Under Current Stressing", Proceedings, ISTFA, P. 303 (1993).

[21] E. I Cole Jr, P. Tangyunyong and D. L. Barton, "Backside Localization of Open and Shorted IC Interconnections", Proceedings, IRPS, P. 129 (1998).

[22] J. Sylvestri, P. McGinnis, "Laser Induced Critical Parameter Analysis of CMOS Devices," US Patent # US 7,038,474 B2, May 2006.

[23] S. Kasapi, W. Lo, J. Liao, B. Cory and H. Marks, "Advanced Scan Chain Analysis Using Laser Modulation Mapping and Continuous Wave Probing", ISTFA 2011 Proceeding, P. 12 (2011).

[24] Y. S. Ng, T. Lundquist, D. Skvortsov, J. Liao, S. Kasapi, H. Marks, "Laser Voltage Imaging: A New Perspective of Laser Voltage Probing", ISTFA 2010 Proceeding, P. 5 (2010).

[25] L. Safran, J. Sylvestri, D. Albert, Z. G. Song and P. McGinnis, "Advanced Fault Localization through the Use of Tester Based Diagnostics with LVI, LVP, CPA, and PEM", ISTFA 2013 Proceedings, P. 313 (2013).

[26] E.I. Cole Jr., P. Tangyunyong, C.F. Hawkins, M.R. Bruce, V.J. Bruce, R.M. Ring, and W.L. Chong, "Resistive Interconnect Localization", ISTFA 2001 Proceedings, P. 43 (2001)

[27] M. R. Bruce, V. J. Bruce, D. H. Eppes, J. Wilcox, E. I. Cole, Jr., P. Tangyunyong, C. F. Hawkins, "Soft Defect Localization (SDL)", ISTFA 2002 Proceedings, P. 21 (2002).

Fast and easy sample preparation with reduced curtaining artifacts using a P-FIB

S. Moreau, D. Bouchu and G. Audoit

Univ. Grenoble Alpes, F-38000 Grenoble, France.
CEA, LETI, MINATEC Campus, F-38054 Grenoble, France.
Phone: +33 4-38-78-06-36 Email: stephane-nico.moreau@cea.fr

Abstract- **The present methodology proposes to reduce curtaining artifacts using a plasma-FIB when milling relatively deep trenches. Finally, the methodology is very fast (< 1 h), simple to set up and can be automated. Its purpose consists in eliminating or reducing the origin of curtaining artifacts by judicious milling. This methodology can advantageously replace the rocking method presented in the literature.**

I. INTRODUCTION

2.5-D and 3-D integration are proposed to address the More Moore's challenge by stacking 2-D dies and connecting them in the 3rd dimension. Such systems promise many significant benefits, including: better electrical performance, lower power consumption, form factor improvement, lower cost, more functionality (heterogeneous integration), and circuit security [1]. For the development of the advanced modules (copper direct bonding, TSV,...), performing an inspection after manufacturing or after reliability tests is required. Dual-beam systems (Focused Ion Beam / Scanning Electron Microscope, FIB/SEM) are important tools for those purpose i.e. morphological inspections. But, the analysis time is very long with standard gallium FIB, because of the large dimensions involved (> 1 000 000 μm^3) and the milling rate of such tools (300 μm^3/min). To keep the sample preparation time acceptable (< 1 h), precise and with a limited heat-affected zone (HAZ), Plasma-FIB (PFIB) is a good candidate with a milling rate up to 300 μm^3/sec. Due to the large milling depth and the differences in milling rate of the various materials present (Cu, Si, SiO_2, TaN/Ta or TiN barrier, voids,...) but also to the difficulty to focalize the high beam currents, curtaining artifacts [2] may complicate the interpretation of the images, notably in the thinnest layers. This paper presents a fast and easy alternative of the rocking method [3, 4] to avoid these curtaining artifacts during sample preparation using a PFIB.

II. PLASMA-FIELD ION BEAM (PFIB) – SAMPLE PREPERATION

The tool used for the fast sample preparation is a FEI Vion™ Plasma FIB (PFIB). Unlike conventional gallium (Ga)-FIB, the ionic source of a PFIB is generated from a plasma of xenon (Xe) ions [5]. Xenon is almost twice as heavy as gallium and the extracted currents are much higher: approximately 1 μA with a PFIB while we obtain 65 nA with a Ga-FIB. Etching with a PFIB is therefore 60 times faster than the etching with a Ga-FIB: for instance, milling a volume of 100 μm×100 μm×100 μm of silicon with a Ga-FIB requires around 50 h while it theoretically only requires 1 h with a PFIB.

This microscope is clearly suitable in the context of characterization of 3D integration devices. Indeed, the thickness of the cross-section can be often higher than 100 μm depending on the size and the depth of the area of interest. The PFIB thus permits to work quickly and effectively on large samples.

The proposed methodology is very easy to set up and fast (< 1 h) to ensure a high throughput as needed by the industry. Roughly, the methodology consists of using a square pattern to etch a given depth of materials. Then, a cleaning cross-section is performed. Through this fast and simplistic method, the sidewall of milled area shows a perfect state allowing a high quality imaging as described hereafter.

Two (2) case studies are presented hereafter:
1. One (1) sample with copper (Cu) direct bonding,
2. One (1) sample with Through Silicon Via (TSV).

In the first case, materials are deposited in a full sheet-like manner, minimizing the risk of curtaining. The second case uses patterned layers with multiple materials, maximizing the potential risk of curtaining.

III. CASE STUDY

Copper direct bonding: simple case study

The studied structure consists of a wafer-to-wafer (WtW) bonding including two copper lines of interest (one per wafer) as described on Fig. 1. Each wafer is processed identically as described hereafter. After trench opening into a deposited SiO_2 layer, TaN/Ta and Cu seed layers are deposited. Cu filling is carried out by electroplating and then annealed at 400 °C. After an optimized damascene-like Chemical and Mechanical Polishing (CMP) surface preparation, 200 mm wafers are bonded at room temperature (alignment precision: ~1 μm), atmospheric pressure and ambient air [6]. A 400 °C post bonding anneal is then applied in order to strengthen the bonding. . To allow a direct probing at the wafer level or a packaging with wire bonding, several technological steps are performed on the bonded wafers: top wafer thinning down to ~55 μm, via etching and fabrication of a redistribution layer (RDL). Finally, in order to protect the copper RDL from further oxidization and to create contact pads for electrical probing and/or wire bonding, an organic passivation and an under-bump metallization (UBM) are processed

Fig. 2 shows the layout of the structure used for electromigration test (reliability test) and the area of interest where the present methodology is applied. The area of interest is

978-1-4799-3911-4/14 $31.00 © 2014 IEEE

located at a 55 μm-depth, at the intersection of one of the tested lines (thickness: 500 nm, width: 3 μm) and a triangular-shaped feed line (width: from 3 to 80 μm). The structure has endured an electromigration test at a temperature of 250 °C during ~3100 h. Due to these harsh conditions, the passivation exhibits cracks which add curtaining effects during PFIB cross-sectioning (see Fig. 3).

N.B.: the first goal of this case study is not to see an electromigration-induced defect but to show the general methodology for a structure buried at a 55 μm-depth.

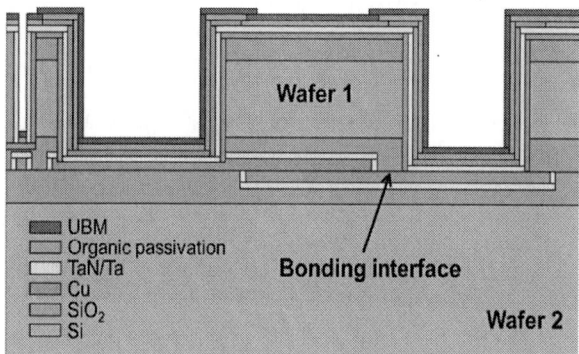

Fig. 1. Cross section of the copper direct bonding integration.

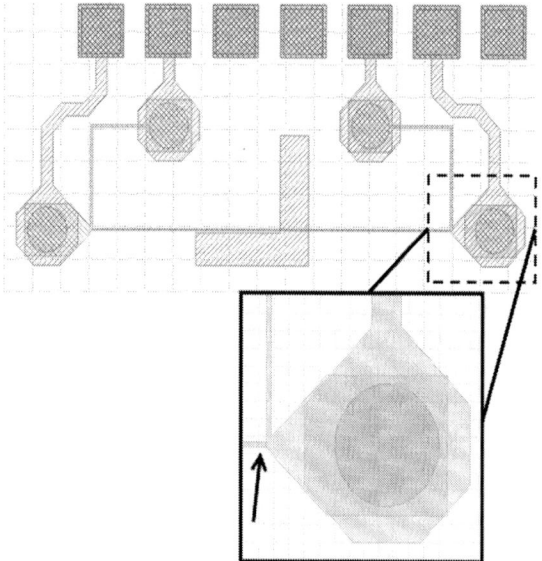

Fig. 2. Layout of the structure used for electromigration tests: global view of the structure plus the area of interest marked by a dotted square and zoom on the region to be investigated and where the present methodology is applied.

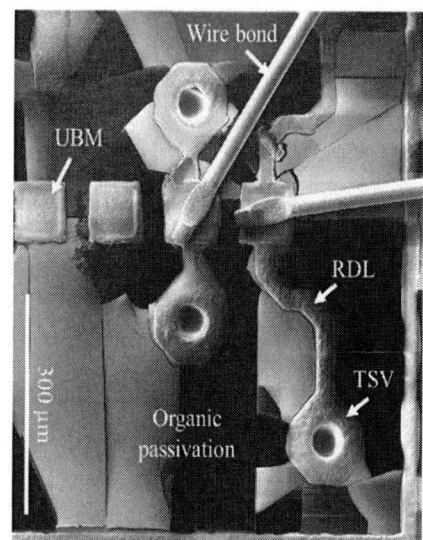

Fig. 3. Ionic micrograph (top view). Flakes are due to the organic passivation's cracking.

For the sake of clarity, the step of parallel etching was split into four (4) sub-steps. In an industrial approach, only two (2) steps (parallel etching + cross-section) can be performed and automated. Table I sums up the patterns, their sizes, the used currents/voltages and the duration of each step. All the millings have been performed with the ion beam normal to the sample surface, a 1 μs dwell time and a beam overlap of 50 %.

Table I. Sample preparation details: patterns, pattern sizes, FIB current and step duration.

Pattern	Dimensions (X μm×Y μm ×Z μm)	Voltage (kV)	Nominal current (μA)	Duration (min)
Flat rect.	80×90×16	30	0.47	~8
	80×90×5			~3
	80×90×20			~11
	80×90×12			~6
Cleaning X-section	30×60×4	30	0.18	~5

The first to the fourth patterns areas were defined larger than the cross section pattern in order to perform an additional *slice & view* - FIB/SEM 3D step with a standard dual beam tool to analyze the 3-D morphology of an EM-induced defect. Indeed, to avoid the problem of redeposition/masking during the slice & view operation, it is required to perform a cross-section larger than the field of view of the zone of interest. Another key-point concerns the total depth reached by a rectangular pattern. The total depth was fixed to ~45 μm (from the silicon top surface) in order to make easier/faster the *slice & view* step performed with a standard dual beam (lower milling rate).

Fig. 4 to 8 illustrate the five (5) above mentioned steps.

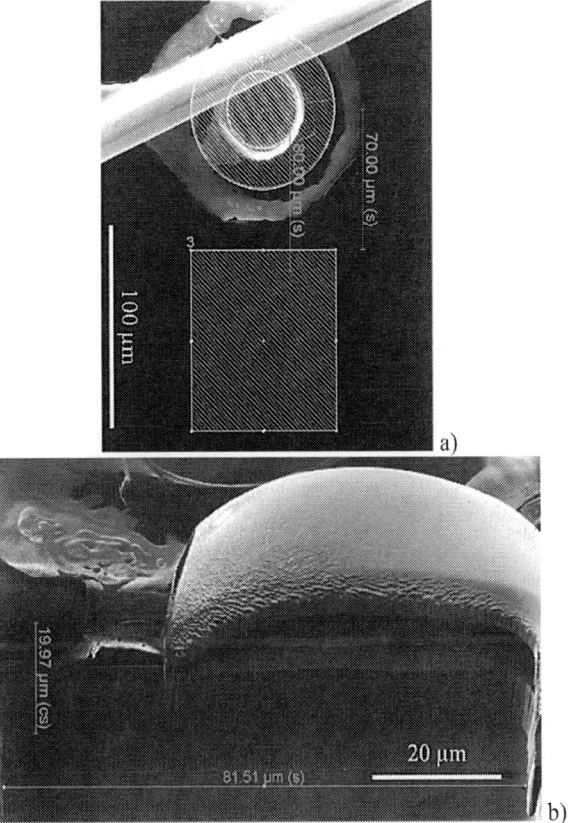

Fig. 4. Ionic micrographs: a) top view + etching pattern and b) tilted view (52°) after the first step of etching.

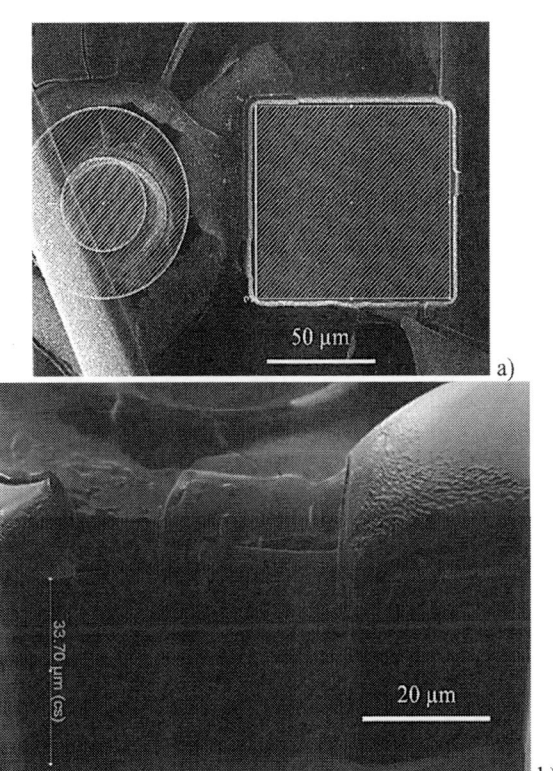

Fig. 6. Ionic micrographs: a) top view + etching pattern and b) tilted view (52°) after the third step of etching.

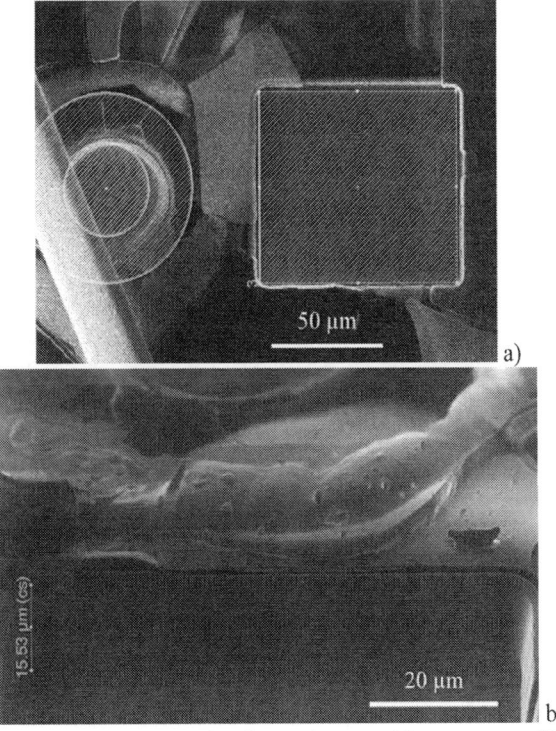

Fig. 5. Ionic micrographs: a) top view + etching pattern and b) tilted view (52°) after the second step of etching.

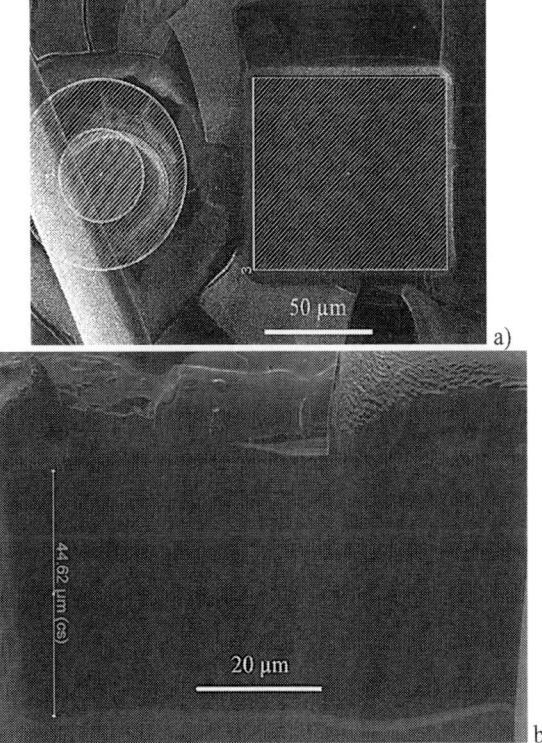

Fig. 7. Ionic micrographs: a) top view + etching pattern and b) tilted view (52°) after the fourth step of etching.

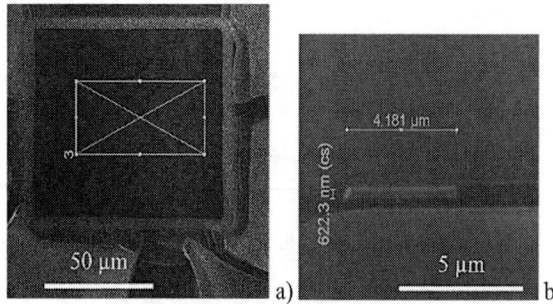

Fig. 8. Ionic micrographs: a) top view + etching pattern, b) tilted view (30°) after the cleaning cross section step of the area of interest.

Fig. 9 shows the benefit of such methodology in case of samples using a Cu direct bonding integration. The benefit is obvious! The Cu line is situated at 55 µm from the silicon top surface.

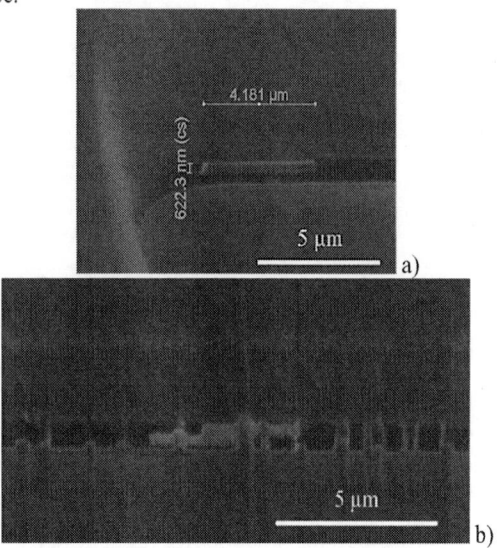

Fig. 9. Comparison of the sample preparation a) with (one-Cu line structure) and b) without (two-Cu line structure) the present method. Curtaining is obvious on b).

Through Silicon Via (TSV): tricky case study

A through-silicon via (TSV) is a vertical electrical connection passing completely through a silicon wafer or die. TSVs are a high performance technique used to create 3-D chip stacking integration circuits, compared to alternatives such as package-on-package, because the density of the vias is substantially higher, and because the length of the connections is shorter.

In that case, we need to observe a whole TSV and not only a buried part like in the previous study. Consequently, the previous methodology requires some modifications by adding, two (2) steps: 1- selective copper etching (optional), 2- selective silicon etching using a xenon difluoride (XeF$_2$) gas. Moreover, to protect the TSV top, two (2) optional deposition steps can be added: 1- SiO$_x$ and 2-platinum (Pt). SiO$_x$ is used in order to have a good contrast with the copper and the second layer is used to

protect the SiO$_x$ layer from the ion beam tail when performing the final milling.

Fig. 10 and 11 illustrate this adapted method for TSV failure/morphological analysis. The TSV use a via-middle integration. The TSV has a diameter of 10 µm and a height of 80 µm.

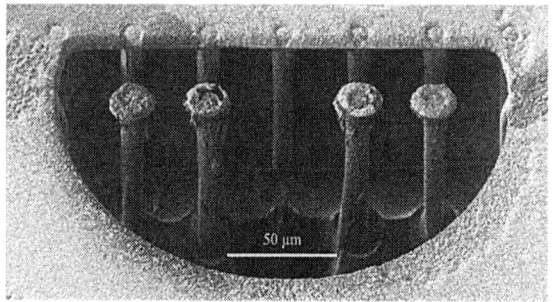

Fig. 10. Ionic micrograph (tilted view, 45°) after selective copper and silicon etching steps.

Fig. 11 shows few curtaining artifacts. The barrier is clearly visible. Thus, the integrity of TSV (copper filling, barrier continuity,…) can be verified easily. More work is in progress to completely remove these undesired curtaining effects. Key parameters such as dwell time and beam overlap should be tuned to further improve the quality of the cut.

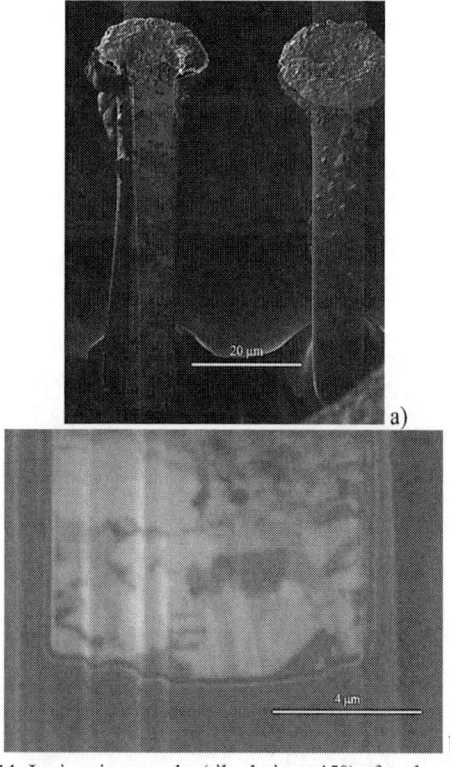

Fig. 11. Ionic micrographs (tilted view, 45°) after the cleaning cross section step: a) whole TSV, b) TSV's bottom. Few curtaining artifacts are present.

IV. SUMMARY

This paper presents a fast (< 1 h) and easy methodology to perform cross-section of 3-D advanced modules and/or ICs with few curtaining artifacts. To reach this high throughput a Plasma-FIB (PFIB) is used. Its purpose consists in eliminating or reducing the origin of curtaining artifacts by judicious milling. This methodology can advantageously replace the rocking method presented in the literature. It can be possible to automate the whole sample preparation thanks to software as FEI iFast™.

ACKNOWLEDGMENT

This work partially supported by the Cluster for Application and Technology Research in Europe on NanoElectronics in the frame of the MASTER_3D project and the French "Recherche Technologique de Base (RTB)" program.

The authors would like to thank the clean room staffs of both LETI and STMicroelectronics (Crolles) for providing us all the samples and the nano-characterization platform (PFNC) people for fruitful discussions on sample preparation and use of SEM/FIB/PFIB tools.

REFERENCES

[1] P. Garrou, C. Bower, and P. Ramm, *Handbook of 3D Integration*: Wiley, 2008.

[2] H. Bender, C. Drijbooms, and A. Radisic, "FIB/SEM Structural Analysis Of Through-Silicon-Vias," *AIP Conference Proceedings,* vol. 1395, pp. 274-278, 2011.

[3] L. Kwakman, M. Straw, G. Coustillier, M. Sentis, J. Schischka, J. Beyersdorfer, and F. Altmann, "Sample Preparation Strategies for Fast and Effective Failure Analysis of 3D Devices," in *International Symposium for Testing and Failure Analysis (ISTFA)*, San Jose (USA, CA), 2013, pp. 17-26.

[4] T. Hrncir and L. Hladik, "Fast and Precise 3D Tomography of TSV by Using Xe Plasma FIB," in *International Symposium for Testing and Failure Analysis (ISTFA)*, San Jose (USA, CA), 2013, pp. 27-32.

[5] R. Young, C. Rue, S. Randolph, C. Chandler, G. Franz, R. Schampers, A. Klumpp, and L. Kwakman, "A Comparison of Xenon Plasma FIB Technology with Conventional Gallium LMIS FIB: Imaging, Milling, and Gas-Assisted Applications," *Microscopy and Microanalysis,* vol. 17, pp. 652-653, 2011.

[6] P. Gueguen, L. Di Cioccio, P. Gergaud, M. Rivoire, D. Scevola, M. Zussy, A. M. Charvet, L. Bally, D. Lafond, and L. Clavelier, "Copper Direct-Bonding Characterization and Its Interests for 3D Integration," *Journal of The Electrochemical Society,* vol. 156, pp. H772-H776, October 1, 2009 2009.

Novel Inverted Sample Thinning Method by Ex-situ Lift-out

Liew Kaeng Nan

United Microelectronics Corporation (Singapore Branch), Ltd.

No. 3, Pasir Ris Drive 12, Singapore 519528

Phone: (65) 62130018 ext 1687 Fax: (65) 62130080 Email: kaeng_nan_liew@umc.com

Abstract- **Curtaining effect and sample thickness constraints are always the key factors of limiting the use of ex-situ lift-out technique in advanced semiconductor device analysis. Over the years, in-situ lift-out technique has gradually replaced ex-situ lift-out because it offers greater advantages that can overcome the mentioned problems. A novel technique has been developed to prepare ultra-thin TEM specimens by inverted FIB thinning without the need of installing FIB chamber-mounted probe.**

I. INTRODUCTION

The advent of advanced semiconductor technology poses a real challenge to device physical characterization. A typical transmission electron microscopy (TEM) specimen is about 100 nm in thickness. However, at sub 28 nm technology node, depending on the structure of interest, the sample needs to be less than 50 nm to prevent structure overlapping. To overcome the problem, low-energy focused ion beam (FIB) is used for preparing ultra-thin TEM specimens. A modern FIB shall be equipped with low-energy ion beam function to reduce the amorphization at the surfaces of TEM specimen induced by high-energy gallium ion beam.

Nevertheless, the curtaining effect could be severe when preparing ultra-thin TEM samples by low-energy FIB. Figure 1 shows a FIB image of TEM sample suffering from curtaining effect. The line-form artifact is created due to uneven milling of various device features across the TEM section and propagates from the denser materials (such as tungsten contacts) vertically to the homogeneous silicon substrate. The artifact does not usually affect the usual low magnification structural analysis, but it seriously impacts the high resolution TEM (HRTEM) analysis, especially on fine structures since the area of interest may be masked by the artifact that renders the sample unusable.

The normal practice of eliminating curtaining effect is to flip the sample upside down by in-situ lift-out technique for FIB milling. When the sample is inverted, the silicon substrate of the sample faces the ion beam source and because silicon substrate is homogeneous, it will not induce curtaining effect. Unlike conventional ex-situ lift-out technique in which the finished TEM lamella is transferred onto a copper grid with carbon support film, in-situ lift-out technique requires manipulation tool to be mounted in the FIB chamber and the TEM lamella transfer is performed inside FIB chamber. Since the micromanipulator can be rotated 360 degrees for flipping the sample upside down, in-situ lift-out technique is widely used for

advanced devices analysis. Figures 2 and 3 show Omniprobe and FEI EasyLift system with in-situ manipulator.

For ex-situ lift-out, the sample can be oriented for inverted FIB milling by performing backside sample polishing from silicon substrate side, prior to loading the sample into the FIB for processing. However, this method can be time-consuming because it is a challenge to localize the small region of interest by backside polishing. The FIB deposited amorphous platinum protective layer at the beginning of TEM sample preparation can be a source of curtaining effect if the platinum is deposited on to a rough sample surface, especially for low-energy milling. Therefore, there are limited options for in standard FIB system to perform in-situ lift-out unless it is equipped with chamber-mounted manipulation tool.

In this article, a novel method has been developed to prepare TEM lamella by inverted FIB thinning without the use of in-situ probe manipulator. The technique can eliminate curtaining effect and produce high quality HRTEM specimens, with the use of ex-situ lift-out technique. Most importantly, the new technique does not require chamber-mounted probe manipulator for sample preparation and therefore cost-effective.

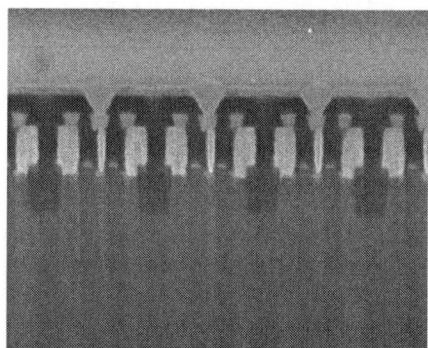

Fig. 1. Curtaining effect.

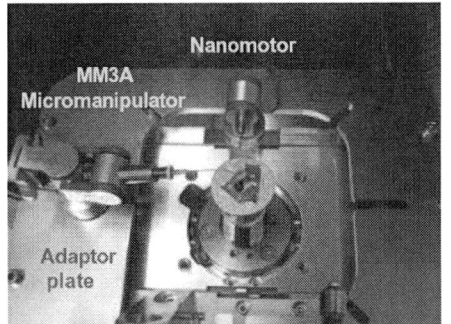

Fig. 2. A micromanipulator and a nanomotor that are installed on a FIB adaptor plate. [1]

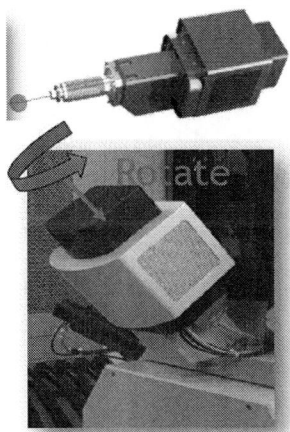

Fig. 3. FEI's EasyLift system. [2]

II. METHODOLOGY

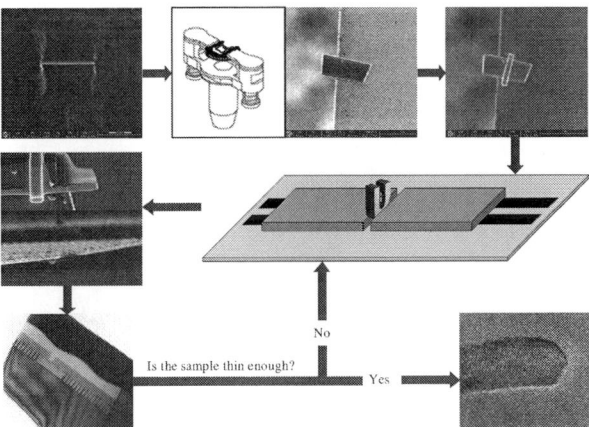

Fig. 4. Flowchart of TEM sample preparation.

Figure 4 outlines a flowchart showing the steps of preparing an ultra-thin TEM sample. The detailed explanation of each step is illustrated below:

1) A 300 nm thick sample is prepared using 30 keV FIB. The sample is then cut free from trench.

2) A silicon-base supported U-shape copper grid is clamped by precision ion polishing system (PIPS) specimen holder. The PIPS specimen holder is then placed onto FIB carrier, as shown in Figure 5.

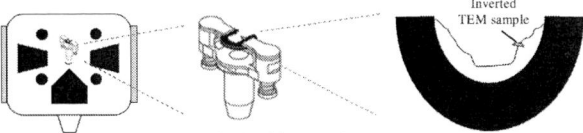

Fig. 5. Lift-out tool set up.

3) The sample is transferred to the copper grid by ex-situ lift out. The sample must be laid upside down on the silicon film. Details of sample orientation are described in results and discussion section.

4) The sample is reloaded into FIB for platinum strip deposition for bonding the lamella to silicon-base support film.

5) The copper grid is removed from the PIPS specimen holder. The configuration of FIB holder is shown in Figure 6. The copper grid is clamped upright by two fixtures, which can be a square wafer slice. The TEM sample is inverted when the copper grid is clamped.

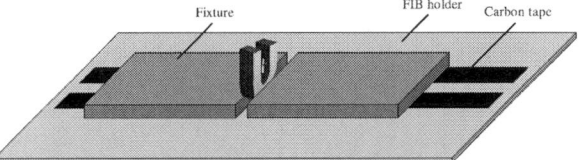

Fig. 6. Pre-FIB tool set up.

6) The sample is reloaded into the FIB. The stage is tilted so that the ion beam is transverse to the vertical axis of the sample. The tilt angle is varies from case to case depending on the geometry of copper grid and can be determined by referring to ion beam image shown in the bottom image of Figure 7. The front side and backside of the sample is then cleaned with low-energy ion beam.

978-1-4799-3911-4/14 $31.00 © 2014 IEEE

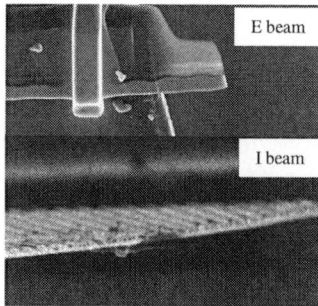

Fig. 7. FIB electron and ion beam image.

7) The sample is the unloaded from FIB and loaded into the TEM for observation. The thickness of the sample can be judged by the diffraction pattern at the region of interest.

8) If the sample thickness is uniform and thin enough for HRTEM, the analysis is performed. Otherwise, repeat Step 5 and 6 until desired sample thickness is achieved.

III. RESULTS AND DISCUSSION

Figure 8A and 8B show high-resolution TEM micrographs from TEM specimens containing advanced semiconductor structures prepared by conventional ex-situ lift-out technique and the novel method developed in this work, respectively. The sample corresponding to Figure 8A is relatively thick and produces a poor phase contrast lattice image, while the sample corresponding to Figure 8B is relatively thin and produces better phase contrast lattice image.

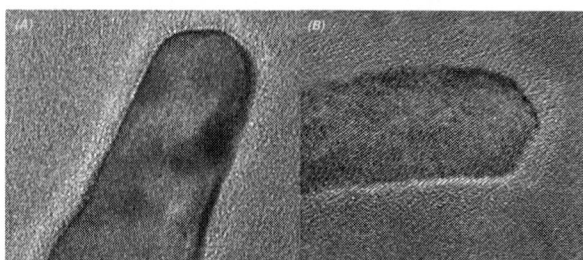

Fig. 8. HRTEM images of lamella prepared by: (A) Conventional ex-situ lift-out technique by 8 keV FIB; (B) Novel ex-situ lift-out inverted thinning by 8 keV FIB.

In spite of the above results, there are also several challenges to sample preparation method developed in this work. Firstly, there is the limitation of TEM sample geometry for lift out. Figure 9 shows the possible sample geometry for the sample to be placed on to silicon-base film so that it can be inverted when the copper grid is subsequently clamped upright on the FIB holder. For the sample to be positioned onto region A, the needle probe must approach the sample from the front/right side or rear/left side of the sample prior to lifting out the sample from trench. This is to ensure that when the lamella is placed on the silicon based film, at least half of the specimen that is not on

silicon film so that it can be re-thinned by FIB. Similarly, for region B, the needle probe shall approach the sample from the front/left side or rear/right side of the sample before picking up the sample. However, region C is not suitable because the sample cannot be inverted at that region.

In practice, the sample transfer process would also be more challenging if it is a site-specific sample. Consider if the region of interest is positioned at the left side of sample as is shown in Figure 10, the probe needle must touch the sample from the right side of sample. The probe needle can either approach the sample from the front side (as shown in Figure 10) or from the rear side of sample (as shown in Figure 11). Different approaches require different sides of silicon based film for landing sample so that the sample can be inverted.

Another limitation of this technique is the restriction of silicon based film geometry for placing TEM lamella. Figure 12 shows the orientation of TEM lamella for bonding to the silicon film to enable sample inversion. Grid A is not a suitable grid to be used for this technique because it is not possible for TEM lamella to be inverted when the grid is subsequently clamped upright. For grid B with silicon film positioned at the left side of the grid, the needle probe must approach the sample from the left side of the sample so that it can be laid onto the silicon film. Similarly, for grid C, the needle probe shall attach to the right side of the sample. Of course, a grid such as the one shown in Figure 9 is the ideal grid since it allows maximum possible sample geometry for sample transfer. Therefore, proper planning must be made in advance to transferring specimen from bulk sample to copper grid.

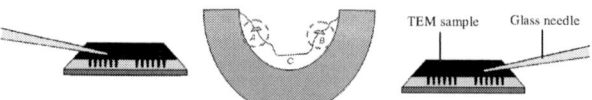

Fig. 9. Limitation of the specimen geometry for lift out.

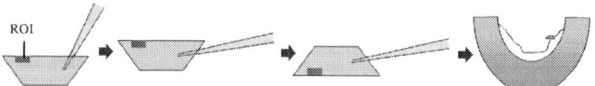

Fig. 10. Needle probe approaches sample from the rear/right side of sample.

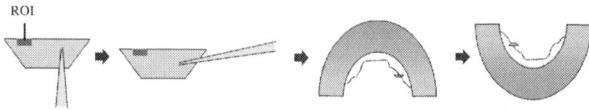

Fig. 11. Needle probe approaches sample from the front/right side of sample.

Fig. 12. Limitation of the position of Si support film at Cu ring for lift out.

IV. CONCLUSION

A method of preparing ultra-thin TEM specimens has been developed by flipping the sample upside down for inverted FIB thinning. This preparation method is compatible with the ex-situ lift-out system and offers high quality TEM specimens without the curtaining effect. Most importantly, this technique does not require FIB chamber-mounted microprobe for the sample preparation and therefore no additional cost is involved. However, this technique is more time-consuming than in-situ lift-out technique.

ACKNOWLEDGMENT

The author would like to thank FA3 for the resources provided in completing this work.

REFERENCES

[1] Jon C. Lee, B.H. Lee, "The Versatile Application for In-situ Lift-out TEM Sample Preparation by Micromanipulator and Nanomotor", Proceedings of ISTFA (2005).

[2] Adapted from FEI Customer Seminar 2012.

FIB-SEM Investigation and Auto-metrology of Polymer-Microlens/CFA Arrays of CMOS Image Sensor

Pradeep Sharma[*], Tai Shan Chiu, Sajal Biring, Te-Fu Chang, Chih-Hsun Chu, and Yong-Fen Hsieh

FIB-TEM and R&D Division, Materials Analysis Technology Inc., 1 A4, 1F, No.1, Lising Road 1,
Science Based Industrial Park, Hsinchu City-300, Taiwan
[*]Email corresponding author: sharma@ma-tek.com
TEL: 00886-3-611-6678, FAX: 00886-3-600-5883

Abstract

We report sample preparation and FIB-SEM investigation of polymer-microlens/CFA arrays of CMOS image sensor for investigating possible nanoscale voids. Polymer staining was employed to delineate boundaries of color filters and microlenses. Newly developed in-house auto-metrology software was used for dimension and uniformity study of SEM images of microlenses.

.

1. Introduction

Due to the increasing requirements of high image resolution, low cost, and low power consumption; CMOS based sensors [1] have gained popularity for consumer electronics. Specifically, CMOS image sensors are commonly used in cameras of cell phones, handy-cams, and webcams. These sensors constitute silicon based signal processing electronics, photo-detectors, color filters, and microlenses. The color filter arrays (CFA) selectively filter particular wavelength of light and they are directly integrated with photo-detectors. CFA are formed by mixing color pigments in photosensitive polymers or resists. Mostly, separate red (R), green (G), and blue (B) color filters are used in CMOS based sensors. These R, G, and B color filters are spatially arranged in special pattern to match with optimum sensitivity of human eyes. Further, for enhancing photosensitivity of CFA, and to increase efficiency of light collection by proper focusing of the light onto photodetectors; polymer based microlenses are fabricated on top of color filters. External light is focused by microlenses onto CFA, which in turn allows particular wavelength of light (either R, or G, or B) to pass through and it is detected by detection by photo-detectors [2]. Then electrical output generated by photo-detectors is processed by integrated circuit electronics for image formation. Presence of nano-scale voids in color filters and non-uniformity of microlenses affects the image quality of sensors. Therefore, it is important to develop protocols for accurate characterization of possible nanoscale defects in CMOS image sensors, and to implement auto-metrology processes for time efficient quality assessment and yield improvement.

Focused ion beam (FIB) cross-sectional sample preparation, and scanning electron microscope (SEM) imaging are widely used to characterize cross-sectional morphology of different biological samples as well as materials and devices. FIB-SEM fabrication and characterization of polymer based nanostructures [3], and FIB-SEM analysis of polymer composites [4] special care. It is essential to employ suitable sample preparation protocols and to operate FIB and SEM under optimized conditions to cause negligible beam induced damages and to prevent formation of artifacts. In this work,

we report sample preparation protocols, and FIB-SEM characterization of polymer-microlens/CFA arrays of CMOS image sensor using optimized electron and ion beam conditions.

2. Experiments

Commercially available electronic device was torn-down to acquire the CMOS image sensors samples for performing this study. To prevent charging by electron or ion beam, about 20 nm gold (Au) films were sputter deposited onto polymer-microlens/CFA arrays. For estimation of suitable electron beam acceleration voltage to be used for scanning electron microscope (SEM) imaging of samples; zero[th] order calculation of electron beam scattering into materials and minimum sample charging were used. Penetration depth of electron beams into thin films of Au was determined using electron beam Monte-Carlo simulator (Win X-ray) [5]. After deposition of Au films, the samples were loaded into dual-beam FIB-SEM system. Thin Au films will not sustain irradiation induced sputtering by focused ion beam (FIB). Therefore, protective layers of platinum (Pt) thin films were deposited. Electron beam (energy = 2 KV) induced Pt (film thickness $\approx$ 25 nm) deposition was conducted onto regions of interest, which was followed by focused ion beam (energy = 8 KV) induced deposition of Pt (thickness $\approx$ 100 nm). Suitable ion beam acceleration voltages and thickness of platinum films were estimated using Transport of Ion in Mater (TRIM) code simulations [5]. After deposition of protective films, FIB cross-sectioning of the samples were performed separately in Fast Mode (30 KV, 250 pA), and Slow Mode (8 KV, 100 pA). Effect of Fast and Slow mode cross-sectioning was imaged using 2 KV electron beam of SEM. For elucidating the effect of electron beam dose, different cross-sectional areas were scanned using dwell time per pixel = 10 μs, and 1 μs. FIB-SEM deposition of Pt, and cross-sectioning-imaging of polymer-microlenses/CFA samples may lead to annealing and thermal deformation/flattening of microlenses. Tapping mode atomic force microscopy (AFM) was utilized for scanning the surface of polymer-microlenses/CFA samples, and to obtain the height of microlenses. Height of microlenses obtained by AFM was compared to height obtained by FIB-SEM characterization. After imaging of cross-sectional view of polymer-microlens/CFA structure, polymer staining was employed to delineate the boundaries of the color filters of polymer-microlens. As an effort for quick and precise assessment of uniformity of microlenses, SEM images of cross-sectioned samples were recorded, and these images were analyzed using in-house newly developed 'Auto-metrology Software'.

978-1-4799-3911-4/14 $31.00 © 2014 IEEE

3. Results and Discussion

Fig. 1(a) shows digital photograph of a CMOS image sensor of a commercial electronics device. Position of sensing part is marked by red colored rectangle. Fig.1(b) shows top view optical micrographs of R-G-B arrays of polymer-microlens/CFA. Polymer-microlenses are transparent. Red, green, and green color shows presence of color filters beneath the microlenses.

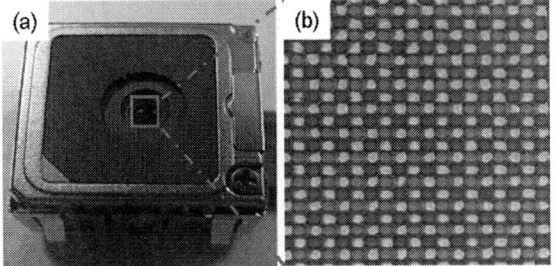

Figure 1. (a) CMOS image sensor of a commercial electronics device. (b) Red-Green-Blue arrays of polymer-microlenses on color filters.

Study of low voltage SEM imaging of different commercial polymers and resists by David et. al. [6] shows that use of about 0.5 to 2 KV electron beam leads to minimum charging of the organic samples. For one time planar, or cross-sectional view of samples, thin coating of metallic (Au, or Pt) or carbon films eliminates the charging issue during SEM imaging. But, FIB-SEM analysis of polymer-microlenses/CFA requires cyclic process of etching of samples by FIB followed by SEM imaging. Therefore, applicability of 1 KV and 2 KV electron beam was investigated for low voltage SEM imaging of polymer-microlenses/CFA.

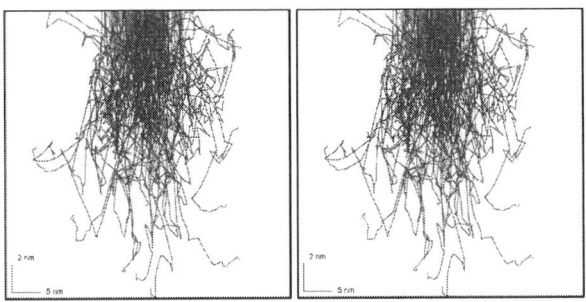

Figure 2. Monte-Carlo simulations of 1 KV (left) and 2 KV (right) electron-beam scattering in protection layer of gold (t ≈ 20 nm). Blue color: forward scattering, Red color: back scattering.

Electron beam Monte-Carlo simulator was to determine penetration depth of 1 KV and 2 KV electron beam into 20 nm gold thin film. As shown in Fig. 2, penetration depth was found to be about 11 nm for 1 KV electron beam, and about 25 nm for 2 KV electron beam. Therefore, about 20 nm thin film of gold was deemed to be fit for imaging plan view of polymer- microlenses/CFA samples without causing charging, or beam induced damage. On the other hand, for cross-

sectional SEM imaging of these samples; 2 KV electron beam provided better resolution than 1 KV beam. Further, as per zero[th] order calculations described by R. F. Egerton et. al. [7], for 2 KV (current = 100 pA) electron beam sputtering rate of polymers/resists was about 0.3 to 1 Å/ms/pixel. Temperature rise due to 2 KV electron beam was estimated to be below 1 K/pixel. For minimizing the impact of electron beam dose [7] induced radiation damage to polymers, the dwell time was kept about 2 to 10 μsec/pixel.

20 nm thin film of gold does not hinder in identification and high resolution inspection of region of interest, which shall be cross-sectioned by FIB and imaged by SEM. This thin film of gold will not sustain ion beam induced sputtering, and deposition of thick metal film will obstruct identification of region of interest. Therefore, selective deposition of Pt by electron and ion beam was performed. FIB can deposit Pt at high speed, but, it will erode Au film. As a solution, 25 nm Pt film is deposited into region by interest using 2 KV electron beam followed by 8 KV FIB induced deposition of 100 nm Pt film. As shown in Fig.3, the penetration depth of 10 KV gallium ions into Pt or Carbon is about 20 to 25 nm only. Therefore, application of 8 KV FIB for deposition of Pt will not damage underlying 25 nm Pt film by SEM/20 nm Au film/polymer- microlenses/CFA structure. Further, as per zero[th] order calculations reported by John et. al. [8], for polymers, the estimated temperature rise of polymer due to irradiation by 8 KV FIB (current = 100 pA) is about 10 K/pixel. Increasing the acceleration voltage, and ion beam current will lead to increase in temperature rise also, and it may yield ion beam induced artifacts in cross-sectioned samples. For biological, or polymer samples, care must be taken to reduce FIB irradiation induced damages [9].

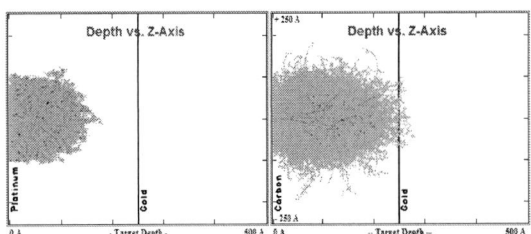

Figure 3. Monte-Carlo simulations of 10 KV gallium ions into Pt/Au and C/Au structures.

Fig. 4(a) shows cross-sectional SEM images of polymer-microlenses/CFA after Slow Mode FIB etching, whereas Fig. 4(b) shows cross-sectional SEM images of polymer-microlenses/CFA after Fast Mode FIB etching. Both images do not show presence of nanoscale voids in color filters, or microlenses. Edges of the color filter and microlenses show a decent edge profile, which indicates absence of voids or non-uniformities on sidewalls. Slow Mode FIB-SEM characterization partially enhances charging of the samples. But, it ensures that if any void is observed, then it is not because of FIB cross-sectioning process. For this particular sample, Fast Mode FIB-SEM characterization provides lower charging of samples, and results are comparable with Slow Mode FIB-SEM characterization. But, Fast Mode can not guarantee that no heat induced damages [8-9] will occur during FIB cross-sectioning. Heat inducted damages are

978-1-4799-3911-4/14 $31.00 © 2014 IEEE   241

primarily controlled by thermal conductivity of polymers, or composite materials, which are to be etched by FIB. Highly conductive materials will show marginal impact of Slow or Fast Mode characterization. Further, Slow Mode FIB-SEM analysis requires 4 to 5 times more time than Fast Mode analysis. Therefore, Fast Mode analysis can be adopted for quick assessment of samples. And Slow Mode can be reserved for verification of nanoscale voids or defects.

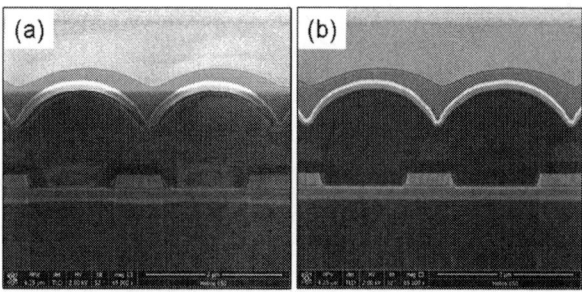

Figure 4. FIB-SEM characterization in (a) Slow Mode, and (b) Fast Mode. Scale Bar = 2 μm.

Cross-sectional SEM images in Fig. 4 show that height of polymer- microlenses is about 700 nm. Height of polymer-microlenses as measured by AFM is found to be about 700 nm also. AFM measurements were performed on pristine region of the same sample. Consistency in height measurement data obtained by SEM and AFM shows that microlenses do not suffer from beam induced flattening during FIB-SEM cross-sectioning and imaging. Typical AFM measurement is shown in Fig.5.

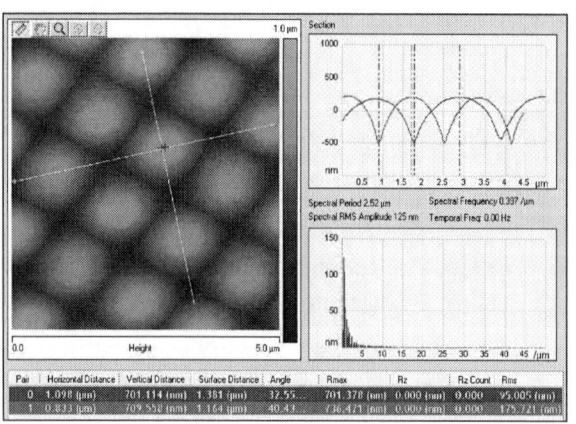

Figure 5. AFM image of polymer- microlenses of CMOS image sensor. Estimated height of microlens is about 700 nm.

As shown in Fig. 4, the boundaries of color filters beneath the microlenses are blurred. In order to delineate these boundaries polymer-staining process was employed. Fig. 6(a) shows post-staining cross-sectional SEM image of sample processed by Slow FIB-SEM characterization, whereas, Fig. 6(b) shows post-staining cross-sectional SEM image of sample processed by Fast FIB-SEM characterization. Fig. 6(a) shows presence of curtain effect on surfaces of microlenses and CFA. This effect can be fine tuned or reduced by

adjusting the percentage of beam overlap during etching process. No curtain effect is observed in Fig. 6(b).

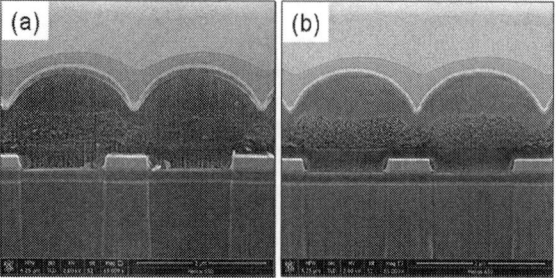

Figure 6. Post-staining SEM images of microlenses/CFA after (a) slow FIB cross-sectioning, and (b) fast FIB cross-sectioning. Scale Bar = 2 μm.

For sample analyzed in Fig. 4, and Fig. 6; no serious impact of electron and ion beam dose was observable. However, color filters, which are mixture of resists and color pigments, or nanocomposite samples, which are mixture of polymers and nanoparticles, are more sensitive to electron and ion beam acceleration voltage and irradiation dose/pixel. This sensitivity arises due to difference in thermal conductivity and glass transition temperature of resist or polymer matrix and pigments or nanoparticles. The higher the difference, the higher is the probability of non-uniform heating/damage by beams. As an example, we show that for some other polymer-microlenses/CFA sample taken from a different device, SEM electron beam dose/pixel can create void like artifacts in CFA. Fig. 7(a) shows example of artificial void formation when for 2 KV electron beam (100 pA current, dwell time/pixel = 10 μs) was used for SEM imaging. These artifacts were removed by reducing the dwell time to 1 μs. Therefore, for some polymer samples, it can be very critical to fine tune electron and ion beam doses to avoid observation of artifacts.

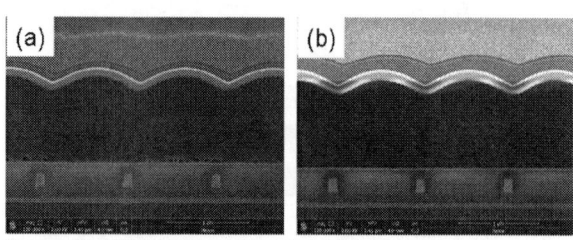

Figure 7. SEM images after fast FIB etching. Electron beam dwell time/pixel = 10 μs for (a), and 1 μs for (b).

Uniformity of polymer- microlenses is required for enhancing photosensitivity of CFA, and to increase efficiency of light collection by proper focusing of the light onto photodetectors. Recently, Materials Analysis Technology, Taiwan developed 'Autometrology Software' for processing and analyzing SEM and TEM images of Fin-FET and LED super-lattice structures. For present study, this software was used to detect edges of the polymer- microlenses, to generate contours of edges, and to record (x,y) coordinates of all points of contours. These coordinates of contours can be easily exported to Microsoft-Excel file, and then used for quick and error free measurement assessment of uniformity of

978-1-4799-3911-4/14 $31.00 © 2014 IEEE

microlenses, radius of curvature of mcirolenses, and pitch of microlenses. Fig. 8 shows a representative image for edge detection of microlenses by software.

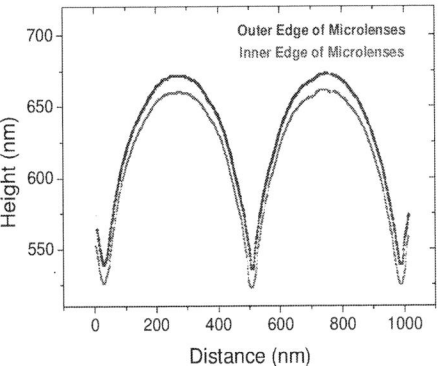

Figure 8. Edge detection of polymer- microlenses by 'Autometrology Software'.

As an independent check for examining uniformity of microlenses, AFM contour images can be analyzed. 2D, and 3D shows AFM contour image of a representative polymer microlenses/CFA sample are shown in Fig. 9(a), and Fig. 9(b).

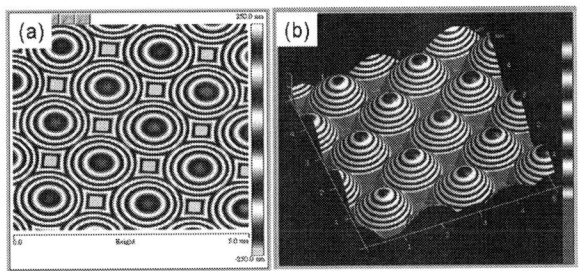

Figure 9. (a) 2D contour map of AFM microlensses/CFA, and (b) 3D contour map of AFM microlenses/CFA.

2D and 3D mapping shows that these lenses are not of identically equal height. As shown in Fig. 10, section of AFM image contour reveals that there is height variation of the order of ± 5 nm. Visible light has wavelength ranging from 400 to 700 nm. Therefore, such minor variations in height may not alter image quality generated by CMOS image sensor. But, monitoring of height of microlenses is crucial to maintain acceptable statistical variation.

AFM images based metrology, and SEM images based metrology can be regarded as complementary to each other. AFM has Å scale height (Z scale) resolution, but, X-Y resolution can be easily affected by interaction of AFM tip, and sample surface morphology. Although, convolution of AFM images due to AFM tip diameter, or aspect ration of AFM tip can be addressed by using ultra-high aspect ration tip, or by using AFM tips, which have carbon nanotubes on their apex. But, utilization of such tips is still limited due to risk of tip damage, and slow scan speed. On the other hand, SEM images provide accurate X-Y resolution at nm scale. If no charging artifacts are present then SEM can provide good

images, which clearly show X-Y dimensions of the objects. But, height resolution of SEM images is limited to precise x-sectioning or tilting of the samples to be characterized. Therefore, combined SEM and AFM metrology provide comprehensive assessment of the micro or nano-scale objects.

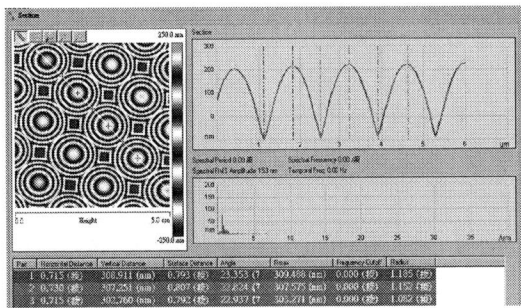

Figure 10. Section of 2D contour map of microlenses/CFA sample.

4. Conclusions

EISS and TRIM code simulations for electron and ion beam scattering into protection films deposited onto polymer samples was performed to estimate suitable acceleration voltages of beams for performing FIB cross-sectioning, and SEM imaging. Simulation results and calculations were used to validate the required thickness of protections films. Methodology for using optimized electron and ion beam acceleration voltages and doses was discussed. Optimized beam conditions were used to mitigate the effects of beam induced sputtering, heating, charging of the polymer microlens samples, and to eliminate formation of artifacts during sample preparation and imaging steps. Further, polymer staining was employed to delineate boundaries of color filters of microlenses. Newly developed auto-metrology software was used for processing SEM images microlenses.

Acknowledgments

Pradeep Sharma, and Sajal Biring acknowledge support by Materials Analysis Technology, Taiwan for supporting this research work.

References

1. Bigas, M. *et al*, "Review of CMOS Image Sensors", *Microelectr J*, Vol. 37, (2006), pp. 433-451.
2. Ohta, J., Smart CMOS Image Sensors and Applications, CRC Press (USA, 2007), pp. 11-55.
3. Lee C. C. *et al*, "Three Dimensional Nanofabrication of Polystyrene by Focused Ion Beam", *J. Microsc*, Vol. 248, (2012), pp. 129-139.
4. Olea-Mejia O. *et al*, Current Microscopy Contributions to Advances in Science and Technology, Formatex Research Center (Spain, 2012), Vol. 2, pp. 1060-1065.
5. Elecetron Beam Trajectories in Materials (Win X-ray) and Ion Beam Trajectories in Solids (TRIM Code) http://www.lehigh.edu/~maw3/link/mssoft/mcsim.html
6. Joy, D. C. *et al*, "Low Voltage Scanning Electron

Microscopy", *Micron*, Vol. 27, No. 3-4 (1996), pp. 247-263.

7. Egerton, R. F. *et al*, "Radiation Damage in TEM and SEM", *Micron*, Vol. 35, (2004), pp. 399-409.

8. Melngailis J., "Focused Ion Beam Technology and Applications", *J. Vac. Sci. Technol. B*, Vol. 5, (1987), pp. 469-495.

9. Bassim, N. D. *et al*, "Minimizing Damage during FIB Sample Preparation of Soft Materials", *J. Microsc*, Vol. 245, (2012), pp. 288-301.

Hourglass concept for RRAM: a dynamic and statistical device model

R. Degraeve, A. Fantini, N. Raghavan[2], L. Goux, S. Clima, Y.Y. Chen[1], A. Belmonte[1], S. Cosemans, B. Govoreanu, D.J. Wouters[1], Ph. Roussel, G.S. Kar, G. Groeseneken[1], M. Jurczak.
Imec, Kapeldreef 75, B3001 Leuven, Belgium, [1] also at ESAT dept, KULeuven, Leuven.[2] Singapore University of Technology & Design (SUTD), Singapore. -- Contact email: Robin.Degraeve@imec.be

Abstract - **In this paper we review a dynamic device model for filamentary RRAM in HfO-based dielectrics. We summarize its transient modeling features and its statistical properties. The model explains with satisfactory quantitative resolution all main features of the RRAM switching, not just the voltage, time and temperature dependence, but also statistical fluctuations resulting from atomistic motion and their resulting LRS and HRS-distributions.**

I. INTRODUCTION

HfO$_2$-based resistive random access memory (RRAM), with demonstrated robust scaling ability down to 10nm and promising performance and reliability, is an interesting alternative memory concept [1,2,3,4]. RRAM operation relies on the voltage-controlled resistance change of a conductive filament in the dielectric of a Metal-Insulator-Metal (MIM) stack. In HfO$_2$, oxygen vacancies (V$_o$), or, complementary, charged oxygen ions, have been identified as the mobile defect species responsible for forming/breaking the filament [5,6,7].

Recently, we demonstrated the concept of the hourglass model that describes the set/reset transient and can capture all the main features of the RRAM device operation [8]. In the present paper, we summarize and review our recent publications related to this work [8-20]. Subsequently, a selection of statistical issues is discussed in the framework of the hourglass model.

II. EXPERIMENT AND DEVICES

N-channel MOS transistors of 0.13µm channel length were processed in a 65nm process technology, with adjusted gate stack definition, so as to allow for operating voltages compatible with RRAM forming and set/reset operation. The resistive switching stack (Fig. 1), consisting of 65nm PVD TiN/5nm ALD HfO$_2$/10nm PVD Hf/30nm PVD TiN, was fabricated afterwards with a back-end-compatible thermal budget, not exceeding 400°C. Cross-bar RRAM elements in a 1 Transistor-1 Resistor (1T1R) configuration are used for which excellent performance down to 10x10nm has been demonstrated [4]. Also 2R-configurations without transistor, but with a load resistor instead are available. Finally, a dual-layer SiO$_2$/SiN passivation scheme has been employed to complete the process. A detailed description of the process conditions and their optimization can be found in [12].

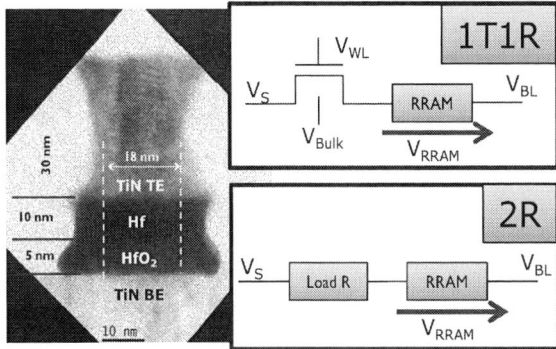

Fig. 1: Sample summary. X-bar RRAM elements with a TiN/HfO2/Hf/TiN stack were used in 2R or 1T1R configuration. Excellent performance down to 10x10nm has been demonstrated [2]. Process details are also in [12].

III. RRAM OPERATION MODELING

Many efforts were already taken in literature to develop a model for understanding the RRAM operation. Some of these approaches start from very fundamental ab-initio calculations, revealing interesting details about the atomistic motion inside and around the filament. Other models describe the device operation at a much higher level of abstraction in order to retain the advantage of an analytical treatment, often resulting in an oversimplification of the problem because the discrete atomistic motion is approximated by a continuum description.

The model in this paper aims at describing the atomistic movement of the ions, while still retaining analytical tractability. In other words, we construct a model that explains with satisfactory quantitative resolution all main features of the RRAM switching, not just the voltage, time and temperature dependence, but also statistical fluctuations resulting from atomistic motion and their resulting distributions of the current-voltage state. In order to reach this goal, a number of well-chosen simplifications and approximations need to be taken. We focus exclusively on the low current regime (I<50uA), since this is the technologically relevant range for RRAM operation.

Our RRAM switching model has five basic ingredients: (i) an electron conduction model for describing the current-voltage characteristics, (ii) a structural model describing the shape of the filament, (iii) a kinetic model describing the vacancy movement inside the filament, (iv) a thermal model describing the heat generation and its catalyzing effect on switching, and (v) a stochastic model describing the statistical variations in the switching behavior.

978-1-4799-3911-4/14 $31.00 © 2014 IEEE   245

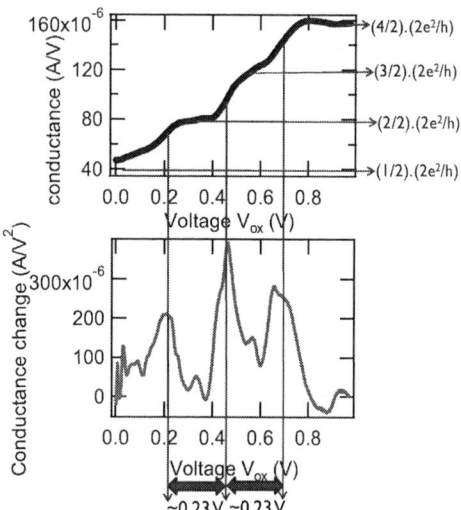

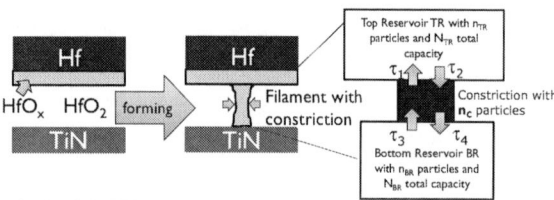

Fig. 2: Top plot shows the first derivative of an IV curve when set fails. Even at room temperature, clear conductance plateaus appear corresponding to the quantized conductances from the QPC model. Bottom plot shows the second derivative, where the peaks indicate the energy levels in the quantized conduction. Approximately equidistant peaks indicate that a parabolic potential well description in the QPC-model suffices. Fig. taken from reference [10].

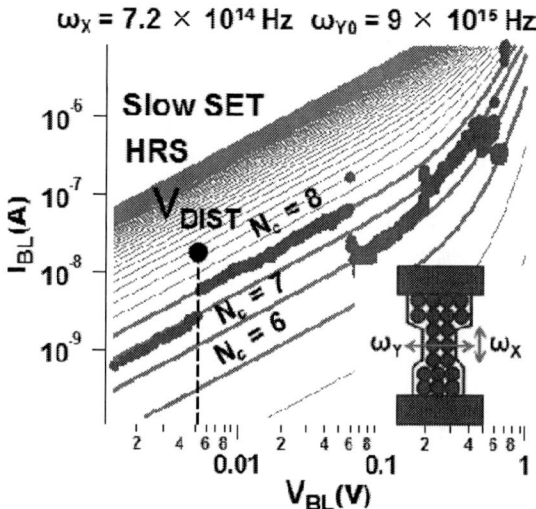

Fig. 3: (a) Stoichiometric HfO_x is present between Hf electrode and HfO_2 dielectric. (b) Forming 'extends' the HfO_x region to the filament. (c) Filament is modeled as a container with a top and a bottom vacancy reservoir (TR and BR), connected by a constriction C with variable width.

A. Electron conduction model.

For describing the current-voltage (I-V) characteristics, the filament is modeled as a *continuous* conduction band inside the dielectric originating from the close-range interaction of many defects. The conductance-controlling discrete energy levels are explained by the spatial confinement of the current to a narrow point in the filament modeled by a *quantum point contact* with saddle-surface potential [9,21,22] (Table 1, Eqs 1). With this approach, the conduction is described by 2 parameters: ω_x and ω_y, which are inversely related to the constriction length and width resp.

In most cases, the conductance quantization is very difficult to measure directly, since physical changes in HfO_2 filament due to oxygen vacancy movements occur even at low voltage. After set *failure*, however, the filament IV curve remains stable and can be measured up to 1V. Clear conductance plateaus become visible in the first derivative of the IV-curve as shown in Fig. 2 [10].

B. Structural model.

Experimental evidence [8,10] suggests that an *abstraction* of the actual filament consists of a top reservoir (TR) of particles connected to a bottom reservoir (BR) by a constriction (C) with variable width (Fig. 3). Subsequent to

forming, we assume no generation or annihilation of vacancies (Eq. 2). Vacancies can only move from TR to C and to BR or the other way around, similar to sand in an hourglass. The number of particles in the constriction n_C is the *integer state variable* of the model. Corresponding to n_C, a set of discrete I-Vs is calculated and after calibration using either Random Telegraph Noise (RTN) or slowly measured I-V data, the QPC conduction model parameters can be uniquely determined (Eq. 3) [16-18]. This is illustrated in Fig. 4.

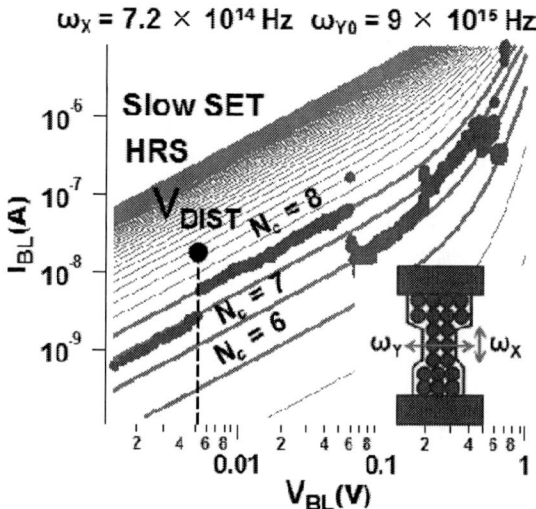

Fig. 4: Solid red lines are a set of IV-curves calculated with the QPC model for discrete values of the number of constriction particles n_c. The experimental data is a slowly measured IV-sweep. Distinct switches from the calculated curve at $n_c=7$ to that at $n_c=8$ and $n_c=6$ are visible, allowing a calibration of the conduction model parameters (top of figure). Figure from reference [18]

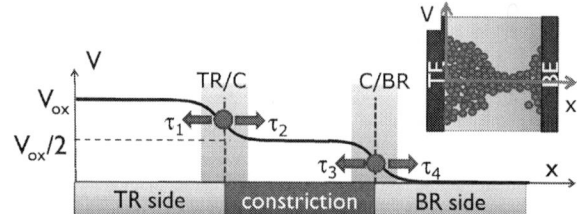

Fig. 5: Potential drop in the filament shows high field at the TR/C and C/BR interfaces. The field depends on the width of C, increasing for smaller n_C. Figure taken form reference [18]

C. Kinetic model

This model describes the *change* of n_C. Four processes need to be considered: particle emission from C to TR and BR (reducing n_C) and emission from TR and BR into C (increasing n_C). For all processes, a time constant can be associated to a single particle hop using an energy barrier E_a (Fig. 3 and Eqs. 4). Note that the details of the bond breaking and ion movement are lumped into one parameter. Since the QPC theory predicts a field enhancement at the C/TR and C/BR interfaces [23], n_C-dependent barrier modulation for particle movement can be considered in these zones (Fig. 5 and Eq. 5), providing extra field enhancement during reset. Recent data show, however, that at low current level the field enhancement saturates to a nearly constant value.

D. Thermal model

The thermal model starts from the classical Joule heating formula, applied in the *maximum field zones* of the filament. (Eq. 6). For very small dimensions, the concept of heating should be more thought of as localized 'power consumption'. Note that the 'unphysical' correction term proposed in [11] has been removed again.

E. Stochastic component

The stochastic component is brought into the model by considering the four time constants defined in Eqs. 4 as the mean values of *exponential time constant distributions*. A Monte Carlo algorithm is used for set/reset transient modeling with inclusion of statistical variations.

Table 1: the equation set of the hourglass model

Current $I = \frac{2e}{h} \int_{-qV_{ox}/2}^{qV_{ox}/2} T(E)dE$ with $E(x,y) = qV_0 - \frac{1}{2}m\omega_x^2 x^2 + \frac{1}{2}m\omega_y^2 y^2$ (Eqs 1)

and T total transmission probability through all energy levels $E_n = eV_0 + \hbar\omega_x\left(n+\frac{1}{2}\right)$ $n=0,1,2,...$

No annihilation/generation: $n_{total} = n_{BR} + n_{TR} + n_C$ (Eq. 2)

Set of IV curves corresponding to integer state variable n_C with $\omega_x(n_C) = \omega_{x,min} \cdot \frac{1}{n_C}$ (Eq. 3)

Kinetic model:

$1/\tau_1 = c.n_C.(1-\frac{n_{TR}}{N_{TR}}).\exp\left(-\frac{E_a - \alpha qV}{kT}\right)$ $1/\tau_2 = c.n_C.\frac{n_{TR}}{N_{TR}}\exp\left(-\frac{E_a + \alpha qV}{kT}\right)$ (Eqs. 4)

$1/\tau_3 = c.n_C.\frac{n_{BR}}{N_{BR}}.\exp\left(-\frac{E_a - \alpha qV}{kT}\right)$ $1/\tau_4 = c.n_C.(1-\frac{n_{BR}}{N_{BR}}).\exp\left(-\frac{E_a + \alpha qV}{kT}\right)$

Constriction-dependent barrier lowering: $\alpha = \alpha_0 + m_n / n_C$ (Eq. 5)

Thermal model: $T = T_{ambient} + \frac{\alpha VI}{n_C}R_{th}$ (Eq. 6)

IV. TRANSIENT SIMULATIONS

A. Voltage sweep SET

After subtracting the voltage drop over the load resistance or transistor, the actual set transient during a voltage sweep is revealed (Fig. 6) [11,14]. Starting from low current, the set voltage first needs to reach V_{trig} to trigger the transient. Then, a fast 'snap back' is seen to a lower voltage followed by a nearly constant voltage current increase (at the 'transient voltage' V_{trans}) until compliance is reached. The trigger voltage is statistically distributed and the transition voltage V_{trans} depends on the ramp rate [14]. These features are reproduced in the simulation with correct V_{trans}-dependence on the sweep ramp rate. The simulated temperature stabilizes to a nearly constant value once V_{trans} is reached [11].

B. Voltage-sweep RESET:

Reset occurs at approximately the same V_{trans} as set [14], as reproduced by the model. Furthermore, very slow high-resolution reset transient measurements reveal huge current fluctuations above V_{trans}, also correctly captured by the model and explained by thermal excitation of the vacancies (not shown here) [11].

C. Pulsed SET/RESET

Similar to sweep simulations, *pulsed* set and reset are modeled. Time transients and cycle-to-cycle variations are shown in Fig. 7. The simulated steep intrinsic distribution of the LRS state and the broadly distributed HRS state are experimentally confirmed [15]. Since HRS and LRS result from a stochastic process, repeated pulses followed by a verify

operation at read condition, also opens the HRS/LRS window. Furthermore, the stochastic nature of set and reset causes extrinsic cell-to-cell variations to be masked by intrinsic cycle-to-cycle variations, in full agreement with literature data [15, 24]. Finally, low voltage RTN, caused by single particle hops, and read current disturb are automatically included in the model without any parameter adaptations [11].

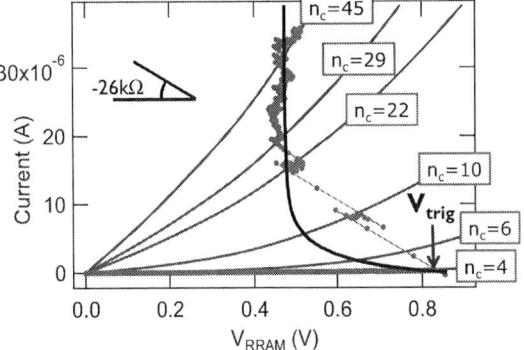

Fig. 6: Example of a slow DC set (0.1V/min in 2R configuration with load resistance=26kOhm). Abrupt current jumps to discrete levels are visible. The dashed curves are QPC calculations for an integer number of defects n_c in the constriction. The blue line corresponds to fixed time constant for the flux in and out of the constriction. Fig. taken from ref. [20].

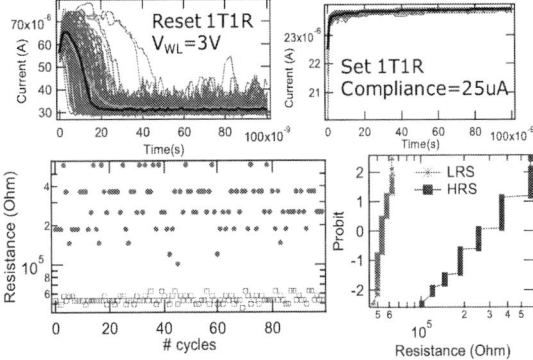

$N_{TR}=100$, $N_{BR}=10$, $n_{total}=109$, $t_{pulse}=100ns@1.5V$

Fig. 7: Simulated time transients and cycle-to-cycle variations of LRS and HRS. Current compliance during set was 25uA, but no compliance was used during reset. Bottom left figure shows 100 simulated cycles, showing the cycle-to-cycle variation. Bottom right figure shows how the HRS distribution is considerably wider than the LRS distribution. Note the discrete levels in the distribution corresponding to integer values of n_c. Fig. taken from ref. [11].

V. STATISTICAL BEHAVIOR: MODEL AND EXPERIMENT

The distribution of the HRS can be largely explained trough a variation of the number of particles in the current-controlling part of the filament (=the constriction) as in Fig. 7. For the LRS-distribution, however, a more complex picture emerges from the experimental observations. Below ~30kOhm, the LRS is lognormally distributed, but at higher resistance a bimodal behavior is observed with the high-resistance mode being wider distributed [15].

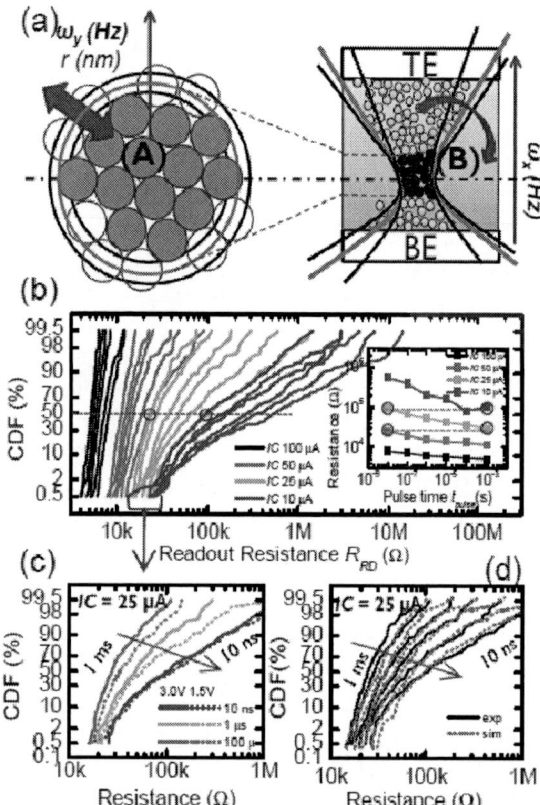

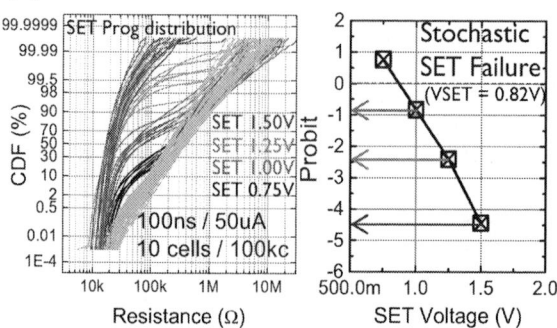

Fig. 9: Left: Single cell SET programing distributions for different V_{set}, starting from the same HRS distribution (grey lines). Right: Extrapolated stochastic SET failure percentages. For $V_{set}>0.82V$, this device has >50% probability for a successful set. Fig. taken from ref. [19].

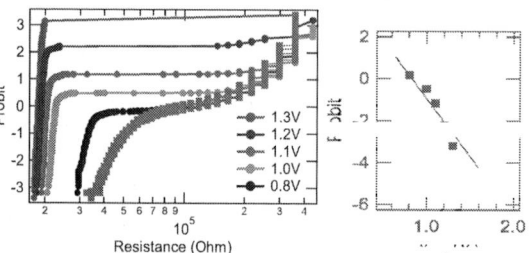

Fig. 10: a simulation of the SET distribution for different V_{set}. On the right is the extracted stochastic set failure probability. No variations of constriction geometry (as in Fig. 8) were included here.

Fig. 8: (a) Physical origin of resistance dispersion: Fluctuation in the number of particles defining radius dominates variability in the range R<~30kOhm, fluctuations in constriction geometry becomes dominant component in the range R>~30kOhm. (b) device-to-device resistance distribution for different compliances and pulse width (10 ns to 1 ms) (c) LRS distribution at current compliance = 25μA for different V_{SET} (d) Measured and simulated distribution for current compliance = 25μA. Fig. taken from ref. [15].

In the QPC model this bimodal behavior cannot be explained exclusively with a fluctuation of the number of defects defining the critical radius as depicted in Fig. 8a, but also changes in constriction geometry (length in particular) should be accounted for. Fig 8b show that the above described broadening of LRS distribution is indeed only a function of the median programmed resistance as it is possible to "trade" current compliance for pulse width and obtain superimposed distributions. As a consequence also relatively high compliance may display broad distribution tails for very short pulses or higher percentiles. Fig. 8c clarifies that, in the described behavior, the impact of delay-to-trigger in SET transition, yet possible, is negligible as pulses with different amplitude (1.5 V and 3 V) yield qualitatively the same distributions shape. Finally in Fig. 8d, Monte Carlo simulations performed with the hourglass model are able to correctly reproduce the LRS distribution as function of both compliance and pulse width when assuming a distribution of ω_x instead of a single value.

Apart from the distribution shape, we can also study the probability for a successful set or reset. Fig. 9 shows the LRS-distribution for V_{set} from 0.75 to 1.5V, starting from the same HRS distribution. The probability for stochastic SET failure can be readily extracted. Exactly the same trend is reproduced by the hourglass model as shown in Fig. 10. Note that in order to limit computation time, the variation of ω_x (as in Fig. 8) was not included. This explains the steep LRS distribution.

Simultaneous matching of the voltage/time relation as well as the stochastic set failure probability demonstrates the strength of our model.

VI. CONCLUSIONS

The statistical SET and RESET transient properties of RRAM under various conditions can be quantitatively predicted by means of the hourglass model. This model consists of an abstraction of the actual filament to a two-reservoir system connected by a constriction. Since RRAM has an intrinsic stochastic behavior, RRAM-controlling circuits should be adapted to deal with this property.

REFERENCES

[1] I.-G. Baek, D. C. Kim, M. J. Lee et al. , "Multi-layer Cross-point Binary Oxide Resistive Memory (OxRRAM) for post-NAND Storage Application", *IEDM Tech Dig.*, pp.750 – 753, 2005.

[2] H. Y. Lee, P. S. Chen, T. Y. Wu et al. , "Low Power and High Speed Bipolar Switching with A Thin Reactive Ti Buffer Layer in Robust HfO₂ Based RRAM", *IEDM Tech Dig.*, pp.297 – 300, 2008.

[3] Joonmyoung Lee, Jungho Shin, Daeseok Lee et al. , "Diode-less Nano-

scale ZrOx/HfOx RRAM Device with Excellent Switching Uniformity and Reliability for High-density Cross-point Memory Applications", *IEDM Tech. Dig.*, pp.452-455, 2010.

[4] B.Govoreanu, G.S.Kar, Y-Y.Chen et al. , "10nmx10nm Hf/HfOx Crossbar Resistive RAM with Excellent Performance, Reliability and Low-Energy Operation", *IEDM Tech. Dig.*, pp.730-732, 2011.

[5] C.H. Wang, Y-H. Tsai, K-C. Lin et al. , "Three-dimensional 4F2 ReRRAM cell with CMOS logic compatible process", *IEDM Techn. Dig.*, pp. 664-667, 2010.

[6] N. Xu, B. Gao. L.F. Liu et al., "A unified physical model of switching behavior in oxide-based RRAM", *Symp. on VLSI Technol. Dig. Of Techn. Papers*, pp. 100-101, 2008.

[7] G. Bersuker, D.C. Gilmer, D. Veksler et al. , "Metal oxide RRAM switching mechanism based on conductive filament microscopic properties", *IEDM Techn. Dig.*, pp. 456-459, 2010.

[8] R. Degraeve, A. Fantini, S. Clima et al. , "Dynamic 'Hour glass' model for SET and RESET in HfO2 RRAM", *Symp. on VLSI Technol. Dig. Of Techn. Papers*, pp. 75-76, 2012.

[9] R. Degraeve, Ph. Roussel, L. Goux et al., "Generic learning of TDDB applied to RRAM for improved understanding of conduction and switching mechanism through multiple filaments", *IEDM Techn. Dig.*, pp. 632-635, 2010.

[10] R. Degraeve, L. Goux, S. Clima et al. , "Modeling and tuning the filament properties in RRAM metal oxide stacks for optimized stable cycling", *VLSI-Technology Systems and Applications, Proceedings of technical papers*, 2012.

[11] R. Degraeve, A. Fantini, N. Raghavan et al., "Modeling RRAM set/reset statistics resulting in guidelines for optimized operation", *Symp. on VLSI Technol. Dig. Of Techn. Papers*, pp. 98-99, 2013.

[12] G.S. Kar, A Fantini, Y.Y. Chen et al., "Process-improved RRAM cell performance and reliability and paving the way for manufacturability and scalability for high density memory application", *Symp. on VLSI Technol. Dig. Of Techn. Papers*, pp. 157-158, 2012.

[13] S. Clima, Y.Y. Chen, R. Degraeve et al. "First-principles simulation of oxygen diffusion in HfO2: role in the resistive switching mechanism", *Appl. Phys. Lett.*, 100, 133102, 2012.

[14] A. Fantini, D.J. Wouters, R. Degraeve et al., "Intrinsic switching behavior in HfO2 RRAM by fast electrical measurements on novel 2R test structures", *presented at IMW 2012*.

[15] A. Fantini, L. Goux, R. Degraeve et al., "Intrinsic switching variability in HfO2 RRAM", *presented at IMW 2013*.

[16] N. Raghavan, R. Degraeve, A. Fantini, et al. , "Modeling the impact of reset depth on vacancy-induce filament perturbations in HfO2 RRAM", *IEEE Electron Dev. Lett.*,Vol. 34, nr. 5, pp. 614-616, 2013.

[17] N. Raghavan, R. Degraeve, A. Fantini et al., "Microscopic Origin of Random Telegraph Noise Fluctuations in Aggressively Scaled RRAM and its Impact on Read Disturb Variability" Proc. IEEE IRPS, 2013.

[18] N. Raghavan, R. Degraeve, A. Fantini et al., "Stochastic Variability of Vacancy Filament Configuration in Ultra-Thin Dielectric RRAM and its Impact on OFF-State Reliability", *IEDM Tech. Dig.*, pp.554-557, 2013.

[19] A. Fantini, L. Goux, A. Redolfi et al. , «Lateral and vertical scaling impact on statistical performances and reliability of 10nm TiN/Hf(Al)O/Hf/TiN RRAM devices", *to be presented at VLSI Technol. Symp.* 2014.

[20] R. Degraeve, A. Fantini, Y.Y. Chen et al. "Reliability of low current filamentary HfO2 RRAM discussed in the framework of the hourglass SET/RESET model", IEEE IIRW Final report, 2012.

[21] E. Miranda, P. Falbo, M. Nafria and F. Crupi, "Electron transport through electrically induced nanoconstrictions in HfSiON gate stacks", *Appl. Phys. Lett.*, vol. 92, 253505, 2008.

[22] M. Büttiker, "Quantized transmission of a saddle-point constriction", *Phys. Rev. B*, vol. 41, no. 11, pp. 7906-7909, 1990.

[23] S. Ulreich and W. Zwerger, "Where is the potential drop in a quantum point contact?", *Superlattices and microstructures*, vol. 23, no. 3/4, pp. 719-730, 1998.

[24] N. Ramaswamy et al., Presented at *IEEE IIRW 2012*.

Study of (correlated) trap sites in SILC, BTI and RTN in SiON and HKMG devices

Erik Bury[1,2], Robin Degraeve[1], Moon Ju Cho[1], Ben Kaczer[1], Wolfgang Goes[3], Tibor Grasser[3], Naoto Horiguchi[1], Guido Groeseneken[1,2]

[1]imec, Kapeldreef 75 – B-3001, Heverlee, Belgium
[2]KU Leuven, Dept. ESAT-MICAS, Kasteelpark Arenberg 10 – B-3001, Heverlee, Belgium
[3]TU Wien - Austria
e-mail: erik.bury@imec.be – phone: +32 16 28 11 08

ABSTRACT

Recently, several experimental groups have found correlations in gate and drain current fluctuations. In this paper, by studying single trap activated leakage paths, both evidence and a refined 4-state defect model are provided, ascribing additional gate tunneling current in nm-FETs to thermally activated defect states. The model is capable of explaining both positive and negative correlations in gate and drain current RTN, but also the mostly uncorrelated nature of these drain and gate RTN signals.

I. INTRODUCTION

Enormous CMOS device improvements have been achieved over the last few decades, both by scaling device dimensions as by reducing the equivalent gate oxide thickness. As a consequence of this, degradation and fluctuations in drain and gate leakage currents become more pronounced, even in a way that they could seriously affect device performance. Phenomena such as bias temperature instabilities and RTN have been extensively studied over the last years [1].

Studying these phenomena in nm-sized FETs can give insight in the underlying physical principles—it is well established that in small devices, charge trapping and de-trapping of single defects can significantly alter the channel current, as shown in Franco et al.'s "ultimate" BTI experiment [2]. Also single leakage paths from the substrate to the gate electrode can be identified and extracted [3], as illustrated in Fig. 1. Finally, it has been shown [4-6] that correlated gate and drain current RTN exists in nFETs. Recently it was shown that even in one single device, both positive as negative correlations can be found [7].

A possible explanation for this latter phenomenon is the electrostatic screening in a direct tunneling model, affected by a discontinuous oxide band banding. Conversely, it has been shown by [8] that electrostatic screening alone cannot alter the gate leakage current more than a few percent. Another theory based on trap-assisted tunneling (TAT) was proposed in [9] and [10], where a trap in the oxide acts as a (thermally activated) stepping stone for tunneling towards the gate electrode, as illustrated in Fig. 2.

This paper shows that the full $I_G V_G$ characteristics of individual trap site-induced leakage paths can be extracted and provides indications that *trap-assisted tunneling can described with a refined 4-state defect model*, in which the *transition rates are crucial* for describing all correlated gate leakage currents in

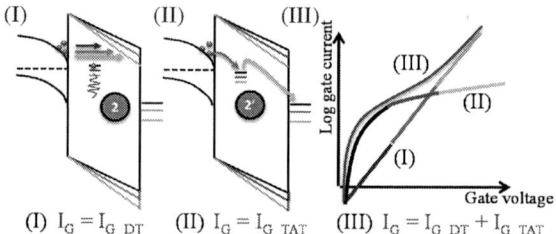

(I) $I_G = I_{G\_DT}$ (II) $I_G = I_{G\_TAT}$ (III) $I_G = I_{G\_DT} + I_{G\_TAT}$

Fig 1. The schematic representation of the components of the gate leakage current. (I) A pre-existing or generated TAT-defect is charged and subsequently relaxed through multi-phonon emission. The trap cannot contribute to the gate leakage current and therefore only a direct tunneling current is measured. (II) The defect can actively contribute by trap-assisted tunneling (TAT). (III) This gate leakage path is superimposed on the direct tunneling gate leakage [9].

nm-size FETs. Moreover, indications are given that most defects responsible for (de-)activating the SILC, also in HKMG devices, should be located in the SiO_2 interfacial layer.

II. MEASUREMENT APPROACH

The measurements in this work were conducted on nanoscale nFETs (dimensions as indicated below) with SiON and HKMG gate stacks respectively. The stacks were specifically selected to have a comparable *physical* thickness. While the lateral dimensions of these devices are chosen small enough to increase the impact of single-defects on the drain current [2], the gate leakage current density has to be sufficiently high to be within measurement resolution at gate biases around the device's threshold voltage. The FET currents were simultaneously measured with a pair of Keithley 2636 units, either with a voltage sweep or with a constant voltage at a rate of 10 samples/s. All the measurements reported here were performed at 25°C unless noted otherwise. The experimental procedures are depicted in Fig. 3. Experiment A reveals the properties of trapped-charge induced BTI-shifts, the voltage dependence of a single leakage path, and the effect of stress on the generation/activation of these gate leakage paths, whereas experiment B reveals the time-dependent characteristics and

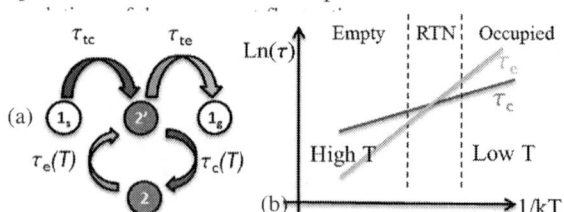

Fig 2. (a) The trap-assisted tunneling model as proposed by [9] and [10], and (b) typical temperature dependence of the time constants characterizing the transition between metastable states 2' and 2.

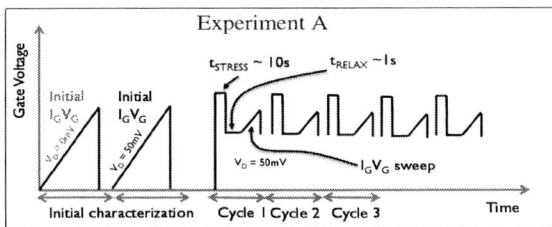
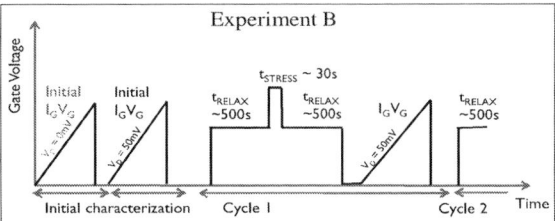

Fig 3. Schematic illustration of the experiments performed. Experiment A provides learning about the *stress dependent evolution* of the SILC as it is focused on measuring I_GV_{GS} after intermittent stress cycles. The cycles are repeated 100 times. Experiment B focusses on the *time evolution and correlation* of both the gate and drain current after one short stress phase. During t_{RELAX} correlations between RTN in I_G and I_D are measured, as well as I_B.

only [Fig. 1(a)]. The latter can be found after scanning multiple devices. Note that the TAT-current is *not necessarily stress-induced*, but it can also be process-induced as shown for the initial I_GV_{GS} in Fig. 4. An inflected gate leakage current (thus with a TAT component) can be observed in both SiON as HKMG devices, but remarkably, *it's not observed substantially more in the latter*, even though it's known that the high-k and SiO2/high-k interface defect density is at least one order of magnitude higher. This is an indication that the enabling trap for the TAT is located in the SiO$_2$ rather than in the high-k material, a conclusion similar to [11] and [12].

The voltage shift of the I_G curves in the linear regime of the MOSFET when increasing V_D from 0 to 0.05V reflects *the lateral position of the TAT path,* as explained in Fig. 5. The linear drop of the channel potential influences the local field in the gate oxide, and thus the leakage current through the TAT-path, as shown in Fig. 1. Therefore, the ratio of voltage shift of this single TAT-path (ΔV_{IG_TAT}) with the applied V_D thus determines the relative position of the trap in the channel:

$$x_{trap} = L_{channel} \frac{\Delta V_{IG\_TAT}}{V_D} \quad (1)$$

This technique is a suitable alternative for the 's-ratio' technique in inversion [13], which relies on the channel resistance differences between the leakage path and drain/source junction, which become unmeasurable in nanoscale and thus ultra-short FETs.

IV. LINK WITH BIAS TEMPERATURE INSTABILITIES

The introduction of HKMG devices also leads to positive BTI, visible as a positive shift of the I_D-V_G characteristic. The *TAT-paths after subsequent stresses are also influenced by the electrostatic charge accumulated in the oxide* due to this BTI-induced charge trapping [Fig. 6(a)]. It is therefore necessary to compensate for this shift. This can be done trace by trace, as the channel current is measured simultaneously.

Our experiments cannot conclusively determine a correlation between the impact of the defects on the drain current (i.e. the position w.r.t. the percolation path) and the screening of the gate current: the trapped BTI-charges will have an electrostatic impact on the TAT-path, but *not* according to their impact on the I_D, i.e. a trap close to a channel percolation

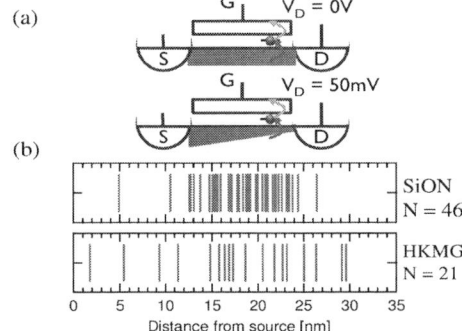

Fig 5. (a) The linear drop over the channel potential allows to extract the lateral position of the TAT-paths, using the V_D dependence of the I_{G_TAT} curves. (b) The SiON traces show a more centered distribution of TAT-paths than the HKMG devices.

path will not necessarily induce a large voltage shift on the TAT-path and vice versa.

However, based on BTI relaxation traces and simple charge sheet approximation (Fig. 6 [a]), we can conclude that our HKMG devices (90x28nm) can easily contain a few tens of trapped charges after stress, and *the 'average' BTI shift and I_G shift are nearly equal*, as shown in Fig. 6 (b). Therefore, in these devices, correcting the I_GV_G traces with a 'BTI' shift measured on the drain, as shown in Fig. 7, is a crude yet useful fix, as it helps to reveal the smaller ΔI_G TAT paths over the shifted background current.

Fig. 7(b) shows that the TAT-paths can both be induced or activated by stress, but *also de-activated* [Fig. 8(a)]. In some cases [Fig 8(d)], the TAT-paths *are activated and de-activated dynamically* during the I_GV_G sweep. It is well known that the generation of defects and thus potential TAT-paths is gate bias dependent (a high stress voltage will induce new defects), but also the activation and de-activation of such a path can show a *gate bias dependence during the sweep*. For the case of Fig. 8

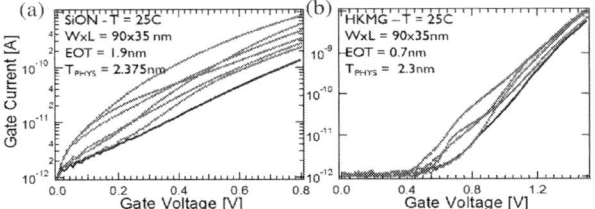

Fig 4. Initial I_DV_G's for different (a) SiON and (b) HKMG devices with similar physical thickness. In both cases, only a few devices show a near-perfect exponential leakage current (black lines). Other devices show a inflected I_DV_G, indicating the presence of at least one TAT-leakage path.

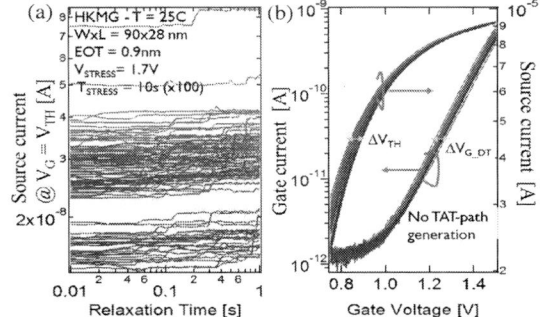

Fig. 6 (a) Experiment A performed on a HKMG device (WxL: 90x28nm, EOT = 0.9nm) for one stress voltage. Relaxation traces of the device after subsequent stresses shows tens of charges being trapped. (b) The I_DV_G and I_GV_G traces after the short relaxation show a comparable total V_{TH} shift for both drain and gate currents, but no TAT-paths have been created.

978-1-4799-3911-4/14 $31.00 © 2014 IEEE   251

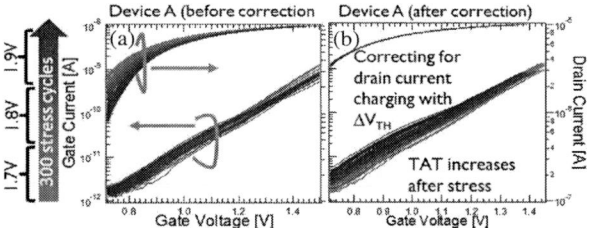

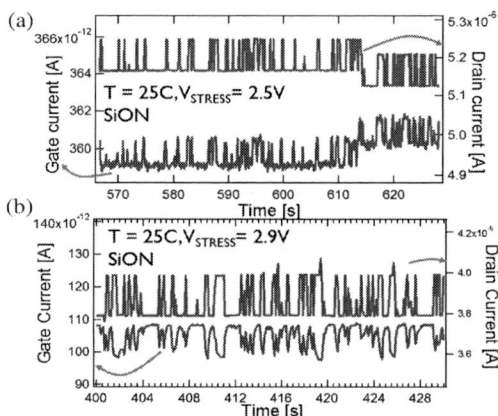

Fig 7. Experiment A performed on a HKMG device (WxL: 90x28nm, EOT = 0.9nm). The V_{TH} is quasi-*continuously* shifting with increasing stress voltage and stress cycles, visible in the I_D, due to the large number of trapped charges. The I_G shows mostly discontinuous steps which are not distinguishable in I_D. In (a) the trapped-charge screening effect on the TAT is apparent, while in (b) correcting for this V_{TH} shift, also illustrated in I_D, makes the TAT paths visible.

(b), the extracted probability for the TAT-path being disabled *increases* with gate bias (Fig 8 [c]).

Switching TAT-paths show RTN-like behavior if the gate leakage current is measured at constant bias, thereby proving that *one trap can activate or de-activate these leakage paths*. As the single path ΔI_G can be $\approx 2.5 \times 10^{-11}$A [Fig. 8(b)], about 1.5×10^8 charges/s should be captured and emitted to the gate by the oxide trap. The *slowest* time constant in the inelastic TAT process (thus either τ_{tc} and τ_{te}) is thus is the order of ns. The (in)elasticity of this tunneling current has no impact on the considerations in the proposed model.

At elevated temperature, the I_G traces become unstable and vary between every sweep. This RTN-like I_G is mostly not visible in I_D. Fig. 8(d) shows experiment A (Fig. 1) performed at a temperature typical for BTI experiments (125°C). More charge trapping—observed as V_{TH} shift in I_D—is apparent. Shifts up to 300mV are observed (not shown here), although the device remains fully functional.

Meanwhile, the gate leakage current is strongly increased to the point where the *individual TAT-paths become indistinguishable*. Their *activation and de-activation time constants are strongly temperature dependent* and decrease below measurement integration time, as predicted in the phonon relaxation model, and illustrated in Fig. 2 (b).

The results of experiment A and B show that PBTI of SiON is much lower than HKMG FETs. Typically, PBTI is attributed to defects in the high-k. In the latter, charge trapping occurs at much lower stress. Due to the defective nature of the high-k layer, phenomena as BTI and I_D RTN are abundantly visible, thereby inhibiting individual characterization, but the TAT-paths remain similar.

V. GATE AND DRAIN CURRENT CORRELATION

I_D and I_G are measured simultaneously and $\Delta I_G/\Delta I_D$ correlations are well within the measurement window. For the devices tested, *most discharging events in the drain show no*

Fig 9 Both (a) positive and (b) negative I_D/I_G correlations are measured on separate SiON nFETs after stress. The negative correlation in (b) also shows signs of a second RTN path in the drain current, and is superimposed both on the low and the high state of the signal, but not visible in the gate current.

effect on the gate current and vice versa. This is remarkable as it is generally assumed that PBTI is primarily caused by traps in the high-k and the SiO₂/high-k interface, far from the channel, so also charging these 'far' traps should have an impact on the drain current.

Apart from the discharging, single and multi-level RTN signals are visible, mostly in I_D. These signals are ideal for studying correlations in TAT-paths and BTI traps. The multi-level RTN traps can be distinguished and treated as separate traps, as long as the ΔI are deviating enough. In some cases, a correlated gate and drain RTN could be found, and remarkably both positively and negatively, as shown in Fig. 9.

Our model, explained schematically in Fig 10, can describe no, positive and negative I_G/I_D correlations for nFET devices, is consistent with the 4-state defect model as proposed by [14]. The no-correlation, which is mostly seen, is explained with an active TAT-path located near the channel. Even in its meta-stable position, the trap will have a net charge, as the time constant for inelastic tunneling from the channel towards the trap ($1s \rightarrow 2'$) is smaller than the time constant to tunnel from the trap towards the gate ($2' \rightarrow 1g$). The net charge is then defined by the product of the occupancy probability and the defect charge state. The trap's de-activated complement state 2 is by definition also a charged defect state [Fig. 10(a)]. Therefore, *no net charging in the oxide occurred whilst activating or de-activating the TAT-path*.

Only the cases where the capture time is dominant (i.e. state $1s \rightarrow 2'$ is slowest, thus the trap located far from the interface) could show a *positive correlation between I_D and I_G*, [Fig. 10

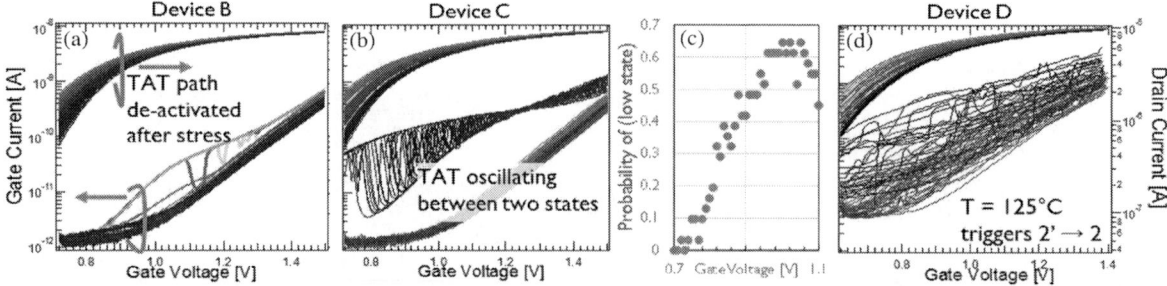

Fig 8. Experiment A performed on other HKMG devices (WxL: 90x28nm, EOT = 0.9nm). The V_{TH} is continuously shifting with stress voltage and cycles for all cases. The I_G shows mostly discontinuous steps which are not distinguishable in I_D. (a) TAT paths are oscillating and then de-activated after stress and (b) vice versa. (c) The gate voltage dependence of occupation probability of the two states in (b). (d) At high temperatures, the $2' \leftrightarrow 2$ transitions are occurring promptly and faster than the measurement integration time.

978-1-4799-3911-4/14 $31.00 © 2014 IEEE

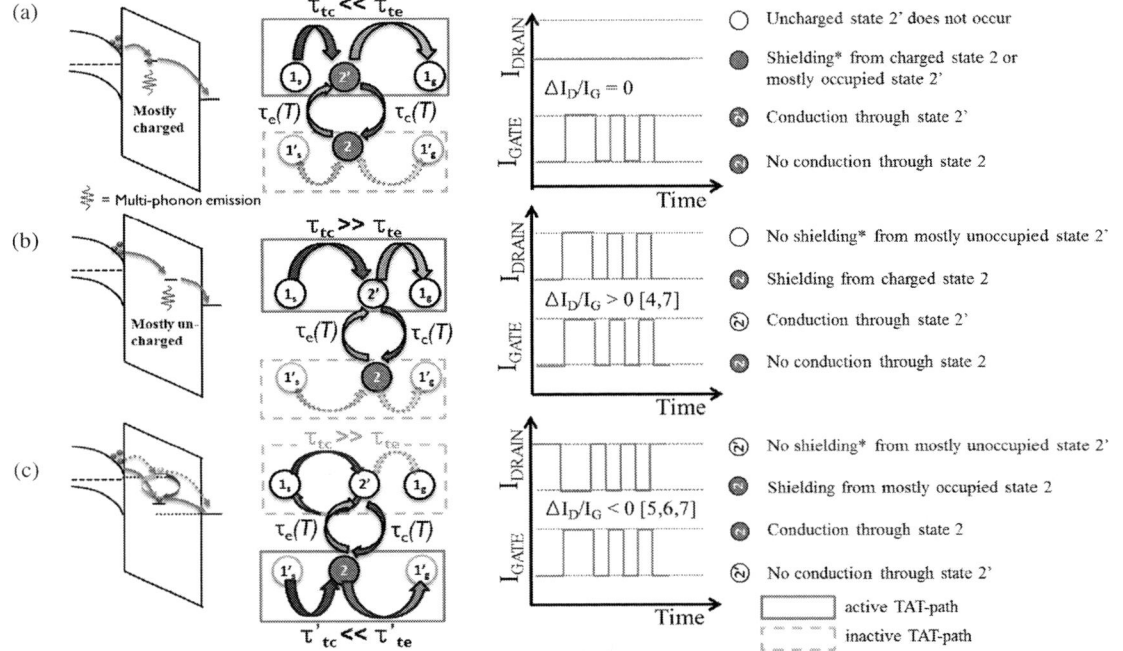

*Shielding is determined by net charge = (occupancy probability) x (defect state charge)

Fig 10. Model for TAT explaining no, positive and negative I_D/I_G correlations for nFET devices. The open symbol indicates the net charge of the defect state, the closed symbol vice versa. (a) The TAT needs an intermediate charged state 2', of which the occupancy probability can vary depending on the time constants τ_{tc} and τ_{te}. When the transition 2'→2 occurs, the TAT path (due to exchange between states 1 and 2') switches off. If $\tau_{tc}<<\tau_{te}$, state 2' is quasi continuously charged, making the 2' → 2 transition invisible for the channel current (= $\Delta I_D/\Delta I_G$ =0). (b) If $\tau_{tc} >> \tau_{te}$ (for states far away from the interface), state 2' is mostly unoccupied. In this case, the phonon relaxation of the TAT defect towards state 2 *will result* in a net charge difference observable in I_D. We therefore find a *positive* RTN correlation ($\Delta I_D/\Delta I_G$ >0). (c) Negative correlations are explained by favorable transitions between the *defect states* 1'$_g$/1'$_s$ and 2. State 2 (the defect state after reconfiguration after electron capturing) *also has to be* a net charged state. Thus, for inverse correlation, the transition rates $\tau'_{tc} << \tau'_{te}$ between the secondary states are 1) much shorter and 2) opposite in their magnitudes: $\tau_{tc} >> \tau_{te}$.

(b)]. These correlations were already shown by [4] and [7] and are confirmed in these measurements.

Also the negative correlation observed earlier in [5-7] and now seen here, cannot just be explained with a TAT scheme and neither with direct electrostatic interaction. A plausible explanation for these correlations is by favorable transitions between the secondary defect states 1'$_g$/1'$_s$ and 2. The transition rates $\tau'_{tc} << \tau'_{te}$ between the secondary states are 1) much shorter and 2) opposite in their magnitudes: $\tau_{tc} >> \tau_{te}$. The transition rate τ_c *towards the secondary defect state* has to be short as state 2' is quasi unoccupied.

In both correlated cases, an indirect gate voltage dependence is expected ascribed to a change in occupancy probability of state 2' influencing the 2'→2 transition, due to typical voltage dependency of BTI parameters τ_c. This gate bias dependence can be further investigated with noise or RTN measurements.

VI. CONCLUSIONS

Extended analysis of currents on all terminals of nanoscaled devices yields large insight in TAT/SILC, RTN and BTI mechanisms we propose a model capable of explaining the measurement observations, and in particular the positive and negative I_D and I_G correlations.

VII. REFERENCES

[1] W. Goes, M. Toledano-Luque, O. Baumgartner, M. Bina, F. Schanovsky, B. Kaczer and T. Grasser, "Understanding correlated drain and gate current fluctuations", in Proc. of IPFA, pp. 51-56, July 2013.

[2] J. Franco, B. Kaczer, B,, M. Toledano-Luque, P.J. Roussel et al., "Impact of Single Charged Gate Oxide Defects on the Performance and Scaling of Nanoscaled FETs", in Proc. IRPS, pp. 5A.4.1 - 5A.4.6, April 2012.

[3] R. Degraeve, B. Govoreanu, B. Kaczer, J. Van Houdt and G. Groeseneken, "Measurement and statistical analysis of single trap current-voltage characteristics in ultrathin SiON", in Proc. IRPS, pp. 360-365, 2005.

[4] M. Toledano-Luque, B. Kaczer, E. Simoen, R. Degraeve, J. Franco et al., "Correlation of single trapping and detrapping effects in drain and gate currents of nanoscaled nFETs and pFETs", in Proc. IRPS, pp. XT. 5.1 – XT. 5.6, April 2012.

[5] C-Y. Chen, Q. Ran, H. Cho, A. Kerber, Y. Liu et al., "Correlation of Id- and Ig-Random Telegraph Noise to Positive Bias Temperature Instability in Scaled High-κ/Metal Gate n-type MOSFETs" in Proc. IRPS, pp. 190 – 195, 2011.

[6] X. Ji, Y. Liao, C. Zhu, J. Chang. F.Yan, Y. Shi, and Q. Guo, "The Physical Mechanisms of Ig Random Telegraph Noise in Deeply Scaled pMOSFETs," in Proc. IRPS, pp. XT.7.1–XT.7.5, 2013.

[7] W . Liu et al, "Analysis of Correlated Gate and Drain Random Telegraph Noise in Post-Soft Breakdown TiN/HfLaO/SiOx nMOSFETs", in Electron Devices Letters, Vol. 35, No. 2, pp. 157-159, 2014.

[8] O. Baumgartner, M. Bina, W. Goes, F. Schanovsky, M. Toledano-Luque, B. Kaczer, H. Kosina and T. Grasser, "Direct Tunneling and Gate Current Fluctuations", in Proc. SISPAD, pp. 17-20, September 2013.

[9] M. O. Andersson, Z. Xiao, S. Norrman, and O. Engstrom, "Model based on trap-assisted tunneling for two-level current fluctuations in submicrometer metal—silicon-dioxide—silicon diodes", Phys. Rev. B., Vol 41, pp. 9836-9842, 1990.

[10] B. Kaczer, M. Toledano-Luque, W. Goes, T. Grasser and G. Groeseneken, "Gate Current Random Telegraph Noise and Single Defect Conduction", in Microelectronic Engineering Volume 109, pp. 123–125, September 2013.

[11] T. Kauerauf, Robin Degraeve, Lars-Åke Ragnarsson, Philippe Roussel, Sahar Sahhaf, Guido Groeseneken, "Methodologies for sub-1nm EOT TDDB evaluation" in Proc. IRPS, pp. 7-16, 2011.

[12] G. Bersuker, D. Heh, C. D. Young, L. Morassi, A. Padovani, L. Larcher and K. S. Yew, "Mechanism of high-k dielectric-induced breakdown of the interfacial SiO2 layer", in Proc. IRPS, pp. 373-378, 2011.

[13] F. Crupi, B. Kaczer, R. Degraeve, A. De Keersgieter and G. Groeseneken, "Location and Hardness of the Oxide Breakdown in Short Channel n- and p- MOSFETs", in Proc. IRPS, pp. 55-59, 2002.

[14] T. Grasser, H. Reisinger, P.-J. Wagner, F. Schanovsky, W. Goes, and B. Kaczer, "The Time Dependent Defect Spectroscopy (TDDS) for the Characterization of the Bias Temperature Instability", in Proc. IRPS, pp. 16-25, 2010.

Transient to Temporarily Permanent and Permanent Hole Trapping Transformation in the Small Area SiON P-MOSFET Subjected to Negative-Bias Temperature Stress

Z. Y. Tung[+] and D.S. Ang

Nanyang Technological University, School of Electrical and Electronic Engineering,
Singapore 639798 (E-mail[+]: TUNG0017@e.ntu.edu.sg)

Abstract—**Examining the drain current recovery traces of a small area SiON p-MOSFET subjected to repeated NBTI stress and relaxation cycling reveals direct evidence of transient to permanent hole trapping transformation inferred from previous studies on big area devices. The results show that the emission times of hole traps are not time-invariant (as normally presumed) but can increase due to evolution of the defect sites into more structurally stable forms. In addition, a new type of switching hole traps, exhibiting intermittent charging during stress and occasional increase in emission time by ~5 orders of magnitude, is observed.**

I. INTRODUCTION

Although the microscopic mechanisms of NBTI remain controversial, it is clear that this phenomenon could severely impact the performance and reliability of advanced p-MOSFETs [1]. One of the major challenges in modeling NBTI lies in its transient nature, which has been recently shown to be mainly linked to the capture and emission of holes by switching oxide traps (SOTs) under pulsed gate stressing [2]. Our recent studies [3]-[6] have suggested that a portion of the SOTs may be progressively transformed into a more permanent form. These studies were made on large area devices and the conclusion was drawn based on the decrease of the recoverable component of NBTI as dynamic stressing progresses. To-date, no direct evidence for the proposed SOT transformation has been presented. Lately, there has been a considerable interest on small area devices which avail themselves to a detailed study of the charge capture/emission events at individual oxide defects from discrete steps in the time dependent NBTI recovery trace [7]-[10]. In the paper, direct evidence for the proposed SOT transformation is presented. After several stressing and relaxing cycles, the emission times of some of the SOTs are clearly increased, indicating that they have been changed into a more permanent form.

II. EXPERIMENTAL DETAILS

Test devices (DUTs) used in this study were p-MOSFETs,

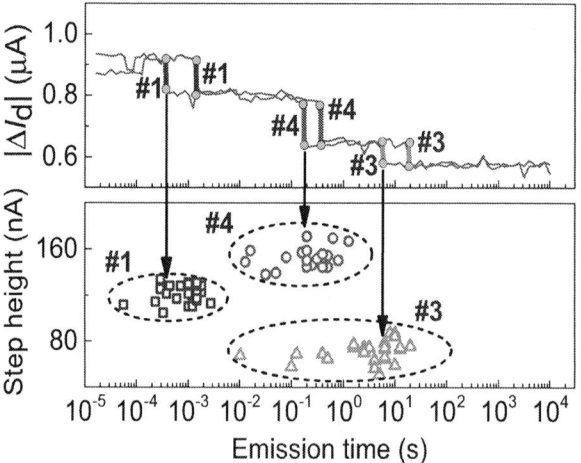

Fig. 1: Construction of a spectral map (bottom) by the extraction of single de-trapping events from two typical representative drain current, I_d, relaxation traces (top) ensuing NBTI stress. The emission time, τ_e, of a defect is obtained by tracking the discrete step height, h from the single de-trapping events, i.e. each defect is identified by a cluster of (τ_e, h) points.

with p^+ polysilicon gate and 1.7-nm SiON gate dielectric prepared by decoupled plasma nitridation. The drawn channel width and length are 120 nm and 60 nm respectively. A DUT was subjected to alternating 10-s NBTI stress and 10-ks relaxation phases each for 30 times at 100 °C. The devices were subjected to a constant gate voltage $V_g = -1.8$ V (oxide field ~8.5 MV/cm) during the stress phase, followed by $V_g = 0$ V during the relaxation phase. During relaxation, linear drain current I_d recovery was measured by pulsing V_g from 0 to −0.7 V, and completing the measurement within a delay of 1 μs (drain bias V_d was constantly set at −0.1 V). In small area devices, only a handful of SOTs are present. They can be individually identified through the different discrete step heights and emission times in the I_d recovery traces using the spectral map obtained from time-dependent defect spectroscopy [7].

III. RESULTS AND DISCUSSION

The degradation of drain current $|\Delta I_d|$, with respect to the

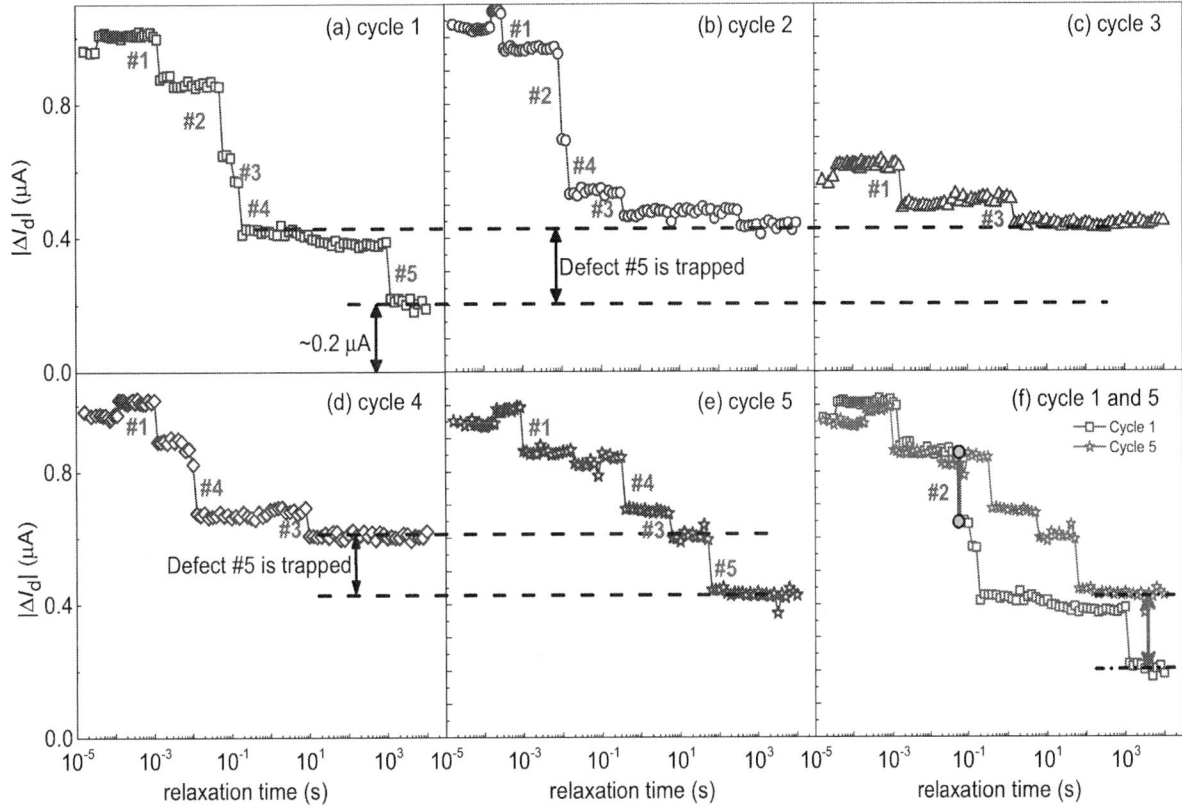

Fig. 2: ΔI_d recovery traces of cycles 1 to 5. **(a)** Recovery trace of cycle 1 shows that $|\Delta I_d| > 0$ after 10-ks of relaxation, implying that the first 10 s of NBTI stress already generates some relatively permanent defects. **(b)** Defect #5 is not recovered in cycle 2 as compared to (a), resulting in a greater $|\Delta I_d|$ degradation at the 10-ks relaxation. **(c)** Defect #2 and #4 are not charged during cycle 3, as can be seen from the much lower $|\Delta I_d|$ degradation at the start of the relaxation phase (compared to that of cycle 1 or 2). The final degradation remains the same indicating that Defect #5 still remains charged after the 3rd relaxation phase. **(d)** Defect #2 and #4 are being charged during NBTI stress in cycle 4 as the initial $|\Delta I_d|$ degradation is similar to those in (a) and (b). Defect #4 recovers but Defect #2 still remains charged since the final degradation after 10-ks relaxation is higher by the Defect #2 step height as compared to those in (b) and (c). Defect #5 still remains charged. **(e)** Defect #5 is finally discharged in cycle 5 since it was charged in cycle 2. However, Defect #2 still remains charged. **(f)** Comparison between cycle 1 and cycle 5 shows that the $|\Delta I_d|$ degradation after 10-ks relaxation in cycle 5 is larger than that in cycle 1. The recovery step height of Defect #2 is not seen and the $|\Delta I_d|$ degradation after 10-ks relaxation in subsequent cycles (starting from cycle 4) is obviously higher by an amount equal to the Defect #2 recovery step height. This may be attributed to the conversion of Defect #2 from a transient hole trap into a more permanent hole trap.

pre-stress value, during each 10-ks recovery was plotted and each SOT was identified through a spectral map made up of discrete I_d step heights, h of the hole-de-trapping events and the times, τ_e corresponding to the de-trapping events (Fig. 1). Clusters of (τ_e, h) points found on the spectral map are believed to be the "fingerprints" of individual SOTs [7].

Recovery traces of cycle 1-5 were shown in Fig 2. It illustrates the identification of individual defects through discrete step height and emission time as well as the way to identify the SOT which has converted to a more permanent form. A non-zero $|\Delta I_d|$ of ~0.2 µA at the end of the first relaxation phase is evident (Fig. 2(a)). This may be attributed to either (1) permanent interface states (e.g. dangling Si orbitals that resulted from dissociated Si-H bonds) that charge up spontaneously every time I_d measurement was made at $V_g =$ −0.7 V; (2) oxide defect(s) that are already being transformed into relatively permanent forms by the prior 10-s stress phase and therefore the trapped hole(s) could not be emitted during the relaxation interval. It should be mentioned that the $|\Delta I_d|$

never returned to ~0 in the remaining 29 recovery traces, indicating that some permanent defects were formed during the very first stress phase.

Detailed examination of I_d recovery traces from all tested devices reveals the following categories of SOTs. Examples of I_d recovery traces obtained on a DUT supporting the different categories are given in Fig. 2, 3 and 4.

- Defects #1 and #3 are typical examples of "cyclical" SOTs (Fig. 3) [2]-[8]. Their presence in every I_d recovery trace (i.e. charged in every stress phase) implies that the capture time constants are much shorter than the applied stress period. The consistent emission behaviors imply that the defect structural changes that occur upon hole-trapping are minimal and fully reversible.

- Defects #4 is coined an "intermittent" SOT. Its behavior is rather complex, comprising periodic absence from the

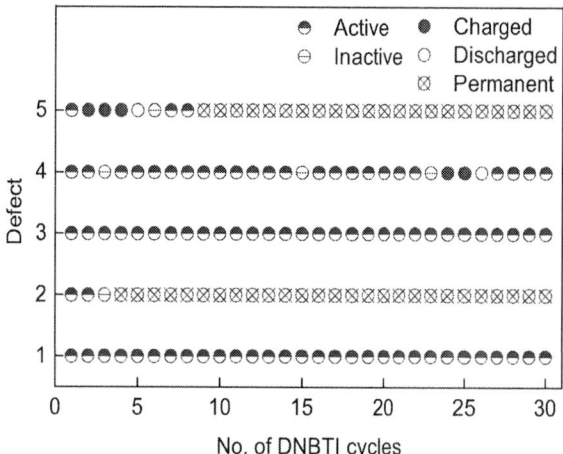

Fig. 3: Evolution of each of the SOTs based on the recovery traces in each cycle as a function of the number of DNBTI cycles. Each dynamic NBTI (DNBTI) cycle consists of a 10-s NBTI stress following by a 10-ks relaxation. A defect is considered to be active when it is charged and discharged in the same DNBTI cycle. It is deemed to be inactive when it is not charged during the stress phase. Some defect, e.g. Defect #5, is charged during stress at cycle 2 but it is not discharged until cycle 5. Defect #2 and Defect #5 transform into a more permanent form after cycle 4 and cycle 9 respectively when they cannot discharge within the experiment window.

I_d recovery trace (i.e. not charged during the prior stress phase) and more importantly temporary but significant increase in emission time (from ~0.01-1s to greater than 10^4 s). Defect #4 was active during the 1st and 2nd stress phase but not during the 3rd. As compared to the 1st and 2nd relaxation phases, the reduced $|\Delta I_d|$ at the beginning and similar $|\Delta I_d|$ at the end of the 3rd relaxation phase confirm that the SOT was not charged during the 3rd stress phase (Figs. 2(a)-(c)). After that, it could be repeatedly charged and discharged in the subsequent cycles except for the 15th and 23rd stress phases in which it was not charged. A temporary increase in emission time occurred in the 24th cycle, in which it was charged during the stress phase but was not discharged until the 26th relaxation phase (i.e. an increase in emission time by ~20 ks). After that, a cyclical behavior again followed until 30th cycle. Part of the behavior of this SOT is similar to the so-called disappearing traps which do not introduce any permanent degradation, as reported in [7]. Such inconsistent charging during stress may be due to the capture time constant being comparable to the stress period. But the temporary yet significant increase of emission time is a new behavior not reported previously.

- Defects #2 and #5 are examples of SOTs whose emission times increase during the course of the experiment. They initially exhibited the behavior of an intermitted SOT. Defect #2 was active in the 1st and 2nd cycles but was not charged in the 3rd cycle (Figs. 2(a)-(c)). After the 4th stress phase, it became permanently charged (i.e. all the way to the last 30th cycle) – Fig. 2(f)

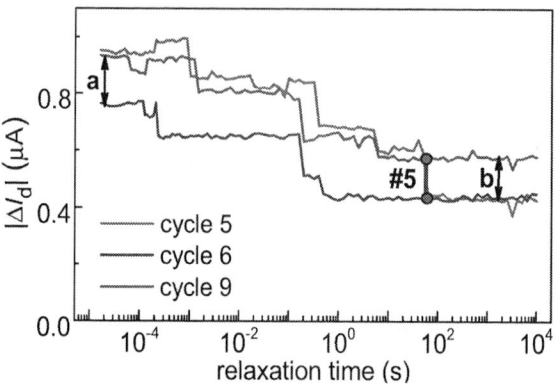

Fig. 4: $|\Delta I_d|$ relaxation traces for cycle 5, 6 and 9. After it was discharged in cycle 5, defect #5 was not activated at cycle 6 as the difference in initial degradation after stress ('a') and final degradation after 10-ks recovery ('b') between cycle 5 and 6 is due to the I_d step height of defect #5. Defect #5 was charged during stress in cycle 9 and became permanently charged thereafter since it was no longer observed in subsequent I_d recovery traces, and each trace exhibited an increase of $|\Delta I_d|$ at the end of 10-ks relaxation equal to the I_d step height of defect #5.

and 3. The emission of defect #5 is observed in the 1st I_d recovery trace, at ~1 ks (Fig. 2(a)). After the 2nd stress phase, it remained charged until after ~100 s into the 5th relaxation phase (Fig. 2(e) and 3). This can be seen from the $|\Delta I_d|$ at the end of the 2nd - 4th relaxation phases, which is higher than of the 1st relaxation phase by an amount similar to the I_d step height caused by Defect #5 (Figs. 2(a)-(d)) After that, it was not charged at 6th cycle but was active again at 7th and 8th cycle (Fig. 3). It became permanently charged after the 9th stress phase. (Fig. 3 and 4)

IV. CONCLUSIONS

The progressive increase in the emission times of Defect #2 and #5 provides an experimental evidence for an initially transient hole trap becoming more permanently charged as the NBTI stress is repeatedly applied. The results confirm our earlier inference of transient to permanent hole trap transformation, derived from the decreasing recovery per cycle of a large area device subjected to NBTI stress/relaxation cycling [2]-[6]. While the nature of such defects remains elusive, the study clearly shows that trap parameters such as the emission time constant are not time-invariant but evolve continuously during stress. The evolution is believed to be linked to phonon-facilitated local defect structural changes which occur upon charging/discharging. In addition, a temporary increase in emission time supports the proposed existence of metastable trap states [11].

ACKNOWLEDGMENT

This work received partial funding support by a Singapore Ministry of Education Research Grant MOE2013-T2-2-099. Z. Y. Tung would like to thank the Singapore Economic

Development Board and GLOBALFOUNDRIES Singapore for a joint Ph.D. scholarship grant.

REFERENCES

[1] D. K. Schroder, "Negative bias temperature instability: What do we understand?," *Microelectron. Reliab.*, vol. 47, no. 6, pp. 841–852, Jun. 2007.

[2] D. S. Ang, Z. Q. Teo, T. J. J. Ho, and C. M. Ng, "Reassessing the Mechanisms of Negative-Bias Temperature Instability by Repetitive Stress/Relaxation Experiments," *Device and Materials Reliability, IEEE Transactions on*, vol. 11, no. 1. pp. 19–34, 2011.

[3] A. A. Boo, D. S. Ang, Z. Q. Teo, and K. C. Leong, "Correlation Between Oxide Trap Generation and Negative-Bias Temperature Instability," *Electron Device Letters, IEEE*, vol. 33, no. 4. pp. 486–488, 2012.

[4] A. A. Boo and D. S. Ang, "Evolution of Hole Trapping in the Oxynitride Gate p-MOSFET Subjected to Negative-Bias Temperature Stressing," *Electron Devices, IEEE Transactions on*, vol. PP, no. 99. pp. 1–4, 2012.

[5] Y. Gao, D. S. Ang, C. D. Young, and G. Bersuker, "Evidence for the transformation of switching hole traps into permanent bulk traps under negative-bias temperature stressing of high-k P-MOSFETs," *Reliability Physics Symposium (IRPS), 2012 IEEE International*. pp. 5A.5.1–5A.5.5, 2012.

[6] Y. Gao, A. A. Boo, Z. Q. Teo, and D. S. Ang, "On the evolution of the recoverable component of the SiON, HfSiON and HfO$_2$ P-MOSFETs under dynamic NBTI," *Reliability Physics Symposium (IRPS), 2011 IEEE International*. p. XT.8.1–XT.8.6, 2011.

[7] T. Grasser, H. Reisinger, P.-J. Wagner, F. Schanovsky, W. Goes, and B. Kaczer, "The time dependent defect spectroscopy (TDDS) for the characterization of the bias temperature instability," *Reliability Physics Symposium (IRPS), 2010 IEEE International*. pp. 16–25, 2010.

[8] T. Grasser, H. Reisinger, P.-J. Wagner, and B. Kaczer, "Time-dependent defect spectroscopy for characterization of border traps in metal-oxide-semiconductor transistors," *Phys. Rev. B*, vol. 82, no. 24, p. 245318, Dec. 2010.

[9] J. Franco, B. Kaczer, M. Toledano-Luque, P. J. Roussel, J. Mitard, L.-A. Ragnarsson, L. Witters, T. Chiarella, M. Togo, N. Horiguchi, G. Groeseneken, M. F. Bukhori, T. Grasser, and A. Asenov, "Impact of single charged gate oxide defects on the performance and scaling of nanoscaled FETs," *Reliability Physics Symposium (IRPS), 2012 IEEE International*. pp. 5A.4.1–5A.4.6, 2012.

[10] B. Kaczer, T. Grasser, J. Franco, M. Toledano-Luque, P. J. Roussel, M. Cho, E. Simoen, and G. Groeseneken, "Recent trends in bias temperature instability," in *J. Vac. Sci. Technol. B*, 2011, vol. 29, no. 1, pp. 01AB01–7.

[11] T. Grasser, "Stochastic charge trapping in oxides: From random telegraph noise to bias temperature instabilities," *Microelectron. Reliab.*, vol. 52, no. 1, pp. 39–70, Jan. 2012.

Evidence for Defect Pairs in SiON pMOSFETs

T. Grasser*, K. Rott†, H. Reisinger†, M. Waltl*, and W. Goes*

*Institute for Microelectronics, TU Wien, Vienna, Austria

†Infineon, Munich, Germany

Abstract—Detailed time-dependent defect spectroscopy (TDDS) studies have recently demonstrated that recovery following negative bias temperature stress in MOSFETs is to good approximation consistent with a collection of independent (effective) first-order reactions. While the data are largely consistent with the first-order picture, several 'anomalies' such as switching traps and disappearing/reappearing traps have already been identified and analyzed. Here, we focus on a newly made observation, namely that emission events apparently belonging to a single defect can in fact be composed of two subsequent emission events if the device is stressed for a long enough time. We analyze this peculiarity as a function of bias and temperature and conclude that it is most likely due to a pair of defects which for some reason have similar configurations and thus similar properties.

I. INTRODUCTION

Using the recently introduced time-dependent defect spectroscopy (TDDS) [1], recovery of the negative bias temperature instability (NBTI) has been extensively studied at the single defect level [2–7]. These studies have shown that NBTI recovery can be well described by assuming a collection of independent (effective) first-order reactions with widely distributed reaction rates, or, alternatively, capture and emission times. At closer inspection, as is the case for many complicated systems, the reactions appear first-order only under certain circumstances, since metastable defect states have a fundamental impact on the dynamics. In particular, these metastable states explain the difference between fixed positive vs. switching traps [8], the decorrelation between capture and emission times as well as their frequency dependence [9]. Also, temporary random telegraph noise (tRTN) stimulated by NBTI stress [1] can only be understood with metastable states [10]. Furthermore, defects have been observed to disappear and reappear during subsequent stress and recovery cycles on a wide range of timescales, suggesting the involvement of hydrogen [11].

Here, we look at a newly observed anomaly, namely the occurrence of two emission events inside a single trace with nearly identical statistics. These statistics will be analyzed and discussed as a function of bias and temperature and then contrasted with three possible explanations.

II. EXPERIMENTAL METHOD

In a TDDS setup, a nanoscale device is repeatedly stressed and recovered (say $N = 100$ times) using fixed stress/recovery times, t_s and t_r. The recovery trace is analyzed for discrete steps of height η occurring at time t_e. Each (η, t_e) pair is then placed into a 2D histogram, which we call spectral

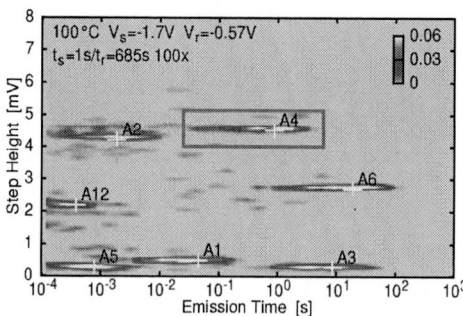

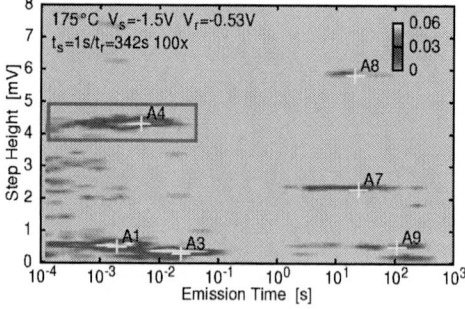

Fig. 1: The spectral map of device A of [1, 8] at 100 °C (top) and 175 °C (bottom). From the visible defects, for longer stress times only defect A4 contains two emission events from a single trace. This feature is maintained at different temperature and different voltages, implying that this is not simply due to a coincidental overlap of two defects. Note how defects have different activation energies, implying a different relative 'movement' on the map with changing temperature. Also note that defect A6 has disappeared from the map in the bottom figure.

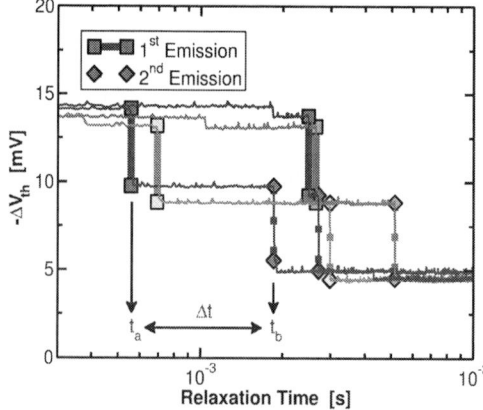

Fig. 2: Typically, in each trace only a single emission event is observed per cluster. However, particularly for larger stress times, two emission events can sometimes be observed. Here, four example traces contributing to the 175 °C spectral maps of Fig. 1 are shown. The step-heights of the second emission event are within 5% of those of the first emission event. Note that the emission events t_a and t_b *do not* directly correspond to the emission times of the defects as discussed below.

978-1-4799-3911-4/14 $31.00 © 2014 IEEE

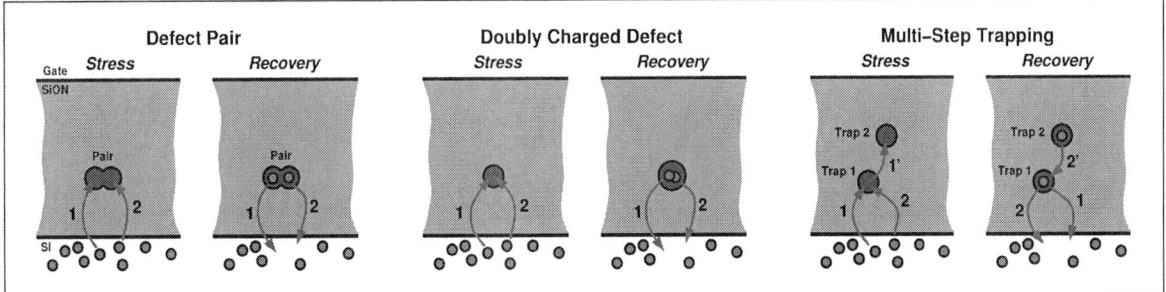

Fig. 3: **Left**: Scenario A: in a certain defective region of the oxide two holes can be trapped subsequently, with the second capture having a larger time constant. A similar configuration is required to explain the same emission time for both emission events. **Middle**: Scenario B: A defect may be able to subsequently capture two holes. **Right**: Scenario C: In a three-step process, a hole is first captured (1), then moves deeper into the oxide (1'), which allows capture of a second hole (2). Emission proceeds in reverse order but always from the same defect, which could explain the similar emission times.

map, see Fig. 1 for examples. The clusters forming in the spectral maps reveal the probability density distribution and thus provide detailed information on the statistical nature of the trap annealing time constant τ_e. We remark that the TDDS only detects the change of the charge state of individual defects, and as such cannot directly differentiate e.g. between the emission of charges from oxide states or the annealing of interface states. Nevertheless, so far, only exponentially distributed emission/annealing events (analyzed on a logarithmic scale) have been observed for the emission events t_e,

$$f(\eta, t_e) = f_\eta(\eta) \frac{t_e}{\tau_e} \exp\left(-\frac{t_e}{\tau_e}\right) \tag{1}$$

which is consistent with independent first-order processes. Due to noise, the step-heights η are typically Gaussian distributed around an exponentially distributed mean $\bar{\eta}$. However, defects may interact electrostatically [1], leading to multiple peaks in the f_η distribution. As such, the extraction of the (η, t_e) pairs is sensitive to noise, particularly RTN, and a certain (hard to quantify) error larger than the typical $\Delta\tau_e = \pm\tau_e/\sqrt{N}$ has to be expected. Furthermore, clusters in the spectral map of the form given by (1) may partially overlap, leading to the erroneous detection of (η, t_e) pairs and assignment to the 'wrong' defect. Still, at not too high stress voltages and not too large stress times, the accuracy is typically very high, and the bias and temperature dependence of $\bar{\eta}$ and τ_e can be easily extracted.

III. REFINED DATA ANALYSIS

At closer inspection of the extraction error in the (η, t_e) pairs we noticed that in some defects two emission events which apparently belong to the same cluster can occur in a single trace. These emission events occurred with a regularity clearly beyond any extraction errors and were investigated in closer detail. Roughly, about 10-20% of the defect clusters in three devices investigated belong to this category. One example, defect A4 (defect #4 of device A previously studied [1]), is shown in Fig. 1. At a first glance, the cluster looks perfectly regular as expected from (1). However, it occasionally contains two emission events from a single recovery trace, inconsistent with (1), see Fig. 2. This phenomenon is observed for all

temperatures studied ($100\,°C - 175\,°C$), not too low biases ($\gtrsim V_{DD} = -1.3\,V$, on a SiON device [12] with EOT=2.2 nm and $V_{th} = -0.7\,V$), starting from stress times about a 1,000 times larger than the capture time of the cluster that produces the first emission event.

The *trivial* explanation for such double emissions, namely that two independent defects by fortuitous coincidence just happen to have the same step-heights and emission times, can be ruled out based on several considerations:

(i) the spectral maps are only scarcely populated and the probability that two defects have the same parameters within experimental resolution is about 10^{-4} (assuming a log-uniform τ_e and an exponential $\bar{\eta}$ distribution).

(ii) All defects have quite a distinct bias and temperature dependence, and as will be shown below, variation of these parameters does not lead to a separation of these clusters.

(iii) Finally, even in the extremely unlikely case that such a pair of defects should exist in one device, the occurrence of such pairs in the other two devices investigated appears highly unlikely.

Therefore, to go beyond the trivial explanation, three possible scenarios to explain such a behavior are considered, see sketches in Fig. 3. For the analysis of the experimental data we have to distinguish between the *physical processes* underlying the double hole emission events from *what is recorded* by TDDS. For example, if we have two defects with identical emission times and step-heights, we will have a 50% probability that the first emission event belongs to defect 1, while the other 50% will belong to defect 2.

A. Scenario A: Defect Pair

In scenario A we assume that we are dealing with two defects which are spatially close but otherwise (nearly) independent. This is based on the assumption that defects may be more likely in certain defective areas of the oxide, as also recently suggested for the case of dielectric breakdown [13]. Then, the two holes are emitted at independent random times t_1 and t_2, while TDDS will see the first event at $t_a = \min(t_1, t_2)$ and the second at $t_b = \max(t_1, t_2)$. If the emission

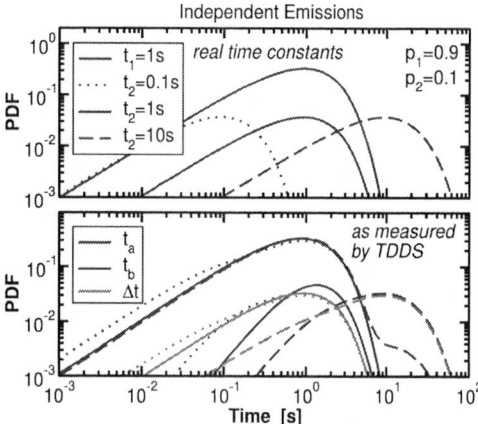

Fig. 4: Scenario A: **Top**: The (real) time constants of the two defects independently emitting a hole. The first emission time is fixed to 1 s while the second is varied from 0.1 s to 10 s (dotted/solid/dashed curves). Typical capture probabilities $p_1 = 0.9$ and $p_2 = 0.1$ are used. **Bottom**: The simulated distribution of the first emission event t_a recorded by TDDS is close to exponential, dominated by $\tau_1 = 1\,\text{s}$, with only a small modulation at short and large times, which is outside our experimental resolution. The distribution t_b of is narrower than exponential. Most importantly, the distribution of Δt is similar to that of the first event t_a.

times t_1 and t_2 are exponentially distributed, the distributions of t_a and t_b can easily be calculated. For this we also need to consider that the probabilities of having either of these emission events are not equal and depend on the probability of having captured a charge in either defect. Experimentally, we typically see mostly emission events related to the first hole for shorter stress times while double emissions are only recorded for larger times. We consider this fact by taking different capture probabilities p_i of the two defects into account, with $i = 1, 2$. Furthermore, the average emission times of each defect are given by τ_i. The p.d.f. of having an emission event from each of these defects is then

$$P_i(t) = p_i f(t; \tau_i) \tag{2}$$

where $f(t; \tau_i)$ is the exponential distribution with characteristic emission time τ_i. Note that $P_i(t)$ is normalized to p_i rather than unity. Under the assumption that both defects emit a hole, the p.d.f. for the event $t_a = \min(t_1, t_2)$ is

$$P_a^0(t) = \int_t^\infty P(t, t_2)\, \mathrm{d}t_2 + \int_t^\infty P(t_1, t)\, \mathrm{d}t_1 = f(t; \tau) \tag{3}$$

using the joint p.d.f. of the independent processes $P(t_1, t_2) = f(t_1; \tau_1)f(t_2; \tau_2)$ and $1/\tau = (1/\tau_1 + 1/\tau_2)$. Then, for the general case of $p_i \leq 1$, we obtain

$$P_a(t) = P_1(t) + P_2(t) - P_b(t), \tag{4}$$

$$P_b(t) = p_2 P_1(t) + p_1 P_2(t) - p_1 p_2 P_a^0(t). \tag{5}$$

The distributions P_a and P_b are shown in Fig. 4 for varying τ_2. As will become clear later, the distribution of the random variable $\Delta t = t_2 - t_1$ is also of interest.

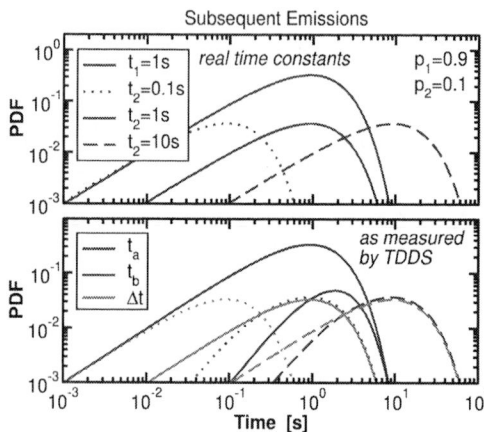

Fig. 5: Scenarios B/C: **Top**: As in Fig. 4, but now for two defects subsequently emitting a hole. **Bottom**: The distribution of the first emission event t_a recorded by TDDS is of course identical to that of t_1, while the distribution t_b is narrower than exponential. Most importantly, however, the distribution of Δt can clearly reveal the case where the second emission event quickly follows the first.

B. Scenario B: Doubly Charged Defect

In scenario B, we assume that a defect can capture two holes, which are then released subsequently. In this case, we would expect $t_a = t_1$ and $\Delta t = t_b - t_a = t_1$ to be independent and exponentially distributed. The resulting distributions are shown in Fig. 5.

C. Scenario C: Multi-Step Trapping

Finally, in scenario C, we assume that after a certain stress time, the holes hop deeper into the oxide, making space for another hole to be trapped in the original defect. Again, from a statistical perspective, emission would then be like in scenario B. However, we would expect the oxide field to have a strong impact on the probability of onward tunneling, similarly to resonant tunneling structures.

Unfortunately, for scenarios B and C where t_1 and Δt are independent, very similar distributions are obtained, which cannot be distinguished using our small sample size of $N = 100$ traces. The most marked difference would be the case where the second emission event would quickly follow the first one, which would result in different distributions for t_b and Δt. So we proceed by extracting τ_1 and τ_2 assuming them to be the original emission times of the independent defects using (4) and (5), and check their plausibility later.

IV. RESULTS AND DISCUSSION

Typical experimental data are shown in Fig. 6, which demonstrates that also τ_2 can be extracted with satisfactory accuracy. The extracted capture and emission times as a function of bias and temperature are shown in Fig. 7 and Fig. 8. As noted before, τ_1 and τ_2 are very similar for all bias conditions and temperatures.

These data can now be compared with the expected predictions of the model scenarios. Before doing so, we briefly summarize general properties of the defects responsible for

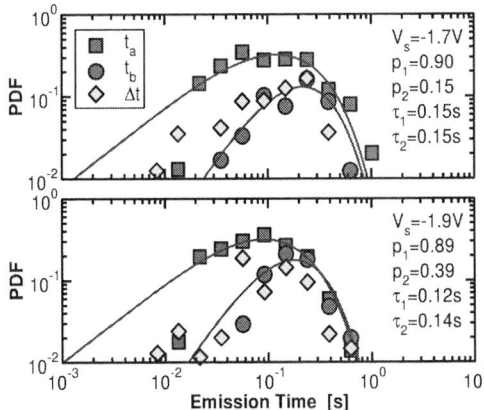

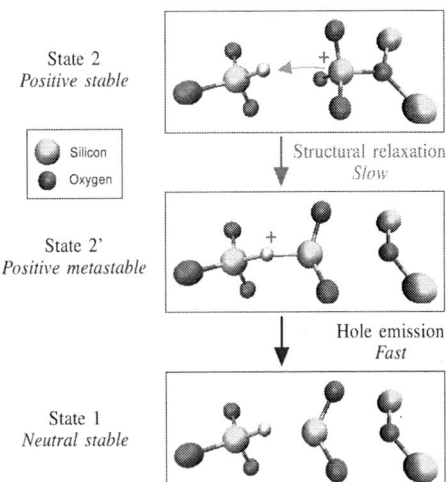

Fig. 9: The emission time constant is dominated by the barrier separating state 2 from state 2', shown above for the hydrogen-bridge in SiO_2 [14]. Once in state 2', hole emission into state 1 is typically fast for lower biases.

Fig. 6: Typical extraction result for defect A4 obtained after a stress time of 10 s at -1.7 V and -1.9 V at 125 °C. From $p_i(t_s)$ the capture time constants can be calculated. Even though only 100 repetitions were available, the noise level in the extracted parameters is satisfactory, showing that the distribution of Δt remains close to the distribution of t_a. While this data do not allow us to distinguish between scenarios A and C, it is inconsistent with scenario B.

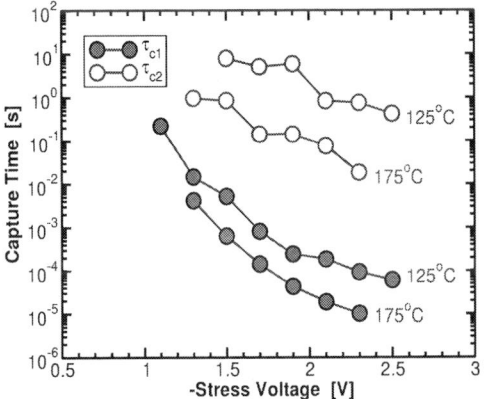

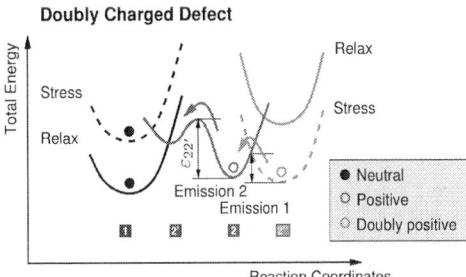

Fig. 10: For a doubly charged defect to produce two emission events with similar emission time constants, the barrier for the non-radiative multiphonon (NMP) [16, 25] transition from the doubly charged state must be similar to the thermal barriers connecting states 2 and 2', $\varepsilon_{22'}$. As the NMP barrier depends on bias, the occurrence of such a scenario is considered unlikely.

Fig. 7: The extracted capture times τ_{c1} and τ_{c2} of defect A4 as a function of bias and temperature. τ_{c2} is typically 3-4 orders of magnitude larger than τ_{c1}.

charge trapping in NBTI [15]. Most importantly, it has been observed that hole emission can take much longer than would be expected from an elastic tunneling process due to significant structural relaxation at the defect site [1, 16, 17]. As such, the emission time is often dominated by a relaxation barrier, in our model written as $\varepsilon_{22'}$, which determines the transition from the stable positive state 2 back to the stable neutral state 1 via a metastable state 2'. While the detailed chemical nature of the defects has not yet been unanimously identified, various forms of the E' centers have been extensively studied in literature [18–24]. In particular, such large relaxation barriers are often observed for the puckered configuration of E' centers [14]. While the prototypical E' center has unfavorable energy levels to be charged during NBTI stress [14], the hydrogen-bridge appears a promising candidate. The three important states for charge emission are shown in Fig. 9, demonstrating the importance of the backward barrier $\varepsilon_{22'}$.

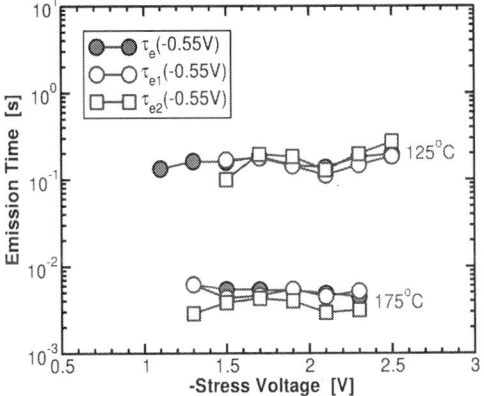

Fig. 8: The extracted emission times τ_1 and τ_2 always track each other closely, independently of bias and temperature.

The main experimental observation is that the two emission times are similar, independently of bias and temperature. This is difficult to reconcile with the doubly charged defect of

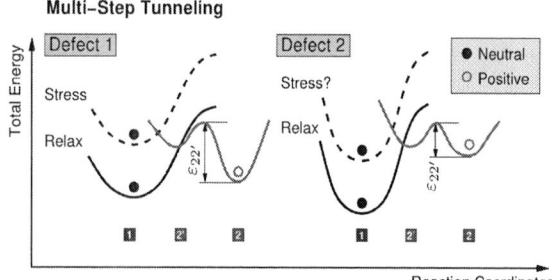

Multi-Step Tunneling

Fig. 11: For a multi-step trapping scenario, the second defect would have to be charged by trapping the hole from the first defect. Assuming a smaller relaxation barrier $\varepsilon_{22'}$ for defect 2, release would then occur quickly into defect 1 once the hole is emitted from there. Under this assumption, emission would be dominated by the properties of defect 1, possibly explaining the similar emission times. However, as the capture times are dominated by the multiphonon processes and the barriers rather than the tunneling times, it is unclear why such a defect would not directly communicate with the channel.

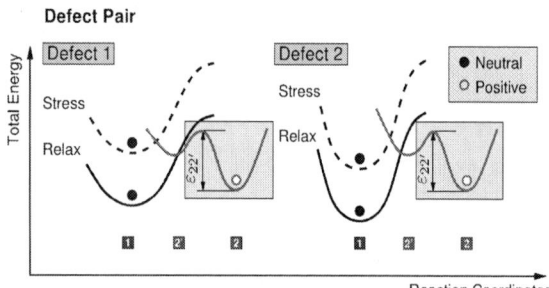

Defect Pair

Fig. 12: For two independent defects to incidentally have similar emission time constants, the thermal barriers connecting states 2 and 2', $\varepsilon_{22'}$, must be similar. The circumstances for this to happen remain to be clarified.

scenario B, see Fig. 10. Here, we would expect the first hole emission to be strongly bias dependent and fast, while the second would be again dominated by the barrier $\varepsilon_{22'}$. This discrepancy could be resolved by a defect which goes through a further structural relaxation with a barrier comparable to $\varepsilon_{22'}$. While this is in principle conceivable, no defect with such properties has been reported to the best of our knowledge.

Regarding scenario C, since onward tunneling deeper into the oxide would be expected to depend strongly on the oxide field, we would expect this second trap to be very close to the first trap for this to be consistent with the data. Also, the emission time of defect 2 into defect 1 would have to be much shorter than the emission time of defect 1, for the second emission to come out with the same time constant, see Fig. 11. However, it is unclear how defect 2 would be prevented from directly communicating with the channel.

Finally, in scenario A the barriers $\varepsilon_{22'}$ would have to be very similar for both defects to explain the observed $\tau_1(V_{\mathrm{G}}, T) \approx \tau_2(V_{\mathrm{G}}, T)$ behavior, see Fig. 12. As discussed initially, given the wide distribution of defect parameters this is unlikely to happen merely by chance but could be the result of two defects residing in the same defective area sharing some part of their configuration.

V. CONCLUSIONS

We have observed that in about 10-20% of the defects studied by TDDS two holes can be captured at larger stress times. The two emission events appear independent and exponentially distributed, with similar emission times for all bias conditions and temperatures. We have discussed three possible scenarios including defect pairs, doubly charged defects, as well as multi-step trapping. Given the experimental evidence, we suggest that we are dealing with a spatially and configurationally closely related pair of defects. The possibility of such configurations must be taken into account in the still unresolved identification of the chemical nature of the defects responsible for NBTI [14].

ACKNOWLEDGMENTS

This work has received funding from the Austrian Science Fund (FWF) project n°23390-N24 and the European Community's FP7 n°261868 (MORDRED).

REFERENCES

[1] T. Grasser, H. Reisinger, P.-J. Wagner, W. Goes, F. Schanovsky, and B. Kaczer, "The Time Dependent Defect Spectroscopy (TDDS) for the Characterization of the Bias Temperature Instability," in *Proc. Intl.Rel.Phys.Symp. (IRPS)*, pp. 16–25, May 2010.

[2] T. Wang, C.-T. Chan, C.-J. Tang, C.-W. Tsai, H. Wang, M.-H. Chi, and D. Tang, "A Novel Transient Characterization Technique to Investigate Trap Properties in HfSiON Gate Dielectric MOSFETs-From Single Electron Emission to PBTI Recovery Transient," *IEEE Trans.Electron Devices*, vol. 53, no. 5, pp. 1073–1079, 2006.

[3] V. Huard, C. Parthasarathy, and M. Denais, "Single-Hole Detrapping Events in pMOSFETs NBTI Degradation," in *Proc. Intl.Integrated Reliability Workshop*, pp. 5–9, 2005.

[4] H. Reisinger, T. Grasser, and C. Schlünder, "A Study of NBTI by the Statistical Analysis of the Properties of Individual Defects in pMOSFETs," in *Proc. Intl.Integrated Reliability Workshop*, pp. 30–35, 2009.

[5] M. Toledano-Luque, B. Kaczer, P. Roussel, T. Grasser, G. Wirth, J. Franco, C. Vrancken, N. Horiguchi, and G. Groeseneken, "Response of a Single Trap to AC Negative Bias Temperature Stress," in *Proc. Intl.Rel.Phys.Symp. (IRPS)*, pp. 364–371, 2011.

[6] J. Zou, J. Zou, C. Liu, R. Wang, X. Xu, J. Liu, H. Wu, Y. Wang, and R. Huang, "On the Statistical Trap-Response (STR) Method for Characterizing Random Trap Occupancy and NBTI Fluctuation," in *IEEE Silicon Nanoelectronics Workshop (SNW)*, pp. 1–2, June 2012.

[7] J. Zou, R. Wang, N. Gong, R. Huang, X. Xu, J. Ou, C. Liu, J. Wang, J. Liu, J. Wu, S. Yu, P. Ren, H. Wu, S. Lee, and Y. Wang, "New Insights into AC RTN in Scaled High-κ/Metal-gate MOSFETs under Digital Circuit Operations," in *IEEE Symposium on VLSI Technology Digest of Technical Papers*, pp. 139–140, 2012.

[8] T. Grasser, K. Rott, H. Reisinger, P.-J. Wagner, W. Goes, F. Schanovsky, M. Waltl, M. Toledano-Luque, and B. Kaczer, "Advanced Characterization of Oxide Traps: The Dynamic Time-Dependent Defect Spectroscopy," in *Proc. Intl.Rel.Phys.Symp. (IRPS)*, pp. 2D.2.1–2D.2.7, Apr. 2013.

[9] T. Grasser, H. Reisinger, K. Rott, M. Toledano-Luque, and B. Kaczer, "On the Microscopic Origin of the Frequency Dependence of Hole Trapping in pMOSFETs," in *Proc. Intl.Electron Devices Meeting (IEDM)*, pp. 19.6.1–19.6.4, Dec. 2012.

[10] M. Uren, M. Kirton, and S. Collins, "Anomalous Telegraph Noise in Small-Area Silicon Metal-Oxide-Semiconductor Field-Effect Transistors," *Physical Review B*, vol. 37, no. 14, pp. 8346–8350, 1988.

[11] T. Grasser, K. Rott, H. Reisinger, M. Waltl, P. Wagner, F. Schanovsky, W. Goes, G. Pobegen, and B. Kaczer, "Hydrogen-Related Volatile Defects as the Possible Cause for the Recoverable Component of NBTI," in *Proc. Intl.Electron Devices Meeting (IEDM)*, Dec. 2013.

[12] H. Reisinger, O. Blank, W. Heinrigs, A. Mühlhoff, W. Gustin, and C. Schlünder, "Analysis of NBTI Degradation- and Recovery-Behavior Based on Ultra Fast V_{th}-Measurements," in *Proc. Intl.Rel.Phys.Symp. (IRPS)*, pp. 448–453, 2006.

[13] E. Wu, B. Li, J. Stathis, R. Achanta, R. Filippi, and P. McLaughlin, "A Time-Dependent Clustering Model for Non-Uniform Dielectric Breakdown," in *Proc. Intl.Electron Devices Meeting (IEDM)*, pp. 401–404, Dec. 2013.

[14] F. Schanovsky, W. Goes, and T. Grasser, "A Detailed Evaluation of Model Defects as Candidates for the Bias Temperature Instability," in *Proc. Simulation of Semiconductor Processes and Devices*, pp. 1–4, 2013.

[15] T. Grasser, "Stochastic Charge Trapping in Oxides: From Random Telegraph Noise to Bias Temperature Instabilities," *Microelectronics Reliability*, vol. 52, pp. 39–70, 2012.

[16] C. Henry and D. Lang, "Nonradiative Capture and Recombination by Multiphonon Emission in GaAs and GaP," *Physical Review B*, vol. 15, no. 2, pp. 989–1016, 1977.

[17] A. Palma, A. Godoy, J. A. Jimenez-Tejada, J. E. Carceller, and J. A. Lopez-Villanueva, "Quantum Two-Dimensional Calculation of Time Constants of Random Telegraph Signals in Metal-Oxide-Semiconductor Structures," *Physical Review B*, vol. 56, no. 15, pp. 9565–9574, 1997.

[18] A. Lelis and T. Oldham, "Time Dependence of Switching Oxide Traps," *IEEE Trans.Nucl.Sci.*, vol. 41, pp. 1835–1843, Dec 1994.

[19] E. Poindexter and W. Warren, "Paramagnetic Point Defects in Amorphous Thin Films of SiO_2 and Si_3N_4: Updates and Additions," *J.Electrochem.Soc.*, vol. 142, no. 7, pp. 2508–2516, 1995.

[20] J. Conley Jr., P. Lenahan, A. Lelis, and T. Oldham, "Electron Spin Resonance Evidence for the Structure of a Switching Oxide Trap: Long Term Structural Change at Silicon Dangling Bond Sites in SiO_2," *Appl.Phys.Lett.*, vol. 67, no. 15, pp. 2179–2181, 1995.

[21] P. Blöchl, "First-Principles Calculations of Defects in Oxygen-Deficient Silica Exposed to Hydrogen," *Physical Review B*, vol. 62, no. 10, pp. 6158–6179, 2000.

[22] D. Fleetwood, H. Xiong, Z.-Y. Lu, C. Nicklaw, J. Felix, R. Schrimpf, and S. Pantelides, "Unified Model of Hole Trapping, $1/f$ Noise, and Thermally Stimulated Current in MOS Devices," *IEEE Trans.Electron Devices*, vol. 49, no. 6, pp. 2674–2683, 2002.

[23] P. Lenahan, "Atomic Scale Defects Involved in MOS Reliability Problems," *Microelectronic Engineering*, vol. 69, pp. 173–181, 2003.

[24] A. Kimmel, P. Sushko, A. Shluger, and G. Bersuker, "Positive and Negative Oxygen Vacancies in Amorphous Silica," in *Silicon Nitride, Silicon Dioxide, and Emerging Dielectrics 10* (R. Sah, J. Zhang, Y. Kamakura, M. Deen, and J. Yota, eds.), vol. 19, pp. 2–17, ECS Transactions, 2009.

[25] S. Makram-Ebeid and M. Lannoo, "Quantum Model for Phonon-Assisted Tunnel Ionization of Deep Levels in a Semiconductor," *Physical Review B*, vol. 25, no. 10, pp. 6406–6424, 1982.

Understanding of Self-Heating Enhanced Degradation in pLDMOSFETs by MR-DCIV Method

Yandong He, Ganggang Zhang and Xing Zhang

Institute of Microelectronics and Key Laboratory of Microelectronic Devices and Circuits, Peking University

Institute of Microelectronics, Peking University, Beijing 100871, P.R. China

Phone: (8610) 62767915   Fax: (8610) 62758331   Email: heyd@pku.edu.cn

Abstract- Self-heating enhanced degradation in pLDMOSFETs was studied by non-destructive MR-DCIV method. Due to self-heating effect in pLDMOSFETs, several times larger MR-DCIV degradation per finger was observed for multi-finger devices with higher temperature rise and less channel edge heat dissipation. Our study has shown that self-heating induced degradation shared the similar trends and mechanism to NBTI.

I. INTRODUCTION

With the advantage of the process compatibility with the mainstream standard CMOS, STI-based LDMOS devices have become popular for its better tradeoff between breakdown voltage and performance[1]. Since LDMOSFET devices are operated under high drain voltage, extensive studies are focused on its breakdown voltage and hot carrier reliability. However, due to its capability of handling high voltage and high current levels, self-heating enhanced degradation of power devices are of prime importance[2]. Higher power dissipation of power device makes it necessary to distinguish the function of the self-heating effect, especially for SOI technology[3]. On the other hand, NBTI in pMOSFETs has become the major reliability concern[4] due to interface states generation under relatively high temperature. In order to characterize the interface state in STI-based LDMOSFETs, a non-destructive MR-DCIV method has been proposed[5] recently to probe the interface

states at channel, accumulation and STI drift region in LDMOSFETs. In this paper, self-heating enhanced degradation in pLDMOSFETs had been investigated by non-destructive MR-DCIV method. The single/multi-finger devices were used as a vehicle to characterize the self-heating effect in pLDMOSFETs. MR-DCIV current degradation per finger was obtained and compared experimentally. The degradation

II. DEVICES AND MR-DCIV METHOD

The STI-based pLDMOSFET with cross section in Fig.1 was fabricated by a 0.18μm SOI CMOS compatible BCD process. SOI wafers with a silicon layer thickness of 1.5μm, a p-type resistivity 8–12Ωcm and with buried-oxide (BOX) thickness of 1μm. The wafers had various doses of implantation to achieve the dose control in p-well and n-drift regions for

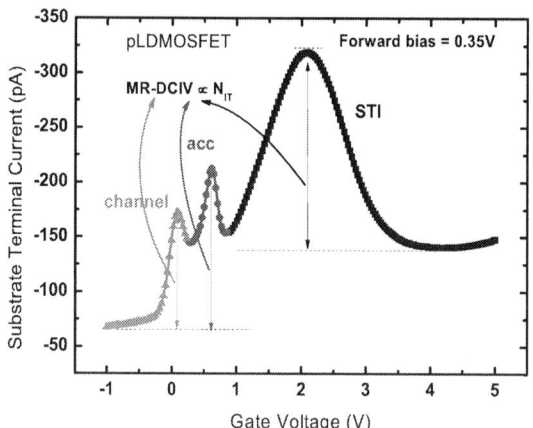

Fig.2. A typical MR-DCIV current of fresh pLDMOSFET

obtaining an optimum reduced surface field (RESURF) effect in order to maximize the breakdown voltage and on-resistance. The off-state breakdown voltage is above 60V. The on-state breakdown voltage is in excess of 40V up to V_{gs}= -5V. A typical MR-DCIV current obtained from a fresh pLDMOSFET was shown in Fig.2. During the MR-DCIV measurement, the source/drain to substrate junction was forward biased, the

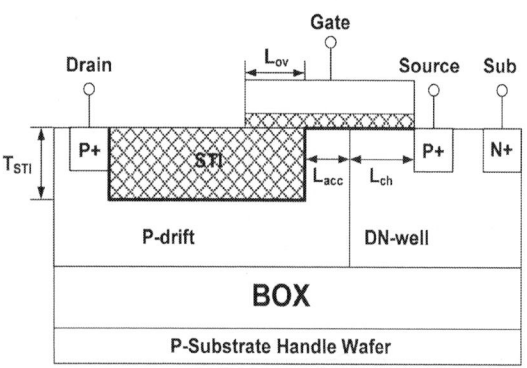

Fig.1. A cross-sectional view of an STI-based pLDMOSFET.

978-1-4799-3911-4/14 $31.00 © 2014 IEEE

substrate current I_{sub} was measured when gate voltage was swept from accumulation to slight inversion. Several distinct MR-DCIV current peaks were corresponding to the interface

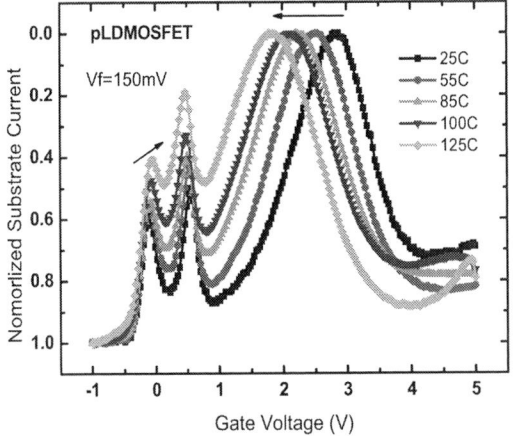

Fig.3. The normalized MR-DCIV current at temperature 25 °C to 125°C

states at channel, accumulation and STI drift region, respectively[5]. The individual MR-DCIV current peak height was proportional to the interface state density at channel, accumulation and STI regions in pLDMOSFETs.

Since the MR-DCIV current would change over several orders magnitude within temperature from 25 °C to 125°C, the normalized MR-DCIV current was shown in Fig.3, demonstrating a group of similar spectra at various temperatures. However, MR-DCIV current peaks were getting closer when the temperature was increased. On the other hand, the extracted MR-DCIV current was summarized in Fig.4. It was easy to find that the MR-DCIV current not only followed the Arrhenius law under various forward biases, but also increased exponentially with source/drain-substrate junction forward bias(V_F) at a constant rate under difference temperatures ranging from 25°C to 150°C. From the good correlation of MR-DCIV peak current with various temperatures and forward biases, the MR-DCIV current at different temperature could be scaled and compared properly.

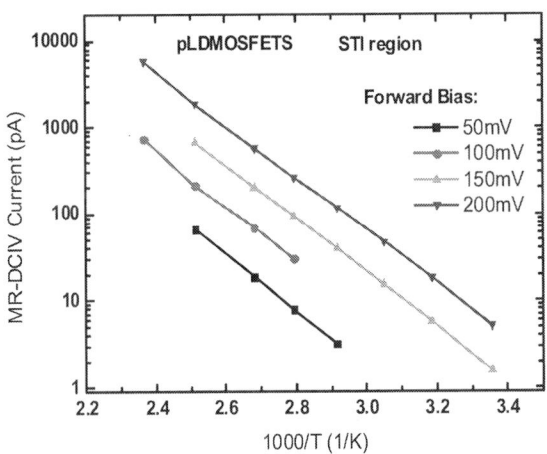

Fig.4. MR-DCIV current at temperature 25 °C to 125°C.

III. RESULTS AND DISCUSSIONS

The typical I_d-V_d characteristics were shown in Fig.5.

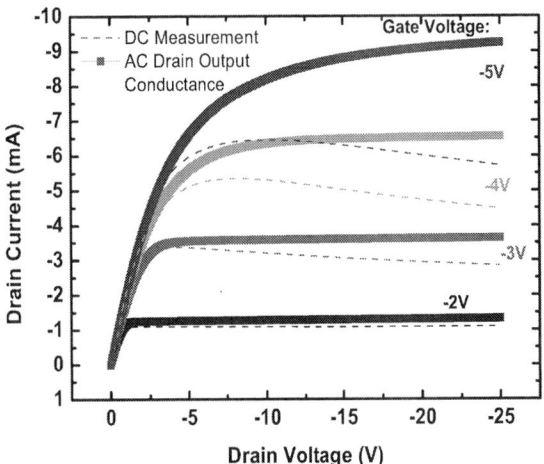

Fig.5. Typical I_d-V_d characteristics of pLDMOSFETs, clearly showing self-heating induced drain current reduction

Compared with DC measurement and AC drain output conductance measurement[6], the self-heating induced drain current reduction was clearly demonstrated especially at higher gate/drain voltage. Under certain a gate voltage bias, the drain current reduction induced by self-heating effect was increased with increasing of drain voltage. The large difference between DC and AC drain output conductance measurement results indicated the occurrence of the severe self-heating effect.

As we know, NBTI degradation was temperature accelerated. For the comparison, NBTI stress degradation was conducted under higher gate voltage with various temperatures. Because the source/drain was ground under NBTI stress, there was no drain current at NBTI stress condition. Without self-heating effect, the relative change of MR-DCIV current under NBTI stresses was shown in Fig.6. It was observed that the major

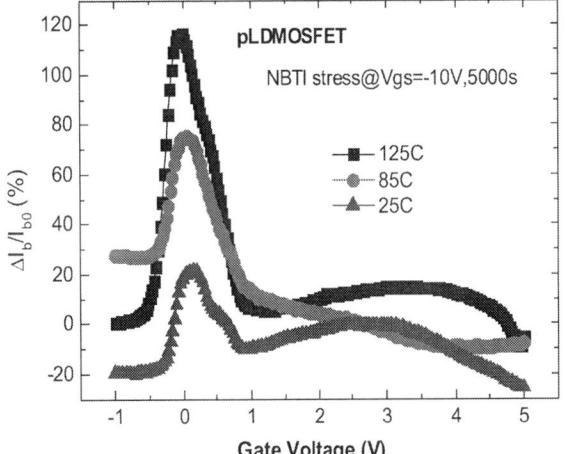

Fig.6. MR-DCIV current degradation after 5000s NBTI stress at various temperatures

change of MR-DCIV current was happened at channel and accumulation region in pLDMOSFETs due to the thinner gate oxide thickness compared to STI region. And the relative variation of MR-DCIV current after 5000s NBTI stress was obviously increased with stress temperature, suggesting that temperature was one of the acceleration factors for interface states generation at channel and accumulation regions in pLDMOSFETs, same as the normal pMOSFETs.

For high power requirement, LDMOSFETs with multi-finger layout were commonly used. In order to investigate the self-heating effect, the V_{gmax} = -5V HCI stress mode was applied to single/multi-finger pLDMOSFETs at constant room temperature of 25°C. The finger number of pLDMOSFETs was varied from 1 to 10. During the stress period, the MR-DCIV current, I_d-V_g and charge pumping current were monitored at predefined stress time interval. In order to predict the power dissipation during the stress period, the drain current was

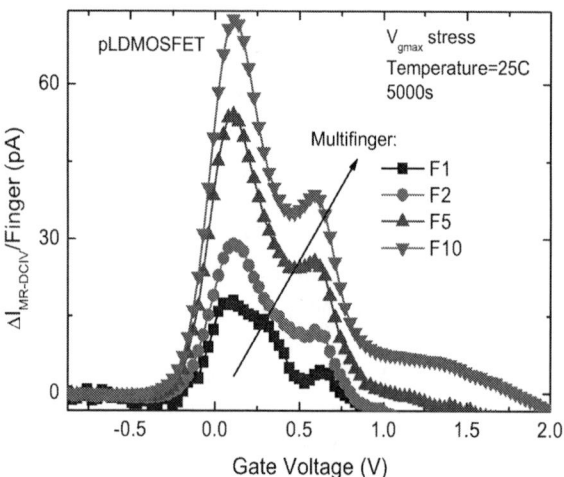

Fig.7. The shift of MR-DCIV current per finger after 5000s V_{gmax} stress in single/multi-finger pLDMOSFETs .

concurrently measured at stress stage.

Because pLDMOSFETs were stressed under the same V_{gmax} stress mode, the electrical stress induced degradation after a certain stress time could be considered at a fixed level. Therefore, for the multi-finger devices the degradation per finger should be a constant value. However, it was found that the incremental of MR-DCIV current per finger was obviously different, illustrated in Fig.7. After 5000s stress, the MR-DCIV current at channel and accumulation region grew up dramatically, indicating the interface states generation during V_{gmax} stress. The charge pumping current result also confirmed these interface states generations(not shown here). It was worthy to be noticed the interface states generation at channel region was 2x larger than that of accumulation region, regardless of finger numbers. This result was consistent with the traditional NBTI effect in standard CMOS devices[4]. The channel and accumulation region was associated with opposite

doping carriers in pLDMOSFETs, therefore the NBTI stress for pLDMOSFETs combined both normal NBTI stress for pMOS and nMOS device.

Based on our experiment with the consistent V_{gmax} stresses, it was found that pLDMOSFETs with various finger numbers yielded quite different MR-DCIV current degradation in channel and accumulation region under gate oxide. The normalized MR-DCIV current shifts after 5000s V_{gmax} stresses were summarized in Fig.8. It could be seen that pLDMOSFET with more fingers may produce larger degradation after constant 5000s stress, indicating more interface states generation for pLDMOSFETs with larger finger numbers. Therefore, considering the electrical bias alone was not sufficient to accurately predict the device degradation, especially for power devices involved with high power dissipation which may lead to device temperature rising.

For understanding this self-heating related degradation enhancement, the on-the-fly drain current during stress time was measured. The correlation between MR-DCIV current change and drain current was shown in Fig.8. It was shown that the MR-DCIV current change per finger was increased with drain current, but the acceleration rate was decreased with the finger number. From the inset of Fig.8, it was noted that the drain current level at stress condition was not proportional to the finger number ideally. The more fingers the device consisted of,

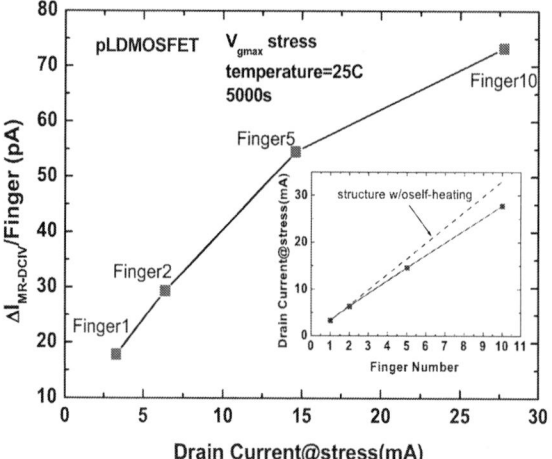

Fig.8. The relationship of MR-DCIV current shift to on-the-fly stress drain current after 5000s V_{gmax} stress

the larger deviation the drain current produced from the non-self-heating device, indicating more serious self-heating effect for multi-finger devices. Although the same bias condition was applied, the normalized power density (dissipated power per finger) for multi-finger devices was not kept constant. The unit DC power dissipation density for multi-finger devices was lower than single finger device. However, the enhanced degradation results suggested that the higher temperature rise was achieved under lower unit DC power dissipation density with multiple-finger layout devices. This behaviour could be attributed to the heat dissipation from channel edge. Due to the

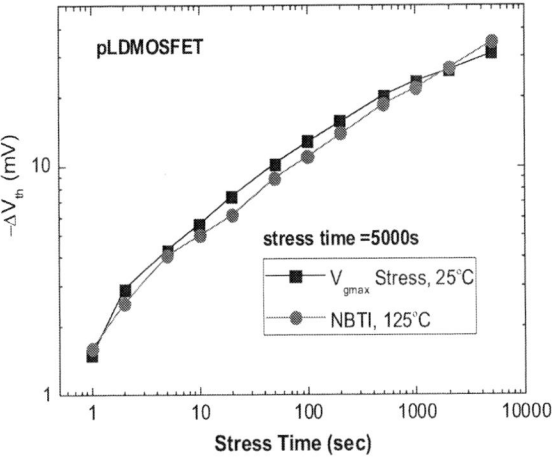

Fig.9. The similar trend for V_{th} shift under 25°C V_{gmax} stress and 125°C NBTI stress

cascaded multi-finger devices with smaller ratio of channel edges, the influence of channel edge heat dissipation was less significant for multi-finger devices compared with single finger device. Thus, under the same stress condition, the internal device temperature became higher for multi-finger device with less channel edge heat dissipation. This self-heating induced temperature rising became the acceleration factor in pLDMOSFETs under V_{gmax} stress, which was also quite similar to normal NBTI.

Moreover, from the linear I_d-V_g characteristics, the similar V_{th} shift was obtained under 25°C V_{gmax} stress and 125°C NBTI stress, shown in Fig.9. Both of them were power law time dependence, which was quite consistent with NBTI of normal pMOSFET. This result could suggest that the self-heating enhanced degradation under V_{gmax} stress may share the same mechanism with NBTI degradation. The high temperature rising due to self-heating effect and the high hole concentration in the channel of pLDMOSFETs under V_{gmax} stress lead to the conclusion that the observed self-enhanced degradation for multi-finger power devices could be explained by NBTI mechanism.

IV. CONCLUSIONS

Self-heating enhanced degradation in pLDMOSFETs with multi-finger layout has been investigated by non-destructive MR-DCIV method. Larger MR-DCIV current degradation per finger was observed for multi-finger devices even though its unit DC power dissipation density was lower than that of single finger device. Compared with NBTI stresses under various temperatures, it was found that due to less channel edge heat dissipation the self-heating induced higher temperature rise became the major degradation factors for interface state

generation at channel and accumulation region in pLDMOSFETs, similar to NBTI degradation in CMOS devices. The evolution of V_{th} shift under self-heating mode was quite consistent with 125°C NBTI. Our results revealed that the thermal management of multi-finger device is of crucial importance to guaranteeing the device performance and reliability.

ACKNOWLEDGMENT

This work is financially supported by the State Key Fundamental Research Project of China (Grant No. 2011CBA00606).

REFERENCES

[1]、 M. Zitouni, F. Morancho, et al., "A new lateral power MOSFET for smart power ICs: the LUDMOS concept", Microelectron. J., Vol.30, (1999), pp.551-561.

[2]、 JF Chen, KS Tian, SY Chen, et al., "On-resistance degradation induced by hot-carrier injection in LDMOS transistors with STI in the drift region", IEEE Trans. EDL, Vol.29 (2008), pp.1071-1073.

[3]、 Dieudonne. F, Haendler S., et al., "Self-heating Eeffects in SOI NLDEMOS Power Devices", Proc. 25th Microelectronics Conf., 2006, pp.191-193

[4]、 D. K. Schroder and J. A. Babcock, "Negative bias temperature instability: Road to cross in deep submicron silicon semiconductor manufacturing", J. Appl. Phys., Vol.94 (2003), pp.1-18

[5]、 Y. He, L. Han, G. Zhang, et al., "Multiregion DCIV: A Sensitive Tool for Characterizing the Si/SiO2 Interfaces in LDMOSFETs", IEEE Trans. EDL, Vol.33 (2012), pp.1435-1437

[6]、 W. Jin, et al., "Self-heating characterization for SOI MOSFET based on AC output conductance", IEDM Tech. Dig., 1999, pp.175-178. M. Zitouni, F. Morancho, et al., "A new lateral power MOSFET for smart power ICs: the LUDMOS concept", Microelectron. J., Vol.30, (1999), pp.551-56

Novel Technique for Deep Vertical Interconnect Access Fault Isolation

T.P. Chua, C.H. Chong and K.N. Liew
United Microelectronics Corporation, Ltd.
No. 3, Pasir Ris Drive 12, Singapore 519528
Phone: (65) 62130018 ext 7688   Fax: (65) 62130004   Email: tze_ping_chua@umc.com

Deep Vertical Interconnect Access (DVIA) was developed in the semiconductor industry for high performance technique which used to create advanced packages and advance integrated circuits. With its physically large diameter (~15um) and depth (~60um), substantial hours will be needed to mill entire DVI using Focosed Ion Beam (FIB) upon locating the failing DVIA. Thermally Induced Voltage Alterations (TIVA) technique has demonstrated significant capability for DVIA fault isolation. We had successfully narrow down failing DVIA inspection area to ~10um and manage to reduce FIB usage time from 4hrs to 2hrs. Save 50% on FIB usage time with novel technique for DVIA fault isolation.

I. INTRODUCTION

DVIA is a vertical electrical connection passing completely through a silicon wafer or die. Trend of establishing DVIA has driven the development of packaging technologies for semiconductor chips for the purpose to achieve higher integration of electronics devices into smaller devices. DVIA is high performance technique which used to connect multiple chips to each other by Cu-filled holes that run directly through the silicon substrate of one of the chips [1] compared to alternatives such as package-on-package, because the density of the vias is substantially higher, and because the length of the connections is shorter. Failure Analysis for DVIA raises more discussion and become more challenging to isolate the fail site of DVIA.

A basic TIVA system for performing fault isolation techniques utilizes a laser scanning (confocal) microscope to sequentially scan a focused laser spot over the integrated circuit. This imaging technique uses a laser beam to pinpoint the location of electrical shorts or electrical high resistant on a device. The laser induces local thermal gradients in the device, which result in changes to the amount of power that the device uses. Scanning can be performed from the front or backside through selection of the laser wavelength. Some device preparation for backside scanning is also necessary [2].

Current Failure Analysis (FA) for DVIA is only able to detect failed DVIA from top view thru TIVA analysis, which provide less information of actual failing site whatever it is from top, middle or bottom of the failed DVIA. So, X-S FIB inspection needs to be carried out on the entire DVIA (Depth ~ 60um). With this method, detailed failure mechanism such as Cu Void or Cu extrusion are difficult to be observed and as mentioned previously, long hours will be spent on FIB machine.

This paper will demonstrate novel technique to isolate failure site of DVIA with the aide of TIVA fault isolation technique. This paper will also show some real failure analysis cases to explain how failed site of DVIA is being isolated and narrow down the analysis zone instead of using FIB to mill whole DVIA to search failure mechanism. This effectiveness of this methodology has been proven in several FA analyses.

III. CONCEPT AND WORKING PRINCIPLES

The novel technique that will be used are known as cross section (X-S) TIVA analysis technique which are effective to detect failure site (short or high resistant) as well as isolate it to an exact failure location of the DVIA. Firstly, failing DVIA will locate by using top view TIVA imaging technique. Failed DVIA was highlighted by TIVA signal as shown in Fig.1 while biasing the DVIA electrical circuit and ground the substrate. After that, the identified failed DVIA thru TIVA is polished to the edge of the samples as shown in Fig.2 in order to perform the X-S TIVA and detect a more precise hot spot along the DVIA from the side view thru TIVA.

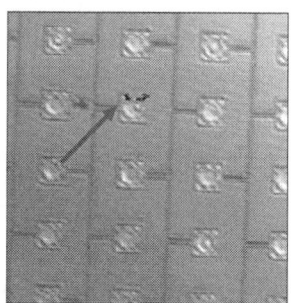

Fig.1: P-V TIVA detection for fail DVIA

Fig.2: O-M image of failed DVIA was polish near to sample edge.

Polished sample is further set-up as X-S analysis to TIVA schematic diagram and physical set-up as shown in Fig.3. The biasing needle is probed to failed DVIA pad and ground needle is probed to silicon substrate. Silver (Ag) gel is to improve conductivity of the biasing needle to the test pad. I-V curve shown in Fig.4 confirmed that this method can provide current to DVIA sample during analysis.

This set-up is able to locate which part of DVIA have problem which will highlight under TIVA analysis which will be discussed in case studies.

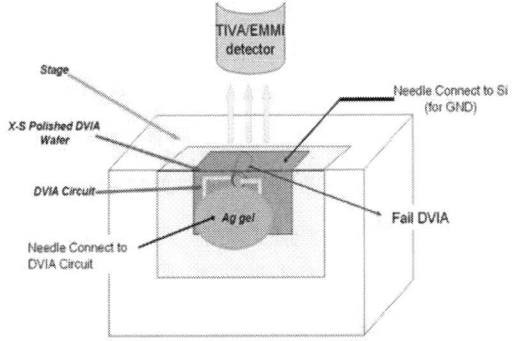

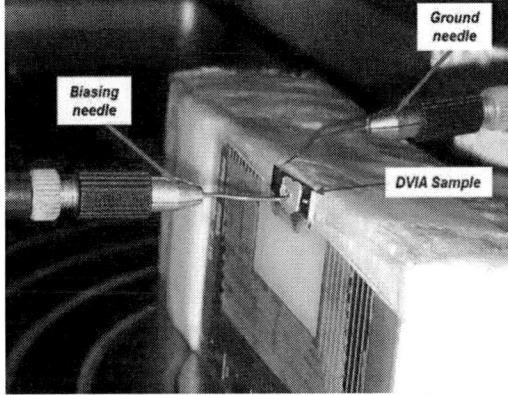

Fig.3: Schematic diagram and physical set-up of X-S TIVA analysis.

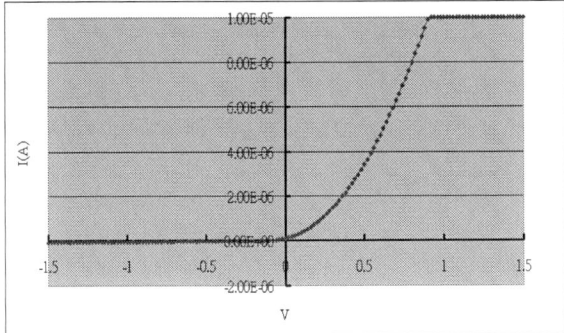

Fig.4: I-V curve of the DVIA Pad vs Ground.

IV. CASE STUDIES

Failed DVIA of Case#1 was located by P-V TIVA (Fig.5). Then the failed DVIA was polished near to the edge of the sample (Fig.6). The polished sample was then subjected to X-S TIVA analysis in order to locate which part of DVIA are highlighted by TIVA signal as shown in Fig.7.

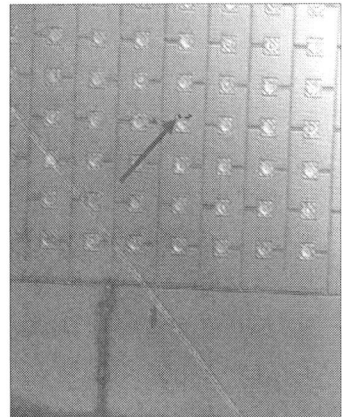

Fig.5: P-V TIVA detection for fail DVIA

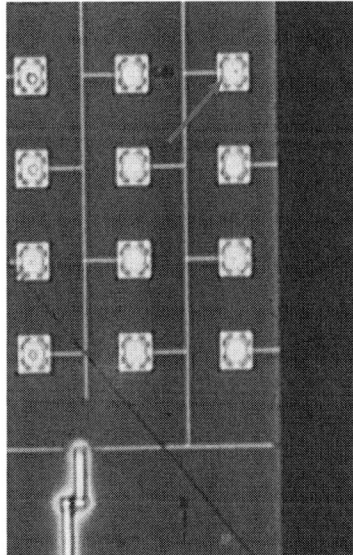

Fig.6: O-M image of failed DVIA was polish near to sample edge.

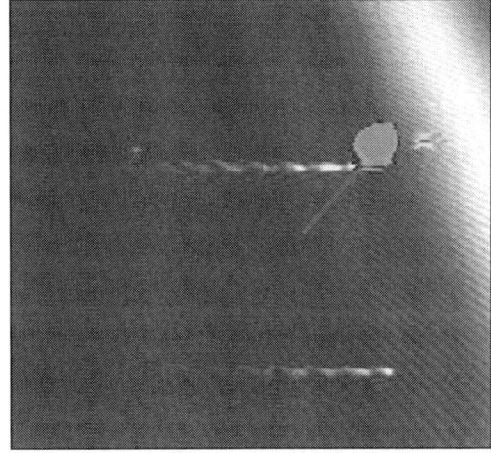

Fig.7: X-S TIVA detection of failed site in DVIA.

Sample which was analyzed by X-S TIVA will be further polished to almost near the failed DVIA and until the failed DVIA is at the corner of the sample as shown in Fig.8.

978-1-4799-3911-4/14 $31.00 © 2014 IEEE   269

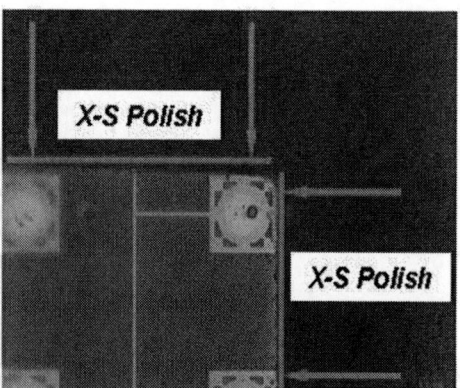

Fig.8: Failed DVIA was polished near to corner of sample.

The completed polished sample will be subjected to X-S FIB milling at the exact site where the TIVA signal was observed previously and results indicated different mode failure mechanism as shown in Fig.9.

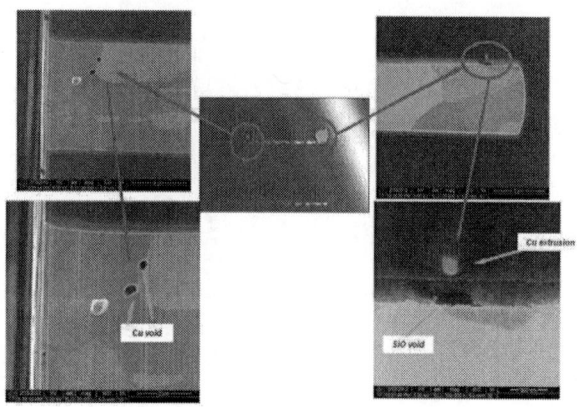

Fig.9: Failure mechanism of (a) Cu Void, (b) SiO void & Cu extrusion were found by X-S FIB milling at the site which highlighted by TIVA.

Case#2 was also performed with X-S TIVA method and abnormal TIVA signal was located from bottom DVIA. Mechanical polishing was again performed on the samples until the failed DVIA is at a corner for easy FIB milling. Subsequent FIB milling analysis at the bottom of the DVIA observed Cu Void and Cu extrusion as shown in Fig. 10.

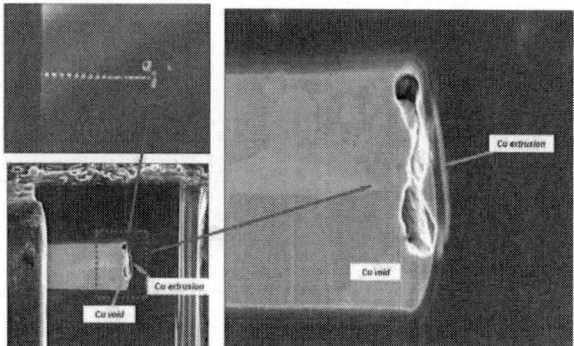

Fig.9: Failure mechanisms of Cu Void & Cu extrusion were found by X-S FIB milling.

V. CONCLUSION

In this paper, a novel technique for DVIA fault isolation has been demonstrated. This X-S TIVA technique able provides more information of DVIA failure site which was highlighted from the abnormal TIVA signal. This method will help to narrow down the region of interest from approximately 60um to 10um as well isolate the area of attention to much smaller proximity. This will help to increase the chances of location defects. This method also reduces FIB machine usage time from 4hrs to 2 hrs and put it to a greater use. This method has demonstrated its capability of increasing the throughput of DVIA failure analysis as compared to conventional way of FIB mill whole DVIA and enhancing the success rate.

ACKNOWLEDGMENT

Authors would like to thank UMC Failure Analysis Department colleagues for discussions and participation during the development of this novel method for DVIA fault isolation.

REFERENCES

[1] X. Gagnard, T. Mourier, "Through silicon via: From the CMOS imager sensor wafer level package to the 3D integration," Microelectronic Engineeing, vol. 87, pp. 470-476, April 2010.
[2] P. Perdu, R. Desplats, and F. Beaudin, "Comparative Study of Sample Preparation Techniques for Backside Analysis" ISTFA, pp. 161-72, 2000.

Back-end Defect Localization for 28nm FPGA

Jack Yi Jie Ng, Liew Chiun Ning, Khoo Khai Ling
Altera Corporation (M) Sdn. Bhd.
Plot 6, Bayan Lepas Technoplex, Medan Bayan Lepas, 11900 Penang, Malaysia.
Tel: (+604)636-8508 Fax: (+604)636-6500 Email: yjng@altera.com

Abstract-This paper presents two case studies, which are based on 28nm Field Programmable Logic Array (FPGA) bulk silicon technology, to highlight the novel approach on locating back-end interconnects and metallization defect by utilizing local software, which are Interconnect Test Generation (ITG) debugger and Functional Interface, then follow by extensive layout study, suspected defect node identification, parallel lapping and Scanning Emission Microscope (SEM) inspection.

I. INTRODUCTION

Semiconductor technology continues to advance to smaller dimensions and additional metal layers with more complex circuit design across the years. The possibility of back-end interconnects and metallization defect increases with the incremental of metal layers as there are 10 or more metal layers for a typical FPGA with 28nm technology node. It also becomes more and more challenging to perform a successful Failure Analysis especially for wafer level functional failures [1]. Die level devices prior to packaging have limitation to run under full functionality for optical based fault isolation like photon emission analysis and dynamic laser stimulation analysis as they are not workable under this circumstance. In the absence of liable fault localization data, physical Failure Analysis is not feasible for these wafer level functional failures.

Therefore, in view on these challenges, the ability to bridge the gap between functional electrical Failure Analysis data from test tools and the identification of the accurate target on layout for physical Failure Analysis becomes increasingly important [2]. Adequate understanding of electrical fault data and further data analysis plays a significant and important role in electrical fault isolation. In this work, a novel approach of combining local debugging software and extensive layout study will be demonstrated based on two case studies which successfully pinpoint to the feasible location for physical Failure Analysis.

II. METHODOLOGY

Interconnect Test Generation (ITG) test is a test which is designed to test the multiple paths or routings in FPGA. Every path is a unique source to sink connection in the chip, where it will pass through several designated resources which are connected by edge to check the continuity of the path or routing as demonstrated in Figure 1[3]. The fault modeling capability includes line break or line shorts to power or ground. It uses the available flops to stitch interconnect into register to register path and automatically test each routing interconnects resource, associated muxes and buffers [2].

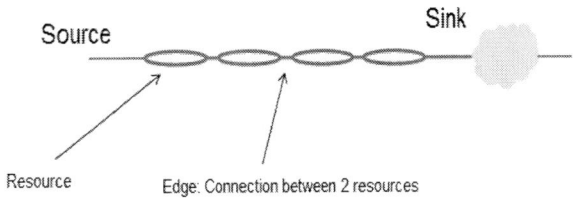

Figure 1. Key concepts of ITG test.

Interconnect Test Generation (ITG) debugger is a test tool that is designed to process test data output from ITG test, which is the log file from tester that consists of all of the failing registers information on a tested device, and generate useful post-analysis data which will help in electrical Failure Analysis. ITG debugger will select the top few failed resources and sort out for further analysis after completing Pass vs Fail analysis as shown in Figure 2. Several important information will be sorted out by the software, such as resource ID, number of path that were tested and failed, and also the failed path ratio, which will identify the path with highest failing %.

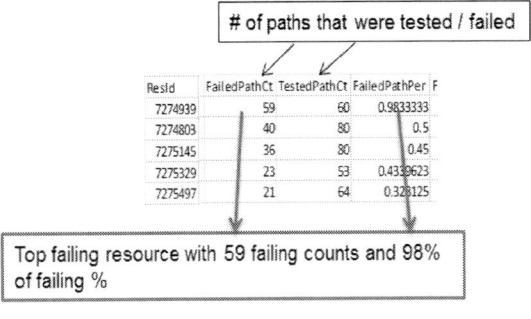

Figure 2. Example of data output from ITG debugger

Functional Interface software will next being used to demonstrate a graphical view of the failing resources, which will help in identifying the most possible failing node for physical Failure Analysis. Figure 3 shows an example of the graphical view results by Functional Interface software. The failing resources will be highlighted in red colour while the passing resources are in black colour. The potential defective node can be easily identified, where it is between fan out 4 and fan out 5 where the path might be broken and cause fan out 5 and fan out 6 to fail.

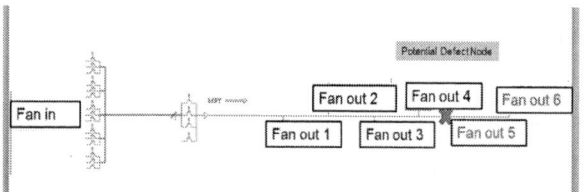

Figure 3. Example of graphical view by using Functional Interface software.

The analysis is followed by extensive layout study to identify the physical location of the potential failing node and physical Failure Analysis is performed on the node by covering all the metal lines and via/contact interconnects that are within the node.

III. CASE STUDY 1

The test data on this 28nm device showed increased failures on ITG test, which suggested increasing defectivity at back-end interconnects and metallization. The goal of this FA is to find the defect and understand the root cause for further process improvement.

There are several dies failing ITG test within the same wafer and the data logs of the failing units are captured for subsequent analysis. ITG debugger is used to analyze the raw data from the logs and generates post-analysis data for candidate selection. All the post-analysis data are reviewed and the candidate with the highest failure rate score on a single path is selected for further analysis.

The post-analysis data from ITG debugger provides important fault information which is the number of failure occurrence on different resources as shown in Figure 4. Top three resources that have the highest failure rate score are further analysed.

ResId	FailedPatl	TestedPat	FailedPathP
5623551	31	81	0.38271605
5623317	13	61	0.21311475
2333810	6	40	0.15

Figure 4. Post-analysis data from ITG debugger

Functional Interface software is then used to open the output file generated from ITG debugger to demonstrate a graphical view on the failing resource as illustrate in Figure 5. The data fan in from the left and the fan out to the right, by traveling through several fan in/fan out resources on a single path. It is observed that several fan out resources to the right including fan_out5, fan_out6 and fan_out7 (highlighted in red) failed while the fan out resources on the left before fan_out5 passed (highlighted in black), which includes fan_out1, fan_out2, fan_out3 and fan_out4 passed. By referring to the data travel direction, which is from left to right, it is quite straight forward

that the metal routing between fan_out4 and fan_out5 is the most possible defective area. The possible defective area is then translated into the physical metal path on the die for further layout study to identify an isolated area for physical Failure Analysis.

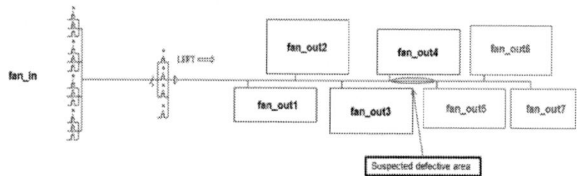

Figure 5. Functional Interface graphical view on the top failing resource suggests that the red circle area is the suspected defect area.

Figure 6 illustrate the physical location of the failing path under layout view which is highlighted in yellow. The direction of the fan outs is to the left as per suggested by the failing path graphical view in Functional Interface software as shown in Figure 5. The suspected defective area is identified at the Metal 4 layer between fan_out4 and fan_out5.

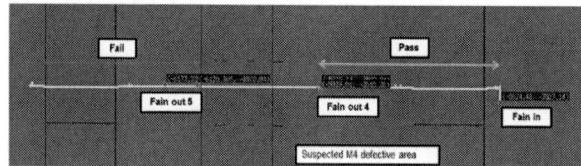

Figure 6. Physical location of the suspected failing metal line under layout view

Top down physical Failure Analysis was carried out on this sample. Layer by layer parallel lapping and Scanning Emission Microscope (SEM) inspection was performed at Metal 4 layer along the defective area. Metal patterning defect was observed on the failing path as illustrate in Figure 7.

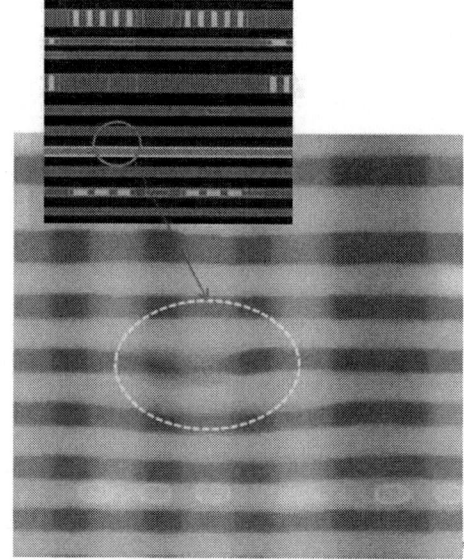

Figure 7. Metal patterning defect was observed on the suspected failing area.

III. CASE STUDY 1

IV. CASE STUDY 2

The sample in case study 1 illustrates a more straight forward situation where the potential defect area can be easily identified from the graphical fan in fan out view of Functional Interface software. However, the electrical Failure Analysis could be more challenging when the graphical results are more complicated and non-straight forward like the sample which is illustrated in case study 2. Figure 8 illustrates the graphical view on the sample which does not have a clear cut suspected defect location. There is only one fan out in this particular failing path and two out of ten of the resources from fan_out1 failed. There is not much clue in identify suspected defect area by only referring to this graphical view. Extensive layout study is needed for this case.

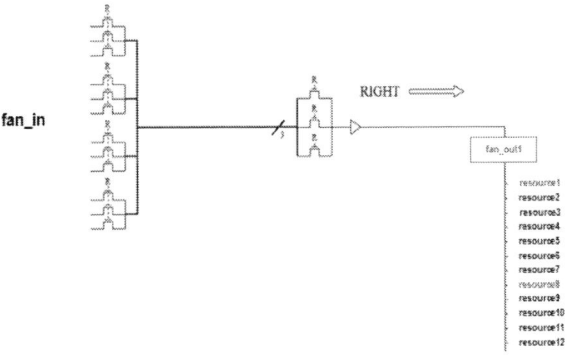

Figure 8. Functional Interface graphical view on for 2nd case study

The resources of the fan out line use muxes to interconnect with path. Extensive layout study was performed along the failing path to identify the common driver/mux that drives the two failing resources in red, which is resource1 and resource8 as illustrate in Figure 8. The red circle area is the common driver area at Metal 2 and below, which is targeted for subsequent physical Failure Analysis.

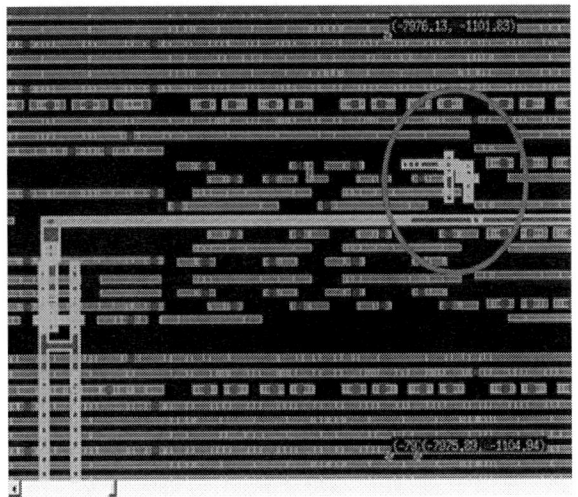

Figure 9. Physical location of the suspected failing driver under layout view at metal 2.

Top down physical Failure Analysis was carried out on this sample. Layer by layer parallel lapping and Scanning Emission Microscope (SEM) inspection was performed at Metal 2 layer at the targeted area. Metal 2 broken and copper losses were observed at the suspected area as illustrate in Figure 10.

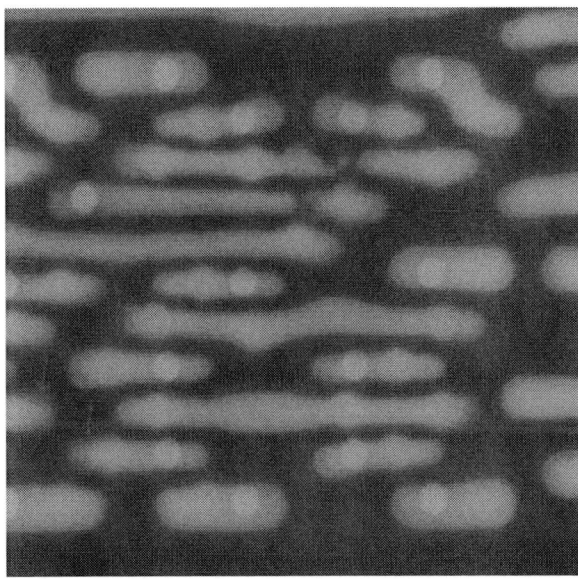

Figure 10. Metal 2 broken and copper lost was observed at the suspected failing area.

V. CONCLUSION

Two successful FA cases are discussed in this paper. ITG debugger with the combination of Functional Interface graphical view on the failing resource and extensive layout study has proven its ability for back-end defect localization. Metal pattern defect was observed on the first case study and metal broken and copper lost was observed on the second case study. This method has significantly improved PFA cycle time and success rates especially for wafer level analysis due to its robustness to identify accurate target for PFA.

ACKNOWLEDGMENT

The authors would like to thank Altera FA team members for their motivation and support.

REFERENCES

[1] Jiang Huang, Ryan Sweeney, Laurent Dumas, Mark Johnston, Pei-Yi Chen, Jeremy Russell. "Effective Defect Localization on Nanoscale Short Failures." *39th International Symposium for Testing and Failure Analysis*, San Jose, California, Nov.,2013, pp. 66-68.

[2] Kit Lau. "Layout Localization in Functional FA". *Altera Technology Topical Technology Review,* San Jose, California, April 2010.

[3] How Whee Yan. "ITG Debugger" *Altera Internal Software Training Material,* March 2013.

40nm NAND Flash Reliability Failure Analysis with Identification Tools Combination

Mei Ying Hsiao, Yi Heng Chen, Ling Kuey Yang
Reliability Testing & Failure Analysis Department, Powerchip Technology Corp.
No. 12 Li-Hsin Rd. 1, Hsin Chu Science Park, Hsin Chu, Taiwan, R.O.C
Phone: 886-3-5795000 EXT.4973   Fax: 886-3-5792142   Email: meiying@powerchip .com

Abstract-In the tradition failure analysis procedure, it is hard to find the physical root cause for 40nm NAND flash reliability testing failure. Because of it has weak micro-leak with repeating and longer metal conducting wire. We used several analysis methods to narrow down the defect location such as OBIRCH (Optical Beam Induced Resistance change) and PVC (Passive Voltage Contrast) to narrow down to 100um. Besides, EBAC (Electron Beam Absorbed Current) combine FIB circuit repair to isolation the failure location to one via. Finally, we preformed the FIB cross section analysis success to find the root cause.

I. INTRODUCTION

With recent shrinking of feature size and increasing number of metal layers in integrated circuits, determination of the physical root cause by failure analysis is becoming extremely difficult especially for the reliability failure samples of 40nm NAND flash product. The reason is weak micro leak and repeating long metal conducting wire in this product. Traditional failure analysis, localization failure site layer by layer analysis is seldom finding physical root cause. To determination the physical root cause, cross section is required. For cross section observation, the area of failure location should be narrow down to 100nm or one via [1].

In this paper, we will perform several analysis methods to narrow down the failure location, and successfully finding the defect root cause for this reliability failure sample.

II. PRODUCT FAILURE ANALYSES

1) Electrical Failure Analysis

In localizing the LSI failure sites, suitable methods are applied by failure modes. In general, DC failure, optical beam induced resistance change (OBIRCH) and an emission microscope (EMMI) are used. Several reasons caused this failure analysis case is difficult. First, we received only one 40nm NAND flash reliability failure sample. The failure item is HTOL (High temperature operating life) 168hrs Isb current high issue and leakage current ~690uA (spec=50uA). Second, EMMI or OBIRCH is difficult to detect the abnormal position for 690uA weak leakage current. Therefore, detail electrical failure analysis is important for this case. From IV curve measurement, the failure sample I_{sb} current is high relativity with Vcc force. After forces over 3.3V, the curve appeared saturated. If we bypass internal voltage control (intVdd=Vcc) Isb will increase with Vcc voltage. We suspected the short portion should be

internal Vdd "Fig. 1," Based on the above results, we try different test condition to find out the EMMI/OBIRCH abnormal hot spot point. Luckily, we found one abnormal hot spot and it is located near by No-1015 IO2/5 area from OBIRCH analysis "Fig. 2," From the layout tracing, the circuit associate to the Vcc and internal Vdd circuit is SAP (Sense PMOS).

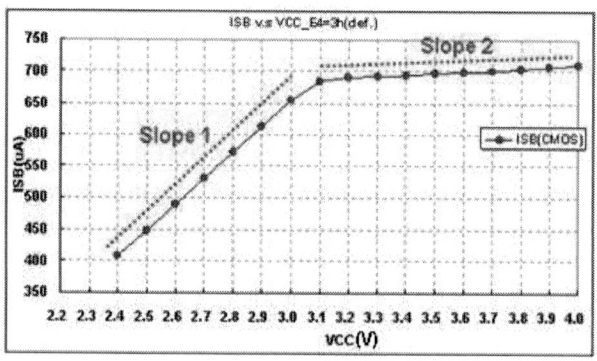

Fig. 1 IV curve of Isb vs. Vcc

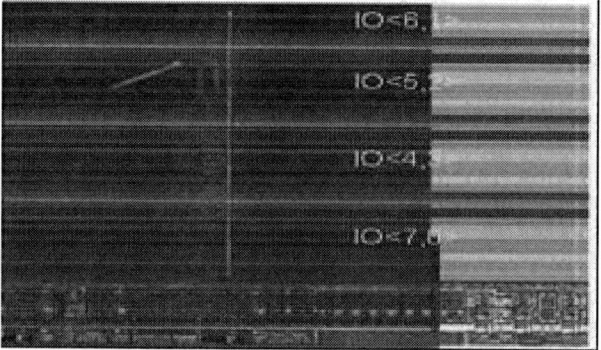

Fig. 2 OBIRCH hot spots

Also checking I-cell Erase, the results showed that the point of failure with the column No-1015 and input/output the circuit of great relevance. Even column and No-1015 shows normal Icell current, but another odd column shows twice of normal I cell current. It means that odd column was selected and NO-1015 also has been selected. The Cell IV curve measurement result also displayed the similar phenomenon. Different input/output circuit display different leakage current. IO6/1 and IO2/5 leakage was significantly higher than other circuits. "Fig. 3," "Fig. 4," Thus, we suspected failure point at circuit number No-1015 (CSL1015) and conclude the following

978-1-4799-3911-4/14 $31.00 © 2014 IEEE   274

model. "Fig. 5," The area is same as the result with OBIRCH detection. Fault occurred in the metal loop possibility. However, to perform a cross-sectional observation the area of localization should be achieved at least down to a 100nm square or specify just one via that connects interconnects fixture in different layers.

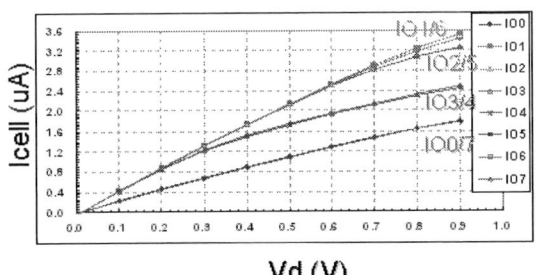

Abnormal I cell of Erase state IO1/6/2/5/3/4

Column address	IO0	IO1	IO2	IO3	IO4	IO5	IO6	IO7
Even column	1.77	1.78	1.77	1.76	1.76	1.76	1.77	1.77
Odd column except 1015	1.79	3.42	3.08	2.37	2.35	3.14	3.36	1.79
1015=col#3F7h	1.77	1.80	1.78	1.80	1.78	1.81	1.79	1.80

unit: uA

Fig. 3 cell program & erase cycle testing

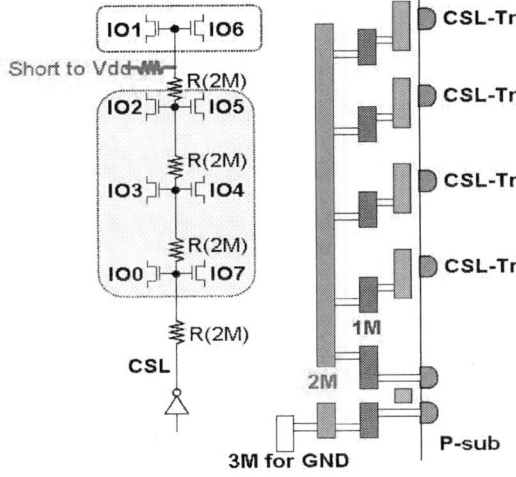

Fig. 4 Icell IV curve

Fig. 5 failure model

2) Physical Failure Analysis

Layer by layer checking in specific circuit is widely used failure analysis technology in metal loop due to resistance

change of OBIRCH result. The rough guideline for the area of localization using this method is about 100um square. PVC (passive voltage contrast) is usual auxiliary judge the failure location. By PVC highlight failure site, it can be reduced failure area around No-CSL1015 this circuit However, Not only IO6/1 and IO2/5 these two circuit but also IO3/4, IO0/7 also found PVC abnormal after compared with other normal circuit. "Fig. 6," We need to use other ways to narrow down the failure area. By the way of layer by layer checking, we found via and metal 2 has shift trouble from SEM images. It means via and metal 2 with very high abnormal risk "Fig. 7,"

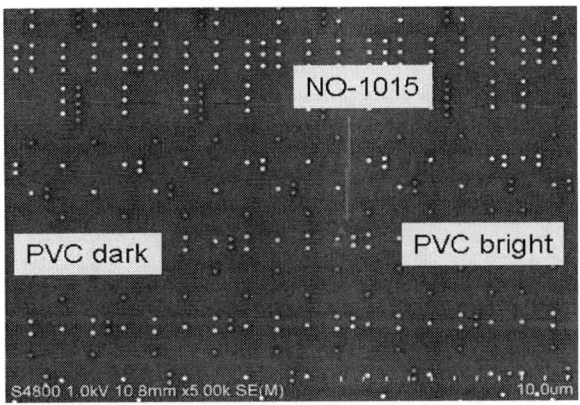

Fig6   PVC bright in NO-1015 circuit

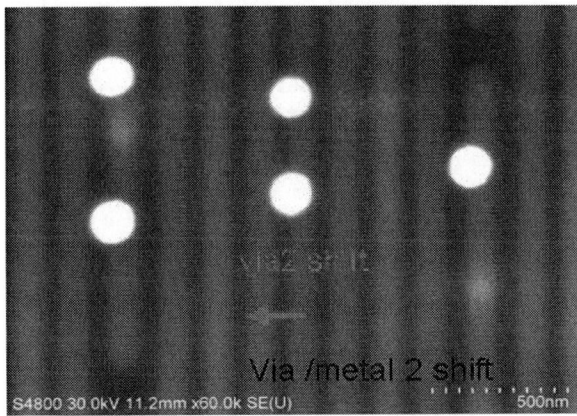

Fig7 Via shift SEM images

For localizing failure sites down to 100nm square range, the electron beam absorbed current method is used. [2] The current image can help to determine whether there is any problem with the metal lines and interconnects in the circuitry of interest. The benefit of absorbed current images is the isolation of buried interconnect defects beneath the surface layer. The SEM micrograph show the EBAC current of one probe to suspected circuit No-CSL1015, IO2/5"Fig. 8,". However, it was disappointing result. We did not find any anomalies.

978-1-4799-3911-4/14 $31.00 © 2014 IEEE   275

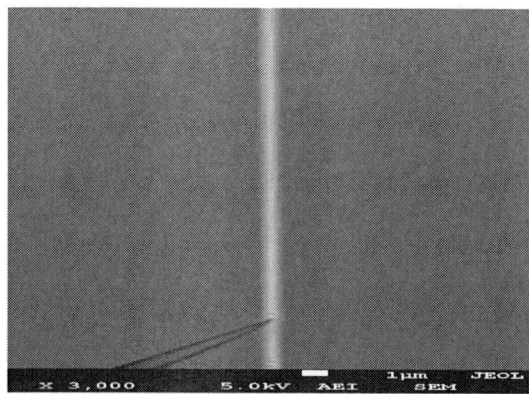

Fig8 one probe SAP & NO-1015 signal

In order find the failure point, we used FIB circuit repair this method and added two probing of electron absorbed current images function try to narrow down the failure site to one via or 100nm square area . From the above EFA result, we identify the failure location at circuit No-1015 IO2/5 and short to SAP this information. We used FIB circuit repaired to isolated SAP path at IO2/5 circuit boundary and added pad for circuit No-CSL1015 on IO3/4 area "Fig. 9,"

Probing to SAP IO6/1 area and circuit No-CSL1015 signal the result looks complicated. From the EBAC and SEM picture, the SAP indicated the white colour and circuit No-1015 indicate black colour. The failure location of IO2/5 still not found the abnormal point but it tells us some things.

(1) SAP signal is cut off on top side, which mean it is isolated from FIB circuit repaired.

(2) SAP IO2/5 this area signal indicated black colour, it is the same as No-1015 this circuit. Short location indeed happened in this area.

(3) The point of failure does not occur in the two metal lines short because of we did not find the potential re-segment phenomenon. Base on the result, we can rule out the abnormal metal short for M2 and M1.

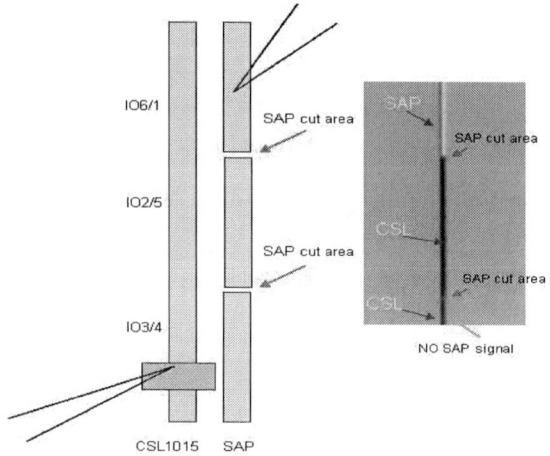

Fig9 Isolate SAP IO2/5 and two probes EBAC

Via may be an exception from SEM layer by layer checking for via and metal 2 shift. From the layout checking, four via contacts on SAP circuit would be the failure points "Fig. 10," To confirm it, Cross section performed one to one checking for the four suspected points. Finally, we found the defect mode is via and metal short and EDX show "W" peak. "Fig. 11," "Fig. 12," We provide the results to fab as a reference for process improvement

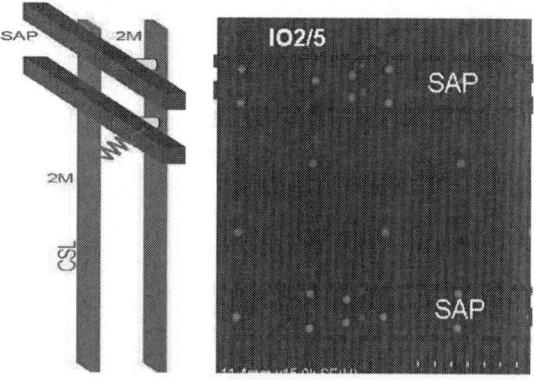

Fig10 Failure mode and SEM images

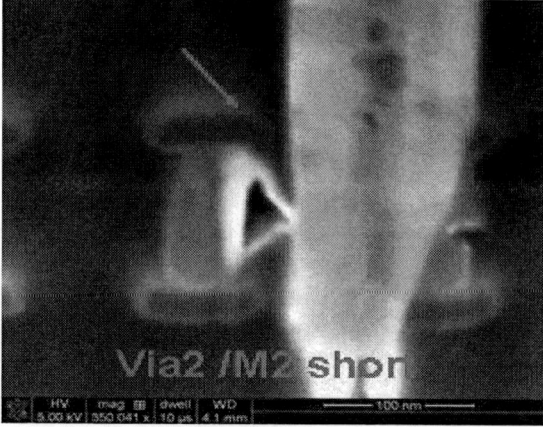

Fig11 Zoom in images

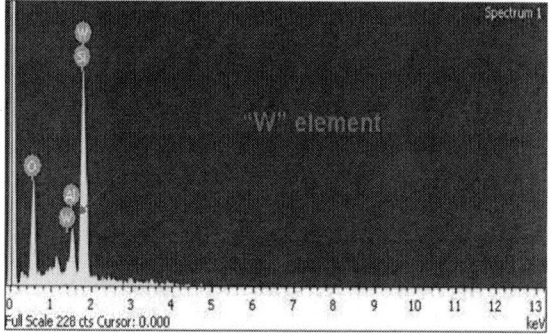

Fig12 EDX result

III. FAILURE MECHANISM

The failure sample behaviour is quite normal in CP testing I_{sb} leak about 6uA before HTOL stress. From FA results showed IMD exit void and via/metal have slightly shift. Through HTOL acceleration test with 125℃ ambient temperature induced via and metal 2 short issue.

IV. CONCLUSION

From the case study, we used several failure analysis methods to find out the root cause because the micro leak of long metal chain and complex circuit. Include OBIRCH, PVC to narrow down to 100um and electron beam absorbed current combined circuit repair methods to isolation the failure location to one via. Finally, we use FIB cross section this method success finding the failure mode is via and metal 2 short trouble.

ACKNOWLEDGMENT

Completion of this literature thanks to a lot of people like Product Development Engineer Division's Director Teng , Power Memory company's Ito san. Your wealth of knowledge grows my knowledge. Executive and partners within the department gave me a lot of forward momentum. .Without your assistance in this literature can not be very smooth finish.

REFERENCES

[1]. Y.Ueki,J.Kinshai,T.Nabeya,T.kawaguchi , M.Todome ,H.Wakamatsu and T.Miyazaki "Case Study on Failure Analysis by Electron Beam Absorbed Current Method" *JEOL news Vol 43 No1 P23-28*

[2].Chen Yaliang, Lin Soon-Huat, Vinod Narang, JM Chin "Absorbed-specimen Current Imaging Implementation and Characterization in Nano-Prober for Resistive Interconnects Isolation in 45-nm Silicon-on-Insulator Microprocessors". *2010. IEEE*

Detailed Package Failure Analysis on Short Failures after High temperature Storage

Z.Y. Oh, F.J. Foo and W. Qiu

Device Analysis Lab, Advanced Micro Devices (Singapore)

508 Chai Chee Lane Singapore 469032

Phone: (65) 67969888 Email: zi-ying.oh@amd.com

Abstract – **This paper describes the failure analysis approach in search of the root cause behind a series of short failures after high-temperature storage (HTS) test at 150°C. Findings revealed that UBM consumption by tin and the eventual disintegration allowed solder diffusion into the die circuitries resulting in the massive short failures.**

I. INTRODUCTION

During field operation, the microprocessor is constantly subjected to a high temperature environment. Therefore, thermal reliability of the flip-chip package is one of the main concerns during qualification. High-temperature storage (HTS) is a useful test for thermal reliability. It involves subjecting test vehicles to elevated temperatures ranging from 125°C to 300°C for a fixed period of time. The conditions are determined from the JEDEC standards for HTS.

A flip-chip package consists of a Silicon die, C4 solder bumps encapsulated by underfill (UF) and an organic substrate, with each component having different intrinsic material properties. The Silicon die and organic substrate have vastly different thermal dissipation rates, thereby resulting in a thermal gradient buildup within the C4 solder bumps, where it is found to be hotter at the Silicon die side [1]. This drives atomic diffusion, where at a sufficiently high thermal gradient, Lead (Pb) atoms can migrate from the hotter Silicon die side towards the colder substrate side in eutectic solder bumps [1]. As a result, Tin (Sn) will be pushed towards the hotter die side. In another study done by Balkan, it was found that during accelerated testing at high temperatures above 150°C, Under-Bump Metallization (UBM) consumption surfaced as a dominant failure mechanism [2]. As such, Sn accumulation at the UBM is believed to be linked to this consumption failure mechnism. This paper dicusses the failure analysis (FA) flow to uncover such failure mechanism in a series of power-supply short failures after HTS.

Common tools utilized in conventional package level FA include, but are not limited to, curve tracer (CT), C-mode scanning acoustic microscope (CSAM), X-ray, optical microscope, scanning electron microscope (SEM) and electron-dispersive x-ray spectroscope (EDX).

Complexity of packages has advanced rapidly with aggressive scaling, finer bump pitch and increase in I/Os. Conventional FA tools are gradually reaching their limitations.

Power-supply shorts usually posed more challenge in FA as numerous chip connections, bumps and package power planes are associated in a single power-supply. Accurate fault isolation (FI) of such failures becomes more critical to ensure a higher success rate during destructive physical failure analysis.

The infrared lock-in thermography (IR-LIT) is a new technique, which in recent years was proven as an effective and non-destructive fault isolation tool for short failures in flip-chip package [3], [4]. This technique is based on the principle that the short defect being a location of high current concentration produces heat which is of similar wavelength to infrared in the electromagnetic spectrum. The heat source and therefore, the defect location, can be easily detected by an infrared thermal microscope. Recent developments of the lock-in frequency function significantly improve the thermal emission detection sensitivity of the infrared thermal microscope [3]. IR-LIT will be extensively used in the FA discussed in this paper.

The use of focused-ion beam (FIB) for milling of package components has been wide-spreading. This is driven by the increased fragility of the dielectrics as ultra-low-k materials are implemented to boost device speed. The use of FIB for cross-section helps to eliminate undesirable artificial damage which may be induced during conventional mechanical cross-sectioning [5]. Despite the drawback of slower milling rate, FIB gives a high success rate when the area of interest is challengingly small. FIB will be employed in the destructive physical analysis section.

II. EXPERIMENTAL SETUP

A. Test Vehicle

Test vehicles built for the high temperature storage test are flip-chip packages assembled with eutectic Sn-Pb solder bumps. The structure of UBM is laddered Titanium/Copper/Nickel (Ti/Cu/Ni). Substrates from two suppliers (denoted as A and B) with two different UFs (denoted as 1 and 2) are coupled to form the matrix below:

Table I. List of test vehicles

Test Vehicle	Substrate	Underfill
A1	A	1
A2	A	2
B1	B	1
B2	B	2

B. Test Condition

The test vehicles are subjected to high temperature storage test according to JESD22-A103D condition B at +150(-0 / +10) °C, for up to a maximum of 1000 hours.

At 500 hours and 1000 hours read-points, the test vehicles are removed from chamber and sent for electrical testing.

C. Electrical Test Results

Results from electrical testing at 500 hours and 1000 hours read-points are listed in Table II below where failures are denoted by shadings.

Table II. Electrical test results

Test Vehicle	HTS 500 hours	HTS 1000 hours
A1		
A2		
B1		
B2		

Power-supply short failures were reported after 500 hours (from A1 and B1) and 1000 hours (from A2 and B2) of HTS from all groups of test vehicles with no commonality in substrate or UF. Failures are removed from the sample pool and submitted for root cause analysis.

III. FAILURE ANALYSIS

Failure analysis began with electrical verification of the reported failure. A curve tracer (CT) is utilized to obtain an I-V curve of the reported failing power-supply. This is done by probing the failing power-supply with respect to ground on the failure unit. The same procedure is repeated on a reference good unit for comparison.

IV curves of the failure and reference unit are displayed in Fig. 1. It clearly verified power-supply short in the failure unit.

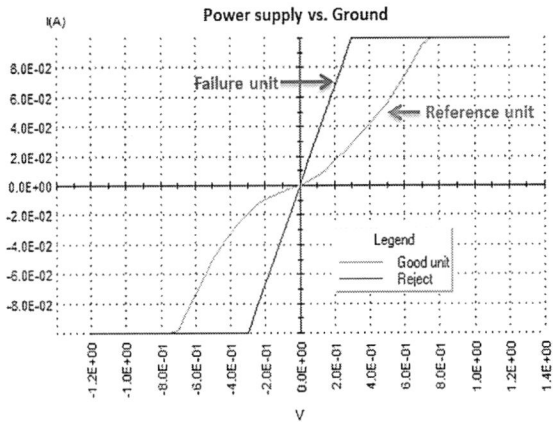

Fig. 1: Short I-V curves of the failing power-supply versus ground

After electrical verification, non-destructive CSAM inspection will follow to check for any C4 bump or interfacial defects within the flip-chip package. The resulting acoustic micrograph is shown in Fig. 2.

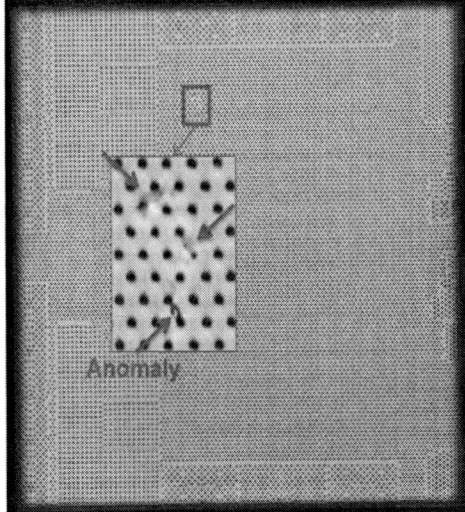

Fig. 2: CSAM micrograph of the failure unit showing anomaly in the polyimide-underfill interface

Fig. 2 displays the polyimide-underfill interface of the failure unit. During a high resolution scan, some form of anomalies can be observed, but the exact signature could not be determined.

Subsequent X-ray inspection also identified several anomalous C4 bumps in the CSAM anomaly region. These bumps are highlighted in Fig. 3.

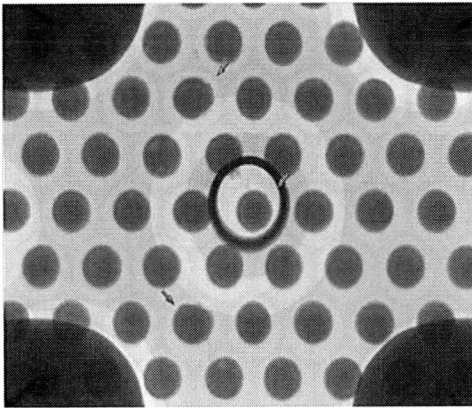

Fig. 3: X-ray image showing anomalous C4 bumps

Although anomalies were identified from CSAM and X-ray inspection, these observations are not sufficiently conclusive to determine the cause of power-supply short. IR-LIT is then employed to isolate the exact location of the short failure. This is achieved by connecting the failing power-supply to ground and supplying a biased voltage of 0.13V to the unit. The resulting current measured was 368mA, with a thermal site observed as shown in Fig. 4.

Fig. 4: IR-LIT image showing thermal site indicating the failure location

This thermal site represents the exact short location. Upon comparison, it is observed that this thermal site location coincides with the region of CSAM and X-ray anomalies. This helps to increase the confidence for destructive physical analysis that follows.

First, parallel lapping was performed to grind away the organic substrate and part of the C4 bumps. Burnt marks were found around the C4 bumps at the thermal site as shown in Fig. 5 which implies that a thermal event has occurred.

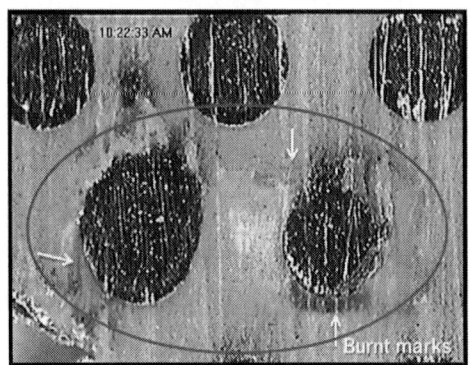

Fig. 5: Optical image of the failure unit after parallel lapping to C4 bump

Cross-section was performed on one of the affected bumps using the focused-ion beam (FIB) to verify the physical defect without inducing any undesirable artificial damage to the die circuitries and inter-layer dielectrics.

The cross-sectional views of a reference good bump and the affected bump are shown in Fig. 6 and Fig. 7 respectively. Corresponding EDX mapping results of the affected bump are displayed in Fig. 8.

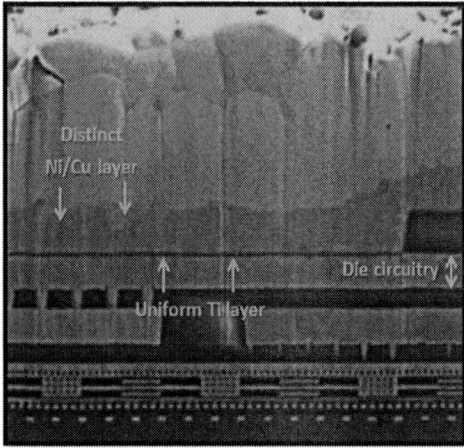

Fig. 6: SEM micrograph of reference bump after FIB cross-section

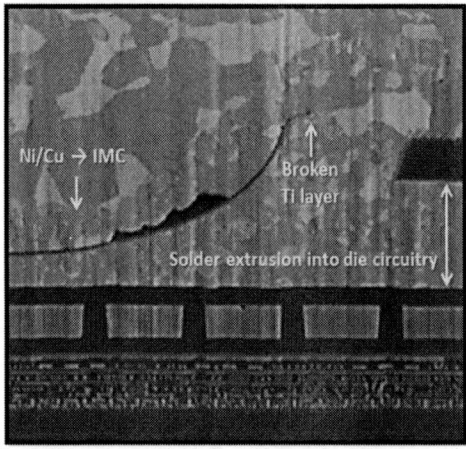

Fig. 7: SEM micrograph of affected bump after FIB cross-section

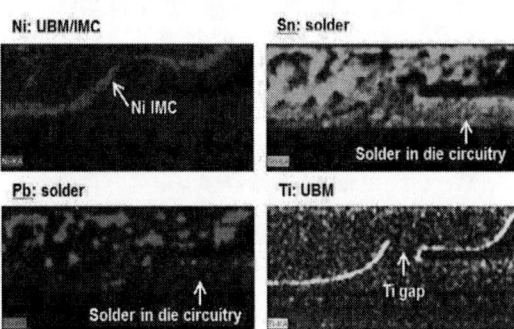

Fig. 8: Elemental maps of the affected bump's UBM

Fig. 6 and Fig. 7 provide comparison between a good UBM in the reference bump and the disintegrated UBM in the affected bump. In addition, solder diffusion into the die circuitry makes it evidently thicker than usual. This allows for quick inspection for similar failure mechanism in other failure units.

From the Ti map in Fig. 8, an obvious gap in the UBM Ti layer can be observed. At the same time, it is also observed that the UBM Ni has all transformed into intermetallic compounds (IMC) and Sn is present in the die circuitry. These are evidences of UBM consumption and disintegration, which has led to solder diffusion into the die circuitry.

Next, further parallel lapping was performed on the same affected bump to remove remaining UF to reach the polyimide layer. Since polyimide appears transparent under visible light, the underlying die circuitries can be inspected under an optical microscope. The resulting image is captured in Fig. 9.

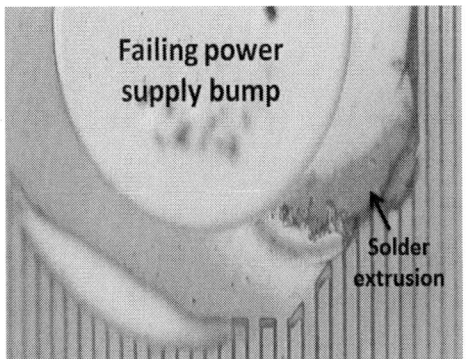

Fig. 9: Optical image of the bump of interest at polyimide layer

The parallel metal lines are alternating power-supply and ground circuitries. Solder is observed to diffuse into the die circuitries and join several metal lines together, hence causing the reported short failure. In order to understand the failure mechanism further, conventional mechanical cross-section was performed on another similar failure unit. A reference bump and an affected failure bump were inspected by SEM and documented in Fig. 10 and Fig. 11 respectively.

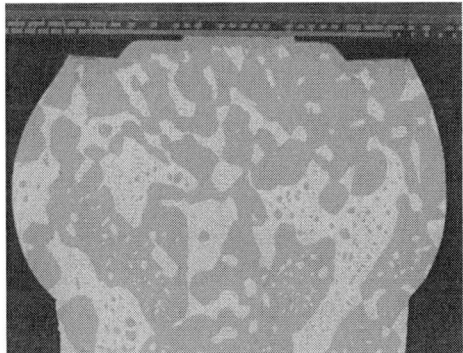

Fig. 10: SEM micrograph of reference bump cross-section

Fig. 11: SEM micrograph of failure bump cross-section

Fig. 12: SEM micrograph of disintegrated UBM

A more refined grain structure is observed in the failure bump (Fig. 11) as compared to the reference bump (Fig. 10). In addition, identical failure signature is seen on this failure bump where disintegrated UBM allowed solder diffusion into the die circuitries, thereby thickening the die metal line, as shown in Fig. 12.

IV. RESULTS AND DISCUSSION

Subsequent FA shows that the same failure mechanism is observed in failures from all 4 groups of test vehicles.

Examining Fig. 11, the refined grain structure is the result of thermally activated solder recrystallization. In addition, signs of thermo-migration can be observed as there is a significantly higher content of Sn near the die side in Fig. 11. This tallied with literature review which states that Pb and Sn atom migration can occur at temperature gradient setup under 150°C [1]. As a result, UBM consumption took place when Sn interacted with the UBM's Ni and Cu to form IMC. The Ti layer eventually broke, allowing solder diffusion into the die circuitries. This phenomenon eventually leads to the short failure.

Since all test vehicles exhibit identical failure mechanism when subjected to 150°C storage temperature, the HTS condition was reviewed.

The exact same experimental setup was repeated with a lower test temperature of 125°C. No electrical failure was reported from all groups of test vehicles, up to the maximum read-point of 1000 hours. Cross-sections performed on the passing units verified UBM integrity.

V. FUTURE WORKS

The homologous temperature of tin-lead solder at 150°C is at 0.93. At this temperature, tin-lead solder is prone to creep. In addition, breakage of UBM Ti layer is observed at the corner in the HTS 150°C failures. Simulation studies can be done to model the stress concentrations at the UBM-solder interface to understand UBM Ti breakage.

VI. CONCLUSION

A detailed FA flow for a series of short failures from HTS test was documented in this paper. Results were then correlated with literature reviews to establish that thermo-migration at 150°C led to Sn accumulation at the hotter Silicon die side. This brought about UBM consumption and disintegration which eventually resulted in short failure due to solder diffusion into die circuitries. Repeating the HTS at 125°C up to 1000 hours observed no failures.

ACKNOWLEDGMENT

The authors would like to thank colleagues from AMD Singapore DA lab for their help in one way or another, and Dr. Daniel Rhee from Institute of Microelectronics for his mentoring efforts in this paper.

REFERENCES

[1] D. Yang, M. Alam, B. Y. Wu and Y. Chan, "Thermomigration in eutectic tin-lead flip chip solder joints," in *Electronics Packaging Technology Conference*, 2006.

[2] H. Balkan, "Flip chip electromigration: impact of test conditions in product life predictions," in *Electronic Components and Technology Conference, 2004. Proceedings. 54th*, 2004.

[3] M.Y. Tay, M.C. Tan, W. Qiu, X.L. Zhao, "Lock-in Thermography Application in Flip-chip Packaging for Short Defect Localization," in *Electronics Packaging Technology Conference*, 2011.

[4] Lihong Cao, Manasa Venkata, Jeffery Huynh, Joseph Tan*, Meng-Yeow Tay*, Wen Qiu*, "Lock-in Thermography for Flip-chip Package Failure Analysis," in *International Symposium for Testing and Failure Analysis*, 2012.

[5] L. T. W. D. Lihong Cao, "Failure Analysis of Flip Chip C4 Package Using Focused Ion Beam Milling Technique," in *37th International Symposium for Testing and Failure Analysis, ISTFA*, 2011.

Structure and Composition of the Cu/Low k Interconnects de-layered with FIB

Dandan Wang[*], Pik Kee Tan, Maggie Yamin Huang, Jeffrey Lam and Zhihong Mai

Technology Development Department, GLOBALFOUNDRIES (Singapore) Pte. Ltd.

60 Woodlands Industrial Park D, Street 2, Singapore 738406

Email: dandan.wang@globalfoundries.com

Abstract-By the gas-assisted focused ion beam (FIB) method, we de-process the device from top layer to bottom layer. It is a highly efficient failure analysis method on the precise location. After removing the dielectric layers under the bombardment of an ion beam, the chemical composition of the top layer was altered with the reduced oxygen content. Further energy-dispersive X-ray spectroscopy and FTIR analysis revealed that the oxygen reduction lead to appreciable silicon suboxide formation. Our findings with structural and composition alteration of dielectric layer after FIB de-layering open up a new insight avenue for the failure analysis in IC devices.

I. INTRODUCTION

With the shrinking dimensions of IC devices, the thickness of each layer becomes thinner. It presents an unprecedented challenge for the rapid and localized de-layering analysis by conventional mechanical polishing method. However, FIB technology compatible with chemically is good choice to enhance milling and deposition processes [1-3]. The precursor gas can greatly enhance the milling rate. Likewise, the selectivity can be significantly enhanced by selecting a gas that reacts with one specific material. FIB to de-layer an integrated circuit (IC) chip offers significant advantages over mechanical polishing methods such as: localized "polishing" without introducing damage to other areas of the die; real-time monitor without withdraw the die and disturbing the milling; in-situ SEM cross-sectioning or TEM lamella preparation without withdraw the die.

Although FIB technology offers many important features, after the bombardment of the ion beam, an inevitable damage will be induced near the surface of materials. For small size device, this damage will lead to the changes of the structure and properties of material and increase the failure risk of device. The structural damage and the ion implantation induced by energetic ion bombardment are main drawbacks of FIB patterning [2-4], especially for electrically active materials [5, 6]. However, to the best of our knowledge, there has been no report on the low-k surface structure and chemical modification during FIB process so far. In this paper, de-layered with a gas-assisted FIB technique, the corresponding structures and the chemical composition on the top layer of an IC device were investigated and discussed in detail.

II. EXPERIMENTAL DETAILS

In the present work, a Helios NanoLab™ 450S DualBeam from FEI Corporation was used for the gas-assisted FIB de-layering. The ion beam used in our work is Ga^+ ion. The FIB accelerating voltage was set to 30 kV and the area for milling is 10×10 μm^2. To mill the copper material, H_2O vapor was injected and the milling current was 0.43 nA. To remove the low k dielectric layer, XeF_2 was used and the milling current was 80 pA. The etching time depends on materials and the layer thickness. Real-time scanning electron microscope (SEM) monitoring was employed to observe the changes of the milling surfaces. The completion of de-layering can be determined with the real-time SEM monitoring and the etching can stop precisely at the layer of interest. IR spectra were recorded on a Nicolet 6700 Analytical FTIR spectrometer coupled with a Nicolet Contiuµm infrared microscope. Sensitive mercury cadmium telluride (MCT/A) detector and KBr beam splitter were used for mid-IR (4000-400 cm^{-1}) data collection with a resolution of 4 cm^{-1}. The IC device was fabricated with a standard Cu/low-k process. Cu was used as the metal interconnects and SiCOH was used as the low-k inter-metal dielectrics (k~2.8). The technology node is 40 nm.

III. RESULTS AND DISCUSSION

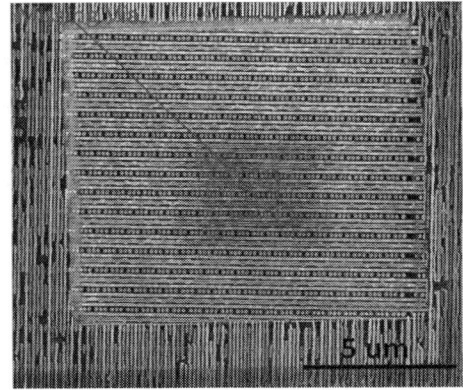

Fig. 1. Planar de-layering SEM image of the FIB window in the failure location.

Fig. 1 shows the SEM images of the gas-assisted FIB planar etching in the failure location. The etching window is 10×10

978-1-4799-3911-4/14 $31.00 © 2014 IEEE

μm^2. It shows missing via after de-layering. The etching gas was introduced and chemisorbed onto the sample surface. Due to the chemical reaction, at moderate ion beam currents, the chemically assisted milling process removes material faster with higher selectivity than that by a conventional ion milling process without gas-assistance. Xenon Difluoride (XeF_2) is used to enhance the material removal rate while milling through inter-layer SiO_2 and Si-based low k dielectrics [4]. It forms the volatile SiF_x compounds. The selective etching of the silicon and silicon-based compounds leaves the underlying metal lines and contacts intact [5].

For the removal of exposed copper lines and contacts, water (H_2O) was used to assist ion beam milling. H_2O contain much oxygen, oxygen offers an inhibitive advantage toward etching the dielectric during the ion beam milling [6]. With water, ion beam etching rate for copper is two times faster than dielectrical materials [6], so copper can be removed selectively and uniformly. As shown in Fig. 1, the silicon dioxide and metal can be milled selectively. In our investigation, the total etching time of the de-layering process is about 3 hours from the layer of metal 6 to the top of the source-drain contacts.

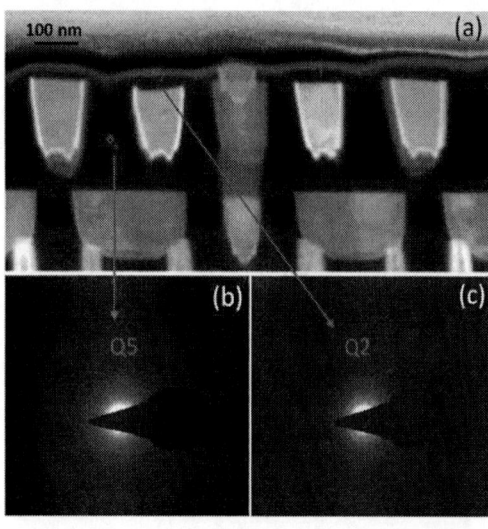

Fig. 2. (a) Dark field TEM image on the defect location. (b) and (c) selected-area electron diffraction patterns on location Q5 and Q2, Q2 is the point located at the FIB induced layer.

In order to further understand the damage of exposed surface under gas-assisted FIB etching, TEM analysis was carried out and the results were summarized in Fig. 2. After expose the metal 2 layer by gas-assisted FIB etching, a thin layer on top was observed with obvious contrast difference to the dielectric layer below. The dark field (DF) TEM image was shown in Fig. 2(a). The highly diffusive ring patterns of selected area electron diffraction (SAED) as shown in Figs. 2(b) and 2(c) reveal that the materials are completely amorphous. As shown in Fig. 3, the energy-dispersive X-ray spectroscopy (EDX) spectra were taken on top layer of samples de-layered by gas-assisted FIB

[Fig. 3 (a)] and mechanical polishing method [Fig. 3 (b)], respectively. Both types of samples contain Si, O, and C elements, but the concentration is different. The top layer of gas-assisted FIB de-layered sample possesses less oxygen than that of a mechanically polished sample. Essentially the Cu concentration is very low from EDX examination, and we mainly attributed the Cu existence to the re-sputtering during the ion beam milling process.

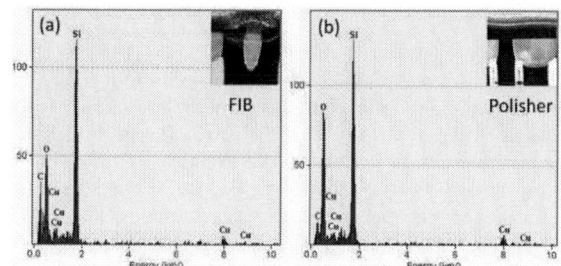

Fig. 3. EDX spectra on top layers for the FIB de-layering sample (a) and the polisher de-layering sample (b).

To have an in-depth understanding on the alternation of the structures and chemical compositions of the amorphous layer, the detailed investigations were carried out using a Fourier Transform Infrared Radiation (FTIR) spectrometer. The model of the FTIR system is Nicolet 6700 from Thermo Fisher Scientific Inc. Totally four samples were top-down de-layered to via 1 and metal 1 by the gas-assisted FIB etch and mechanical polishing method, labeled as FIB-v, FIB-m, polisher-v, and polisher-m, respectively. In the FTIR spectra in Fig. 4, in addition to the H_2O and CO_2 peaks [7], there are three peaks corresponding to stretching vibrations of C-H_2 (2850 cm^{-1}), Si–O bond (1250–950 cm^{-1}) [8, 9], and Si-C vibration (800 cm^{-1}), respectively. Clearly, the peaks corresponding to the Si–O bond exhibit a slight different for the four samples.

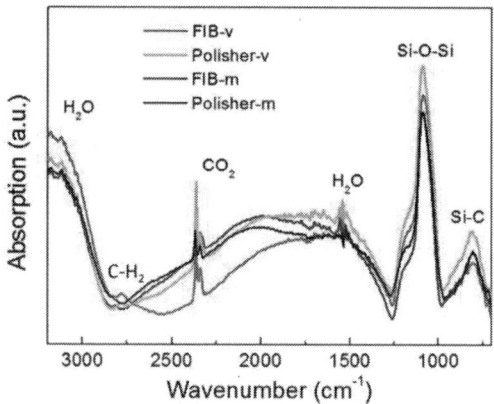

Fig. 4. FTIR spectra on the metal 1 and via 1 layer with FIB de-layering and mechanical polishing.

978-1-4799-3911-4/14 $31.00 © 2014 IEEE

In SiCOH low k dielectric, the Si–O–Si asymmetric stretching band frequency is related to the Si–O–Si bonding angle. After zooming in the spectra of Si–O bonds, the detailed results are shown in Fig. 5. The Si–O–Si asymmetric stretching band at 1250–980 cm^{-1} can be de-convoluted into three peaks centering at ~ 1020-1400 cm^{-1} (P1), ~1065 cm^{-1} (P2) and ~1145 cm^{-1} (P3), respectively. P3 at 1145 cm^{-1} is attributed to the Si–O–Si bonds in a cage structure with a bond angle of approximately 150° [10-13]. P2 at 1065 cm^{-1} is attributed to the stretching of smaller angle Si–O–Si bonds in a network structure. In a fully relaxed stoichiometric thermal silicon oxide grown at temperatures >1000 °C, the bond angle is reported to be ~144° [7, 14], appears around 1080 cm^{-1} in some reports. P1 at 1020~1400 cm^{-1} is ascribed to the stretching of an even smaller Si–O–Si bond angle. It is reported that the frequency of the Si–O–Si asymmetric stretching vibration decrease in silicon suboxides due to the silicon atoms having one or more non-oxygen neighbors with a higher probability [14, 15]. A bonding structure model based on the distortion of SiO$_3$X tetrahedral due to electronegativity differences in sub-oxide has been proposed to explain the change in bonding angle. In the sub-oxide, the substitute atom, such as C or Si, has less electronegativity than that of oxygen. Such atom attracts the bonding electrons in the Si–X bond less strongly than oxygen in the Si–O bond thus distorting the tetrahedra and reducing the bonding angle to < 144° [16]. Obviously, P1 in FIB samples appear with higher wavenumber in comparison with polisher samples. The location difference of P1 between FIB and polisher samples is because of that the different atoms replace O in Si–O bonds in two types of samples. In FIB samples, X'-Si bond has higher electronegativity than that of X-Si bond in polisher samples, thus it gives rise to higher wavenumber in FTIR results.

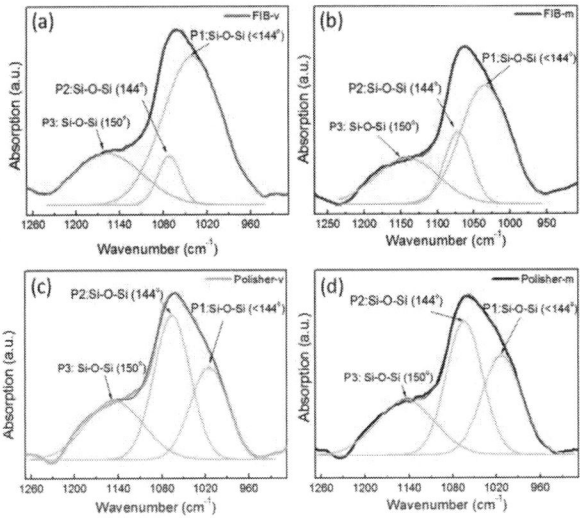

Fig. 5. Deconvolution of the enlarged Si-O-Si FTIR absorption bands.

After fitting and integrated area calculation, the de-convolution results are summarized in Table I. We found that the peak center of P1 for the FIB de-layered samples moving to the higher wavenumber side. According to the bonding energy theory mentioned above, it means the substitute atom in FIB de-layering and polisher samples are different, and the Si-substitute atom bonding energy in FIB de-layering sample is a little bit higher. Comparing the integrated area of each peak among four samples, the P3 is same and it means the Si-O bonding angle >150° is the same for both FIB delayering and polisher samples. For the P1, the integrated area is very big for the FIB de-layered samples, suggesting that there are more suboxide bonds in FIB de-layered samples. During the FIB de-layering, the higher voltage energy was used for the ion milling; consequently the bombardment caused temperature increasing will enhance the vibration of the band. Hence it will break the Si-O bonds and lead the formation of suboxide bonds, which is also in agreement with the EDX results aforementioned. Grill [17] and Jaeyeong Heo [18] et al. reported that the ratio of the suboxide bonds can cause the changes of k value. The sub-oxide bonds has less angle (<144°), and it occupy the small volume, so it lead to the reducing of air aperture in the film and increase the k value. In the FIB de-layered samples, there is much sub-oxide bonds with angle (<144°), and the k may be higher than polisher samples.

Table I De-convoluted Si-O peak position and the corresponding area ratio for FIB and mechanical polishing de-layered samples.

	P1 (silicon suboxide, Si-O-Si, angle <144°)	P2 (network Si-O-Si, angle ~144°)	P3 (Si-O-Si, angle ~150°)	Integrated area percentage		
				P1	P2	P3
FIB-v	1040	1072	1156	64%	8.4%	27.6%
FIB-m	1034	1071	1141	54.7%	18.5%	26.8%
Polisher-v	1019	1065	1145	28.8%	43.2%	28%
Polisher-m	1018	1063	1144	33.3%	40.7%	26%

IV. SUMMARY

The gas-assisted FIB milling method makes it possible to de-layer an IC chip at a highly precise location efficiently with in situ image capturing on the area of interest throughout the delayering process. The structures and chemical compositions at the de-layered area were investigated with TEM and FTIR analysis. After gas-assisted FIB de-layering, there was an amorphous SiCOH layer on top of the de-layered area. FTIR spectra and EDX results show that this layer possesses less oxygen with Si–O–Si bonds angle <144°. The k value after the gas-assisted FIB de-layering became higher than polisher sample.

ACKNOWLEDGMENT

We would like to thank the Failure Analysis Lab in Globalfoundries for providing the excellent TEM support.

REFERENCES

[1] Datta, Yuh-Renn Wu, and Y. L. Wang, *Appl. Phys. Lett.*, 75, 2677 (1999).

[2] J. H. Kim, J. H. Boo, Y. J. Kim, *Thin Solid Films*, 516, 6710 (2008).

[3] M. K. Lee and K. K. Kuo, *Jpn. J. Appl. Phys.*, 45, 2447 (2006).

[4] R. Mikhail Baklanov, J. Marneffe, D. Shamiryan, A. M. Urbanowicz, H. L. Shi, V. T. Rakhimova, H. Huang, and P. S. Ho, *J. Appl. Phys.*, 113, 041101 (2013).

[5] X. Li, E. A. Delenia, R. M. Ring, U.S. Patent 7029595 B1, (2002).

[6] D. L. Scott, U.S. Patent 6,407,001, (2002).

[7] G. Herzberg, *Infrared and Raman Spectra of Polyatomic Molecules*; D. Van Nostrand Co.: New York, 1945.

[8] C. T. Kirk, *Phys. Rev. B*, 28, 3225 (1983).

[9] Simon, in *Modern Aspects of the Vitreous Silica Gordon and Breach*, New York, 1975.

[10] P. Bornhauser and G. Calzaferri, Spectrochim. *Acta*, Part A 46, 1045 (1990).

[11] P. Bornhauser and G. Calzaferri, *J. Phys. Chem.*, 100, 2035 (1996).

[12] C. Marcolli and G. Calzaferri, *J. Phys. Chem.*, 101, 4925 (1997).

[13] L. H. Lee, W. C. Chen, and W.-C. Liu, *J. Polym. Sci., Part A: Polym. Chem.*, 40A, 1560 (2002).

[14] G. Lucovsky, M. J. Manitini, J. K. Srivastava, and E. A. Irene, *J. Vac. Sci.Technol. B*, 5, 530 (1987).

[15] P. G. Pai, S. S. Chao, Y. Takagi, and G. Lucovsky, *J. Vac. Sci. Technol. A*, 4, 689 (1986).

[16] Y.H. Kim, M. S. Hwang, H. J. Kim, J. Y. Kim, and Y. Lee, *J. Appl. Phys.*, 90, 3367 (2001).

[17] Grill and D. A. Neumayer, *J. Appl. Phys.*, 94, 6697 (2003).

[18] Jaeyeong Heo and Hyeong Joon Kim, *J. Electrochem. Soc.*, 153, 228 (2006).

Identification of Cu-Al Intermetallic Phases of Copper Wire Bonding Using TEM Nano Beam Diffraction Indexing Technique

F.K. Yong

Infineon Technologies (Kulim) Sdn. Bhd.
Lot 10 & 11, Industrial Zone Phase II, Kulim Hi-Tech Park,
09000, Kedah Darul Aman, Malaysia
Phone: +60 (4) 4278741 Email: fookhong.yong@infineon.com

Abstract -The paper shares the application of electron diffraction spot indexing to identify intermetallic phases in copper wire bond. Standard Scanning Transmission Electron Microscopy (STEM) technique with a converged electron probe was modified to produce parallel nano beam for forming electron diffraction spot patterns, CuAl(η_2) and CuAl$_2$(θ) phases in copper wire bond were identified by indexing the diffraction spots.

I. INTRODUCTION

Copper wire bonding on aluminum metallization is cost-effective and, offers better thermal and electrical conductivity with improved mechanical robustness of Cu-Al intermetallic compounds (IMC). In order to meet market demand for miniaturized products, semiconductor packages have started to shrink and the use of fine pitch wires with smaller diameter have increased, especially in high pin count devices. Development and optimization of copper wire bonding process requires in-depth study of various aspect, which includes characterization of Cu-Al intermetallic compounds after isothermal aging.

Currently various techniques have been deployed to study the IMC phases. Among those techniques, the most popular and cost effective is Scanning Electron Microscopy (SEM) based Energy Dispersive Spectroscopy (EDS) analysis after mechanical cross section of the ball bond. For improved spatial resolution, Scanning Transmission Electron Microscopy (STEM) based EDS analysis of a thin specimen is preferred. The Cu-Al intermetallic phases are identified through the atomic or weight percent quantification of the Cu and Al content derived from the X-ray counts. Thin intermetallic layers Cu wire bonding down to 50nm in thickness have been characterized with STEM-EDS technique, at zero hour before the aging process [1]. However, without proper standards, the STEM-EDS technique produces an empirical composition data with lots of challengeable ambiguities. Parallel nano beam diffraction (NBD) analysis to overcome the EDS analysis ambiguities has been proposed with minor modification of existing microscope in STEM mode.

Diffraction is a phenomenon in which travelling waves bend around the small obstacles and the spreading out of waves when passing through small opening or slit. Typical example of diffraction visible to naked eye us depicted in Figure 1; a coherent wave of laser diffracts at multiple parallel slits and forms diffraction pattern as series of sharp spots resulting from wave interference [2]. Spot intensity is higher when it is close to the center undiffracted beam. When the slits are rotated, the diffraction spots will also rotate accordingly. Typically, distance between spots is inversely proportional to the size of slit and spacing between the slits.

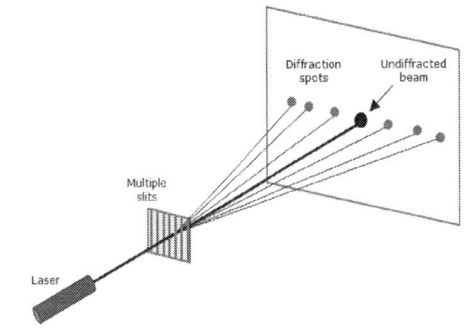

Fig. 1. Typical diffraction pattern of laser waves when passing through multiple slits.

Electron diffraction is similar to that of laser diffraction but a CCD camera is needed to record the diffraction pattern. Due to its short wave length and the uniqueness of diffraction behavior, electron diffraction is very often used to determine lattice properties of a crystal system, i.e. symmetry and position of the spots in the diffraction pattern contains information about the symmetry of the crystal and the spacing of atoms in the crystal structure. Figure 2 shows an example of high resolution transmission electron microscope (TEM) image of silicon crystal structure in which the electron beam is aligned to parallel <110> zone axis of silicon; various interplanar spacing of silicon crystal planes are identified to satisfy Bragg condition:

$$n\lambda = 2d\sin\theta$$

where *n* is an integer, λ is the wavelength of incident wave, *d* is the spacing between the planes in the atomic lattice, and θ is the angle between the incident ray and the scattering planes. As such these sets of crystallographic planes will be responsible for respective rows of spots in a diffraction pattern. These rows of diffraction spots from various sets of crystallographic planes will super impose to form complete electron diffraction spot pattern in the reciprocal space. The distance between the transmitted beam (center spot) and the diffracted spot is inversely proportional to the interplanar spacing of specific set of crystallographic planes corresponding to that spot (Figure 3).

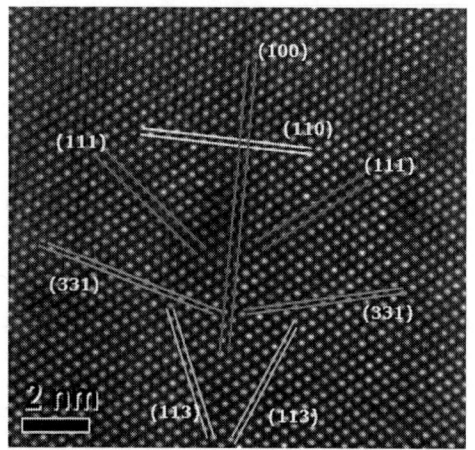

Fig. 2. High resolution TEM image of silicon lattice structure at <110> zone axis. Lattice spaces satisfying the Bragg Law are indicated by parallel lines.

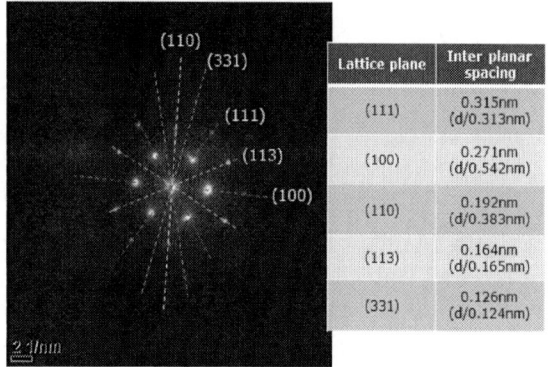

Fig. 3. Complete electron diffraction spot pattern of silicon at <110> zone axis. Interplanar spacing can be derived from the reciprocal of the distance between the transmitted (center) and diffracted spots.

Previous section demonstrated the relationship between the electron diffraction spot and the crystal structure parameters. Such electron diffraction pattern from unknown intermetallic phase can be used to determine its crystal structure by indexing the diffraction spots and matching the interplanar spacing to a crystallographic database of intermetallic compound. In this article, the application of nano beam diffraction (NBD) technique in STEM mode is demonstrated to characterize the IMC layers of copper wire bonding. The sample preparation, STEM-NBD mode set up and electron diffraction spot indexing are discussed in the following sections.

II. MOTIVATION

A microelectronic chip with 8 mils copper wire bonding on aluminum metallization was selected for analysis. The unit had

been subjected to 5000 hours of 175°C thermal stress for the purpose of IMC characterization using STEM-EDS technique. The STEM-EDS elemental profile across the IMC is shown in Figure 4. Presence of $CuAl(\eta_2)$ could be easily predicted at area B of the elemental line profile because it matched closely with composition information on the phase diagram; however it was difficult to predict presence of $CuAl_2(\theta)$ at area A due to a difference of 10 weight percent, when compared to phase diagram. The NBD technique in STEM mode was attempted to produce more conclusive results.

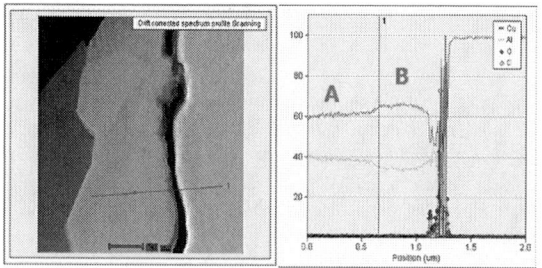

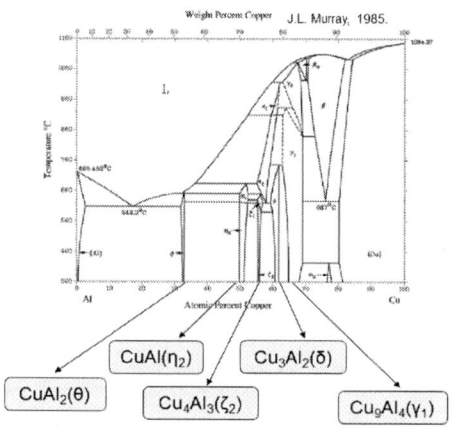

Fig. 4. Example of EDS elemental line profile across Cu-Al IMC; prediction of $CuAl_2(\theta)$ at area A is challenging due to 10 wt% difference when compared to phase diagram [3].

III. MATERIALS AND METHOD

TEM/STEM Specimen Preparation

The TEM sample was prepared on a cross section surface of wire ball bonding; sample thickness was typically ~80nm. It is important to note that the sample did not lose its crystallinity due to the ion beam milling during FIB preparation. For NBD analysis, existing sample was used after STEM-EDS analysis.

TEM/STEM Tool Setup

A 200kV Tecnai TEM system (FEI company) was used for STEM-EDS and NBD analysis described in this work. The TEM was originally set up with the standard STEM illumination whereby a converged beam is incident on specimen with spot size of less than 1nm. For NBD analysis purpose, this STEM mode was slightly modified to produce a near parallel beam of

~5nm diameter (Table I), i.e. the extraction voltage (EV) was lowered and a smaller condenser-2 (C2) aperture was used to generate small beam size; small spot size was also used to prevent extensive exposure of CCD camera to high intensity diffraction spots; the mini-condenser was switched on along with the upper objective lens to generate a near parallel nano beam. A schematic of representation of the illumination system is depicted in Figure 5. It's essential to note that the C2 is typically switched off in STEM mode; centering of small C2 aperture is rather challenging and pivot point alignment is required to improve beam positioning accuracy. A dark room environment is normally needed as the beam intensity is too low for naked eye.

Table I
Comparison of Standard and Modified STEM Mode
Illumination for NBD Analysis

	Standard STEM Illumination	Modified STEM Illumination
Acceleration Voltage	200kV	200kV
Extraction Voltage	4500	4000
Gun Lens	5	5
Spot Size	7/8	10/11
C2 Aperture	75/100um	10um
Mini Condenser	Off (Nano Probe)	On (Micro probe)
Beam condition	Focus Convergent	Near Parallel
Beam diameter	< 1nm	~5nm
C1 cross over	at specimen	at front focal plane

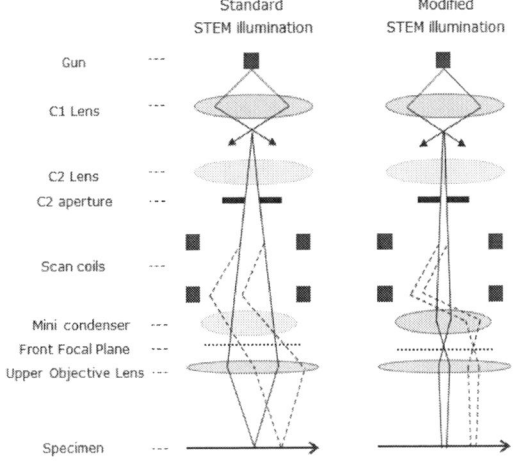

Fig. 5. Comparison of standard STEM mode and modified STEM mode for NBD analysis.

STEM And Diffraction Spots Imaging

In STEM-NBD mode, image of specimen is captured with high angle annular dark field (HAADF) detector (which is placed at the conjugate plane of back focal plane), similar to standard STEM imaging. For recording diffraction spot pattern, the back focal plane is projected onto the viewing screen or the CCD camera. It's important to identify a camera length with optimum diffraction spots coverage on CCD camera and perform diffraction image calibration using a known specimen, typically Si at <110> zone axis.

Diffraction Spots Indexing and Orientation Identification

The IMC is identified by indexing diffraction spots using Single Crystal^{TM} diffraction software (CrystalMaker Software Ltd). This program was used to access crystallographic data files from libraries and subsequently simulate electron diffraction spot pattern of identified IMC layers. In this case study, crystallographic data files were obtained from American Mineralogist Crystal Structure Database [4]. The grid tool in Single Crystal^{TM} diffraction software was then used to determine the orientation of recorded diffraction pattern, with respect to a simulated pattern, which is automatically oriented and indexed (Figure 6).

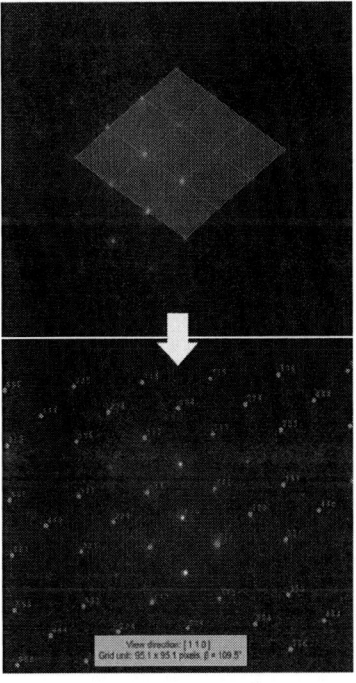

Fig. 6. Application of grid tool to find the orientation of diffraction pattern and to index various spots.

IV. RESULT

For Cu-Al IMC characterization, the camera length was set at 200mm and sample was aligned in such a way the IMC grain of interest was at near zone axis for easy diffraction indexing. Finding the zone axis base on the diffraction pattern is challenging, especially when the crystallographic structure is unknown. This problem was overcome by increasing the camera length to ~2m, which resulted in enhanced diffraction contrast in STEM image. The sample was tilted until the IMC grain exhibited relatively dark contrast (Figure 7), which suggested that the incident beam was closely oriented to a zone axis.

With such sample zone axis alignment technique established, high symmetry diffraction spot patterns were successfully captured from various IMC grains. Using Single Crystal^{TM} software, areas A and B as referenced in Figure 4 were confirmed to be $CuAl_2(\theta)$ and $CuAl(\eta_2)$ respectively; the recorded diffraction patterns were matched perfectly to the simulated diffraction patterns, with each spots indexed and crystal orientation identified; various reciprocal lattice spacings were measured and matched with crystallographic database (Figure 8 & 9).

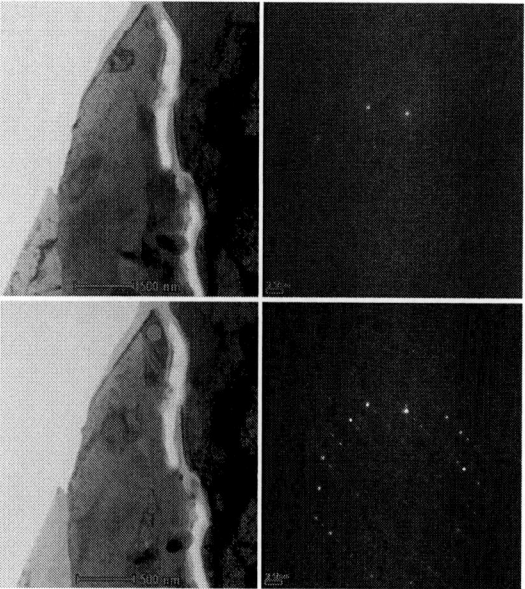

Fig. 7. STEM image with NBD illumination at 2m camera length and respective diffraction pattern at different specimen orientation with respect to electron beam. The higher symmetry diffraction pattern was recorded after tilting the sample until it exhibited darker contrast.

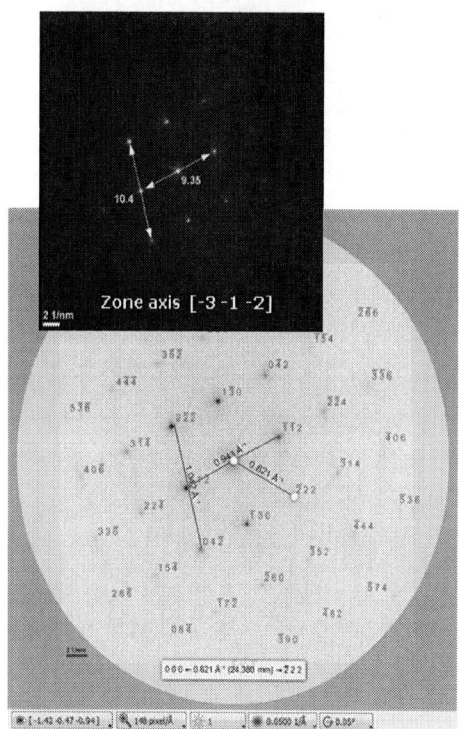

Fig. 8. NBD indexing result from Area A confirming the presence of $CuAl_2(\theta)$ phase. Spot (-2,2,2) has reciprocal distance of $0.621A^{-1}$ from the center spot, this is equivalent to half the d-spacing for <111> crystal plane of $CuAl_2(\theta)$.

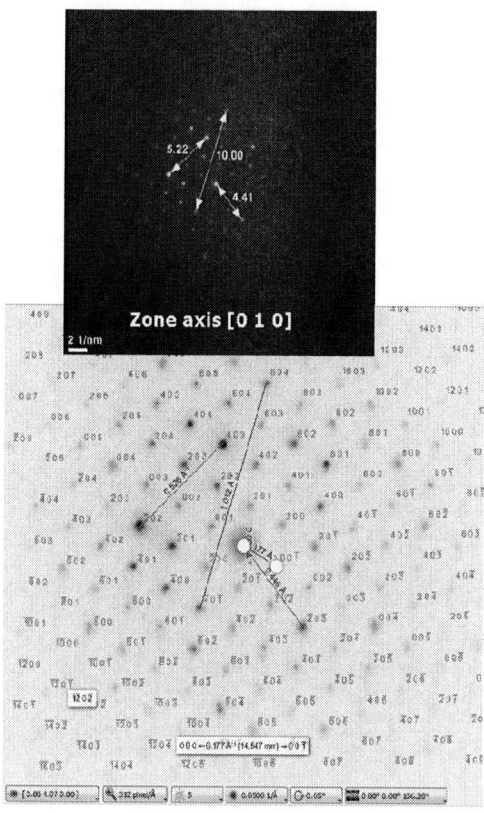

Fig. 9. NBD indexing result from Area A confirming the presence of $CuAl(\eta_2)$ phase. Spot (0,0,-1) has reciprocal distance of $0.177A^{-1}$ from the center spot, this is equivalent to the d-spacing for <001> crystal plane of $CuAl(\eta_2)$.

IV. CONCLUSION

In summary, the NBD indexing to identify the Cu-Al IMC has been demonstrated in STEM-NBD mode. The utilization of parallel nano beam in STEM micro probe mode is essential for producing electron diffraction spot patterns. The Single Crystal[TM] software was used for indexing the diffraction patterns. Using this technique, the uncertainties of phase identification based on empirical composition derived from EDS data has been eliminated.

REFERENCES

[1] Y.Y Tan, F.K. Yong, "Cu-Al IMC Micro Structure Study in Cu wire Bonding With TEM", IEEE Proceedings of 17th IPFA - 2010, Singapore.

[2] HyperPhysics (©C.R. Nave, 2012), "Diffraction Grating" Department of Physics and Astronomy, Georgia State University, Atlanta, Georgia 30302-4106.

[3] J.L. Murray, "The Aluminium-Copper System", International Metals Review 1985 Vol. 30 No.5, pg.212

[4] Robert T.Downs, Michelle Hall-Wallace, "The American Mineralogist Crystal Structure Database", Department of Geosciences, University of Arizona, Tucson, Arizona 85721-0077, U.S.A.

Experiments and Results of Raman and FTIR Complementary Vibrational Spectroscopy for IC Reliability Failure Analysis

Huang Yamin, Hao Tan, Dandan Wang, Jeffrey Lam and Zhihong Mai
GLOBALFOUNDRIES Singapore Pte. Ltd.
60 Woodlands Industrial Park D, Street 2, Singapore 738406
Email: maggie.huang@globalfoundries.com

Abstract-Time-dependent dielectric breakdown (TDDB) of ultra-low-k materials is one of the most critical reliability issues in leading edge Cu/low-k technology due to the weak intrinsic breakdown strength of ultra-low-k materials as compared to that of SiO_2 dielectrics. With continuous device dimension scaling, this problem is further exacerbated for Cu/ultra-low-k interconnects. There are different TDDB models proposed to address this issue, however, there is no direct evidence to get into the failure mechanism. The key technical reason is that the damage to the dielectric material properties is not able to be monitored during the TDDB test. In this paper, we will describe the experiments and the setup used to capture the dielectric bonding damage during the reliability test. Raman and FTIR complimentary vibrational spectroscopy were used to detect the dielectric bonding on the pattern wafer, which has historically been a challenge for current leading edge Cu/low k or ultra-low-k technologies due to the influence of the metal interconnects and the thin dielectric layer. From our experiments, we successfully detected the TDDB degradation behavior of ultra-low-k dielectric in Cu/ultra-low-k interconnects and found the intrinsic degradation of the ultra-low-k dielectric. Further study on the damaged structures with TEM analysis revealed that the Ta ions migrated from the Ta/TaN barrier bi-layer into the ultra-low-k dielectrics. In addition, no out-diffusion of Cu ions was observed in our TEM investigation on Cu/Ta/TaN/SiCOH structures.

I. INTRODUCTION

With the shrinking in dimensions of IC devices, Cu and low-k dielectrics have been introduced into IC devices to reduce the RC delay. In the past decade, a lot of research and application efforts have made Cu and low-k dielectric technology the main stream of the back end of line of leading edge IC process. In reliability studies of low-k dielectric IMD, leakage current will increase due to the degradation of low-k/Cu interconnects, which will finally result in dielectric related failure. Vibrational spectroscopy is a rapid and easily used technique for the identification of the chemical bonds of materials. The most common characterization tools for the complementary vibrational spectroscopy are Raman and FTIR. Many of the spectral patterns for the organic groups attached to silicon are highly specific and identification of organosilicon materials can be made more rapidly from Raman and FTIR spectroscopy than by any other techniques.

While there are many studies on low-k dielectric characterization, few are on real patterned wafers and IC

devices. Due to the small thickness of the low-k IMD layer and complicated mixed structures and materials in IC device, the vibrational spectroscopy detection is very challenging and a detailed analysis is needed to extract low-k dielectric signals from the mixed structures in spectroscopy. In this paper, Raman with high resolution and various laser sources were employed intensively to characterize patterned wafers with SiCOH as the low-k/ultra-low-k dielectrics. FTIR was also used as a complement to investigate the chemical composition of SiCOH low-k dielectrics. The influence of different SiCOH densities on spectroscopy was studied. Furthermore, the analysis of spectroscopy on SiCOH/Cu with different pattern layouts was studied in details. The experiments and the setup used to capture the dielectric bonding damage during the reliability test were described.

II. EXPERIMENTAL DETAILS

In leading edge technology, an IC device has multiple layered structures with interconnects formed by different dielectrics and metals. In our experiments, patterned low-k/Cu wafers were used for the Raman spectroscopy study. The patterned wafers were fabricated with a standard CMOS process.

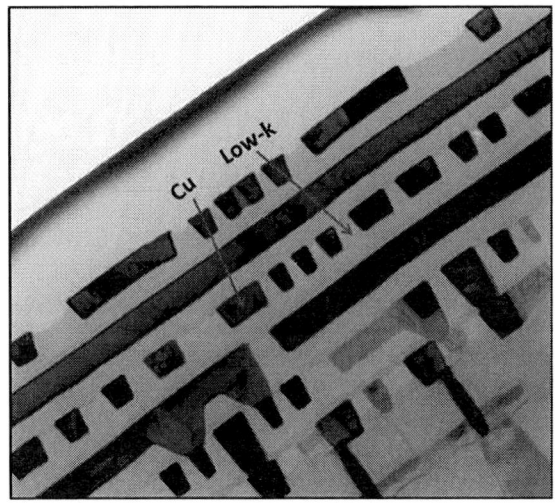

Fig. 1. A typical cross-section TEM of an
IC device with low-k/Cu CMOS process.

Fig. 1 is a typical TEM image of an IC device with low-k/Cu CMOS process. SiCOH was used as low-k IMD deposited with a CVD process and Cu was used as the metal interconnects fabricated with conventional dual damascene process. With physical sputtering, a Ta layer was deposited as the barrier layer between Cu interconnects and the surrounding low-k IMD. A SiN_x layer between the SiCOH low-k dielectrics was used to provide mechanical support to the low-k/Cu interconnects construction. The SiCOH low-k dielectric, Ta barrier layer and SiN_x layer have a thickness of 200, 10, and 50 nm, respectively.

In reliability test on the low k material, breakdown properties were assessed in many different types of reliability tests, e.g., breakdown test, infant mortality test (IM), TDDB test and electromigration (EM) test. Fig. 2 shows a cross section image of the low-k dielectric breakdown during a reliability test, the 168 hours IM test. Recent research has shown that the current leakage issue could be due to Cu atoms diffuse into the low-k dielectric. However, there is no detailed research to address the methodology of the analysis on the chemical property degradation of the low-k dielectric during the reliability test on an IC device.

Fig. 2. Low-k dielectric breakdown (indicated by the arrow) during the reliability test, IM 168 hours.

In our experiments, an electrical test (ET) comb structure was used, as shown in Fig. 3. The comb structure was biased at given voltages at the two pads. Our TDDB test was setup on a MPS150 cascade probe station. It is a very cost-effective and easy to use solution for TDDB test. It can support microprobing on a substrate or a whole wafer up to 8 inch in diameter. The platen can support up to sixteen positioners. A semiconductor parameter analyzer, Keithley 4200-SCS, is used to electrically stress the device in order to simulate the TDDB test. Leakage current and voltage applied were measured during the TDDB test. The Keithley 4200-SCS performs electrical characterization and measurement of semiconductor devices, materials and processes. It can be applied to basic I-V and C-V measurement, including advanced ultra-fast pulsed I-V, transient I-V measurements and waveform capture. DC I-V measurements are the most critical to device and material characterization. The Source Measure Units (SMU) of the 4200-SCS are very precise and accurate to source current or voltage and, at the same time, to measure current and voltage. The 4200-SCS can also be applied for the measurements of

sub-pA leakage and $\mu\Omega$ resistance. Raman spectra were captured by a T64000 Raman system from Horiba. The T64000 system is integrated with a triple spectrometer design to achieve better optical stability. Its confocal LabRAM Raman microscope provides rigid and stable mechanical coupling. It has efficient optical coupling and fast throughput. For Raman spectroscopy experiments on low k material, green to deep UV laser sources were used to get maximum signal detection. The laser sources in our system are listed in Table I. A Nicolet 6700 FTIR system with Continuum microscope was used to capture the FTIR spectra from the low k material for TDDB study.

Table I
LASER SOURCES FOR RAMAN SPECTROSCOPY

	Visible 532nm	Near Ultra Violet 325 nm	Deep Ultra Violet 266 nm
Spatial Resolution	1 um	Less than 500 nm	
Application	Bulk Property	Ultra Surface Property	

The vibrational spectroscopy study on conventional SiCOH thin films was carried out on two completed CMOS IC device samples. One sample has dense low-k SiCOH while the other has porous ultra-low-k SiCOH. The dielectric constants and optical properties of the low-k and ultra-low-k SiCOH are listed in Table II. The samples were polished to the interested layers and the low-k/ultra-low-k micro regions were characterized by vibrational spectroscopy. The conventional mechanical polishing method used in the study can avoid chemical reaction. This method can prevent damaging or changing the chemical bondings of the low-k/ultra-low-k dielectrics.

Table II
K VALUE AND OPTICAL PROPERTIES OF
LOW-K AND ULTRA-LOW-K SICOH

	K value	Refractive index at 532 nm	Extinction coefficient at 532 nm	Refractive index at 325 nm	Extinction coefficient at 325 nm
Low-k	2.8	1.46	0.010	1.49	0.013
Ultra-low-k	2.3	1.19	0.001	1.23	0.005

III. RESULTS AND DISCUSSION

The typical curve of leakage current versus stress time in TDDB is shown in Fig. 3. The early stage of the TDDB test on the comb capacitor structures is defined by point A and point B. All our ultra-low-k dielectric samples were found to follow this general conduction behavior, that is, upon the application of the electric field, the current flow was initially high but decayed exponentially until it reached the saturation level, as indicated in the first stress period of curve AB in Fig. 3. The quick initial current decrease is believed to be caused by electron trapping, which thickens the electron tunneling barrier from the cathode into dielectric conduction band. Generally, this conduction mechanism of dielectric is independent of extrinsic factors, such as out-diffusion of Cu ions. However, as for the increasing current with increased stressing time during the second stress period of curve BC, the dependence of leakage on Cu diffusion is still being debated in the published research papers.

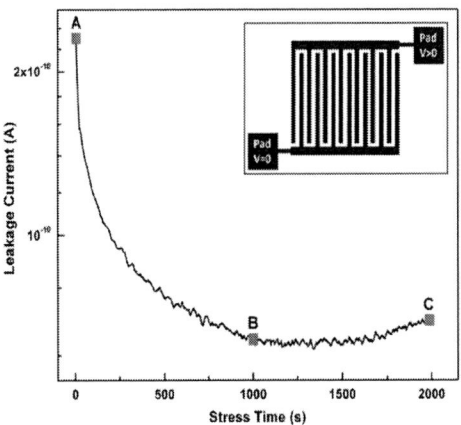

Fig. 3. A plot showing the leakage current as a function of time in Cu/ultra-low-k comb structure. The insert shows the top-down schematics of the comb structure used in the study.

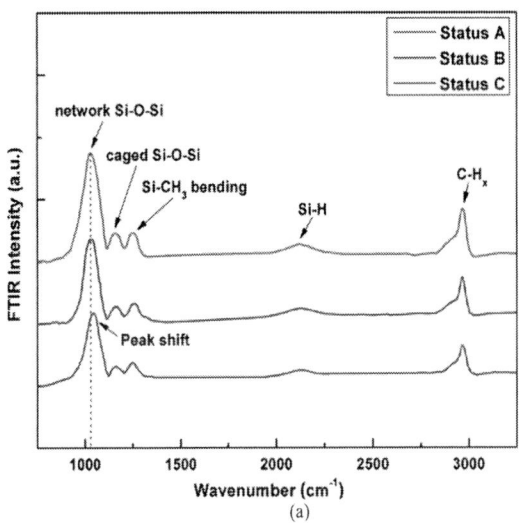

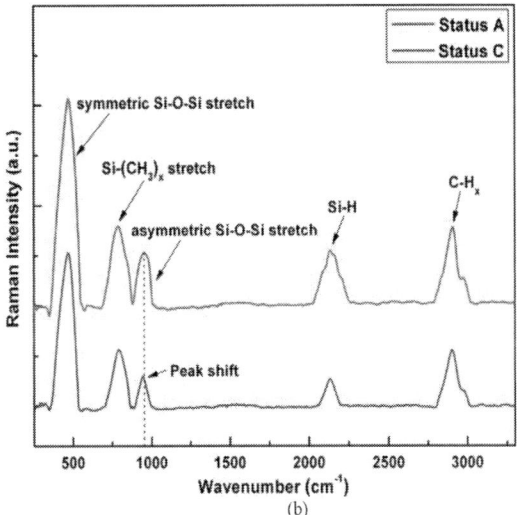

Fig. 4. In-situ FTIR (a) and Raman (b) spectroscopy on the Cu/ultra-low-k comb structure at different stress status.

Fig. 4 (a) is the result of in-situ FTIR spectra captured on the comb capacitor structure during TDDB test. It shows the chemical bonding evolution in ultra-low-k dielectric from the stress status A to C as indicated in Fig. 3. The original sample (status A) shows clearly the ultra-low-k chemical bondings including the network Si-O-Si band, the caged Si-O-Si band, the bending mode of Si-CH$_3$ band, the Si-H band and the C-H group bands consisting of the symmetric and asymmetric stretching modes of C-H$_3$ and C-H$_2$ bands. The initial period of leakage decay of the ultra-low-k dielectric is because the ion in the dielectric initially moves following the electrical field applied. In status B, the chemical bonding spectra signal intensity has decreased, indicating an occurrence of intrinsic degradation in ultra-low-k dielectric, since no extrinsic actors were introduced in this stress period. The result suggests that the injected electrons from cathode have enough energy to break the chemical bonding of ultra-low-k dielectric and the degradation is more severe for the network Si-O-Si bands at 1027 cm^{-1}. When the leakage current begins to increase due to continuous stress at status C, the intensity of the FTIR spectrum of the chemical bonding of ultra-low-k dielectrics decreases further. Additionally, a peak shift towards the higher wave frequency vibration is observed for the Si-O-Si network band. It is a combination of effects from both stress and strain on the ultra-low-k dielectric. The formation of strain is due to an additional force on the ultra-low-k bonding breakage from bigger size species (compared to electrons) such as metal ions, injected from the anode. The metal ions migration is expected to have caused an increase in the leakage current in the dielectric.

Raman measurements were carried out before and after the TDDB test for verification of degradation in the ultra-low-k structure, as shown in Fig. 4 (b). The chemical bondings include Si-O-Si symmetric stretching band, Si-O-Si asymmetric stretching band, Si(CH)$_x$ stretching bands, Si-H band, C-H$_3$ symmetric stretching band and C-H$_3$ asymmetric stretching band. Compared to the original sample, all the ultra-low-k bands (especially the Si-O-Si stretching bands) are obviously degraded after stress. More importantly, a shift towards the lower wave frequency for the asymmetric stretching band of Si-O-Si is observed. This left band shift in the Raman spectra is generally caused by strain, confirming that bonding damage has occurred in the dielectric during and after the TDDB test.

It is found that the intrinsic degradation of the ultra-low-k dielectric firstly occurs under the applied electrical field and Ta ions then migrate into ultra-low-k along the weakened interface of Cu/Ta/TaN/SiCOH, causing a more severe damage to the ultra-low-k dielectric, as shown in Fig. 5. The Ta ions inside the ultralow- k can induce an increased local electrical field between Cu electrodes and thus accelerate the ultra-low-k degradation to final breakdown. In our investigation on the Cu/Ta/TaN/SiCOH structures, no out-diffusion of Cu ions was observed.

In order to have a clear understanding of the effect of the

leakage current on the ultra-low-k dielectric degradation in TDDB test, high resolution TEM and EDX were used to investigate the possible damage information within ultra-low-k dielectric.

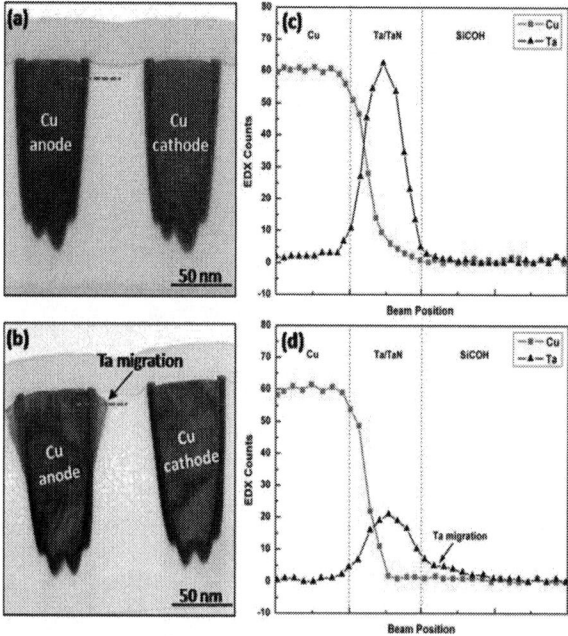

Fig. 5. TEM cross-section images of the Cu/ultra-low-k comb structure for the original sample before stress (a) and the sample with a certain time of stress with leakage current (b). EDX line profile of the Ta migration along the interface of Cu/Ta/TaN/SiCOH before stress (c) and after stress (d), respectively.

Fig. 5(a) shows the cross-section image of the comb structure before the application of stress and no abnormality is observed in the original sample. However, Fig. 5(b) shows a severe degradation in ultra-low-k of the comb structure at the stress status C indicated in Fig. 3. It is seen that the Ta/TaN barrier bi-layer has decomposed and only a weakened Ta layer is left to cover the Cu lines.

EDX line scans along the interface of Cu/Ta/TaN/SiCOH at the anode for both two samples were conducted. Fig.s 5(c) and (d) illustrate that there is no Cu diffusion out of Ta/TaN bilayer in both positions. Compared with the results from original sample in Fig. 5(c), the Ta/TaN liner peak maximum is much reduced and the width is increased in Fig. 5(d), which confirms that the Ta ions out migration has occurred during TDDB test, however, the Ta/TaN bi-layer is sufficient to stop Cu out diffusion in this condition. This finding further proves that the strain measured in vibrational spectra is induced by out diffused metal ions, which subsequently results in more damage in the ultra-low-k dielectric, as manifested by the increased leakage current. Ta ions drifted into ultra-low-k dielectric at a location which is a distance away beneath the capping layer, as shown in Fig. 5(b). This is because the capping layer has a stronger dielectric strength than that of the ultra-low-k material, so that

the capping layer can push the out-migrating Ta ions into the ultra-low-k volume. Meanwhile, the ultra-low-k dielectric has degraded due to the effect of the electrical field. As a result, migrating Ta ions prefer the path along the weakened ultra-low-k dielectric. The Ta ions diffusion inside ultra-low-k dielectric shows a non-uniform distribution due to the gradient local electrical field formed between Cu lines from upper to lower interface of dielectric and barrier, as shown in Fig. 5(b). The highest local electrical field is formed at the upper interface with the smallest Cu line to line spacing. Ta ions diffusion reduces the dielectric gap between Cu lines and thus enhances the local electrical field, which results in an acceleration of ultra-low-k dielectric degradation to the final breakdown.

IV. CONCLUSIONS

In conclusion, we have set up the vibrational spectroscopy with Raman and FTIR to detect the degradation of the low-k dielectrics during the TDDB test. Based on the TEM and EDX analysis, it is found that the intrinsic degradation of the ultra-low-k dielectric firstly occurs under the applied electrical field and Ta ions then migrate into ultra-low-k dielectric along the weakened Cu/Ta/TaN/SiCOH interface, causing a more severe damage to the ultra-low-k dielectric. The diffused Ta ions inside the ultra-low-k dielectric will induce an increased local electrical field between Cu electrodes and thus accelerate the ultra-low-k dielectric degradation to the final breakdown. In our investigations on the Cu/Ta/TaN/SiCOH structures, no out-diffusion of Cu ions is observed.

ACKNOWLEDGMENTS

We would like to thank the Failure Analysis Lab in Globalfoundries for providing the excellent TEM support.

REFERENCES

[1] Maex K., Baklanov M. R., Shamiryan D., Iacopi F., Brongersma S. H., and Yanovitskaya Z. S., "Low Dielectric Constant Materials for Microelectronics," *J. Appl. Phys.*, Vol. 93, No. 11 (2003), pp. 8793-8841.

[2] Chen L. S., Bang W. H., Park Y. J., Ryan E. T., King S., and Kim C.U., "Observation of Space Charge Limited Current by Cu Ion Drift in Porous Low-k/Cu Interconnects," *Appl. Phys. Lett.*, Vol. 96, No. 9 (2010), pp. 0919031-09190314.

[3] Grill A. and Neumayer D. A., "Structure of Low Dielectric Constant to Extreme Low Dielectric Constant SiCOH Films: Fourier Transform Infrared Spectroscopy Characterization," *J. Appl. Phys.*, Vol. 94, No. 10 (2003), pp. 6697-6707.

[4] Cartereta C. and Labrosseb A., "Vibrational Properties of Polysiloxanes: from Dimer to Oligomers and Polymers. 1. Structural and Vibrational Properties of Hexamethyldisiloxane $(CH_3)_3SiOSi(CH_3)_3$" *J. Raman Spectrosc.*, Vol. 41, No. 9 (2010), pp. 996-1004.

[5] Trujillo N. J., Wu Q., and Gleason K. K.," Ultralow Dielectric Constant Tetravinyltetramethylcyclotetrasiloxane Films Deposited by Initiated Chemical Vapor Deposition (iCVD)," *Adv. Funct. Mater.*, Vol. 20, No. 4 (2010), pp. 607-616.

978-1-4799-3911-4/14 $31.00 © 2014 IEEE

3D EBSD Characterizations on Copper TSV for 3D Interconnections

W.N. Putra[1,2] , A.D. Trigg[2] , H.Y. Li[2] , C.L. Gan[1]
[1]School of Materials Science & Engineering,
Nanyang Technological University, 50 Nanyang Avenue, Singapore 639798
[2]Institute of Microelectronics, Agency for Science, Technology and Research (A*STAR),
11 Science Park Road, Singapore 117685
E-mail: wahy0008@e.ntu.edu.sg

Abstract - **Microstructure analysis plays an important role in the reliability study of copper Through-Silicon Vias (TSVs). While conventional 2-dimensional (2D) Electron Back-Scatter Diffraction (EBSD) is a useful technique, 3-dimensional (3D) EBSD characterization provides a more accurate picture of the TSV microstructure. Information that is missing in 2D observations, such as grain shape and volume, can be obtained from the 3D technique. In this study, we did 3D characterizations by serial sectioning of the TSV samples and mapped the microstructure on each slice. These maps were then reconstructed into 3D images. From the result, it showed that the increase in Cu grain volume after thermal annealing can be up to 99%, as compared with 55% and 67% increase in calculated grain volume as determined from single and averaged 2D EBSD maps, respectively.**

I. INTRODUCTION

Copper Through-Silicon Via (TSV) is a key technology for three-dimensional integrated circuit applications [1], and its microstructure is being studied for a better understanding on its electrical and reliability characteristics. Observing the copper microstructure is crucial as it will provide information on how the copper TSV will behave under high temperature processes [2]. The shape of TSV itself will also affect the grain growth, and the resulting stress within the TSV as well as the surrounding silicon [3].

Electron backscatter diffraction (EBSD) technique is commonly used for microstructure characterizations. For TSV, there have been studies on microstructure using two-dimensional (2D) EBSD to obtain information such as grain size and orientation [4,5]. Our previous work showed that the grain size increases by about 20% in diameter after thermal annealing at 400°C, which indicated grain growth during the thermal process [5]. This paper will focus on the three dimensional (3D) EBSD characterization and analysis, and its advantages over 2D observation.

II. TSV SAMPLE PREPARATION

For this experiment, we used TSV with 5 μm diameter and 50 μm depth (aspect ratio of 1:10). The via depth was formed by a deep-reactive-ion-etching (DRIE) process on 12-inch silicon wafers. Before being filled with metal, the etched vias were lined with insulating oxide layer by plasma-enhanced chemical vapor deposition (PECVD) and also Titanium based diffusion barrier by physical vapor deposition (PVD) process. After that, conformal copper seed layer was deposited followed by multiple steps electroplating (ECP) of the via with copper for TSV solid filling. Figure 1 shows the cross section of TSV sample used in this experiment.

The grain growth of the Cu TSV was studied, in which each sample was annealed from 200 - 400°C and maintained for 60 minutes in a tube furnace which has temperature stability of ±1°C. Nitrogen gas was used in the furnace to prevent surface oxidation of the samples.

III. 3D FIB SERIAL SECTIONING AND EBSD MAPPING

Three-dimensional characterizations can be performed by physical sectioning or by transmissive radiation e.g. 3D x-ray tomography. Although transmissive radiation technique is non-destructive, it will not provide details of the microstructure. On the other hand, serial sectioning technique coupled with EBSD is a simple and useful 3D characterization method. It consists of serial cutting or sectioning of the sample, and recording or mapping the structure of each slice. This series of maps are then stacked and reconstructed into a single 3D structure. Serial sectioning technique can be applied for a wide range of materials with the only drawback is that this technique is destructive [6].

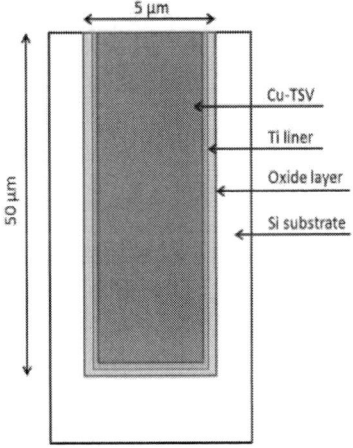

Fig. 1. Cross section view of TSV sample

978-1-4799-3911-4/14 $31.00 © 2014 IEEE

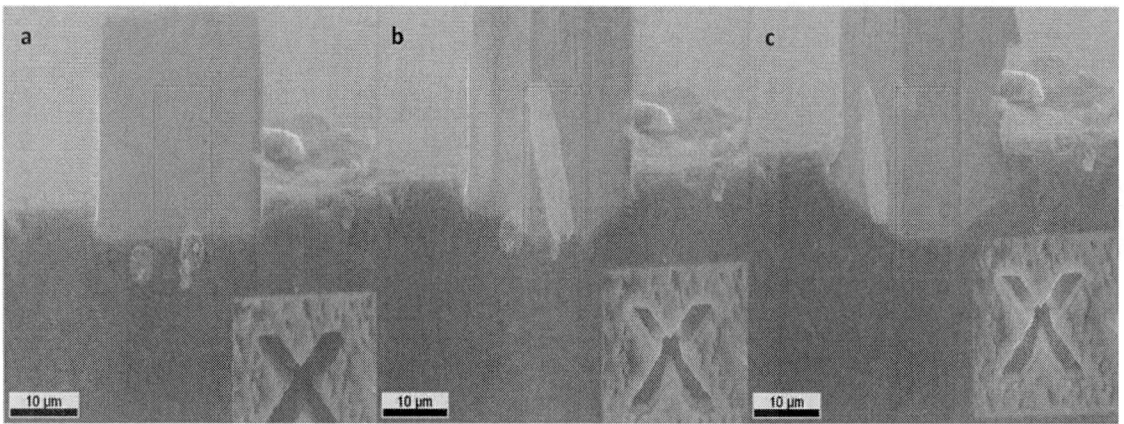

Fig. 2. SEM image of (a) before; (b) during; and (c) after automated FIB serial sectioning. The green rectangle shows the EBSD map area. The X-mark for fiducial can be seen as well.

The equipment used for this study was FEI NovaNanolab DualBeam 600i with integrated EDAX Digiview IV EBSD detector. For 3D sectioning by Focused Ion Beam (FIB) and EBSD mapping, the sample needs to be mounted on a special holder that is pre-tilted at 54°. This special sample holder will allow the sample to be rotated and tilted at 52° and 70° for FIB sectioning and EBSD mapping respectively, with minimum stage movement to maintain accuracy.

To avoid drifting of the stage during long duration mapping, the TSV sample was mounted to the holder using conductive double-sided copper tape. Generous amount of silver paint was also applied to ensure better conductivity from sample to holder. Stage movement for serial sectioning and mapping was controlled automatically using FEI EBS3 software.

Considering the large size of the TSV, the ion beam current used for FIB milling was set to 9.5 nA and step size for EBSD mapping was set at 0.12 μm to optimize the characterization duration. Using this setting, one single slice of 2D EBSD map could be obtained in roughly 25 - 30 minutes. In general, 40 slices were made to reconstruct the whole part of TSV that was sectioned.

The FIB serial sectioning started before the TSV was exposed and completed after the TSV was completely sliced through (Fig. 2). The total time needed for one set of 3D EBSD data is approximately 17 hours.

Automated FIB serial sectioning method features movement of the sample back and forth from the FIB milling position to EBSD mapping position. Despite the special sample holder which was designed to maintain accuracy, it still can be seen in Figure 2 that the fiducial mark shifted over time. This problem occurred from compucentric rotation inaccuracy of the equipment stage. Best effort had been done to adjust and calibrate the concentric rotation movement of the stage to minimize the problem.

Depending on the sample placement, hence TSV position, a minimum of 30 slices is usually usable. The 3D TSV map was then constructed from 30 2D EBSD images which were obtained by the FIB sectioning with 200 nm gap between slices along the TSV diameter. Figure 3 shows examples of a single slice 2D map and a 3D EBSD map. Auto alignment, data clean up and data extractions were done using TSL OIM^{TM} Analysis software.

IV. GRAIN SIZE ANALYSIS

As mentioned previously, the 3D structure was obtained by reconstructing several 2D maps into a single 3D map. To ensure that the serial sectioning method is valid, we performed 2D EBSD analysis using single maps from 3D data set to verify the results. For conventional 2D grain diameter study, a single map from the center of TSV was used. Four samples of TSV were taken, two samples each for before and after annealing. The grain diameter obtained before annealing was 0.99 μm whereas after annealing, it was 1.146 μm. This means the increase in grain diameter is around 16%. Our previous work showed that grain size increment is in this range as well (Fig. 4) [5,7].

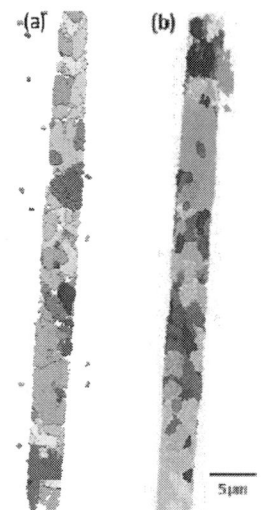

Fig. 3. (a) 2D map. (b) 3D map

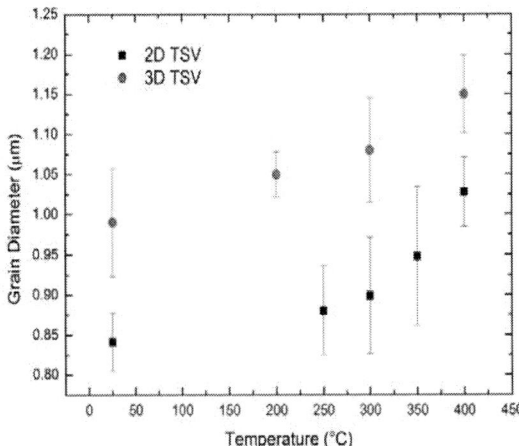

Fig. 4. Grain size (diameter) comparison, taken from previous 2D EBSD work [5] and current 3D EBSD data

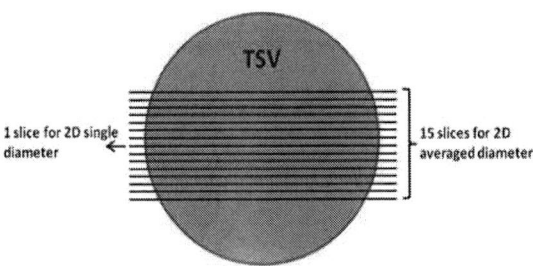

Fig. 5. Schematic illustration on maps taken for 2D single and averaged diameter calculation

The 2D grain diameter data can also be used to calculate a rough grain area and volume by using simple mathematic formula. The calculated grain areas are 0.769 μm^2 and 1.031 μm^2 before and after annealing respectively, which means the increase in calculated grain area is around 34%. This area increment is bigger than diameter increment.

Using the sphere formula, it is found that the calculated grain volumes are 0.507 μm^3 and 0.788 μm^3 before and after annealing, respectively. This means the increase in calculated grain volume is even larger, which is around 55%. This calculation assumes several conditions, i.e. sphere grain shape and isotropic grain growth.

For further analysis, we averaged the 2D grain size calculation using several 2D maps from the same data set of the 3D EBSD sample. Fifteen maps were used from the center area of TSV and the grain diameter for each map was calculated and then averaged. Schematic illustration of the maps taken is shown in Figure 5.

The 2D average grain diameters are 0.936 μm and 1.114 μm before and after annealing, respectively, which shows only a 19% increment. By calculation, however, this translates to a grain volume of 0.4353 μm^3 and 0.7302 μm^3 before and after annealing, respectively. The volume increase is around 68%, which is bigger than the previous single map result.

It has been shown that from 2D characterization data, the 3D grain volume can be calculated using mathematic formula. Statistical approach from several maps was performed as well to get better prediction on the grain volume. However, the grain volume information may not be accurate because the calculation still assumes ideal spherical grain shape as mentioned before. The TSV grain in real samples may have irregular shape instead of sphere, thus 3D characterization is needed.

After 3D reconstruction, information on real grain volume can be obtained. Grain volume analysis shows an increase from 1.2818 μm^3 to 2.5547 μm^3 before and after annealing, respectively. This means that the volume has increased by about 99% after annealing, which is almost twice the calculated 2D single map volume. Comparison between 2D single, averaged and 3D volume are shown in Figure 6. The result shows the importance of 3D characterization as 2D characterization is not able to give accurate grain volume information. From 2D averaged grain size, it is clear that the grain volume cannot be derived from statistical approach. True size and shape of grains can only be determined by 3D technique.

The significant difference in area/volume between the 2D and 3D results is likely due to the TSV grain shape and the limited planar 2D mapping. From our analysis, TSV microstructure can be characterized by the core and peripheral grains. In a two-dimensional map, it is difficult to distinguish between these grains. Frequently, the larger size core grain is partially blocked by the smaller peripheral grains, which makes the large core grains look much smaller. The other possibility is where a large grain is "split" by several smaller grains, which made the software incorrectly identify the single large grain as two (or more) smaller grains.

These two examples will give a lower average grain size. 3D map, however, is able to clearly distinguish between medium and large core grains and smaller peripheral grains, thus making grain size calculations more accurate. Grain shape visualization and identification will be discussed in the next section.

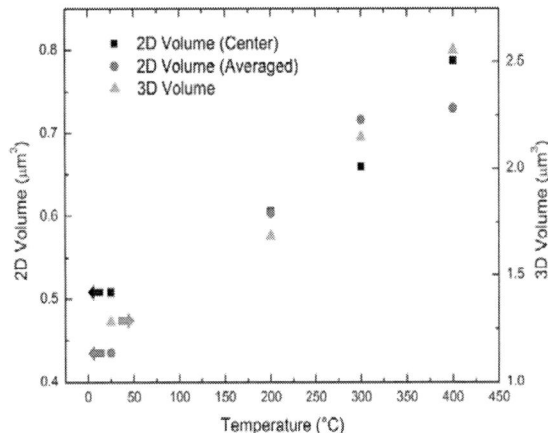

Fig. 6. Grain size comparison from single and averaged 2D calculated volume and 3D volume

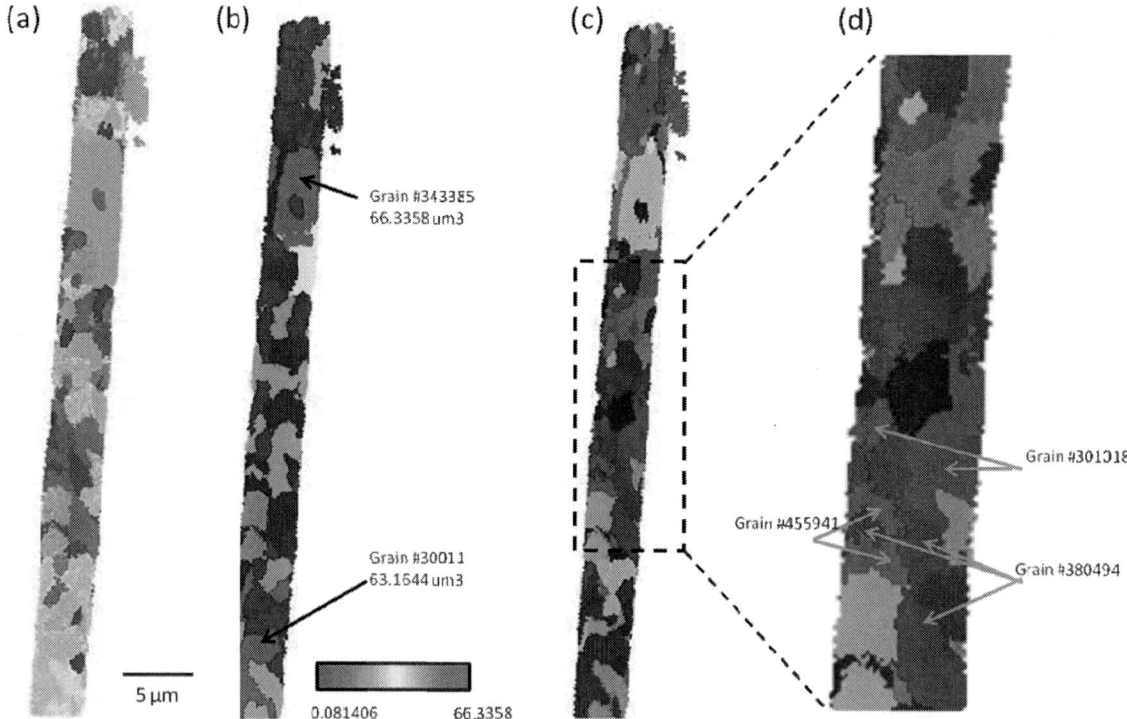

Fig. 7. (a) Inverse Pole Figure (IPF) map; (b) Grain size map; (c) Normal view and (d) Zoomed view of unique grain map

It is important to have better accuracy on the grain size from microstructural characterization. From the simulation done by Wu *et al.* [8], it was known that stress at area with larger grains is smaller than area with smaller grains. Furthermore, morphology of grains inside a TSV will affect the stress distribution. Large core grains mean they occupy significant amount of area in the center of TSV while smaller grains are spread through the peripheral of TSV. This might create stress concentration inside the TSV [8]. Stress concentration is not preferred due to its tendency for mechanical failure.

Another impact of this significant change in grain size is the grain boundary area. In TSV, copper is usually plated under different conditions and will result in a residual stress concentrated at grain boundaries [9]. This residual stress within the TSV can be very high to cause failures such as delamination or fracture. It is also well known that the stress inside TSV will affect the mobility in MOSFET devices up to 7% [10]. Understanding the grain size and grain boundary in TSV will be helpful to minimize these problems.

V. GRAIN SHAPE VISUALIZATION AND IDENTIFICATION

Using EBSD analysis software, we can choose the type of map to show. The most common EBSD map is an Inversed Pole Figure (IPF) map which creates a grain orientation map. IPF map is useful to observe any preferred orientation in TSV. However, it is difficult to observe individual grain and identify

its size from IPF map. Different maps are required for individual grain observation as shown in Figure 7. For easier comparison, previous Figure 3b is inserted again as Figure 7a, and it can be seen that it is difficult to differentiate between core grain and peripheral grain. It is also almost impossible to point out large or small sized grains. For this study, we used two different types of map to visualize and identify grains in TSV.

For better separation between large and small grains, the grain size map was used. Grain size map (Fig 7b) will mark clearly large grains (red colored) from the other small and medium grains surrounding them. From the figure, it can be seen that the two large grains are core grain with several smaller peripheral grain blocking the view. If the analysis was done in 2D characterization, these two large grains will be identified as two or more different grains with smaller size, which will affect the grain size calculation.

Another different map that is useful to identify individual grain is the unique grain color map (Fig 7c). This map is useful to separate small and medium sized grains. The map will show individual grains marked with different color. Zoomed map (Fig 7d) shows an example where one grain can be mistakenly identified as two or more different grain if the analysis was done in 2D characterization. In the figures, there are three examples of grains (purple, magenta and blue colored grain) with this condition. After checking by software using grain number ID, they have the same ID number which confirms that those grains with the same color are actually one single grain.

VI. CHALLENGE AND POSSIBILITIES OF 3D EBSD CHARACTERIZATION FOR TSV OBSERVATION

3D EBSD characterization has many advantages over 2D method for TSV microstructure analysis. There are, however, several observations which can be done easily in 2D mode but quite challenging in 3D method. In 2D EBSD map, we can easily obtained information on grain boundary length and type. It is known that in copper, there are normal grain boundaries and twin grain boundaries. These two types of grain boundaries were also observed in TSV [7]. Grain boundary type information in TSV is important as they may affect the resistivity of TSV itself [11, 12].

The information on grain boundary length in 3D characterization will not be "actual" but more "calculated" as the 3D map itself is a reconstruction of several 2D maps. However, even though it will be tedious, the grain boundary type is still possible to be checked by tracking through the stack of 2D maps.

Twin boundary in copper could serve as a void nucleation site [13]. Void formation in copper TSV has been observed and one of the possible causes is stress concentration [14]. It is possible for 3D EBSD characterization to help further analyze the origin of void and its relationship with twin boundary. The main problem is the void dimension. A very small void dimension needs smaller EBSD step size and narrower FIB slice gap for reasonable quality of 3D map. This setting will cause longer characterization duration, but we can shorten the duration by focusing the mapping only on the area of void and not necessarily the whole TSV. That is provided the location of void is known, which is another challenge.

To locate the void, FIB slice must be done carefully, and only after void is located, 3D EBSD characterization can be carried out. Because of this procedure, it will be impossible to get a perfect fully intact 3D void map because small portion of the void will be cut by FIB during locating the void itself. However, the data from EBSD will be more important as compared with transmissive radiation technique which is only able to show position of void without further information.

VII. CONCLUSIONS

Three-dimensional characterization by FIB serial sectioning and EBSD mapping is a simple yet useful technique for microstructure analysis on copper TSV. Compared with conventional 2D EBSD technique, 3D characterization provides more information such as grain shape and volume.

From the 3D EBSD experiments, we learnt that the Cu TSV grain size increased more significantly after annealing due to grain growth than what was determined from 2D grain area analysis. The increase in grain volume can be up to 99%, as compared with 55% and 67% increase from single and averaged 2D maps, respectively. Several 3D EBSD maps could also be used to provide better visualizations on copper grain morphology inside TSV.

This study on microstructure could give a more accurate understanding on the copper grain characteristics through different thermal processes, and thus contribute to better structural or process design to minimize any potential reliability issues.

ACKNOWLEDGMENTS

The authors would like to thank Dr. Liu Qing (NTU), Dr. Stuart I. Wright (EDAX) and Dr. Rene de Kloe (EDAX) for the help on 3D EBSD experimental work and analysis.

REFERENCES

[1] International Technology Roadmap for Semiconductors Report Update 2012

[2] Okoro, C. et al., "A Detailed Failure Analysis Examination of the Effect of Thermal Cycling on Cu TSV Reliability." *IEEE Transactions on Electron Devices* 61, no. 1 (2014): 15–22. doi:10.1109/TED.2013.2291297.

[3] Jiang, Tengfei, et al., "Impact of Material and Microstructure on Thermal Stresses and Reliability of Through-silicon via (TSV) Structures." In *Interconnect Technology Conference (IITC), 2013 IEEE International*, 1–3, 2013. doi:10.1109/IITC.2013.6615584.

[4] C. Okoro et al., "Elimination of the axial deformation problem Of Cu-TSV In 3D Integration," *AIP Conference Proceedings*, vol. 1300, pp. 214-220, 2010.

[5] Heryanto, A et al. "Effect of Copper TSV Annealing on Via Protrusion for TSV Wafer Fabrication." *Journal of Electronic Materials 41*, no. 9 (September 1, 2012): 2533–2542. doi:10.1007/s11664-012-2117-3.

[6] Zaefferer, S., S. I. Wright, and D. Raabe. "Three-Dimensional Orientation Microscopy in a Focused Ion Beam–Scanning Electron Microscope: A New Dimension of Microstructure Characterization." *Metallurgical and Materials Transactions A* 39, no. 2 (February 1, 2008): 374–89. doi:10.1007/s11661-007-9418-9.

[7] Putra, W.N., H.Y. Li, A.D. Trigg, and C.L. Gan. "Microstructure Investigation of TSV Copper Film." In *Electronic Components and Technology Conference (ECTC), 2013 IEEE 63rd*, 1414–19, 2013. doi:10.1109/ECTC.2013.6575758.

[8] Wu, Zhiyong, Zhiheng Huang, Yucheng Ma, Hua Xiong, and Paul P. Conway. "Effects of the Microstructure of Copper through-Silicon Vias on Their Thermally Induced Linear Elastic Mechanical Behavior." *Electronic Materials Letters* 10, no. 1 (January 1, 2014): 281–92. doi:10.1007/s13391-013-3053-y.

[9] W Lee, Gyujei, Ho Young Son, Joon Ki Hong, Kwang-Yoo Byun, and Dongil Kwon. "Quantification of Micropartial Residual Stress for Mechanical Characterization of TSV through Nanoinstrumented Indentation Testing." In *Electronic Components and Technology Conference (ECTC), 2010 Proceedings 60th*, 200–205, 2010. doi:10.1109/ECTC.2010.5490902.

[10] S. E. Thompson, S. Guangyu, C. Youn Sung, and T. Nishida, "Uniaxial-process-induced strained-Si: extending the CMOS roadmap," *IEEE Transactions on Electron Devices*, vol. 53, pp. 1010-20, 2006.

[11] J. M. E. Harper et al., "Mechanisms for microstructure evolution in electroplated copper thin films near room temperature," *Journal of Applied Physics* 86, no. 5 (September 1, 1999): 2516-2525.

[12] W. Steinhögl et al., "Comprehensive study of the resistivity of copper wires with lateral dimensions of 100 nm and smaller," *Journal of Applied Physics* 97, no. 2 (December 27, 2004): 023706-023706-7.

[13] Sekiguchi, A., J. Koike, S. Kamiya, M. Saka, and K. Maruyama. "Void Formation by Thermal Stress Concentration at Twin Interfaces in Cu Thin Films." *Applied Physics Letters* 79 (2001): 1264. doi:10.1063/1.1399021.

[14] Shin, Hae-A-Seul, Byoung-Joon Kim, Ju-Heon Kim, Sung-Hwan Hwang, Arief Suriadi Budiman, Ho-Young Son, Kwang-Yoo Byun, et al. "Microstructure Evolution and Defect Formation in Cu Through-Silicon Vias (TSVs) During Thermal Annealing." *Journal of Electronic Materials* 41, no. 4 (February 17, 2012): 712–19. doi:10.1007/s11664-012-1943-7.

Case Study of Wet Chemical Stain to Identify Implant Related Low Yield Issue

Yi- Chen Lin, Sheng-Min Chen

Reliability Testing & Failure Analysis Department, Powerchip Technology Corp.
No. 12 Li-Hsin Rd. 1, Hsin Chu Science Park, Hsin Chu, Taiwan, R.O.C
Phone: 886-3-5795000 EXT.4856 Fax: 886-3-5792142 Email: yijen@powerchip .com

Abstract- **Ion implant is very important process in semiconductor manufacturing. In this study, we discuss a problem of low yield caused by an implant related defect on a specific location and structure in the device. The paper explains how general Failure Analysis (FA) techniques such as top view analysis by Scanning Electron Microscope (SEM), Passive Voltage Contrast (PVC) and cross section by Focused Ion Beam (FIB) coupled with Transmission Electron Microscopy (TEM) are unable to identify the defect which causes the gate driver failure which in turn leads to the implantation related low yield issue. It was found that Emission Microscopy (EMMI) analysis for global isolation, followed by nano-probing for electrical characterization of the gate driver was needed. Cross section wet chemical stain technique was then used to identify the localized implant junction failure.**

I. Introduction

To select a suitable method and tool for Physical Failure Analysis in wafer FAB is very important as it will help us to rapidly identify possible root causes of failures and find the solutions. Generally, failure isolation approaches flow as Fig1[1]. The typical failure analysis approaches to work on these failures are using general I-V curve check for any leakage or high resistance issue. In cases where I-V curve tracing of leakage or high resistance first, then electrical fault isolation by EMMI/OBIRCH(Optical Beam Resistance Change) analysis will be performed. EMMI/OBIRCH is considered as a powerful electrical fault isolation tool that has been widely used in I-V leakage related failure analysis. PVC technique is commonly using to localize characterization contract. Another advanced technique is nano-probing for device electrical characterization. In this paper we will discuss in implant related current failure result in low yield, the EMMI, nano-probing, PVC, that was able to isolate the fault location but no defect was observed after de-processing of failed sample.

Cross section wet chemical stain is necessary in this implant related current failure. For the junction failure low yield issue, the wet chemical stain is more effective FA method than the SIMS(Secondary Ion Mass Spectrometry) analysis because SIMS analysis just for blank wafer and it is large ion sputtering area, it cannot measure junction profile of a real device in the circuit. So follow "Convenient FA approach Flow"(Fig. 1) rapidly identify the presence of localized abnormal device and wet chemical junction stain technique helping us to successfully find out the root cause. Failure mechanism and root cause will be discussed in this paper.

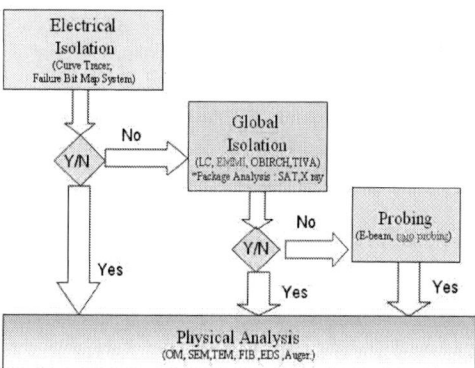

Fig. 1: Convenient FA approach Flow (Flow between failure site isolation tool)[1].

II. Results and Discussion

A. Physical Failure Analysis

Particular device of a logic product suffered the current and gate driver failure that are several lots impacted. The yield lost about 8% per lot. The failure CP(Chip probing) map show up mainly at 3o'clock area of wafers. (Fig. 2).

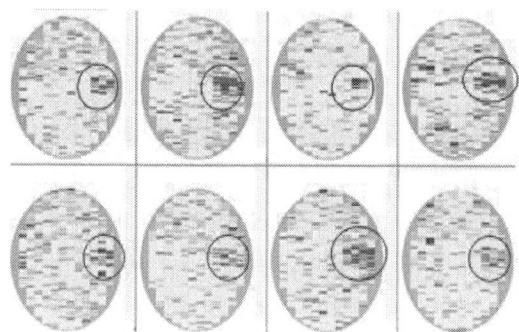

Fig. 2: Wafer location at 3o'clock signature fail.

After electric testing, the fail mode was current and gate driver soft bin fail. We also confirm the WAT (Wafer Acceptance Test) electric characteristic data and process in-line lot history, not found specific tool and process step have strong commonality after correlation in lot history(because too many tools and process steps related). It is very difficult to identify problem step and tool. We just know the front-end process problem. How to find out the root cause? We follow convenient FA approach flow (Flow between failure site isolation tool) [1] to analysis.

978-1-4799-3911-4/14 $31.00 © 2014 IEEE

First, global isolation by EMMI tool was able to isolate the failure location on hot spot area (Fig. 3). Then we focus EMMI abnormal hot spot layer by layer de-processing for top view inspection and PVC result are normal, preliminary FA with no obvious defect found.

According to I-V current test for the forward and reverse voltage I-V test result show lower Id(Fig. 6) and reverse voltage test must be using higher voltage about 8-9 V that can be turned on(Fig. 7) the transistor. It explained the localization have abnormal higher resistance issue.

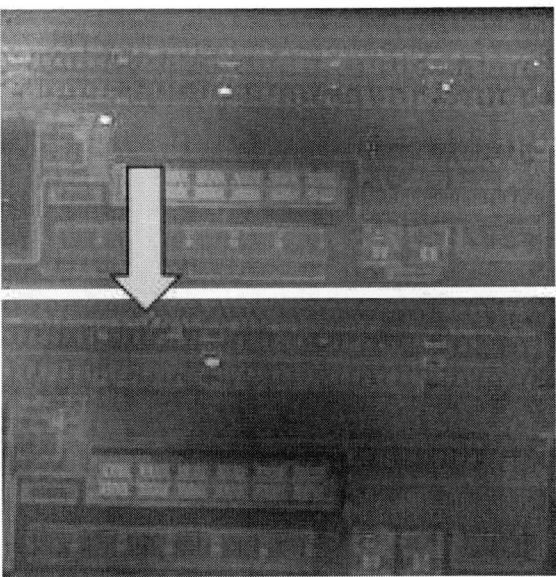

Fig. 3: Find some abnormal hot spot

Second, we prepare sample for the nano-probing analysis. By studying the layout – HV (high voltage) NMOS devices with suspected weak point were analyzed by nano-probing (Fig. 4. 5). The purpose is to check the device electrical characterization. After analysis, we found the Id - forward of HV NMOS of failure sample lower than good sample about 2mA(Fig. 6) and the Id - reverse of HV NMOS device was abnormal also.

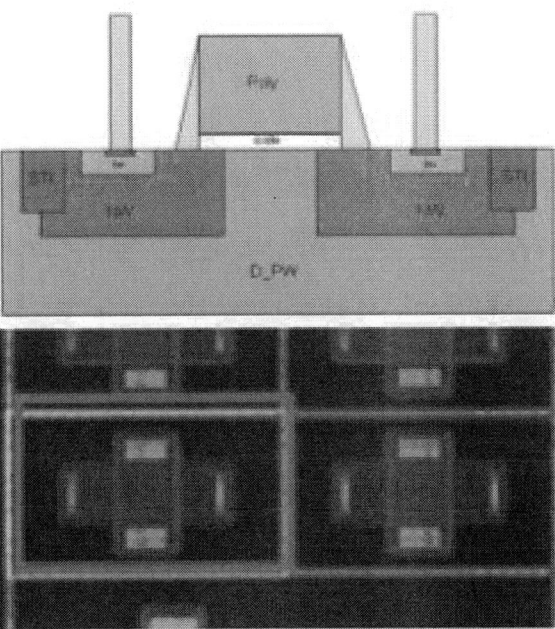

Fig. 5. HV NMOS device structure

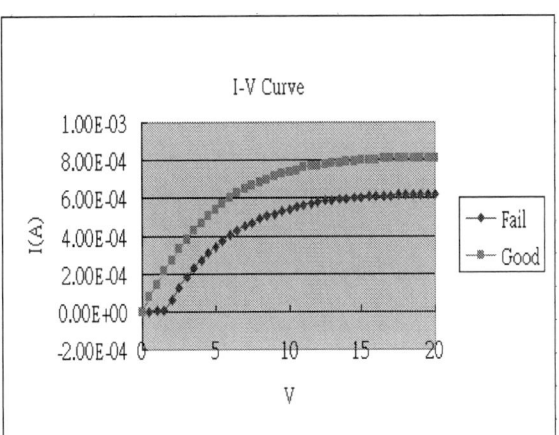

Fig. 6. Nano probe Good/Fail main die HVNMOS device :Forward(20V): I-V curve

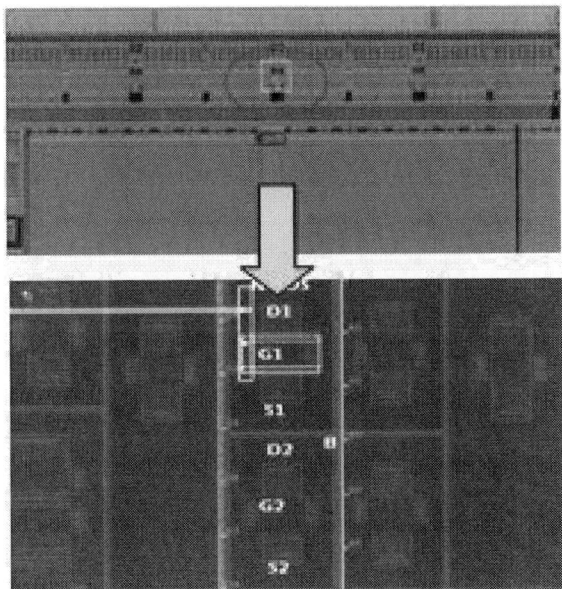

Figure 4: layout tracing –HV MOS device was weak point

978-1-4799-3911-4/14 $31.00 © 2014 IEEE   301

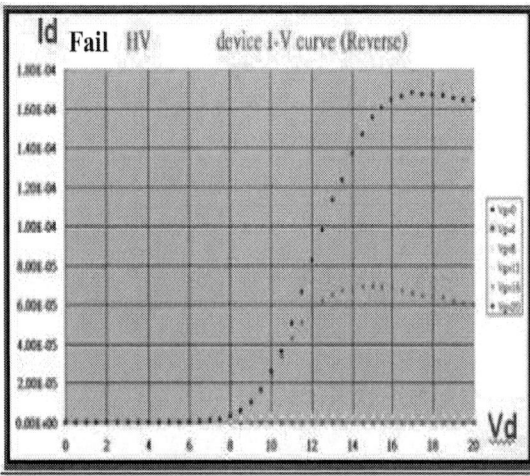

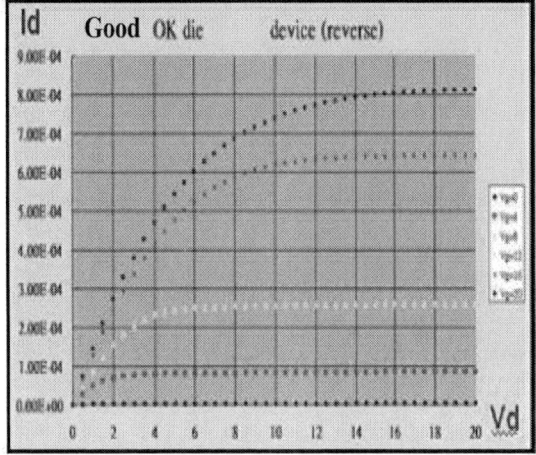

Fig. 7. Nano probe Good/Fail die HV device I-V curve Reverse voltage(0v,4v,8v,12v16v,20v): abnormal I-V curve

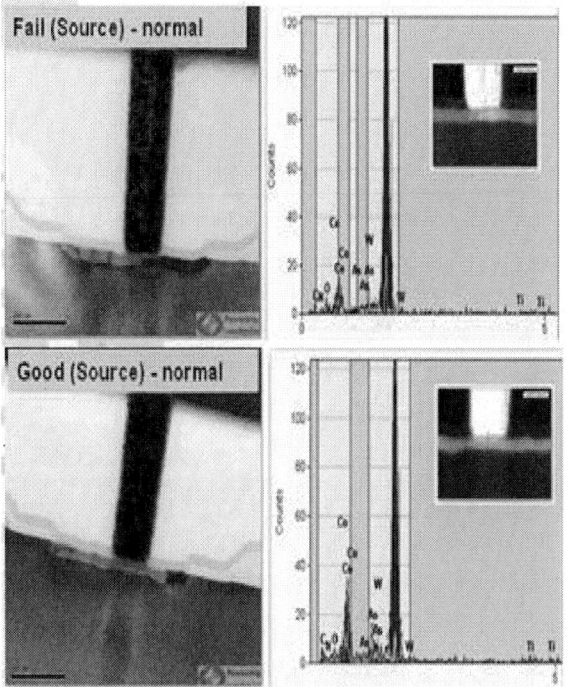

Fig. 8. TEM and EDX Analysis for contact bottom

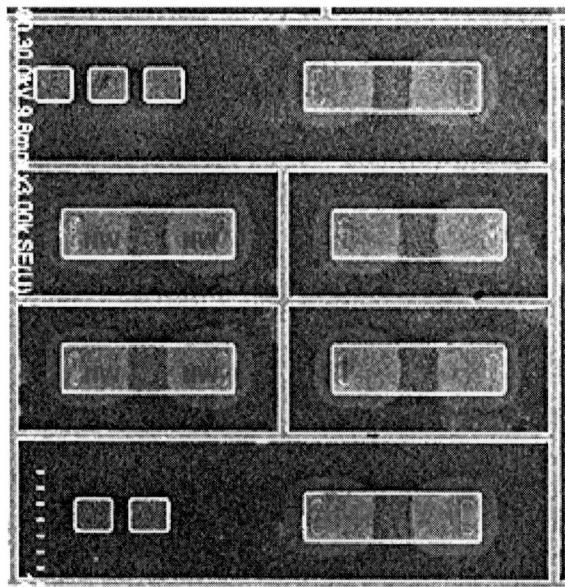

Fig. 8-1. Top view well implant stain result –normal

Thus, the breakout by nano-probe after EMMI analysis that is able to stimulate the failure condition and localization. But it can not identify the failure type. To identify the HV device Reverse I-V curve fail for higher resistance issue, we suspected failure defect maybe on contact bottom of CoSi or implant related problem. Focus these issues, we check (a) contact profile and bottom CoSi by Cross section TEM(Transmission Electron Microscope) and EDX(Energy Dispersive Spectrometer) analysis comparable normal sample(Fig. 8). (b) Top view wet chemical Well implant stain the result is normal too. (Fig. 8-1) (c) Effective Cross section wet chemical junction stain: the procedure is (c.1)cross section by FIB and smaller current ion miller be carefully to avoid the surface damaged then (c.2)flushing the sample with solution and take Ga+ ion out the substrate surface,(c.3) wet chemical implant stain by n^{+} acid (CH3COOH+HNO3+HF mix acid) and control the stain time in few seconds is very important. We are successful to find the implant junction abnormal (Fig. 9) by (c) method. The result is no N/P junction implant the shock barrier is higher and Id low that induced the Gate driver failure.

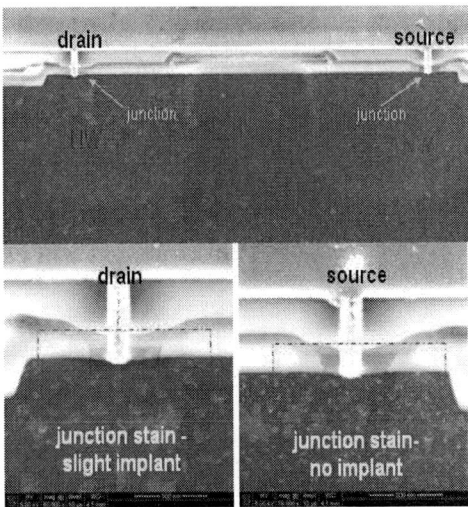

Fig. 9: Failure sample – Wet chemical junction stain to find slight and no implant problem, its comparable with nano probe analysis.

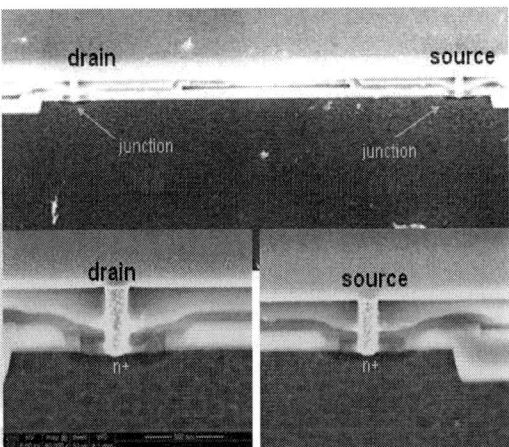

Fig. 10. Good sample - Wet chemical junction stain normal profile

B. Failure mechanism and Root cause

After Physical Failure Analysis, we found the particular structure at the HV NMOS junction implant abnormal induce the higher resistance issue. And to find out the root cause

(I)From WAT(HV NMOS) related implant items vs Gate fail trend chart to show HV NMOS Id just unstable but still in spec, and Photo tool was key issue. The root cause was Photo CD marginal induce the junction implant trouble then caused the Id low issue.

(II)The other hand the CP Map show to appear mainly at 3o'clock wafer specific location issue – Double confirm Photo tool related parameter and found that was development nozzle and wafer rotating speed parameter shift and marginal caused the specific location fail.

C. Investigation and Solution

Investigation this gate driver fail case CD marginal is key point that could result PR residues caused the Gate drive failure. For this event, experiment lot on Photo tool of risk higher

condition and confirm the in-line PR(Photo Resist) profile image by SEM to find the experiment lot PR bridge defect. (Fig. 11). From the experiment lot result – CD smaller could cause the PR residue worse and Id lower risk.

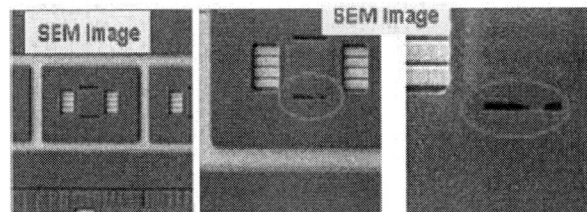

Fig. 11: In-line check PR profile image by SEM to find the PR micro bridge defect.

We aim the PR bridge marginal up action: (I) The solution is setup best parameter condition of photo tool and (II) 3o'clock CP map issue - optimized develop PR remove ability to improve and solute it.

III. CONCLUSION

Selection of a suitable method and tools is very important for successful failure analysis. In this implant related low yield case, it was found that following a convenient failure analysis flow could rapidly find the failure location. This was followed by an effective cross-section, wet chemical staining technique which was necessary to determine the root cause quickly

REFERENCES

[1] Lawrened C Wagner, "Failure Analysis Challenges",8th *IPFA Singapore* , p.36-p.41.2001.
[2] Poo Khai Yee, Lim Saw Sing."Application of Atomic Force Microscope in IC/Dscrete Failure Analysis " *IEEE* , p.457-462, 2013
[3] J.Colvin and K. Jarnsh, "Atomic Force microscopy and Analystical Techmiques with Scanning Probe Microscopy " *Microelectric Failure Analysis Desk Reference Edition ASM International, Chapter 13,January 2004.*
[4] Ang G him Boon, Chen Changqung, Zhao Si Ping, Neo Soh Ping, Yip Kim Hong, Loh Scok Khim, Ng Hui Peng, Angela Teo, Ng Peng Tiong, "Failure Analysis on non-visible front – end defects in deep NWELL implantation related process." 20th *IEEE*, p.181-185, 2013.
[5] Miao Wu, Li Tian, Diwei,Chunlei Wu, Gaojie Wen," Failure Mechanism of Induced by Pattern – Dependent Photo Resist Distortion" 20th *IEEE*, p.678-681, 2013

Comprehensive Study and Corresponding Improvements on the ESD Robustness of Different nLDMOS Devices

Yuan Wang[1,2]*, Guangyi Lu[1], Lizhong Zhang[1], Jian Cao[1], Song Jia[1,2], and Xing Zhang[1,2]

[1]Institute of Microelectronics, Peking University, Beijing 100871, China

[2]Innovation Center for MicroNanoelectronics and Integrated System, Beijing 100871, China

*E-mail: wangyuan@pku.edu.cn

Abstract-**Four-terminal and three-terminal asymmetrical n-type LDMOS (asym-nLDMOS) devices are investigated in 0.18μm 40V SOI BCD technology. To improve normal asym-nLDMOS devices ESD robustness, an additional p-sink implant is added beneath their source/drain diffusion regions. Transmission line pulse measured results show that the novel asym-nLDMOS devices have a suitable triggering voltage and 30-48% improvement of second breakdown current.**

I. INTRODUCTION

The lateral diffused MOS (LDMOS) transistor is widely used as an output driver in high-voltage smart power technologies. For typical integrated circuits (ICs), electrostatic discharge (ESD) failure has been considered as a profound reliability threat [1]. The LDMOS is usually seemed as an electrostatic discharge (ESD) self-protecting device [2-3]. Actually, the LDMOS has not enough ESD robustness because of the harsher environment in which it is generally used [4]. The base push-out effect in its parasitic bipolar junction transistor (BJT) under ESD events degrades the LDMOS ESD self-protection ability [5]. Moreover, due to the good isolation ability and layout-saving advantage, silicon-on-isolation (SOI) technique has been extensively used to fulfill the LDMOS. For the SOI process, the self-heating effect of the buried oxide layer will also worsen the ESD performance [6]. The LDMOS with an embedded silicon controlled rectifier (SCR), named the LDMOS-SCR is an effective solution for the LDMOS ESD protection [7-14]. The LDMOS-SCR combines the characteristics of both the LDMOS and SCR. Under ESD stress, the existence of the drift region enhances the triggering voltage similarly as with the LDMOS, while the SCR with positive-feedback characteristic possesses excellent ESD protection. But, as is well-known, the SCR is susceptible to latch-up danger during normal circuit operations in a noisy environment [15]. Therefore, some additional ESD protection design is needed to protect the LDMOS device, especially in the SOI technology.

In this paper, the ESD performance of a four-terminal (4-T) asymmetrical n-type LDMOS (asym-nLDMOS) device is investigated in CSMC 0.18μm SOI 40V (BV 60V) BCD technology. And then, a modified device, in which an additional p-sinker implant is added beneath the source/drain diffusion regions, is proposed to improve the ESD robustness of the original nLDMOS. Moreover, the ESD characteristic of a layout-saving three-terminal (3-T) asym-nLDMOS is also presented in this work. 3-T asym-nLDMOS device saves footprint as it has a source/substrate abutted design which also results in its ESD robustness difference with 4-T asym-nLDMOS. The paper is arranged as follows: Section II describes the structure of the proposed asym-nLDMOS devices and analyzes their ESD characteristics by TCAD simulation. Silicon-based transmission line pulse (TLP) measured results of 4-T asym-nLDMOS are discussed in section III. The ESD robustness of 3-T asym-nLDMOS and its improvement are described in section IV. Finally, conclusion is drawn in section V.

II. FOUR-TERMINAL ASYM-nLDMOS AND TCAD ANALYSIS

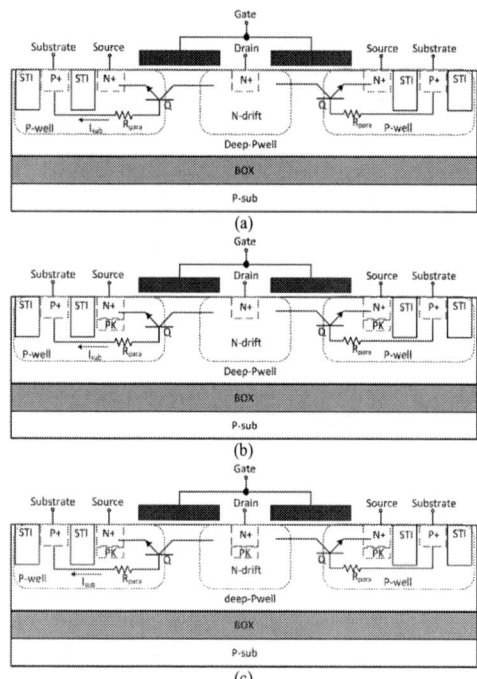

Fig. 1. Cross-section views of (a) standard 4-T asym-nLDMOS, and novel 4-T asym-nLDMOS (b) with only source PK layer, (c) with both source and drain PK layer

Fig.1a shows the cross-section of a standard 4-T asym-nLDMOS in a CSMC 0.18μm 40V (BV 60V) SOI BCD process. The drain-source operation voltage V_{DS} is 40V, and the minimum endurable breakdown voltage of high-voltage trench isolation is 60V. The thickness of the silicon film is 1.5μm, and the buried oxide (BOX) is 1μm. This asym-nLDMOS has following characteristics: the lateral drift region is just resided in the drain region, and the source and substrate are separated by shallow trench isolation (STI) region. Fig.1b and Fig1c show the cross-section of the novel asym-nLDMOS. An additional p-sinker implant, named PK layer, is added beneath the source/drain diffusion regions to inspect the ability of adjusting and controlling the triggering voltage (V_{t1}) and second breakdown current (I_{t2}) of 4-T asym-nLDMOS.

In order to predict the proposed device performances and analyze the principle of operation, a TCAD simulation for asym-nLDMOS devices is implemented.

Firstly, Fig.2 depicts the TCAD simulation results of (a) normal asym-nLDMOS and (b) asym-nLDMOS with PK layers only beneath source diffusion regions. It is obvious that the triggering voltage V_{t1} of (b) is bigger than (a). It is because the difference of the source pn junction in the current discharge path. The source pn junction is formed by P-well/N+ diffusion in (a), but by PK/N+ diffusion in (b). As is known [16], the built-up potential of pn junction V_D is decided by

$$V_D = \frac{KT}{q}\ln\left(\frac{N_A * N_D}{n_i{}^2}\right).$$
(1)

Where K is Boltzmann constant, T is absolute temperature, q is electron charge, n_i is the intrinsic carrier concentration, and N_A and N_D is the p-type and n-type doping concentration, respectively. In this case, N_A in (b) PK region are larger than that directly in (a) P-well region, that is to say, V_D of (b) is larger than that of (a). Hence, the pn junction turn-on voltage V_{on} of (b) is also larger than that of (a). Whereas V_{on} is equal to the product of the P-well parasitic resistor R_{para} multiplied by the current I_{sub} that flows through the P-well region as shown in Fig.1. For a fixed device length, R_{para} is constant. Hence, to trigger the device, I_{sub} required by (b) should be larger than (a). And then, the value of I_{sub} is determined by the avalanche breakdown multiplication factor M, M [16] can be expressed by

$$M = \frac{1}{1-(V/V_{BD})^n}.$$
(2)

Where n is constant, V is the drain biasing voltage, and V_{BD} is the avalanche breakdown voltage of the drain pn junction formed by N-drift/deep P-well same in (a) and (b). Larger Isub of (b) requires larger M of (b), which results in V of (b) is larger than that of (a).

Secondly, the position of PK layer is investigated. Fig.3 depicts the TLP simulation results of 4-T asym-nLDMOS devices (b) only with source PK layer, and (c) with source and drain PK layer. It is shown that the triggering voltage V_{t1} is reduced when PK layer is inserted into the drain region. In this case, the decrease of V_{t1} is decided by the avalanche breakdown voltage V_{BD} of the drain pn junction. The drain pn junction is formed by N-drift/deep P-well in (b), but by N+ diffusion/PK in

(c). Because the p-type doping density of PK region is higher than deep P-well region and the n-type doping density in the N+ diffusion is also higher than N-drift region, V_{BD} of (b) is higher than that of (c). Moreover, with the drain's PK layer in (c), an additional embedded SCR device is inserted into the intrinsic asym-nLDMOS device. The drain's N+ diffusion region serves as the SCR's anode, the substrate's P+ diffusion region serves as its cathode, and the drain's N+ diffusion/PK/N-drift/P-well forms the npnp structure. It is predictable that the novel device shown in Fig.1c combines the characteristics of both the asym-nLDMOS and SCR device. Under an ESD stress, the existence of the light-doping N-drift region determines its triggering voltage V_{t1} similarly as with the normal asym-nLDMOS, while the additional SCR structure with positive feedback provides excellent ESD protection ability.

To clearly understand this characteristic, Fig.4 depicts the distribution of the hole current density in these two device structures. The drain biasing voltage is 50V, which is lower than V_{t1} of (b) and bigger than V_{t1} of (c). The hole current comes from the avalanche breakdown of the drain pn junction. It is obvious that that main hole carrier is generated in the N+ diffusion/PK pn junction region in (c) different with (b).

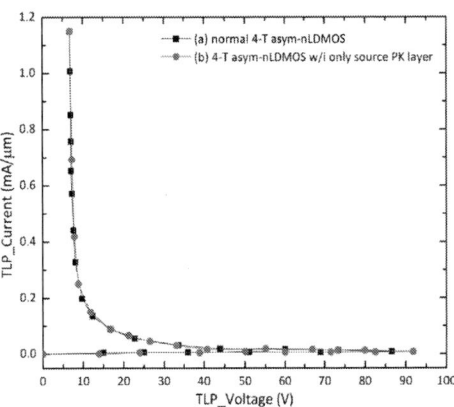

Fig.2 TLP Simulations of (a) normal 4-T asym-nLDMOS (b) 4-T asym-nLDMOS with only source PK layer

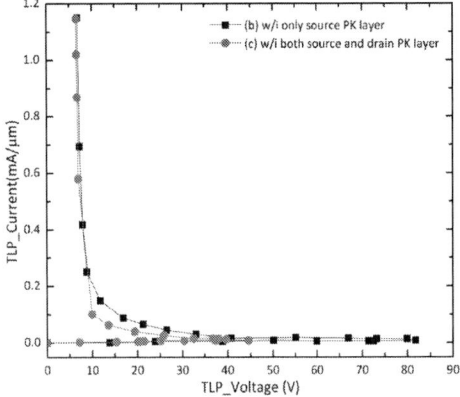

Fig.3. TLP Simulations of 4-T asym-nLDMOS with (b) only source PK layer and (c) both source and drain PK layer.

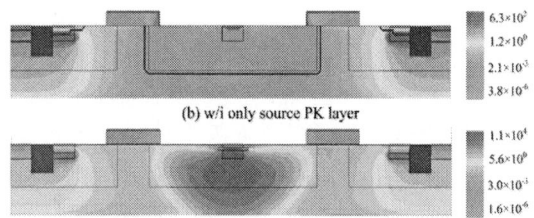

(b) w/i only source PK layer

(c) w/i both source and drain PK layers

Fig.4 Hole current density of modified 4-T asym-nLDMOS

III. SILICON-BASED TLP MEASURED EXPERIMENT OF FOUR-TERMINAL ASYM-NLDMOS

These three asym-nLDMOS is implemented in CSMC 0.18μm 40V (BV 60V) SOI BCD technology. The width of each device is 40μm. TLP measurement with its pulse width as short as 100 ns is used to investigate the device's characteristics (i.e., triggering voltage, holding voltage, secondary breakdown current, etc.) under ESD stress [17]. For all of these asym-nLDMOS, the gate, source, and substrate terminals are connected together as the cathode, and the drain terminal is served as the anode under ESD TLP testing.

Fig.5 depicts the TLP measured results. The triggering voltages V_{t1} of (a) normal asym-nLDMOS and (b) asym-nLDMOS only with source PK layer (both above 80V) are much bigger than that of (c) asym-nLDMOS with both source and drain PK layers (about 45V), which is in good agreement with simulation results in Fig.2 and Fig.3. As mentioned earlier, the minimum endurable breakdown voltage of this process is 60V, which is determined by the anti-high-voltage capacity of the trench isolation. Hence, the value of V_{t1} in (a) and (b) is so high as to threaten the endurance of high-voltage isolation. By comparison, the value of V_{t1} in (c) is about 45V, which is suitable to 40V VDS and 60V breakdown voltage of the trench isolation. The acute decrease of V_{t1} in (c) is derived by the pn junction occurring avalanche breakdown is changed. The heavy-doping N+-diffusion/PK junction in (c) takes place of the light-doping N-drift/deep-P-well junction in (a) and (b).

After triggering, all of three devices have strong snapback and low holding voltage. In snapback stage, the strong source electron injection into the bulk leads to a push-out of the depletion region into the low-doping region of the deep P-well region. In Fig.5, the secondary breakdown in (a) and (b) takes place immediately after the snapback. The second breakdown current I_{t2} of (a) is about 1.35A, and I_{t2} of (b) is about 1.5A. The improvement of I_{t2} from (a) to (b) come from the fact that the additional PK layer beneath source diffusion region result in fewer electrons to induce the base push-out effect of the parasitic BJT, labeled Q in Fig.1, under ESD events. Furthermore, the device (c) with both drain and source PK layer has a highest I_{t2} about 1.75A, about 30% improvement. The reason for (c) having a better ESD performance could be attributed to the better current discharging ability of the additional embedded SCR's cross coupled structure. Once the device is turned on, the embedded SCR will replace the intrinsic asym-nLDMOS to become the primary current discharge device. The on-state resistance of the embedded SCR is much smaller

than that of the intrinsic asym-nLDMOS, which dramatically reduces the power dissipation and results in an improved ESD performance.

IV. THREE-TERMINAL ASYM-NLDMOS AND IMPROVEMENT

To save layout area, the source and substrate of 4-T asym-nLDMOS are abutted to form a 3-T asym-nLDMOS. The cross-section and layout top view of 3-T asym-nLDMOS are shown in Fig.6. Intersected N+ and P+ implants are placed along the width direction of the source/substrate stripes to form source and substrate abutted pick-ups. The source/substrate abutted design results in its ESD robustness difference with the 4-T asym-nLDMOS.

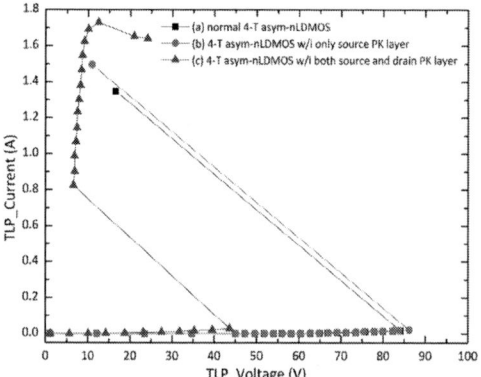

Fig. 5 TLP testing results of three 4-T asym-nLDMOS

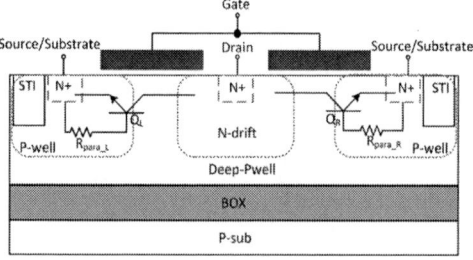

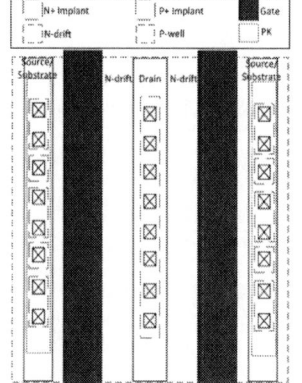

Fig.6. Cross-section and layout top view of 3-T asym-nLDMOS

978-1-4799-3911-4/14 $31.00 © 2014 IEEE

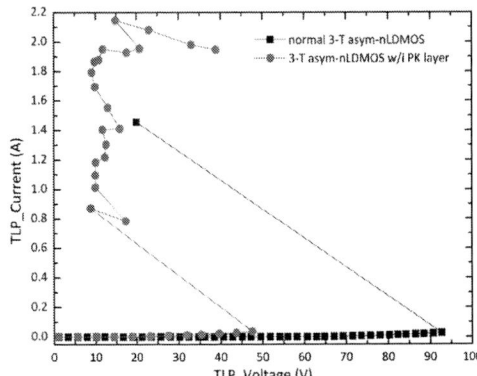

Fig.7 TLP test results for different 3-T asym-nLDMOS

Fig.7 depicts TLP testing result of normal 3-T asym-nLDMOS. Compared with normal 4-T asym-nLDMOS as shown in Fig.5, V_{t1} (92V) and I_{t2} (1.45A) of normal 3-T asym-nLDMOS are both larger. 3-T and 4-T normal asym-nLDMOS have the same the turn-on voltage Von of the source pn junction, which is equal to the product of the P-well parasitic resistor R_{para} multiplied by the substrate current Isub. Due to a source/substrate abutted design, R_{para} of 3-T device is smaller than that of 4-T device. To achieve a same Von, Isub required by 3-T device should be larger than 4-T device. The value of Isub is determined by the drain biasing voltage as mentioned earlier. Hence, V_{t1} (92V) of normal 3-T asym-nLDMOS is larger than that of normal 4-T device. The higher I_{t2} of 3-T asym-nLDMOS comes from the fact that source and substrate abutted design results in fewer electrons to induce base push-out effect of parasitic BJT under ESD events. Fig.7 also shows TLP testing result of 3-T asym-nLDMOS with PK layer beneath both source and drain region. The novel 3-T device has an improved ESD robustness with a suitable V_{t1} (47V) and the highest I_{t2} (2.14A, about 48% improvement). The decrease of V_{t1} and the increase of I_{t2} for 3-T asym-nLDMOS are attributed to the inserted PK layer, which is as same as 4-T asym-nLDMOS. Here we will not go further on this issue.

V. CONCLUSION

The different asym-nLDMOS devices are investigated in CSMC 0.18μm SOI BCD technology. Through additional PK layers beneath source and drain diffusion regions, we can improve the ESD characteristic of asym-nLDMOS devices with a suitable triggering voltage V_{t1} and a high second breakdown current I_{t2}. The TCAD simulation and Silicon-based TLP testing experiment are used to verify these improved methods. V_{t1} of the novel asym-nLDMOS devices is around 47V, which falls in between the drain-source operation voltage and the trench isolation breakdown voltage. And I_{t2} of the novel devices has 30%-48% improvement.

ACKNOWLEDGEMENTS

This work is supported by the National Basic Research Program of China (Grant No. 2011CBA00606) and the Young Scientists Fund of the National Natural Science Foundation of China (Grant No. 61106101). The authors would like to thank Prof. Shurong Dong at Zhejiang University for help with TLP testing.

REFERENCES

[1]. Duvvury, C. "ESD Protection Device Issues for IC Designs," *IEEE Conf. Custom Integrated Circuits,* San Diego, CA, May 2010, pp.41-48.

[2]. Parthasarathy, V. *et al*, "A Double RESURF LDMOS With Drain Profile Engineering for Improved ESD Robustness." *IEEE Electron Device Lett.,* Vol.23, No.4, (2002), pp.212-214.

[3]. Duvvury, C. *et al*, "Lateral DMOS Design for ESD Robustness," *International Electron Devices Meeting Tech. Dig.,* Washington, DC, Dec. 1997, pp.375-378

[4]. Vashchenko, V. A. *et al*, "Improving The ESD Self-Protection Capability of Integrated Power NLDMOS Arrays," in *Proc. 32nd Electrical Overstress/Electrostatic Discharge Symp.,* Reno, NV, Oct. 2010, pp.1-8.

[5]. Ker, M. D. *et al*, "The Impact of Low-Holding-Voltage Issue in High-Voltage CMOS Technology And The Design of Latchup-Free Power-Rail ESD Clamp Circuit For LCD Driver ICs," *IEEE J. Solid-State Circuits,* Vol. 40, No.8, (2005), pp.1751-1759.

[6]. Raha, P. *et al*, "Heat Flow Analysis For EOS/ESD Protection Device Design in SOI Technology," *IEEE Tran. Electron Devices,* Vol.44, No.3, (1997), pp.464-471.

[7]. Wang, Y. *et al*, "A Novel ESD Self-Protecting Symmetric nLDMOS for 60V SOI BCD Process," *Proc. IEEE Int. Conf. EDSSC,* Hongkong, June 2013, pp.6628102

[8]. Zhang, P. *et al*, "LDMOS–SCR: a replacement for LDMOS with high ESD self-protection ability for HV application," *Semiconductor Science and Technology,* Vol.27, No.3, (2012), pp.035006

[9]. Lee, J.-H. *et al*, "Novel ESD Protection Structure with Embedded SCR LDMOS for Smart Power Technology," *Proc. 40th Int. Reliab. Phys. Symp.,* Apr. 2002, pp 156–161

[10]. Walker, A. J. *et al*, "Novel Robust High Voltage ESD Clamps for LDMOS Protection," *Proc. 45th Int. Reliab. Phys. Symp.* Phoenix, AZ, Apr. 2007, pp 596–597

[11]. Griffoni A. *et al*, "Off-State Degradation of High-Voltage-Tolerant nLDMOS–SCR ESD Devices," *IEEE Trans. Electron Dev.,* Vol.58, No.7, (2011), pp.2061-2071

[12]. Zhang P. *et al*, "Study of LDMOS–SCR: A High Voltage ESD Protection Device," *Proc. 10th IEEE Int. Conf. ICSICT,* Shanghai, Nov. 2010, pp 1722-1724.

[13]. Zhang P. *et al*, "Analysis of LDMOS–SCR ESD Protection Device for 60V SOI BCD Technology," *Proc. IEEE Int. Conf. EDSSC,* Hongkong, Dec. 2010, pp 1-4

[14]. Shrivastava, M. *et al*, "A Drain-Extended MOS Device With Spreading Filament Under ESD Stress," *IEEE Electron Device Lett.,* Vol.33, No.9, (2012), pp.1294-1296.

[15]. Ker, M. D. "ESD Protection for CMOS ASIC in Noisy Environments With High-Current Low-Voltage Triggering SCR Devices," *Proc. IEEE Int. ASIC Conf. and Exhibits,* Portland, OR, Sep. 1997, pp. 283–286.

[16]. Sze, S. M. *et al*, *Physics of semiconductor devices,* Wiley (New York, 2006).

[17]. Piatek Z. *et al*, "Transmission line pulsing tester for on-chip ESD protection testing," *Proc. Mixed Design of Integrated Circuits and System Symp.,* Gdynia, 2006, pp 595-599.

978-1-4799-3911-4/14 $31.00 © 2014 IEEE

Systematic Methods to Identify and Verify Non-visible Defects in Silicon Substrate

Hongwei Huang, Winnie Wei, JJ Xin, Candy Liu, Luke Wu, Clieve Dai, Pinglung Liao, Wei Xu

HH-Grace Semiconductor Manufacturing Corporation

1399 Zuchongzhi road, Zhangjiang Hi-tech Park, Shanghai, China

Hongwei.Huang@hhgrace.com

Abstract

For failure analysis, most of defects are visible to imaging tools, such as OM, SEM, FIB, TEM etc. However, there are still lots of non-visible defects which cannot be caught by these tools. As complexity for such non-visible defect failure analysis is much high, FA engineers were often puzzled where to begin from. Two such cases were presented in this paper with solutions. The systematic methods for these cases include electrical data mining, brainstorming or fish-bone diagram method to list all failure possibilities, and then proper characterization tools or methods were used to identify and verify the hypotheses. Finally DOE (design of experiments) was used to verify the root cause. As a result, phosphorus contamination was found for embedded Flash products' MOS threshold voltage shift issue, and higher substrate oxygen concentration for Power MOS products source to drain low breakdown voltage issue.

Introduction:

General procedure of failure analysis includes failure verification, defect/fault isolation, sample preparation, imaging with chemical analysis until root cause was found. Most of researches or efforts were put on the defect isolation to pinpoint defect location. After that, optical microscopy (OM), SEM, FIB, and TEM (EDS) are used to observe these defects, such as profile/critical dimension shift, metal/poly bridging, CT/VIA open/short, dislocation, ACT/poly/metal block etch, defect, ESD/EOS, design issue, etc. These visible defects provided clearly evidence and gave much confidence for process debug and corrective actions implementation. After all, the defect picture won a thousand words.

However, no anomalies were always found even with specific defect location. Such defects include, but not limited to, metal ion contamination,[1] little doping dosage shift,[2] electron trap center caused by dangling bonds,[3] etc. For these non-visible defects, complexities of failure analysis grow greatly. Simply increasing analysis sampling size is not helpful to case analysis success rate. Systematic methods, from electrical data analysis, brainstorming or fish-bone diagram analysis to list all and high possibility hypotheses; then proper analysis tools or methods were chosen to identify root cause, finally DOE (design of experiment) should be conducted to verify the fail mode and provide direction for corrective actions.

In this paper, we presented two such cases of non-visible defects in silicon substrate. There are no obvious defects found by image tools even after Wright or wet treatment of samples. With the above systematic methods, root causes were found and corrective actions were implemented to eliminate the failure.

Case one: device Vt shift caused by Phosphorous contamination

1. Fail mode, Analysis and Hypothesis

Several hundred lots of Flash and Embedded Flash memory products in 8-inch FAB suffered MOS transistor threshold voltage (Vt) shift issue at WAT (wafer acceptance test) sites, more than 300 pieces wafers were scrapped. Three WAT parameters, Vt of NMOS, PMOS and ZMOS shifted simultaneously in the same period. ZMOS is NMOS with zero threshold voltage, but actually value of Vt target is slighter larger than zero. Figure 1 showed trend chart of those three parameters of one embedded Flash product, divided by each parameter Vt target. In the figure, each dot represents one production lot (25 pieces).The X-axis is process time, from time B, three parameters all shifted away from baseline(A to B). N type MOS Vt (Vt-NMOS and Vt-ZMOS) trend down but P type MOS Vt(absolute value) trend up at the same time.

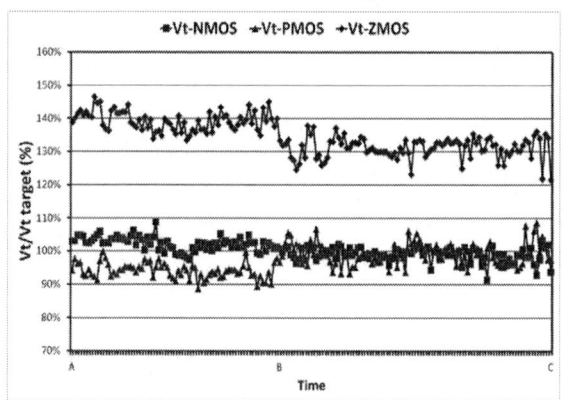

Figure 1: Trend chart of three Vt parameters by HTO process time. For ZMOS, Vt value decreased 16% from 138.5% (time A-B) to 122.5 %(time B-C). For NMOS, Vt decreased 3.2% from 102.4% to 99.2%, while for PMOS, Vt increased 5% of target from 94.2% to 99.2%.

As shown in figure 2, TEM cross-section of one Flash failed bit showed no visible defect, all the CD and profiles are within spec, no dislocation was found. Therefore, non-visible defect was highly suspected.

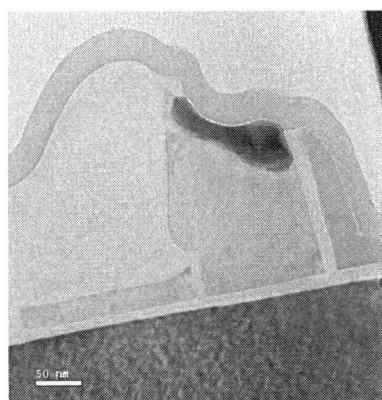

Figure 2: TEM cross-section picture of one failed bit.

After lot and tool history review, wafer source issue or implant tool excursion were excluded. Therefore, electrical data was studied in-depth to infer cause of failure. Also, brainstorming was conducted and fish bone diagram was drawn. Finally, failure was suspected by phosphorous or arsenic counter-doping caused by contamination after ruling out other possibilities. The reason was that Vt is a function of substrate doping concentration. If P or As was unintentional doped in the NMOS and PMOS at the same time, the Vt value of NMOS will shift lower because P or As will act as counter-dopants while absolute value of PMOS Vt will shift higher. Also, implantation dosage of ZMOS is much lower than that of NMOS, therefore, same level contamination can cause larger impact on Vt of ZMOS than that of NMOS.

2. Root Cause Identification

To verify the hypothesis, one typical failed wafer was scrapped and one 120μm*90μm area in scribe line was chosen for SIMS analysis. SIMS analysis was made with 10keV O2+ primary beam on CAMECA IMS6F. Before analysis, sample was treated by HF solution to gate poly silicon layer, keeping below oxide and silicon substrate untouched, which ensure the SIMS data accuracy.

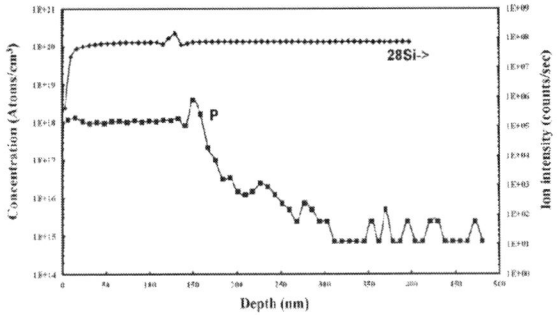

Figure 3: SIMS result of P profile in layers from poly, gate oxide to substrate of failed wafer.

As seen from the figure 3, one phosphorous element peak was observed in the thin gate oxide layer, maximum data is 3.9E+18 atoms/cm³. Then, P was driven into the substrate gradually. Arsenic (As) was also measured but its concentration is below detection limit; therefore it was not shown in the figure. To double confirm this result, two logic product wafers were chose to do SIMS test, one is baseline wafer and another is failed wafer with same fail mode Vt shift. SIMS results were shown in figure 4. One phosphorous peak was observed in the oxide layer for the failed wafer while no such P peak for baseline wafer. These SIMS results confirmed the hypothesis of P counter-doping.

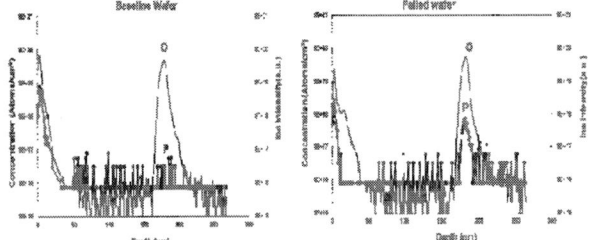

Figure 4: SIMS result of P profile for logic product wafers, left is for baseline wafer, right is for failed wafer.

Wafer ID	Wafer layers before HTO deposition	Idle time in HTO stocker	HTO deposition thickness	P Con. (E+10 atoms/cm²)
#1	Monitor wafer Si Sub+300Å oxide	168hour	No	539
#2	Monitor wafer Si-sub	No	145Å, run with 168hrs idled dummy wafers	<Detection Limit
#3	Monitor wafer Si-sub+145Å oxide	No	145Å, run with 168hrs idled dummy wafers	7.45

Table 1: Different wafer condition and corresponding P concentration

3. Root cause verification:

Based on above analysis result, process loops with both silicon surface at cell and periphery were exposed at the same time were highly suspected. Finally dog process at the HTO (High Temperature Oxidation) deposition step was pinpointed. If P element was mixed in the HTO deposition tool, it will be absorbed at the surface of substrate and was driven into the substrate during the following thermal processes. To find the source of P element and verify the above suspicion, below DOE experiment was designed, as shown in table 1, Phosphorus concentration in dummy wafers and witness wafers were analyzed on ICP-MS (Inductively Coupled Plasma Mass Spectrometry) which Model is Agilent 7500CS.

A fresh monitor wafers, #1, with 300Å oxide above the silicon substrate were put in the stocker in HTO tools for 168hours. The oxide layers on substrate, including HTO oxide layer, were dissolved by HF solution firstly, and then loaded in ICP-MS tool for P concentration analysis. The result showed that this wafer has a very high P concentration level, as much as 5.39E+12 atoms/cm², which means that HTO stocker was contaminated by P element, and contaminated dummy wafers which were stored in the stocker before mixed running with production wafers.

Another two fresh monitor wafers, one is bare wafer, #2, another is with 145Å oxide on substrate, #3, were selected to deposit 145Å HTO oxide with dummy wafers which were already idled in HTO stocker for 168hrs. After HTO deposition, both wafers were taken out to measure P concentration by ICP-MS. For #2, the P concentration is below detection limit but for #3, the result is 7.45E+10 atoms/cm², which is higher than spec 5E+10 atoms/cm². The difference of P concentration between wafer #2 and #3 showed that some P was driven in the silicon substrate and weren't dissolved by HF solution during sampling process for #2. Again, this experiment confirmed root cause of Vt shift.

Further engineering wafers verification was conducted, two witness wafers from 1 lot(25 pieces) were deposited HTO oxide with fresh dummy wafers while other 23 wafers with normal dummy wafers stocked in HTO stocker. The Vt values of the two wafers didn't show any drop, compared to other 23 wafers'.

4. Corrective Actions:

Since root cause of P contamination was confirmed, corrective actions were implemented accordingly. HEPA (High Efficiency Particulate Air) filter was installed for HTO stockers. After that, Vt of thousand production lots were pulled back to baseline.

Case II: Power MOS BVDss fail

1. Fail mode, Analysis and Hypothesis

Recently, two Power MOS products suffered WAT BVDss fail issue and 250 pieces of wafers were scrapped. BVDss was parameter of breakdown voltage of source to drain, which are 8% lower than spec for these suffered wafers. These lots history was checked, however no abnormity or common tools were found.

Physical failure analysis was performed and cross-section was observed by FIB and TEM for several samples, but no physical defect evidence was found. Figure 5 was FIB cross section picture of the failed test structure together with the doping types were marked. Process critical dimensions were within spec, no dislocation, no breakdown or burnt out were found. Junction staining was also performed without any finding. Therefore, the defect was also attributed as non-visible one.

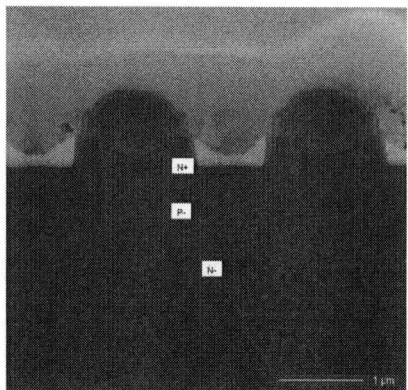

Figure 5: FIB cross-section picture

For these products, 5µm N-EPI layer was deposited on the substrate of source wafers. But EPI tools had no shift during the production, EPI film thickness of monitor wafer was stable and within specification. Therefore suspicion of tool shift can be rule out.

Further WAT electrical test data analysis suggested that the breakdown occur at the junction of P+ to N- sub. Fish bone diagram was drawn for all the possibilities for this breakdown fail. Hypotheses included substrate defect, dislocation, deep P+ dosage change cause P+ to N- junction change, deeper trench depth, and ring IMP profile was driven in because of over thermal budget, etc. All except one were denied based on failure mode, WAT data and PFA result, that is: the fail was caused by high sub oxygen concentration? If the oxygen concentration was high, it would diffuse into EPI drift layer (about 2µm at the bottom of the 5-µm EPI layer) and acted as recombination center which would increase the current in reverse-biased p-n junctions. The higher oxygen concentration in substrate, the deeper it would diffuse into EPI drift layer, which eventually would cause effective depletion region of above junction would shift more into substrate doping drift layer. As a result, junction breakdown voltage (BVDss) would decrease rapidly.

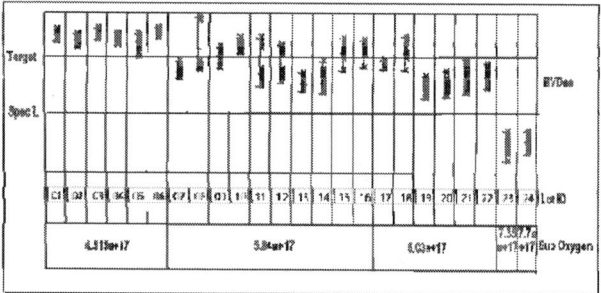

Figure 6: Different substrate oxygen concentration levels and corresponding BVDss

2. Root cause verification:

For this case, the P-N junction was too deep for SIMS to measure oxygen profile (more than 5 µm) from top to down, but FTIR (Fourier Transform Infrared Spectroscopy) was widely used to measure oxygen concentration in substrate. [4] Therefore, DOE experiment was designed to verify oxygen effect on the breakdown voltage, and oxygen concentrations of these wafers were measured by FTIR before EPI layer deposition. The DOE results were shown in figure 6. There are four oxygen concentration levels and 24 lots were run for data collection.

Strong correlation was found between BVDss & oxygen concentration of substrate. Higher oxygen concentration leads to lower BVDss voltage. This split results confirmed our hypothesis that oxygen play an important role on junction breakdown voltage. For mass production, source wafer oxygen concentration should be controlled.

Stage	Condition	Thickness	#13	#14	#15	#16	#17
EPI DEP	Worst	4.75μm	V	V			
	Best	5.25 μm			V	V	
	POR	5 μm					V
WAT	Yield (%)		6.45	0	100	99.19	99.19
	BVDss (%)		87.90	99.19	0	0.81	0

Table 2: DOE of EPI thickness with yield and BVDss fail rate

Because oxygen diffusion caused shorter effective EPI layer thickness, increasing EPI thickness should improve the yield and gain more process margin. Table 2 is the DOE table of EPI thickness with yield and BVDss fail rate. It was clearly shown that thick EPI layer can bring high yield and low BVDss fail rate. This result matched with our expectation and suggested that 5μm EPI layer thickness is too marginal to cover substrate oxygen variation.

3. Corrective actions:

Based on the DOE results, source wafer oxygen concentration was control with specification. EPI thickness was re-targeted from 0.5μm to 0.525μm and WAT spec re-targeted from 33V to 35.5V, with no side effect on others parameters.

Conclusions

As shown in the two examples, for non-visible defect, directly imaging method cannot provide physical evidence of fail mechanism and directions for corrective actions. Systematic methods should be employed, which included in-depth electrical data mining to find clues, followed by brainstorming and fish-bone diagram to list all possibilities. Then proper tools or methods were used to characterize root cause. Finally, DOE experiment should be conducted to confirm the hypotheses and to verify the root cause. The key process for such non-visible defect analysis should be "bold hypothesis, cautious verification".

Acknowledgments

Great supports from HHGrace FAB 3 Diffusion department, integration department and FA chemical lab were highly appreciated.

References

1. Masataka Hourai, *et al*, "A Method of Quantitative Contamination with Metallic Impurities of the Surface of a Silicon Wafer," Jpn. J. Appl. Phys. 27 (1988) pp. L2361-L2363.

2. Kun Lin, *et al*, "Scanning Capacitance Microscopy Application for Bipolar and CMOS Doping Issues in Semiconductor Failure Analysis," *Proc 31st International Symposium for Testing and Failure Analysis*, San Jose, 2005, pp. 307-310.

3. C.Q. Chen, *et al*, "Non-Visible Defect Analysis of OTP Device," *Proc 37th International Symposium for Testing and Failure Analysis*, San Jose, 2011, pp. 198-201.

4. A. Baghdadi1, *et al*, "Interlaboratory Determination of the Calibration Factor for the Measurement of the Interstitial Oxygen Content of Silicon by Infrared Absorption," J. Electrochem. Soc. 1989 136(7), 2015-2024.

Material Characterization and Failure Analysis of Through-Silicon Vias

Chenglin Wu, Tengfei Jiang, Jay Im, Kenneth M. Liechti, Rui Huang and Paul S. Ho
Department of Aerospace Engineering and Engineering Mechanics, University of Texas, Austin, TX 78712, USA
Microelectronics Research Center and Texas Materials Institute, University of Texas, Austin, TX 78712, USA
Email: ruihuang@mail.utexas.edu; paulho@mail.utexas.edu

Abstract - **In this paper, the effects of Cu microstructure on the mechanical properties of TSV and via extrusion are studied using two types of though-silicon vias (TSVs) with different grain size distributions. A direct correlation is found between the Cu grain size and the mechanical properties of the TSVs. An analytical model is used to explore the relationship between the mechanical properties and via extrusion. The results show that small and uniform grains in the Cu vias led to smaller via extrusion. Such grain structures are effective for reducing via extrusion failure to improve TSV reliability.**

I. INTRODUCTION

Copper (Cu) through-silicon via (TSV) is a critical element in three-dimensional (3D) integrated circuits. Typically, fabrication of TSVs involves etching of via holes, deposition of liner and Cu seed layers, electroplating of Cu, post-electroplating annealing, and CMP removal of Cu overburden. In the via-middle scheme widely adopted for 3D integration, the back-end-of-the-line (BEOL) layers are deposited on top of the wafer after the fabrication of TSVs. As one of the reliability issues, extrusion of the Cu vias occurs primarily during the BEOL processing, which can cause the BEOL layers to deform, leading to mechanical and electrical failures of the interconnect structures (Fig. 1) [1-5]. Thus via extrusion has been a major concern for yield and reliability of 3D integration. Previous studies have suggested that the stress and mechanical properties of the Cu via directly affect via extrusion [4-6]. The underlying mechanism of via extrusion has been examined by considering plastic deformation in Cu and via/Si interfacial sliding [7,8]. For the purpose of process optimization, it is important to establish a correlation between the microstructures of the Cu via and the mechanical properties, which in turn can be correlated to via extrusion. In this work, we characterized the microstructures and mechanical properties of Cu TSVs, followed by measurements and modeling of via extrusion.

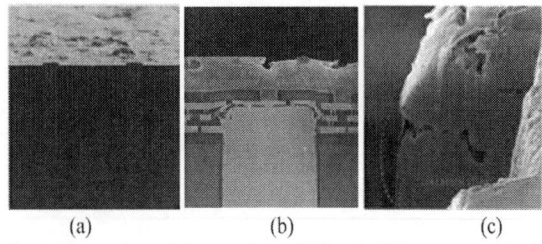

(a) (b) (c)

Fig. 1. Observations of via extrusion and failure in TSV structures: (a) Via extrusion after thermal cycling [4]; (b) and (c) Fracture and delamination due to via extrusion [3].

II. MICROSTRUCTURES OF COPPER VIAS

The TSV samples used in this study were fabricated using the standard via-middle scheme. The via diameter was $D = 5.5$ μm and the via depth was $H = 55$ μm. The total thickness of the wafer was 780 μm. Two different processing conditions were used to fabricate two types of TSVs with different microstructures, referred to as TSV-A and TSV-B. A number of vias from each type were cross-sectioned by focused ion beam (FIB) and their microstructures were measured by electron backscatter diffraction (EBSD). Both types of vias were found to have essentially random grain orientations (Fig. 2a). However, the grain size distributions were different (Fig. 2b): in TSV-A, the grain sizes were relatively uniform while the distribution in TSV-B was rather polarized, with several large grains mixing with small grains. The large grains in TSV-B sometimes spanned across the entire via diameter. This can be seen from the grain mapping and grain size distribution in Figure 2. Quantitatively, the average grain size was found to be 2.83 μm for TSV-A and 3.82 μm for TSV-B.

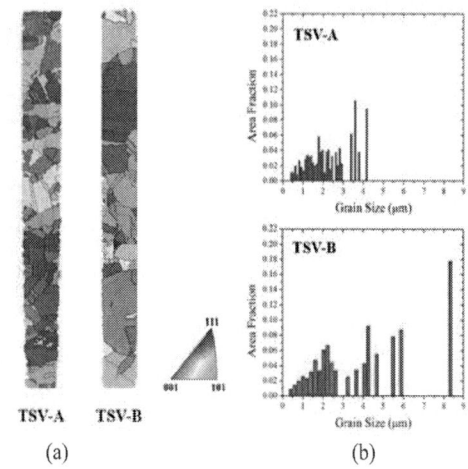

TSV-A   TSV-B

(a) (b)

Fig. 2. EBSD measurement of Cu microstructures for TSV-A and TSV-B: (a) grain mapping; (b) grain size distributions.

III. MECHANICAL PROPERTIES OF COPPER VIAS

To determine the elastic and plastic properties of the Cu vias, nanoindentation measurements were carried out. For several vias of each type, the BEOL layers were removed by FIB, and then quasi-static indentations were conducted on the top of the vias using Hysitron TI 950 TriboIndenter® equipped with a

Berkvovich diamond tip. A two-segment load versus time profile was applied with a loading/unloading rate of 100 nN/s and a peak load of 800 μN. Figure 3 shows the measured load-displacement responses. The elastic moduli of the Cu vias were deduced from the unloading curves based on the Oliver-Pharr method [9], as shown in Figure 4. The average elastic modulus was found to be 117 GPa for TSV-A and 93 GPa for TSV-B. The elastic modulus for TSV-B is lower than typically expected for Cu (~110 GPa) based on previous studies on electroplated Cu thin films [10], which may be related to the surface roughness or the grain textures near the top surface of the via.

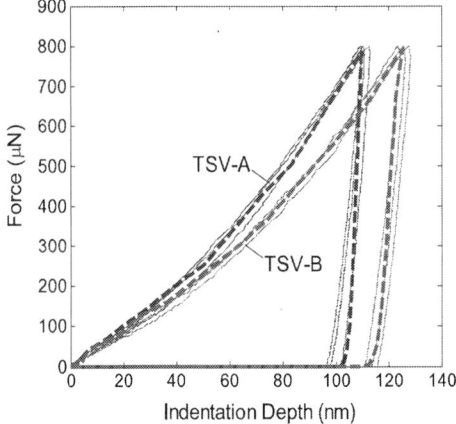

Fig. 3. Nanoindentation measurements for TSV-A (red) and TSV-B (blue), in comparison with FEA simulations (dashed lines).

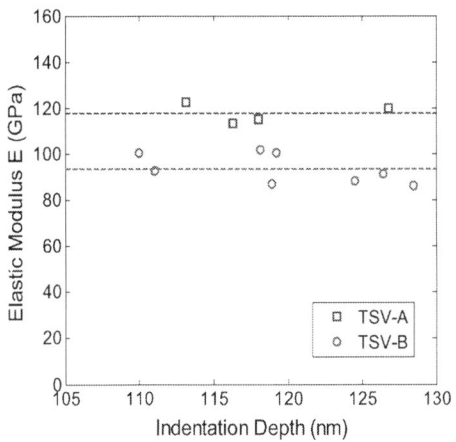

Fig. 4. Elastic modulus of Cu vias extracted from nanoindentation measurements.

To determine the plastic properties of the Cu vias, an axisymmetric finite element analysis (FEA) model was constructed to simulate the nanoindentation experiments (Fig. 5). An elastic diamond indenter with the cono-spherical shape was used with a tip radius of 100 nm. The material properties used for the diamond tip were: $E_i = 1222$ GPa and $v_i = 0.2$. For the Cu vias, the average elastic moduli obtained from the nanoindentation measurements were used for TSV-A and TSV-B along with Poisson's ratio $v_{Cu} = 0.35$. The elastic

properties for Si were: $E_{Si} = 130$ GPa and $v_{Si} = 0.28$. The interface between Cu and Si are assumed to be perfectly bonded.

To model plasticity in the Cu vias, we used a classical metal plasticity model in ABAQUS [11] with Mises yield surface and isotropic hardening in the FEA simulations. The yield stress was specified as a function of plastic strain:

$$\sigma_y\left(\bar{\varepsilon}_p\right) = \sigma_{y0}\left[1 + \frac{7E\bar{\varepsilon}_p}{3\sigma_{y0}}\right]^{1/n}. \tag{1}$$

where σ_{y0} is the initial yield strength, $\bar{\varepsilon}_p$ is the equivalent plastic strain, and n is the hardening exponent. The initial yield strength σ_{y0} and the hardening exponent n were deduced using an iterative approach based on comparison between the FEA simulations and the nanoindentation experiments. For a given set of yield strength and hardening exponent, the indentation response of the TSV was simulated and compared to the force-displacement curve obtained from the experiment. The hardening exponent was fixed while adjusting the yield strength until the peak displacement was within 10% of the experiment value. The hardening exponent was then adjusted to achieve a better fitting to the experimental curve. This process was repeated until a reasonable fitting was obtained (see Fig. 3). The hardening exponent was found to be $n = 9.5$ for both vias, while the initial yield strength was 250 MPa for TSV-A and 190 MPa for TSV-B. The lower yield strength for TSV-B is qualitatively consistent with the theoretical expectation based on the Hall-Petch relation, namely, the yield strength decreasing with increasing grain size due to the grain boundary strengthening mechanism.

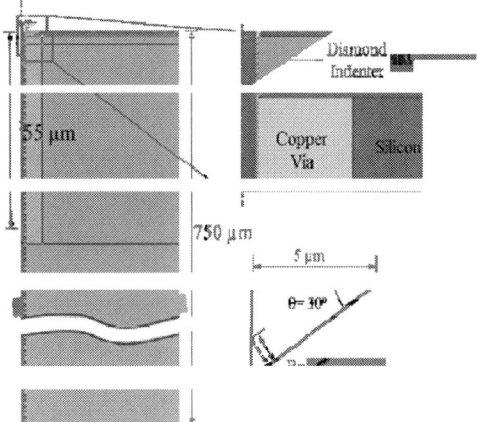

Fig. 5. An axisymmetric model for finite element simulations of nanoindentaiton using a cono-spherical diamond indenter.

IV. MEASUREMENT OF VIA EXTRUSION

For both types of TSVs, via extrusion and damage of the BEOL layers were observed. The extent of the via extrusion was measured at the via cross-section by high resolution scanning electron microscope (SEM), as shown in Figure 6. The average extrusion was 117 nm for TSV-A and 147 nm for TSV-B, showing the amount of via extrusion for TSV-B about 25%

978-1-4799-3911-4/14 $31.00 © 2014 IEEE   313

larger than for TSV-A. A clear correlation appeared to exist between the average grain size of the Cu via and the amount of via extrusion (Fig. 6). For TSV-A which has smaller and more uniform grains, the amount of via extrusion is smaller than TSV-B. The difference can be traced to the different mechanical properties of Cu due to different grain structures resulting from different process conditions. Based on the observed correlation, TSVs with uniform small grains would be more favorable for reducing via extrusion.

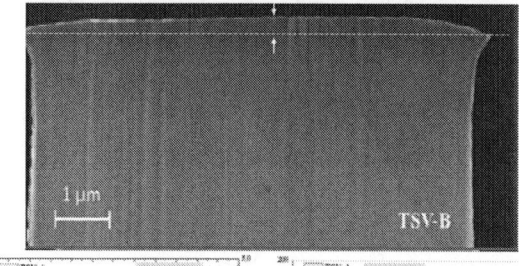

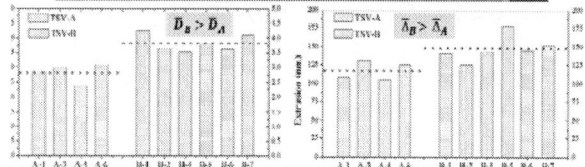

Fig. 6. SEM images of via extrusion (upper panel), and correlation between grain size and via extrusion for TSV-A and TSV-B.

V. MODELING OF VIA EXTRUSION

To elucidate the effects of mechanical properties on via extrusion, a simple analytical model was formulated taking into account Cu Plasticity [7], followed by FEA simulations. Both the analytical model and FEA considered TSVs subject to a thermal cycle from the room temperature (T_R) to a high process temperature (T_H) and then back to the room temperature, with a thermal load $\Delta T = T_H - T_R$. First, assuming a free sliding interface, the mismatch of thermal expansion between the Cu via and Si induces a biaxial compressive stress in Cu upon heating:

$$\sigma_r = \sigma_\theta = -\Delta T \left(\alpha_{Cu} - \alpha_{Si} \right) \left(\frac{1-\nu_{Cu}}{E_{Cu}} + \frac{1+\nu_{Si}}{E_{Si}} \right)^{-1} \tag{2}$$

where α_{Cu} and α_{Si} are the coefficients of thermal expansion (CTEs) for Cu and Si, respectively. The stress induces an elastic strain in the axial direction of the via, which result in an elastic extrusion at the high temperature T_H:

$$\frac{\Delta H_e}{H} = \varepsilon_{z,Cu} - \varepsilon_{z,Si}$$

$$= \Delta T \left(\alpha_{Cu} - \alpha_{Si} \right) \left(1 + \frac{2\nu_{Cu}}{E_{Cu}} \left(\frac{1-\nu_{Cu}}{E_{Cu}} + \frac{1+\nu_{Si}}{E_{Si}} \right)^{-1} \right) \tag{3}$$

where H is the via height and ΔH_e is elastic extrusion. The elastic extrusion increases linearly with temperature as, $\Delta H_e = \beta_e H \Delta T$, with $\beta_e = 20.64$ ppm/°C by using the typical values for the thermomechanical properties of Cu and Si (α_{Cu} = 17 ppm/°C, α_{Si} = 2.3 ppm/°C, E_{Cu} = 110 GPa, E_{Si} = 130 GPa, ν_{Cu} = 0.35, and ν_{Si} = 0.28).

If no plastic yielding in Cu, the elastic via extrusion would decrease with the same rate upon cooling and vanish at the room temperature after a full thermal cycle. On the other hand, assuming prefect plasticity with a yield stress σ_y for the Cu via, plastic yielding of Cu is predicted when heating above a critical temperature

$$\Delta T_y = \frac{\sigma_y}{\alpha_{Cu} - \alpha_{Si}} \left(\frac{1-\nu_{Cu}}{E_{Cu}} + \frac{1+\nu_{Si}}{E_{Si}} \right) \tag{4}$$

which is proportional to the yield strength of Cu. Beyond the critical temperature ($\Delta T > \Delta T_y$), the Cu via deforms plastically, resulting in a plastic extrusion [7]:

$$\frac{\Delta H_p}{H} = \left(3\alpha_{Cu} - 2\alpha_{Si} \right) \left(\Delta T - \Delta T_y \right) \tag{5}$$

The plastic extrusion also increases linearly with temperature, but with a higher rate as, $\Delta H_p = \beta_p H \left(\Delta T - \Delta T_y \right)$, with $\beta_p = 46.4$ ppm. The plastic extrusion rate is over twice of the elastic extrusion rate, leading to more significant via extrusion at the high temperature. More importantly, the plastic extrusion does not vanish after cooling, resulting in a non-zero residual extrusion after a full thermal cycle [7]:

$$\Delta H_r = H \left(\beta_p - \beta_e \right) \left(\Delta T - \Delta T_y \right) \tag{6}$$

Thus, the magnitude of the residual extrusion depends on the highest temperature during the thermal cycle and the plastic yield strength of the Cu via. Increasing the yield strength of Cu would increase the yield temperature ΔT_y and thus decrease the residual extrusion for the same thermal load ΔT.

Using the elastic-plastic properties extracted from the nanoindentation experiments, the magnitude of via extrusion versus the maximum process temperature are plotted in Figure 7 for both TSV-A and TSV-B. Since the yield strength is lower for TSV-B, the critical thermal load for via extrusion is lower. Subject to same thermal load ($\Delta T > \Delta T_y$), the analytical model predicts that the amount of via extrusion for TSV-B is higher than TSV-A, consistent with the experimental observations (Fig. 6). Using the average extrusion measured for TSV-A and

TSV-B, we found that the corresponding thermal load is around 350°C for both TSVs, although the thermal load for TSV-B is slightly lower. The deduced thermal load is in reasonable agreement with typical process temperatures (~400°C), although the exact thermal processes are not available for these TSV samples. Incidentally, the predicted via extrusion for TSV-B compare closely with reported data in a previous study [5].

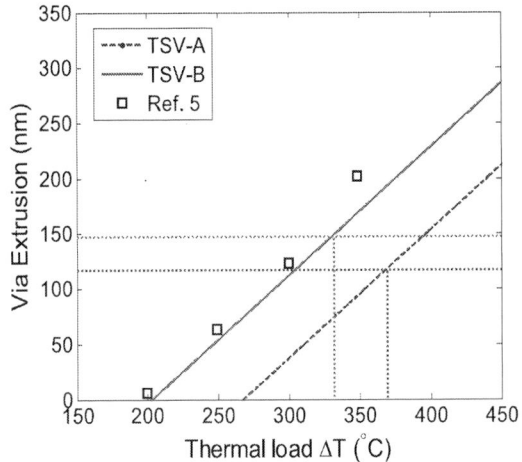

Fig. 7. Via extrusion versus maximum process temperature predicted by the analytical model, in comparison with experiments.

Next we used an axisymmetric FEA model similar to Figure 5 to simulate via extrusion during a thermal cycle, using the elastic-plastic properties extracted for TSV-A and TSV-B. As shown in Fig. 8, the numerical results are consistent with the predictions by the analytical model, where the residual via extrusion after the thermal cycle was higher for TSV-B than for TSV-A, both subject to the same thermal load $\Delta T = 350$°C. It was found that the analytical model slightly overestimated the via extrusion due to the assumption of perfect plasticity (no strain hardening) and uniform stress field in the Cu vias.

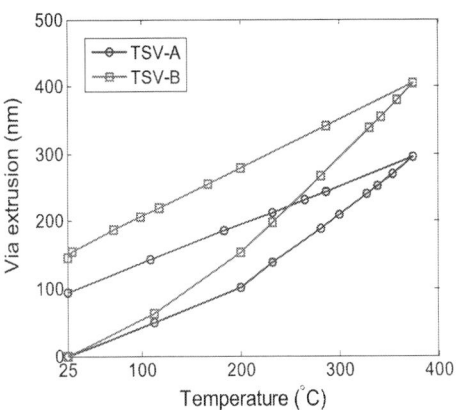

Fig. 8. Numerical simulations of via extrusion during a thermal cycle for TSV-A and TSV-B.

We further apply the FEA model to investigate the effect of interfacial properties on Cu extrusion. The development of via extrusion during a thermal cycle is evaluated for two cases with different bonding behaviors and the results are shown in Figure 9 for comparison with the analytical model. First, a perfectly bonded interface is assumed between the copper via and silicon, for which the residual extrusion is significantly reduced, by ~3x, at room temperature. The results indicate that the contribution of plasticity to extrusion is reduced by the interfacial bonding between via and silicon. In the second case, a cohesive via/Si interface is assumed, which is represented by a bilinear traction-separation relationship with an adhesion energy of 2.5 J/m² and a shear strength of 50 MPa. The via extrusions at both room and maximum temperatures are considerably higher than the bonded case although they are still lower than the analytical model. Hence, the via extrusion depends on both Cu plasticity and interfacial adhesion.

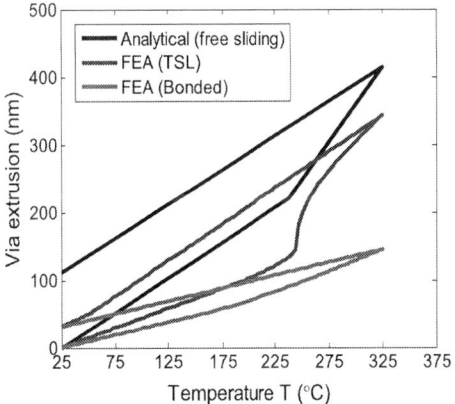

Fig. 9. Comparison of via extrusion calculated from the analytical model and FEA simulations with different interfacial properties: free sliding, perfect bonding, and cohesive with a bilinear traction-separation law (TSL).

VI. CONCLUSIONS

In summary, the effect of the average grain size on the elastic-plastic properties of Cu TSV has been examined and correlated to via extrusion. The effect was investigated for two types of grain structures. It was found that smaller and more uniform grains resulted in higher yield strength and therefore is more favorable for reducing the via extrusion. The findings in this study suggest that in the fabrication of TSVs, it will be effective to control the processing parameters in order to achieve more uniform and small grains to improve via extrusion reliability.

ACKNOWLEDGMENT

This work was supported by Semiconductor Research Corp.

REFERENCES

[1] I. De Wolf, K. Croes, O. V. Pedreira, R. Labie, A. Redolfi, M. Van De Peer, K. Vanstreels, C. Okoro, B. Vandevelde, and E. Beyne, "Cu pumping in TSVs: Effect of pre-CMP thermal budget," Microelectron. Reliab. 51, 1856–1859 (2011).

[2] A. Heryanto, W. N. Putra, A. Trigg, S. Gao, W. S. Kwon, F. X. Che, X. F. Ang, J. Wei, R. I Made, C. L. Gan, and K. L. Pey, "Effect of copper TSV

annealing on via protrusion for TSV wafer fabrication," J. Electron. Mater. 41 (9), 2533–2542 (2012).

[3] S. Kang, S. Cho, K. Yun, S. Ji, K. Bae, W. Lee, E. Kim, J. Kim, J. Cho, H. Mun, and Y. L. Park, "TSV optimization for BEOL interconnection in logic process," in Proc. IEEE Int. 3DIC, Osaka, Japan, Jan. 31/Feb. 2, 2012, pp. 1–4.

[4] S. K. Ryu, T. Jiang, K. H. Lu, J. Im, H.-Y. Son, K.-Y. Byun, R. Huang, and P. S. Ho, "Characterization of thermal stresses in through-silicon vias for three-dimensional interconnects by bending beam technique," Appl. Phys. Lett. 100, 041901 (2012).

[5] D. Zhang, K. Hummler, L. Smith, and J.-Q. Lu, "Backside TSV protrusion induced by thermal shock and thermal cycling," Proceedings of IEEE Electronic Components and Technology Conference (2013), pp. 1407–1413.

[6] T. Jiang, S. K. Ryu, Q. Zhao, J. Im, R. Huang, and P. S. Ho, "Measurement and analysis of thermal stresses in 3D integrated structures containing through-silicon-vias," Microelectron. Reliab. 53, 53–62 (2013).

[7] T. Jiang, C. Wu, L. Spinella, J. Im, N. Tamura, M. Kunz, H.-Y. Son, B.G. Kim, R. Huang, P.S. Ho, "Plasticity mechanism for copper extrusion in through-silicon vias for three-dimensional interconnects", Appl. Phys. Lett. 103, 211906 (2013).

[8] S.-K. Ryu, T. Jiang, J. Im, P.S. Ho, R. Huang, "Thermo-mechanical failure analysis of through-silicon via interface using a shear-lag model with cohesive zone", IEEE Trans. on Device and Materials Reliability 14, 318-326 (2014).

[9] W. C. Oliver and G. M. Pharr, "An improved technique for determining hardness and elastic modulus using load and displacement sensing indentation experiments", J. Mater. Res. 7, 1564-1583 (1992).

[10] D. W. Gan, R. Huang, P.S. Ho, J. Leu, J. Maiz, T. Scherban, "Isothermal stress relaxation in electroplated Cu films. Part I: Mass transport measurements", J. Applied Physics 97, 103531 (2005).

[11] ABAQUS Theory and Analysis User's Manuals (Version 6.13), Dassault Systèmes Simulia Corp., Providence, RI, USA (2013).

Electromigration Reliability of Open TSV Structures

Wolfhard H. Zisser[a], Hajdin Ceric[a,b], Josef Weinbub[a] and Siegfried Selberherr[a]

[a]Institute for Microelectronics, Technische Universität Wien, Gußhausstraße 27-29/E360, 1040 Wien, Austria

[b]Christian Doppler Laboratory for Reliability Issues in Microelectronics

Email: {zisser|ceric|weinbub|selberherr}@iue.tuwien.ac.at

Abstract—**A study of electromigration in open through silicon vias is presented. The calculations are based on the drift-diffusion model for electromigration combined with mechanical simulations. The results show that the highest stresses are located at the aluminium/tungsten interfaces, near the region where the electrical current is introduced into the open through silicon vias, which happens to be the location of the highest current density at the interface. There, the electromigration induced degradation, e.g. void nucleation, is most probable to occur.**

I. Introduction

Three-dimensional (3D) integration is a promising approach for the development of systems with higher performance. Interconnections for 3D integrated circuits, though, include components not used in planar 2D architectures, such as through silicon vias (TSVs). Open through silicon vias introduced in [1] are a TSV concept in which the cylindrical structure is coated, rather than entirely filled with the conducting metal. The advantage of this technology, is that it can reduce the stress originating from the mismatched thermal expansion coefficients between the substrate and the TSV.

Electromigration (EM) addresses the material transport due to microscopic forces acting on mobile defects. These forces originate from the electric current and the electric field and cause therefore the flow of vacancies in current direction. The flow of material builds up stress in the interconnects, especially in the ones which carry high current densities, and leads to mechanical degradation, e.g. void nucleation or delamination [2].

In this work we investigate the possible EM reliability issues associated with this particular TSV technology. We use the drift-diffusion model for mass transport, in the aluminium part, which accurately treats the effects of EM. The results give the initial strain for a mechanical calculation to treat the stress in the whole structure. We simulate a segment of the geometry of the open TSVs as shown in Fig. 1. The result show that the highest stress is located at the outer surface of the aluminium cylinder along the interface between aluminium and tungsten. The highest vacancy concentration is found in the aluminium, in the region covered by the tungsten, which is therefore the region most sensitive to reliability issues.

II. Approach

The TSV geometry considered is shown in Fig. 1. Here, the tungsten, shown in red, forms an empty cylinder closed on the bottom side. Below that (not shown in figure) an aluminium plate is placed on which a solder pump is mounted to connect to other wafers. On the top side, an aluminium layer (shown in blue) forms a second empty cylinder, which overlaps with the inside, upper part of the tungsten cylinder wall. The upper side of the aluminium connects to the planar interconnect structure by a round plate as shown in Fig. 1. These open TSVs are different compared to the traditional copper TSVs which have their cylinders completely filled.

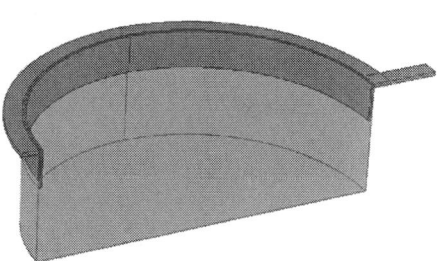

Fig. 1. TSV structure: Aluminium in blue and tungsten in red. The tungsten cylinder is shortened to 10% of the real length. For the simulated TSV the upper plate is removed.

In order to address EM in materials, two important microscopic forces must be considered to determine the material transport. The first is the so called direct force ($\vec{F}_{direct}$), caused by the the local electric field acting on the ionic atoms in the metal. The second is called the wind force ($\vec{F}_{wind}$), which is caused by the electrons scattered by the atoms in the metal [3]. The sum of these two forces determines the total force, as

$$\vec{F} = \vec{F}_{direct} + \vec{F}_{wind} = (Z_d + Z_w)e\vec{E} = Z^*e\vec{E}, \quad (1)$$

where Z_d and Z_w are the so called direct valence and wind valence, respectively, and Z^* is the effective valence, which describes the sensitivity to EM. $\vec{E}$ is the electrical field and e is the elementary electron charge.

For macroscopic modelling of the time evolution of the vacancy distribution C_v in a bulk material, a drift-diffusion model [4] with an additional generation/annihilation term G is used as

$$\frac{\partial C_v}{\partial t} = -\nabla \cdot \vec{J}_v + G. \quad (2)$$

The generation/annihilation term G, usually called Rosenberg-Ohring term [5], [6], is computed by

$$G = \frac{C_{v,eq} - C_v}{\tau}, \quad (3)$$

where $C_{v,eq}$ is the equilibrium concentration and τ is the characteristic relaxation time of the vacancy concentration.

The vacancy flux $\vec{J}_v$ is driven by three main forces, all of which are included in the bracket of the following equation

$$\vec{J}_v = -D_v \left(\nabla C_v - \frac{|Z^*|}{k_B T} C_v \vec{E} + \frac{f\Omega}{k_B T} C_v \nabla \sigma \right), \quad (4)$$

where k_B is the *Boltzmann* constant, T is the temperature, D_v is the diffusion coefficient of the vacancies, Ω is the atomic volume, and f is the relaxation factor. The first term in the bracket (first force) is a typical diffusion flux term due to different concentrations of vacancies. The second flux term is caused by the EM as described above, which is determined by the electric field in the structure. The third term is the flux due to different stresses in the material. A fourth flux term due to temperature gradients in the material could also be included, but is neglectable in this study, due to the homogenous temperature distribution in the aluminium and the tungsten.

For the stress term a solid mechanics simulation is required. The initial strain, which serves as an input to the solid mechanics simulation, is obtained by the following equation [7].

$$\frac{\partial \epsilon^v}{\partial t} = \Omega[(1 - f) \nabla \cdot \vec{J}_v + fG] \qquad (5)$$

The mechanical constraints were chosen as follows: The outside of the cylinder is surrounded by a silicon oxide/silicon substrate. Therefore, the position of the outer surface of the material is considered to be fixed. In the actual structures, inside the cylinder there is a thin silicon oxide layer, which is also taken into account in the calculations.

III. Results

The TSV structure includes surfaces connecting tungsten and aluminium. Since studies have shown that tungsten has a much lower sensitivity to EM [8], we focus our EM study on the aluminium structure on the top of the TSV, and include the tungsten only in the mechanical and electrical simulations.

The open TSV structure, which geometrically forms an arc, is shown on the left side of Fig. 2. On the top, an aluminium layer (shown in yellow) forms a second arc, which overlaps with the inside, upper part of the tungsten wall. The geometry considered in our calculations is a segment of this open TSV and shown on the right side in Fig. 2. The upper side of the aluminium (yellow) connects to the rectangular supply line aluminium interconnect, which is also included in the simulation domain. In the inner cylindrical surface of the TSV the aluminium and the tungsten are coated by a silicon oxide film (red).

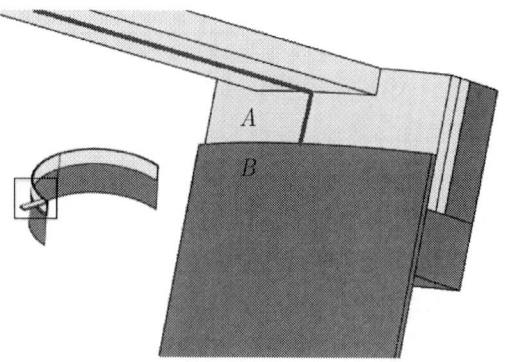

Fig. 2. TSV structure showing aluminum in yellow, tungsten in green, and silicon oxide in red. The blue line indicates a cut through the TSV for use in the subsequent figures.

For the mechanical stress simulations we use fixed boundary conditions on the aluminium surface labelled A and for the tungsten surface labelled B in Fig. 2. On the open surface of the oxide (red segment in Fig. 2), in the inner region of the arc, we employ open boundary conditions.

First we calculate the electrical current density in the considered geometrical structure, as shown in Fig. 3. The current flows through the aluminium interconnect (see arrow in Fig. 3), into the aluminium part of the TSV structure, and from there into the tungsten. The regions with the highest current density are shown in red. In the TSV part, a high current density is observed in the tungsten due to its thinner size compared to that of the aluminium section. At the corner of the interface of the two materials, however, a high current density is also observed (see zoomed-in inset). This current crowding at the corner at the very end of the interface is a result of the higher conductivity of aluminium and its larger thickness compared to tungsten, which provides a low resistance path for the current to flow to the corner of the interface, before it passes into the tungsten.

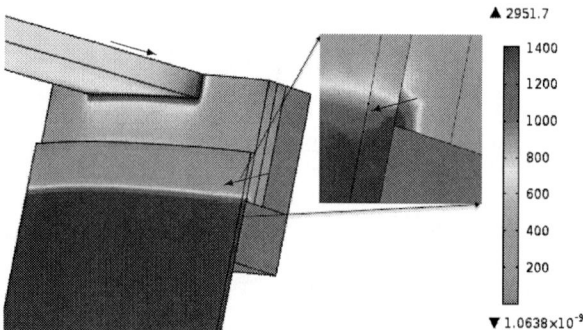

Fig. 3. Current density in the structure (A/cm^2). The inset is a zoom-in at the corner of the aluminium/tungsten interface.

After calculating the electrical current, we solve (2) together with the solid mechanics equations (5) to obtain the vacancy concentration in the aluminium, in order to determine the stress in the whole structure. Fig. 4 shows the relative change of the vacancy concentration ($c = (C - C_{eq})/C_{eq}$) compared to the equilibrium concentration C_{eq}, at the beginning of switching on the current. The figure is a cut through the blue line of Fig. 2. Most vacancies accumulate in the aluminium region, especially at the corner of the interface between aluminium and tungsten, where the highest current density is observed. The interface is blocking the vacancies and vacancy accumulation is created.

The corresponding tensile stress at $t = 1$ s inside the TSV structure is shown in Fig. 5. In the aluminium region, where the vacancies are highly concentrated, higher tensile stress values are observed compared to the stress values in the tungsten region due to the presence of excess vacancies. At the interface of the two materials, the highest stress values are observed, attributed to the tendency of the aluminium to shrink compared to tungsten due to the difference in the vacancy concentration (c.f. Fig. 6).

An important observation, is that the results show a localized EM behaviour, extending only a few μm into the interconnect. Therefore EM studies can be restricted to these critical parts of the structures and can then be used for the prediction of the entire structure's resistance against EM degradation.

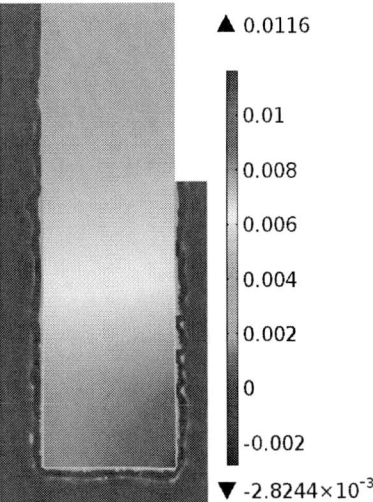

Fig. 4. Relative vacancy concentration change compared to the vacancy equilibrium concentration after $t = 1$s of current flow. A surface cut along the red line of Fig. 1 is shown. Red colour indicates the peak concentration.

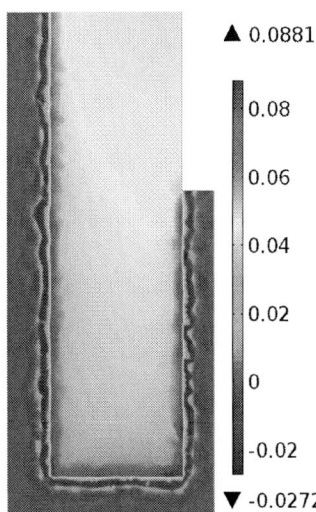

Fig. 6. Relative vacancy concentration change compared to the vacancy equilibrium concentration after $t = 10^6$s of current flow. A surface cut along the red line of Fig. 1 is shown. Red colour indicates the peak concentration.

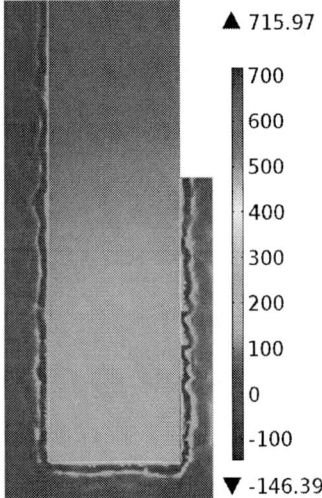

Fig. 5. Tensile stress (in Pa) after $t = 1$s of current flow. A surface cut along the red line of Fig. 1 is shown. Red colour indicates the peak tensile stress.

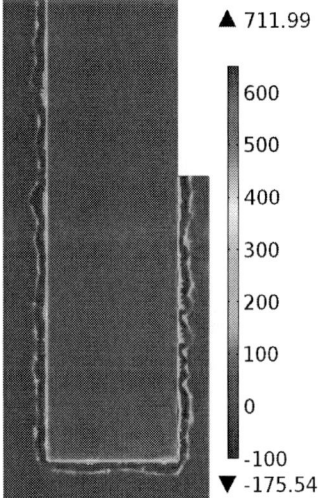

Fig. 7. Tensile stress (in MPa) after $t = 10^6$s of current flow. A surface cut along the red line of Fig. 1 is shown. Red colour indicates the stress peak.

For short times (e.g. 1 s) the stress build up is only driven by the displacement of vacancies due to the high current densities. After 10^5 s, however, the vacancies are less driven by the current and more by the tensile stress gradients. Fig. 6 shows the vacancy distribution after a longer time interval, again along the cut through the blue line of Fig. 2. Interestingly, the vacancies are more concentrated along the interfaces, both the aluminium/tungsten and aluminium/oxide interfaces. The highest tensile stress is also located along the aluminium/tungsten interfaces as shown by the red regions in Fig. 7.

The time evolution of the maximum stress in the structure (which is observed along the aluminium/tungsten interface) is shown in Fig. 8. Initially, the stress grows linearly in time proportional to the number of vacancies transported into the structure due to EM and annihilation in the aluminium. At longer times the stress growth rate decreases. The reason is that more stress is built up in the structure, the more it opposes the current density (see third term of (4)), which reduces the vacancy flow and therefore the stress build-up. Fig. 9 shows the maximum vacancy concentration versus time, which confirms the typical EM behaviour observed by Kirchheim [9]. In the beginning, the vacancy concentration increases, as vacancies accumulate in the aluminium. A quasi-steady state concentration is reached after a fairly short time [9]. As the vacancies recombine, the aluminium shrinks, which creates the stress accumulation observed in Fig. 9. At larger times, however, the vacancy concentration increases rapidly. This increase happens after a high stress magnitude is developed. In reality the tensile stress in the structure increases in such a way that void nucleation is triggered.

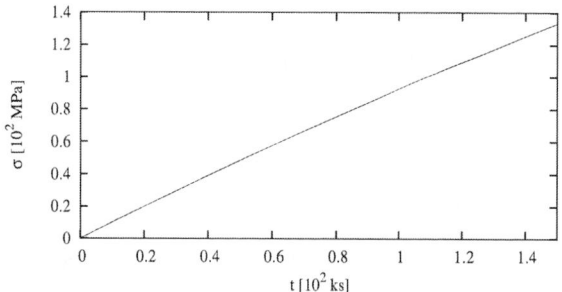

Fig. 8.   Maximum stress versus time at the aluminium/tungsten interface.

Fig. 10 shows the stress development in time for different current densities. The stress build-up rises almost linear with the current density. When these results are fitted to Black's equation [10], with an appropriately chosen stress threshold as a failure condition, a current exponent of 0.74 is obtained shown in Fig. 11.

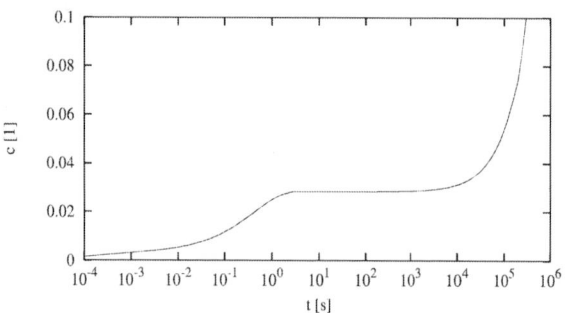

Fig. 9.   Maximum relative vacancy density change versus time. Located at the aluminium/tungsten interface.

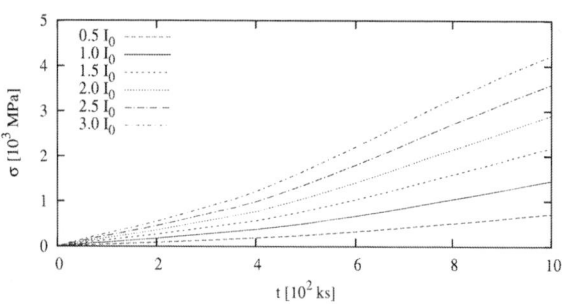

Fig. 10.   Stress build up in the TSV for different current densities. I_0 is the current used for Fig. 8 and Fig. 9.

IV.   Conclusions

Electromigration simulations in open TSVs were performed using the drift-diffusion model for mass transport, including solid mechanics simulations for stress and charge transport simulations for the current densities. We show that the largest stress in the material occurs along the aluminium/tungsten interface. The largest vacancy concentration in the via is observed in the aluminium, also near the regions surrounded by tungsten, where the stress is high. Finally, we

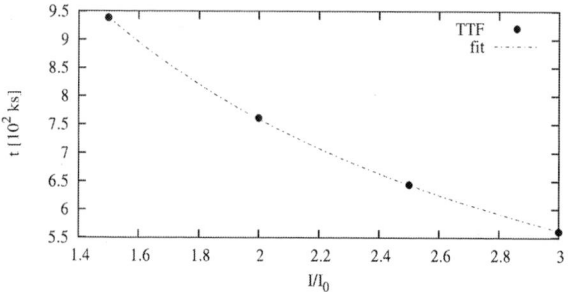

Fig. 11.   Simulated TTF for different current densities. Broken line shows the exponential fitting of the results.

show that EM has a localized behaviour, extending only a few μm into the interconnect, which indicates that EM studies can be restricted to these critical parts of the structures rather than the entire TSV.

Acknowledgment

This work was supported by the Austrian Science Fund FWF, project P23296-N13.

References

1. J. Kraft, F. Schrank, J. Teva, J. Siegert, G. Koppitsch, C. Cassidy, E. Wachmann, F. Altmann, S. Brand, C. Schmidt, and M. Petzold, "3D sensor application with open through silicon via technology," *Proc. ECTC*, pp. 560–566, 2011.

2. I. A. Blech and C. Herring, "Stress generation by electromigration," *Appl. Phys. Lett.*, Vol. 29, No. 3, pp. 131–133, 1976.

3. R. S. Sorbello, "Microscopic driving forces for electromigration," *Proc. Mater. Research Soc. Symp.*, Vol. 427, pp. 73–81, 1996.

4. R. L. de Orio, "Electromigration modeling and simulation," Dissertation, Technische Universität Wien, June 2010.

5. R. Rosenberg and M. Ohring, "Void formation and growth during electromigration in thin films," *J. Appl. Phys.*, Vol. 42, No. 13, pp. 5671–5679, 1971.

6. H. Ceric, R. L. de Orio, J. Cervenka, and S. Selberherr, "A comprehensive TCAD approach for assessing electromigration reliability of modern interconnects," *IEEE Trans. Device Mat. Rel.*, Vol. 9, No. 1, pp. 9–19, 2009.

7. H. Ceric, R. Heinzl, C. Hollauer, T. Grasser, and S. Selberherr, "Microstructure and stress aspects of electromigration modeling," *Proc. AIP*, Vol. 817, No. 1, pp. 262–268, 2006.

8. J. Tao, K. Young, N. W. Cheung, and C. Hu, "Electromigration reliability of tungsten and aluminum vias and improvements under AC current stress," *IEEE Trans. Electron Devices*, Vol. 40, No. 8, 1993.

9. R. Kirchheim, "Stress and electromigration in Al-lines of integrated circuits," *ACTA Metall. Mater.*, Vol. 40, No. 2, pp. 309–323, 1992.

10. J. R. Black, "Electromigration: A brief survey and some recent results," *IEEE Trans. Electron Devices*, Vol. 16, No. 4, pp. 338–347, Apr 1969.

Effects of Sidewall Scallops on the Performance and Reliability of Filled Copper and Open Tungsten TSVs

Lado Filipovic, Roberto Lacerda de Orio, and Siegfried Selberherr

Institute for Microelectronics, Technische Universität Wien

Gußhausstraße 27-29/E360, 1040 Wien, Austria

{filipovic|orio|selberherr}@iue.tuwien.ac.at

Abstract

The effects of the presence of scallops along the sidewalls of filled (copper) and open (tungsten) TSVs are studied. The Bosch process is used in order to generate highly vertical deep trenches; however, the process results in scallops along the etched sidewalls. A model for the Bosch process is implemented in an in-house level set simulator in order to generate various TSV structures with small and large sidewall scallops. The resulting geometries are imported into a finite element tool in order to analyze the performance and reliability of the devices. The electrical parameters of the TSVs are shown to vary when scallops are present for both types of TSVs. In addition, the maximum thermo-mechanical stress increases in the presence of scallops, while the average stress along the interfaces remains relatively unchanged. Electromigration analyses were also performed on the structures in order to determine stress development during the early stages of operation. It was found that the filled TSV with scalloped sidewalls experiences a higher current density and suffers from increased stress, while the sidewall scallops do not cause variation in the stress of open tungsten TSVs. The open tungsten TSVs experience most Electromigration-induced stress in the connecting metal layers and not along the sidewall.

I. Introduction

The microelectronics manufacturing industry has aggressively scaled devices with "more Moore" integration over the last decades. The increased process equipment and factory costs for scaling are expected to limit scaling at the 6nm node [1]. Recently, a great amount of effort has been directed towards adding more functionality to applications beyond memory and logic, deemed "more than Moore" integration. A major development in this direction is the through-silicon via (TSV), a three-dimensional integration technology which allows for the fabrication of systems connecting various technologies, dense device packing, lower power consumption, and reduced RC delay [2]. The two main methods to etch the silicon layer for TSV implementation are the Bosch process and plasma etching [1].

Each silicon etching method has its own flaws and reliability concerns. Problems specific to the Bosch process are a rough, scalloped TSV sidewall, notch formation at the TSV bottom, and potential step coverage issues relating to depositing layers on a scalloped wall [3]. The etching of deep trenches using an ion-enhanced plasma, such as SF_6/O_2, results in significant sidewall tapering, making the formation of deep vertical trenches a challenge [4]. This work compares, through simulations, the electrical and reliability properties of filled copper TSVs with an aspect ratio of 1:11 as well as

open tungsten TSVs with a geometric aspect ratio of 1:3. The effects of the scallops along the length of the TSV sidewalls for each type of via is analyzed using simulations of electrical performance, thermo-mechanical stress after a 300°C temperature drop, and the electromigration (EM) induced stress while operating at a current of 1A for an extended period of time. Non-scalloped structures with the same aspect ratios have been used in order to extract model parameters which successfully replicate experimental measurements.

II. Silicon Etching using the Bosch Process

In order to generate the TSV profiles, an in-house process simulator is used. The simulator is implemented using the level set framework and it is capable of simulating a sequence of processing steps, including etching and deposition [5]. The Bosch process is performed in order to etch deep trenches in silicon using multiple cycles of polymer deposition and polymer/silicon etching.

The first step of a Bosch process cycle involves the deposition of a thin chemically inert polymer layer, usually in a C_xF_x gas environment. The subsequent etching step is performed in an ion-enhanced plasma environment, usually using SF_6 gas. The polymer protects the structure from the chemical etching, while the ions attack the polymer layer at the trench bottom. This results in an exposure of the substrate at the bottom, where chemical etching can then proceed, while the sidewalls are still protected.

TABLE I.   ETCH PARAMETERS FOR THE TWO TYPES OF TSVS IN ORDER TO GENERATE SMALL AND LARGE SCALLOPS ALONG THE SIDEWALL

	Rate	Etch ratio	Cycle time (sec) Small	Cycle time (sec) Large
Deposition	10nm/sec	-	4	12
Si etch isotropic	39nm/sec	Si:mask 80:1 Si:poly 13:1	11.2	33.6
Si etch directional	20nm/sec	Si:mask 80:1 Si:poly 2:1		
Resulting scallop height:			480nm	1.45μm
Total number of dep/etch cycles (Cu):			120	40
Total number of dep/etch cycles (W):			520	173

The parameters used to investigate the etched profile of the TSVs are listed in Table I. For both types of TSVs two simulations are performed in order to generate structures with small and large sidewall scallops, with scallop heights of 48nm and 1.45μm, respectively. The filled copper TSVs have a depth of 58μm, requiring 120 and 40 cycles for a complete etch, while the tungsten TSVs have a 250μm depth, requiring 520 and 173 cycles, in order to generate the small-scalloped and large-scalloped structures, respectively. Each structure is imported into a finite element simulator, where its performance is compared to that of an ideal, flat-sidewalled TSV.

A. Generating the Filled Copper TSVs

After the silicon etching step, several deposition steps are necessary to generate the full TSV. For the filled copper TSVs, an isotropic model is implemented in order to deposit a 500nm layer of SiO_2 along the etched walls for electrical isolation. In addition, a 100nm layer of tantalum is deposited to serve as a barrier to potential copper diffusion into the substrate. The resulting size of the TSV is approximately $5\mu m \times 58\mu m$, which can be filled without appearing seam voids, using electrochemical deposition of Cu with chemical vapor deposition of tungsten and a sputter TiW/Cu seed layer [6]. A profile of the top of the TSV structure is shown in Fig. 1, while an enlarged section showing the mesh used for finite element analysis is shown in Fig. 2. The meshed structure is imported into a finite element tool for electrical and reliability analysis.

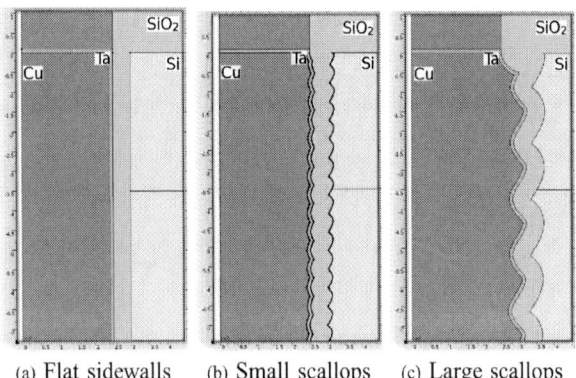

(a) Flat sidewalls   (b) Small scallops   (c) Large scallops

Fig. 1. Materials at the top of the three filled copper TSVs. The backside (bottom) of the structure is also connected to a copper layer, through a tantalum liner.

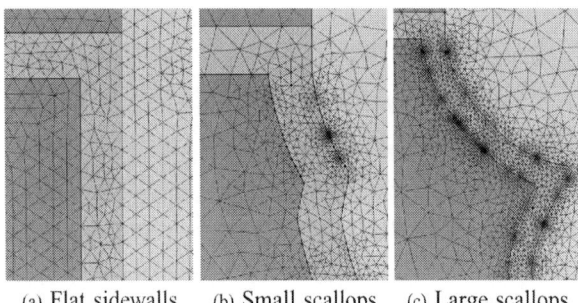

(a) Flat sidewalls   (b) Small scallops   (c) Large scallops

Fig. 2. Mesh at the top of the three filled copper TSVs. The image is an enhanced section from Fig. 1 where the oxide, copper, and tantalum layers are visible.

B. Generating the Open Tungsten TSVs

The open TSV does not require an electrochemical deposition step, but several material depositions are needed along the trench surface in order to generate the final TSV structure.

For the tungsten TSVs a 500nm layer of SiO_2 is deposited along the etched walls for isolation, followed by a 100nm layer of tungsten, an additional 100nm layer of SiO_2, and a 100nm Si_3N_4 liner. An isotropic model is once again used for the deposition steps. The resulting sizes of the TSVs are approximately $80\mu m \times 250\mu m$. The bottom profile of the tungsten TSV structure is shown in Fig. 3.

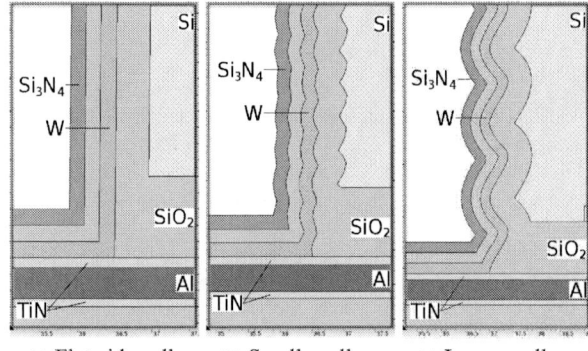

(a) Flat sidewalls   (b) Small scallops   (c) Large scallops

Fig. 3. Bottom of the three open tungsten TSVs with materials labeled. A variation in sidewall scallop height and width is evident.

A very fine mesh is required in order to properly simulate the device performance using finite element methods. An enhanced section of the TSV bottom, where the thin material layers of silicon nitride, oxide, tungsten, titanium nitride, and aluminum can be seen with their meshed elements, is shown in Fig. 4.

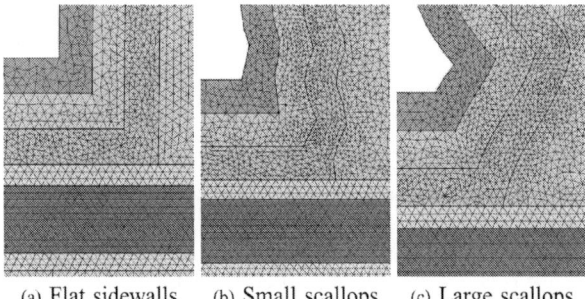

(a) Flat sidewalls   (b) Small scallops   (c) Large scallops

Fig. 4. Mesh at the bottom of the three filled copper TSVs. The image is an enhanced section from Fig. 3 where the oxide, tungsten, titanium nitride, aluminum, and silicon nitride layers are visible.

It is worthwhile noting that the structures generated with small scallops experience a slight sidewall tapering (89%) while the large-scalloped structures show almost perfectly vertical walls. Therefore, when comparing the structures to ideal flat-sidewalled TSVs, it is important to differentiate which characteristics are influenced by the sidewall tapering and which are influenced by the sidewall scallops themselves. Additionally, the large-scalloped structures experience more lateral etching, making the trench wider than the flat, ideal, cylindrical trench.

978-1-4799-3911-4/14 $31.00 © 2014 IEEE

C. Electromigration - Void Nucleation Model

The model used in order to calculate the electromigration-induced stress through the TSV metal layers is given in [7]. The stress is used to detect early failures in metal lines, when compared to a critical stress, which is a material-dependent property. The total vacancy flux is given by

$$\vec{J}_v = -D_v \left(\nabla C_v + \frac{eZ^*}{kT} C_v \rho \vec{j} - \frac{Q^*}{kT^2} C_v \nabla T + \frac{f\Omega}{kT} C_v \nabla \sigma \right),$$

(1)

where D_v is the vacancy diffusivity, C_v is the vacancy concentration, e is the elementary charge, Z^* is the effective charge, ρ is the metal resistivity, $\vec{j}$ is the current density, Q^* is the heat of transport, f is the vacancy relaxation ration, Ω is the atomic volume, and σ is the hydrostatic stress.

The accumulation and depletion of vacancies is found according to the continuity equation

$$\frac{\partial C_v}{\partial t} = -\nabla \cdot \vec{J}_v + G,$$

(2)

where G is a given surface function which models vacancy annihilation and generation. The vacancy transport results in the creation of mechanical strain

$$\frac{\partial \varepsilon}{\partial t} = \Omega \left[(1 - f) \nabla \cdot \vec{J}_v + fG \right],$$

(3)

where ε is the trace of the strain tensor, which is applied to a mechanical simulation using a linear elastic model for copper. The stress resulting from the mechanical simulation, with the strain applied, gives the EM-induced stress.

III. Performance and Reliability of the Copper TSVs

Using finite element tools, the performance and reliability of the simulated TSVs have been analyzed. The parasitic capacitance between the metal layer and the bulk silicon for each TSV is shown in Fig. 5. Even though the TSVs' depths are identical and the deposited oxide thickness is 500nm for each structure, there is some variation of the low-frequency capacitance values. However, at high frequencies, as the device enters the resistive mode of operation, the capacitance between the TSVs does not vary significantly.

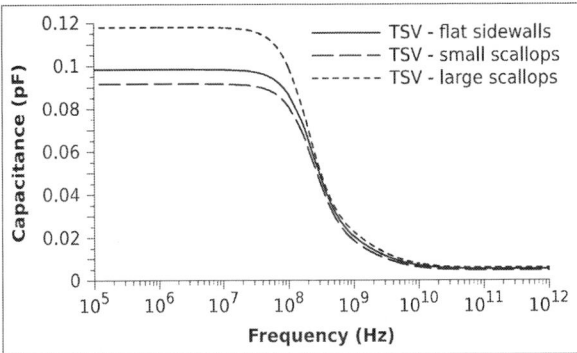

Fig. 5. Frequency-dependent capacitance (pF) for the three filled copper TSVs shown in Fig. 1.

The extracted capacitance, inductance, and TSV resistance are given in Table II. A copper electrical conductivity of 5.9987×10^7 S/m (at $20°C$) with a temperature-dependent resistivity of $0.0043°C^{-1}$ is used [8]. Relative permittivities of oxide and silicon, used for the capacitance simulation, are 4.2 and 11.7, respectively.

TABLE II.   ELECTRICAL RESULTS INCLUDING RESISTANCE, CAPACITANCE, AND INDUCTANCE FOR THE FILLED COPPER TSV

	Flat	Small scallops	Large scallops
Resistance	$116\mu\Omega$	$135\mu\Omega$	$98.8\mu\Omega$
Capacitance	98.35pF	91.65pF	117.92pF
Inductance	3.88fH	3.25fH	5.06fH

The resistance appears to increase for the small-scalloped structure, which is due to the tapered sidewalls during etching. The tapering results in a smaller copper volume and less area through which the current flux can propagate. The structure with large scallops shows a reduced resistance, while capacitance is significantly increased. The reduced resistance is due to the longer lateral etching time during processing, which resulted in a larger TSV diameter. The increased capacitance can be attributed to the thinning which occurs during oxide deposition around the large scallops. As scallops switch between convex and concave during material deposition, there is not a uniform thickness throughout the oxide, but rather regions of slightly thinner and slightly thicker oxide around each scallop.

A. Thermo-Mechanical Stress

The thermo-mechanical stress was analyzed for each structure by applying a $\Delta T = 300°C$ (temperature drop from $320°C$ to $20°C$). Assuming a stress-free temperature of $320°$, this analysis is meant to simulate the structure cooling after a thermal processing step. The stress builds up in the structures due to the variation in the coefficient of thermal expansion (CTE) between adjoining materials. The variations in the CTEs between different materials relevant for both the filled copper TSV and open tungsten TSV are shown in Table III. A linear elastic model is used for the metal layers during the mechanical simulation.

TABLE III.   COEFFICIENTS OF THERMAL EXPANSION FOR ALL RELEVANT MATERIALS

Material	Cu	W	Ta	SiO_2	Si	Si_3N_4
CTE (10^{-6}/K)	16.5	4.5	6.3	0.5	2.6	2.3

The maximum stress increases in the presence of scallops, while the average stress along the interfaces remains relatively unchanged. Fig. 6 shows the stress distribution along one-dimensional radial cut lines through the middle depth of the TSVs. The peak stresses observed in the scalloped structures correspond to the pinched area between two scallops. The pinched region between scallops absorbs the stress from the surrounding area, so that the stress is not uniform, as in the case of the flat TSV. The top of Fig. 6 shows the stress through a one-dimensional cut line going through a location where the scallops at the copper/tantalum and tantalum/oxide interfaces are pinched. Therefore, the Cu/Ta and Ta/SiO2 interfaces experience a spike in the thermo-mechanical stress. The bottom of Fig. 6 shows the stress where the scallops at the oxide and silicon interface are pinched. Similarly, the spike in thermo-mechanical stress is noted at the SiO2/Si interface.

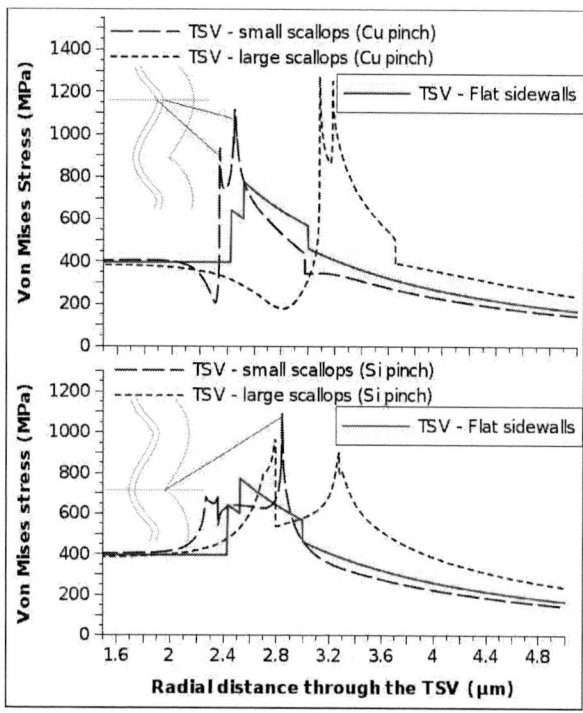

Fig. 6. Von Mises stress through the vertical middle of the filled copper TSVs assuming a drop from a stress-free temperature of 320°C to 20°C.

B. Electromigration-Induced Stress

Electromigration analyses were performed on the structures using a model presented in [7], with the resulting current density and EM-induced stress through the structures shown in Fig. 7 and Fig. 8, respectively. The increased current density along the scallop edges can be attributed to the thinning of the metal structure at the sidewalls due to tapering during processing, forcing more current to flow around the scallops.

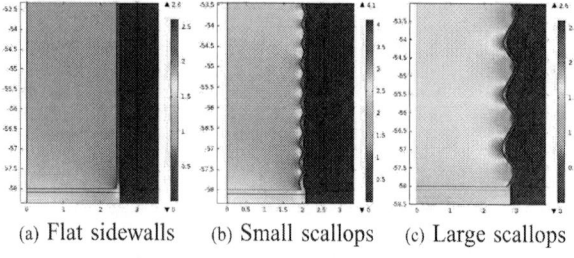

(a) Flat sidewalls (b) Small scallops (c) Large scallops

Fig. 7. Current density distribution (MA/cm²) in the three filled copper TSVs when a 1A current is applied through the top of the structure.

The stress in Fig. 8 is generated after operating the device at 1A for approximately 700hrs. The growth of the maximum EM-induced stress during the time-dependent simulation is shown in Fig. 9. The EM weakness at the bottom of the structure is due to the current flow being directed from the top, while the ground node is at the bottom. The vacancies build up around the anode end of the TSV at Cu/capping layer

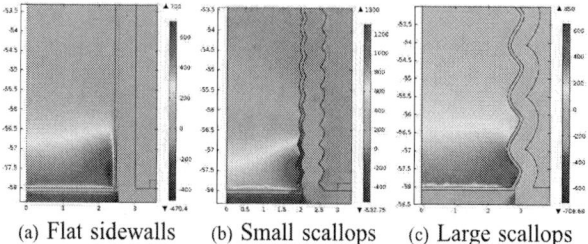

(a) Flat sidewalls (b) Small scallops (c) Large scallops

Fig. 8. Electromigration-induced stress (MPa) accumulation at the bottom of the three filled copper TSVs when a 1A current is applied through the top of the structure for 700hrs.

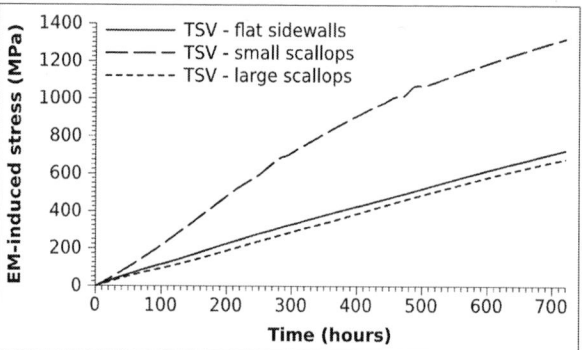

Fig. 9. Maximum EM-induced stress (MPa) through the filled copper TSVs during operation at a 1A current.

interface [9], which is at the TSV bottom in the given example.

The TSV with small-scalloped sidewalls experiences the highest current density through the copper because of the tapered sidewalls, resulting in a thinning metal layer. The increased current density leads to an increase in the vacancy concentration as given in (1), while the increased vacancy concentration results in an increased stress as given in (3). The flat TSV and the TSV with large-scalloped sidewalls display a very similar EM response. The large-scalloped TSV experiences the lowest stress due to its increased width, discussed earlier. The increased width results in a reduction in the current density, which in turn reduces the vacancy flux and stress generation.

IV. Performance and Reliability of the Tungsten TSVs

A similar analyses to that given for the performance and stress generation for filled copper TSVs is performed for open tungsten TSVs. The parasitic capacitance between the metal layer and the bulk silicon for each TSV is shown in Fig. 10. Once again, variation is observed in the low-frequency capacitance values, especially in the case of the large-scalloped TSV structure, while the high-frequency capacitance does not vary significantly.

The extracted capacitance, inductance, and TSV resistance are given in Table IV. The presence of scallops results in an increased resistance and inductance, with the values increasing further when the scallop size is increased. This increase occurs because the tungsten, which is deposited on top of a scalloped sidewall, will have a longer effective length through which

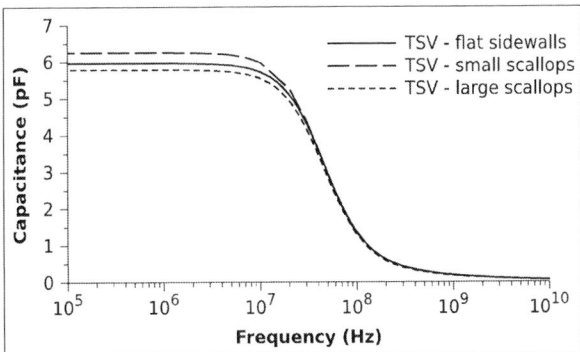

Fig. 10. Frequency-dependent capacitance (pF) for the three open tungsten TSVs shown in Fig. 3.

TABLE IV. ELECTRICAL RESULTS INCLUDING RESISTANCE, CAPACITANCE, AND INDUCTANCE FOR THE OPEN TUNGSTEN TSV

	Flat	Small scallops	Large scallops
Resistance	409mΩ	426mΩ	473mΩ
Capacitance	5.96pF	6.26pF	5.79pF
Inductance	3.49pH	3.61pH	3.86pH

current flows. The increased current path leads to an increased overall TSV resistance. The electrical conductivities of tungsten and aluminum used in this simulation are 1.3×10^7S/m and 3.5×10^7S/m, respectively. These values resulted in a good agreement with measured results for similar structures in [10]. The signal loss through the structure is also shown in Fig. 11, where no major variation can be observed between the tested structures as well as the structure measured in [10].

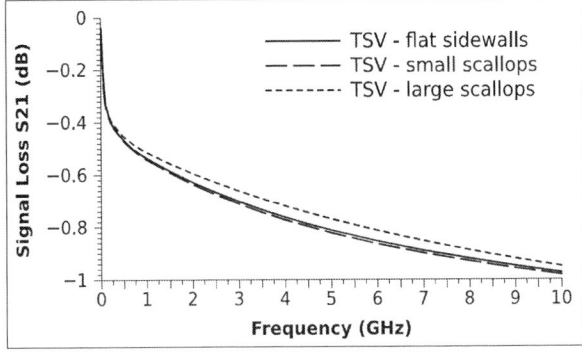

Fig. 11. Frequency dependence on the signal loss S21 (dB) for the three open tungsten TSVs. A higher loss is noted at high frequencies for the large-scalloped structure.

A. Thermo-Mechanical Stress

The thermo-mechanical stress build-up in the open tungsten TSV was also analyzed with a cool-down after a thermal processing step at a stress-free temperature of 320°C to 20°C. The CTEs for the relevant materials are given in Table III. The maximum stress once again increases in the presence of scallops, peaking at locations where two scallops meet, as shown in Fig. 12 by investigating the stress distributions along

one-dimensional radial cut lines through the TSV middle. The highest stress is seen in the tungsten layer, while the lowest is observed through the oxide. The TSV structure with flat sidewalls experiences a homogeneous stress through each material layer. However, the scalloped structures show a peak, which is higher than that noted in the flat TSV, but also a minimum which is below that noted in the flat TSV. The scalloped structures in the top graph in Fig. 12 show the stress peak at the location where two scallops at the SiO_2 liner/W interface meet. The stress peaks shown in the bottom graph in Fig. 12 occur where a scallop has its longest width, at the W/SiO_2 isolation interface.

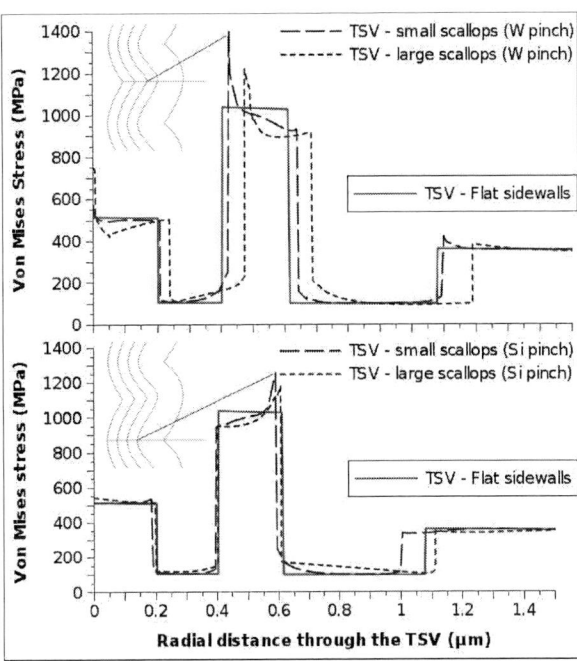

Fig. 12. Von Mises stress through the vertical middle of the open tungsten TSVs assuming a drop from a stress-free temperature of 320°C to 20°C.

B. Electromigration-Induced Stress

Electromigration analyses were performed on the structures using a model presented in [7], with the resulting current density and EM-induced stress through the structures shown in Fig. 13 and Fig. 14, respectively. There appears to be no major change in the current density distribution between the three structures, except for slight increases at locations where two scallops meet.

The stress in Fig. 14 is generated after operating the devices at 1A for approximately one year. The stress generated in the TSV is not affected by the scalloped sidewalls, because the aluminum layer is where EM effects are critical. The stress level after one year of operation is shown to be approximately 30MPa, which is significantly smaller than the levels seen for copper TSVs. This is mainly due to the large volume of aluminum used, which covers the entire bottom of the TSV. This ensures a low current density and low vacancy build-up in this layer. The thickness of the bottom aluminum layer plays a more significant role in determining the EM-induced stress,

978-1-4799-3911-4/14 $31.00 © 2014 IEEE

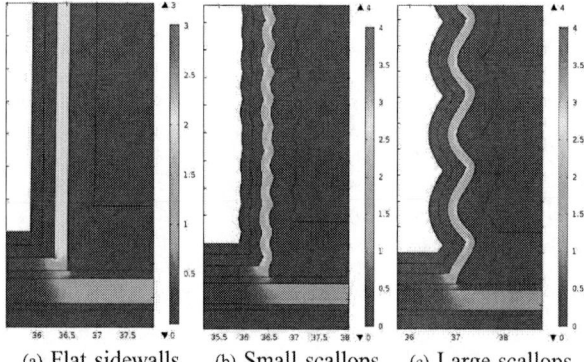

(a) Flat sidewalls (b) Small scallops (c) Large scallops

Fig. 13. Current density distribution (MA/cm^2) in the aluminum layer of the three open tungsten TSVs when a 1A current is applied through the structure.

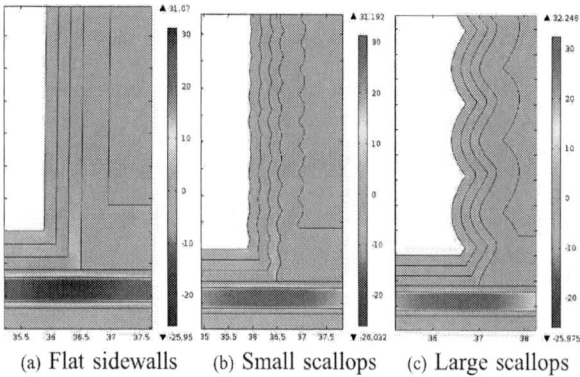

(a) Flat sidewalls (b) Small scallops (c) Large scallops

Fig. 14. Electromigration-induced stress (MPa) in the aluminum layer of the three open tungsten TSVs when a 1A current is applied through the structure for one year.

as shown in Fig. 15, where the stress build-up over time is shown. The use of a 20% thinner aluminum layer results in a 67% increase in the observed stress.

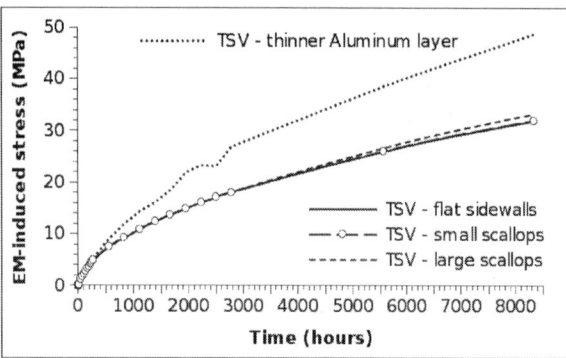

Fig. 15. Maximum EM-induced stress (MPa) through the aluminum layer in the open tungsten TSVs, while operating with a 1A current applied through the structure.

V. Conclusion

The performance of several TSVs has been tested through simulations. Two different TSV structures have been analyzed

and the effects of sidewall scallops which are generated during TSV processing on the TSVs' performance and reliability have been shown. The simulated topographies of the structures were imported into a finite element simulator in order to compare the performances of the different devices. Electrical parameter extraction showed that the TSV resistance, inductance, and capacitance are affected by the scalloped walls. The thermo-mechanical stress is also shown to be influenced by the process, while the EM-induced stress for the open tungsten TSV is independent of the sidewall structure, due to the stress developing in the bottom aluminum layer. Filled copper TSVs with scalloped sidewalls only experience an incresed EM-induced stress due to sidewall tapering, suggesting that the scallops themselves do not influence the EM response significantly.

Acknowledgment

The research leading to these results has received funding from the European Union Seventh Framework Programme under grant agreement no. 318458 SUPERTHEME.

References

1. "The International Technology Roadmap for Semiconductors (ITRS)," http://www.itrs.net/.

2. Li, R. *et al.*, "Continuous deep reactive ion etching of tapered via holes for three-dimensional integration," *J Micromech Microeng*, Vol. 18, No. 12 (2008) p. 125023(8pp).

3. Ranganathan, N. *et al.*, "Influence of Bosch etch process on electrical isolation of TSV structures," *IEEE Trans-CPMT*, Vol. 1, No. 10 (2011) pp. 1497–1507.

4. Filipovic, L., de Orio, R. L., and Selberherr, S., "Process and reliability of SF6/O2 plasma etched copper TSVs," *Proc 15th International Conference on Thermal, Mechanical and Multi-Physics Simulation and Experiments in Microelectronics and Micro-Systems Conf.*, Gent, Belgium, Apr. 2014, 4pp.

5. Ertl, O. and Selberherr, S., "A fast level set framework for large three-dimensional topography simulations," *Comput Phys Commun*, Vol. 180, No. 8 (2009), pp. 1242–1250.

6. Wolf, M.J. *et al.*, "High aspect ratio TSV copper filling with different seed layers," *Proc 58th Electronic Components and Technology Conf.*, Buena Vista, FL, May. 2008, pp. 563–570.

7. de Orio, R.L., Ceric, H., and Selberherr, S., "A compact model for early electromigration failures of copper dual-damascene interconnects," *Microelectron Reliab*, Vol. 51, No. 9 (2011) pp. 1573–1577.

8. Bai, G. *et al.*, "Copper interconnection deposition techniques and integration," *1996 Symposium on VLSI Technology. Digest of Technical Papers*, Honolulu, HI, June. 1996, pp. 48–49.

9. Lane, M.W., Liniger, E.G., and Lloyd, J.R., "Relationship between interfacial adhesion and electromigration in Cu metallization," *J Appl Phys*, Vol. 93, No. 3 (2003) pp. 1417–1421.

10. Roger, F. *et al.*, "TCAD electrical parameters extraction on through silicon via (TSV) structures in a 0.35 μm analog mixed-signal CMOS," *Proc International Conference on Simulations of Semiconductor Processes and Devices Conf.*, Denver, CO, Sept. 2012, pp. 380–383.

978-1-4799-3911-4/14 $31.00 © 2014 IEEE

Imaging of Through-Silicon Vias using X-Ray Computed Tomography

J.P. Gambino, W. Bowe, D. M. Bronson, S. A. Adderly

IBM Microelectronics, 1000 River Street, Essex Junction, VT, 05452, USA

Phone: (802) 769-1438   Fax: (802) 769-9452   Email: shawn.adderly@us.ibm.com

Abstract- X-Ray computed tomography (CT) can be useful in evaluating defects in through-silicon vias (TSVs). X-Ray CT images of two different TSV processes are presented; copper TSVs used for stacked memory on logic and tungsten TSVs used for power amplifiers. It is found that TSVs in the edge exclusion region are susceptible to defects from the TSV etch and TSV metallization processes.

Keywords – TSV, X-Ray Tomography, TSV Failure Analysis

I. INTRODUCTION

Through-silicon vias (TSV) are used in many applications, including CMOS image sensors, Si interposers, and stacked memory. These applications can be broadly classified into three types [1]. The first type is a simple vertical connection to back of the wafer, with no die stacking. Some examples are CMOS image sensors and SiGe power amplifiers. The second type is 2.5D integration, where die are attached to a Si interposer, with TSVs in the interposer. An example is a multiple field programmable gate arrays (FPGAs) on a silicon interposer. The third type is 3D integration, where die are stacked and TSVs are in active die. Examples of this approach are memory on memory or memory on logic (Fig. 1).

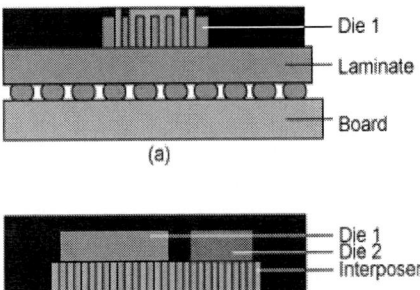

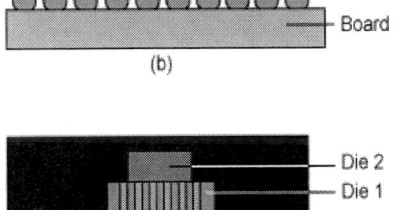

Figure 1: Applications for TSVs; (a) TSV is in an active die to make a connection to a bond pad on backside of wafer; (b) TSV is used in an interposer; (c) TSV is in an active die to allow die stacking.

Failure analysis and construction analysis of TSVs is challenging because of the large dimensions. Fails can occur at the top of the TSV, the bottom of the TSV, or in the middle of the TSV. A typical TSV height is 50 to 100 μm and may include an array of TSVs connected together. Thus requiring that many different parts of the TSV must be examined when there is a fail.

Novel failure analysis techniques are required to ensure that TSVs are etched to the proper depth and filled properly. Traditional techniques 2-D cannot provide information about the formation of the TSV across the device wafer, nor do they provide topographic information about the TSV. Recently it has been shown that X-ray computed tomography (CT) can be used to characterize voids in copper-filled TSVs. In this report, we show that X-ray CT is useful for defects in TSVs in the edge exclusion region of the wafer.

A. Tungsten TSVs for SiGe power amplifiers

Mobile phones consist of a number of building blocks, including the front-end module, the baseband processor, and the applications processor (Fig. 2a) [2]. The front end module connects the baseband processor to the antenna. Incoming signals are amplified by the low noise amplifier (LNA) and sent to the baseband processor. Outgoing signals are amplified by the power amplifier (PA) and transmitted by the antenna (Fig. 2b).

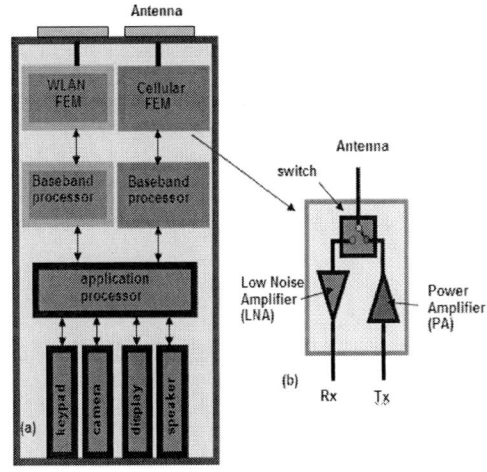

Figure 2: (a) Block diagram of mobile phone, showing front-end modules (FEM), processors, and other components. (b) Schematic of front-end module showing low noise amplifier (LNA) in the receiver circuit and power amplifier (PA) in the transmitter circuit [2].

Historically, the front end module consisted of multiple die, such as GaAs for the power amplifier and CMOS for the control circuits and the switches [3]. However, improvements in SiGe bipolar technology have allowed Si-based bipolar transistors to replace GaAs for the some applications, such as wireless local area network (WLAN) transceivers. As a result, the front end module can be fabricated in one BiCMOS die rather than in multiple die, resulting in cost savings and a smaller form factor. One way to improve the performance of SiGe bipolar transistors is to use through silicon vias (TSVs), to form the ground connection for the emitter of the npn devices [3]. An important parameter for power amplifiers is the power gain of the transistor. Power gain is important for providing the maximum transmitted signal (for example, when the cell phone tower is far away), while minimizing power consumption (i.e., to allow long battery life). The power gain of Si bipolar transistors in common-emitter circuits is given by equation 1 [4].

$$\text{Gain} \sim 1 / \omega^2 C_{CB} L_E \qquad [1]$$

Where ω is the operating frequency, C_{CB} is the collector-base capacitance, and L_E is the emitter inductance to ground. Note that gain is inversely proportional to the emitter inductance; hence gain can be increased by reducing the emitter inductance. TSVs allow fabrication of power amplifiers with low inductance, while taking advantage of the low cost and good thermal properties of a wirebond package. The TSVs in this application are very simple compared to those used in 3D-IC technology, since they only provide a ground connection (i.e. the TSVs are not used for power or signal lines). Hence, the TSVs can be grounded to the substrate, eliminating the need for a dielectric liner inside the vias [5,6].

B. Copper TSVs for interposers and 3D integration

A common example of 3D integration using copper TSVs is a stack of one or more memory die connected to a processor (Fig. 3) [7,8]. Stacking the die provides increased bandwidth between the processor and the memory, resulting in improved system performance.

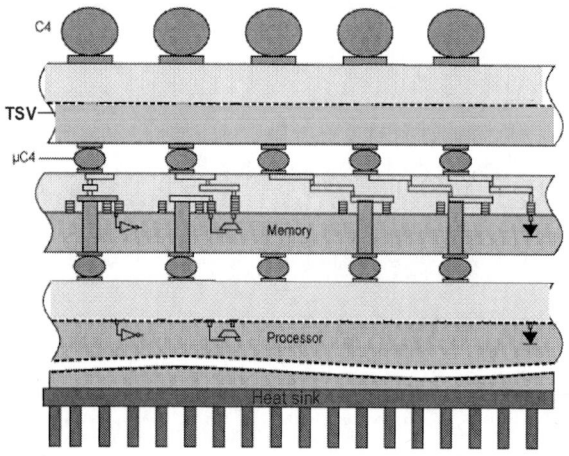

Figure 3: Schematic of memory die stacked on a processor [7].

C. Edge exclusion

Reducing the wafer edge exclusion results in an increase in usual die per wafer, and can enhance the yield of the "former" edge-most region [9]. For example, for a 200mm wafer, moving from a 3mm edge exclusion to a 2mm edge exclusion, results in an increased die area per wafer of ~ 2% (Fig. 4) [10].

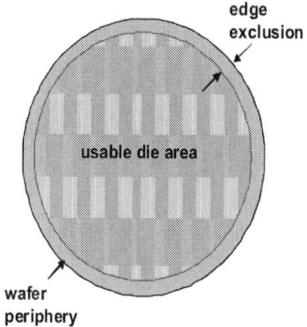

Figure 4: Schematic of Si wafer showing useable die area and edge exclusion region.

Of course, there are many challenges to reducing the size of the edge exclusion regions, including wafer edge flatness, lithography, and CMP. For TSVs, some of the challenges of reducing the edge exclusion region are metal fill and TSV RIE. Metal deposition is undesirable on the bevel region of the wafer, because of the possibility of cross-contamination on chamber surfaces and robot handlers. Hence, metal deposition is typically prevented at the edge of the wafer. To reduce the size of the edge exclusion region requires improved control of the diameter where metal deposition is prevented.

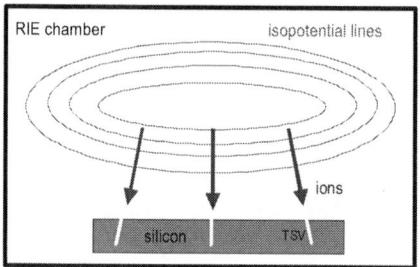

Figure 5: Schematic of RIE chamber used for TSV RIE. The non-vertical ion trajectory at the edge of the wafer results in tilted TSVs.

TSV etch is challenging at the edge of the wafer because of the shape of the isopotential lines in the RIE tool (Fig. 5) [11]. Ions impinge at a non-vertical angle at the edge of the wafer, resulting in tilting of TSVs.

D. X-ray computed tomography

X-Ray Computed Tomography (CT) has been widely used for failure analysis of packaged microelectronic devices [12-14]. X-Ray CT provides a non-destructive method to characterize submicron defects, and has been used to characterize C4 bumps,

978-1-4799-3911-4/14 $31.00 © 2014 IEEE

BGA solder, wirebonds, and die cracking [12]. Recently, x-ray CT has been used to characterize voids in TSVs [15,16].

X-ray CT consists of an x-ray source, a rotating stage, and a detector [12]. The stage rotates the sample through 180° in equally spaced angles and the detector collects a two dimensional image at each angle (Fig. 6). The 2D images are then superimposed and processed to form a 3D image of the sample.

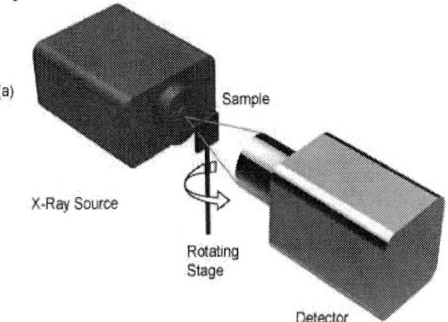

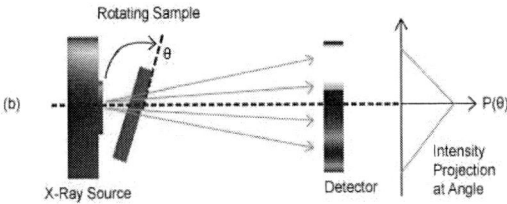

Figure 6: (a) Schematic of x-ray CT system. (b) The projection of 2D images are collected at each angle and superimposed into a 3D image [12].

II. EXPERIMENT

A. Tungsten TSVs for power amplifiers

Tungsten-filled, grounded TSVs were fabricated in 200 mm silicon wafers using a standard BiCMOS process flow (Fig. 7) [5,6]. Tungsten was chosen as the TSV conductor because it

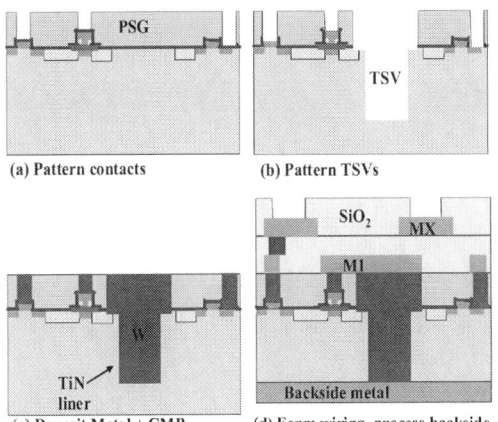

Figure 7: Schematic of process flow showing (a) contact patterning, (b) TSV patterning, (c) contact and TSV metallization, and (d) backside grind and metallization.

can fill high aspect ratio (50:1) vias sufficiently and it has low chemical reactivity with Si (i.e., compared to Cu). The TSVs are formed as part of the contact module, after the BiCMOS devices are fabricated. The TSV process consists of contact patterning (Fig. 7a), TSV patterning (Fig. 7b), Ti/TiN liner deposition followed by W deposition, and finally W chemical mechanical polishing (CMP) (Fig. 7c). The TSV lateral dimensions are ~ 4 μm x 51 μm, with a nominal depth of 100 or 150 μm. Note that the TSVs are metallized simultaneously with the contacts. The TiN liner and bulk W are deposited by chemical vapor deposition (CVD), to minimize void formation in the TSV (Fig. 8). After the completion of back-end-of-the-line (BEOL) wafer processing, the wafer backside is ground and polished, to either 100 μm or 150 μm, and the backside is metallized prior to dicing (Fig. 7d).

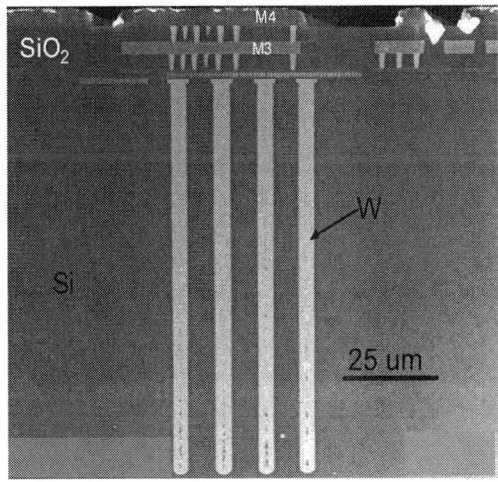

Figure 8: SEM cross-section of an array of tungsten-filled TSVs prior to wafer thinning.

B. Copper TSV for memory stacked on logic

The copper TSVs were fabricated at the fatwire levels in a 45nm CMOS process [8]. The process consists of TSV patterning, deposition of a conformal oxide, sputter

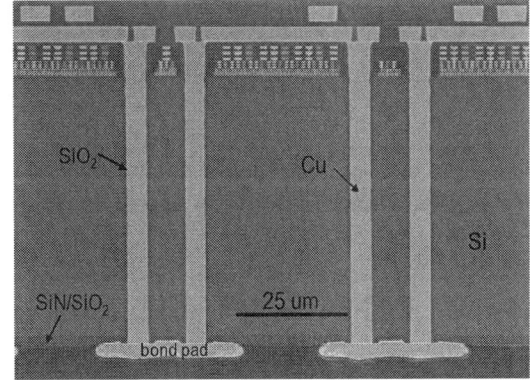

Figure 9: SEM cross-section of copper-filled TSVs after wafer thinning and backside bond pad formation.

978-1-4799-3911-4/14 $31.00 © 2014 IEEE

deposition of the barrier and seed layers, Cu plating, and Cu CMP. Additional wiring layers are built after TSV processing. After frontside processing, wafers are bonded to glass handlers, and thinned by grinding, polishing, and etching. Backside oxide/nitride are deposited, followed by TSV metal exposure. The final steps are fabrication of a backside redistribution level and bond pads (Fig. 9).

C. X-ray CT measurements

The X-ray CT images were obtained using an Xradia 3D tomographic x-ray microscope, with a 60KeV x-ray beam at 6W power. The resolution is ~ 1 μm. Analysis of a TSV using X-ray CT is the similar to using it for packaging analysis. However, in the case of a TSV, the material of interest was known, and it was possible to filter the x-ray source to get the best material contrast of the interconnect. 2-D X-rays were used to look for changes in contrast in the TSVs. When the TSVs of a different brightness were detected, X-ray CT was used to obtain 3D images. Sample preparation for 3-D X-ray tomography is done by cleaving a piece of the substrate in the area of interest, and mounting it to a silicon slide using wax.

III. RESULTS AND DISCUSSION

X-ray CT images of tungsten TSVs in the center of the wafer show well formed TSV arrays (Fig. 10). In contrast, images from the extreme edge of the wafer show partially filled TSVs (Fig. 11). SEM cross-sections from the edge exclusion region confirm that the tungsten fill is inadequate (Fig. 12).

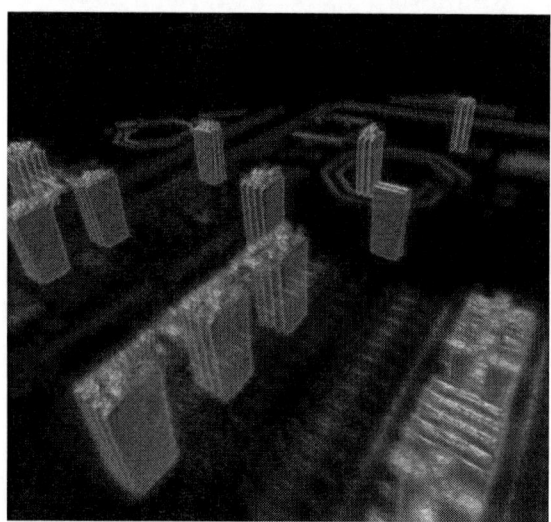

Figure 10: X-ray CT image of tungsten-filled TSVs at the center of the wafer.

Optimization of the tungsten CVD process is required to ensure good fill of the TSV is achieved at the extreme edge of the wafer.

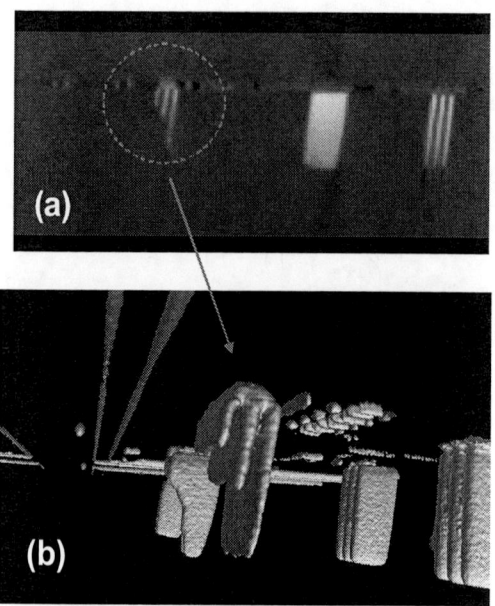

Figure 11: X-Ray CT image of tungsten-filled TSVs in the edge exclusion region of the wafer, (a) cross-section and (b) 3D image.

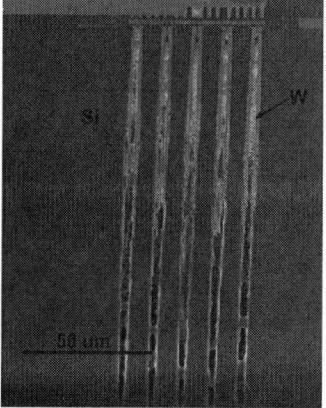

Figure 12: SEM cross-section of tungsten-filled TSVs in the edge exclusion region of the wafer.

Figure 13: X-Ray CT image of copper-filled TSVs at the center of the wafer.

X-ray CT images of copper-filled TSVs in the center of the wafer show well formed TSV arrays (Fig. 13). In contrast, images from the extreme edge of the wafer show partially filled TSVs (Fig. 14).

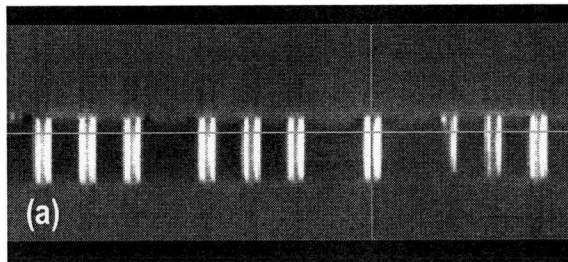

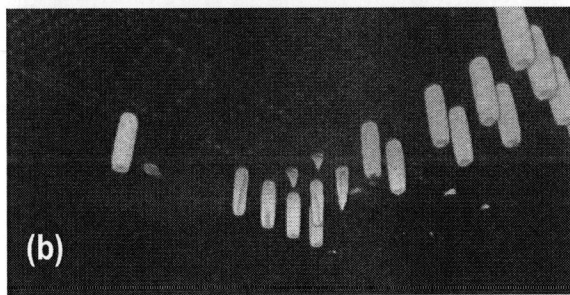

Figure 14: X-ray CT image of copper-filled TSVs in the edge exclusion region of the wafer. (a) cross-section and (b) 3D image.

These results suggest that X-ray CT measurements can be a useful tool for yield learning as well as for failure analysis

IV. CONCLUSION

X-Ray computed tomography (CT) is useful in evaluating defects in through-silicon vias (TSVs). X-Ray CT images of two different TSV processes are presented; copper TSVs used for stacked memory on logic and tungsten TSVs used for power amplifiers. It is found that TSVs in the edge exclusion region are susceptible to defects from the TSV etch and TSV metallization processes.

ACKNOWLEDGMENT

The authors acknowledge the staff of the IBM Microelectronics manufacturing facilities in Essex Junction, VT and East Fishkill, NY, for their assistance with processing the wafers. We would also like to thank the IBM failure analysis team and for their assistance. Additionally we would like to thank Sue Feng for her assistance in rendering several of the figures used in this paper.

REFERENCES

[1] J.P. Gambino, S.A. Adderly, J.U. Knickerbocker, "A Review of Through Silicon Via Manufacturing Challenges for Vertical Chip Integration", *to be published in Microelectronic Engineering*, 2014.

[2] P. Li, P. DeCarlo, "WiFi Front End Module Design for Smart Phone Applications", *Solid State and Integrated Circuit Technology (ICSICT)*, 2010, pp. 1-4.

[3] P. Zampardi, "Performance and Modeling of Si and SiGe for Power Amplifiers", *Silicon Monolithic Integrated Circuits in RF Systems*, 2007, pp. 13-17.

[4] P.H.C. Magnee, F. van Rijs, R. Dekker, D. M. H. Hartskeerl, A. L. A. M. Kemmeren, R. Koste, H. G. A. Huizing, "Enhanced RF Power Gain by Eliminating the Emitter Bondwire Inductance in Emitter Plug Grounded Mounted Bipolar Transistors", *Bipolar/BiCMOS Circuits and Technology Meeting (BCTM)*, 2000, pp. 199-202.

[5] A.K. Stamper, P. Andry, M. Erturk, R. Groves, A. Joseph, P. Lindgren, P. Mclaughlin, R. Previti-Kelly, E. Sprogis, K. Stein, C.K. Tsang, P. Wang,, J. Dunn, "Through Silicon Via Integration in CMOS and BiCMOS Technologies", *Advanced Metallization Conference (AMC 2008)*, MRS 2009, pp. 495-500.

[6] R. Malladi, A. Joseph, P. Lindgren, W. Ni, D. Wang, Hanyi Ding, M. Erturk, R. Previti-Kelly, "3D Integration Techniques Applied to SiGe Power Amplifiers", *ECS Trans.*, 16(10), pp. 1053-1067.

[7] M. Wordeman, J. Silberman, G. Maier, M. Scheuermann, "A 3D System Prototype of an eDRAM Cache Stacked Over Processor-Like Logic Using Through-Silicon Vias", *Proc. IEEE Int. Solid-State Circ. Conf.*, 2012, pp. 186-187.

[8] M. G. Farooq, T. L. Graves-Abe, W. F. Landers, C. Kothandaraman, B. A. Himmel, P. S. Andry, C. K.Tsang, E. Sprogis, R. P. Volant, K. S. Petrarca, K. R. Winstel, J. M. Safran, T. D. Sullivan, F. Chen, M. J. Shapiro, R. Hannon, R. Liptak, D. Berger, S. S. Iyer, "3D Copper TSV Integration, Testing and Reliability", *Proc. IEEE Int. Elec. Dev. Meeting*, 2011, pp. 143-146.

[9] T. Tran, W. Roberts, J. Tiffany, I. Jekauc, N. Clements, P. Jowett, R. Ferguson, D. Mattson, C. Demmert, M. Richmond, C. Wiendl, M. Bruno, A. Brock, T. Taylor, "Extreme Edge Engineering – 2mm Edge Exclusion Challenges and Cost-Effective Solutions for Yield Enhancement in High Volume Manufacturing for 200 and 300mm Wafer Fabs", *Proc, IEEE/SEMI Advanced Semicond. Manuf. Conf.*, 2004, pp. 453-460.

[10] K.W. Lee, " New Edge Exclusion Proposal", presented at Semicon Taiwan, 2013.

[11] B. Wu, A. Kumar, and S. Pamarthy, "High aspect ratio silicon etch: A review", *J. Appl. Phys.* 108, 2010, p. 051101.

[12] M. Pacheco, D. Goyal, "New Developments in High-Resolution X-ray Computed Tomography for Non-Destructive Defect Detection in Next Generation Package Technologies", *Proc. ISTFA*, 2008, pp. 30-35.

[13] M. Oppermann, T. Zerna, K. J. Wolter, "X-ray Computed Tomography on Miniaturized Solder Joints for Nano Packaging", *Proc. IEEE Elec. Packaging Tech. Conf.*, 2009, pp. 70-75.

[14] M. Oppermann, T. Zerna, K.J. Wolter, "X-ray Computed Tomography for Nano Packaging – A Progressive NDE Method", *Proc. IEEE Elec. Packaging Tech. Conf.*, 2010, pp. 853-858.

[15] V.N. Sekhar, S. Neo, L.H. Yu, A.D. Trigg, C. C. Kuo, "Non-Destructive Testing of a High Dense Small Dimension Through Silicon Via (TSV) Array Structures by Using 3D X-ray Computed Tomography Method (CT scan)", *Proc. IEEE Elec. Packaging Tech. Conf.*, 2010, pp. 462-466.

[16] L.W. Kong, J.R. Lloyd, M. Liehr, A.C. Rudack, S. Arkalgud, A.C. Diebold, "Measuring Thermally Induced Void Growth in Conformally Filled Through-Silicon Vias (TSVs) by Laboratory X-ray Microscope", SPIE Proc., vol. 8324, 2012, p. 832412.

The Observation of Mobile Ion of 40nm node by Triangular Voltage Sweep

Clement Huang, James W. Liang, Alex Juan and K. C. Su

Reliability Technology & Assurance Division, UMC Inc.

No.3, Li-Hsin Rd. II, Hsinchu Science Park, Taiwan 300, ROC

Phone: +886-935336621; Fax: +886-3-5630602; e-mail: clement_huang@umc.com

Abstract—Using the different test structures, we investigated the Triangular Voltage Sweep (TVS) variables, e.g. temperature, capacitor area & voltage sweep rate to observe the mobile ion in dielectric layer for Back-end process. We found that temperature 125°C could activate mobile ions. The amount of mobile ion is strongly correlated with tested topology. The amount of mobile ion is also dependent on the voltage sweep rate. Time-dependent dielectric breakdown (TDDB) lifetime will reduce 1 order when mobile ion concentration raise 1 order. It is extremely important to specify the reasonable dielectric area of test structure (intra-metal comb length) for both of TDDB and TVS.

Keywords: cupper, Triangular Voltage Sweep, Time-dependent dielectric breakdown, voltage breakdown, Bias temperature stress, mobile ion, voltage sweep rate, capacitor area, temperature, intra-metal, comb structure.

I. INTRODUCTION

With future interconnect size reduction, time-dependent dielectric breakdown (TDDB) will become the most serious problem of copper (Cu) interconnect reliability. Many people believe that the ion drift is the root cause of TDDB failures because Cu ions are fast diffusers. However, some researchers have the opinion that since the optimized barrier structures can completely suppress Cu ion drift, the intrinsic breakdown occurs by thermochemical Si-O bond breakage under high electric fields even for the backend [1]. Table1 summaries the two different familiar models for TDDB.

Model	Factor	Equation
E	thermochemical *Si-O bond breakage*	$TF \propto Exp(-\gamma E)$
√E	1. Schottky or Poole-Frenkel conduction 2. *ion drift (Cu diffusion)*	$TF \propto Exp(-\alpha\sqrt{E})$

Table.1. Two familiar models for TDDB.

To investigate the TDDB mechanism, Bias Temperature Stress (BTS) & Triangular Voltage Sweep (TVS) has been frequently used, and TVS is better than BTS. TVS has four advantages over the BTS method: 1. It determines the mobile

charges without interference from the interface trap charges. 2. It can determine the type of ion (such as Cu+ or H+ [2,3]) that is contaminating the oxide, because the peak current for different ions occurs at different biases. 3. It provides measurements an order of magnitude more sensitive than BTS. 4. It is faster than the BTS method, since the device only needs heating once.

II. EXPERIMENTAL

The wafers were fabricated using a 40nm node standard damascene process and copper metallization process based on low-k technology with dielectric constant 2.5 material. The cap material on top of the dielectric is Nitride-based etch stop layer SiCN. The "Intra metal-to-metal" comb structures contain narrow metal lines was introduced to investigate the mobile ion effect on dielectric TDDB reliability (Fig.1). The variables of this experiment are:

(a) Metal layer: Metal1, Metal2; (b) Temperature: 25~300°C; (c) Voltage sweep step: 0.05 ~ 1.00V; (d) Capacitor area : 1X ~ 1000X (X is the normalized multiple).

Fig.1. Top view of "Intra metal-to-metal" comb structures.

TVS experiments were performed on the comb structures using an HP4070 meter. A TVS scan from -15V to +15V was performed at different temperatures ranging from 25 to 300°C. The different scan rates from 0.05 to 1V/s were also evaluated. Fig.2 illustrates the TVS measurement method. A negative (or positive) bias was applied to one side of the capacitor and hold for a few seconds, then sweeping the voltage at slow linear rates. The Voltage break-down (BVD) and TDDB test was also performed by the same test pattern for various capacitor area from 1X to 1000X. After BVD finished, TDDB would be executed under a constant electric field.

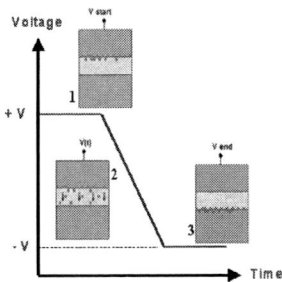

Fig.2. TVS measurement method.

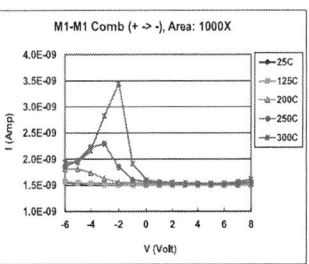

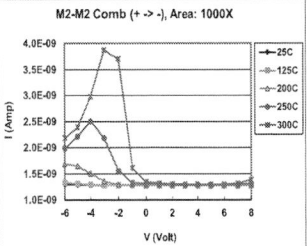

Fig. 3. TVS test results for the intra-metal comb structure capacitor

III. RESULTS AND DISCUSSION

The temperature effect

In the TVS test procedure, a negative voltage (v) was applied to the capacitor, and swept the voltage with a slow linear rate a=dv/dt from a negative to a positive voltage. When the extra ions existed, the coherent motion of accumulated ions contributed to the distortion of the sensed leakage current. The sensed leakage current (Amp) can be described as Eq(1):

$$I_{measured} = I_{displacement} + I_{leak} + I_{ionic} \qquad Eq(1)$$

The ion concentration Qm (ions/cm^2) measured using TVS can be described as Eq(2):

$$Qm = \int \frac{I(v)dv}{aqA} \qquad Eq(2)$$

Where v is the sweep voltage; I is the measured current; a is the sweep rate; A is capacitor area; and q is the electron charge (1.6E-19 coulomb). Suppose the kinetics of the Cu ion movement is described as either an ionization or de-trapping process (determined by the numbers of ions evaluated), activation energy (Ea) can be extracted from the Arrhenius equation [4].

In terms of temperature, Fig.3 shows the difference of M1 & M2 comb structure capacitor for +/-15V TVS test. Higher temperature has higher mobile ion peak, and the peaks disappear when the temperature is below 125°C. That means the activation temperature that drives the free ion movement is larger than 125°C. The peaks are reversible when the temperature decreased. The peak area represented the free ion concentration. At the same temperature (e.g. 300C), the mobile ion peak of M1 comb is a little higher than M2. Because IMD2 thickness is thicker than IMD2 (IMD1:1250A; IMD2:1400A), so M2 comb structure capacitor could contains more mobile ion and induce higher electric current during TVS test. The Ea value, 0.435ev and 0.433eV, individually obtained from the M1 comb and M2 comb capacitor, are similar, as shown in Fig.4.

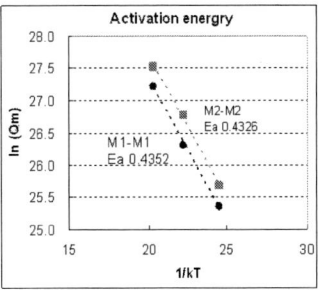

Fig. 4. Activation energy extracted from M1, M2 comb capacitor.

The voltage sweep step effect

Fig.5 shows the TVS results for different sweep rates for M1 comb structure. Higher sweep rate has higher mobile ion peak. As the sweep rate decreased, the measured low-level excess current decreased and shifted about 1V (from -3V to -2V). It implies if there are 2 peaks overlap at one specific voltage, we can change sweep rate to separate the peaks [3]. Though the high sweep rate could obtain higher mobile ion peak, it could also speed up the dielectric breakdown [5].

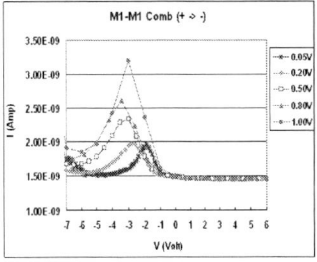

Fig. 5. TVS test by different voltage sweep rate

The capacitor area effect

Regarding to the mobile ion amount should be proportional to the volume of the intra-metal dielectric, we use metal comb structure with different length to check TVS

978-1-4799-3911-4/14 $31.00 © 2014 IEEE   333

performance with various capacitor area. Fig.6 indicates that TVS peak only detected at 1000X capacitor area. Fig.7 shows the correlation between capacitor area and the mobile ion amount inside the dielectric. The Qm was calculated from the Eq.2 with different areas - 1X ~ 10000X. The results showed that total mobile ion concentration was strongly affected by the capacitor area, especially above a critical value. From critical capacitor area experiment we conclude that if area > 100X, the detect sensitivity of mobile ion is higher.

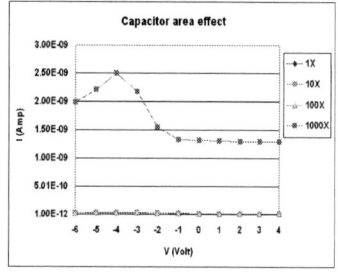

Fig.6. TVS test by different metal comb structure capacitor area.

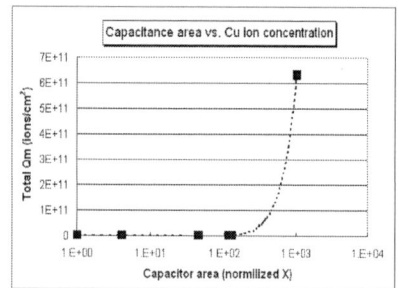

Fig.7. The capacitor area correlated with the ion amounts

The correlation between TVS and IMD TDDB

Fig.8 shows the relationship between IMD BVD/TDDB and capacitor area. I-V curve of BVD for 1000X capacitor is 15% worse than 1X, and the TDDB mean-time-to-failure (MTTF) is 1 order lower than 1X. Bigger capacitor area has more dielectric material, so it could contains more mobile ions and induce higher electric current during the test. This means that while the capacitor is under TDDB testing, the free movable ions traveled toward one side of the capacitor and decreased the effective capacitor spacing [5]. The mobile ion remaining on the dielectric interface (such as Cu residue) degraded the IMD TDDB performance. The higher total mobile ions concentration, the lower BVD/TDDB performance. While the performance of IMD TDDB is inversely proportional to capacitor area, but on the contrary TVS (total Qm) is proportional to capacitor area (if Cu dual-damascene process is well control and no excursion occur). From Fig.9, it is obvious that TDDB lifetime soon degraded with the total mobile ion concentration (TDDB reduce 1 order if mobile ion concentration raise 1 order). So it is extremely important to specify the reasonable dielectric area of test structure (intra-metal comb length) for both of TDDB and TVS.

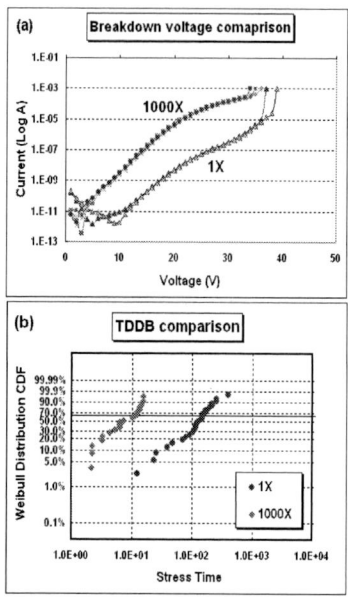

Fig.8. The IMD reliability for different capacitor area: (a)I-V curve of BVD; (b) TDDB distribution.

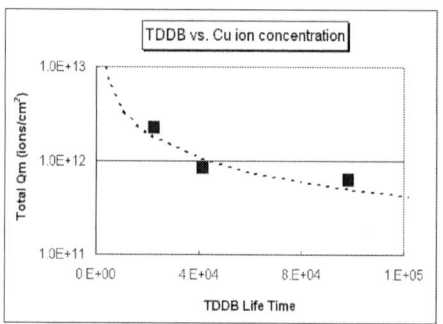

Fig.9. The correlation between TVS and IMD TDDB life time

IV. CONCLUSION

We investigated the TVS variables, e.g. temperature, capacitor area & voltage sweep rate: T>125°C to activate mobile ions. The amount of mobile ion is strongly correlated with tested topology (metal spacing ↓ or capacitor area ↑, mobile ion ↑). The amount of mobile ion is also dependent on the voltage sweep rate (sweep rate ↑, mobile ion ↑). TDDB lifetime reduce 1 order when mobile ion concentration raise 1 order. It is extremely important to specify the reasonable dielectric area of test structure (intra-metal comb length) for both of TDDB and TVS.

REFERENCES

[1] H. Miyazaki et al., "The Observation of Stress-Induced Leakage Current of Damascene Interconnects after Bias Temperature Aging", International Reliability Physics Symposium, pp. 150- 197, 2008.

[2] I. Ciofi et al., "Water and copper contamination in SiOC:H damascene: novel characterization methodology based on

triangular voltage sweep measurements", IEEE-IITC Proceedings, pp. 181-183, 2006.

[3] L. Anderson et al., "A case study in a 100-spl times-reduction in sodium ions in a 0.8 -spl mu-m BiCMOS process using triangular voltage sweep", IEEE-IRW, pp. 45-48, 1996.

[4] A. MALLIKARJUNAN et al., "Mobile Ion Detection in Organosiloxane Polymer Using Triangular Voltage Sweep", Journal of The Electrochemical Society, pp 155-159, 2002.

[5] L. Anderson et al., "An efficient approach to quantify the impact of Cu residue on ELK TDDB", International Reliability Physics Symposium, pp. 619-623, 2009.

Electromigration Reliability of Solder Bumps

Hajdin Ceric[a,b] and Siegfried Selberherr[a]

[a]Institute for Microelectronics, Technische Universität Wien, Gußhausstraße 27-29/E360, 1040 Wien, Austria
[b]Christian Doppler Laboratory for Reliability Issues in Microelectronics

Abstract

Characteristic of solder bumps is that during technology processing and usage their material composition changes. We present a model for describing a growth of intermetallic compound inside a solder bump under the influence of electromigration. Simulation results based on our model are discussed in conjunction with corresponding experimental findings.

I. Introduction

For the realization of modern three-dimensional (3D) integrated circuits new interconnect structures such as through-silicon-vias (TSVs) and solder bumps, together with complex multi-level 3D interconnect structures are gaining importance. The application of new structures and materials inevitably introduces new reliability issues. The interconnect reliability is affected by degradation processes induced by thermal gradients, electromigration (EM), and stressmigration. Solder bumps are important components for 3D integration, because they enable vertical stacking of wafers. Pure Sn has been identified as the best Pb-free solder for ultra fine pitch solder bumps for advanced 3D interconnect applications due to its baseline advantages of being electrodeposited and exhibiting a low melting temperature. Failure analyses have shown that

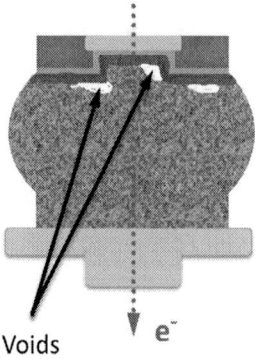

Fig. 1. EM failure is caused by voids which are formed between Ni under bump metalization (UBM) and Sn solder bump.

failures in Sn bumps occur by EM induced voiding at the interface between the intermetallic compound (IMC) and the solder (cf. Fig. 1). EM in Sn-based solder bumps is much more complicated than EM in copper due to the presence of impurities. The development of a failure in a copper interconnect takes place in two distinctive phases: a void nucleation phase and a void evolution phase. During the first phase practically no resistance increase can be measured. The situation is quite different in the case of EM failure development in a Sn bump, where an IMC layer is also present [1]. From the beginning of EM stressing a continuous growth of the bump resistance (cf. Fig. 2) is observed. After a certain period of EM stressing, the bump resistance starts to rise with a significantly steeper slope. Chen *et al.* [1] assume that the two slopes of

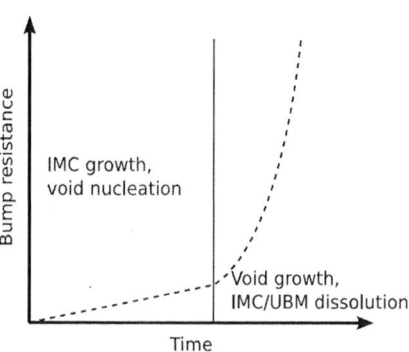

Fig. 2. Resistance change due to IMC growth and voiding with two different slopes.

the resistance growth represent two different stages of failure development: void nucleation combined with IMC growth and void propagation with IMC dissolution. The investigation of the physical mechanisms behind such a failure behavior is the main subject of this work.

Important for the layouts attached to the solder bumps is an under bump metallization (UBM), which separates the Sn bump from the surrounding metallization. The solder bump with UBM has a lower maximum current density and peak temperature in the solder, which contributes to longer EM lifetimes.

II. Physics of Intermetallic Compound Growth

The solder bump itself is usually designed and realized as an alloy, for example as SnAg, SAC405 (Sn-4% Ag-0.5% Cu) [2]. A solder bump interface to the UBM, which is usually made of Ni, is also a reason for the development of alloys in solder bumps. At this interface an IMC is formed, which is a thin layer consisting of alloys with Sn as the principal component.

Until now several attempts have been published to model EM induced IMC development [3], [4], however, none of them is applicable for numerical simulation. The growth of the IMC is determined by atomic fluxes of different material components ($\vec{J}_{Sn}$, $\vec{J}_{Cu}$, and $\vec{J}_{Ni}$). The fluxes are driven by gradients of chemical potentials (μ_{Sn}, μ_{Cu}, and μ_{Ni}) and EM. In each of the chemical potentials an impact of the local mechanical stress and the local concentration of the corresponding atom type is contained [5]:

$$\vec{J}_{Sn} = -\frac{C_{Sn}}{k_B T} \mathbf{D}_{Sn}(\nabla \mu_{Sn} - |Z_{Sn}^*|e\nabla\varphi), \quad (1)$$

$$\vec{J}_{Cu} = -\frac{C_{Cu}}{k_B T} \mathbf{D}_{Cu}(\nabla \mu_{Cu} - |Z_{Cu}^*|e\nabla\varphi), \quad (2)$$

$$\vec{J}_{Ni} = -\frac{C_{Ni}}{k_B T} \mathbf{D}_{Ni}(\nabla \mu_{Ni} - |Z_{Ni}^*|e\nabla\varphi). \quad (3)$$

$\mathbf{D}_{Sn}$, $\mathbf{D}_{Cu}$, and $\mathbf{D}_{Ni}$ are the tensorial diffusivities [6]. After Cu and Ni atoms are injected in a Sn bump, the chemical reactions described by the following equations take place:

$$\frac{\partial C_{Sn}}{\partial t} = -\nabla \cdot \vec{J}_{Sn} - \kappa_1 C_{Sn} C_{Cu} - \kappa_2 C_{Sn} C_{Ni} \qquad (4)$$

$$\frac{\partial C_{Cu}}{\partial t} = -\nabla \cdot \vec{J}_{Cu} - \kappa_1 C_{Sn} C_{Cu} \qquad (5)$$

$$\frac{\partial C_{Ni}}{\partial t} = -\nabla \cdot \vec{J}_{Ni} - \kappa_2 C_{Sn} C_{Ni} \qquad (6)$$

$$\frac{\partial C_{\mathrm{IMC1}}}{\partial t} = \kappa_1 C_{Sn} C_{Cu} \qquad (7)$$

$$\frac{\partial C_{\mathrm{IMC2}}}{\partial t} = \kappa_2 C_{Sn} C_{Ni} \qquad (8)$$

(4)-(8) describe a consummation of Sn, Cu, and Ni, respectively, and a production of two types of IMCs (IMC1 = Cu_6Sn_5 and IMC2 = Ni_3Sn_4). This process is schematically illustrated in Fig. 3. The rates of the chemical reactions κ_1 and κ_2 are thermally activated parameters according to an Arrhenius law. The model assumes that all transport processes take place

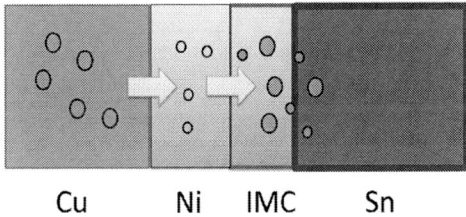

Fig. 3. EM of Cu and Ni atoms into Sn bump and the forming of IMC.

in Sn, i.e. Sn is the only transport medium in the model. However, when a thin layer of IMC is formed, impurities migrate through this layer in order to reach the Sn region, where the chemical reaction which produces the IMC occurs. Migration through the IMC for both Cu and Ni is characterized by specific diffusion coefficients and effective valences.

A. Model Parameterization and Properties of Sn

The Sn solder bump microstructure plays an important role in interconnect reliability. Compared to Cu, Sn crystallization produces 100-1000 times larger grains. Correspondingly, the role of grain boundaries as fast diffusivity paths is much more pronounced. Sn solder bumps often consist of several large Sn grains, such that most solder bumps exhibit one or at most a few Sn grain orientations [7].

Sn has a bulk tetragonal crystal structure which exhibits highly anisotropic diffusional, electrical, mechanical, thermal, and electrical properties [8].

A clear dependence of the thermo-mechanical response of a Sn solder bump on microstructure and Sn grain orientation was also observed [7]. The coefficient of thermal expansion is higher in the c-axis direction than in the a- or b-axis directions.

The isotropic diffusivity coefficient in Ni is [9]

$$D_{Ni} = 2.9 \exp\left(-\frac{2.88\,eV}{k_B T}\right) \frac{\mathrm{cm}^2}{s}. \qquad (9)$$

The effective valence for self-EM in Ni is, to the best of our knowledge, not available in the literature. Therefore, we use a

simple estimate from the ballistic EM model [10]

$$Z^* = -n_0 l \sigma_{tr}, \qquad (10)$$

where n_0 is the electron concentration, l is the electronic mean free path, and σ_{tr} is the transport cross section of the atom evaluated at the Fermi energy. By taking the values for n_0, l, and σ_{tr} from [11] we obtain an estimated $Z^*_{Ni} \approx -10$.

Measurements of self-diffusion in Sn clearly show an anisotropic atomistic transport. From [12] we have

$$D_{c,Sn} = 3.7\,10^{-8} \exp\left(-\frac{0.25\,eV}{k_B T}\right) \frac{\mathrm{cm}^2}{s}, \qquad (11)$$

$$D_{a,Sn} = D_{b,Sn} = 8.4\,10^{-4} \exp\left(-\frac{0.45\,eV}{k_B T}\right) \frac{\mathrm{cm}^2}{s}. \qquad (12)$$

The value of the self-EM effective valence in Sn has been obtained experimentally in [12]: $Z^*_{Sn} \approx -79$.

Diffusion of Ni in Sn is studied in [13] and the following values are obtained:

$$D_{c,Ni(Sn)} = 1.99\,10^{-2} \exp\left(-\frac{0.19\,eV}{k_B T}\right) \frac{\mathrm{cm}^2}{s}, \qquad (13)$$

$$D_{a,Ni(Sn)} = D_{b,Ni(Sn)} = 1.87\,10^{-2} \exp\left(-\frac{0.56\,eV}{k_B T}\right) \frac{\mathrm{cm}^2}{s}. \qquad (14)$$

For effective valence of Ni in Sn we have $Z^*_{Ni(Sn)} \approx -67$ [12].

The diffusivity coefficients (11)-(14) build diffusivity tensors which are used in (1), (2), and (3). The basic structures of the diffusivity tensors are given by

$$\bar{\bar{D}}_{Sn} = \begin{bmatrix} D^a_{Sn} & & \\ & D^b_{Sn} & \\ & & D^c_{Sn} \end{bmatrix}, \qquad (15)$$

$$\bar{\bar{D}}_{Ni(Sn)} = \begin{bmatrix} D^a_{Ni(Sn)} & & \\ & D^b_{Ni(Sn)} & \\ & & D^c_{Ni(Sn)} \end{bmatrix}. \qquad (16)$$

From (11)-(14) at $T = 150°C$ we have

$$\frac{D_{c,Sn}}{D_{a,Sn}} \approx 174, \qquad \frac{D_{c,Ni(Sn)}}{D_{a,Ni(Sn)}} \approx 10^4. \qquad (17)$$

The anisotropy of Ni diffusion in Sn is much more pronounced than the anisotropy of Sn self-diffusion.

The configurable parameters of the model are the rates of chemical reactions κ_1 and κ_2. All others parameters are obtained from measurements or theoretical calculations. The model itself allows to compute the time-dependent change of bump resistance and the thickness of the IMC layer which can also be experimentally determined and used for model calibration.

B. Intermetallic Compound Resistivity

The bump resistance increases due to void formation and microstructure changes during EM can be precisely measured with Kelvin bump probes [1]. IMCs such as Cu_6Sn_5 and Ni_3Sn_4 have a higher resistivity than pure Sn. In the case of Cu_6Sn_5, the resistivity is approximately 60% higher and in the case of Ni_3Sn_4 it is even 160% higher than the resistivity of Sn at room temperature (see Table I).

TABLE I.	MATERIAL PROPERTIES AT 20 °C	

Materials	Resistivity (nΩm)	Thermal conductivity (WK/m)
Cu	16.8	403.0
Ni	69.3	76.0
Sn	110.0	67.0
Cu_6Sn_5	175.0	34.1
Ni_3Sn_4	285.0	19.6

III. Simulation Results and Discussion

Model (1)-(8) has been implemented in an in-house three-dimensional simulation tool. For the EM reliability analysis

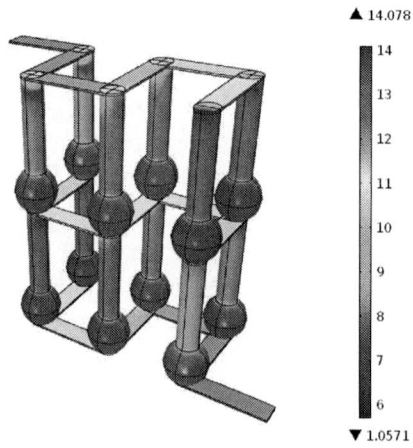

Fig. 4. Multilevel interconnect for 3D integration. Presented is the current density distribution on the logarithmic scale ($\times 10^3$ A/cm^3). Wafers are removed from the picture. The copper TSV length is 100μm and the diameter chosen 25μm. The UBM thickness is 2.5μm and the bump diameter is 40μm. Due to the chosen geometry the lowest calculated current density is inside the bumps (2.66 10^3 A/cm^2).

we use a multilayer interconnect structure, similar to the one usually used for wafers' stacking (cf. Fig. 4). The first step

Fig. 5. Structure of the solder bump used for the study. On the top of the Sn bump a Ni UBM is placed.

is the determination of the maximal current density which is reached in some particular bump for given operating conditions (voltage, temperature) of the whole interconnect structure. In the second step the EM in a single solder bump is analyzed. In

our study we have considered Ni_3Sn_4 as the primary IMC. The bump structure used in our simulation is sketched in Fig. 5.

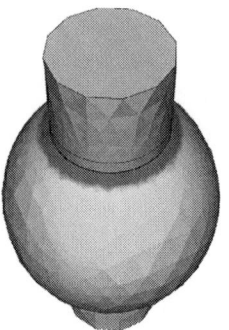

Fig. 6. IMC layer formed at the interface between a nickel UBM and a Sn solder bump.

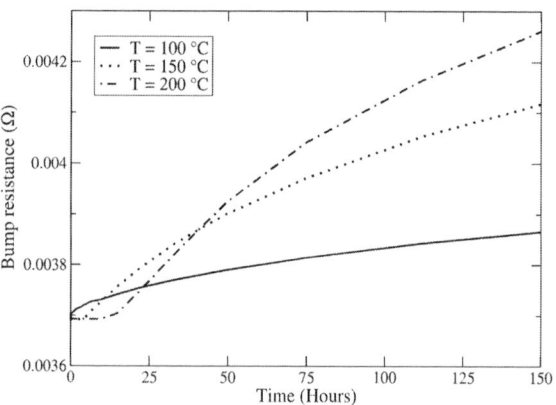

Fig. 7. Initial phase of solder bump resistance growth for three different temperatures.

The formation and growth of the IMC at the interface between the UBM and the Sn (cf. Fig. 6) is caused by several physical mechanisms. In the initial phase both Cu and Ni penetrate into the Sn bump and segregate just below the UBM/bump contact surface. Corresponding to this initial phase of EM stressing, there is an increase of the IMC's resistance as shown in Fig. 7. As expected the dynamics of IMC growth is enhanced at elevated temperatures and it is determined by the following mechanisms:

- Diffusion and EM of Cu and Ni in Sn

- Chemical reactions which convert Cu, Ni, and Sn into IMC (e.g. Cu_6Sn_5 and Ni_3Sn_4)

- Diffusion and EM of IMC in Sn

All these processes are thermally activated following Arrhenius law. Simulation allows to observe and study the interplay between the above mechanisms. As we can see in Fig. 7, in the first hours of EM stressing the resistance increase at 100°C is higher than at 150°C and 200°C. Both, migration and chemical reaction are enhanced at elevated temperature, but it seems obvious that in the early phase, migration keeps the concentration of impurities below the threshold necessary

978-1-4799-3911-4/14 $31.00 © 2014 IEEE

for an IMC production which would cause an observable resistance increase. The delay in the IMC formation process can also be observed in Fig. 8.

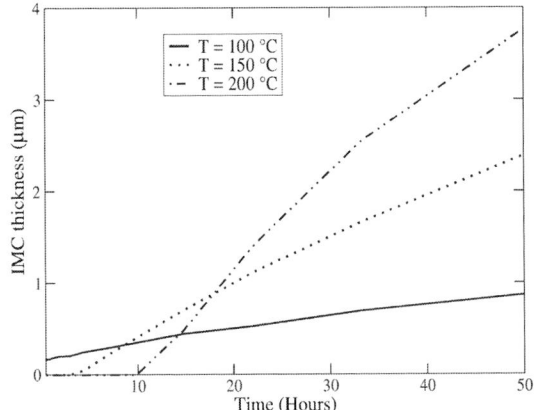

Fig. 8. Growth of IMC thickness in time. Fast migration of Cu and Ni at 150°C and 200°C prevents IMC emergence for several hours of stressing.

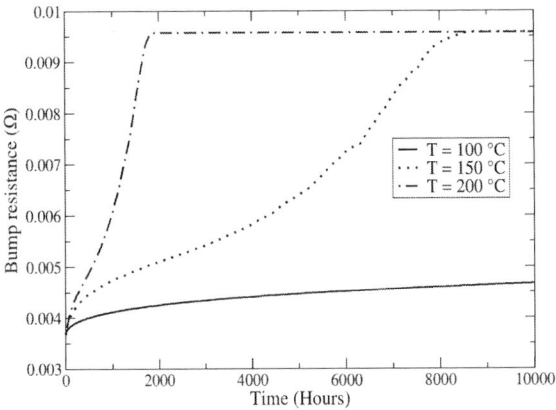

Fig. 9. Late phase of the resistance growth. The resistance of the solder bump increases, until the whole Sn is converted into IMC.

EM of vacancies ultimately leads to void formation and failure of the bump, a scenario which is most commonly observed in EM experiments [1]. However, simulation also permits to study a situation, where no void nucleation takes place but EM stressing proceeds, until the whole Sn of the solder bump is converted into an IMC. As we can see in Fig. 9 the bump resistance rises with a gradually increased slope, until the whole bump consists only of IMC (Ni_3Sn_4). From our simulation, we conclude that a much steeper second slope (cf. Fig. 2) observed by Chen $et\ al.$ [1], which appears abruptly after approximately 100h of stressing, can only be caused by an emergence of the new phase between the IMC and the Sn layer. We support the findings of Chen $et\ al.$ [1] that this "second phase" is actually the beginning of void evolution.

IV. Conclusion

The development of an IMC phase inside of Sn-based solders bump represents a reliability risk for interconnect structures used for realization of 3D ICs. In this work we have presented a model for IMC growth in Sn solder bumps. The model includes a description of the Cu and Ni migration into the Sn bump and the chemical reaction which produces two different types of IMCs. Simulations based on our model predict 3D profiles of the IMC, time-dependent resistance change, and time dependent change of the IMC thickness. The obtained results are utilized for explanation and discussion of experimental observations and measurements.

References

1. H.-Y. Chen, M.-F. Ku, and C. Chen, "Effect of Under-Bump-Metallization Structure on Electromigration of Sn-Ag Solder Joints," *Advances in Materials Research*, vol. 1, no. 1, pp. 83-92, 2012.

2. Ch. Hau-Riege, R. Zang, Y.-W. Yau, P. Yadav, B. Keser, and J.-K. Lin, "Electromigration Studies of Lead-Free Solder Balls used for Wafer-Level Packaging," *Proc. Electronic Components and Technology Conference*, pp. 717-721, 2011.

3. H. Yu, V. Vuorinen, and J. Kivilahti, "Effect of Ni on the Formation of Cu_6Sn_5 and Cu_3Sn Intermetallics," *Proc. Electronic Components and Packaging Technology Conference*, pp. 1-12, 2006.

4. L. Meinshausen, H. Frémont, and K. Weide-Zaage, "Migration Induced IMC Formation in SAC305 Solder Joints on Cu, NiAu and NiP Metal Layers," *Microelectronics Reliability*, vol. 52, pp. 1827-1832, 2012.

5. H. Ceric, R. Heinzl, Ch. Hollauer, T. Grasser, and S. Selberherr, "Microstructure and Stress Aspects of Electromigration Modeling," *Stress-Induced Phenomena in Metallization, AIP*, pp. 262-268, 2006.

6. R. L. de Orio, H. Ceric and S. Selberherr, "Strain-Induced Anisotropy of Electromigration in Copper Interconnect,"*Proc. International Semiconductor Device Research Symposium*, 2 pages, 2007.

7. T. R. Bieler, H Jiang, L. P. Lehman, T. Kirkpatrick, E. J. Cotts, and B. Nandagopa, "Influence of Sn Grain Size and Orientation on the Thermomechanical Response and Reliability of Pb-free Solder Joints,"*IEEE Trans. Comp. Pack. Techn.*, vol. 31, nr. 2, pp. 370-381, 2008.

8. M. Lu, "Effect of Microstructure on Electromigration in Pb-free Solder Interconnect," *Stress-Induced Phenomena in Metallization, AIP*, pp. 229-234, 2010.

9. A. R. Wazzan and J. E. Dorn, "Analysis of Enhanced Diffusivity in Nickel," *J. Appl. Phys.*, vol. 36, no. 1, pp. 222-228, 1965.

10. H. B. Huntington and A. R. Grone, "Current-Induced Marker Motion in Gold Wires," *J. Phys. Chem. Solids*, vol. 20, no. 1/2, pp. 76-87 1961.

11. N. A. Ashcroft and N. D. Mermin, Solid State Physics, *Holt, Rinehart and Winston*, 2003.

12. P. H. Sun and M. Ohring, "Tracer Self-Diffusion and Electromigration in Thin Tin Films," *J. Appl. Phys.*, vol. 47, no. 2, pp. 478-485, 1976.

13. D. C. Yeh and H. B. Huntington, "Extreme Fast-Diffusion System: Nickel in Single-Crystal Tin," *Phys. Rev. Lett.*, vol. 53, no. 15, pp. 1469-1472, 1984.

14. H. Ceric, R. Orio, and S. Selberherr, "TCAD Study of Electromigration Failure Modes in Sn-Based Solder Bumps," *Proc. International Conference on Simulation of Semiconductor Processes and Devices*, pp. 264-267, 2012.

New Technique for Acquiring Dead Pixel Free and Fine Inspection Image of Advanced LSI Package with Rough Surface Using Scanning Acoustic Tomograph

Kaoru Kitami, Masakatsu Murai, Natsuki Sugaya, Osamu Kikuchi and Shigeru Ohno

Hitachi Power Solutions Co., Ltd.

NS Bldg. 4F, 730, Horiguchi, Hitachi-naka, 312-8504, Japan.

Phone: +81-50-3139-0305 Fax: +81-29-271-2292 Email: shigeru.ono.nq@hitachi.com

Abstract-We have developed three new gate tracking functions to acquire dead-pixel-free and fine inspection images for advanced LSI packages with rough surface using a scanning acoustic tomograph. These are predicted surface gate tracking, double surface gate tracking and predicted S2-gate tracking methods. The advantages of these functions are demonstrated by using various test samples.

I. Introduction

As the integration density of large-scale integration circuits (LSIs) increases, means for quality control and failure analysis have become important in order to ensure reliability of products. Scanning acoustic tomograph (SAT) has been established as a nondestructive testing method for LSI packages [1].

For the nondestructive testing of LSI packages, the SAT is usually operated in reflection method and generates a transverse cross-sectional image. Fig. 1 shows the principle of generating inspection images by the SAT. An ultrasound transducer irradiates ultrasound to a sample, receives the series of ultrasound echoes reflected at each boundary between different materials and transduces into a time-domain electrical waveform. A surface detection gate (S-gate) having an appropriate duration is set to the first echo (surface echo) in order to obtain a reference signal (S-trigger) for data acquisition for one pixel on an inspection image. The S-trigger is generated at the reference point of time when the first echo signal exceeds an appropriate signal level (S-trigger level) during the S-gate duration. A focused boundary surface detection gate (F-gate) is set to the echo signal

appearing with an appropriate delay against the reference point. The maximum amplitude within the F-gate duration is converted to color tone and displayed to one pixel on the inspection image. A monochrome image can be obtained by repetition of this operation all over the observation area of the sample, since the amplitude of the echo signal depends on materials of the focused layer. If the first echo drifts because of slight inclination of the sample surface, the S-gate can track the surface echo in the S-gate duration.

However the inspection image of a sample having a rough surface or a complicated structure may include dead pixels because the amplitude of the surface echo becomes unstable. At these pixels, the surface echo is weaker than the S-trigger level, and the reference point for the F-gate setting cannot be generated. Therefore the dead pixels appear on the inspection image because the F-gate cannot be set. The inspection image with many dead pixels causes false detection of flaws. In order to suppress such effect, one can choose a low frequency transducer or a transducer with long focal depth. These transducers are insensitive to surface roughness. On the other hand, the resolution of acquired image is decreased because such transducers have large beam spot size.

Therefore, it is necessary to develop a new technique to acquire dead-pixel-free and fine inspection images for advanced LSI packages with rough surface using the SAT.

II. Experimental

In this paper, flaw inspection was carried out with FineSAT

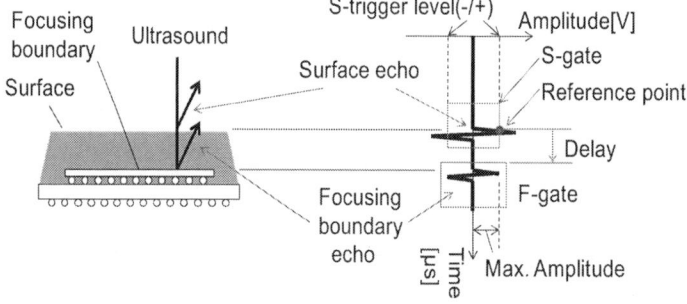

Fig.1. Imaging principle of the SAT.

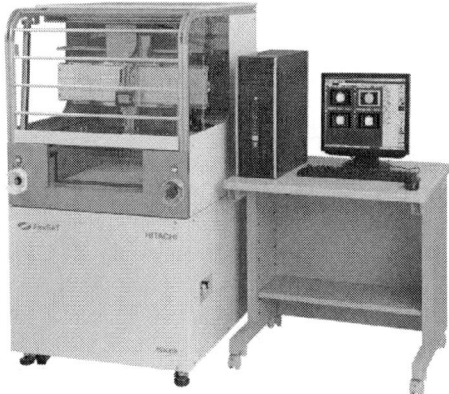

Fig.2 Exterior of FineSAT III.

III (Hitachi Power Solutions) as shown in Fig. 2. The system is equipped with a high-speed and high-precision scanner with a maximum speed of 1000 mm/s and a minimum pitch of 0.5 μm. The maximum received frequency is 500 MHz. The system can be utilized with a wide variety of transducers with frequency ranging from 5 to 300 MHz.

In order to obtain dead-pixel-free and fine inspection images of samples with unstable surface echo, three kinds of newly developed gate tracking method were adopted in this study. Fig. 3 shows the schematic diagram of the new gate tracking methods.

The first one is predicted surface gate tracking (PSGT) method (Fig.3 (a)). This function generates a virtual reference point at the same timing as the reference point was obtained for the last pixel data acquisition in the case of the surface echo being weaker than the S-trigger level. An active pixel can be acquired by setting of the F-gate against the virtual reference point. This function may be effective when the amplitude of the surface echo is unstable because of surface roughness of the sample.

The second one is double surface gate tracking (DSGT) method (Fig.3 (b)). This function uses two S-gates being set to the top surface and another reference boundary surface existing between the top and the focused surfaces. The first reference point which is obtained from the first S-gate (S1-gate) is used to set the second S-gate (S2-gate). In addition, the reference point

which is obtained from the S2-gate is used to set the F-gate. Both reference points are obtained at the time when the echoes from, respectively, the top and reference boundary surfaces exceed the S1 and S2-trigger levels. This function may be effective when the distance between the top and focused surfaces fluctuates depending on the sample location.

The last one is predicted S2-gate tracking (PS2GT) method (Fig.3 (c)). This method is the mixture of PSGT and DSGT methods. This function generates a virtual reference point at the same timing as the reference point was obtained for the last pixel data acquisition in the case of the surface echo being weaker than the S2-trigger level. The amplitude of the reference boundary surface often becomes unstable from a reason such as fluctuation of filler density or the longitudinal structure. Therefore this function may be preferred to be used for complicated and state-of-the-art LSI packages. As the other variation, the virtual reference points obtained with both S1- and S2-gates can be used in order to set S2- and F-gates, respectively.

III. Results and Discussion

Fig. 4 demonstrates the advantage of PSGT method. The test sample was a resin-molded IC having a rough surface (Fig.4 (a)). The inspection was carried out by using an in-house transducer with a frequency of 90 MHz and a focal length of 7 mm. The S-gate and the F-gate were set to the resin surface (surface (1)) and the top surface of a silicon chip (surface (2)). The surface around the location (A) was relatively flat compared to the surface around the location (B). Since the rough surface scattered the incident ultrasound wave, the intensity of received echo signal at the location (B) was smaller than location (A) as shown in Fig. 4(b) and (c). In this case, the S-trigger level has been set between the echo signal intensity obtained at the locations (A) and (B) intentionally. Since the conventional S-gate could not generate the S-trigger at the location (B), the F-gate was lost at the location (B) and dead-pixels were displayed on the inspection image (Fig.4 (d)). The dead-pixels were appeared as black dots in the case that the conventional S-gate was used. On the other hand, a dead-pixel free and fine inspection image of the silicon chip surface was obtained in the case that the PSGT function was activated, as shown in Fig.4 (e).

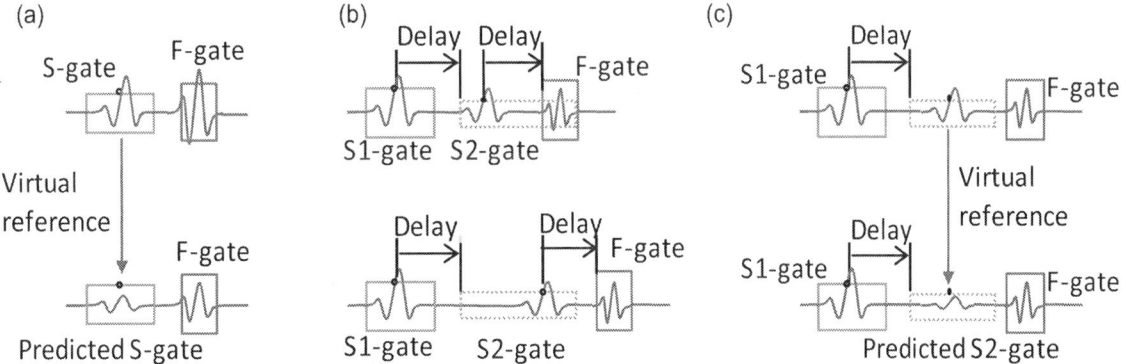

Fig.3 Imaging principle of the newly developed gate tracking functions. (a) Predicted surface gate tracking (PSGT), (b) double surface tracking (DSGT) and (c) predicted S2-gate tracking (PS2GT) methods

Fig. 5 indicates another example of PSGT method. The test sample was also a resin-molded IC having locally rough surface. The sample was removed from a substrate of a malfunctioned smartphone. The resin surface was relatively rough and character marking was made on it. The inspection was carried out by using the in-house transducer with a frequency of 90 MHz and a focal length of 7 mm. The S-gate and the F-gate were set to the resin surface and the top surface of a silicon chip. The inspection was made with switching from the conventional S-gate method to PSGT method. There were many dead pixels appeared on the upper half of the inspection image using the conventional S-gate because of the rough surface and character marking. On the other hand, there were no dead pixel appeared on bottom half of the inspection image using PSGT method.

Fig. 6 indicates the effectiveness of DSGT method by using a dummy test sample. The longitudinal cross-sectional structure of the dummy sample was shown in Fig.6 (a) and had an acrylic plate with a thickness of 1 mm on the top, an aluminum heat sink on the bottom and an epoxy resin adhesive filled between them. A ring shaped spacer with a thickness of 1 mm was inserted on the left side of the sample in order to incline the top plate to the heat sink. There were four boundary surfaces denoted as (1)-(4) in Fig.6 (a). The inspection was carried out by using the in-house transducer with a frequency of 25 MHz and a focal length of 17 mm. Four echoes reflected on the each boundary surface were detected as denoted by (1)-(4) as shown in Figs.6 (b) and (c). The ultrasound waveform obtained by using the conventional S-gate and DSGT methods were indicated in Figs. 6 (b) and (c), respectively. In Fig.6 (b), the S-gate and F-gate were set to the echo from the surface (1) and the boundary (4). The conventional S-gate obtained the F-gate at the location (B). However, at the location (A), the conventional S-gate could not generate the reference point and lost the F-gate because the surface (1) is inclined to the boundary (4). Therefore

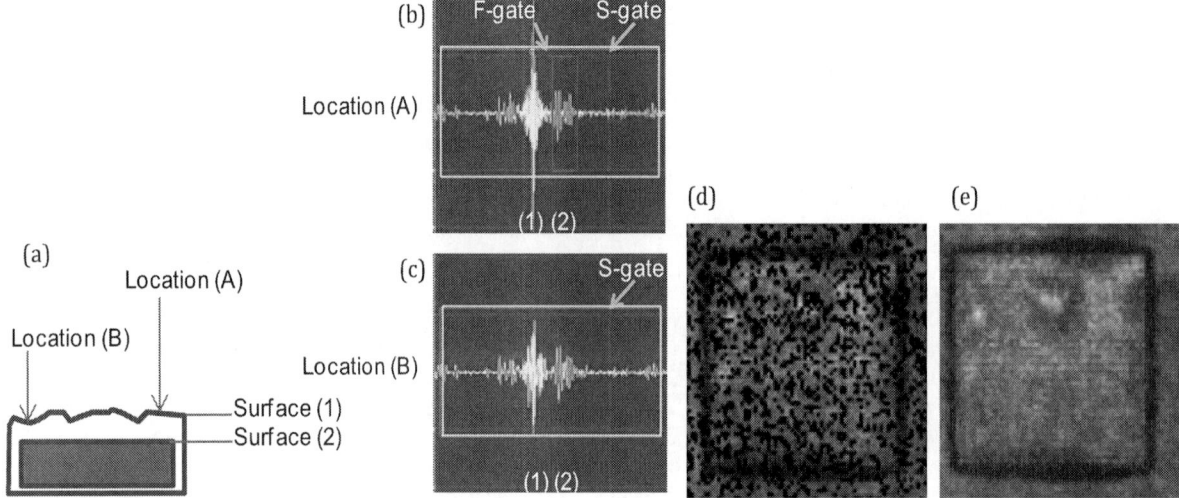

Fig.4 Example of the application of PSGT method. (a) A schematic illustration of a test sample, (b) and (c) examples of received echoes obtained from the points having smooth and rough surfaces, (d) and (e) comparison of obtained images between the cases activated and inactivated PSGT function.

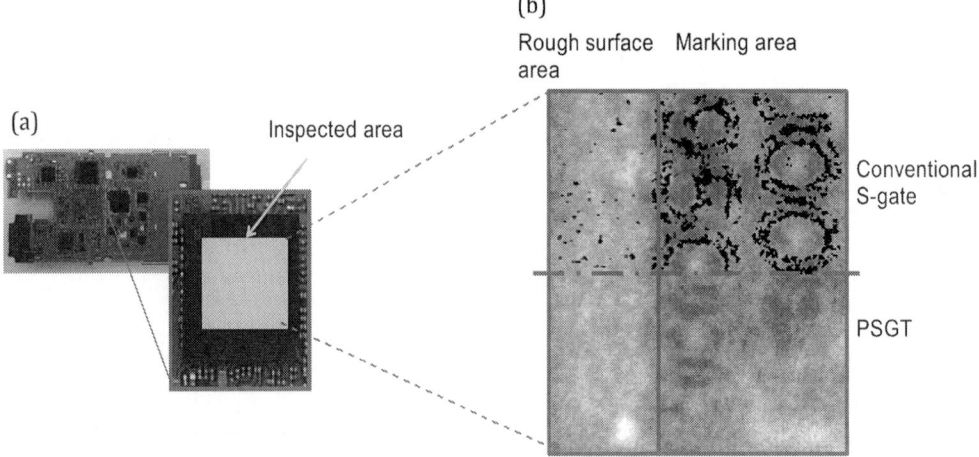

Fig.5 The other example of the application of PSGT method. (a) A test sample mounted on a substrate of a smartphone and (b) obtained image with the conventional S-gate and the newly developed PSGT methods.

the echo from the boundary (4) was appeared with a different delay from the surface echo (1) and was out of the S-gate duration. As a result, dead pixel (black) area was indicated on the bottom half of the inspection image as shown in Fig.6 (d). In order to obtain a dead-pixel free and fine inspection image, DSGT method was applied. The S1-gate, the S2-gate and the F-gate are set to the top and the boundary surfaces, (1), (3) and (4). The boundary surface (3) was parallel to the surface (4) so that F-gate could obtained with a certain delay from the reference point generated by S2-gate all over the test sample as shown in Fig.6 (c). Finally, the dead-pixel area is observed on the image acquired with the DSGT method (Fig.4 (e)).

As for the PS2GT method, we did not demonstrate an example here, because we could not an appropriate test sample. However, this function was prepared for more complicated LSI packages in the near future.

IV. Conclusions

We have developed three new gate tracking functions to acquire dead-pixel-free and fine inspection images for state-of-the-art and complicated LSI packages using SAT.

The first one is predicted surface gate tracking (PSGT) method. This function generates a virtual reference point when the S-trigger cannot be obtained. This function has been demonstrated to be effective when the surface echo is unstable because of the surface roughness of the sample.

The second one is double surface gate tracking (DSGT) method. This function uses two S-gates (S1- and S2-gates). The first reference point obtained from S1-gate is used to set the second S-gate (S2-gate) and the second reference point is used to set the F-gate. This function has been demonstrated to be effective when the distance between the top and focused surfaces fluctuates depending on the sample location.

The last one is predicted S2-gate tracking (PS2GT) method. This method is the mixture of PSGT and DSGT methods. This function generates a virtual reference point when the S2-trigger cannot be obtained. This function has the other variation that generates virtual reference points when both S1- and S2-trigger cannot be obtained. This function may be effective for more complicated LSI packages in the near future.

This paper demonstrated that these functions have advantages to obtain dead-pixel free inspection images of state-of-the-art and complicated LSI packages.

References

[1] M. Tanaka, M. Sakimoto, K. Nishi, K. Ohtsuka, "A Novel Test Method of Resistance to Soldering Heat of Plastic Encapsulated Surface Mount LSIs", NIKKEI ELECTRONICS 1990. 12. 24, p.143 (written in Japanese).

[2] K. Kitami, M. Takada, O. Kikuchi and S. Ohno, "Development of High Resolution Scanning Acoustic Tomograph for Advanced LSI Packages," Proc., 20th IEEE International Symposium on the Physical and Failure Analysis of Integrated Circuits (IPFA), p.530 (2013).

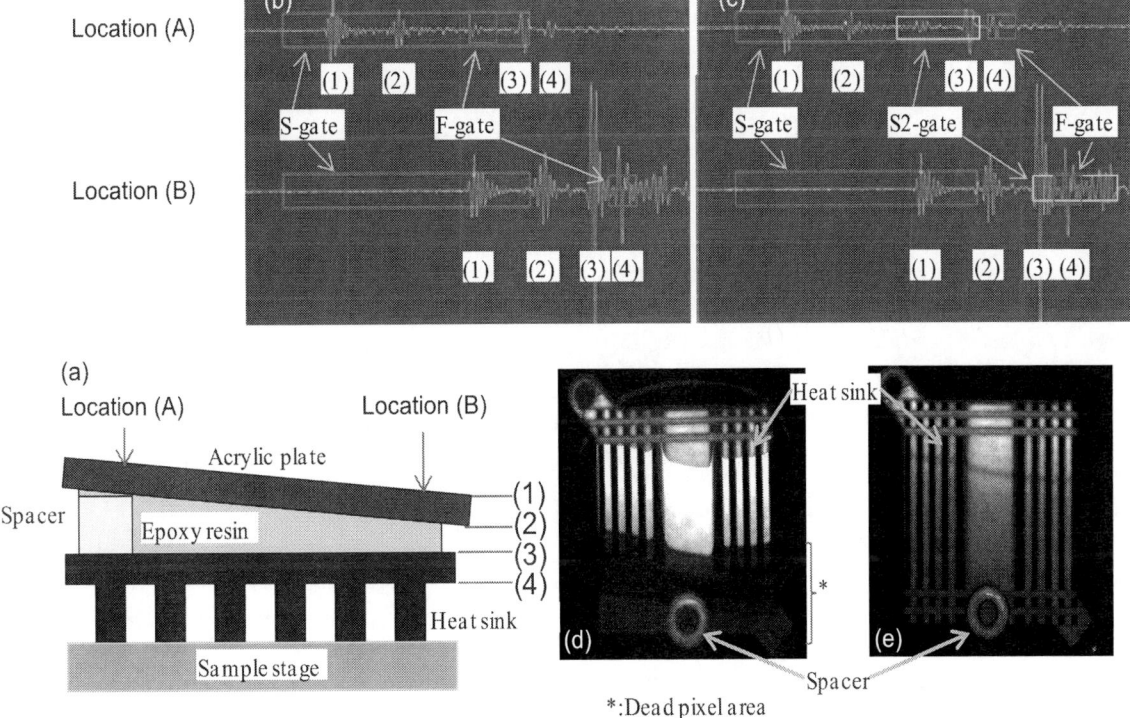

Fig.6 Result of double surface gate tracking method (a) The cross-sectional view of the sample, reflected ultrasound echoes obtained by using (b) the conventional S-gate method, (c) DSGT method, transversal cross-sectional images obtained by using (d) the conventional S-gate method and (e) DSGT method.

Reliability Analysis From Field Data And Prediction Models For Customer Risk Assessments: Case Studies And Strategy

Corinne Bergès, Yves Chandon and Pierre Soufflet

Freescale Semiconductor SAS

134 Avenue du General Eisenhower - BP 72329, 31023 Toulouse Cedex 1, France

Email : corinne.berges@freescale.com

Abstract-In the automotive industry, risk analysis are performed as soon as quality incidents with the same defect signature are reported from the field. The tools used may be the defect distribution modeling, used for the failure rate estimations required by ISO26262 standard, but adapted to the stronger constraints of quality issues. One of these constraints is the low number of failing parts on which a modeling has to be performed. Some case studies seemed to indicate that it is possible to assert the assumption of a constant failure rate when there are a small number of defects. A novel strategy to assess risk has been developed from this number of defects.

I. INTRODUCTION

In automotive component industry, ISO26262 standard requires an estimation of the residual technology-dependant failure rate: this failure rate is observed during the component lifetime, between the youthful and wearout periods, in the typical bathtub curve. When the components are in their design or qualification phases, this failure rate estimation can be performed by standard industrial methods like IEC/TR 62380, IEC 61709, MIL HDBK 217 or FIDES, or using results from qualification reliability tests described in the automotive AECQ100 standard, like the High Temperature Operating Life tests. For components that have already been produced and are currently in the field, and for which a risk assessment is required, most of the time, the only data that can be used for this estimation are quality incidents from the field, reported by automotive customers. The methods for failure rate estimation from field data are more and more studied, and field data are more and more taken care of since estimation accuracy relies on their quality [3, 4, 5]. The new interest for these methods and an improved quality of field data have a beneficial effect on the risk assessments required by customers, especially when, alerted by some field failures with a same signature, they fear more failures

at short term, and ask for accurate quantification of the failures that may happen.

This article deals with the findings in terms of risk analysis, from several field case studies. In the first part, the method for failure rate estimation from field data will be reviewed and its theoretical application for customer risk assessment will be presented. Some field case studies will show the practical application. Finally, the complete strategy to write a customer risk assessment will be detailed.

II. METHOD FOR FAILURE RATE ESTIMATION FROM FIELD DATA AND APPLICATION FOR CUSTOMER RISK ASSESSMENT

A failure rate estimation from field data, required by ISO26262 standard, may be performed by modeling them according to the typical Weibull model [1, 2]. Failure rate typically fits with the hazard function h(t) whose formula is the following one:

$$h(t) = (Beta/t) (t/Alpha)^{Beta}$$

where Alpha and Beta respectively are the scale and shape Weibull parameters.

Failure rate may be also expressed in FIT (Failure In Time):

$$\text{Failure rate in FIT at 1 year} = h(t) \times 10^9$$

By this field modeling and from the found Alpha and Beta parameters, cumulative distribution function F(t) may be deduced: it allows to give the failure quantity expressed in ppm (part per million) from the lifetime start to any instant in the life:

$$F(t) = 1 - e^{-(\frac{t}{Alpha})^{Beta}}$$

$$\text{Failure quantity (ppm) at } t = 10^6 \times [F(t) - F(0)]$$

For a field failure, customers are expected to give a failure mileage, more accurate than a failure time for which variability is much larger: variability on car usage and therefore on mileage also exists but it is easier to take it into account. So, the variable in all the previous formulas will be the mileage. The unit FIT will be replaced by a FIM unit (Failure in Mileage), with a possible conversion from one figure to the other, using the hour usage average per months:

Failure rate in FIT = Usage average per months (hours) x Failure rate in FIM

A difficulty may become apparent when customers do not report this failure mileage data for all or part of the quality incidents: an extrapolation from the incidents with mileage data to all the incidents with and without mileage data will be performed thanks to a proportional coefficient between the data with mileage and the data without mileage: this coefficient is defined by:

$$\text{Coefficient} = [\ (NW+OW) \ / \ NW \ ]$$

where:

Total number of FF with failure mileage data = NW

Total number of FF without failure mileage data = OW

Therefore, on a quality crisis, broken out from several failures with a same signature, customers will require data such as failure rate and quantity of potential failures, which corresponds to a risk assessment. This field modeling will be able to provide these results. It may be difficult when a field modeling is requested on only a few field returns although Weibull curve modeling is more accurate for a significant quantity of returns. Making the things worse is the fact that, voluntarily, field modeling is performed only on the first two or three years of the product life, which corresponds to the guarantee period (quality car makers offer a three years/100 000 km warranty): indeed, beyond this period, some quality incidents are not reported to car dealers and this lack of input data may distort the field modeling, failure rate and quantity results [6]. In extreme cases, for only one or two defects, modeling is no longer possible: the extra assumption of a constant failure rate may be asserted, allowing to make a failure quantity estimation by a simple calculation. This assumption fits with an exponential model. In case of an important quantity of defects with the same signature but without mileage data, using an exponential model makes possible to directly take into account these data, which is never possible with field modeling that needs a mileage data for each known incident.

Using an exponential model, the formula to estimate the failure rate λ is the following one:

$$\lambda_{\text{failure/hour}} = r/hf$$
$$\lambda_{\text{FIT}} = r/hf \text{ x } 10^9$$

where: r is the number of field failures, and hf the total hours spent for the shipped parts in the field. With a λ confidence level, failure rate is estimated according to the following formula:

$$\lambda_{\text{failure/hour}} = \chi_{[2(r+1);100(1-\alpha)]}^2 / (2 \text{ x hf})$$
$$\lambda_{\text{FIT}} = 10^9 \text{ x } \chi_{[2(r+1);100(1-\alpha)]}^2 / (2 \text{ x hf})$$

When the number of rejects is null (r = 0), the previous formula becomes:

$$\lambda_{\text{failure/hour}} = -\ln(\alpha) \ / \ hf$$
$$\lambda_{\text{FIT}} = -10^9 \text{ x } \ln(\alpha) \ / \ hf$$

Besides the particular difficulty of a field modeling on a small number of events, a specificity of a customer risk assessment is that the results from field data will be applied for other parts than parts in field. An estimation of potential failure quantity will be performed on the parts which are currently in the customer assembly lines or still stored before shipment to customers.

Before any estimation of quantity and failure rate, facing a quality incident, the first request from the customers will be a complete characterization of the defect obtained after a failure analysis. To some extent, field modeling will be able to confirm this characterization with regard of the Weibull parameters found by this field modeling.

III. CASE STUDIES

The method to assess a risk depends essentially on the number of defects in the quality crisis. Several case studies are presented according to this number.

A. Field modeling for more than twenty field incidents

The simpler case in term of failure rate and quantity estimation corresponds with a crisis of more than twenty quality incidents, for two similar products dedicated to power switching in automotive applications. Risk assessment started by a full failure analysis that revealed a latent defectivity issue on a capacitor: after a spatial localization of the defect with an appropriate technique like the light emission microscopy technique, a SEM inspection showed melted polysilicon going through the Nitride dielectric layer and shorting the Metal1 plate with the Polysilicon plate [Fig. 1.].

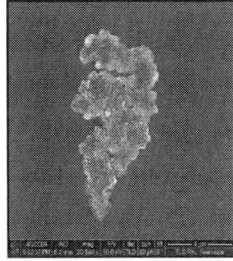

Fig. 1. SEM picture of the defective particle on the capacitor

Twenty five failure cases from the field were counted: customers reported mileage data for all of them. Fifteen other similar cases from customer assembly line were included in the risk assessment since they presented the same defectivity defect on the capacitor. For this latent failure, and the failure in the assembly lines may be interpreted like a failure in the very early mileage. For these ten and five extra cases coming from assembly, mileage failure was set to the null value for the field modeling. Assumption of a mean 1670 km mileage covered per month is asserted and the typical variability on car usage is taken into account, by applying a standard deviation of 350 km. With these two data of mean and standard deviation on car usage, the mileage distribution of the surviving parts in field will be generated: this surviving population with their current mileage will be compared to the failing parts with their failure mileage. Another parameter in the field modeling is the delay between the shipment date from the component manufacturing site and the lifetime start: a statistical study of this delay can be launched in order to make an estimate. Distribution of this delay shows a mean value about 166 days: this figure will be used in field modeling. A statistical fit study between this delay and the manufacturing date at component manufacturer's shows that the distribution is wider for an older manufacturing date: this may be linked with a larger assembly duration at customer's when the product is newly introduced into their assembly lines [Fig. 2.].

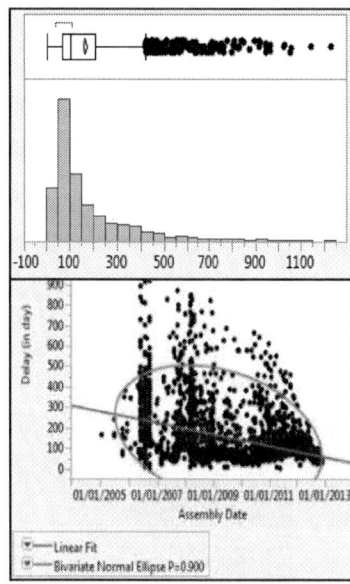

Fig. 2. Delay (in days) between shipment date and lifetime start: distribution and bivariate Fit by assembly date

Finally, field modeling enables to find a Weibull model with a very low Beta parameter equal to 0.09 [Fig. 3.]. Besides the failure quantity and rate estimation for the parts in the field, failure probability is applied to the quantity of parts stored by the component manufacturer or by the customers. A defect with a same root cause that would not been taken into account in the field modeling, because of a poor characterization for example, could impact the results. A study consisted in removing three of the field failures among the forty cases, at the mileages 236 km,

2596 km and 26336 km. The failure quantity decreased in the same proportion than the number of defects (-7.5%) although the failure rate decreases twice as much (-14%). Impact on the Beta parameter and on its standard deviation is low, but very high on the Alpha parameter. In the same way, adding a failure incident without any relationship into the field modeling would increase the failure quantity and rate results.

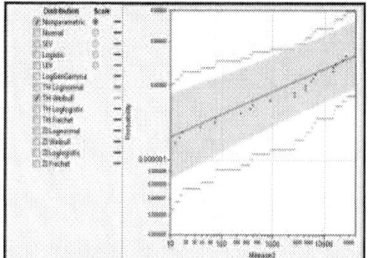

Fig. 3. Field modeling for a large number of defects

B. Field modeling and Weibull model for a few quality incidents with failure mileage data

When only a few quality incidents are reported with failure mileage data, field modeling allows to calculate model parameters but the confidence intervals are very large or even are hard to estimate. Two quality issues are in this case: the first one fits with only five failing components. On these components dedicated to automotive valve switching, a leakage was observed on the switch. An Optical Beam Induced Resistance Change (OBIRCh) analysis, performed on the leakage, allowed localizing the defect. Then, deprocessing and FIB cross-sections performed on the defect helped to characterize the root cause: a polysilicon residue deposited on top of polysilicon and organic flakes were observed, corresponding to a manufacturing defectivity issue done in a single process step [Fig. 4.].

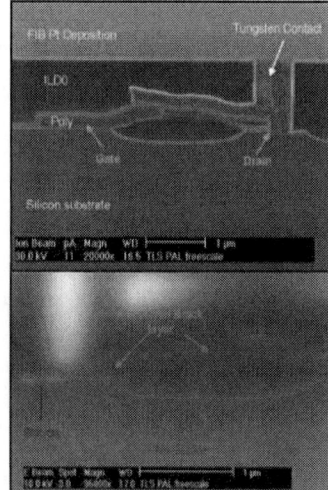

Fig. 4. Polysilicon residue and cross section through the flake

Among the reported five cases for this issue, four of them came from the field but only three of them had a failure mileage data. One incident observed in the assembly line at the customer's was integrated into the field modeling, with a mileage data set to zero. Field modeling gave a Weibull model and a beta parameter estimate equal to 0.163 but no upper values for the model parameters. This low value for Beta confirmed a defectivity issue, without confidence that this Beta stays inferior to one in the confidence interval.

A second case pertains to the same type of component and the same electrical signature for only four customer returns from the field: a valve switch is leaky. An OBIRCh technique localized precisely the defect but the physical analysis did not allow observing the particle [Fig. 5.]. An intermetallic defectivity issue was suspected. A field modeling found a Weibull model with merely estimates for the parameters but no upper values. A Beta equal to 0.228 confirmed a manufacturing defectivity issue.

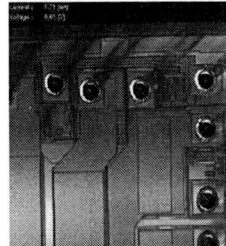

Fig. 5. OBIRCh localization of the defect

C. Field modeling with an exponential model

A field modeling performed on two quality issues, highlighted an exponential field model or a model very similar to an exponential one.

The first issue was linked to ten field failures with mileage data. For this automotive component, several pins were shorted to ground, or open. Several Electrical Overstress Signatures (EOS) were observed on the die [Fig. 6.]. A field modeling found a Weibull model with a Beta parameter equal to 0.897. This Beta close to one indicates a failure rate almost constant, that is to say an event independent of the moment in the product life. These EOS are suspected to be related with the customer application.

Fig. 6. Electrical Overstress Signatures: constant failure rate

For an automotive component dedicated to drive electronic braking, six failing parts revealed delamination at the ILD0/polysilicon, polysilicon/field oxide and ILD0/silicon interfaces [Fig. 7.]: all these six cases came from a same wafer lot of two hundred and fifty parts. Field modeling was performed with two different models, Weibull and exponential. The Akaïke or Bayesian information criterions (AIC, BIC) were calculated in order to determine the most appropriate model, balancing relevance of defect explanation and accuracy prediction for the risk analysis:

Bayesian information criterion = BIC = -2loglikelihood + k ln(n)
Akaïke information criterion = AICc = -2loglikelihood + 2k[n/(n-k-1)]
AICc = AIC + [2k(k+1)]/(n-k-1)

where k is the number of estimated parameters in the model, and n, the number of observations in the set of data.

In that case, the model with the lower AIC criterion, the exponential one, was chosen, since an estimation of the defect probability was required within the modeling period. At that point, this exponential model showed that it was not a defectivity problem, even it did not appear like a wearout phenomena with a Weibull Beta parameter superior to one.

After three new months of mileage and no more failing parts, this risk assessment was updated: then, the best fit was a defective subpopulation Fréchet or Weibull model, which allowed to provide a more optimistic failure quantity prediction.

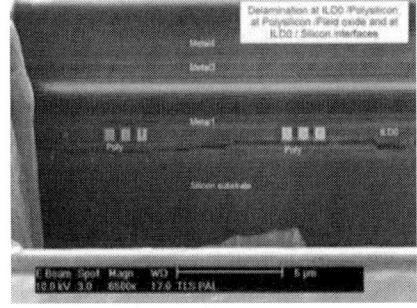

Fig. 7. Delamination on six failing parts

D. Calculation when field modeling is no longer possible

When field modeling is not possible with only one failing part or even with no failing part, the new assumption of a constant failure rate is asserted.

For one case of suspected metal defectivity, where a cross-section on the OBIRCh localization did not reveal the defect, distribution of the hours spent in field for all the surviving parts is generated in order to directly calculate a failure rate expressed in FIT [Fig. 8 and 9].

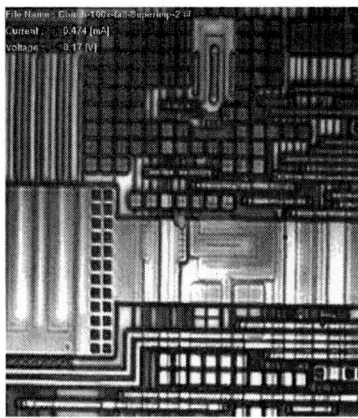

Fig. 8. Cross-section on the OBIRCh localization was failing

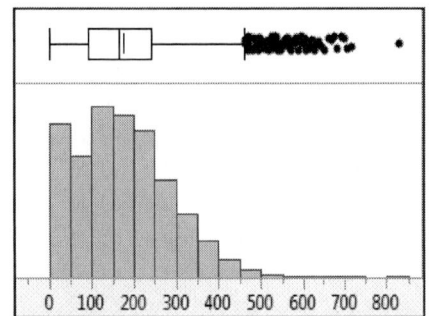

Fig. 9. Distribution of the hours spent in the field for the surviving parts

This same method is applied for a case of suspected pad corrosion occurred during manufacturing or assembly process. Chlorine traces, the corrosive agent, were revealed by Energy Dispersive X-Ray (EDX) spectroscopy analysis [Fig. 10.].

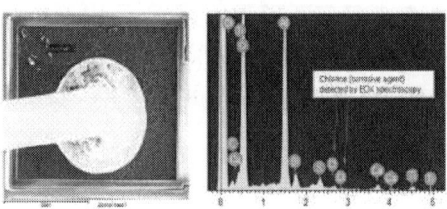

Fig. 10. Corroded pad and Energy Dispersive X-Ray (EDX) spectroscopy analysis

The extreme case is the one with zero failure. Application of the appropriate formula, for a component type that represented 17 billions of hours in field, gave a failure rate equal to 0.053 FIT at a 60% confidence level.

IV. STRATEGY

From these case studies, a full strategy in five steps may be implemented in order to perform a customer risk assessment [Fig. 11.].

The first step is the determination of the customer return root cause. Typically, it requires a precise electrical characterization of the failing part and a full failure analysis, performed by all the available tools for defect localization and physical characterization. It is a major phase in the risk assessment since it will allow to bring together the cases which are similar. Its result is qualitative but it finally guarantees accuracy in failure rate and probability estimations. It is always possible to take into account some suspected cases, since field modeling will give an indication of the defect type through the Beta value in the Weibull model.

In the second step, a precise counting of the failure parts is performed: it is important to know the rate between the failing parts for which a failure mileage is reported by the customers and the ones for which any indication about the failure event is given. The larger the ratio with/without, the more precise the failure rate and probability are. In order to increase the number of failing parts with mileage data, it is possible, in some cases, to bring together the returned parts from customer assembly line, when the defect is proven to be exactly the same one as the one from field.

Then, field modeling or exponential calculation is performed, according to the number of failures reported, and the result is applied on the stored parts and the parts which are not in the field yet. Typically, a field modeling may demonstrate four main root causes for a defect. A latent die defect may have a Weibull Beta parameter inferior to one, a latent package defect will show it superior to one. Application defects and such events as EOS, have a failure rate close to one, and therefore are modeled by an exponential model. Inside the latent die defect type, defect-controlled failures (extrinsic) are different than the defect-free ones (intrinsic). And this distinction is important when corrective actions have to be implemented: determination of these corrective actions fits with the fourth step of a customer risk assessment. Lastly, the found failure rate and probability will be compared with the results of the stress tests implemented in qualification phases, since the Early Life Failure Rate (ELFR) test targets the defects within the first life year, as the defect quantity at one year is estimated from field for a customer risk assessment, and the High Temperature Operating Life (HTOL) test is focused on the life failure rate, similar to the estimation performed from field. Since the returns from field may demonstrate all defect types (package and die defects), defects highlighted by other tests such as the Temperature Cycling for example, have to be taken into account. Let's note that at this phase of comparison between on one hand the failure rate (expressed in FIM or in FIT) found by field monitoring and on the other hand qualification stress accelerated test results (always expressed in FIT), it is always possible to take several assumptions around the typical monthly-covered mileage or monthly-spent hours values in field, in order to challenge the comparison conclusions.

Finally, this first risk assessment will be regularly updated, even when there is no new field return: that will reduce the confidence intervals of the results, above all for the crisis with only a few quality incidents.

References:

1. Bergès C, Chandon Y, Gubian R, "Innovative methodology for failure rate estimation from quality incidents, for ISO26262 standard requirements", *IPFA 2012*

2. Bergès C, Goxe J, "Benefits of field failure distribution modeling to the failure analysis", *ESREF 2013*

3. Wu M, Wang W, Tian L, Wu C, Fan D, "A novel data analysis methodology in failure analysis", *ISDEA 2012*

4. Barnett TS, Grady M, Purdy KG, Singh AD, Combining negative binomial and weibull distributions for yield and reliability prediction. *IEEE 2006*

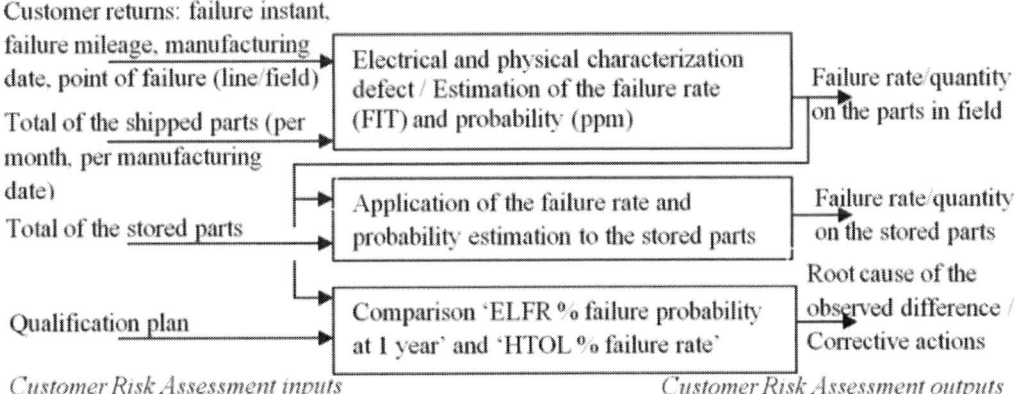

Fig. 11. Scheme of the customer risk assessment process

V. CONCLUSION

A new strategy to analyze risk from field returns has been implemented. Thanks to this strategy, a risk assessment can be performed whatever the number of returns reported by customers. Besides the full electrical and physical characterization of the defect, main result from a risk assessment is a failure rate and probability estimation, for the parts already in field as the ones stored by the customer. A risk assessment has to lead to corrective actions according to the observed defect. It also has to lead to a comparison with the specifications in terms of failure rate and probability for which the product has been sized and that have been checked by accelerated stress tests in qualification phase. This comparison may be very difficult since field potentially reveals any type of defects, although qualification tests can target only one type of defect. It may be the subject of further studies.

5. Haggag A, Barr A, Walker K, Winemberg L, "Realistic 55nm IC Failure In Time (FIT) estimates from automotive field returns", *IRPS 2013*

6. Alam MM, Suzuki K, "Lifetime estimation using only failure information from warranty database", *IEEE Trans Reliab 2009;58(4)*

Defect Localization by Lock-in IR-OBIRCH on Some Recovered Cases

Chunlei Wu[1,2], Grace Song[2], Suying Yao[1]

[1]School of Electronic Information Engineering, Tianjin University, Tianjin 300072, China
[2]Freescale Semiconductor (China) Limited, Tianjin, China
Phone: (+86) No.22 85684614/15922127484   Email: b09059@freescale.com

Abstract- **There are some recovered cases during failure analysis (FA) process, although every FA step is performed right and very carefully. Sometimes the failure root cause still need to be identified after recovering, because the failed IC is unique and the failure root cause is very important to improve the products' quality. Sometimes the defect could be localized by Lock-in IR-OBIRCH, although the failure has disappeared. In this paper, two cases are demonstrated to show how to locate defects by Lock-in IR-OBIRCH after the failed ICs have recovered.**

I. INTRODUCTION

Some recovered cases are encountered during FA process, although every FA step is performed very carefully by FA engineers. These recovered cases challenge FA tremendously and result in a low success rate. It is hard for a FA engineer to finding the failure root cause after recovering. Recovering could happen in every FA step, such as failure verification, de-capsulation, fault localization such as photon emission microscopy (PEM) [1, 2] analysis, making probe pads by Focused Ion Beam (FIB) and Micro-probing and so on. There are some possibilities that could induce recovering, such as a leakage path is opened by forcing excessive power (current and voltage) during electrical test, charging or discharging and so on.

However, sometimes the failure root cause still need to be identified after recovering, because the failed IC is unique and the failure root cause is very important to improve the products' quality. Sometimes the defect could be localized by Lock-in IR-OBIRCH, although the failure has disappeared.

In this paper, two cases are demonstrated to show how to locate defects by Lock-in IR-OBIRCH after the failed ICs have recovered.

II. Lock-in IR-OBIRCH principle

IR-OBIRCH [3, 4] techniques employ a laser with a wavelength around 1340nm to locally heat the device to avoid creating electron-hole pairs. Compared with the reference, the localized laser heating induces different resistance change in the circuit due to a defect in a failed IC. The effect of the resistance change is converted to a current variation which is monitored by the IR-OBIRCH amplifier for defect localization. In the IR-OBIRCH system, there are two modes of laser signal, one is "lock off" mode, and another one is "lock in" mode. The laser signal is constant in the "lock off" mode. However, the laser signal is modulated by 5 kHz or 50 kHz signals in the "lock in"

mode. Comparing with the "lock off" mode, the noise signal whose frequency is not 5 kHz or 50 kHz is filtered under the "lock in" mode and signal to noise ratio (SNR) is improved. Therefore, the "lock in" mode laser is better than the "lock off" mode laser when Lock-in IR-OBIRCH is performed to localize a leakage site.

Lock-in IR-OBIRCH is very sensitive to the current variation and it can detect minimum 10pA current. The defect is still right there, although the failed IC has recovered. Therefore, the defect maybe induce a tiny current variation in the circuit while laser is locally heating it, and then this tiny current variation may be detected by Lock-in IR-OBIRCH and be marked on the die image synchronously. So it is possible to locating the defect by Lock-in IR-OBIRCH, although the failed IC has recovered.

III. CASE STUDY 1

This functional failure case involved a 0.28μm SMARTMOS8 MV analog and mixed-mode IC with four metal layers. The failure mode was confirmed in the FA Lab that "over_current_lo" fault of OUT3 was reported incorrectly during enabling CSNS function of OUT3 for the failed IC. But for the reference one, no "over_current_lo" fault of OUT3 was reported during enabling CSNS function of OUT3. The "over_current_lo" fault means the current of OUT3 is higher than the threshold. CSNS function enable means the current of OUT3 is output to CSNS pin in term of a defined proportion for monitoring whether it is normal or not. Generally, for the reference IC, CSNS function enable does not trigger "over_current_lo" fault. Thus, there is a wrong relevancy between CSNS function enable and "over_current_lo" fault on the failed one.

PEM was performed and an abnormal emission spot was found in a NMOS in the logic circuit area whose gate connects "oclo_deglitch_3" signal (Fig. 1). The abnormal emission spot revealed the voltage of "oclo_deglitch_3" maybe in intermediate level between VDD and GND. So, a hypothesis was proposed according to the special failure mode and the abnormal emission spot that "csnc_en3" signal maybe have a leakage with "oclo_deglitch_3" signal, which resulted in the voltage of "oclo_deglitch_3" maybe in intermediate level between VDD and GND (Fig. 2). However, the failed IC recovered after depositing 2 probe pads on these 2 signals metal lines by Focused Ion Beam (FIB). The I-V characteristic between "csnc_en3" signal and "oclo_deglitch_3" signal was examined by probing these two signals but no difference was found between the failed IC and the good one (Fig. 3). Therefore,

978-1-4799-3911-4/14 $31.00 © 2014 IEEE

there was no any leakage between these two signals after the failed IC has recovered.

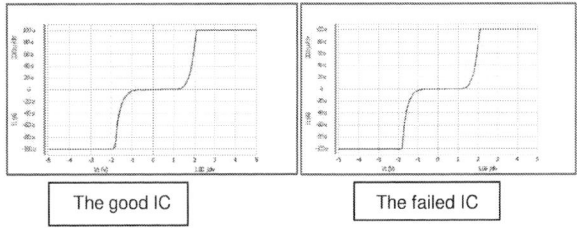

| The good IC | The failed IC |

Fig. 3 The I-V curve of "csns_en3" - "oclo_deglitch_3"

The layout examination revealed both of these two signals have very long path in the layout, so it was impossible to catching the defect by physical FA without an accurate leakage location. As mentioned above, Lock-in IR-OBIRCH is very sensitive to the current variation and it can detect minimum 10pA current. The defect is still right there, although the failed IC has recovered already. Therefore, the defect may induce a tiny current variation in the circuit while laser is locally heating it, and then this tiny current variation may be detected by Lock-in IR-OBIRCH and be marked on the die image synchronously. So it is possible to locating the defect by Lock-in IR-OBIRCH, although the failed IC has recovered. Thus, Lock-in IR-OBIRCH was tried to locate the suspected leakage spot while 2V was forced between "csns_en3" and "oclo_deglitch_3". And amazingly, an abnormal OBIRCH spot was just found between "csns_en3" metal line and "oclo_deglitch_3" metal line on metal3 layer on the failed IC, although it has already recovered (Fig. 4). This abnormal OBIRCH spot proved the hypothesis proposed above was right.

Thus, as Lock-in IR-OBIRCH analysis has pinpointed the exact position of the leakage current, FIB cross section was performed and a metal bridge defect was revealed at the IR-OBIRCH spot site (Fig. 5).

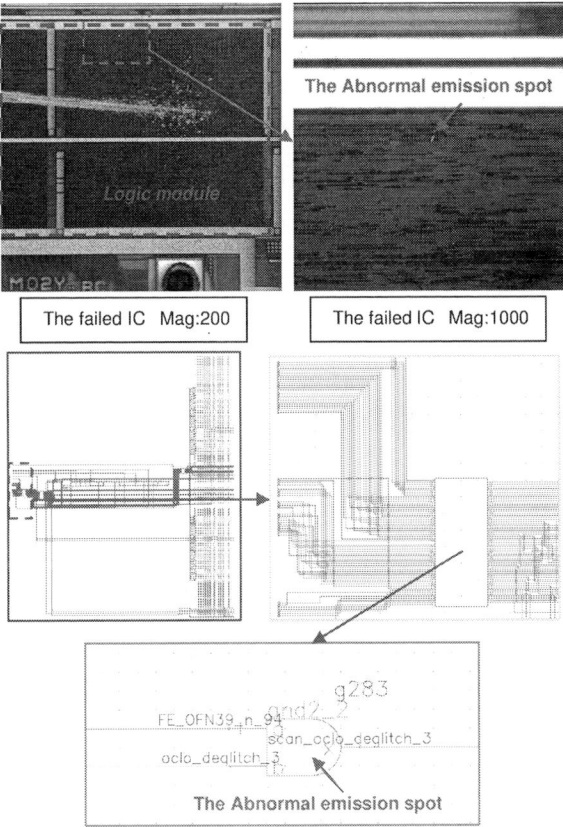

Fig. 1 PEM images of the failed IC and its location in the schematic

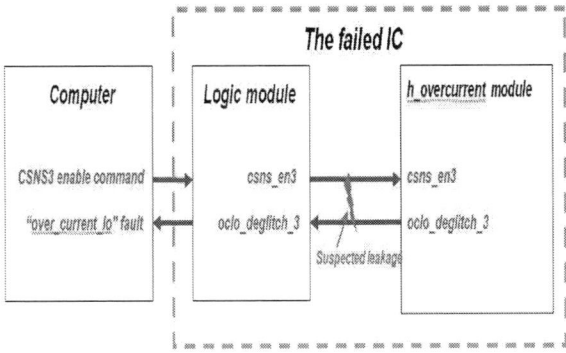

Fig. 2 The sketch of the hypothesis

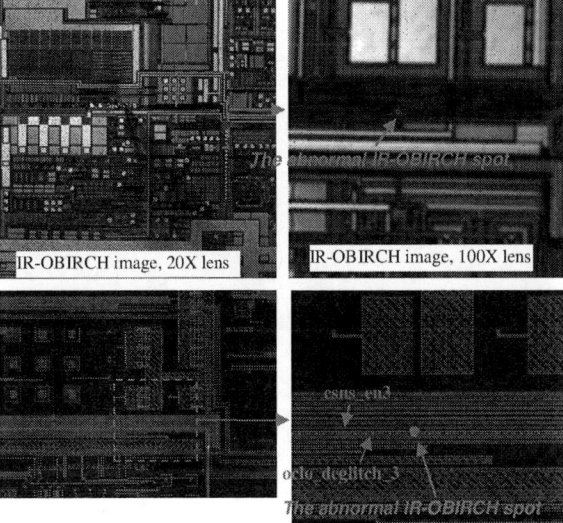

Fig. 4 The IR-OBIRCH spot images and its location in the layout

978-1-4799-3911-4/14 $31.00 © 2014 IEEE   351

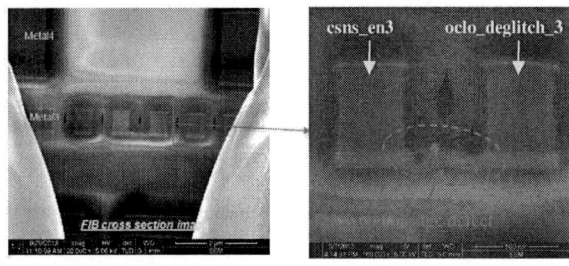

Fig. 5 The metal bridge defect at the abnormal IR-OBIRCH spot

IV. CASE STUDY 2

This functional failure case involved a 0.8μm SMARTMOS5AP analog and mixed-mode IC with two metal layers. The functional failure mode was confirmed that its interrupt function was abnormal for the failed IC. PEM was performed under the functional set up and a normal emission spot was missed at a zener DZ1 in "dsat_thresh" block compared with the good one (Fig. 6), which indicated abnormal voltage on "desat_level" signal. Microprobe analysis based on the emission result found "desar_level" signal was 13.91V instead of 12.48V for the failed IC (abnormal higher than the reference). Further microprobe analysis found "bias" signal inside this block was about 46mV instead of 1.26V for the failed IC (Fig. 7). The abnormal lower "bias" signal could turn off NMOS Mn0 and then resulted in the abnormal higher "desar_level" signal. So I-V characteristic was performed between "bias" and "VSS" and a resistive short (~9kohm) was found. But the I-V curve changed to normal after repeated measurement (Fig. 8). And the failed IC was also confirmed functionally recovered at Bench and ATE.

Even so, Lock-in IR-OBIRCH was still tried to locate the suspected leakage spot while 1V was forced between "bias" and "VSS". An abnormal IR-OBIRCH spot was detected on NMOS Mn0 in "dsat_thresh" block (Fig. 9). Fig. 6 shows its location in the schematic.

At last, physical analysis revealed a gate oxide rupture at the IR-OBIRCH spot location (Fig. 10).

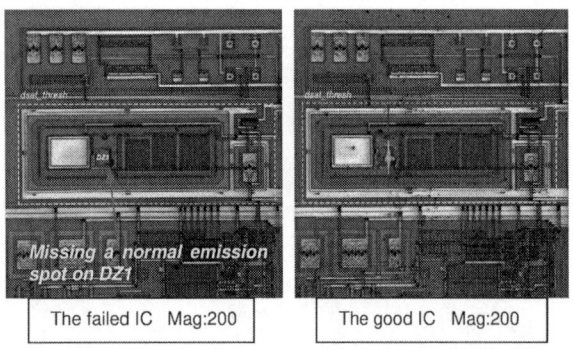

Fig. 6 The emission result of the good IC and failed IC

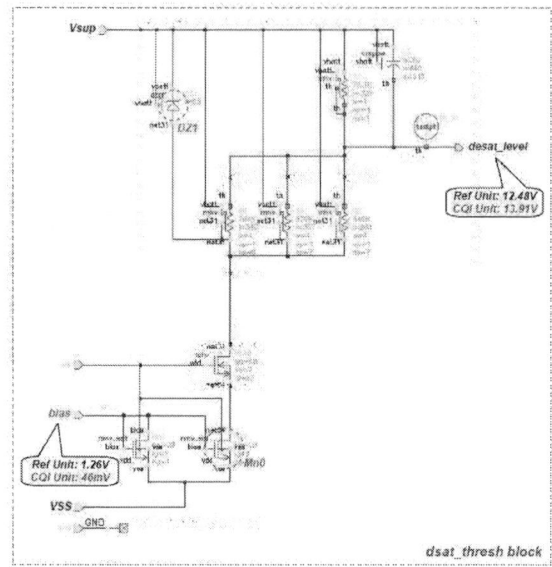

Fig. 7 The schematic of "dsat_thresh" block and probe result

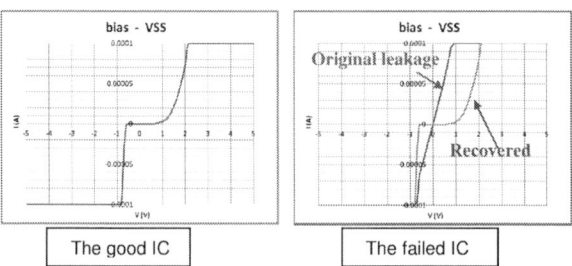

Fig. 8 The I-V curve of "bias" and "VSS"

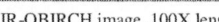
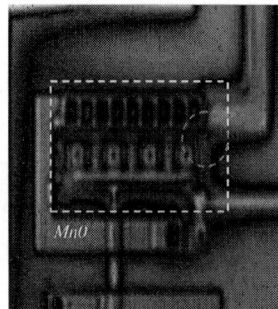

IR-OBIRCH image, 100X lens   Its superimposed image, 100X lens

Fig. 9 An abnormal IR-OBIRCH spot was detected on Mn0

978-1-4799-3911-4/14 $31.00 © 2014 IEEE   352

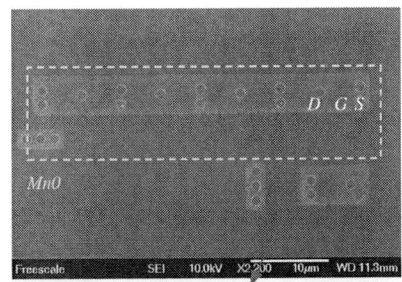

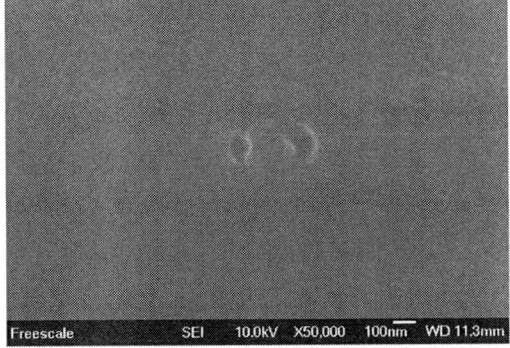

Fig. 10 The gate oxide rupture defect at the abnormal IR-OBIRCH spot

V. SUMMARY

In this paper, different types of defects were located by Lock-in IR-OBIRCH with a normal I-V characteristic, although these failed ICs have recovered. In case 1, the possible reason of recovering was the leakage path triggered by the metal bridge defect was opened because the defect's nature was changed by locally electrical field at the defect area due to Ga$^+$ ions accumulation under FIB. This kind of recovering maybe avoided if more advanced FIB is employed; In case 2, the possible reason of recovering was the leakage path triggered by the gate oxide rupture defect was opened by forcing excessive power (current and voltage) during I-V curve test. The excessive power resulted in a higher temperature within the gate oxide rupture defect and then changed its nature, so the leakage path was opened. Thus, an appropriate power (as less as possible) should be chosen during electrical test for avoiding this kind of recovering.

The two cases revealed a possibility that sometimes the failure root cause could still be localized precisely by Lock-in IR-OBIRCH even on some recovered cases. Because Lock-in IR-OBIRCH is very sensitive to the current variation and it can detect minimum 10pA current. The defect is still right there, although the failed IC has recovered. Therefore, the defect maybe induce a tiny current variation in the circuit while laser is locally heating it, and then this tiny current variation may be detected by Lock-in IR-OBIRCH and be marked on the die image synchronously. We can try to perform Lock-in IR-OBIRCH to pinpoint the defect, so long as enough clues are gotten and a reasonable hypothesis is proposed. That was confirmed by these two functional failure cases demonstrated in this paper.

REFERENCES

[1] Bruce MR, Bruce VJ, "ABCs of Photon Emission Microscopy," *Electronic Device Failure Analysis,* vol. 5, 2003, pp. 13-20.
[2] Phang JCH, Chan DSH, Tan SL, Len WB, Yim KH, Koh LS, Chua CM, Balk LJ, "A Review of Near Infrared Photon Emission Microscopy and Spectroscopy," *IPFA 2005,* pp. 275-281.
[3] K. Nikawa, S. Tozaki, "Novel OBIC observation method for detecting defects in Al stripes under current stressing," *ISTFA 1993,* pp. 303–310.
[4] Clifford Howard, Anusha Weerakoon, Diana Mitro, Dawn Glaeser, "Topside Defect Localization Using OBIRCH Analysis," *ISTFA 2005,* pp. 231-236.

Single bit cell SRAM failure due to titanium particle

Rachel Siew and W.F. Kho

Freescale Semiconductor Malaysia

2 Jalan SS8/2, Free Industrial Zone Sungei Way

47300, Petaling Jaya, Selangor Malaysia

Phone: (603) 76692406 Fax: (603) 76692936 Email: rachelsiew@freecale.com

Abstract

A single bit cell SRAM failure with shorted bit line (BL) and bit line bar (BLB) storage nodes were analyzed. Instead of using top down deprocessing to localize the defect, a modified approach incorporating Atomic Force Probe (AFP) and STEM dark field imaging of a thick sample were used. Using the modified approach the single bit cell failure was found to be due to a titanium particle on the spacer of a transistor that shorted the polysilicon gate and active silicon.

Introduction

SRAM bit cell characterization has proven to be a reliable approach to identify defects in single bit cell failures. By comparing measured I-V curves with I-V curves that were simulated using Spice, defects within the bit cell can be localized with high accuracy [1]. An exception is when the bit line and bit line bar storage nodes are resistively shorted. In this case, the location of the short is much more difficult to localize as the short may be anywhere within the entire bit cell. In the past a top down deprocessing approach (Figure 1) has been used to localize the defect. However this approach is still risky even in the hands of an experienced analyst as a stringer like defect can float away during top down deprocessing. Furthermore chemical identification of the small defect with EDX is difficult due to a large X-ray interaction volume. In this paper we report a modified approach that include AFP and STEM dark field imaging of a thick sample to address these issues.

Electrical Analysis

A device failed memory BIST test and was verified on bench to fail (stuck low) at a single bit of SRAM. Photon

Emission Microscopy (PEM) was performed but no anomalous emission spots were noted. Micro probing the bit line and bit line bar of the failing bit while writing and reading from the bit also showed no faulty signals. Bit cell characterization was then performed [1].

Figure 2 shows a schematic of a 6T SRAM bit cell. The word line of the bit cell is set at VDD. The bit line (BL) and bit line bar (BLB) are then set to conditions as shown in Table 1 to generate four sets of current(I) versus voltage (V) curves. For simplicity only the I-V curves of conditions I and II are shown. Figure 3 and 4 show the I-V curves of a good and the failing bit respectively. On the failing bit, the drain current of NFET1 is suppressed and start at negative current. The set of I-V curves were compared with Spice simulated curves and a best match was noted for a resistive short defect between the bit line (BL) and bit line bar (BLB)storage nodes (Figure 5 and 6).

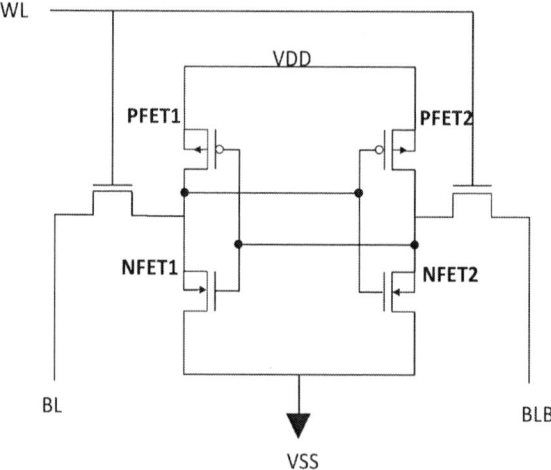

Figure 2. Schematic of 6T SRAM bit cell

Condition	BL	BLB
I	Sweep 0V-VDD	VDD
II	Sweep VDD-0V	0V
III	VDD	Sweep 0V-VDD
IV	0V	Sweep VDD-0V

Table 1: The four conditions for performing electrical analysis on 6T Bit Cells shown in Figure 2

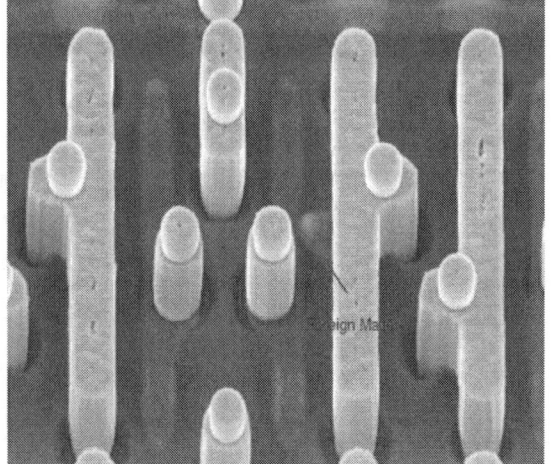

Figure 1. Top down deprocessing to contact and local interconnect to expose shorting particle

978-1-4799-3911-4/14 $31.00 © 2014 IEEE   354

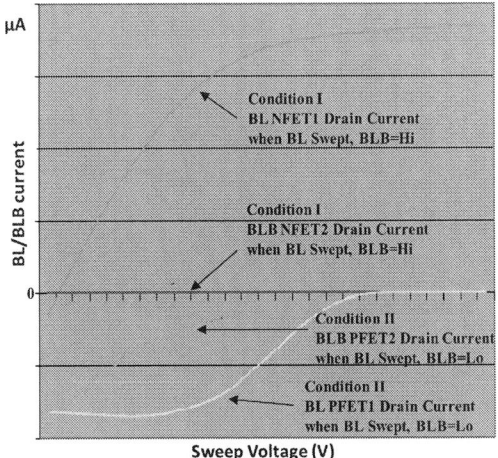

Figure 3. I-V curves of a good bit under conditions I and II

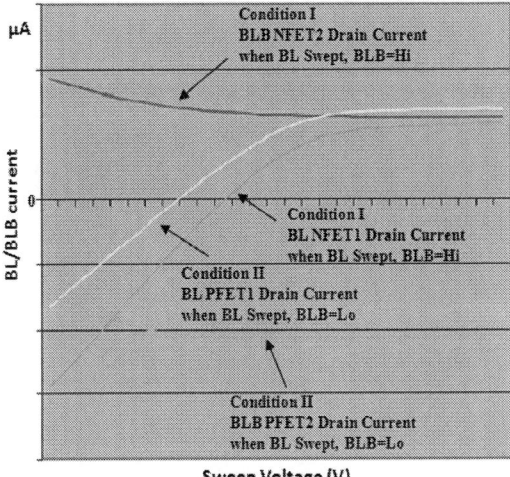

Figure 4. I-V curves of the failing bit under conditions I and II

A review of the bit cell layout indicate the only possible way for a resistive short between BL and BLB storage nodes is at or below Via 1 (Figure 7). In order to check the hypothesis that the BL and BLB storage nodes are resistively shorted, probing was performed on the bit cell with Atomic Force Probe (AFP) after the die was parallel lapped to Via 1 level. As the BL storage node is not accessible at Via 1 level an indirect method was used for confirmation. This is done by turning on PFET1 by applying a negative bias on the gate of this transistor. Two other probes are then placed at SD2 and S1 (Figure 8). If the BL and BLB storage nodes are shorted a resistive trace is expected when a voltage of -1V to 1V is swept across SD2 and S1. Conversely if the BL and BLB storage nodes are not shorted a diode curve on both nodes is expected. AFP probing results verified the hypothesis that the BL and BLB storage nodes are shorted with the presence of a resistive trace (Figure 9).

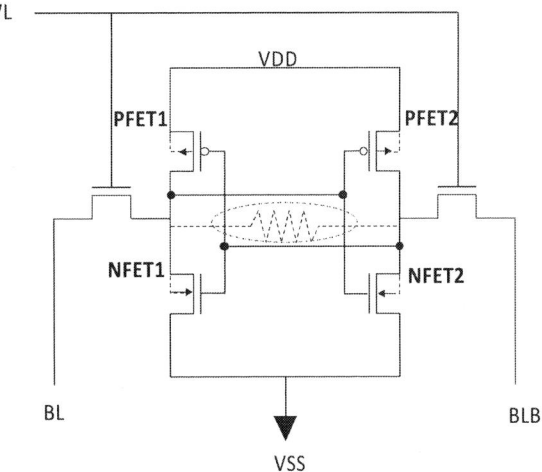

Figure 5. Schematic of a resistive short between the Bit Line Bar(BLB) and Bit Line(BL) storage nodes

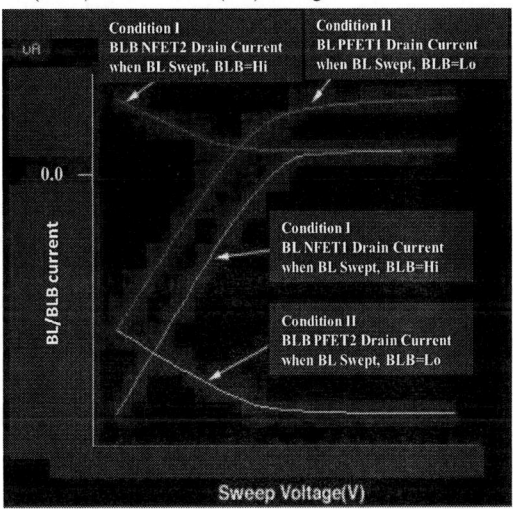

Figure 6. SPICE simulated I-V curves for a resistive short between BL and BLB storage nodes

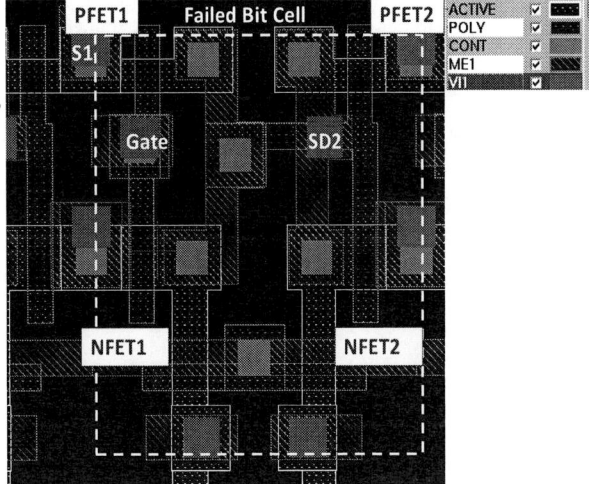

Figure 7. Layout view of bit cell

978-1-4799-3911-4/14 $31.00 © 2014 IEEE   355

Current path through resistive defect
shorting BLB to BL storage node

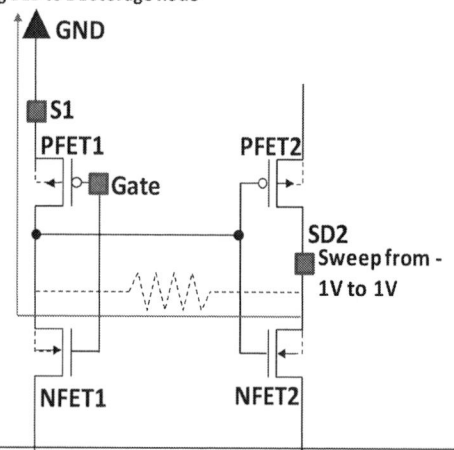

Figure 8. Schematic at Via 1 level showing location of AFP probes at gate of PFET1, location S1 and SD2

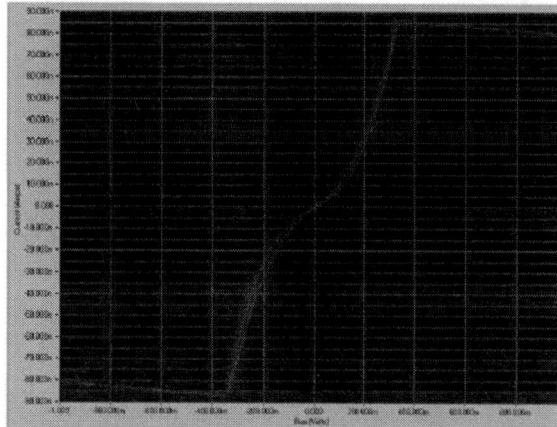

Figure 9. Resistive I-V trace between probes placed at SD2 and S1 indicate a short between BL and BLB storage nodes

Physical Analysis

In order to locate the defect site, a sequential Focused Ion Beam (FIB) cross section on the sample was considered. However as the area that can potentially short the BL/BLB storage node is large there is a risk of missing a small defect. In view of this, a thick TEM sample encompassing the entire bit cell was instead prepared and inspected under Dark Field Scanning Transmission Electron Microscope (STEM) mode. STEM can be used to image much thicker samples compared to conventional TEM (up to 10x thicker) as there is no objective lens below the sample to cause chromatic aberration [2]. In addition, the high contrast of metallic related shorts in STEM dark field images is also easier to recognize [3]. Figure 10 to 12 shows STEM dark field images of progressively thinner samples as the sample is rethinned to further isolate the location of the defect. A metallic material was found on the spacer of NFET1 shorting the polysilicon gate and silicon active area. Spot mode EDX analysis detected presence of titanium on the material (Figure 13 and 14).

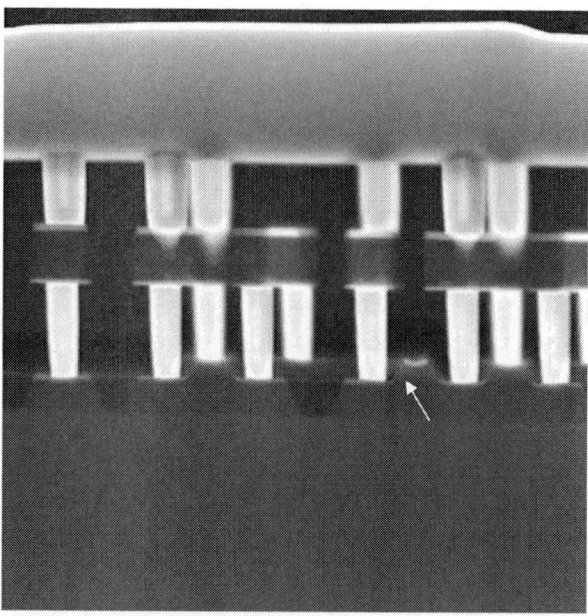

Figure 10. STEM dark field image of the entire bit cell. Anomaly is noted at spacer of NFET1

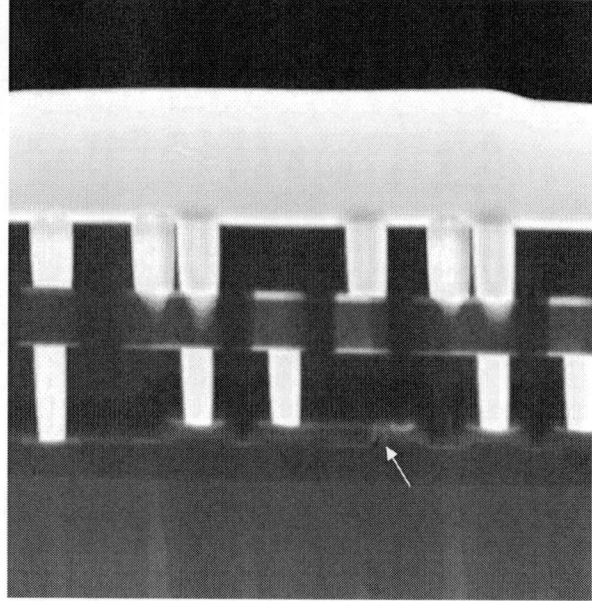

Figure 11. Metallic element more apparent on spacer of NFET1 with further thinning

Conclusions

A single bit BL to BLB bar storage node short SRAM failure was analyzed using a modified approach instead of using top down deprocessing. Following SRAM bit cell characterization that showed a resistive BL to BLB storage node short, AFP was used to confirm the location of the defect is at Via 1 level or lower. STEM Dark Field mode imaging of a thick sample encompassing the entire bit quickly identified a potential defect. Subsequent rethinning and STEM Dark Field imaging identified a titanium particle shorting the BL and BLB storage nodes.

978-1-4799-3911-4/14 $31.00 © 2014 IEEE

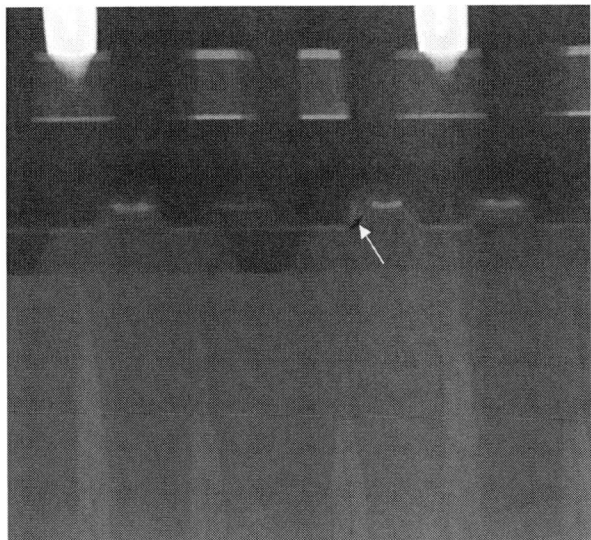

Figure 12. Thin sample with minimal overlapping features clearly shows metallic element on spacer of NFET1

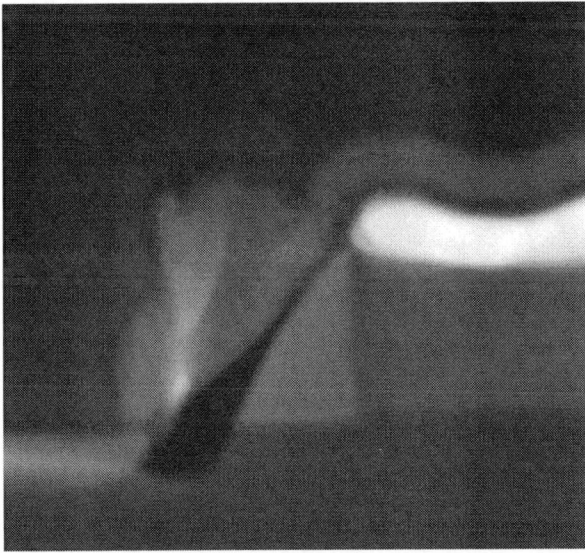

Figure 13. Closer view of metallic element on the spacer of NFET1

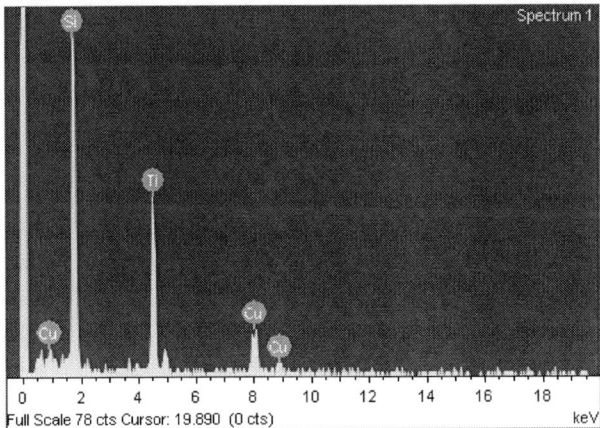

Figure 14. EDX of metallic element on the spacer (Spot 1 in Figure 13) show presence of titanium

Acknowledgments

The authors would like to thank Mr. Gary Chan for his advice and helpful comments.

References

1. R. Mulder, S. Subramanian, E. Widener, T. Chrastecky, "Improved SRAM 6T Bit cell Failure Analysis using MCSpice Bit Cell Defect Modelling, in *ISTFA Symposium*, 2003, pp. 363-370
2. Wang. N et. al., "Use of STEM in nanometer level defect analysis of SRAM devices", in *ISTFA Symposium*, 2001, pp. 313-318
3. S. Subramanian *et. al.*, "Advanced Defect Characterization by STEM analysis", in *Electronic Device Failure Analysis*, Volume 10, Issue 2, May 2008, pp. 20-28

Gate Oxide Rupture Localization by Photon Emission Microscopy with the Combination of Lock-in IR-OBIRCH

Chunlei Wu[1,2], Suying Yao[1]

[1]School of Electronic Information Engineering, Tianjin University, Tianjin 300072, China
[2]Freescale Semiconductor (China) Limited, Tianjin, China
Phone: (+86) No.22 85684614/15922127484   Email: b09059@freescale.com

Abstract-There are many failure analysis cases are induced by the gate oxide rupture. It is a common and important failure mechanism in failure analysis. Photon emission microscopy with the combination of Lock-in IR-OBIRCH are very effective to localize the gate oxide rupture in MOS transistor, which can decrease analysis cycle time and improve success rates remarkably. In this paper, some different cases are presented to show how to locate the gate oxide rupture in MOS transistor accurately and quickly by photon emission microscopy with the combination of Lock-in IR-OBIRCH.

I. INTRODUCTION

Gate oxide breakdown is one of the key integration and reliability issues for advanced semiconductor technology [1]. The abrupt loss of the insulating property of the thin gate oxide layer used in metal-oxide-semiconductor (MOS) structure is one of its most important failure mechanisms [2]. A gate oxide rupture is generated in a MOS transistor when the gate oxide breakdown happened and the MOS was stressed continuously after that. The job of failure analysis is to localize the gate oxide rupture which is a proof to reveal the gate oxide breakdown in MOS transistor.

Photon emission microscopy (PEM) [3, 4] is a nondestructive, fast tool for locating emission spots on an IC where light is emitted. Sometimes, this technique can directly pinpoint the gate oxide rupture location on gate oxide layer prior to destructive physical analysis when the voltage drop and leakage current in the breakdown spot of gate oxide layer are sufficient to the sensor of PEM. If PEM could not locate that directly, schematic/layout study and microprobe have to be performed to isolate the breakdown MOS. And then Lock-in Infrared Optical Beam Induced Resistance Change (IR-OBIRCH) [5, 6] should be employed to localize the gate oxide rupture accurately as it is good at locating the leakage current spot. Thus, the complementary combination of PEM, Lock-in IR-OBIRCH and schematic/layout study and microprobe is more effective to localize the gate oxide rupture in MOS transistor than only one FA technique is employed reported in the literature.

In this paper, some different cases are presented to show how to locate the gate oxide rupture in MOS transistor quickly by PEM with the combination of Lock-in IR-OBIRCH.

II. CASE STUDY 1

This functional failure case involved a 0.8μm SMARTMOS5AP analog and mixed-mode IC with two metal layers. The functional failure mode was confirmed that the charge pump voltage was about 12.6V instead of 26V under 13V power supply for the failed IC.

PEM was performed under functional set up and an abnormal emission spot was found (Fig. 1). The abnormal emission spot was examined in the layout/schematic and its location was at a PMOS in the charge pump module. Further layout study revealed its location was just in the middle of the channel (Fig. 2). Generally, a good PMOS transistor does not emit lights under PEM if it only works at ON/OFF state. It may emit lights under PEM if it works at saturated state, but the saturated emission spot mainly focuses on the drain side of the PMOS and the shape of the saturated emission spot is same as the shape of active area in the drain side. However, in this case, the abnormal emission spot was just in the middle of the channel of the PMOS, which revealed a gate oxide breakdown spot definitely according to above analysis.

Schematic study and microprobe analysis were not mandatory for this case. However, for double confirming the PEM result, I-V curve test was performed on this PMOS and a resistive leakage was found between gate and source of this PMOS (Fig. 3).

Lock-in IR-OBIRCH was employed to localize the gate oxide breakdown point again, while 200mV was being forced between gate and source by microprobe needles. Two abnormal OBIRCH spots were found. The abnormal OBIRCH spot 1 was found just at the same location as the abnormal emission spot. And the abnormal OBIRCH spot 2 was found at a contact which revealed the leakage current flowed from PMOS source to PMOS gate through the gate oxide breakdown point and the contact due to the gate oxide breakdown issue in the PMOS (Fig. 4). The abnormal OBIRCH spot 2 only revealed the leakage current path, and the abnormal OBIRCH spot 1 revealed the gate oxide breakdown point. So in this case, the gate oxide breakdown point was localized by both PEM analysis and Lock-in IR-OBIRCH analysis directly.

At last, de-process analysis revealed a gate oxide rupture at the abnormal emission/ OBIRCH spot area (Fig. 5).

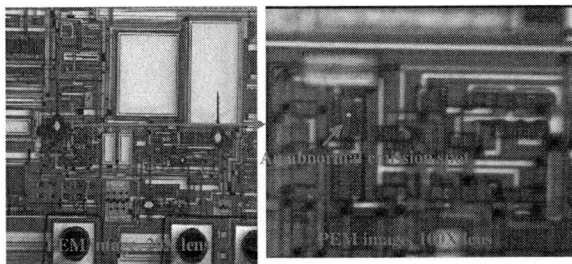

Fig. 1 PEM images of the failed IC

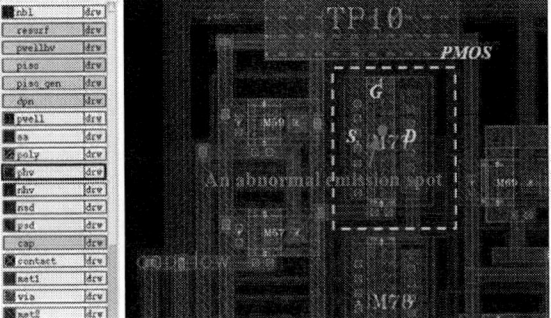

Fig. 2 The abnormal emission spot in the layout

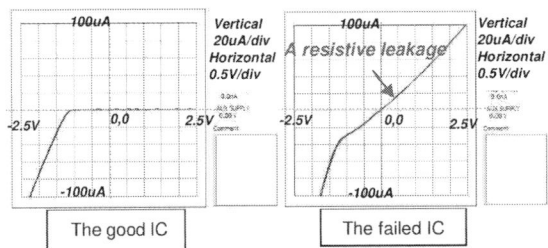

Fig. 3 The I-V characteristics of gate–source of the PMOS

Fig. 4 The abnormal OBIRCH spot images

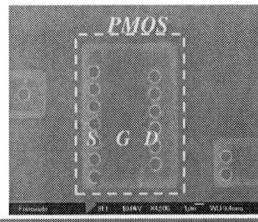

Fig. 5 The gate oxide rupture at the abnormal emission/OBIRCH spot area

III. CASE STUDY 2

This functional failure case also involved a 0.8μm SMARTMOS5AP analog and mixed-mode IC with two metal layers. The functional failure was verified that the Pull-up Current at pin SP2 (Pin7) was 0mA instead of 2mA for the failed IC.

PEM was performed under functional set up but no abnormal emission spot was found for the failed IC compared with the good IC. Schematic study and microprobe have to be performed for isolating the failure root cause. After that, a leakage I-V curve between net103 and GND was found (Fig. 6). Therefore, Lock-in IR-OBIRCH was employed to pinpoint the leakage spot in the circuit, while 500mV was being forced between net103 and GND by microprobe needles. An abnormal OBIRCH spot was found at the gate area of a NMOS whose poly gate connects to net103 and its source connects to GND. And another OBIRCH spot was found at the contact which revealed the leakage current flowed from metal 1 to poly gate via the contact due to the gate oxide breakdown issue in the NMOS. Fig. 7 shows the abnormal OBIRCH spot images and Fig. 8 shows its location in the layout and schematic.

At last, de-process analysis revealed a gate oxide rupture at the abnormal OBIRCH spot area (Fig. 9).

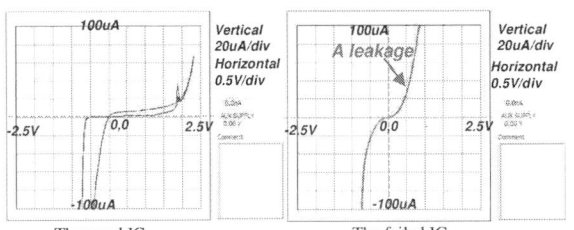

Fig. 6 The I-V characteristics of net103-GND

978-1-4799-3911-4/14 $31.00 © 2014 IEEE   359

Fig. 7 The abnormal OBIRCH spot images

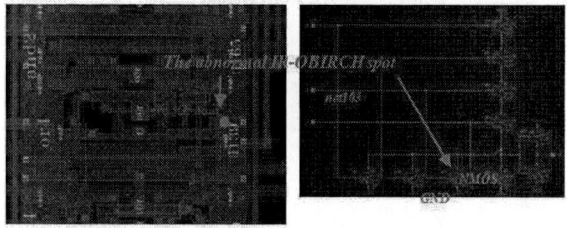

Fig. 8 The abnormal OBIRCH spot in the layout and schematic

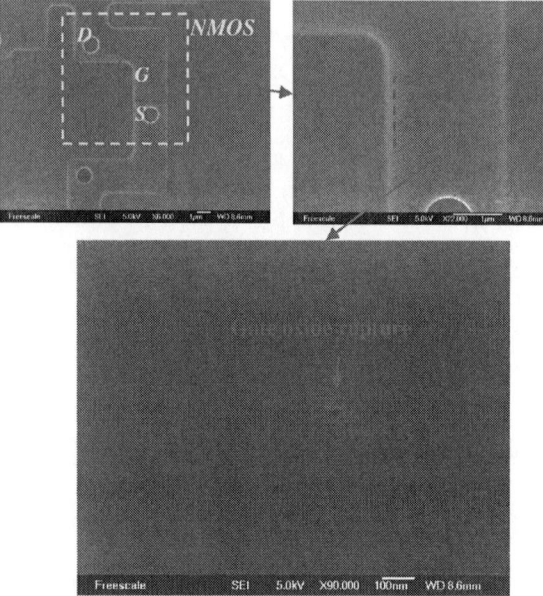

Fig. 9 The gate oxide rupture at the abnormal OBIRCH spot area

IV. CASE STUDY 3

This functional failure case involved a 0.8μm SMARTMOS5AP analog and mixed-mode IC with two metal layers. The functional failure mode was confirmed that OUT1 could not turn on when 10K ohm resistor loaded to GND.

PEM was performed under functional set up and an abnormal emission spot was found (Fig. 10). The abnormal emission spot was examined in the layout/schematic and its location was at a NMOS in I261 module (Fig. 11). Further layout study revealed its location was on the drain side of the NMOS and the shape of the abnormal emission spot was same as the shape of active area in the drain side. So this was a typical NMOS saturated emission phenomena. It revealed the voltage of the NMOS gate (net38) should be under intermediate level, so there was a leakage current triggering that. Thus, those signals and devices relevant to the NMOS gate (net38) in the schematic were studied. And microprobe analysis was performed focused on the area relevant to net38. A leakage I-V curve of net38-VDD was found quickly (Fig. 12).

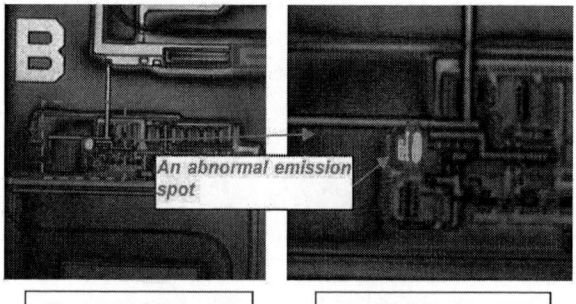

Fig. 10 The emission result of failed IC

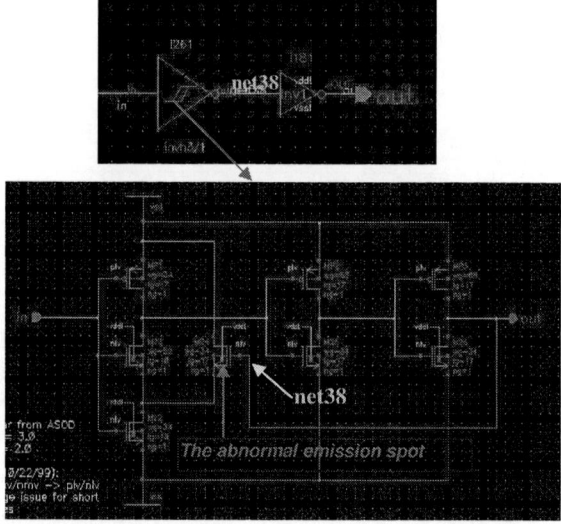

Fig. 11 The abnormal emission spot in the schematic

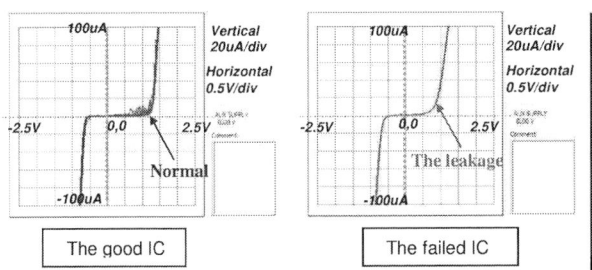

Fig. 12 The I-V curve of net38 and VDD

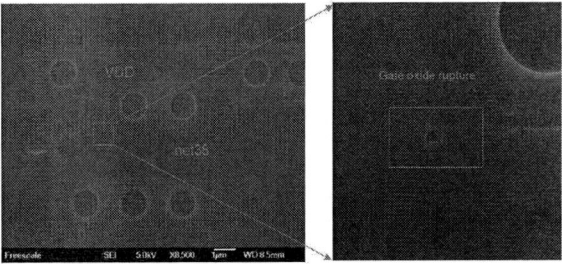

Fig. 14 The gate oxide rupture at the abnormal OBIRCH spot area

Lock-in IR-OBIRCH was employed again to pinpoint the leakage spot in the circuit, while 1V was being forced between net38 and VDD by microprobe needles. An abnormal OBIRCH spot was found at the gate area of a PMOS within an inverter I181 (Fig. 13) whose poly gate connects to net38 and its source connects to VDD. As Fig. 11 shows, the net38 signal connects the NMOS gate of I261 and the PMOS gate of I181. The leakage between net38 (gate) and VDD (source) in the PMOS of I181 resulted in net38 voltage was at an intermediate voltage level, which triggered the NMOS of I261 to be in saturated status and emit lights.

At last, de-process analysis revealed a gate oxide rupture at the abnormal OBIRCH spot area (Fig. 14).

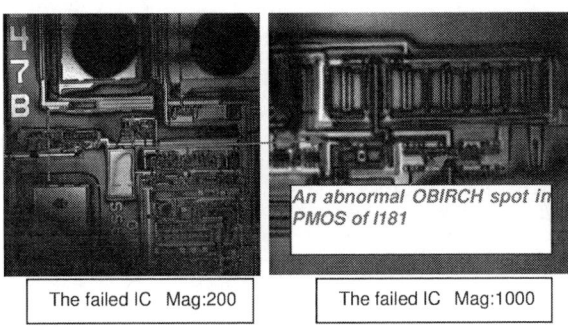

Fig. 13 The abnormal OBIRCH spot images

V. CONCLUSIONS

In this paper, three different functional failure cases were presented to show how to locate the gate oxide rupture in MOS transistor quickly by PEM with the combination of Lock-in IR-OBIRCH. In case 1, PEM located the gate oxide rupture directly. So schematic (microprobe) analysis was not mandatory for this case. In case 2, PEM did not pinpoint the gate oxide breakdown point. Therefore, schematic study and microprobe have to be performed for isolating the failure root cause. Then the gate oxide rupture was localized accurately by Lock-in IR-OBIRCH. In case 3, PEM revealed a valuable clue (the saturated emission NMOS) for further isolating the fault quickly by microprobe, although it did not pinpoint the gate oxide rupture directly. At last, the gate oxide rupture was localized accurately again by Lock-in IR-OBIRCH.

These three successful cases indicated the gate oxide rupture could be located precisely and quickly by PEM and Lock-in IR-OBIRCH techniques.

REFERENCES

[1] Yung-Huei Lee, Neal R. Mielke, William McMahon, Yin-Lung Ryan Lu, and Sangwoo Pae, "Thin-Gate-Oxide Breakdown and CPU Failure-Rate Estimation," *IEEE transactions on device and materials reliability, vol. 7, NO. 1, 2007*, pp. 74-83.
[2] Enrique Miranda, Jordi Sune, Rosana Rodriguez, and Montserrat Nafria, "Soft Breakdown Conduction in Ultrathin (3-5nm) Gate Dielectrics," *IEEE transactions on electron device, vol. 47, NO. 1, 2000*, pp. 82-89.
[3] Bruce MR, Bruce VJ, "ABCs of Photon Emission Microscopy," *Electronic Device Failure Analysis,* vol. 5, 2003, pp. 13-20.
[4] Phang JCH, Chan DSH, Tan SL, Len WB, Yim KH, Koh LS, Chua CM, Balk LJ, "A Review of Near Infrared Photon Emission Microscopy and Spectroscopy," *IPFA 2005,* pp. 275-281.
[5] K. Nikawa, S. Tozaki, "Novel OBIC observation method for detecting defects in Al stripes under current stressing," *ISTFA 1993,* pp. 303–310.
[6] Wu Chunlei, Linda Zhai, Winter Wang, Grace Song, Li Jinglong, Yu Joe et al., "Defect Localization Using Photon Emission Microscopy Analysis with the Combination of OBIRCH Analysis," *IPFA 2010,* pp. 63-68.

Non-Destructive Techniques for Internal Solder Bump Inspection of Chip Scale Package-Ball Grid Array Package

Jason H. Lagar, Rudolf A. Sia and Marlyn C. Grancapal
Analog Devices, Inc., Philippines
Gateway Business Park, Javalera, Gen. Trias, Cavite, Philippines
jason.lagar@analog.com, rudolf.sia@analog.com, marlyn.grancapal@analog.com

Abstract

Non-destructive inspection of Chip Scale Package-Ball Grid Array (CSP-BGA) package for anomalies related to continuity test failures specifically on the internal solder bumps, which connect the die to the Printed Circuit Board (PCB) substrate, is a challenge. Curve trace analysis can trace which internal solder bumps are involved but confirming its physical status needs more reliable and advanced non-destructive techniques. C-mode Scanning Acoustic Microscopy (CSAM) and Micro-Computed Tomography (μCT) scan were evaluated. Results of this paper showed that depending on the physical attribute of the bump anomaly, it could be seen either in μCT scan or CSAM. μCT scan will show those solder bumps with abnormal size or formation and CSAM using a 100 MHz transducer will show those bumps which fractured from its die pad connection. μCT scan can also be utilized for inspecting the metal traces, through hole vias and external solder balls of the PCB substrate. With these two non-destructive techniques, conventional destructive physical analysis techniques like mechanical cross-section, delayering and deprocessing are no longer required saving cycle time and cost. The samples are also saved for further electrical verification, fault isolation and destructive die-level physical analysis, if needed.

Introduction

CSP-BGA technology has received tremendous attention in the electronic packaging area due to its good features such as high I/O capability, short interconnects, miniature size and high performance. Its package assembly is composed of the silicon die, internal solder bumps, PCB substrate, underfill epoxy and external solder balls. See Figure 1 for the top view optical photo and cross-section of a CSP-BGA package. One of its critical components is the internal solder bump which directly connects the silicon die to the PCB substrate. These solder bumps are composed of tin, silver and copper with a diameter of 100μm and protected by an underfill epoxy material. Recent issues for this package related to continuity test failures call for non-destructive techniques to be evaluated. Curve trace analysis can confirm the bumps with open connection or anomalous I/V curve, but determining the failure mechanism using conventional destructive physical analysis techniques such as mechanical cross-section, parallel lapping, laser/chemical deprocessing, decapsulation or FIB cross-section limits further analysis whenever necessary.

In an effort to check the internal bump anomalies non-destructively, Micro-Computed Tomography (μCT) scan and C-mode Scanning Acoustic Microscopy (CSAM) were evaluated. μCT scan, also known as 3D X-ray, is an imaging procedure that utilizes computer-processed X-rays to produce tomographic images or slices of specific areas of a material.

Series of 2D X-ray projections are acquired while sample is progressively rotated to 360 degrees with an increment of around 1 degree per step. These projections contain information on the position and density of absorbing object features within the sample. Digital geometry processing is used to generate a three-dimensional image of the inside of an object from these large series of 2D X-ray projections (see Figure 2) [1]. Since the internal solder bumps have enough density to absorb X-rays, it can easily be seen in a μCT scan. Any inhomogeneity or discontinuity on the bumps can also be detected. Figure 3 shows different μCT scan images of the internal solder bumps and external solder balls of a CSP-BGA package.

Figure 1. Top view optical photo (left) and package cross-section (right) of a CSP-BGA package

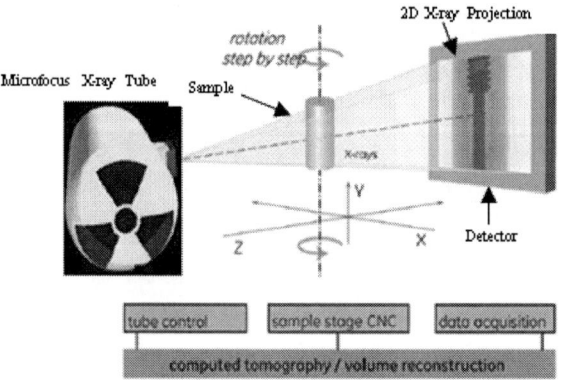

Figure 2. Main components and working principle of a μCT system [1]

CSAM, on the other hand, is based on the interaction between ultrasonic waves and sample material. The amplitude and polarity of ultrasonic waves can be modified along its propagation paths as it interacts with inhomogeneities and discontinuities on the sample material interfaces. As the pulsed ultrasonic waves of fixed frequency from the transducer travel through the sample, reflections occur at discontinuities and disturbances from the interfaces within the sample due to difference in acoustic impedance. Sample and

978-1-4799-3911-4/14 $31.00 © 2014 IEEE

transducer are submerged in deionized water. The reflected pulses are received by the same transducer that functions also as a detector. The amplitude and polarity of the detected signals are analyzed and processed and displayed as pixels with defined gray values thereby creating an image [2]. Figure 4 shows an example of a CSAM image using a 100 MHz transducer focused on the interface of the internal solder bumps and its silicon die pad connection. The bumps, which have uniform dark contrast in the CSAM image, are not showing any sign of disturbance or discontinuity. These are electrically good bumps. The use of a high frequency transducer during CSAM can give high resolution imaging due to its small spot size.

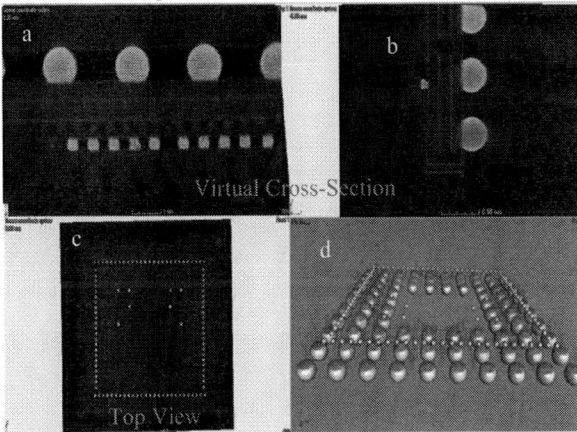

Figure 3. Virtual cross-section views (a, b, c) and 3D reconstruction (d) of μCT scan of the solder bumps and balls in a CSP-BGA package

Figure 4. CSAM image focused on the interface of the internal solder bumps and its silicon die pad connection

These two non-destructive techniques offer some advantages. If the defective bumps can be seen using these techniques then it would be easy to identify which bumps will be subjected for mechanical cross-section or other destructive physical analysis, if needed, to reveal the physical status of the affected bumps. Since these techniques are non-destructive, additional electrical verification, localization analysis and die-level physical analysis will still be possible since the sample and bumps are still intact [3]. This paper will demonstrate the effectiveness of Micro-Computed Tomography (μCT) scan and C-mode Scanning Acoustic Microscopy (CSAM) at non-

destructively analyzing defects of internal solder bumps of a CSP-BGA package.

Experimental Section

CSP-BGA samples failing continuity-related test parameters were screened out using an Automatic Test Equipment (ATE). Curve trace analysis was conducted to verify which of the internal solder bumps were causing the continuity failures. Those bumps with anomalous I/V curves or open connections were identified. CSAM using a 100 MHz transducer was conducted focusing on the internal solder bump-to-die pad connection interface. Those bumps with anomalous I/V curve or open connections but did not show any sign of anomaly or was inconclusive during CSAM were subjected for μCT scan. Those samples showing bump anomalies in either of the two techniques were subjected to mechanical cross-section for verification of the failure mechanism.

Results and Discussion

After screening out those CSP-BGA samples with continuity-related test failures using an ATE and identifying those internal solder bumps with anomalous I/V curve or open connection, CSAM was conducted. CSAM (using a 100 MHz transducer) results showed less signals reflected from those suspected bumps which are shown as bright contrast. Figure 5 shows the difference under CSAM imaging between suspected bumps (bright contrast) and bumps tested to be electrically good (dark contrast).

Figure 5. CSAM images showing the difference in contrast between the suspected internal solder bumps (red markings) and electrically good solder bumps (green markings)

Some of the identified bumps with open connection or anomalous I/V curves were not observed to have any anomaly or sometimes inconclusive under CSAM. Those bumps were subjected to μCT scan. Since solder bumps have enough density to absorb X-rays, those bumps with abnormal formation or size can easily be detected in this technique. See Figure 6 for the different μCT scan views of a solder bump with anomaly. It was also observed that those suspected bumps under CSAM showed no signs of anomaly under μCT scan.

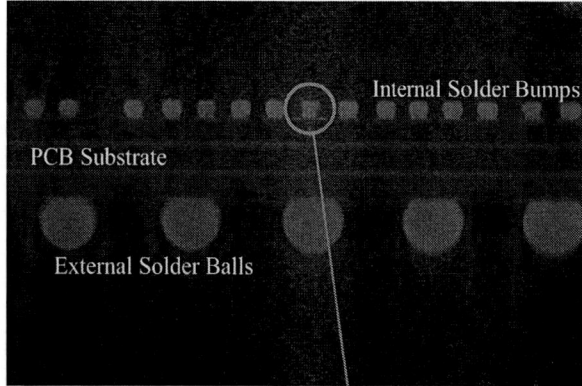

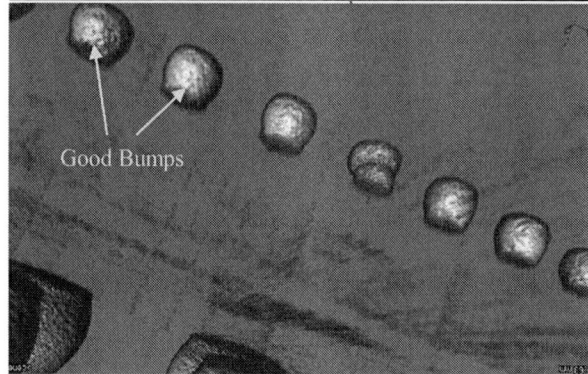

Figure 6. µCT scan virtual cross-section view (top) and 3D reconstructed image (bottom) showing the bump with abnormal size or formation

Samples were then subjected for mechanical cross-sectioning. Results showed that those anomalous bumps under CSAM imaging were fractured from its die pad connection (see Figure 7). The gap between the bumps and its die pad connection caused the distortion of the pulsed ultrasonic waves being reflected to the transducer of the CSAM system. This fracture could not be detected under µCT scan.

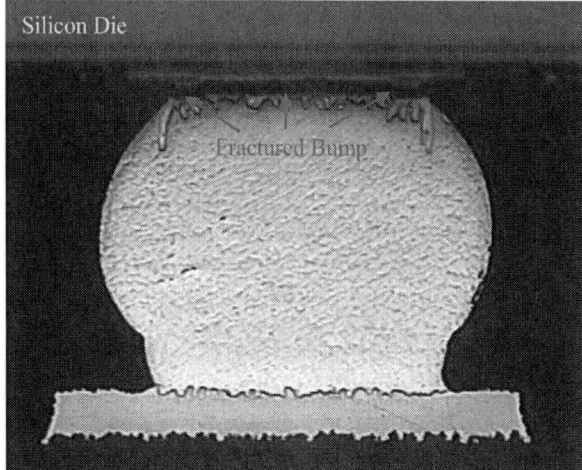

Figure 7. Optical photo post mechanical cross-section showing the fractured bump

Figure 8 shows the bump with abnormal size or formation seen during µCT scan (Figure 6). Similar details can be seen in the virtual cross-section of µCT scan as is apparent in the mechanical cross-section. The advantages of the µCT scan is that it is non-destructive and is relatively easy and fast to take the virtual cross-section views versus the difficulty and time consuming process in preparing a mechanical cross-section prior to inspection. A cross-section of a known good internal solder bump is also shown in Figure 9 for comparison.

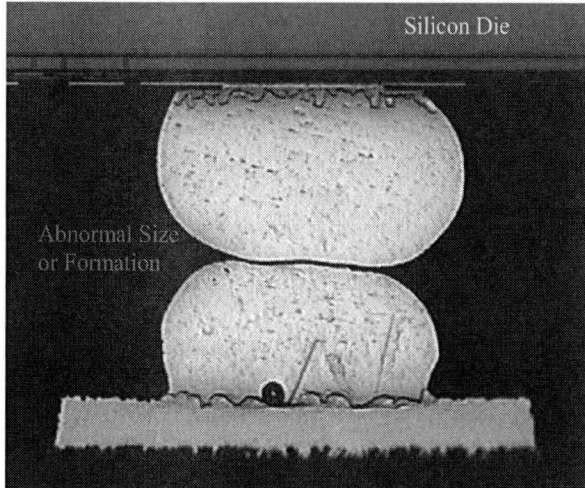

Figure 8. Optical photo post mechanical cross-section of a bump with abnormal size or formation

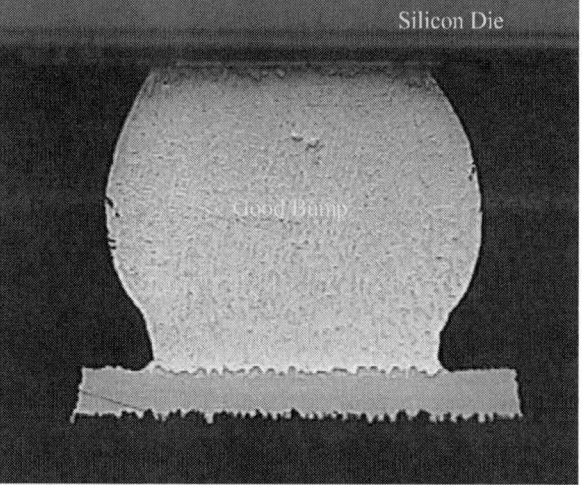

Figure 9. Optical photo post mechanical cross-section of a good solder bump

µCT scan can also be used to inspect the through hole vias and metal traces in the PCB substrate and also the external solder balls if no anomaly can be seen in the internal solder bumps (see Figure 10). If no anomaly can be observed on these parts, then the sample can still be subjected to further fault isolation and die-level physical analyses.

978-1-4799-3911-4/14 $31.00 © 2014 IEEE

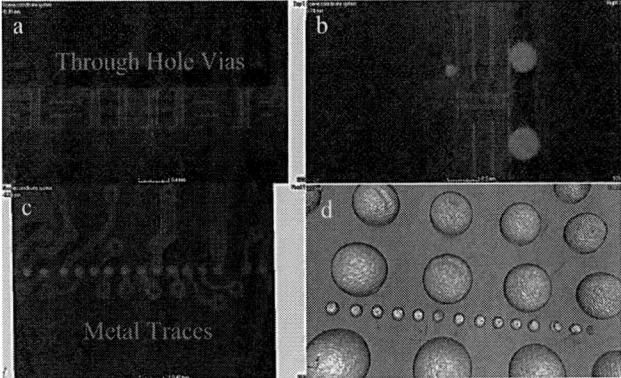

Figure 10. μCT scan virtual cross-section views (a, b, c) and 3D reconstructed image (d) of the through hole vias, metal traces, internal solder bumps and external solder balls in a CSP-BGA package

Conclusions

Micro-Computed Tomography (μCT) scan and C-mode Scanning Acoustic Microscopy (CSAM) were both very effective in revealing the physical characteristics of internal solder bumps with electrical continuity anomalies in a CSP-BGA package. CSAM can detect bumps which fractured from its die pad connections while μCT scan can detect bumps with abnormal size or formation. μCT scan can also be used to inspect the metal traces, through hole vias and external solder balls of the PCB substrate. Both techniques contributed as non-destructive physical analysis, reducing the probability of failure mechanism undetermined and inconclusive analysis. Mechanical cross-section and deprocessing are no longer needed so some process time and cost will be saved and the units will be spared for further electrical evaluations, fault isolation and destructive die-level physical analysis.

Acknowledgments

The authors wish to express their gratitude to Cindy Cabezas, Melanie Cajita and Raymond Mendaros of Analog Devices, Inc., Philippines for their invaluable support towards the completion of this paper. Special thanks also to Mikah Zabaljauregui and Bernardino Mazon for providing the samples to be used for the evaluation.

References

1. Holger Roth and Tobias Neubrand, "Non-destructive Inspection of through Mould vias in Stacked Embedded Packages by Micro-CT", *Proceedings of the 13th Electronics Packaging Technology Conference, 2011.*
2. Lili Ma *et al*, "Application of C-mode Scanning Acoustic Microscopy in Packaging," *Proceedings of the 2007 IEEE.*
3. Morgan Cason and Raleigh Estrada, "Application of X-ray MicroCT for Non-Destructive Failure Analysis and Package Construction Characterization", *Proceedings of the 18th EEE International Symposium on the Physical and Failure Analysis of Integrated Circuits, 2011.*

One Die Logic Analysis through the Backside

M.R. Bruce, L.K. Ross, and C.M. Chua
SEMICAPS Pte Ltd.
28 Ayer Rajah Crescent, #03-01
Singapore 139959
Email: mike.bruce@semicaps.com

Abstract

On Die Logic Analysis (ODLA) uses a scanning optical microscope (SOM) to quickly determine logic timing patterns, and then uses this information to identify logic pattern matches/mismatches on-the-fly from the backside. In this paper, the ODLA system and methodology will be described along with how, in one universal method, it can replace a slew of techniques such as Laser Timing Probe (LTP), Frequency Mapping (FM), and Phase Imaging (PI). It will be demonstrated on a chain of scan cells.

Introduction

Historically, logic analysis from the backside on integrated circuits has been inferred from waveforms using well established techniques like Laser Voltage Probing (LVP) and Laser Timing Probe (LTP) [1-6]; however, these techniques are time consuming as they are based on manual point-by-point probing. More recent laser scanning techniques such as Frequency Mapping (FM) [6] and Phase Imaging (PI) [7] greatly speed up the logic analysis process but they too suffer from limitations: 1) the signal frequency must be precisely known, and 2) the waveform must be highly periodic like a clock otherwise the sensitivity decreases rapidly. For this reason, it is not possible to map more complex patterns typical of automated test pattern generation (ATPG) in a timely fashion.

To overcome these limitations as well as the complexity required by multiple analysis techniques, On Die Logic Analysis (ODLA) was developed. ODLA quickly identifies logic timing states and uses this information to identify logic pattern matches/mismatches (i.e., pass/fail of a pattern) in an automated fashion and, furthermore, represents a more universal method to collect all the information that previously required a combination of techniques (e.g. LTP, FM, and PI) and instruments (e.g. oscilloscope, multiple spectrum analyzers (SA), phase analyzer). In the following sections, the ODLA system and methodology will be described, and then exemplified on a chain of scan cells.

System Description

A schematic of the system is shown in Fig. 1. Compared to standard backside SOM's [6] for probing, the oscilloscope, spectrum analyzers (multiple SA's are required to monitor more than one frequency), and phase analyzers have been replaced by a single high speed digitizer. The rest is processed with software. As described in [6], a scanning optical microscope directs a continuous wave (CW) 1319 nm laser onto the device-under-test (DUT) in a region where a logic cell of interest is expected to be, the modulated return light is then detected by a high speed photo-diode detector (8 GHz bandwidth (BW)).

The high speed digitizer (5 GS/s, 2 GHz BW) receives the amplified signal from the photo-diode and outputs the data to an outboard computer. A reference trigger (such as from the beginning of a loop or clock) initiates the digitization process. A couple of unique features of the digitizer are: 1) it can internally sum 65,000 acquisitions (trigger events) in real time, and 2) it has a dual-banked memory buffer. The net result is there is no data loss or communication latency for acquisition or trigger rates up to 10 MHz. For comparison, a high end digitizing oscilloscope can only collect "burst" of waveform data for about 1,000 acquisitions; after the "burst" the scope must process the data adding 10's to 100's of microseconds in latency. Since it is necessary to average 10,000's to 100,000's acquisitions for a decent waveform (signal-to-noise ratio (SNR) ≥ 3), this latency can cause as much as 95% loss of data for a 1 μsec loop.

The outboard computer continues additional averaging if needed and extracts the logic. Logic is extracted *in-situ* from a live waveform acquisition using a combination of box car averaging, edge enhancement, and thresholding. The digitizer/computer combination can continue acquiring data until the desired SNR is achieved (typically SNR~3) or times out after a predetermined amount of time in the case that no signal is present.

As will be shown in the next section, cell locations for probing are determined from GDSII (Graphic Data Stream) or other Electronic Design Automation (EDA) format. Virtual Layer (VL) files can be created to represent cell areas that are overlaid and aligned to the image. However, instead of conventional scanning, the laser is controlled pixel-by-pixel within the Area-of-interest (AOI) as defined by the logic cell or VL. It is important to note that layers are no longer "dummy" polygons overlaid on the image as is typical in other SOM systems, instead the "polygons" represent active SOM AOI's that have been mapped onto the image where the laser positioning can be controlled pixel-by-pixel for probing or scanning. This can further speed up the analysis by avoiding sampling areas of no interest as is currently practiced in conventional laser scanning.

As is illustrated in Fig. 2, a reference pattern is defined and is compared to an extracted logic pattern to determine if they match ("pass") or mismatch ("fail"). During the test as logic cells are automatically probed pixel-by-pixel, the logic cell polygons are highlighted on the image in green for "pass" or red for "fail" (see Fig. 3 for illustration). It should be noted the raw waveforms for each pixel are also available for inspection if necessary.

Within a logic cell boundary or AOI, it is not necessary to probe every pixel location. For passing cells, only a single positive match is required since the odds of a pass by accident is extremely small; in which case the probe will then move to

978-1-4799-3911-4/14 $31.00 © 2014 IEEE

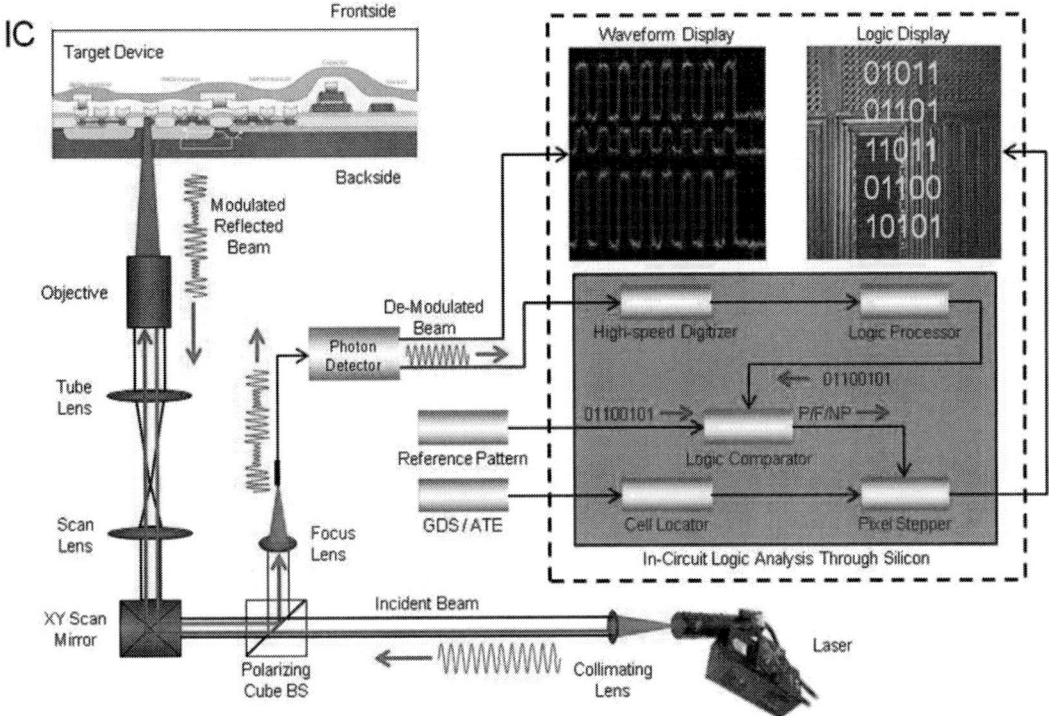

Fig. 1. Schematic of ODLA logic analyzer system with pixel by pixel sampling.

the next cell. For failing cells, a larger sample of pixels is probed; however, not every pixel is probed within the cell boundary (typically about 10-15 pixels) since if the algorithm missed a passing cell (e.g., signal too weak or search grid too sparse) the odds are it will determine the logic chain "recovered" downstream if the so-called failing cell was really passing. Once a passing/failing transition boundary has been found (e.g. see Fig. 3) further focus on the boundary cells can determine the actual failure down to the gate level. Note that it is necessary to index the order of the logic elements in the chain to take full advantage of the efficiencies offered by selective pixel probing. Further efficiencies are obtained when searching in a binary pattern.

Results and Discussion

In this section, ODLA will be demonstrated on a chain of scan cells (28 nm node) in shift mode running at 5 MHz. Fig. 4 shows a backside Solid Immersion Lens (SIL) image of the areas to be probed. The overlaid blue virtual layers (extracted from GDSII) represent unprobed scan cell areas. The overlay has been adjusted to accurately align with the image. Each blue VL cell defines an AOI that can be scanned or probed automatically within the boundaries by the laser. In Fig. 4 there are a total of about 100 cells as represented by the overlaid blue rectangles mapped onto the image.

A laser probe is automatically directed to each cell location (blue rectangles in Fig. 4) and acquires a logic waveform. A sample waveform, logic extracted from the waveform, and reference logic is shown in Fig. 5 for one of

1. Logic State Waveform Display:

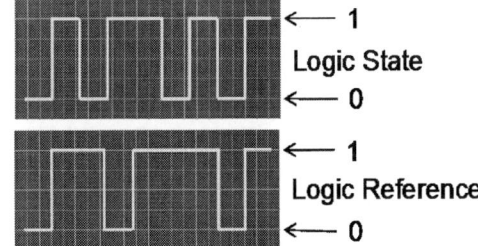

2. Logic State Binary Display:

0 1 0 1 1 0 1 0 1 Logic State

0 1 0 1 1 0 1 0 1 Logic Compare

Fig. 2. (1) Logic state (upper) and logic reference (lower) waveforms, (2) Binary logic pattern comparison between the measured logic state and the reference logic state shown above in (1). Green indicates where specific logic states match and red indicates where specific logic states do not match. In this case, a single failed state renders the entire pattern a "fail" and is highlighted in "red" on the image (see Fig. 3).

the pixels with an active signal. A 5MHz 50/50 duty cycle pattern with the phase adjusted to the first cell is set up as the reference pattern. As mentioned, up to 15 pixels are probed in each single cell in a grid like pattern (user definable). If the logic pattern matches (i.e., passes) the reference at any one

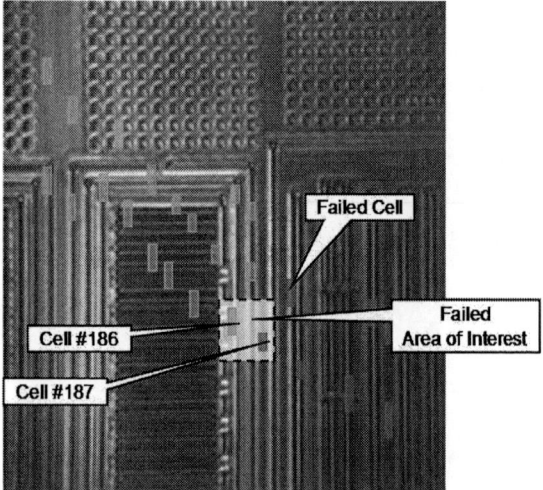

Fig. 3. Illustration of cell-by-cell analysis. Green indicates passing pattern matches and red where failing patterns are observed. The transition from green (passing) to red (failing) indicates where the fail first occurred and represents the true failure area.

pixel within the boundary of the cell, the cell is outlined green and the laser is moved to the next cell. If the pixel fails, the laser steps to the next pixel within the cell until either a pass is achieved or all 15 points in the cell have been probed. Only if all the pixels fail to match the reference pattern within the cell element will the cell be outlined as a fail (red) (note: unprobed cells remain blue until probed). Up to 1500 pixels were probed (not counting the reduction from "passes"). Since each pixel is probed for about 100 msec (about 100,000 waveform acquisitions), the entire chain is probed in less than 3 minutes (note: with a digitizing scope, this would take about 1 hour).

Fig. 6 represents the cells from Fig. 4 after logic probing. Passing (5 MHz reference pattern) cells are outlined in green while the failing cells are outlined in red. The transition boundary between passing and failing represents the region where the chain is broken. In Fig. 6, a clear boundary can be seen between the green passing (lower half of image) and red failing regions (upper half of image). In this particular case, the chain is broken between cells C065 (green) and C066 (red). Further probing within or between cells C065 and C066 can be conducted to isolate the exact failure to gate level; however, that was not done for this case. Note that in this example every cell was probed at least once; however, for much larger chains a binary search would be more efficient and is planned to be implemented in the future. For instance, for a chain with a million cells, only about 20 cells would be required to be probed in a binary search, much less than was done in this study.

In this case study, assuming the frequency is precisely known, it is true that FM and PI together would also be capable of finding the failure; however, with ODLA the complexity is greatly mitigated without compromising analysis time. Hardware wise, FM and PI require multiple SA's (e.g., first harmonic, clock, second harmonic, etc.) and phase information (e.g., network analyzer or lock-in) while ODLA requires only a single digitizer. It is important to note

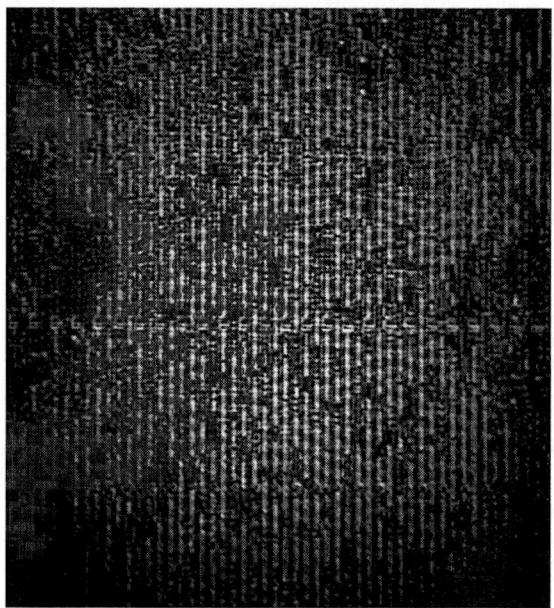

Fig. 4. SIL image of a scan chain area to be probed. The blue rectangles were extracted from GDSII as a virtual layer representing the location and boundaries of the scan cells. The blue rectangles also represent multiple AOI locations for laser probing and scanning.

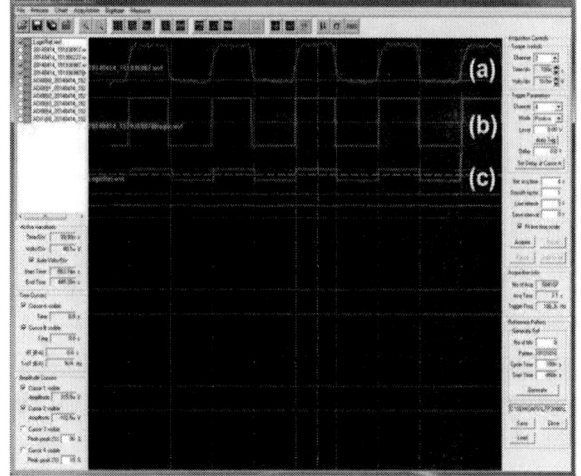

Fig. 5. Example logic extraction from a single pixel of a probed scan chain : a) waveform, b) logic extracted from waveform, and c) reference logic pattern. This case represents a passing condition.

that the purpose of FM and PI is not to measure "frequency" or "phase", but to infer correctness of "logic"; in other words, FM/PI is just trying to determine where the logic went wrong in an *ad hoc* fashion. That is exactly what ODLA does by working directly with the logic itself and mapping "correctness" or "incorrectness" of logic onto the spatial image/layout in an intuitive and timely fashion.

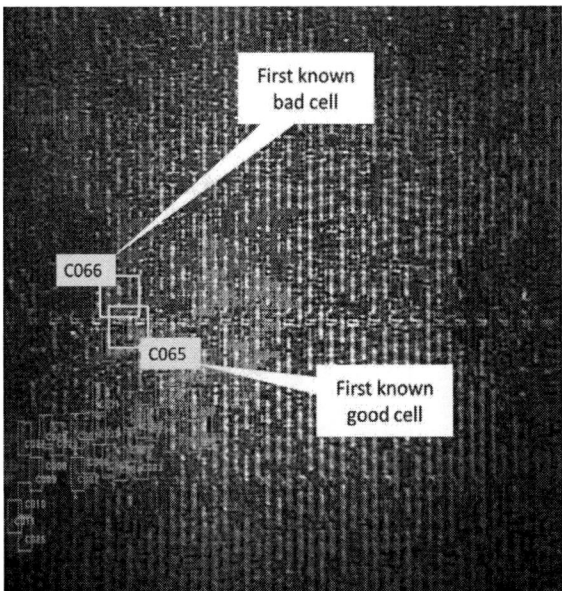

Fig. 6. ODLA probe results from the scan chain in Fig. 4. Probed cells are outlined in green (passing) or red (failing). The blue rectangles on the far left edge of the field of view (FOV) remain unprobed as their boundary locations are undefined (not closed). In all about 100 cells were probed, up to 15 points each. The failure location region is indicated within or between cells C065 and C066.

Conclusions

On Die Logic Analysis (ODLA) has been described and applied to scan chain analysis. ODLA quickly extracts logic timing states which are used to determine "pass/fail" conditions against a reference pattern. The "pass/fail" results can be displayed on the device image and/or layout to determine where a fail occurred. The main limitation of ODLA is the bandwidth and loss of timing edge information; however, in cases where only the logic is desired, the enhancement in acquisition speed is well worth the trade-off in edge accuracy.

For the first time, we can begin to think of probing as more like an ATE (Automated Test Equipment). ODLA is very similar to the ATE method of comparing measured patterns to an expected reference pattern for pass/fail determination; however, ATE only has direct access to the I/O pins so internal logic must be inferred by applying multiple patterns using ATPG. Whereas ATPG diagnosis will try to "guess" where candidate failures can occur, ODLA directly measures the internal logic within the die leaving nothing to "guess work". In the future, it is expected that when ODLA is combined directly with ATPG diagnosis, all candidate failure locations can be automatically probed, thus eliminating any uncertainty as to where the logic failed in the die.

Acknowledgments

The authors acknowledge support of the staff of SEMICAPS Pte Ltd. and GLOBALFOUNDARIES Singapore Pte Ltd.

References

1. Heinrich, H. K. *et al,* "Noninvasive Sheet Charge Density Probe for Integrated Circuits", *Appl. Phys. Lett.* Vol. 48, p. 1811 (1986).

2. Koskowich, G. N. *et al,* "Optical Charge Density Modulation as an Internal Probe for CMOS IC's", *IEEE J. Quant. Electron.* Vol. QE-24, p. 1981 (1988)

3. Pannaccia, M. *et al,* "Novel Optical Probing Technique for flip-chip Mounted Integrated Circuits*", Proceedings of the IEEE International Test Conference (ITC)* Washington, DC 1998, p. 740.

4. Bruce, M. R. *et al,* "Waveform Acquisition from the Backside of Silicon using Electro-optic probing", *Proceedings of the 25th International Symposium for Testing and Failure Analysis (ISTFA)* Santa Clara, CA 1999, p. 19.

5. Kasapi, S. *et al,* "Laser Beam Backside Probing of CMOS Integrated Circuits", *Microelectron. Reliab.* Vol. 39, p. 39 (1999).

6. Koh, L. S. *et al,* "Laser Timing Probe with Frequency Mapping for locating Signal Maxima", *Proceedings of the 33rd International Symposium for Testing and Failure Analysis (ISTFA)* San Jose, CA 2007, p. 33.

7. Ng, Y. S. *et al,* "Scan-shift debug using LVI Phase Imaging", *Proceedings of the 39th International Symposium for Testing and Failure Analysis (ISTFA)* San Jose, CA 2013, p. 322.

8. Bruce, M. R. *et al,* "Through Silicon In-circuit Logic Analysis for localizing Logic Pattern Failures", *Microelectron. Reliab.* Vol. 52, p. 2043 (2012).

Temperature Effect on Reflected Laser Probing Signal of Multiple Elementary Substructures.

M. M. Rebaï [a,b], F. Darracq [a], J-P. Guillet [a], D. Lewis [a], P. Perdu [b], K. Sanchez [b].

a- IMS Laboratory, CNRS UMR5218, University of Bordeaux, 351 Cours de la libération, - 33405 Talence – France.

b- CNES DCT/AQ/LE, 18 Avenue Edouard Belin - 31401 Toulouse Cedex 9 Fance.

Phone No.: +33 54 000 28 14, Fax No.: +33 55 637 15 45, email: mohamedmehdi.rebai@ims-bordeaux.fr

Abstract- Electro-Optical Probing (EOP) has shown its efficiency in the world of failure analysis. The different external physical parameters effects, especially the temperature, on the EOP signals are not well known and not that much described in the literature. In addition to thermoreflectance, the temperature is a parameter that affects directly the free carrier's distribution and carrier mobilities inside the semiconductor. Temperature also modifies the absorption coefficient and not only the refractive index as known in the thermo-reflectance domain. All the physical and environmental parameters contribute to the modulation of the reflected laser probing beam onto structures under test.

In this paper we will expose the origins of the reflected laser beam and the impact of the temperature on the EOP signal. For the first time, all the parameters, including temperature, have been taken into account. It opens the door of laser probing techniques improvements in failure analysis of submicron devices.

1. INTRODUCTION

Laser probing techniques are non-destructive powerful tools in the failure analysis domain. In up to date technologies, the laser probing techniques are performed from the backside of the studied IC's. One of the challenges is related to the spot size that can cover a cluster of multiple elementary structures in different electrical states. Therefore, the reflected signal is a complex contribution of each cluster covered by the laser spot.

When we probe an active region of an elementary structure, the incident and reflected laser beam are affected by the internal physical and electrical parameters of each layer of the IC [1]. These internal parameters are themselves sensitive to the external parameters such the used wavelength of the laser probe beam and the temperature. To determine the impact of the temperature on the intensity of the reflected beam we present in this paper some of the obtained simulations and experimental results. It characterizes the influence of the temperature on EOP techniques.

2. LASER BEAM – SEMICONDUCTOR INTERACTION

In laser-semiconductor interaction, many physical and electrical parameters have to be taken into account. Depending on the ICs technology and the probing parameters, the reflected beam is function of the absorption coefficient and the refractive index. The latter are themselves functions of doping density, temperature, wavelength and electrical field. In previous paper [1], we presented multiple models that we used in this study.

For this study, we used the following equations to predict the ratio Y(V) and normalized ratio $Y_{Normalized}(V)$ between the intensity of the reflected I_{ref} and the intensity of the incident I_{inc} laser beams when our laser spot covers multiple regions of a cluster of elementary structures.

$$Y(V) = \frac{I_{ref}}{I_{inc}}$$

(1)

$$Y_{Normalyzed}(V) = 100 * (Y(V) - Y_{min})/(Y_{max} - Y_{min})$$

(2)

3. SIMULATION AND EXPERIMENTAL RESULTS

3.1. EOP simulation and experimentation on an integrated PMOSFET

Using previous models presented in [1], we simulate the intensity of a reflected beam on a PMOSFET for a varying temperature from 290K to 340K with a 1350 nm laser beam wavelength (the temperatures are chosen in function of the experimentation capacity). We consider a laser $1/e^2$ radius value equal to 3µm. Such value has been experimentally obtained using a high magnification and long working distance microscope objective (100X, N.A=0.7).

The technological data of the simulated structures are from the commercial 0.35 µm BiCMOS technology used to create the ISLAND (Integrated Structures for Laser ANalysis Demonstrations) test vehicle.

The considered PMOS transistor is a reproduction of the PMOSFETs of the ISLAND device. The structure is simulated with a TCAD tool for different temperatures to have the information on the density profile distribution under the source, gate and the drain. The following figure represents a schematic profile of the simulated PMOSFET.

Using all TCAD information, we calculate the ratio Y between the reflected and the incident laser beam intensity under the S, G and D cuts and their contribution in the total EOP signal.

978-1-4799-3911-4/14 $31.00 © 2014 IEEE

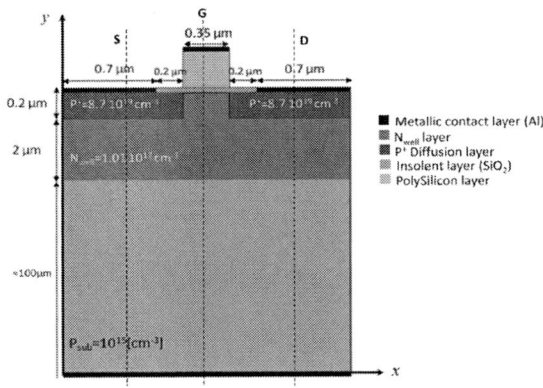

Fig. 1: Schematic of the simulated PMOSFET

Fig. 2-a- represents the normalized variation of the total ratio Y for a varying V_{sg} and a V_{sd} = 3.3V for different temperatures and Fig. 2-b- represents a comparison between simulation and experimental results when the PMOS is biased for two cases: V_{sd} = 3.3V and Vs_g = 1.5V ; V_{sd} = 3.3V and V_{sg} = 2.5V.

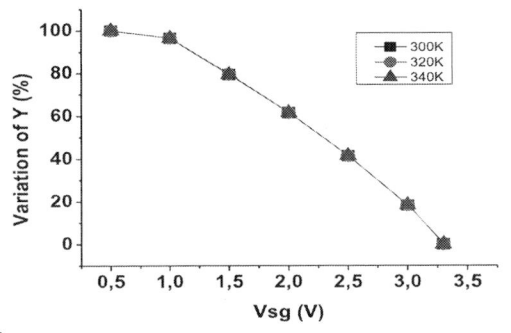

-a-

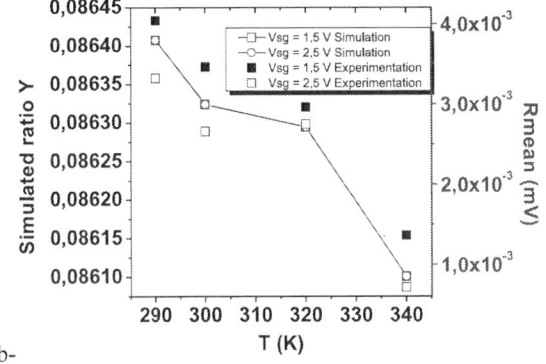

-b-

Fig. 2: Simulation of the ratio Y of the PMOSFET when V_{sd} = 3.3V and a varying V_{sg} for different temperatures.
-a- Variation of the normalized ratio Y.
-b- Variation of the ratio Y.

We can see on Fig. 2-a- that the normalized ratio Y decrease when the V_{gs} increase for a V_{sd} = 3.3V this is due to

the increase of the free carriers density during the polarization of the PMOS that causes an increase of the absorption coefficient and a decrease of the refractive index. We can also see that the normalized variation for the different temperatures is almost the same but on Fig. 2-b- we can see that the ratio Y decreases when the temperature increases. The decrease due to the temperature is more important than the variation due to the biased voltage. This is due to the impact that has the temperature on the different parameters such the free carriers' distribution, the carrier mobilities [2], the absorption coefficient [3] and the refractive index [4].

The following figure show the impact of the temperature on the free carrier absorption coefficient in the P^+ diffusion regions (y=102.18μm), under the gate (y=102,18μm) and the well region (y=101.11μm) when V_{sg} = 3V and V_{sd} = 3V for λ=1350nm.

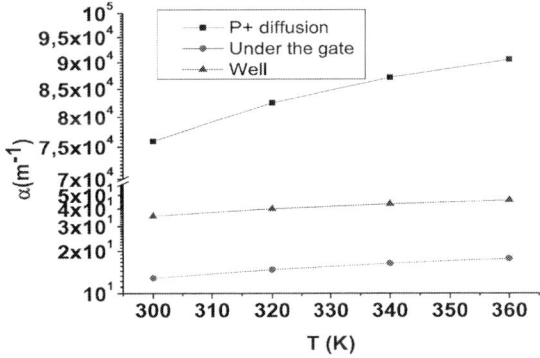

Fig. 3: Absorption coefficient α in the different regions of the PMOS when Vsg = 3V and Vsd=3V as function of the temperature

One can see on Fig. 3 that the free carrier absorption increases with the temperature. This is due to the thermal agitation of the free carriers that increases their distribution density inside the silicon. We can also see that the temperature effect on the free carrier absorption is more important in the highly doped regions (P^+ diffusion regions). All along the structure the mean value of this increase is 28% between 300 and 360 K.

The temperature effect on the refractive index is presented on the following figure for the same previous regions.

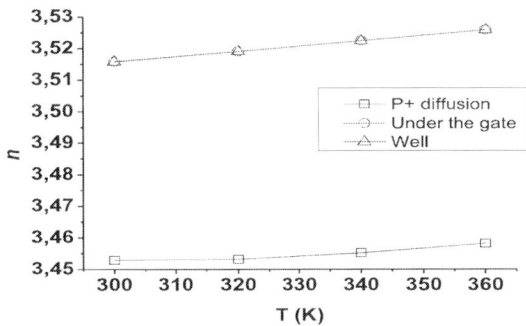

Fig. 4: Refractive index n in the different regions of the PMOS when Vsg = 3V and Vsd=3V as function of the temperature

As for the absorption coefficient, Fig. 4 shows that the refractive index is also sensitive to the increase of the temperature and increases with it. But the refractive index is less sensitive to the temperature than the free carrier absorption coefficient. All along the structure the mean value of this increase is just 0.24% between 300 and 360 K. The variation of the reflected beam with the temperature seems to be dominated by the variation of the absorption coefficient.

3.2. Simulation and experimentation on a simple inverter

The experimentations on the simple inverter were done on the ATLAS platform of the IMS laboratory of Bordeaux. The used optical and electrical setup is presented in Fig. 5. The laser source produces a continuous beam at a wavelength of 1350 nm and a maximum power of 80 mW. The incident power on the Device Under Test (DUT) is about 9.6 mW. The minimum laser spot diameter ($1/e^2$) is 6μm. We used a PIN InGaAs photodiode (PD1 in Fig. 5) to monitor the incident laser beam power and an InGaAs avalanche photodiode (APD in Fig. 5) to measure the reflected beam power. The APD and PD1 are plugged to a lock-in amplifier for a synchronous detection. The waveform generator and the lock-in amplifier are monitored by the computer to automate the experimentation. The probed structure is an integrated inverter of the ISLAND vehicle (see Fig. 6). An InGaAs camera has been used to localize the inverter.

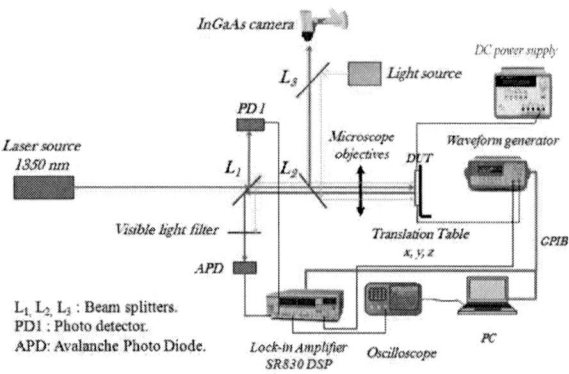

L₁, L₂, L₃ : Beam splitters.
PD1 : Photo detector.
APD: Avalanche Photo Diode.

Fig. 5: Optical and electrical setup for the experimentations.

During the experiment, the inverter bias (V_{DD}) was 3V and a pulsed square signal (V_{in}) was applied to the input of the chain. The parameters of the input signal are a pulse duration of 500 μs, a frequency of 480 Hz and an amplitude varying from 0.1 to 3V. The pulse duration and the frequency have been chosen in order limit the self-heating of the chip and to optimize the signal to noise ratio. The lock-in amplifier is used to extract the signal from noise at the input frequency of the biased DUT. The laser spot was centered on the middle of the inverter as shown in Fig. 6.

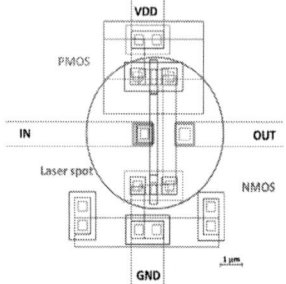

Fig. 6: Illustration of a laser beam probing a simple inverter.

Each measurement R_{mean} is a mean value of 100 acquisitions. The Fig. 7 presents the variation of R_{mean} as a function of V_{in} for different temperatures.

One can see on Fig. 7 that we have a peak of the modulated signal for $V_{in} = 1.6V$, whatever the temperature. This corresponds to the switching voltage of the inverter. Both transistors are then passing in this configuration of our probed structure.

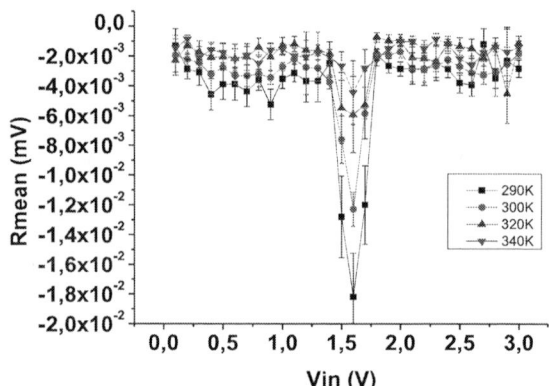

Fig. 7: Experimentation results of a reflectometric measurement on a simple inverter for different temperatures.

The amplitude of the peak decreases with the increase of the temperature (see Fig. 8) and the loss of the measured amplitude is about 76.6% between 290K and 340K (Fig.7).

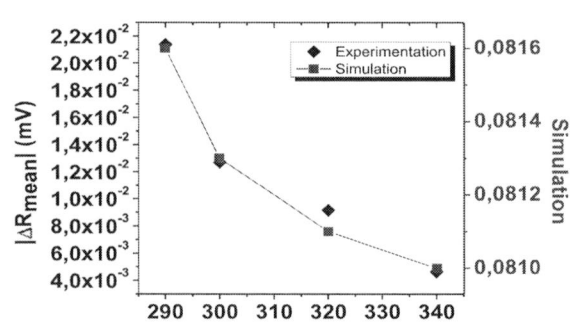

Fig. 8: Absolute value of the peak's amplitude as a function of the temperature (measured and simulated) on a simple inverter, when $V_{in} = 1.5V$.

Fig. 9 represents the EOP signal when the laser spot is focused on the NMOS for $V_{ds} = 3V$ and a varying V_{gs} ($V_{in} = V_{gs}$) for different temperatures.

We can see on Fig. 9 that EOP signal (R_{mean}) decreases with the increase of V_{gs} bias whatever the temperature. But the decrease is less pronounced for the highest temperature values. As it has been observed in fig.7, an increase of the temperature induce a decrease of the EOP signal

A similar experimentation was held for a laser focused on the PMOS transistor of the simple inverter. Fig. 10 represents the measured EOP signal from the PMOS.

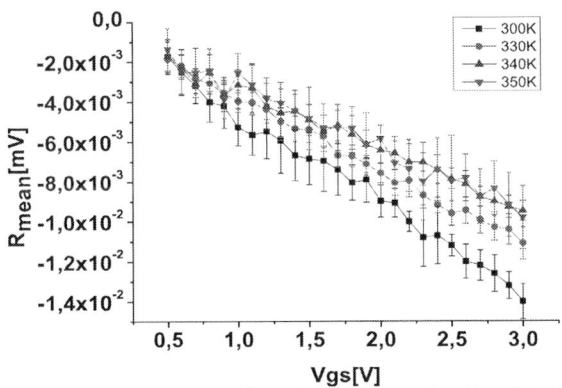

Fig. 9: Measured EOP signal from the NMOS as a function of V_{gs} for different temperatures.

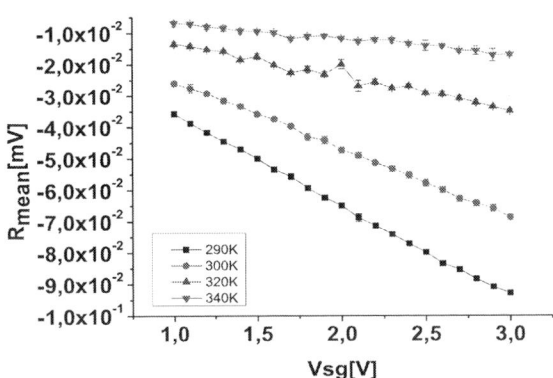

Fig. 10: Measured EOP signal from the PMOS as a function of V_{sg} when $V_{sd}=3V$ for different temperatures.

We can see on Fig. 10 that the measured signal from the PMOS transistor decreases with the applied voltage. This was expected as shown previously in the simulation part. We can also see that the EOP signal is more important for the PMOS than for the NMOS. It may be explained by differences in doping levels.

Moreover, the variation of the EOP signal between 300 and 340 K is greater in the case of the PMOS transistor than in the case of the NMOS transistor.

3.3. Experimentation on a NAND gates chain

The two inputs NAND gates chain is constituted of 100 gates wired as inverters. The same optical setup and measurement configuration were used. We biased the structure with a constant $V_{dd}=3V$ and a square pulsed input voltage (V_{in}) varying from 0.5 to 3.1V with a frequency of 480Hz. Each gate is composed of four MOS transistors (2 PMOSFET and 2 NMOSFET). The gate size is 6x8 μm. The laser beam is focused in the middle of the gate, thus partly covering all the transistors during the probing of the chain.

Figure 11 presents the EOP signal from the NAND chain as a function of V_{in} for different temperatures. An increase of the input voltage results in a decrease of the EOP signal. We also have a strongest absolute value of the signal for cooler temperatures. V_{in} is the gate potential of the MOSFETs.

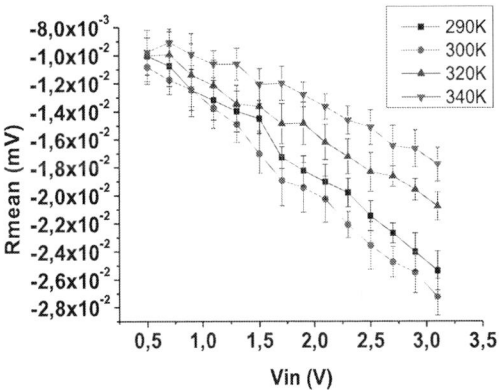

Fig. 11: EOP signal from the probed region of the XOR chain as a function of V_{in} for different temperatures.

When V_{in} is low, the two PMOSFET are ON and the two NMOSFET are OFF. The situation is reversed when V_{in} goes high. From the EOP signal levels from Fig. 9 and Fig. 10 we can suppose that the main contribution to the total EOP signal is from the PMOS transistors than the NMOS ones.

3.4. Experimentation on a XOR gates chain

This chain is constituted of 2-inputs XOR gates. Each XOR gate has one input is connected to the ground. We biased the structure with a constant $V_{dd}=3V$ and a square pulsed input voltage (V_{in}) varying from 0.5V to 3.1V with a frequency of 480Hz has been applied to the other input.

The XOR gate is composed of ten transistors (5 PMOSFETs and 5 NMOSFETs). During the experiment, the laser was centered on three PMOSFETs. The variation of the measured EOP signal as function of V_{in} and for different temperatures is given in Fig. 12.

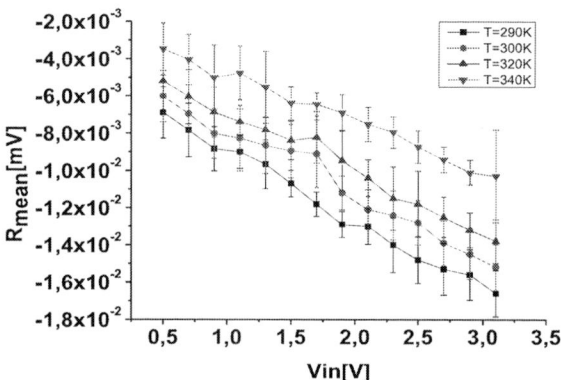

Fig. 12: EOP signal from the probed region of the XOR chain as a function of V_{in} for different temperatures.

The EOP signal decreases with the increase of V_{in} as shown on Fig. 12. When the input is low or high, two of the three probed PMOSFETs are ON. We also see on Fig. 12 the temperature effects on the EOP signal for this case of study. As we saw on the previous study cases, the measured signal decreases with the increase of the temperature.

4. DISCUSSION

All the experiments done on various structures of the ISLAND test vehicle indicate a non-negligible effect of the temperature on EOP measurements. First simulations tend to indicate a major role of the sensitivity of free carrier absorption to the temperature. Of course, these results have to be confirmed by testing other designs and technologies. However, they open two questions.

The first is the use of external temperature for the EOP testing of integrated circuits. Would it be easier to obtain a good contrast when the device temperature is different from the room temperature? The use of the EOP technique on deep submicron devices is not straightforward. Cooling the device may facilitate it.

The second is relative to the devices' self-heating. In our experiment we used short square pulses to bias the DUT in order to minimize the effect of self-heating, but some tests requires bias conditions which would induce an inhomogeneous temperature by self-heating. Our study shows that different temperature of the device active areas may result in a significant EOP contrast which cannot be directly related to any electrical activity.

5. CONCLUSION

Probing submicron elementary structures is getting more and more difficult with their size reduction. In this paper we studied the impact that has the temperature in a complex reflectometric measurement to understand its influence on different parameters that affect the reflected probing laser beam onto an Integrated Circuit (IC). Using our previous mathematical models obtained from the analysis of the physical and optical path of the laser beam onto the ICs and following the simulation methodology proposed in previous

paper [1], we presented some simulation and experimental results of the impact of the temperature on an EOP technique for some structures. We presented the impact of the temperature on the physical parameters such as the absorption coefficient and the refractive index in elementary structures by simulations and saw that the absorption coefficient and the refractive index increase with the increase of the temperature. Experimentation results on the influence of the temperature on the EOP signal from the test of some integrated structures of the ISLAND devices are given. We saw that our EOP signal is better for cooler temperatures. This work is significant for devices that heat up during power up and cooling can be a solution for EOP improvement.

REFERENCES

[1] Mehdi M. REBAÏ et al, "How to interpret the reflected laser probe signal of multiple elementary substructures in very deep submicron technologies" Conference Proceedings of ISTFA, , (San Jose, CA Nov 2013), pp 471-480.

[2] S. Noor Mohammad et al, "Temperature, Electric Field and Doping Dependent Mobilities of Electrons and Holes in Semiconductors", Solid State Electronics, Vol. 36, No. 12, (1993), 1667-1683.

[3] R. A. Soref and B. Bennett, "Electro-optical Effects in Silicon", IEEE Journal of Quantum Electronics, Vol.QE-23, (1987), 123-129.

[4] H. H. Li, "Refractive Index of Silicon and Germanium and Its Wavelength and Temperature derivatives", J.Phys. Chem. Ref. Data, Vol 9, No 3 (1980) 561-658.

Defect Localization Enhancement using Light Induced CI-AFP

N. Dayanand, A.C.T Quah, C.Q.Chen, S.P.Neo, G.B.Ang, M. Gunawardana, Z.H.Mai, J.C.Lam

Product, Test, Failure Analysis Group, GLOBALFOUNDRIES Singapore

60 Woodlands Industrial Park D, Street 2, Singapore 738406

Phone No:(65) 6670-1508, Email: Dayanand.NAGALINGAM@globalfoundries.com

Abstract

This paper describes the effectiveness of using light induced Current Imaging – Atomic Force Microscopy (CI-AFP) to localize defects that are not easily detected through conventional CI-AFP. Defect localization enhancement for both memory and logic failures has been demonstrated. For advanced technology nodes memory failures, current imaging from photovoltaic effects enhanced the detection of bridging between similar types of junctions. Light induced effects also helped to improve the distinction between gated and non-gated diode, as a result enhanced localization of gate to source/drain short.

I. INTRODUCTION

As technology node shrinks, traditional SEM based Passive Voltage Contrast (PVC) technique has become less effective to pick up front-end of line defects as junction leakages and resistive open defects become more subtle. Thus, marginal PVC contrast difference between good and bad samples requires further verification and Current Imaging – Atomic Force Probing (CI-AFP) [1] has become an effective technique used at tungsten contact level to localize and characterize transistor level defects. CI-AFP is typically performed at contact mode with tip biased at a small positive or negative voltage and a scanned current image is created by measuring the leakage current flowing from the tip to sample or sample to tip. The measured current image reflects the source/drain/well junction leakage and gate oxide leakage at a particular biased voltage.

Conventional current imaging was done in the dark to minimize the interferences of the junction leakage current with the photo-generated current. This technique has been very effective in localizing memory and logic hard failures such as high resistance contact, junction leakage, gate-S/D short, bridge between NMOS and PMOS transistors, [1], [2] for devices in bulk process. However, there are certain defects that are not easily detectable by conventional CI-AFP. One such failure mode is a bridge between two or more similar type junctions (p+/nwell - p+/nwell or n+/pwell - n+/pwell). In this paper, we described 2 case studies on the use of light induced effects to improve the effectiveness of localizing a short between 2 similar junction types in the memory array and also a case study where light induced effects help to distinguish between gated and non-gated diode in a logic gate to S/D short scenario.

In this paper, the CI-AFP analysis and nanoprobing measurements were conducted using Multiprobe AFP MPIIb system. Fig.1 shows a snapshot of the AFP DUT stage, 2 out of a total of 4 probe heads in position and a 40X optical lens [3]. The DUT stage is able to rotate and move in X, Y and Z directions. The DUT stage is also fixed onto a precision snap stage for precise sample movement within +/- 200um in X & Y directions. The 40X optical lens with 100X digital zoom is mounted above the stage for gross tip to tip positioning and also for navigation to the region of interest (ROI). The optical lens is illuminated with a tungsten light source which is turned off during current imaging or nanoprobing for accurate junction leakage and transistor characterization. However, in this work, the light source is deliberately turned on during current imaging to generate photo-carriers for fault localization.

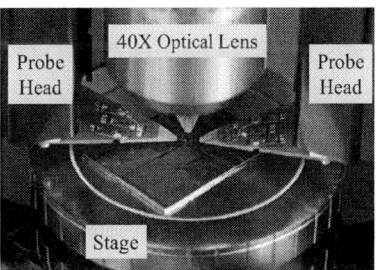

Fig.1 AFP stage and housing.

II. RESULTS AND DISCUSSION

SCENARIO1: SRAM Pull-Up Gate–S/D leakage detection using PV effect.

Scenario 1 illustrates two cases where light induced CI-AFP was used to effectively localize SRAM Pull-Up (PU) transistors Gate-Source/Drain short which is otherwise not easily detectable with conventional current imaging analysis. Fig.2 shows the layout of a single 6T SRAM cell which is commonly used in advanced technology node memory array. While the NMOS pull-down (PD) and pass-gate (PG) transistors have isolated S/D contacts, the PU transistors have rectangular contacts (CAREC) to locally connect the gate of one PU to the drain of the other PU transistor. Thus, gate of the PU exhibits a P+/N diode junction characteristic.

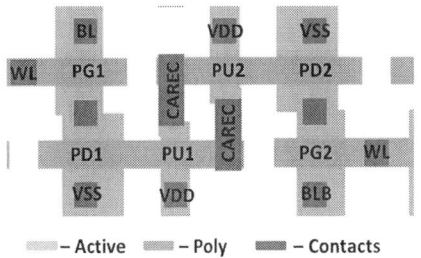

Fig.2. 6T SRAM cell layout

978-1-4799-3911-4/14 $31.00 © 2014 IEEE

SRAM single bit failure was identified on a 40nm device and physical failure analysis was performed at the failed bitmap address. Both SEM inspection at via1/metal1 and PVC analysis at via 1 and contact levels, as shown in fig 3(b), observed no anomaly. Conventional CI-AFP was employed and Fig 4(a) and 4(b) shows the current image with a small positive and negative bias to the tip (±0.5V) respectively. In the positive bias scan, PU S/D junctions are forward biased resulting in diffusion current flowing from tip to sample denoted by white contrast as shown in fig 4(a). The PD & PG S/D contacts were in reverse bias with very small leakage current. For the negative biased scan in fig 4(b), the NMOS S/D junctions were in turn forward biased resulting in current flow from sample to tip denoted by black contrast. From these results, one could deduce that there is no resistive active contacts and CARECs, no NMOS gate-S/D short, no gate oxide breakdown and no junction leakages. However, we could not identify if there was a bridge among PU contacts, resistive open on PG gate contact, transistor S/D bridge or if the failure is due to transistor mismatch. Thus, conventionally, an extensive 6 transistors characterization, which is time consuming, would usually have to be performed to further localize the defect location.

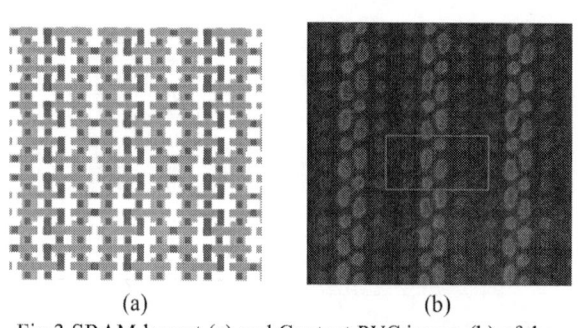

(a)   (b)

Fig.3 SRAM layout (a) and Contact PVC image (b) of the failed cell (highlighted in red box).

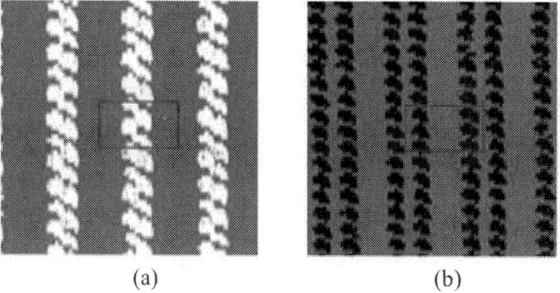

(a)   (b)

Fig.4 CI-AFP with +0.5V (a) and −0.5V bias (b). Failed cell is highlighted in red box

On the other hand, Light induced CI-AFP was used with tip at -0.5V negative bias as shown in fig 5(a). It was observed that the 2 PU CARECs are showing a darker contrast, indicating a larger reverse bias leakage as compared to the CARECs from neighboring good cells. A simple two point nanoprobing between the two CARECs indeed revealed an electrical short, as shown in fig 5(b), that would cause the cell to fail. Further SEM inspection after wet etch observed a silicided polysilicon residue at the STI region resulting in a bridge between the 2 CARECs as shown in fig 6.

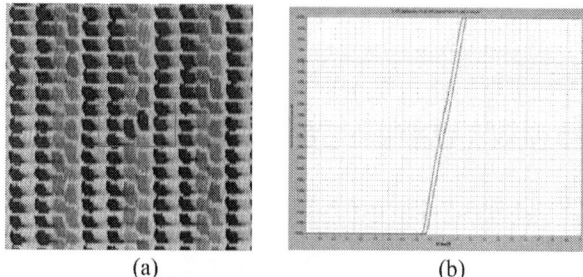

(a)   (b)

Fig.5 Light induced CI-AFP at -0.5V (a) showed distinct contrast at 2CAREC. Nanoprobing confirmed short (b).

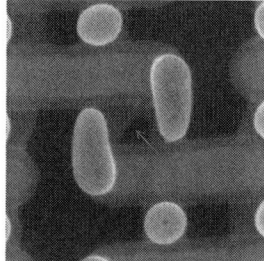

Fig.6 SEM image showing silicided polysilicon residue in STI region causing bridge between 2 CARECs.

The same light induced current imaging was applied on another similar SRAM failure where no abnormal contrast was observed at contact level from conventional CI-AFP analysis. Fig 7 showed the light induced CI-AFP scan with small negative bias to the tip. It was evident that there was a distinct darker contrast observed on both the CAREC and VDD node as compared with other PU contacts. Further 2 point nano-probing verified CAREC-VDD short and SEM inspection at poly and active layer revealed a minute poly residue shorted the poly that was connected to the abnormal CAREC to the VDD contact as shown in fig.8.

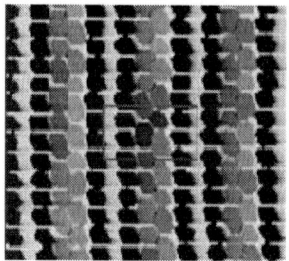 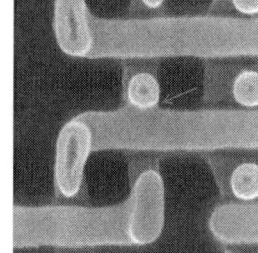

Fig.7 Light induced CI-AFP at -0.5V showed distinct contrast between CAREC and VDD

Fig.8 Poly residue resulting in short between CAREC and VDD.

The mechanism behind the contrast difference can be attributed to the difference in the photovoltaic effect on the shorted and non-shorted junctions. Fig.9 showed the current density vs voltage (J-V) characteristics of a p-n junction in the dark (black curve) and under illumination (blue curve) [4]. As visible light impinges on the p-n junction, electron-hole pairs (E-H) are generated. These minority carriers flow across the depletion region, resulting in open circuit potential difference, V_{OC}, built up between the junction where p-type electrode is positively charged while the n-type electrode is negatively charged. The short circuit current density, J_{SC}, can be

978-1-4799-3911-4/14 $31.00 © 2014 IEEE   376

measured during AFP scanning when the tip comes in contact with the junction and closes the circuit. Effectively, the illuminated *J-V* characteristic of the junction is shifted down by photo-generated current density, J_{Ph}, given by eqn (1) [5]

$$J_{Ph} = qG(L_n + W + L_p) \qquad (1)$$

Where, *G* is the *E-H* pair generation rate dependent on the light intensity, *Ln* and *Lp* is the minority carrier diffusion length for electrons and holes respectively and *W* is the width of the depletion region.

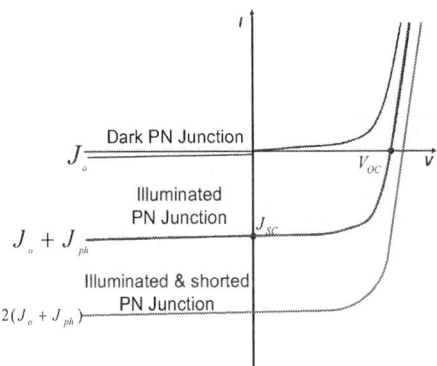

Fig.9 *J-V* characteristics of shorted and non shorted junction with and without illumination

For the 2 memory failure cases described above, the bridge due to the conductive defects would result in parallel connections between 2 p+/nwell junctions as illustrated in schematic diagram in fig 10.

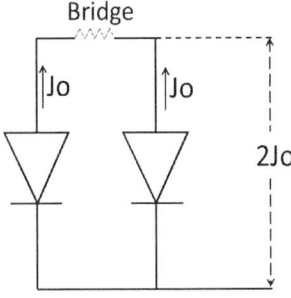

Fig.10 Schematic of 2 p+/nwell junctions in parallel.

In typical reverse biased current imaging, the reverse saturation current, $2J_o$, at a small reverse biased voltage of the shorted junctions remain too small to be detectable by the system. In the forward bias scan, the large diffusion current could easily mask out the difference between shorted and non-shorted junctions. Thus, by comparing the difference in the photo-generated current in the reverse bias condition, the contrast difference between shorted and non-shorted p+/nwell junctions can be easily observed from the additional J_{Ph} on the contacts of the shorted p+/nwell junctions as shown by the red curve in fig 9 . This methodology can also be applied to logic array but one has to take note of the complications from the irregularities of the junction widths.

SCENARIO2: PV effect on gated and non-gated diode to precisely localize logic Gate – S/D short

Following is a 28nm case where light induced CI-AFP precisely localized gate – S/D short in a transistor array structure. Electrical fault isolation identified an exclusive spot at a transistor array in SRAM decoder which is shown in fig11.

Fig.11 EMMI image showing spot at SRAM decoder.

Standard PFA approach was carried out with layer by layer de-processing with SEM inspection and PVC analysis at CA level, however no anomaly was observed. The EMMI spot was identified to be on a PMOS array and hence conventional CI-AFP was performed with a positive bias (+0.5V ~ +1V) to the AFP tip. Strong gate leakage was observed at the structure, shown in fig.13; however the leaky S/D had to be identified through nano-probing. Light induced CI-AFP scan at zero bias surfaced out the leaky S/D with distinctive contrast as shown in fig.14. Further two point nano-probing between gate and leaky node confirmed short. STEM image of a defect in core logic with similar failure mode is shown in fig.15.

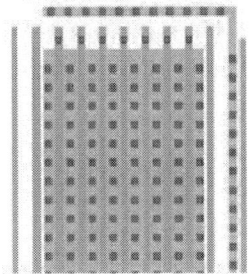

Fig.12. PMOS Transistor array structure

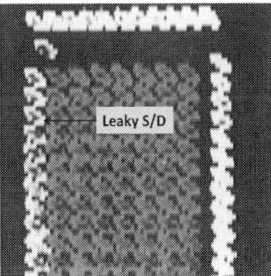

Fig.13. CI-AFP showing leaky gate at +0.5V bias.

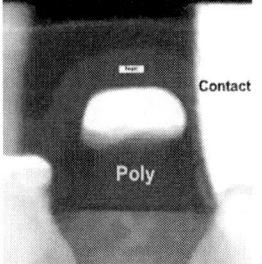

Fig.14. Light induced CI-AFP at zero bias surfaced out the leaky S/D node

Fig.15. STEM image showing Poly shorted to contact

The contrast difference between shorted and non-shorted nodes can be achieved due to photovoltaic effect on gated and non-gated junction. When visible light is illuminated on the sample, E-H pairs are generated in both p+/nwell and nwell-p-sub junctions. Since the area of nwell–psub junction is bigger than p+/nwell junction, the excess carrier(light induced carrier) will be decided by nwell-psub junction. Therefore, excess electron will be induced in the nwell region. p+/well junction can be treated as normal junction. While performing an AFP scan across a normal S/D contact, it creates a path for the excess electron to diffuse from nwell to p+ active as shown in fig.16(a). This photo-generated current is reflected as white contrast in current image.

Alternatively for a gated junction (gate-S/D short), the ionized impurities in the nwell underneath the gate, will be shared between p+ active and gate. This causes some field lines from the nwell to terminate on the ionized P+ Active and some to terminate on the gate, thereby reducing the junction barrier [7]. While performing AFP scan on such a gated junction, the amount of excess electron diffusing from nwell to p+ active will increase in this region due to reduced junction barrier as shown in fig.16(b). Therefore contrast difference is achieved due the difference in amount of electrons diffusion between a gated and non-gated junction.

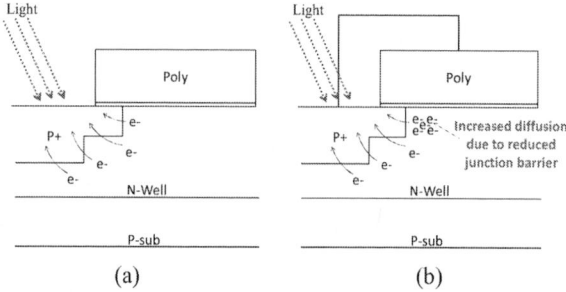

(a)   (b)

Fig.16 Electron diffusion in an ordinary junction (a) and gated junction (b)

Although the case study describes the scenario of single poly gate transistor, we believe that the same method could be used on multiple finger transistors, commonly found in logic circuitry as shown in fig.17, to distinguish the defective gated junction without going through tedious nanoprobing process.

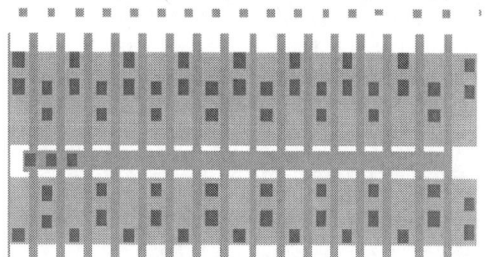

Fig.17 Typical Transistor Array structure in logic circuitry

III CONCLUSION

The effectiveness of light induced CI-AFP to enhance defect localization had been demonstrated in two different scenarios where leakages could not be initially detected or precisely localized through conventional CI-AFP. With the results in hand, a successful effort was made to theoretically understand the mechanism involved in achieving a significant contrast difference at the leaky nodes with CI-AFP. A detailed explanation on the mechanisms has been illustrated for each case.

ACKNOWLEDGMENT

The authors would like to thank Product, Test and Failure Analysis group for the constant support and encouragement to experiment the technique with real time samples and actively participate in results discussion to understand the mechanism.

REFERENCES

1. Tom X. Tong et al, "Current Image Atomic Force Microscopy (CI-AFM) combined with Atomic Force Probing (AFP) for location and characterization of advanced technology node", *Proc Int Symp Testing & Failure Analysis (ISTFA)*, pp 42-46, 2004.
2. C.Q. Chen et al, "Application of AFP on failure analysis of 65nm technology SRAM", *Proc. Intl. Symp. Physical & Failure Analysis of Integrated Circuits (IPFA)*, pp148-151, 2011.
3. Multiprobe Inc., AFP Manual, P-EM-05 V4.0, July 2008.
4. Purnomo Sidi Priambodo et al, "Electric Energy Management and Engineering in Solar Cell System, Solar Cells - Research and Application Perspectives",
5. P. Würfel, "Physics of Solar Cells: From Principles to New Concepts", Wiley-WCH, Weinheim, 2005.
6. Roy. K et al, "Leakage current mechanisms and leakage reduction techniques in deep-submicrometer CMOS circuits", *Proc. of the IEEE*, Vol.91, No.2 (2003), pp305-327.
7. Badih El-Kareh et al, "Silicon Devices and Process Integration-Deep Submicron and Nano-Scale Technologies", pp277-286.

Cluster matching in
Time Resolved Imaging for VLSI analysis

S. Chef[1,2], S. Jacquir[2], P. Perdu[1], K. Sanchez[1] and S. Binczak[2]

[1]CNES, DCT/AQ/LE, Bpi 1414, 18 Avenue Edouard Belin, 31401 Toulouse, France
[2]Le2i UMR CNRS 6306, University of Burgundy, 9 Avenue Alain Savary, 21000 Dijon, France
Phone: +33380399035   Email: samuel.chef@u-bourgogne.fr

Abstract- **If scaling has the benefit of enabling manufacturers to design tomorrow's integrated circuits, from the failure analyst point of view it also has the drawback of making devices more complex. The test sequence for modern VLSI can be quite long, with thousands of vector. Dynamic photon emission databases can contain millions of photons representing thousands of state changes in the region of interest. Finding a candidate location where to perform physical analysis is quite challenging, especially if the fault occurs on a single vector. In this paper, we suggest a new methodology to find single vector fault in dynamic photon emission database. The process is applied at the post-acquisition level and is based on clustering algorithm and nearest neighbor research.**

I. INTRODUCTION

As reported in [1], the acquisition of photons emitted by switching transistors in integrated circuits enables the failure analysts to monitor in time and space the activity at circuit's nodes. The use of time correlated single photon counting sensor such as micro-channels plates (MCPs) or single photon avalanche diodes (SPADs) makes the characterization of each acquired photons at the picoseconds scale possible. Over the years this technique, also known as Time Resolved Imaging (TRI), has become a key tool in failure analysis laboratories for the defect localization in faulty devices.

In order to retrieve electrical parameters from these 3D optical data, several post-acquisition processes have been reported over literature. Usually, the first step is to separate noise from signal photons. Techniques like spatio-temporal photon correlation (STPC) [1], Positive Photon Discrimination (PPD) [2] or iterative binarizations [3] can be used to highlight signal photons, or at least signal spots at a coarser level, among background noise. Once it has been done, signal propagation through the area [4] or emission frequency of the spots [5] can be deduced from the filtered database. These are just two examples of electrical parameters among many others. It must be noticed that most of these processes require a good signal over noise ratio to give optimal results.

In 2013, a new approach based on unsupervised classification has been proposed [6]. Its aim is not to extract electrical parameters but to directly find in 3 dimensions the groups of photons - or clusters - that can be linked to a suspicious node. Once these clusters have been found, the potential defect can be characterized in time (when or how often emission from this node occurs) and in space (giving a candidate location to start deeper analysis). The time characterization can be as important as the space information because both can be correlated with the design and the electrical test pattern to have a better understanding of what is happening.

Nevertheless, in [6], the suspicious clusters were found by analysis of their statistical properties. In the case where the defect leads to an extra or missing emission, this kind of study should be irrelevant as no abnormal statistic should stand out. Another way of highlighting these clusters among these huge databases should be found. In this paper a method to find this kind of photon emission event is reported. As a follow-up to the work started in [6], the process relies on clustering and the comparison of clusters.

The next section of the paper explains in details the clusters matching procedure. In the following section, the method is applied on real databases acquired on a 8 bit microcontroller. A conclusion finishes this paper.

II. METHOD DESCRIPTION

Photons are randomly generated in saturated MOSFET during switching. Because of the stochastic nature of the process, it would be nonsense to directly compare a single photon from a database A to another one from the database B, The probability to spot a photon at the exact same time and position during both acquisitions is really low. On the other hand, if single test vector and node are considered, for a sufficient number of repetition of the electrical pattern, a group of photons occupying a certain volume in the (x,y,t) space will be acquired. Because of the implied physical phenomena, the photon emission in dynamically stimulated MOSFET is a stationary process. If the operation is repeated for a second acquisition, then another group of photons will be acquired at almost the same location and time and occupying a similar volume. It means that these two groups might be comparable because they came from the same node and test vector.

There are various metrics to characterize a group of points. One of the first that comes to mind is the centroid coordinates. Even if photon emission is random, it can be expected that for one local emission event, the coordinates of the centroid do not change much between the two samples. Ensuring that in both databases, a centroid is found at equivalent position is a way to grant that a node is properly working on this test vector. On the contrary, if it is not the case, then it should send a warning: something is happening for one sample but not for the other. Of course, if it is a systematic thing, that is to say for every test vector, no emission happens in one the sample, then there is no need to start such process as a simple spatial projection should enable the analyst to find this node. But, if the fault occurs only from time to time, or worst on the single test vector level, then it is worth it. The association of one cluster from database A to another from the database B is the cluster matching procedure.

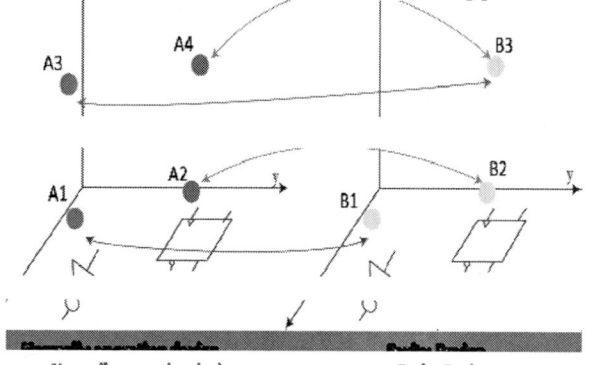

Fig. 1. Cluster matching schematic.

To have a better understanding of cluster matching, a schematic is available in Fig. 1. Two hypothetical samples of the same integrated circuit are considered, A and B. The first one plays the part of the normally operating device whereas B is the faulty integrated circuit. At this step, clustering has already been performed and clusters are pictured by their centroid. Each one is labeled. The label A1 indicates the first cluster from the database A, A2 the second one and so on. The NND computation gives the association marked by the arrows of different colors. It must specified hat in order to make sense, the centroids must be expressed in the same referential system. The cluster A3 is associated to B3, so as A4. If everything was fine, the association relation should grant unity, which is not the case here. A4 occurs at the same location as B3, so it is its true match. A3 is related to the inverter gate whereas B3 is on the flip-flop gate.

A simple way to isolate an extra or missing event is to compute the distance between a cluster centroid in A to all of the centroids in B. The minimum distance indicates its match. If everything is normal, the match distance should be quite small. If there is an extra or a missing cluster, then the nearest neighbor distance (abbreviated hereafter NND) should be bigger than the average one. As a conclusion, this metric is a mean to isolate the extra or missing emission events or clusters. To find outliers, a plot of all of the nearest neighbor distances like a histogram or a box plot helps to have a synthetic view of the results and immediately visualize it.

1. The main steps of the process are the following:
2. Clustering.
3. Centroid computation.
4. Nearest neighbor research (matching).
5. NND analysis.
6. NND threshold setting.
7. Visualization of photons belonging to clusters of NND above threshold.

The first step is to get the photons cluster from the databases acquired on the two devices. In [6], it has been shown that DBSCAN algorithm (Density-Based Clustering for Application with Noise) [7] fits to this application. Indeed, it has been

designed to deal with noise and does not require the number of output class (number of clusters) as a parameter, compared to other clustering algorithms. Predicting this number is quite difficult with TRI databases. After clustering, each cluster centroid position is computed. For each one in the database A, its nearest neighbor in the database B is sought. As the space used for photon is a 3D Euclidian space, the Euclidian distance (or L2-norm) is enough to perform the nearest neighbor search. The final step of the procedure is the computation of the NND histogram. Outliers are found by choosing the centroids which NND are bigger the rest of the distances.

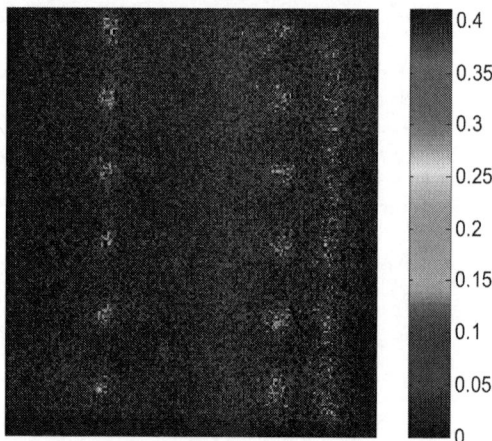

Fig. 2. Static difference of the two emission images

III. APPLICATION TO RAM ANALYSIS

The cluster matching process is applied to analyze databases acquired on RAM embedded in a 8-bit microcontroller. The test sequence consists in reading different data stored in memory at the same address. The sequences last 22 µs and 24 µs. The static difference between the two emission image (usual database comparison process) is reported in Fig. 2. The color bar indicates the difference in arbitrary unit (a.u.). The difference does not have highlighted one spot in particular, meaning that when an emission event happens in only one of the database, it is not always at the same place. It also does not occur on every test vector.

For this analysis a variation of DBSCAN is used, STDBSCAN [8]. The benefit of this algorithm over classic DBSCAN is an easier parameters choice for TRI data. The dxy and dt parameters can be set according to hardware. For instance, if the sensor has a precision in time of 100 ps, then there is no use to choose a smaller value for the first shot. If the photon density is really high and clusters seem to have merged, it can be useful to try with more selective values. The parameters has been set to $dxy = 2$ pixels and $dt = 100$ ps. On the first attempt, the minimum number of points in the neighborhood to consider a photon as signal, i.e. r, has been defined to 5 photons. As a counterpart, STDBSCAN favors the splitting of clusters, which can lead to an over segmentation.

This is not a major concern for this application. If a group of photons is split in two, as long as there is enough photons to ensure that it is signal, the NND for both subgroups should quite high if there no matching event in the other set. Eventually, the photons for both subgroups will be displayed and it might be possible to merge them after the process.

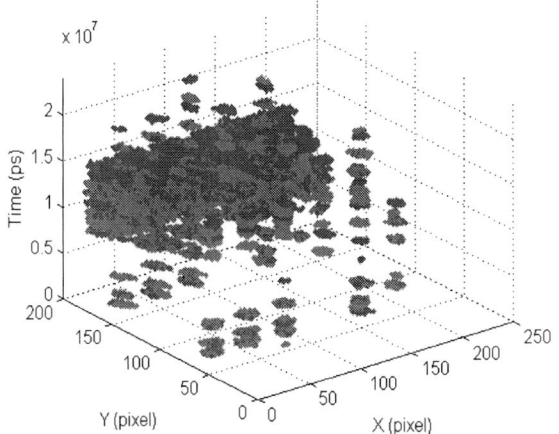

Fig. 3. Photons remaining after clustering for both databases. Set A in red and set B in blue

After STDBSCAN, 638 clusters have been discovered in database A against 705 clusters in database B. As the sets have been acquired on the same device with the same bias conditions, there are good chances that the difference in the number of clusters can be due to emission events. Although, several cases should be considered. First, it remains possible that some clusters have been split when they should be gathered. It can be the case when the local density is low and in one dataset, only a single photon makes the link. In the second database, because there is no photon at this location, or because it is a little further from the clusters, the association cannot be made. Second, it is not because B has more clusters than A that it means more emission events happens in B and only the NND from B to A should be studied. It is also possible that something happening in A does not exist in B. Both NND histogram shall be analyzed.

The scatter plot of photons remaining after clustering is reported in Fig. 3. The groups of photons validate the second hypothesis. For both sets, there are things happening only in one of them. There is a high photon density in the area defined over the interval [150; 200] in Y, between 5 and 10 µs. It is a bit concerning because there is a higher probability of merging, as stated above, leading to inaccurate results. For other locations, clusters are well defined, meaning that the cluster matching strategy is more likely to succeed with this kind of configurations.

The box plots of the NND from B to A and from A to B are reported in Fig. 5. This kind of plot helps to have a synthetic view of data repartition by showing where are located the first order statistic. The boundaries of the blue box indicate the first quartile and the third quartile. The red line is the median value. The black lines at the end of whiskers are set according to the lowest and highest datum inside 1,5 times the inter quartile range (respectively the lower and higher limit). Points outside of these values are considered as outliers.

The two box plots exhibits a difference of repartition for the two NND sets. The distribution of the NND from A to B is more located than from B to A. The distribution "A to B" seems to be closer to a normal distribution than the "B to A", which in turns is more likely to follow a khi-square law. This is just some observations and only statistical tests can confirm and reject them.

In B, there are 167 clusters which NND is larger than the higher whisker end, located at 5930 a.u. For the second box plot, the top whisker value is 538 a.u. and there 132 outliers. These two values are chosen as thresholds for photons isolation. In Fig. 5 and 6, photons belonging to clusters of NND greater than thresholds are isolated for both databases. In overall, results seem to be correlated with what can be observed on Fig. 3, even though some photons are clearly missing in Fig. 5 compared to Fig. 3 because the threshold was set too high.

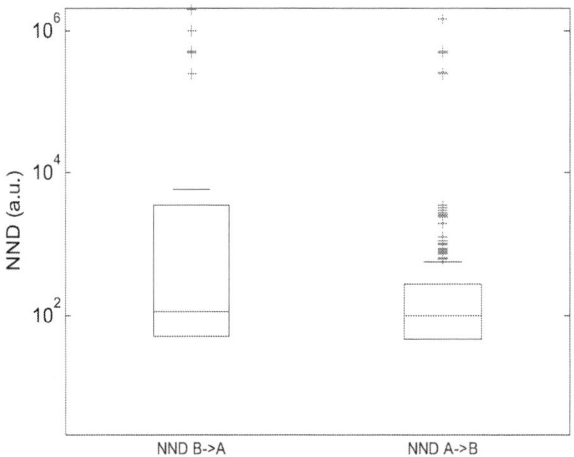

Fig. 4. Box plot of the NND repartition

IV. DISCUSSION : APPLICATION BOUNDARIES

There are several critical steps in this application. First, the success of the cluster matching procedure is determined by the success of clustering. If the parameters are not well defined, for instance if the threshold to identify a photon as a core object is set too high, an emission event can be completely missed. This issue is similar to one of the concern met with STPC. Although, the real issue here is the signal over noise ratio problem. A node that emits only few photons during the whole acquisition time will be difficult to be identified as signal, because the photon distribution in its neighborhood will be closed to the noise neighborhood distribution. Currently, the only way to ensure that these groups of photons can be identified by clustering is to play with the acquisition parameters. To be more specific, a longer duration can increase the number of collected photons and remains the best choice.

It is known that photon emission is dependant of bias voltage and operating frequency of the device. It can be tempting to increase these parameters in order to solve the above mentioned issue, but it would bring another one. Changing voltage has an impact on MOSFET switching timing. If the timing are not the same, then the hypothesis on which the cluster matching procedure is built, something normal should happen at the same

978-1-4799-3911-4/14 $31.00 © 2014 IEEE   381

place and time in both devices, is no longer true. Only comparable things can be compared. In the case of a deterministic delay, some preprocessing strategies can be applied. As an example, the cross-correlation between the optical waveforms of the whole area for the two acquisition can help to find an average delay. If there is an asynchronous signal that commands the operations, the timing difference appears as a random phenomenon. The cross correlation can still indicate an average value but it is worthless to precisely correct each and every one of the switching timings. The cluster matching procedure has a really low probability of succeed.

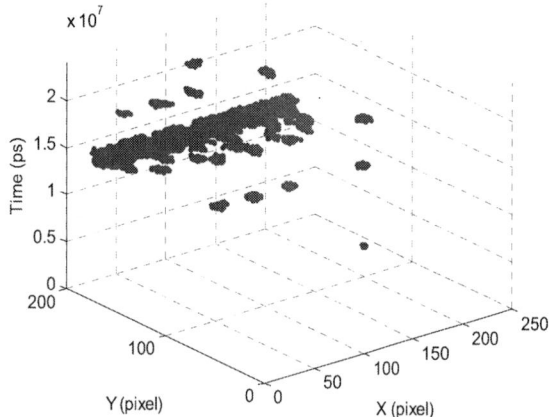

Fig. 5. Photons belonging to clusters of high NND in B

A spatial difference between the two acquisition scenes, for instance the positioning of the device is not same, is less problematic. Different strategies can be considered. Still, the active part of the sensor remains the same between acquisitions (for instance, 200x200 pixels) so if the positioning is not the same, even after correction, some part of the acquired area for one of the sample do not have a match in the second database. A part signal can be unusable for the cluster matching analysis and there is a loss of information.

The last critical point is the threshold definition. In the case of a large value, there are big chances that the isolated photons are true differences between databases. In the application reported in the previous section, the threshold was set according to the repartition of the NND, based on the first order statistics. Still, there were some isolated photons seen in Fig. 3 that have not been found in Fig. 5. The definition of outliers used here is more relevant in the case of a normal distribution, which is not exactly what have been observed in this case. It is a starting point for threshold setting but in near future, this part of the process should be investigated more precisely.

V. CONCLUSION

Finding logic fault when it occurs at the single vector level can be quite challenging. In this paper, a method for the analysis of dynamic photon emission by cluster matching has been reported. It has been applied on RAM activity analysis, which can be seen as equivalent to logic fault analysis. Nevertheless, the process is still at its early stages of development and several points remain to be improved in order to have a robust and reliable procedure.

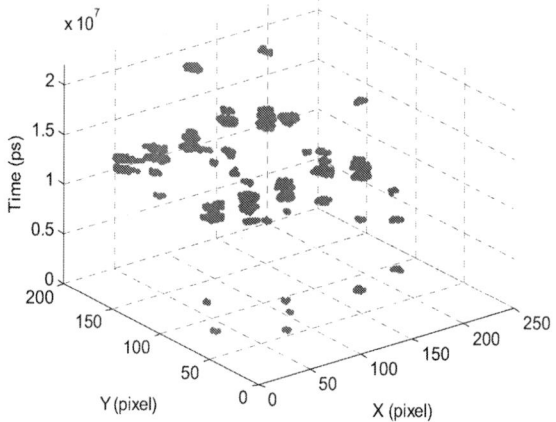

Fig. 6. Photons belonging to clusters of high NND in A

ACKNOWLEDGMENT

Authors would like to thank Renesas Electronics (databases), Hamamatsu Hpks for their technical support (TriPhemos) and the Regional Council of Burgundy for its financial support.

REFERENCES

[1] J.C. Tsang, J.A. Kash, D.P. Vallett, "Picosecond imaging circuit analysis," *IBM Journal of Research and Development*, Vol. 44, No. 4, pp. 583-603, 2000.

[2] R. Desplats *et al.*, "A new approach for faster IC analysis with PICA: STPC-3D," *Proceedings of the 10th IEEE International Symposium on the Physical and Failure Analysis of Integrated Circuits (IPFA)*, pp. 45-53, 2000.

[3] S. Chef, S. Jacquir, K. Sanchez, P. Perdu, S. Binczak "Filtering and emission area identification in the time resolved imaging data," *Proceedings of the 38th International Symposium for the Test and Failure Analysis of Integrated Circuits (ISTFA)*, pp. 264-272, 2012.

[4] G. Bascoul, P. Perdu, D. Lewis, "Signal propagation analysis by digital lock-in Time Resolved Imaging," *Proceedings of the 19th IEEE International Symposium on the Physical and Failure Analysis of Integrated Circuits (IPFA)*, pp. 1-5, 2012.

[5] S. Chef, S. Jacquir, K. Sanchez, P. Perdu, S. Binczak, "Frequency mapping in dynamic light emission with wavelet transform," *Microelectronic reliability*, Vol. 53, No. 9, pp. 1387-1392, 2013.

[6] S. Chef, P. Perdu, G. Bascoul, S. Jacquir, K. Sanchez, S. Binczak "News statistical post-processing approach for precise fault and defect localization in TRI database acquired on complex VLSI," *Proceedings of the 20th IEEE International Symposium on the Physical and Failure Analysis of Integrated Circuits*, pp. 136-141, 2013.

[7] M. Ester, H.P. Kriegel, J. Sanders, X. Xu, "A density-based algorithm for discovering clusters in large spatial database with noise," *Proceedings of the 2nd International conference on Knowledge Discovery and Data Mining (SIGKDD)*, pp. 226-231, 1996.

[8] D. Birant, A. Kurt, "ST-DBSCAN: An algorithm for clustering spatial -temporal data," *Data & Knowledge Engineering*, Vol. 60, No. 1, pp. 208-221, 2007.

Fusion Prognostics-based Qualification of Microelectronic Devices

Michael Pecht, Elviz George, and Arvind Vasan
Center for Advanced Life Cycle Engineering (CALCE)
University of Maryland
College Park, Maryland, USA
Phone: (301) 405 5323. Fax: (301) 314 9269. Email: pecht@calce.umd.edu

Preeti Chauhan
Intel Corporation
Chandler, Arizona, USA

Abstract- The rapid evolution of electronic products has resulted in numerous choices for customers. This has made for intense competition between manufacturers to reduce costs and minimize the time to market for their products. One bottle-neck in getting products to market is the qualification process, which has traditionally been time-consuming and often inadequate to prevent failures in field. In particular, in the past decade, there have been significant numbers of microelectronic devices that have passed qualification tests but failed in the field. The resulting costs of these failures have been in the billions of dollars. Thus, there is a need to develop approaches to qualification methodologies that quicken the development time but also prevent product failures in the field.

This paper discusses the current state of qualification practices in the electronics industry. Then, an alternative approach, called fusion prognostics, for qualification is presented that can make the process more efficient and cost-effective. This approach involves an *in-situ* qualification process that incorporates a fusion of machine learning techniques and physics-of-failure based prognostics. The machine learning techniques are used to monitor the degradation behavior during testing. On the other hand, the physics-of-failure techniques identify critical failure mechanisms and the acceleration factors.

I. BACKGROUND

There are numerous examples where failure to adequately qualify a microelectronic device resulted in costly field failures, exorbitant warranty costs, and loss of reputation and revenue for the companies. For example, between 1983 and 1995, 22 million Ford vehicles were affected by defective microelectronic ignition modules that could cause the vehicle to stall abruptly while driving [1]. The ignition modules were found to intermittently fail when the engine was hot and function again when cooled, without leaving any physical evidence of the failure. Ford projected warranty returns of 10 per 100 modules (10%) before five years or 50,000 miles based on its previous ignition system. Actual field returns were 40% on average, and as high as 99% in some cases, but because many of the ignition modules exhibited intermittent failures, dealerships had high no-fault-found rates. Ford settled the ignition lawsuit by agreeing to pay for

repairs in vehicles with the flawed ignition modules at a cost of approximately $2.7 billion [1], [2].

In 2010, Toyota recalled 1.33 million vehicles due to issues in the engine control microelectronics module, which caused the vehicle to fail to start or to stall while driving [3]. This issue was due to cracks that developed at certain solder points or on the electronic devices used to protect circuits against excessive voltage on the engine control unit's (ECU's) printed circuit board [4].

A family of mold compounds with red phosphorous flame retardant was introduced as the encapsulant for microelectronic devices in the 1996. Sumitomo Bakelite introduced these mold compounds as an environmentally friendly alternative to bromide and antimony oxide flame-retardants. However, microelectronic devices packaged with these mold compounds began to fail after a few months in field. These compounds led to cause current leakage and resistive shorts between adjacent leads inside lead-frame devices. Hundreds of millions of dollars were lost in field returns of final products, and resulted in the discontinuation of these mold compounds in 2002 [5].

As of May 2014, General Motors had recalled 12.8 million vehicles, primarily due to faulty ignition switches that could unintentionally shut off the engine and disable airbags, power steering, and anti-lock brakes during driving. These faulty ignition switches resulted in at least 13 deaths and 31 crashes [6]. The faulty device did not meet the specifications through the life of the product under all the use conditions [7]. Several lawsuits were filed against the company, including at least two class action suits. GM agreed to pay $35 million in fines to settle the federal probe associated with the recall [8], but litigation costs are expected to be significantly higher.

These are but a few examples that show how ineffective qualification processes can result in microelectronic devices that can fail in the field. And, once in the field, the costs for the root cause investigation, finding solutions, litigation, recalls, and warranty service, not to mention loss of reputation and market share, can be staggering.

978-1-4799-3911-4/14 $31.00 © 2014 IEEE

II. QUALIFICATION

The continuous evolution of consumer electronic products, multitude of choices for customers and increased competition between manufacturers has resulted in a demand for continually new products at affordable costs. Thus, manufacturers are forced to shorten the development cycles to stay competitive in the market; a challenging endeavor for manufacturers trying to ensure product reliability while offering competitive pricing. Regardless, manufacturers need to qualify a product to "demonstrate that it can reliably operate under use conditions [9]" before the mass production of their products. As noted, improper qualification can result in failures in the field, resulting in financial and reputation loss for the companies.

Qualification is carried out to ensure that a product will meet or exceed the reliability targets in field use [10]. Qualification must be implemented prior to manufacturing the product for shipment. That is, mass manufacturing should not begin, until the device is qualified. Conducting qualification testing products after manufacturing can result in increased costs and delayed shipments if problems with the products are uncovered during the tests. Furthermore, on products that have undergone "significant" design or manufacturing changes, qualification tests should also be performed [10]. JEDEC JESD 46 [11] defines significant changes as "changes that result in impact to form, fit, function, or reliability of a product". For example, changes in die structure, packaging materials or the fabrication process warrant requalification of the product by manufacturers.

Two main approaches for qualification testing are standards-based testing and knowledge-based testing. Standards-based testing is based on a predefined suite of reliability tests for a device. Knowledge-based (use condition based testing) is a relatively new approach where the device is qualified based on the use conditions the product will encounter during its operating life. These two approaches are discussed in detail in Section III.

III. THE QUALIFICATION CHALLENGE

The most common approach for qualification testing is standards-based testing. The most compelling reason for the use of standards-based testing is the industry consensus on the use of standards due to the lack of development and history of any other qualification practice. Standards-based testing provides a way to assess the device reliability based on a set of test conditions. In standards-based testing, if the product survives a certain specified pass criterion mentioned in the standard, the product is considered to have passed qualification testing [12]. Unfortunately, this approach fails to keep up with the ever changing use conditions of the microelectronic devices, and changing environmental regulations. The proliferation of numerous microelectronic device types and materials makes it almost impossible to develop specific standards based on geometry and material type; which are needed to determine acceleration factors. As new technologies are introduced in the market, companies who qualify their new products based on standards that were meant for their products' predecessors or the closest technology, may find themselves having field failures. Also, as more materials are banned from use in electronics, the industry resorts to new materials and processes to adapt. However, in many cases, the industry uses new materials and processes without complete understanding of the reliability risks involved and use standards meant for their predecessors. For instance, although many low-silver lead-free SAC solders are already being used by many companies [13], the applicability of existing standards developed for tin-lead and SAC305 solders on the new low-silver SAC solders has not been evaluated.

Standards-based testing might provide a less costly means to qualify a product, but not necessarily provide meaningful information, since the results from a standards-based qualification test might not relate to the actual field condition (the use condition that standard was based on might be no longer applicable). Examples of standards-based qualification test procedures include JESD 22 series [14] and MIL STD 883 [15]. Qualification test plans provides information on the selection of test conditions, definition of pass/fail criteria, and other specifics on how to conduct a generic qualification test. Examples of qualification test plans include JEDEC JESD 47 [16] and AEC Q100 [17].

Most original equipment manufacturers (OEMs) and original device manufacturers (ODMs) conduct a suite of tests to qualify their products based on the test conditions provided in standards. For instance, International Rectifier qualifies their insulated gate polar transistors (IGBTs) meant for industrial applications based on the JESD-47 standard and for automotive applications based on the AEC-Q101 standard. Qualification tests conducted by International Rectifier require their IGBTs to survive for a specific amount of time in order for them to have passed the qualification process. International Rectifier defines failure to occur when the device parameters exceed the specifications provided in the datasheets. For example, IGBTs meant for industrial applications are required to survive at least 1000 hours at 150°C or 175°C in order to pass the high temperature reverse bias test [18]. However, the selection of the test temperature or time to survive is not based on the actual conditions to which the device will be subjected to.

Pass/fail results from a standards-based qualification testing utilizing PoF techniques may provide information on whether the product would actually survive in the end use condition. However, with the changing use conditions, package architecture, and materials, the test conditions listed in the standards might be no longer valid. In fact, a product that has

978-1-4799-3911-4/14 $31.00 © 2014 IEEE

failed the required standards-based qualification test may survive satisfactorily in the actual use conditions. This leads to rejection of reliable products that could have otherwise been shipped, thereby resulting in loss of resources and money [19]. On the other hand, a product that has passed the standards-based test may fail prematurely in field [20] thus leading to a false acceptance of a defective product. The premature failure may be due to the fact that the qualification test condition is not stringent enough or the failure to capture intermittent failures during the qualification process.

Another reason for the use of standards-based testing is their wide acceptability in the industry which helps to provide some confidence into qualification testing of products where there is a lack of understanding of end-use conditions of the product. The development of most electronics today involves complex global supply chains for design, materials, manufacturing, assembly and test. Cost and schedule (availability) are key metrics used to create the supply chains. The segmentation of supply chains most often has microelectronic device suppliers providing parts to a variety of industries, for a variety of end products with different use conditions. Knowledge of the use conditions of any given microelectronic device in a final product is often not known by the suppliers. Hence, it might not be feasible (practically and economically) for suppliers to qualify a microelectronic device for different use conditions in order to cater the requirements of original equipment and design manufacturers. At the same time, since standards are still widely accepted in the industry, it provides an easy way to qualify the products.

Owing to the above disadvantages of standards based testing, more and more semiconductor companies are leaning towards use conditions (knowledge-based) testing. SEMATECH guidelines [12] on knowledge-based qualification recommend that "qualification must take into account the most severe use conditions that are likely to be encountered". Semiconductor companies conduct use condition based qualification testing wherein they identify the key failure mechanism and modes for a given use condition and tailor the qualification tests to that. The parameters to be monitored during reliability testing (relating to the failure mechanisms and modes) are identified. Reliability model is developed by running multiple reliability stresses for a given failure mechanism and measuring the parameters at regular intervals of the test. Based on the acceleration factor with the use conditions, the testing duration and stresses are selected for qualification. Although, major semiconductor companies do qualify their products against the worst case use conditions, this might not be a universally accepted practice yet. There is a great resistance in the industry to deviate from the standards-based testing since there is not enough credibility around any other qualification testing approach. Hence the companies are often "forced" to follow standards-based testing instead of knowledge-based testing.

However, both standards- and use condition- based testing fail to capture intermittent failures, or to determine a device 'issue' no-fault-found (NFF) or retest OK (RTOK) cases [21]. Intermittent behaviors result in a loss of functionality for a specified duration of time; however, the functionality may recover when the stress condition that caused the intermittent behavior is removed. Intermittent failures can be very expensive for manufacturers. According to a research report by Accenture in 2011, 68% of the returned consumer electronic products are characterized as NFF by the manufacturer. The NFF processing cost was $2.5 billion in 2011, which was 15% of the total cost of consumer electronics returns attributed to U.S.-based consumers. A 1% reduction in the number of NFF cases could translate to annual savings of 4% in return and repair costs, or $21 million for a typical large consumer electronics manufacturer and $16 million for an average consumer electronics retailer [22]. If the intermittent behavior is not captured during qualification testing, then this behavior may manifest itself as intermittent failures or NFFs in the field. Periodic monitoring of system parameters is also susceptible to missing intermittent behavior, but superior to only identifying failure after the test has been conducted. For example, the intermittent failures in the Ford ignition modules were due to damage from a high temperature condition, which recovered when the vehicle was not accelerating. Clearly, *in-situ* monitoring of the product parameters increases the likelihood of capturing intermittent behavior. To successfully utilize *in-situ* monitored data, parameter selection and sampling rate (frequency of monitoring) determination are critical.

Majority of the industry still strongly supports standards-based testing due to lack of a better qualification testing approach. However, continued occurrence of costly field failures warrants the need to transition from a standards-based qualification approach to a more comprehensive qualification process that utilizes a variety of tools and techniques. This paper presents an approach to fusion-based qualification utilizing a combination of physics-of-failure (PoF) and machine learning concepts. The objective is to develop a rapid, cost effective and efficient approach to qualification.

IV. SOLUTIONS

The qualification of electronic products can be carried out using various approaches, including PoF-based and data-driven-based. The PoF-based approach helps to obtain application-based accelerated test conditions. The data-driven-based approach provides anomaly detection by in-situ monitoring of operating and environmental parameters thereby detecting intermittent faults. In this paper, a fusion prognostics-based qualification approach is proposed that combines the PoF- and data-driven-based approaches. It involves an in-situ qualification process that incorporates a fusion of machine learning techniques and physics-of-failure based prognostics. Fig. 1 shows the various approaches for product qualification: the PoF-based, prognostics-based, and fusion prognostics-based approaches.

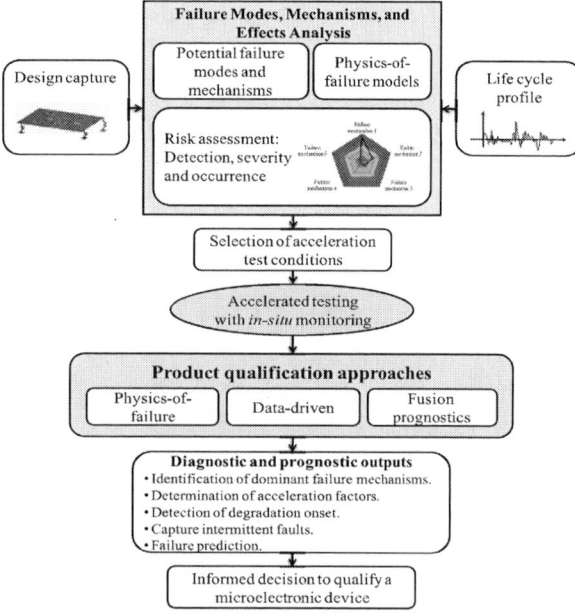

Fig. 1. Approaches to product qualification (modified using concepts from Ref. [23]).

A. Physics of Failure (PoF)-based Qualification

PoF-based qualification approaches have been discussed at length in the literature [24], [25], as well as in JEDEC [16], [20], such as JESD 94 and JESD 47, and SEMATECH publications [12], [26]. PoF-based qualification is based on the knowledge of field use conditions, expected failure mechanisms, and the related failure models [9], [12]. Failure modes, mechanisms, and effects analysis (FMMEA) uses knowledge of expected life-cycle conditions to identify potential weaknesses in the product design. Then, based on the anticipated targeted application conditions (e.g., load type, level, and frequency), combined with the most severe failure locations, failure mechanisms can be prioritized according to the level of their severity and likelihood of occurrence [27]. Virtual qualification can be conducted using simulation software and helps to determine the probability of the product meeting its life requirements [28].

The first step towards PoF-based qualification is to obtain critical failure mechanisms for a given device architecture and usage environment. Once the failure mechanisms are identified, appropriate failure models are applied to obtain time-to-failure estimates. Acceleration factors are obtained from the time to failures under the qualification test conditions and use conditions obtained from the failure models. The qualification test conditions can hence be chosen based

on the acceleration transforms to assess the product reliability in short test duration. Without identifying an appropriate acceleration factor, there is no way to relate times to failure under accelerated conditions to the expected operating conditions of the product. An understanding of the anticipated use conditions of the product is needed to translate results from a qualification test to useful information.

Although failure mechanisms and models for electronic components have been developed for a wide range of electronics and are widely documented in the literature [10], [29]-[36], variations in materials and the device architecture can significantly affect the model results. Whenever a new failure mechanism is identified, appropriate failure models need to be developed. The JEDEC standard JESD 91A [35] describes a method for developing PoF models for failure mechanisms in electronic components. Qualification methodologies for specific applications are prescribed in JEDEC JESD 94 [20], JEP 148 [9], and in SEMATECH documents [12], [26], [37], [38].

B. Data-Driven Qualification

In-situ monitoring of operating and environmental parameters, such as power, current, voltage, temperature, humidity, vibration, and acoustic signal, during qualification tests can be used to make informed decisions on the device reliability. Employing pattern recognition methods on these monitored data during qualification tests can help in identifying the onset of degradation in the device.

Patil et al. [39], [40], through in-situ monitoring of IGBT during accelerated testing (power cycling under constant frequency), identified that gate oxide and die attach degradation affects the quasi static capacitance-voltage (C-V) and ON state collector-emitter voltage (VCE) measurements respectively. Using the identified precursor parameters, Patil et al. [41] developed a Mahalanobis distance-based anomaly detection method to detect the onset of degradation in IGBT packages. Sutrisno et al. [42] further extended this approach by employing k-Nearest Neighboring (k-NN) method to detect the onset of degradation in IGBT packages during power cycling under different frequencies. Furthermore, Sutrisno et al. [42] developed a health indicator using the anomaly detection results obtained from k-NN to estimate the health of an IGBT during qualification testing. Thus, monitoring appropriate precursor parameters and employing appropriate data-driven methods aids in understanding the life consumed and time at which the device started to degrade during qualification test. This information can be used to (1) anticipate the onset of degradation in use condition and (2) estimate remaining life of the devices during the time they are shipped.

In addition to these advantages, data-driven prognostic methods can also be used to reduce the time of qualification tests which requires 'test-until-failure'. Once a failure prediction method is established using the data collected under a batch qualification test, the future qualification tests' time can be shortened to a time at which reliable failure time estimates can be obtained. One such failure prediction method was developed by Patil *et al.* [43], where an empirical degradation model was developed based on the trend exhibited by the V_{CE} parameter after the onset of die attach degradation. By integrating the degradation model with a statistical filter, Patil *et al.* [43] was able to predict the time at which V_{CE} crossed the failure threshold i.e. the time of failure, once anomalies were detected.

In-situ monitoring of product parameters also enables determination of intermittent faults that are often missed in standards-based qualification. Data-driven methods that could capture outliers in the monitored data can aid in capturing such faults. The need to capture intermittent failures has prompted manufacturers to develop integrated sensors for placement on products to continuously monitor system parameters. For example, applications of lithium-ion secondary batteries, such as electric vehicles, smart phones, personal digital assistants, notebooks, and electric cars, require batteries to charge and discharge rapidly. This can cause the interior temperature of a battery to rise quickly, raising a safety issue. The *in-situ* measurement of battery temperature helps to keep the battery temperature in check, since overcharging may lead to unstable voltage and current, which in turn may cause potential safety problems, such as thermal runaway and explosions. Lee *et al.* [44] developed micro temperature sensors integrated onto a lithium-ion secondary battery for *in-situ* monitoring of battery temperature.

C. Fusion Prognostics-based Approach

The fusion prognostics-based approach exploits the advantages of PoF and data-driven approaches to estimate the remaining useful life of the product, detect intermittent faults or anomalous behavior of the product [27]. In the PoF approach, the RUL is estimated based on an understanding of the failure mechanisms that lead to the degradation and eventual failure of the product. However, the PoF approach cannot capture the intermittent failures, since the PoF models do not account for any unexpected change in the system parameters that characterize intermittent failures.

Data-driven approaches look for patterns in the *in-situ* data obtained from the device during qualification tests to detect the onset of degradation, and identify the underlying failure mechanism. Since data driven techniques involve "learning" from the data and help in making decisions, it is possible to detect any unexpected changes in the system parameters. This in turn enables the detection and analysis of intermittent failures in the product. Anomaly detection can be carried out by comparing the *in-situ* data against a healthy baseline data. Baseline data is collected under several modes and loading conditions under which a product is expected to operate [45], [46]. For example, Jaai *et al.* [46] used Multivariate State Estimation Technique (MSET) and Sequential Probability Ratio Test (SPRT) to detect the onset of failure in Ball Grid Arrays (BGA) subjected to Accelerated Temperature Cycling (ATC) tests. The resistance data monitored during ATC was compared with the baseline data to detect anomalies.

The benefits of PoF-based and data-driven based approaches were combined to develop a fusion-prognostics based qualification approach. To conduct a fusion-prognostics qualification process, accelerated tests are conducted till an anomaly is first detected instead of testing till failure. In-situ monitoring and machine learning techniques are employed to detect anomalies, which could potentially indicate the onset of failure. After the identification of an anomaly, the parameters contributing to the anomaly can be isolated using appropriate techniques (such as PCA, LDA, SVM, SOM, and more). From the existing literature on PoF models and data-driven approaches, potential failure mechanisms that may arise as a result of the contributing parameters to anomalies are listed. Using the relevant PoF models corresponding to the potential failure mechanisms and acceleration factors, the time to failure time distribution can be estimated.

The distance between the time to anomalous behaviour and time to failure needs to be characterized for each of the listed failure mechanisms. An understanding of this time difference will prevent the prediction of failure too early in the actual use conditions. For instance, if the detection of anomalous behaviour is at 90% of time to failure in the qualification test, the predicted behaviour using PoF models will be provide an estimate of anomalous behaviour in field at 90% of time to failure. This may be acceptable as the product can be deployed in field for most of its useful life. However, if the a particular anomalous behaviour is not an intermittent and occurs at 50% of time to failure in qualification test, then the use of that particular anomaly for fusion-based qualification will result in early withdrawal of the product from the field.

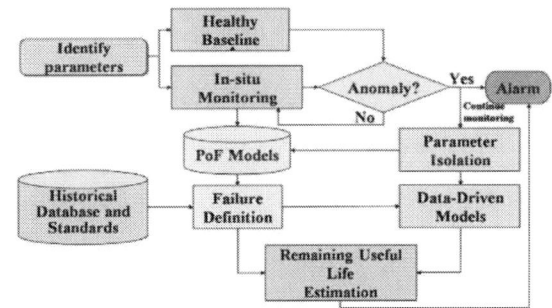

Fig. 2. Fusion prognostics approach.

As demonstrated, qualification based on fusion prognostics enables the health monitoring of the product instead of pass/fail decisions. This enables the detection of intermittent behavior that might have been otherwise missed. Since the fusion prognostics approach can effectively capture anomalous behavior and intermittent faults, root cause analysis of NFF errors can be carried out. The fusion prognostics-based approach can result in the significant reduction in qualification test times. It also provides the benefit of having an estimation of RUL, and, hence, catastrophic failure prevention instead of failure identification.

V. CONCLUSIONS AND RECOMMENDATIONS

Companies who make electronic products face several challenges, including shorter development cycles, the drive to reduce costs, and a more diffuse and complex supply chain over which often have little control. Due to these challenges, companies often adopt standards-based testing to qualify their products. However, standards-based qualification in the absence of the understanding of the correlation between the end use and test conditions can lead to ineffective qualification. This is especially true for those companies whose product lifetimes are long (e.g., more than eight years) and where the environments are harsher than for most commercial applications (high temperature electronics), or where the consequences of failure are severe (medical electronics).

PoF qualification approaches can be utilized to focus on the critical failure mechanisms for a given use condition. Using the failure models for the critical failure mechanisms, test conditions can be selected so that passing the qualification test will help ensure that the product performs reliably in the field. On the other hand, the use of prognostics techniques detects intermittent failures, and provides early warning of failure. The detection of intermittent behavior is enabled by the identification of anomalous behavior and the identification of environmental and product performance parameters that contribute to the anomaly. Anomaly identification prior to the failure can help to provide a warning of an impending failure and also allow root cause analysis of the failure.

It is not feasible for device manufacturers to qualify their devices to the specifications and targeted use conditions of each product manufacturer to whom they sell. For example, a manufacturer of a small commodity item, such as a common resistor, probably has little incentive to qualify to the requirements of the small-volume medical device community. However, if a device manufacturer does not qualify to a product manufacturer's specifications, the onus lies on the product manufacturer to make sure that the devices used in their products work reliably under the targeted use conditions of the end product. A more viable and acceptable approach may be to have a single qualification process that maximizes resources, but this may be too demanding and unnecessary for a large number of customers.

Sometimes the product manufacturers may need to conduct additional testing at their end. Medical electronics manufacturers, for instance, often "requalify" components when they believe the qualification practices of the device manufacturer to be insufficient for the use conditions of their products.

Significant risks are entailed in the use of qualification approaches in scenarios where the application conditions are not well known, acceleration models are non-existent or incorrect, and the tests and evaluation conditions have not been properly set. The Ford, Toyota, and General Motors examples mentioned in the introduction have shown that in addition to huge financial losses, unreliable products can result in the loss of goodwill and tarnish the reputation of the company. Since the potential for damage from failures and subsequent litigation is likely to be higher for those closest to the final customer in the supply chain, the incentives to improve current qualification processes will likely be from the top down in the supply chain. However, microelectronics companies can also suffer severe financial losses and should also consider changing their qualification practices to adapt to the ever-changing product life conditions, reduced costs, and shorter time to market.

No device level qualification method will capture the scenarios of unexpected use environments, and interactions of all hardware and software within a system. This requires qualification to be conducted at each level in the product development. Conducting board-level and product-level qualification tests will enable OEMs and ODMs to capture unexpected failures due to device–device and device–product interactions.

REFERENCES

[1] Class Action Lawsuits and Recalls, "Defective Ford ignition modules", July 30, 2007, Available online: http://classactionlawsuit.blogspot.com/2007/07/defective -ford-ignition-modules.html, Last accessed: May 18, 2014.

[2] D. Thomas, K. Ayers, and M. Pecht, "The 'trouble not identified' phenomenon in automotive electronics", Microelectronics Reliability, Vol. 42, No. 4/5, pp. 641–651, April/May, 2002.

[3] Toyota, "Toyota announces voluntary safety recall on certain Toyota Corolla and Corolla Matrix models", 2010, Available online: http://www.toyota.com/recall/corolla-matrix.html, Last accessed: January 9, 2013.

[4] CBS News, "Latest Toyota recall one of its largest", August 30, 2010, Available online: http://www.cbsnews.com/2100-500395_162-6808992.ht ml, Last accessed: January 9, 2013.

[5] M. Pecht and Y. Deng, "Electronic device encapsulation using red phosphorus flame retardants", Microelectronics Reliability, Vol. 46, No. 1, pp. 53–62, January, 2006.

[6] A. Carrns, "Understanding particulars of GM's safety recall", NY Times, April 4, 2014, Available online: http://www.nytimes.com/2014/04/05/your-money/unders tanding-gms-safety-recall.html, Last accessed: May 17, 2014.

[7] C. Isidore, "The 57-cent part at the center of GM's recall crisis", CNN Money, April 2, 2014, Available online:

http://money.cnn.com/2014/04/02/news/companies/gm-r
ecall-part/, Last accessed: May 17, 2014.

[8] C. Isidore and R. Marsh, "GM to pay $35 million over delayed recall", CNN Money, May 16, 2014, Available online: http://money.cnn.com/2014/05/16/news/companies/gm-n htsa/, Last accessed: May 17, 2014.

[9] JEDEC, "Reliability qualification of semiconductor devices based on physics of failure risk and opportunity assessment", JEP 148A, December, 2008.

[10] A. Dasgupta, J. Evans, and M. Pecht, "Quality conformance and qualification of microelectronic packages and interconnects", New York, Wiley-Interscience, 1994.

[11] JEDEC, "Customer notification of product/process changes by semiconductor suppliers", JESD 46C, October, 2006.

[12] D. Eaton, N. Durrant, S. Huber, R. Blish, and N. Lycoudes, "Knowledge-based reliability qualification testing of silicon devices", Austin, Texas, Sematech, 2000.

[13] E. George, M. Osterman, M. Pecht, R. Coyle, R. Parker, and E. Benedetto, "Thermal cycling reliability of alternative low-silver tin-based solders," 46th International Symposium on Microelectronics, Orlando, Florida, US, September 29 –October 3, 2013.

[14] JEDEC, "Reliability test methods for packaged devices", JESD 22 Series.

[15] Department of Defense, "Test method standard circuits", MIL STD 883 H, February, 2010.

[16] JEDEC, "Stress test driven qualification of integrated circuits", JESD 47 H, February, 2011.

[17] AEC, "Stress test qualification for integrated circuits, AEC Q100 Rev F, 2003.

[18] International Rectifier, "Qualification of discrete devices: MOSFET's, IGBT's, and diodes", January 01, 2012, Available online: http://www.irf.com/product-info/reliability/qrdiscrete.pd f, Last accessed: June 3, 2014.

[19] E. Suhir, C. Wong, and Y. Lee, "How to make a device into a product: Accelerated life testing its role, attributes, challenges, pitfalls, and interaction with qualification testing", Micro- and Opto-Electronic Materials and Structures: Physics, Mechanics, Design, Packaging, Reliability, New York, Springer-Verlag, 2007.

[20] JEDEC, "Application specific qualification using knowledge based test methodology, JESD 94A, 2008.

[21] H. Qi, S. Ganesan, and M. Pecht, "No-fault-found and intermittent failures in electronic products", Microelectronics Reliability, Vol. 48, No. 5, pp. 663–674, May, 2008.

[22] D. Douthit, M. Flach, and V. Agarwal, A returning problem, Reducing the quantity and cost of product returns in consumer electronics, Accenture report, 2011, Available online: http://www.accenture.com/SiteCollectionDocuments/PD F/Accenture-Reducing-the-Quantity-and-Cost-of-Custo merReturns.pdf, Last accessed: May 28, 2014.

[23] M. Pecht, and J. Gu, "Prognostics-based product qualification", IEEE Aerospace Conference, pp.1-11, March 7-14, 2009.

[24] K. Upadhyayula and A. Dasgupta, "Physics-of-failure guidelines for accelerated qualification of electronic systems", Quality and Reliability Engineering International, Vol. 14, pp. 433–447, November/December, 1998.

[25] M. Pecht and A. Dasgupta, "Physics-of-failure: an approach to reliable product development", International Integrated Reliability Workshop, Final Report, pp. 1-4, October 22-25, 1995.

[26] R. Blish, S. Huber, and N. Durrant, "Use condition based reliability evaluation of new semiconductor technologies", Sematech Technology, Transfer # 99083810A-XFR, August 31, 1999.

[27] R. Jaai and M. Pecht, "A prognostics and health management roadmap for information and electronics-rich systems", Microelectronics Reliability, Vol. 50, No. 3, pp. 317–323, March, 2010.

[28] M. Osterman, A. Dasgupta, and T. Stadterman, "Virtual qualification of electronic hardware," Case Studies in Reliability and Maintenance, W. R. Blischke and D. N. P. Murthy, Wiley, Hoboken, New Jersey, 2003.

[29] J. Hu, D. Barker, A. Dasgupta, and A. Arora, "Role of failure mechanism identification in accelerated testing", Journal of Illuminating Engineering Society, Vol. 36, No. 4, pp. 39–45, July/August, 1993.

[30] A. Dasgupta and J. Hu, "Failure mechanism model for brittle fracture", IEEE Transactions on Reliability, Vol. 41, No. 3, pp. 328–335, September, 1992.

[31] A. Dasgupta and J. Hu, "Failure mechanism model for ductile fracture", IEEE Transactions on Reliability, Vol. 41, No. 4, pp. 489–495, December, 1992.

[32] A. Dasgupta and J. Hu, "Failure mechanism models for plastic deformation", IEEE Transactions on Reliability, Vol. 41, No. 2, pp. 168–174, June, 1992.

[33] J. Li and A. Dasgupta, "Failure mechanism models for creep and creep rupture", Transactions of the American Society of Mechanical Engineers, Vol. 115, pp. 416–423, 1993.

[34] A. Dasgupta, "Failure mechanism models for cyclic fatigue", IEEE Transactions on Reliability, Vol. 42, No. 4, pp. 548–555, December, 1993.

[35] JEDEC, "Method for developing acceleration models for electronic component failure mechanisms", JESD91A, August, 2003.

[36] D. Searls, T. Dishongh, and P. Dujari, "A strategy for enabling data driven product decisions through a comprehensive understanding of the usage environment", Proceedings of Pacific RIM/ASME International Intersociety Electronic and Photonic Packaging Conference, Kauai, Hawaii, July, 2001.

[37] J. Veshinfsky, B. Purvee, R. Susko, J. McCullen, and N. Durrant, "Use condition based reliability evaluation: An example applied to ball grid array (BGA) packages", Sematech, 99083813A-XFR, August, 1999.

[38] JEDEC, "Failure mechanisms and models for semiconductor devices", JEP122F, November, 2010.

[39] N. Patil, D. Das, K. Goebel, and M. Pecht, "Identification of failure precursor parameters for Insulated Gate Bipolar Transistors (IGBTs)", International Conference on Prognostics and Health Management, pp. 1-5, October 6-9, 2008.

[40] N. Patil, J. Celaya, D. Das, K. Goebel, and M. Pecht, Precursor parameter identification for insulated gate bipolar transistor (IGBT) prognostics, IEEE Transactions on Reliability, Vol. 58, No. 2, pp. 271-276, June, 2009.

[41] N. Patil, D. Das and M. Pecht, Mahalanobis distance approach to field stop IGBT diagnostics, Proceedings of the 10th International Seminar on Power Semiconductors, Prague, pp.79-84, 1-3 September, 2010.

[42] E. Sutrisno, "Fault detection and prognostics of insulated gate bipolar transistor (IGBT) using a k-nearest neighbor classification algorithm", M.S. Thesis, University of Maryland, College Park, 2013.

[43] N. Patil, D. Das, K. Goebel, and M. Pecht, "A prognostic approach for non-punch through and field stop IGBTs", Microelectronics Reliability, Vol. 52, pp. 482–488, 2012.

[44] C. Lee, S. Lee, M. Tang, and P. Chen, "In-situ monitoring of temperature inside lithium-ion batteries by flexible micro temperature sensors", Sensors, Vol. 11, No. 10, pp. 9942-9950, October, 2011.

[45] S. Kumar, E. Dolev, and M. Pecht, "Parameter selection for health monitoring of electronic products", Microelectronics Reliability, Vol. 50, No. 2, pp. 161–168, February, 2010.

[46] R. Jaai, M. Pecht, and J. Cook, "Detecting failure precursors in BGA solder joints using prognostic algorithms", IEEE Reliability and Maintainability Symposium, 2009.

978-1-4799-3911-4/14 $31.00 © 2014 IEEE

9781479939114